OXFORD STATISTICAL SCIENCE SERIES

OXFORD STATISTICAL SCIENCE SERIES

For a full list of titles please visit
http://www.oup.co.uk/academic/science/maths/series/osss/

10. J.K. Lindsey: *Models for Repeated Measurements*
11. N.T. Longford: *Random Coefficient Models*
12. P.J. Brown: *Measurement, Regression, and Calibration*
13. Peter J. Diggle, Kung-Yee Liang, and Scott L. Zeger: *Analysis of Longitudinal Data*
14. J.I. Ansell and M.J. Phillips: *Practical Methods for Reliability Data Analysis*
15. J.K. Lindsey: *Modelling Frequency and Count Data*
16. J.L. Jensen: *Saddlepoint Approximations*
17. Steffen L. Lauritzen: *Graphical Models*
18. A.W. Bowman and A. Azzalini: *Applied Smoothing Techniques for Data Analysis*
19. J.K. Lindsey: *Models for Repeated Measurements, Second Edition*
20. Michael Evans and Tim Swartz: *Approximating Integrals via Monte Carlo and Deterministic Methods*
21. D.F. Andrews and J.E. Stafford: *Symbolic Computation for Statistical Inference*
22. T.A. Severini: *Likelihood Methods in Statistics*
23. W.J. Krzanowski: *Principles of Multivariate Analysis: A User's Perspective, Revised Edition*
24. J. Durbin and S.J. Koopman: *Time Series Analysis by State Space Methods*
25. Peter J. Diggle, Patrick Heagerty, Kung-Yee Liang, and Scott L. Zeger: *Analysis of Longitudinal Data, Second Edition*
26. J.K. Lindsey: *Nonlinear Models in Medical Statistics*
27. Peter J. Green, Nils L. Hjort, and Sylvia Richardson: *Highly Structured Stochastic Systems*
28. Margaret Sullivan Pepe: *The Statistical Evaluation of Medical Tests for Classification and Prediction*
29. Christopher G. Small and Jinfang Wang: *Numerical Methods for Nonlinear Estimating Equations*
30. John C. Gower and Garmt B. Dijksterhuis: *Procrustes Problems*
31. Margaret Sullivan Pepe: *The Statistical Evaluation of Medical Tests for Classification and Prediction, Paperback*
32. Murray Aitkin, Brian Francis, and John Hinde: *Statistical Modelling in GLIM4, Second Edition*
33. Anthony C. Davison, Yadolah Dodge, N. Wermuth: *Celebrating Statistics: Papers in Honour of Sir David Cox on his 80th Birthday*
34. Anthony Atkinson, Alexander Donev, and Randall Tobias: *Optimum Experimental Designs, with SAS*
35. M. Aitkin, B. Francis, J. Hinde, and R. Darnell: *Statistical Modelling in R*
36. Ludwig Fahrmeir and Thomas Kneib: *Bayesian Smoothing and Regression for Longitudinal, Spatial and Event History Data*

Bayesian Smoothing and Regression for Longitudinal, Spatial and Event History Data

Ludwig Fahrmeir

Department of Statistics
Ludwig-Maximilians-University Munich

Thomas Kneib

Department of Mathematics
Carl von Ossietzky University Oldenburg

OXFORD
UNIVERSITY PRESS

OXFORD
UNIVERSITY PRESS

Great Clarendon Street, Oxford OX2 6DP

Oxford University Press is a department of the University of Oxford.
It furthers the University's objective of excellence in research, scholarship,
and education by publishing worldwide in

Oxford New York

Auckland Cape Town Dar es Salaam Hong Kong Karachi
Kuala Lumpur Madrid Melbourne Mexico City Nairobi
New Delhi Shanghai Taipei Toronto

With offices in

Argentina Austria Brazil Chile Czech Republic France Greece
Guatemala Hungary Italy Japan Poland Portugal Singapore
South Korea Switzerland Thailand Turkey Ukraine Vietnam

Oxford is a registered trade mark of Oxford University Press
in the UK and in certain other countries

Published in the United States
by Oxford University Press Inc., New York

First published 2011

British Library Cataloguing in Publication Data
Data available

Library of Congress Cataloguing in Publication Data
Data available

Typeset by SPI Publisher Services, Pondicherry, India
Printed in Great Britain
on acid-free paper by
CPI Antony Rowe, Chippenham, Wiltshire

ISBN 978–0–19–953302–2

1 3 5 7 9 10 8 6 4 2

Preface

The last two decades have seen a tremendous increase in semiparametric regression modelling where the traditional assumptions underlying generalized linear models (linear impact of covariates, independence of observations, univariate exponential family for the responses) can be relaxed. A starting point was the monograph by Hastie and Tibshirani (1990) that introduced the unifying concept of generalized additive models (GAM) to a broader audience and combined several diverse approaches in one general framework. Since then, semiparametric regression models have been extended in several directions, for example with the inclusion of spatial effects that account for spatial correlations, the possibility of analysing longitudinal data or the consideration of more general response types such as survival times or categorical responses, see Ruppert, Wand and Carroll (2003) and Wood (2006) for detailed and up-to-date references on such extended generalized additive models.

With this book, we make an attempt to collect several recent advances in semiparametric regression and present them in a unifying, Bayesian perspective. We will consider simulation-based full Bayesian Markov chain Monte Carlo (MCMC) inference as well as empirical Bayes procedures, closely related to penalized likelihood estimation and mixed models. In general, we will take a pragmatic Bayesian perspective and will therefore also discuss issues close to (or even beyond) the boundary of Bayesian statistics if this allows to shed more light on the properties of the inferential procedures. In the main text and in the applications, we focus on semiparametric regression and smoothing based on basis expansions of unknown functions and effects in combination with smoothness priors for the basis coefficients. In a penalized likelihood approach, such smoothness priors correspond to roughness penalties. Other approaches for semiparametric regression, in particular concepts based on Bayesian variable selection (see for example Denison, Holmes, Mallick and Smith, 2002), or nonparametric Bayesian approaches based on Dirichlet process mixtures, are also described but in less detail.

We assume a thorough background knowledge on generalized linear models as provided for example in Fahrmeir and Tutz (2001, Chs. 2 and 3). Some basic knowledge on (penalized spline) smoothing and mixed models is a plus but Chapters 2 and 3 concisely present the essential tools of smoothing and mixed models to bring all readers to the same level. Chapter 2 deals with basic concepts of (Bayesian) smoothing and semiparametric regression for cross-sectional data. It provides information on penalized spline smoothing in generalized additive models, also beyond the usual scope of the univariate exponential family by allowing for categorical responses. It summarizes ideas for penalized likelihood/empirical Bayes inference as well as as for Markov chain Monte Carlo simulation algorithms.

Chapter 3 comprises an introduction to generalized linear mixed models (GLMM). Beyond what is usually covered in textbooks on GLMMs, we present flexible random effects distributions, in particular based on Dirichlet process mixture priors. Much of what is presented in Chapter 3 will reappear in Chapter 4 as building blocks for inferential procedures in semiparametric mixed models that combine the ideas of GLMMs for longitudinal data and GAMs for modelling flexible, nonlinear covariate effects. Full and empirical Bayes inference for GLMMs will allow for a data-driven determination of smoothing parameters and variance components in semiparametric mixed models. Dirichlet process modelling can be used to relax the normality assumption of random effects models in a full Bayesian formulation. In addition, Chapter 4 discusses general issues of imposing prior structures to incorporate appropriate regularization properties, the relation between smoothing temporal effects and modelling of temporal correlations, and model choice and model checking questions.

Chapter 5 adds spatial components and bivariate surface smoothing to the class of semiparametric mixed models to account for spatial correlations or interactions between continuous covariates in the data. The spatial component can take different forms depending on the type of spatial information. The main differentiation will be between continuous and discrete spatial data, representing either coordinate information or the presence of a spatial grouping variable. Much of Chapter 5 is self-contained so that it can also serve as a first introduction to spatial statistics.

The final chapter introduces models that describe transitions of a process with categorical state space in continuous or discrete time. Hazard regression models for survival times are an important special case and a large proportion of Chapter 6 is devoted to this specific class of models. As an alternative, we will consider semiparametric extensions of accelerated failure time models. Chapter 6 also covers extensions to multi-state models where multiple transitions are analysed simultaneously. The regression specification of the transition intensities will reuse ideas from previous chapters and will consist of a combination of, for example, nonlinear effects, random effects and spatial effects. In addition, time-dependence is included explicitly in the predictor of the regression model.

All methods considered in this book are illustrated through worked examples from various fields such as forestry, development economics, medicine, or marketing. We expect the methods to be of interest in all fields of applied statistics and therefore collected different examples that emphasize the generality of the statistical theory. Most of the examples have been analysed using implementations in the Bayesian software BayesX available from http://www.stat.uni-muenchen.de/~bayesx/ and codes as well as some of the data sets are made publicly available on the website http://www.smoothingbook.org/ accompanying this book.

Although, in principle, this book is a research monograph, we believe that it can (and actually should) also serve as the basis of advanced regression courses at the masters level and that it also contains material valuable to many applied researchers. The mathematical level is moderate but requires familiarity with matrix algebra and also assumes basic knowledge in Bayesian statistics and Markov chain Monte Carlo simulation techniques (as provided for example in Harville (1997) and Gelman, Carlin, Stern and Rubin (1995)).

München & Oldenburg,
December 2010

Ludwig Fahrmeir
Thomas Kneib

Acknowledgements

This book would not have been possible without the help of:

- Andreas Bayerstadler, Christian Seiler and Roman Hornung who contributed significantly to the realization of the book in LaTeX and the computations related to several of the examples.
- Christiane Dargatz, Nora Fenske, Sylvia Frühwirth-Schnatter, Michael Höhle, Stefanie Kalus, Susanne Konrath, Fabian Scheipl, Volker Schmid and the students participating in the seminar on 'Semiparametric Regression for Longitudinal, Spatial and Event History Data' during the winter term 2009/10 who proofread parts of the book at various stages of its production.
- Several colleagues and collaborators who kindly allowed us to use their data sets or material emerging from joint work. In particular, Jan Beyersmann, Andreas Brezger, Nora Fenske, Axel Göttlein, Felix Heinzl, Andrea Hennerfeind, Stefanie Kalus, Susanne Konrath, Stefan Lang, Daniel Sabanés Bové, Fabian Scheipl, Heinz-Erich Wichmann and Alexander Yassouridis had significant impact on several pieces of work that turned into parts of this book.
- The BayesX development team: Christiane Belitz, Andreas Brezger, Stefan Lang, Eva-Maria Fronk, Felix Heinzl, Andrea Hennerfeind, Manuela Hummel, Alexander Jerak, Susanne Konrath, Petra Kragler, Cornelia Oberhauser, Leyre Estíbaliz Osuna Echavarría and Daniel Sabanés Bové.
- The German Science Foundation providing financial support in terms of the Collaborative Research Center 386 'Statistical Analysis of Discrete Structures' and the research project 'Bayesian regularisation for regression models with high-dimensional predictors' (Grants FA 128/5-1, FA 128/5-2).
- Raymond J. Carroll who initially proposed this book project to us and Oxford University Press for the smooth collaboration at all stages of the production (in particular Helen Eaton, Dewi Jackson, Alison Jones, Keith Mansfield, Caroline Broughton, Victoria Mortimer and Abhirami Ravikumar).

Contents

List of Algorithms

List of Examples

1 Introduction: Scope of the Book and Applications

1.1 Semiparametric regression

Generalized linear models (GLM) form by far the most popular class of regression models for investigating the relation between a (possibly non-normal) response and a set of covariates. A generalized linear model comprises two basic assumptions. The first assumption states conditions for the response distribution: The response variables y_i, $i = 1, \ldots, n$, are following a distribution belonging to the univariate exponential family class and are conditionally independent given the covariates. The second assumption specifies the structure of the regression model and assumes that the expectations of response variables are related to a linear predictor formed of the covariates via a one-to-one transformation, that is,

$$\mathbb{E}(y_i) = h(\eta_i), \qquad \eta_i = \boldsymbol{u}_i'\boldsymbol{\beta}$$

where h is the response function and $\eta_i = \boldsymbol{u}_i'\boldsymbol{\beta}$ is a linear combination of covariates $\boldsymbol{u}_i$ supplemented with regression coefficients $\boldsymbol{\beta}$. Popular special cases of generalized linear models are

- *linear models* where $y_i \sim \mathrm{N}(\eta_i, \sigma^2)$ and h is the identity,
- *binary regression models* where $y_i \sim \mathrm{B}(1, \pi_i)$ and $\pi = \exp(\eta_i)/(1 + \exp(\eta_i)$ for logit models or $\pi_i = \Phi(\eta_i)$ for probit models,
- *log-linear Poisson regression* where $y_i \sim \mathrm{Po}(\lambda_i)$ and $\lambda_i = \exp(\eta_i)$.

Extensive introductions to generalized linear models can, for example, be found in McCullagh and Nelder (1989), Dey, Gosh and Mallick (2000) or Fahrmeir and Tutz (2001).

Generalized additive models (GAM) extend generalized linear models by replacing the purely parametric predictor $\boldsymbol{u}_i'\boldsymbol{\beta}$ by a semiparametric predictor

$$\eta_i = \boldsymbol{u}_i'\boldsymbol{\beta} + f_1(x_{i1}) + \ldots + f_p(x_{ip})$$

that acknowledges the possible nonlinearity of some of the covariate effects by including the (smooth) functions $f_1, \ldots, f_p$ of continuous covariates $x_1, \ldots, x_p$. Several ideas for representing the unknown functions are in use, ranging from purely nonparametric kernel-based methods or local regression techniques to semiparametric basis function approaches. We will focus on the latter in this book since basis function approaches fit well into the modular, unifying framework for semiparametric regression considered here. In particular, the functions can be represented parsimoniously in terms of a relatively moderate number of parameters and the resulting models can easily be extended to include, for example, random effects or spatial effects. Chapter 2 reviews the methodology for generalized additive models that will be utilized in later parts of the book and may therefore also serve as a first introduction to this model class. More extensive reviews can be found in Ruppert, Wand and Carroll (2003) and Wood (2006) or in the classic textbook by Hastie and Tibshirani (1990). However, Chapter 2 also goes beyond these books in considering Bayesian formulations of semiparametric regression in more detail and in extending generalized additive models beyond the univariate exponential family class. More specifically, we will also consider additive regression models for categorical responses with either ordered or unordered categories.

Generalized additive models and extensions of this class have become one of the most popular tools for data analysis in the regression context. The main subject of this book are exactly such extensions for some non-standard regression situations, that is,

- semiparametric mixed models for longitudinal data comprising both random effects and nonlinear effects (Chapter 4),
- spatial regression models, further extending the predictor of semiparametric mixed models by including spatial effects or interaction surfaces (Chapter 5),
- semiparametric (mixed) models for event history data (Chapter 6).

When longitudinal data y_{ij} have been observed for subjects $i = 1, \dots, n$, at time points $t_{i1} < \dots < t_{ij} < \dots < t_{in_i}$, the additional inclusion of subject-specific random effects in generalized additive models may be required. This yields additive mixed predictors of the form

$$\eta_{ij} = \boldsymbol{u}_{ij}'\boldsymbol{\beta} + f_1(x_{ij1}) + \dots + f_p(x_{ijp}) + \boldsymbol{z}_{ij}'\boldsymbol{\alpha}_i, \quad i = 1, \dots, n, \; j = 1 \dots, n_i,$$

where $\boldsymbol{\alpha}_i$ refers to random effects of covariates $\boldsymbol{z}$ that are specific for subject i and are supplemented with appropriate random effects distributional assumptions. From a Bayesian perspective, this corresponds to the specification of a prior distribution for the subject-specific coefficients. If time t is considered as one of the continuous covariates, say $t = x_1$, then $f_1(t)$ represents a nonparametric population trend. The Bayesian perspective on longitudinal regression models will considerably facilitate the formulation of priors that relax the common assumption of a Gaussian distribution for the random effects. Chapter 3 gives a brief introduction to linear and generalized linear mixed models with linear mixed predictors $\eta_{ij} = \boldsymbol{u}_{ij}'\boldsymbol{\beta} + \boldsymbol{z}_{ij}'\boldsymbol{\alpha}_i$. It provides the necessary background knowledge for the rest of the book and also discusses some recent advances in Bayesian generalized linear mixed models, such as the consideration of finite mixture or Dirichlet process mixture priors for random effects.

Chapter 4 will bring together the modelling ideas from generalized additive models and random effects models in the class of semiparametric mixed models (SPMM), including models with additive mixed predictors as an important subclass. This chapter may be considered the central part of the book since it discusses several general questions and models that will remain important throughout the book. For example, it considers issues related to model choice and model checking, discusses relations between smoothing and correlation and details empirical Bayes and full Bayes inference.

Chapter 5 considers situations where spatial information on the collected observations is available. When observations have an explicit spatial alignment, it is quite likely that observations that are spatially close by are correlated and cannot be treated as independent. This can be accounted for by including spatial random effects or, in a Bayesian interpretation, a spatial function with prior structure chosen such that spatial correlations are induced. For cross-sectional data, this yields a geoadditive predictor structure such as

$$\eta_i = \boldsymbol{u}_i'\boldsymbol{\beta} + f_1(x_{i1}) + \ldots + f_p(x_{ip}) + f_{\text{spat}}(s_i)$$

but of course spatial effects can also be included in longitudinal extensions, leading for example to

$$\eta_{ij} = \boldsymbol{u}_{ij}'\boldsymbol{\beta} + f_1(x_{ij1}) + \ldots + f_p(x_{ijp}) + f_{\text{spat}}(s_i) + \boldsymbol{z}_{ij}'\boldsymbol{\alpha}_i.$$

The exact form of the spatial function $f_{\text{spat}}(s_i)$ depends on the form of the spatial information s_i. This information may either come in the form of point references (i.e. coordinates of the points in the observation area) or in terms of a geographical grouping variable that could, for example, indicate that a specific observation has been collected within a geographical region or district. In the latter case, additional information on the arrangements of geographical regions is required to enable the construction of a smooth spatial surface based on the discrete spatial information. Chapter 5 also discusses the estimation of interaction surfaces for flexible interaction effects between continuous covariates. Actually, this problem is closely related to spatial smoothing in the case of point-referenced observations. Much of Chapter 5 is a self-contained introduction to spatial statistics that can also be read without too much knowledge about the rest of the book. Still, of course the chapter, also discusses how spatial effects can be embedded in semiparametric mixed models.

While Chapters 2 to 5 deal with smoothing and regression for responses from either univariate or multivariate exponential families, the last chapter on event history data considers a different model structure. Here, we are interested in analysing duration times between transitions of individuals or objects within a discrete state space. The most common example is the analysis of survival times where the state space only consists of two states: 'alive' and 'dead'. Already in this situation a new difficulty arises that hinders the application of regression models considered in previous chapters. Due to the data collection process, individual trajectories are often only partially observed, that is, parts of the observation history are censored. A proper statistical analysis has to take this special data structure into account to yield valid conclusions.

A class of models that has turned out to be very useful when censoring is present in the data, relies on the specification of transition intensities, that is, regression models for hazard rates in case of continuous-time data or transition probabilities in case of discrete-time data. In both cases, time-dependence of the quantities describing the transitions has to be taken explicitly into account. For example, a hazard regression model comprising nonlinear and spatial effects may be specified as

$$\begin{aligned}\lambda_i(t) &= \exp(\eta_i(t)),\\ \eta_i(t) &= g_0(t) + \boldsymbol{u}_i'\boldsymbol{\beta} + f_1(x_{i1}) + \ldots + f_p(x_{ip}) + f_{\text{spat}}(s_i)\end{aligned}$$

where the only additional quantity is the baseline effect $g_0(t)$ that models the time-dependence of the hazard rate $\lambda_i(t)$. In extended versions, additional time-dependent effects can be added, such as time-varying effects $u_{ij}g_j(t)$ of a covariate u_{ij} or effects of time-dependent covariates $u_{ij}(t)$ and $x_{ij}(t)$ in either linear form $u_{ij}(t)\beta_j$ or as nonparametric effects $f_j(x_{ij}(t))$. Discrete-time models reduce to models for transition probabilities and therefore can be based on methods from previous chapters, at least after appropriate data augmentation. Binary regression models are appropriate to model transitions to a single absorbing state while multicategorical regression models can be used to model several competing transitions simultaneously.

1.2 Applications

In this section, we will introduce a number of examples that will serve as illustrations throughout the book. Together with the description of some of the key features of different data sets, we will also relate the predictors and model classes from the previous section to specific applications.

We start with an example on *longitudinal data* where repeated measurements are collected on the same response variable (and possibly also the covariates). Such data typically require the inclusion of correlations between repeated measures since these can hardly be considered independent. Moreover, temporal trends and/or time-varying effects may be required.

Example 1.1: Childhood obesity

The prevalence of childhood obesity has increased continuously in industrialized countries during the last decades. Assessment of childhood obesity is typically based on the body mass index (BMI) of children. The usual approach is to classify children as obese using reference charts that are constructed from age-specific upper quantiles of the BMI distribution in reference populations. To assess the influence of a child's age and risk factors on its BMI, we will apply linear mixed models (Chapter 3) and semiparametric extensions (Chapter 4) to longitudinal data from the LISA study, kindly provided by Heinz-Erich Wichmann (Helmholtz-Zentrum München, German Research Center for

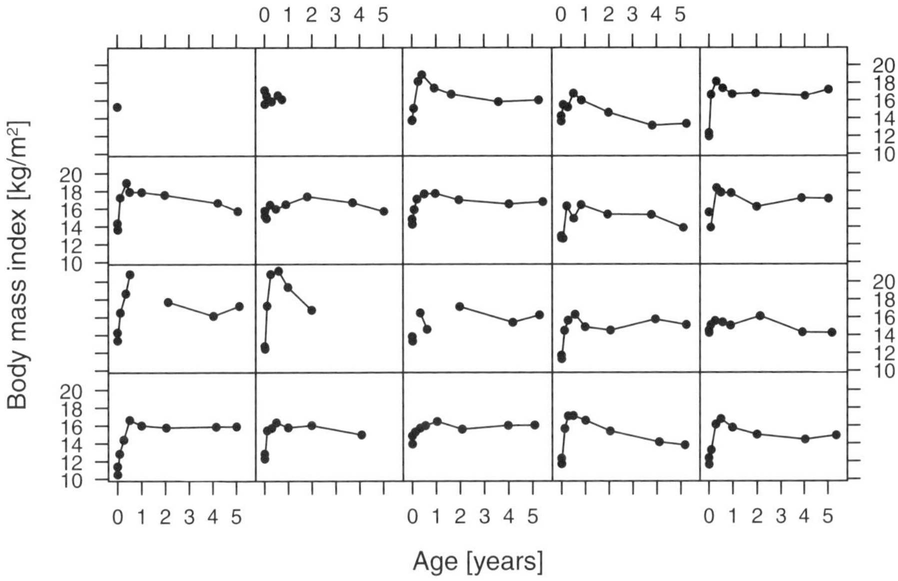

Fig. 1.1 Childhood obesity: Longitudinal BMI patterns by age of some randomly selected individuals.

Environmental Health). This is a prospective birth cohort study conducted in four cities in Germany (Bad Honnef, Leipzig, Munich, Wesel) to investigate influences of *L*ife-style factors on the development of the *I*mmune *S*ystem and *A*llergies in East and West Germany. The data set includes information on 3,097 healthy neonates born between November 1997 and January 1999. Follow-up time was until the age of six by questionnaires in connection with the nine mandatory examinations at birth and around the age of 2 weeks, 1, 3, 6, 12, 24, 48 and 60 months. Thus, the maximum number of observations per child is nine. Figure 1.1 shows individual BMI patterns by age for some randomly selected children. Figure 1.2 displays empirical BMI distributions stratified with respect to age together with empirical quantiles. For all analyses, we will restrict the data set to complete cases, resulting in a total of 17,316 observations for 2,043 children.

Table 1.1 and Table 1.2 give an overview of the continuous and categorical covariates that are available for the analyses. They cover most of the risk factors discussed in the literature, such as parental weight (maternal BMI), socioeconomic factors (sex, area and maternal education), physical activity (hours spent watching TV and playing on the computer as well as hours spent outdoors), nutrition (breastfeeding) and rapid growth factors, such as weight gain until the age of two years, maternal weight gain during pregnancy and maternal prenatal smoking. All covariates except for age are considered as being time constant.

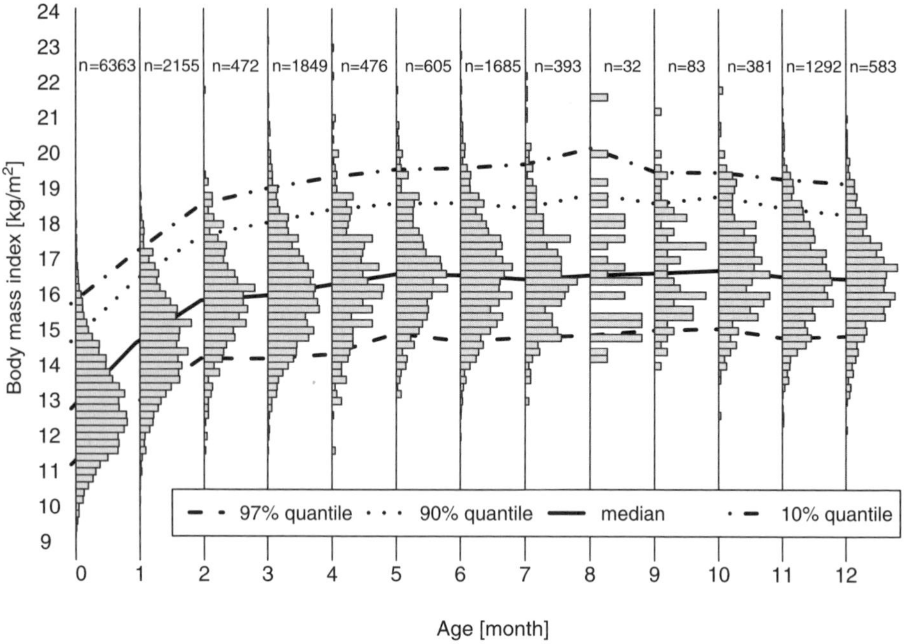

Fig. 1.2 Childhood obesity: Empirical BMI distributions stratified with respect to age together with empirical quantiles.

Table 1.1 Childhood obesity: Description of continuous covariates for 17,316 complete cases from 2,043 children. Age is a time-dependent covariate, all other covariates are considered as being time constant.

Covariate	Unit	Median	Mean	Sd.
age	years	0.52	1.39	1.76
weight gain until the age of 2 years	kg	8.84	8.91	1.29
hours spent outdoors at the age of 4 years	hours per day	3.50	3.46	1.15
maternal BMI at pregnancy begin	kg/m^2	21.72	22.58	3.73
maternal BMI gain during pregnancy	kg/m^2	4.96	5.12	1.63

Analyses of this longitudinal data will be based on linear or additive mixed models of the form

$$y_{ij} = \boldsymbol{u}'_{ij}\boldsymbol{\beta} + f(t_{ij}) + \boldsymbol{z}'_{ij}\boldsymbol{\alpha}_i + \varepsilon_{ij}, \; i = 1, \dots, n, \; j = 1, \dots, n_i,$$

where y_{ij} is the body mass index of child i at age t_{ij}, $\boldsymbol{u}_{ij}$ are the values of (a subset of) the covariates in Tables 1.1 and 1.2, and ε_{ij} are i. i. d. Gaussian errors. Figures 1.1 and 1.2 suggest that age t has a nonlinear effect on the response y.

Table 1.2 Childhood obesity: Description of categorical covariates. Frequencies and percentages are related to complete cases of 2,043 children.

Covariate	Categories	Frequency(%)	No.
sex	0 = female	52.8	1,079
	1 = male	47.2	964
nutrition until the age of 4 months	0 = bottle- and/or breastfeeding	40.5	828
	1 = breastfeeding only	59.5	1,215
hours spent watching TV and playing computer at 4 years	1 = less than 1 hour	72.1	1,472
	2 = 1 up to 2 hours	26.6	543
	3 = more than 3 hours	1.3	28
maternal smoking during pregnancy	0 = no	86.0	1,756
	1 = yes	14.0	287
highest level of maternal education	1 = no degree or Hauptschule (CSE)	6.5	133
	2 = Realschule (secondary school)	36.1	737
	3 = Fachabitur (subject specific high school diploma)	19.0	389
	4 = Abitur (high school diploma)	38.4	784
area	0 = rural (Bad Honnef, Wesel)	21.5	439
	1 = urban (Leipzig, Munich)	78.5	1,604

We will therefore model the nonlinear population trend $f(t)$ parametrically in Section 3.3 and semiparametrically (as a P-spline) in Chapter 4. The random effects term $\boldsymbol{z}'_{ij}\boldsymbol{\alpha}_i$ will be used to model individual-specific deviations of individual age effects from the population effect $f(t_{ij})$. Previous analyses with such models are described, for example, in Fenske *et al.* (2008) and Heinzl, Kneib and Fahrmeir (2009). △

A further stage of complexity arises when, additionally, longitudinal data are spatially aligned in addition. In this case, both temporal and spatial correlations may be present and have to be taken into account simultaneously.

Example 1.2: Forest health

Vital forests are an important prerequisite for a working ecosystem, in particular due to their regulatory impact on climate and the water cycle. Data and statistical analyses focussing on forest health are therefore required to gain more information on the determinants of forest health. Here we are considering data collected in annual visual forest health inventories carried out between 1983 and 2004 in a northern Bavarian forest district (exhibiting 10 km from south to north and 15 km from west to east). The data have both a longitudinal and a spatial dimension since the health status of beeches at the same observation plots is assessed repeatedly over time.

In this example, forest health is measured in terms of a categorical response variable reflecting the degree of defoliation ranging from healthy trees (0% defoliation) to trees with 100% defoliation in 12.5% steps. The frequencies of the resulting nine categories are visualized in Fig. 1.3 and indicate that very high amounts of defoliation are actually quite rare. We will therefore consider only aggregated variants of the original response, coded either into a binary response variable for the presence of any defoliation or a categorical response variable with three ordered levels corresponding to no defoliation (0%), medium defoliation (12.5–37.5%) and severe defoliation (more than 37.5%). Figure 1.3 visualizes the temporal trends of the frequencies of the three categories. These trends indicate a long-term increase in the number of severely defoliated trees, although the corresponding frequency is generally low. For medium defoliation, the trend function has a maximum at the end of the 1980s followed by a more or less constant period until the end of the observation period.

Figure 1.4 visualizes both the spatial distribution of the plots within the observation area and the percentage of time points for which a tree was classified to be damaged. Although it is hard to identify a clear spatial pattern, it still seems to be recommended to include the spatial information to account for possible spatial correlation and for unobserved spatial heterogeneity in the data.

In addition to temporal and spatial information, a number of covariates have been collected in each year, see Table 1.3 for a summary. Several of these covariates also account for the temporal and spatial alignment of the data and we are particularly interested in the question of whether temporal and spatial correlations remain unexplained after accounting for the effects of covariates. To take full advantage of the information contained in the covariates, we will use

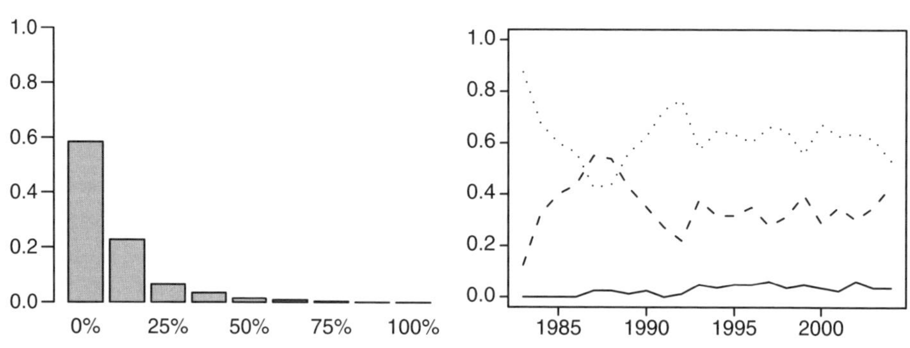

Fig. 1.3 Forest health: Barplot of the distribution of the original defoliation degree in nine categories (left panel) and empirical time trends for the three response categories (right panel, dotted line = no defoliation, dashed line = medium defoliation, solid line = severe defoliation).

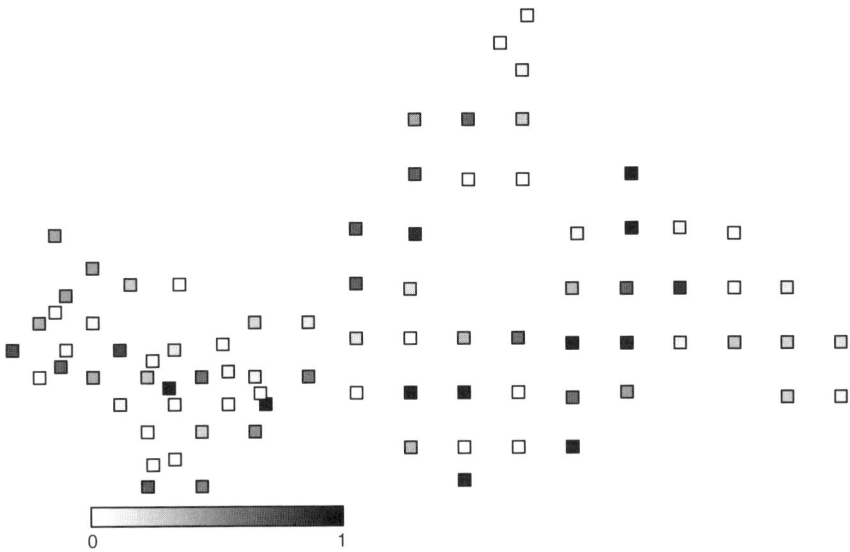

Fig. 1.4 Forest health: Fraction of time points for which an observation plot was classified to be damaged. Dark colour corresponds to trees that were damaged for most of the time points.

Table 1.3 Forest health: Description of covariates.

Covariate	Description
age	age of the stand in years (continuous, $7 \leq \text{age} \leq 234$)
time	calendar time (continuous, $1983 \leq \text{time} \leq 2004$)
elevation	elevation above sea level in metres (continuous, $250 \leq \text{elevation} \leq 480$)
inclination	inclination of slope in per cent (continuous, $0 \leq \text{inclination} \leq 46$)
soil	depth of soil layer in centimetres (continuous, $9 \leq \text{soil} \leq 51$)
ph	ph-value in 0–2 cm depth (continuous, $3.28 \leq \text{ph} \leq 6.05$)
canopy	density of forest canopy in per cent (continuous, $0 \leq \text{canopy} \leq 1$)
stand	type of stand (categorical, 1=deciduous forest, -1=mixed forest).
fertilization	fertilization (categorical, 1=yes, -1=no).
humus	thickness of humus layer in 5 categories (ordinal, higher categories represent higher proportions).
moisture	level of soil moisture (categorical, 1=moderately dry, 2=moderately moist, 3=moist or temporarily wet).
saturation	base saturation (ordinal, higher categories indicate higher base saturation).

non- and semiparametric modelling techniques to overcome the limitations of strictly linear models. In summary, this yields predictor structures like

$$\eta_{it} = \boldsymbol{u}_{it}'\boldsymbol{\beta} + f_1(age_{it}) + f_2(t) + f_{1,2}(age_{it}, t) + f_{\text{spat}}(s_i) + \alpha_{s_i}$$

where f_1 and f_2 are smooth functions of age and calendar time, $f_{1,2}$ captures an additional interaction effect between age and time, f_{spat} is a spatial function based on the location information and α_s is an additional plot-specific random effect that accounts for the longitudinal structure of the data. In analyses of the categorical response, additional category-specific effects may (and will) be considered.

The forest health data will serve as a running example throughout almost the whole book. In particular, we will illustrate smoothing approaches and random effects models for categorical data in Chapters 2, 3 and 4. In Chapter 5, we will further investigate the presence of spatial correlation in the forest health data. Previous analyses of the same data are, for example, contained in Kneib and Fahrmeir (2010), Kneib and Fahrmeir (2006), or Fahrmeir and Lang (2001b). △

Spatial information can of course not only be present in longitudinal data but also in cross-sectional data structures. As in the case of longitudinal data, combinations of spatial effects and flexible nonlinear effects may still be required to obtain an adequate description of the relation between covariates and responses.

Example 1.3: Childhood malnutrition in developing countries
Malnutrition of wide parts of the population, and in particular of children, is one of the most urgent public health challenges in developing and transition countries. It is not only related to the growth of children but also has severe long term impacts. For example, Caulfield *et al.* (2004) estimated that malnutrition is an underlying cause for about 53% of child deaths worldwide. As a consequence, both the reduction of (childhood) malnutrition and childhood mortality are among the United Nations Millennium Development Goals.

A rich database on health and population trends in developing countries is provided by the Demographic and Health Surveys (DHS, www.measuredhs.com) consisting of more than 200 surveys conducted in 75 countries that are available free of charge for research purposes. DHS data sets provide nationally representative information on fertility, family planning, maternal and child health, as well as child survival, HIV/AIDS, malaria and nutrition. In this book, we will use information from the 1998/99 survey for India and the 2003 survey for Nigeria. The Indian data will be analysed with respect to childhood malnutrition, while we will focus on childhood mortality and in particular survival times in the Nigerian data.

Childhood malnutrition is commonly assessed in terms of a Z-score relating an anthropometric indicator of the child to values obtained from a reference population. Consequently, the Z-score is defined as

$$Z_i = \frac{\mathrm{AI}_i - \mu_{\mathrm{AI}}}{\sigma_{\mathrm{AI}}}$$

where AI_i denotes the value of the anthropometric measurement for child i, while μ_{AI} and σ_{AI} correspond to median and standard deviation in the reference population (stratified with respect to some covariates such as age and gender). Depending on the choice of the anthropometric indicator, different aspects of malnutrition can be assessed. Insufficient weight for height captures acute malnutrition (wasting) while insufficient height for age is an indicator for chronic malnutrition (stunting). We will focus on the latter in this book.

A large number of potential covariates is contained in the database available from the DHS. Table 1.4 gives a summary of variables extracted from this database that will be investigated for their impact on childhood malnutrition. The India data will serve as an illustration of additive models in Chapter 2 and geoadditive models in Chapter 5. In the latter case, the predictor will be of the form

$$\begin{aligned}\eta_i = \boldsymbol{u}_i'\boldsymbol{\beta} &+ f_1(cage_i) + f_2(breastfeeding_i) + f_3(mbmi_i) + f_4(mage_i) \\ &+ f_5(medu_i) + f_6(edupartner_i) + f_{\mathrm{spat}}(s_i) + \alpha_{s_i}\end{aligned}$$

where f_1 to f_6 are smooth functions of the covariates age of the child (cage), body mass index of the mother (mbmi), age of the mother (mage), education years of the mother (medu), and education years of the mother's partner (edupartner). The spatial function f_{spat} and the additional district-specific random effect α_s represent different types of spatial effects of the district the individual observation pertains to. While f_{spat} captures spatially correlated effects, α_s captures localized, small scale spatial variation. An analysis of a similar data set obtained from the 2006 India DHS (that does not contain spatial information and is therefore not considered in this book) can be found in Fenske, Kneib and Hothorn (2009) who use quantile regression techniques to go beyond usual mean regression models when investigating childhood malnutrition. △

Example 1.4: Munich rental guide

According to German law, increases in rents for flats can be justified based on 'average rents' paid for flats that are comparable in size, equipment, quality and location. Therefore, most larger German cities publish rental guides that provide information on average rents obtained from regression models with net rents or net rents per square metre as dependent variables. In this example, we analyse data from the 2007 rental guide for Munich, consisting of information on approximately 3,000 flats in the city of Munich collected by Infratest Sozialforschung. From the roughly 250 variables contained in the questionnaire, 238 categorical covariates can be extracted that describe characteristics of the flats, such as the quality of bathroom equipment, whether the flat is rented for the first time, the

Table 1.4 Malnutrition data: Description of covariates.

Covariate	Description
cweight	weight of the child in kg (continuous, $2.5 \leq$ cweight ≤ 20.5)
cheight	height of the child in cm (continuous, $49.0 \leq$ cheight ≤ 112.5)
stunting	stunting (height-for-age) according to the WHO definition (continuous, $-599 \leq$ stunting ≤ 597)
cbmi	BMI of the child in kg/m^2 (continuous, $10 \leq$ cbmi ≤ 25.98)
cage	age of the child in months (continuous, $0 \leq$ cage ≤ 35)
breastfeeding	months of breastfeeding (continuous, $0 \leq$ breastfeeding ≤ 35)
csex	sex of the child (categorical, 1=male, 2=female)
ctwin	child is twin (categorical, 0=no, 1=yes)
cbirthorder	birth order number (ordinal, 1=1, 2=2, 3=3, 4=4, 5=5 or higher)
mbmi	BMI of the mother in kg/m^2 (continuous, $12.45 \leq$ mbmi ≤ 39.85)
mweight	weight of the mother in kg (continuous, $27.9 \leq$ mweight ≤ 106.9)
mheight	height of the mother in cm (continuous, $127.2 \leq$ mheight ≤ 184.5)
mage	age of the mother in years (continuous, $13 \leq$ mage ≤ 49)
medu	education of the mother in single years (continuous, $0 \leq$ medu ≤ 22)
edupartner	education of the mother's partner in single years (continuous, $0 \leq$ edupartner ≤ 22)
munemployed	mother is unemployed (categorical, 0=yes, 1=no)
mreligion	religion of the mother (categorical, 1=Hindu, 2=Muslim, 3= Christian, 4=Sikh, 5=Other)
mresidence	type of place of residence (categorical, 1=urban, 2=rural)
deadchildren	number of dead children (ordinal, 0=0, 1=1, 2=2, 3=3 or more)
electricity	availability of electricity (categorical, 0=no, 1=yes)
radio	ownership of a radio (categorical, 0=no, 1=yes)
television	ownership of a television (categorical, 0=no, 1=yes)
refrigerator	ownership of a refrigerator (categorical, 0=no, 1=yes)
bicycle	ownership of a bicycle (categorical, 0=no, 1=yes)
motorcycle	ownership of a motorcycle (categorical, 0=no, 1=yes)
car	ownership of a car (categorical, 0=no, 1=yes)
distH	district (categorical, $5 \leq$ distH ≤ 482, 438 districts)

presence of a garden or a balcony, etc. In addition, the continuous covariates, size of the flat and age of the building, in which the flat is located have been recorded. Moreover, spatial information on the subquarter of Munich where the flat is located is available (see Fig. 1.5 for a map of the subquarters).

Earlier analyses have shown that both size and year of construction have a nonlinear impact on the net rent and that spatial variation remains unexplained

Fig. 1.5 Munich rental guide: Map of the subquarters in Munich.

even after accounting for covariate information describing the quality of a flat. Hence, we will use a geoadditive regression model with predictor

$$\eta_i = \boldsymbol{u}_i'\boldsymbol{\beta} + f_1(size_i) + f_2(year_i) + f_{\text{spat}}(s_i)$$

to predict net rents in Chapter 5. As before, f_1 and f_2 are smooth functions of the covariates size of the flat (size) and year of construction (year) while f_{spat} is a spatial function defined on the subquarters of Munich. However, in contrast to previous analyses, the vector $\boldsymbol{u}$ formed of the categorical covariates is high-dimensional, and we will therefore also supplement the corresponding regression coefficient with informative regularization priors to ensure stable estimation of their effects. The analysis presented in Chapter 5 is based on results reported in Kneib, Konrath and Fahrmeir (2010b) and Fahrmeir, Kneib and Konrath (2010). △

A different data structure arises when the response variable of interest corresponds to a duration time. New types of models are then required for the following reasons: First of all, duration times are intrinsically nonnegative, which should be accounted for in the model formulation. Secondly, time-dependence should be included at different stages of the model formulation. For example, covariates may be time-varying or effects of covariates may change over time. Thirdly, duration times can often not be completely observed, leading to the problem of censored data. A class of models that allows to one to incorporate all these three issues is hazard regression models, where the hazard rate or intensity for an event of interest is specified in terms of a regression model. Again, nonlinear and spatial effects as well as a number of extensions may be required in these models.

Fig. 1.6 Coronary artery bypass graft: Observation area in London and Essex.

Example 1.5: Coronary artery bypass graft
This data set contains information on 3,015 patients with definite coronary artery bypass graft (CABG) who were referred to a single cardiothoracic unit from five contiguous health authorities. Waiting times from angiography to CABG are available in days such that a continuous time analysis is appropriate. Given covariate information includes the gender of the patient, the age in years and the number of diseased vessels (1, 2 or 3). In addition, information on the residence of the patients is available in terms of one of 488 electoral wards (see Fig. 1.6 for a graphical representation of the observation area). In addition to nonlinear effects of the continuous covariate age and spatial effects, particular interest in this application is on the time-varying effect of the number of diseased vessels. Hence, a suitable hazard rate regression model may be formulated as

$$\lambda_i(t) = \exp(\eta_i(t)),$$
$$\eta_i(t) = \beta_0 + \beta_1 \text{sex}_i + g_0(t) + \text{DV}_{i2} g_1(t) + \text{DV}_{i3} g_2(t) + f_1(\text{age}_i) + f_{\text{spat}}(s_i) + \alpha_{s_i}$$

where $\lambda_i(t)$ is the hazard rate corresponding to the waiting time for CABG, $g_0(t)$ represents the baseline hazard, $g_1(t)$ and $g_2(t)$ are time-varying effects corresponding to the indicators DV_2 and DV_3 for the number of diseased vessels (two for DV_2 and three for DV_3), $f_1(\text{age})$ is a nonparametric effect of age, $f_{\text{spat}}(s_i)$ and α_{s_i} correspond to spatially structured and unstructured effects and β_1 is the gender difference.

We use the CABG data to illustrate Bayesian inference in continuous-time geoadditive survival models in Chapter 6. The data were originally analysed by Crook, Knorr-Held and Hemingway (2003) based on discrete-time methodology and in Hennerfeind, Brezger and Fahrmeir (2006) using geoadditive hazard regression. △

Example 1.6: Childhood mortality in developing countries
Similarly to the Indian data in Example 1.3, this data set comprises information obtained from the DHS database on Nigeria collected in 2003. A similar set of covariates as in Example 1.3 will be related to the survival time of children in the applications presented in Chapter 6, utilizing both discrete-time and continuous-time models. For the continuous-time model, we consider the hazard rate specification

$$\begin{aligned}\lambda_i(t) = \exp\big[&g_0(t) + f_1(\text{bmi}_i) + f_2(\text{age}_i) + f_3(\text{bord}_i) + f_4(\text{size}_i)\\ &+f_{\text{spat}}(s_i) + \boldsymbol{u}_i(t)'\boldsymbol{\beta}\big]\end{aligned}$$

where g_0 represents the baseline intensity, f_1 to f_4 are smooth functions of the body mass index of the mother (bmi), the age of the mother (age), the position of the child in the birth order (bord) and the size of the household (size) and f_{spat} represents a smooth spatial effect based on district information. Previous analyses of the Nigeria data are for example contained in Adebayo and Fahrmeir (2005) and Kneib (2006). △

Example 1.7: Human sleep
Human sleep is a complex biological process and its statistical analysis poses several challenging difficulties. In this example, we consider data collected at the Max-Planck Institute for Psychiatry in Munich as a part of a larger study on sleep withdrawal. We will restrict our attention to a part of the data collected to obtain a reference standard of the sleeping behaviour of the study participants at the beginning of the study. For 70 participants, the sleep process is recorded by electroencephalographic (EEG) measurements which are afterwards classified into a discrete set of sleep stages. For example, in a simple three-state classification, we may differentiate between the states awake, Non-REM and REM. In addition to sleep states obtained from the EEG measurements every 30 seconds throughout the night, blood samples are taken from the patients approximately every 10 minutes, providing measurements on the nocturnal secretion of certain hormones such as cortisol. For each of the participants, there is information available for exactly one night.

Figure 1.7 shows the realizations of sleep processes for two exemplary participants together with the nocturnal secretion of cortisol. Questions of interest are, for example, as follows:

- What is the temporal development of transition intensities between the three considered states?

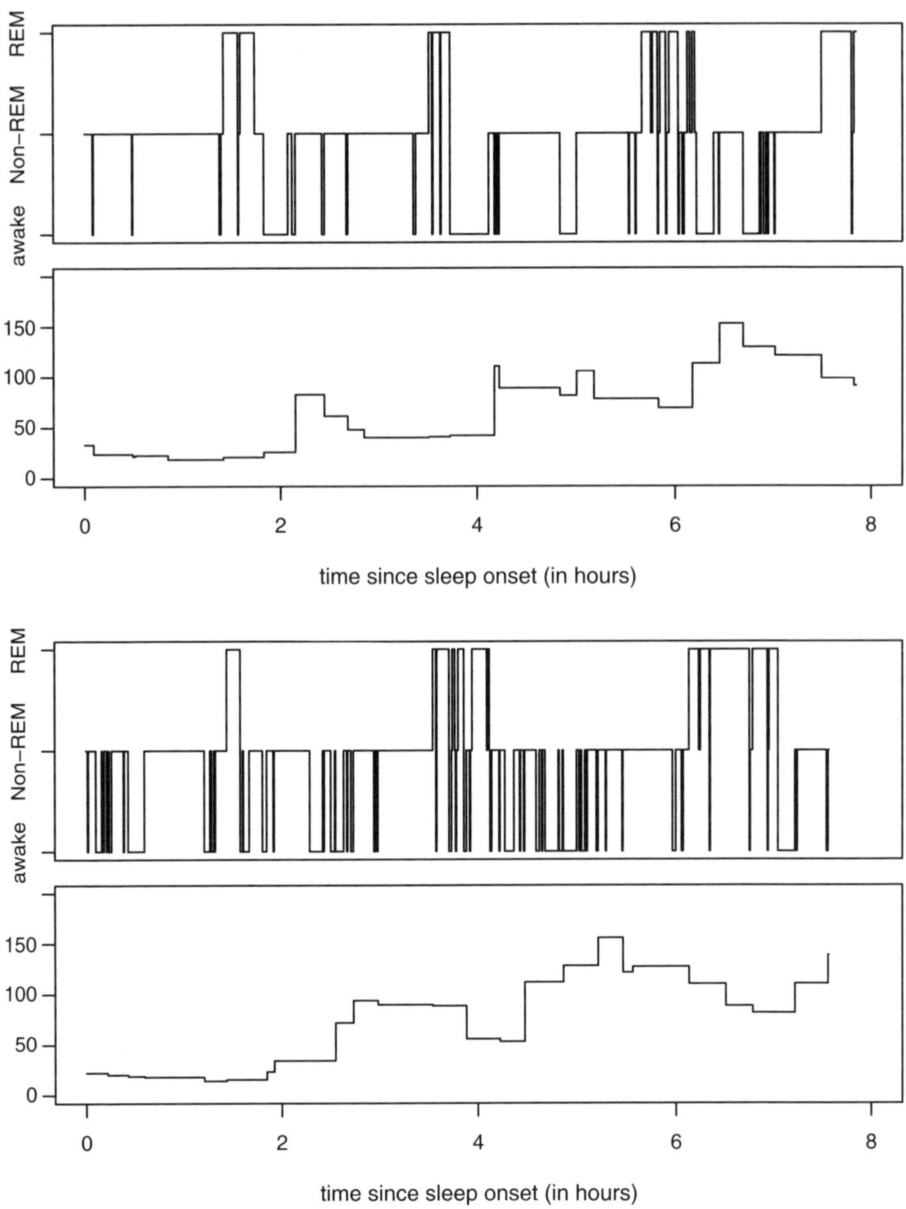

Fig. 1.7 Human sleep: Realization of two sleep processes (first and third panel) together with the corresponding nocturnal secretion of cortisol (second and fourth panel).

- Are these transition intensities impacted by the nocturnal secretion of hormones? For example, it has been hypothesized that increased secretion of cortisol is associated with a higher propensity for the transition from Non-REM to REM state.
- Are there individual-specific differences between the transition intensities?

These and some further questions will briefly be investigated in the context of time series smoothing in Chapter 2 and more extensively based on transition intensity models in Chapter 6. For example, we will consider the following model for the transition intensity between REM and Non-REM sleep phases:

$$\lambda_{RN,i}(t) = \exp\left[g_0^{(RN)}(t) + c_i(t)g_1^{(RN)}(t) + \alpha_i^{(RN)}\right].$$

In this model, g_0 corresponds to a baseline intensity for this transition while g_1 captures a time-varying effect due to the level of cortisol reflected in the binary variable c that takes the value one if the level is high and zero otherwise. The additional individual-specific random effect α captures individual sleeping behaviour such as an overall increased tendency for the transition of interest. Our analyses in Chapter 6 will be based on Kneib and Hennerfeind (2008) and Kalus *et al.* (2009). △

2 Basic Concepts for Smoothing and Semiparametric Regression

This chapter reviews basic concepts for smoothing and semiparametric regression based on roughness penalties or—from a Bayesian perspective—corresponding smoothness priors. In particular, we will introduce several tools for statistical modelling and inference that will be utilized in later chapters. Moreover, we will highlight the close relation between frequentist penalized likelihood approaches and Bayesian inference based on smoothness priors.

Section 2.1 considers the classical smoothing problem for time series of Gaussian and non-Gaussian observations. We start with Whittaker's (1923) 'method of graduation' for estimating trends in Gaussian time-series data based on appropriate difference penalties and show its close relation to the Bayesian posterior mean estimate based on corresponding random-walk smoothness priors. Several extensions and modifications of this basic model, such as correlated errors, nonequidistant observations or locally adaptive approaches are also briefly discussed. For non-Gaussian time series, penalized likelihood estimates can still be derived as posterior mode estimates based on difference penalties or random-walk smoothness priors, but the correspondence between posterior mean and penalized likelihood estimates is lost.

Based on the considerations from Section 2.1, Section 2.2 introduces penalized splines and their Bayesian counterpart as a computationally and conceptually attractive alternative to random-walk priors. Penalized least squares and penalized likelihood inference for fixed smoothing parameters will be discussed extensively for Gaussian data as well as data from univariate and categorical non-Gaussian distributions. Fundamental concepts of Markov chain Monte Carlo simulations, including the determination of smoothing parameters, will be introduced to be available for later consideration. Moreover, the relations to other smoothing approaches will be sketched.

Section 2.3 extends the univariate smoothing approaches to additive and generalized additive models. While classical approaches such as backfitting are mentioned, we will focus on low-rank smoothers, in particular P-splines, that enable penalized likelihood or empirical Bayes estimation and Markov chain Monte Carlo estimation techniques for full Bayes inference.

2.1 Time series smoothing

2.1.1 Gaussian observation models

In the classical smoothing problem treated by Whittaker (1923), equidistant time series observations $\boldsymbol{y} = (y_1, \ldots, y_T)'$ are assumed to be the sum

$$y_t = \gamma_t + \varepsilon_t, \qquad t = 1, \ldots, T, \tag{2.1}$$

of a smooth trend $\boldsymbol{\gamma} = (\gamma_1, \ldots, \gamma_T)'$ and an irregular component of independent noise variables $\boldsymbol{\varepsilon} = (\varepsilon_1, \ldots, \varepsilon_T)'$ with $\mathbb{E}(\varepsilon_t) = 0$.

Whittaker suggested to estimate $\boldsymbol{\gamma}$ by minimizing the *penalized least squares* (PLS) criterion

$$\text{PLS}(\boldsymbol{\gamma}) = \sum_{t=1}^{T} (y_t - \gamma_t)^2 + \lambda \sum_{t=3}^{T} (\gamma_t - 2\gamma_{t-1} + \gamma_{t-2})^2 \tag{2.2}$$

with respect to $\boldsymbol{\gamma}$. Minimization of the PLS criterion balances two desirable criteria: a close fit to the data, as measured by the first term in (2.2), and smoothness of the trend estimate, as measured by the second term, a roughness penalty

$$\text{pen}(\boldsymbol{\gamma}) = \sum_{t=3}^{T} (\gamma_t - 2\gamma_{t-1} + \gamma_{t-2})^2 = \sum_{t=3}^{T} (\Delta^2 \gamma_t)^2$$

given by the sum of squared second order differences of the parameter sequence $\Delta^2 \gamma_t = (\gamma_t - 2\gamma_{t-1} + \gamma_{t-2})$. The smoothing parameter λ controls the trade-off between these two goals. Tacitly it is assumed that the errors are independent and have identical variance $\text{Var}(\varepsilon_t) = \sigma^2$; otherwise the criterion has to be modified appropriately.

Because $\text{pen}(\boldsymbol{\gamma}) = 0$ for a global linear trend $\gamma_t = a + bt$, $t = 1, \ldots, T$, the roughness penalty does not penalize a straight line but it rather penalizes local deviations from such a linear trend. Instead of second order differences, we might also use first or higher order differences

$$\begin{aligned} \Delta^1 \gamma_t &= \gamma_t - \gamma_{t-1}, \\ \Delta^d \gamma_t &= \Delta^{d-1} \gamma_t - \Delta^{d-1} \gamma_{t-1}, \qquad d = 2, 3, \ldots \end{aligned}$$

to define general roughness penalties

$$\text{pen}(\boldsymbol{\gamma}) = \sum_{t=d+1}^{T} (\Delta^d \gamma_t)^2.$$

For $d = 1$, $\text{pen}(\boldsymbol{\gamma})$ penalizes deviations from a horizontal line $\gamma_t = a$, and, generally, a penalty with squared differences of order d penalizes deviations from a global polynomial of degree $d - 1$.

Example 2.1: Global warming

To investigate the presence and magnitude of global temperature changes during the last decades and centuries, several climate reconstructions have been carried out to enable the comparison of the currently increasing temperatures to historic natural variability in average temperatures. Such climate reconstructions are typically based on one or more proxy variables, such as tree-ring data or lake and ocean sediments, and reconstruct annual average temperatures either globally or for parts of the globe such as the Northern hemisphere.

As an example for equally-spaced time series, we consider the climate reconstruction for yearly average Northern hemisphere temperatures of Moberg *et al.* (2005) visualized in Fig. 2.1. The data shows a large-scale trend that makes it suitable for time-series smoothing (although the assumption of independent errors maybe debatable). Note that our aim is not to answer questions related to climate change, which probably is more a question of the methods employed to reconstruct previous temperatures than a question of statistical methods applied to the resulting time series afterwards. Instead, we aim at recovering the long-term trend, separating it from short-term random fluctuations. △

Penalized least-squares smoothing

To derive the PLS estimate of γ, it is convenient to rewrite the PLS criterion in matrix notation. Defining the first order difference matrix of dimension $(T-1)\times T$

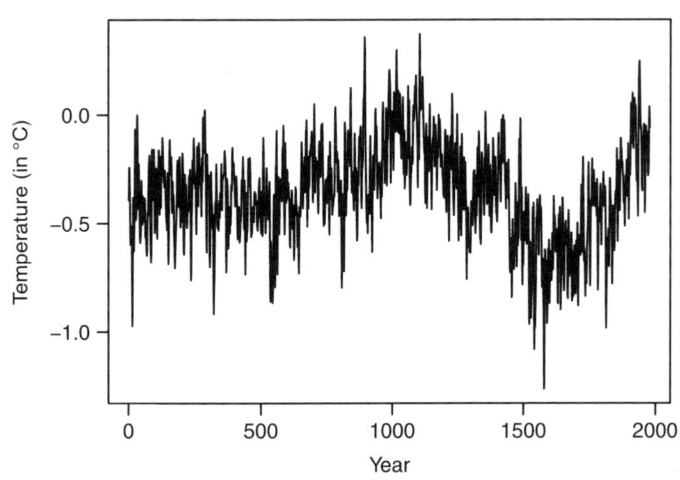

Fig. 2.1 Global warming: Yearly averaged northern hemisphere temperature reconstruction for the last 2000 years.

$$D_1^{(T-1)\times T} = \begin{pmatrix} -1 & 1 & & & 0 \\ & -1 & 1 & & \\ & & \ddots & \ddots & \\ 0 & & & -1 & 1 \end{pmatrix}. \tag{2.3}$$

we obtain the vector of first order differences as

$$D_1^{(T-1)\times T}\gamma = \begin{pmatrix} \gamma_2 - \gamma_1 \\ \vdots \\ \gamma_T - \gamma_{T-1} \end{pmatrix}.$$

Higher order differences can be expressed through higher order difference matrices which are defined recursively as

$$D_d^{(T-d)\times T} = D_1^{(T-d)\times(T-d+1)} D_{d-1}^{(T-d+1)\times T}.$$

For example, for second order differences ($d = 2$), we obtain the $(T-2)\times T$ difference matrix

$$D_2^{(T-2)\times T} = \begin{pmatrix} 1 & -2 & 1 & & & 0 \\ & 1 & -2 & 1 & & \\ & & \ddots & \ddots & \ddots & \\ 0 & & & 1 & -2 & 1 \end{pmatrix}. \tag{2.4}$$

In the following, we will suppress the explicit notation of dimensions on the matrices in order to keep the expressions simple.

Roughness penalties of order d can now be expressed as quadratic forms

$$\text{pen}(\gamma) = \sum_{t=d+1}^{T} (\Delta^d \gamma_t)^2 = \gamma' D_d' D_d \gamma = \gamma' K_d \gamma$$

with $T \times T$-penalty matrices $K_d = D_d' D_d$. Such quadratic penalties will be of high relevance throughout the book and will reappear in several different situations, such as the estimation of random effects, functional effects or spatial smoothing. Appropriate smoothness conditions are then expressed in terms of different penalty matrices K.

For first and second order difference penalties, the penalty matrices are given by

$$K_1 = \begin{pmatrix} 1 & -1 & & & 0 \\ -1 & 2 & -1 & & \\ & \ddots & \ddots & \ddots & \\ & & -1 & 2 & -1 \\ 0 & & & -1 & 1 \end{pmatrix}, \quad K_2 = \begin{pmatrix} 1 & -2 & 1 & & & & 0 \\ -2 & 5 & -4 & 1 & & & \\ 1 & -4 & 6 & -4 & 1 & & \\ & \ddots & \ddots & \ddots & \ddots & \ddots & \\ & & 1 & -4 & 6 & -4 & 1 \\ & & & 1 & -4 & 5 & -2 \\ 0 & & & & 1 & -2 & 1 \end{pmatrix}. \tag{2.5}$$

Note that the penalty matrices do not have full rank: It is easily seen that

$$\operatorname{rank}(\boldsymbol{K}_d) = T - d,$$

reflecting the fact that unpenalized polynomials of degree $d-1$ form the null space of $\boldsymbol{K}_d$. Omitting the index d on $\boldsymbol{K}$ for notational simplicity, the PLS criterion (2.2) becomes

$$\begin{aligned} \operatorname{PLS}(\boldsymbol{\gamma}) &= (\boldsymbol{y}-\boldsymbol{\gamma})'(\boldsymbol{y}-\boldsymbol{\gamma}) + \lambda\boldsymbol{\gamma}'\boldsymbol{K}\boldsymbol{\gamma} \\ &= \boldsymbol{y}'\boldsymbol{y} - 2\boldsymbol{\gamma}'\boldsymbol{y} + \boldsymbol{\gamma}'(\boldsymbol{I}+\lambda\boldsymbol{K})\boldsymbol{\gamma}. \end{aligned} \tag{2.6}$$

Setting the first derivative

$$\frac{\partial \operatorname{PLS}(\boldsymbol{\gamma})}{\partial \boldsymbol{\gamma}} = -2\boldsymbol{y} + 2(\boldsymbol{I}+\lambda\boldsymbol{K})\boldsymbol{\gamma}$$

to zero and solving for $\boldsymbol{\gamma}$ yields the PLS smoother

$$\hat{\boldsymbol{\gamma}}_{\text{PLS}} = (\boldsymbol{I}+\lambda\boldsymbol{K})^{-1}\boldsymbol{y}. \tag{2.7}$$

It has the structure of a ridge or quadratic shrinkage estimator, shrinking observations $\boldsymbol{y}$ towards a smooth trend $\boldsymbol{\gamma}$. In the frequentist setting of PLS smoothing, this trend is unknown but deterministic or 'fixed'. Since $\mathbb{E}(\boldsymbol{\varepsilon}) = \boldsymbol{0}$ and $\mathbb{E}(\boldsymbol{y}) = \boldsymbol{\gamma}$, we have

$$\mathbb{E}(\hat{\boldsymbol{\gamma}}_{\text{PLS}}) = (\boldsymbol{I}+\lambda\boldsymbol{K})^{-1}\boldsymbol{\gamma} \tag{2.8}$$

so that $\hat{\boldsymbol{\gamma}}_{\text{PLS}}$ is biased with

$$\operatorname{Bias}(\hat{\boldsymbol{\gamma}}_{\text{PLS}}) = (\boldsymbol{I}+\lambda\boldsymbol{K})^{-1}\boldsymbol{\gamma} - \boldsymbol{\gamma} = [(\boldsymbol{I}+\lambda\boldsymbol{K})^{-1} - \boldsymbol{I}]\boldsymbol{\gamma}$$

which is non-zero unless $\lambda = 0$.

Assuming i.i.d. errors with covariance matrix $\operatorname{Cov}(\boldsymbol{\varepsilon}) = \sigma^2\boldsymbol{I} = \operatorname{Cov}(\boldsymbol{y})$, we obtain

$$\begin{aligned} \operatorname{Cov}(\hat{\boldsymbol{\gamma}}_{\text{PLS}}) &= (\boldsymbol{I}+\lambda\boldsymbol{K})^{-1}\operatorname{Cov}(\boldsymbol{y})(\boldsymbol{I}+\lambda\boldsymbol{K})^{-1} \\ &= \sigma^2(\boldsymbol{I}+\lambda\boldsymbol{K})^{-2} \end{aligned} \tag{2.9}$$

as the covariance matrix of the PLS smoother. Though being biased, PLS estimates are typically considered with the hope of reducing the MSE due to a smaller variance of the estimate. Of course, this conjecture crucially depends on a suitably determined value for the smoothing parameter λ, which therefore is an integral part of the model-fitting procedure. More generally, we observe a common problem in statistics, the so-called *bias variance trade-off*: Choosing small values of the smoothing parameter leads to unbiased but typically highly variable estimates for the trend. To be more specific, in the situation of time-series smoothing a smoothing parameter $\lambda = 0$ would lead to a simple interpolation of the observed data, which is of course not a desirable trend estimate. On the other hand, choosing large values of the smoothing parameter induces very little variation of the estimated trend but may result in high bias.

Under the additional assumption of Gaussian i.i.d. errors $\varepsilon_t \sim \mathrm{N}(0, \sigma^2)$, model (2.1) becomes

$$\boldsymbol{y} = \boldsymbol{\gamma} + \boldsymbol{\varepsilon}, \qquad \boldsymbol{\varepsilon} \sim \mathrm{N}(\boldsymbol{0}, \sigma^2 \boldsymbol{I}), \tag{2.10}$$

and the PLS smoother is also Gaussian, that is

$$\hat{\boldsymbol{\gamma}}_{\mathrm{PLS}} \sim \mathrm{N}\left((\boldsymbol{I} + \lambda \boldsymbol{K})^{-1} \boldsymbol{\gamma}, \sigma^2 (\boldsymbol{I} + \lambda \boldsymbol{K})^{-2}\right). \tag{2.11}$$

Bayesian smoothing

In a corresponding Bayesian model, the unknown trend is considered a random vector $\boldsymbol{\gamma}$ with some prior $p(\boldsymbol{\gamma})$, and the observation or data model is specified through a conditional density $p(\boldsymbol{y}|\boldsymbol{\gamma})$. The observation model corresponding to (2.10) is

$$\boldsymbol{y}|\boldsymbol{\gamma} \sim \mathrm{N}(\boldsymbol{\gamma}, \sigma^2 \boldsymbol{I}).$$

Assuming σ^2 to be known, we have

$$p(\boldsymbol{y}|\boldsymbol{\gamma}) \propto \exp\left(-\frac{1}{2\sigma^2}(\boldsymbol{y} - \boldsymbol{\gamma})'(\boldsymbol{y} - \boldsymbol{\gamma})\right).$$

A Bayesian smoothness prior corresponding to the roughness penalty term in (2.2) is given by the stochastic difference equations

$$\gamma_t - 2\gamma_{t-1} + \gamma_{t-2} = u_t, \qquad t = 3, \ldots, T$$

with Gaussian i.i.d. errors $u_t \sim \mathrm{N}(0, \tau^2)$, or in other words, by a random-walk prior of order $d = 2$ (RW(2) for short)

$$\gamma_t = 2\gamma_{t-1} - \gamma_{t-2} + u_t, \qquad t = 3, \ldots, T.$$

To obtain a *proper* joint Gaussian prior for the vector $\boldsymbol{\gamma}$ (i.e. a prior that integrates to one), we have to specify a *proper* prior for initial values γ_1, γ_2, for example

$$(\gamma_1, \gamma_2)' \sim \mathrm{N}(0, c\boldsymbol{I}_2),$$

independent from the RW(2) errors. For $c \to \infty$ we obtain *flat improper* priors

$$p(\gamma_1) \propto \text{const}, \quad p(\gamma_2) \propto \text{const}. \tag{2.12}$$

In this case, the joint prior for γ is a *partially improper* Gaussian prior. As can be easily seen, it can be written in the form

$$p(\gamma) \propto \exp\left(-\frac{1}{2\tau^2}\gamma'\boldsymbol{K}\gamma\right), \tag{2.13}$$

where $\boldsymbol{K} = \boldsymbol{K}_2$ is the penalty matrix in (2.5). We therefore obtain the important relationship between frequentist and Bayesian smoothing, that the penalty term equals the negative log-prior in a Bayesian setting (up to additive constants). This equivalence will be exploited at several points throughout this and the following chapters.

Since $\boldsymbol{K}_2$ has rank $T-2$, the distribution in (2.13) is a singular (*partially improper*) Gaussian smoothness prior with *precision matrix* $\boldsymbol{K}_2/\tau^2$. As for the variance σ^2 in the observation model, we assume the variance parameter τ^2 to be known or given and can therefore safely ignore the normalizing constant in (2.13). Because $p(\gamma|\boldsymbol{y})$ and $p(\gamma)$ are Gaussian, the posterior

$$p(\gamma|\boldsymbol{y}) = \frac{p(\boldsymbol{y}|\gamma)p(\gamma)}{p(\boldsymbol{y})} \propto p(\boldsymbol{y}|\gamma)p(\gamma) \tag{2.14}$$

is also Gaussian and is characterized by its posterior expectation $\mathbb{E}(\gamma|\boldsymbol{y})$ and its posterior covariance matrix $\text{Cov}(\gamma|\boldsymbol{y})$. Because of normality, the posterior expectation and the posterior mode, that is, the maximizer of (2.14) or, ignoring constants, of the log-posterior

$$\log p(\gamma|\boldsymbol{y}) = \log p(\boldsymbol{y}|\gamma) + \log p(\gamma)$$

coincide. Using the (conditional) independence assumptions, and assuming flat priors (2.12), the posterior mode can be calculated by minimizing

$$\frac{1}{\sigma^2}\sum_{t=1}^{T}(y_t-\gamma_t)^2 + \frac{1}{\tau^2}\sum_{t=3}^{T}(\gamma_t - 2\gamma_{t-1}+\gamma_{t-2})^2 \tag{2.15}$$

with respect to γ. Defining

$$\lambda = \frac{\sigma^2}{\tau^2}$$

as the variance ratio reveals that the criteria (2.2) and (2.15) are identical for matching choices of the smoothing and variance parameters. In particular, the penalty term turns out to be derived from the negative log-prior of the Bayesian approach and vice versa.

As a result, the PLS (point) estimate $\hat{\gamma}_{\text{PLS}}$ (2.7) is identical to the posterior mode and (because of posterior normality) to the posterior mean estimate $\mathbb{E}(\gamma|\boldsymbol{y})$. A more thorough look at this close relationship is possible if we derive the complete posterior $p(\gamma|\boldsymbol{y})$ explicitly. Assuming flat priors for initial values and ignoring constants we get

$$\begin{aligned} p(\gamma|\boldsymbol{y}) &\propto p(\boldsymbol{y}|\gamma)p(\gamma) \\ &\propto \exp\left(-\frac{1}{2\sigma^2}(\boldsymbol{y}-\gamma)'(\boldsymbol{y}-\gamma)\right)\exp\left(-\frac{1}{2\tau^2}\gamma'\boldsymbol{K}\gamma\right) \\ &\propto \exp\left(-\frac{1}{2}(\gamma-(\boldsymbol{I}+\lambda\boldsymbol{K})^{-1}\boldsymbol{y})'\frac{1}{\sigma^2}(\boldsymbol{I}+\lambda\boldsymbol{K})(\gamma-(\boldsymbol{I}+\lambda\boldsymbol{K})^{-1}\boldsymbol{y})\right) \end{aligned}$$

after completing squares in the quadratic form. Thus,

$$\gamma|\boldsymbol{y} \sim \mathrm{N}\left(\mathbb{E}(\gamma|\boldsymbol{y}), \mathrm{Cov}(\gamma|\boldsymbol{y})\right) \tag{2.16}$$

with posterior mean

$$\mathbb{E}(\gamma|\boldsymbol{y}) = (\boldsymbol{I}+\lambda\boldsymbol{K})^{-1}\boldsymbol{y} \tag{2.17}$$

and posterior covariance matrix

$$\mathrm{Cov}(\gamma|\boldsymbol{y}) = \sigma^2(\boldsymbol{I}+\lambda\boldsymbol{K})^{-1}. \tag{2.18}$$

This result remains true if we assume random-walk smoothness priors of any order d. For example, first order difference penalties correspond to a random-walk model of order $d = 1$ (RW(1)), that is,

$$\gamma_t = \gamma_{t-1} + u_t, \qquad t = 2, \ldots, T,$$

with u_t i.i.d. $\mathrm{N}(0, \tau^2)$. With a flat prior $p(\gamma_1) \propto$ const for the initial value, this can be written as a *partially improper* Gaussian smoothness prior

$$p(\gamma) \propto \exp\left(-\frac{1}{2\tau^2}\gamma'\boldsymbol{K}\gamma\right), \tag{2.19}$$

where $\boldsymbol{K} = \boldsymbol{K}_1$ is the penalty matrix in (2.5).

Thus, posterior normality as in (2.16) holds for any penalty matrix $\boldsymbol{K}$. Comparing with the corresponding results in (2.7), (2.9) and (2.11) we see that the PLS smoother $\hat{\gamma}_{\mathrm{PLS}}$ coincides with the posterior mode and posterior mean $\mathbb{E}(\gamma|\boldsymbol{y})$ as a point estimate, but

$$\mathrm{Cov}(\hat{\gamma}_{\mathrm{PLS}}) = \sigma^2(\boldsymbol{I}+\lambda\boldsymbol{K})^{-2} \neq \sigma^2(\boldsymbol{I}+\lambda\boldsymbol{K})^{-1} = \mathrm{Cov}(\gamma|\boldsymbol{y}).$$

The posterior covariance matrix is 'larger' in the sense that the difference

$$\mathrm{Cov}(\gamma|\boldsymbol{y}) - \mathrm{Cov}(\hat{\gamma}_{\mathrm{PLS}}) = \mathrm{Cov}(\gamma|\boldsymbol{y})(\boldsymbol{I} - (\boldsymbol{I}+\lambda\boldsymbol{K})^{-1})$$

is positive semidefinite. Thus, the equivalence between the penalized least squares and the Bayesian approach holds numerically for the point estimates, but in a strict sense they are not the same as statistical estimators.

Example 2.2: Global warming

Figure 2.2 displays estimated smooth time trends obtained for the climate reconstruction data with either first or second order random-walk prior and (relatively) small or large values of the smoothing parameter. Note that the values of the smoothing parameter are generally large because we are considering a very long time series. Computation of the PLS estimates is quite demanding since approximately 2,000 parameters are involved and a $2{,}000 \times 2{,}000$ system of equations has to be solved for each fit. Of course, a natural strategy to overcome this difficulty would be to group adjacent time points and to assign a single constant parameter value to such groups. This already points to the penalized spline smoothing approaches that will be considered in Section 2.2.

Qualitatively, all four plots in Fig. 2.2 yield comparable results with the expected impact of the smoothing parameter, resulting in smoother estimates

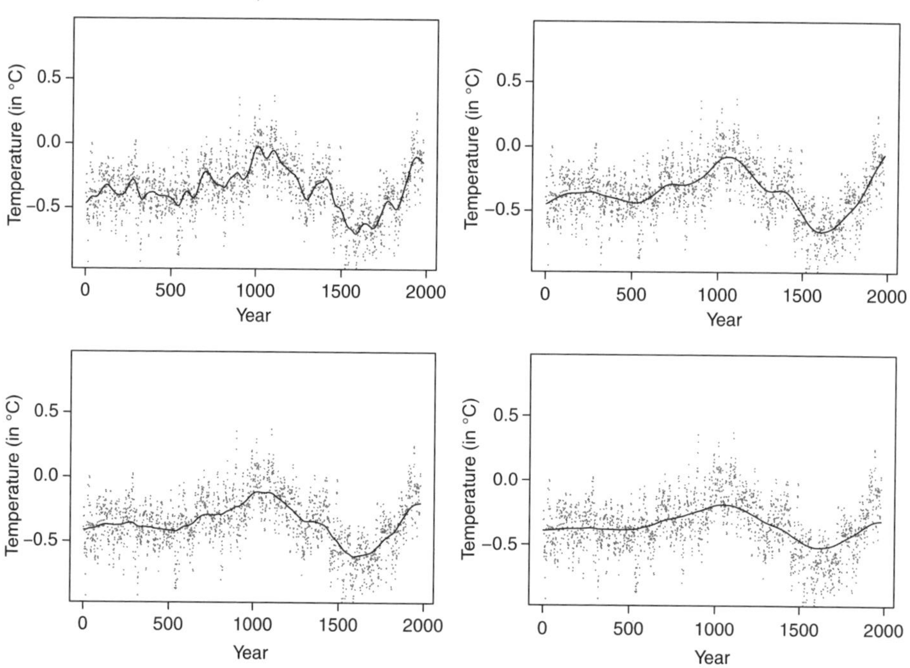

Fig. 2.2 Global warming: First (upper row) and second order (lower row) random-walk fits based on different values for the smoothing parameter λ ($\lambda = 4{,}000$ or $20{,}000$ for the first order random-walk and $\lambda = 100{,}000$ or $5{,}000{,}000$ for the second order random-walk).

for larger smoothing parameter values. Note, that the impact of the difference order is somewhat unintuitive and also not typical. Usually, a smaller difference order implies more wiggly function estimates whereas we observe the contrary behaviour in our example. This may, however, be caused by the smaller null space that remains unpenalized with first order differences, forcing the estimate to a straight line for large values of the smoothing parameter. △

2.1.2 Some modifications and extensions

In the following, we mention and briefly discuss some additional issues which are of relevance not only for the simple scatterplot smoothing setting of Section 2.1.1 but also in the context of more complex data structures and models. We will revisit these issues in later chapters and consider them there in more detail (see in particular Section 4.2 for details on estimating smoothing parameters and Section 4.4.3 for locally adaptive function estimation).

Estimation of smoothing parameters and variances

Up to now, we have assumed that the observational variance σ^2 and the smoothing parameter λ or—from the Bayesian perspective—the smoothing variance τ^2 of the random-walk priors are known. In practice, they have to be estimated from the data. From a frequentist point of view, (generalized) cross validation is the most popular technique for estimating smoothing parameters. In the Bayesian framework, we distinguish two concepts:

- In a full Bayes (FB) approach, the variances σ^2 and τ^2 are considered as additional (random) parameters obeying appropriate priors. They are estimated jointly with the trend γ as characteristics of the posterior $p(\gamma, \sigma^2, \tau^2|\boldsymbol{y})$, usually through Markov chain Monte Carlo (MCMC) simulation. The basic concepts of MCMC based estimation of smoothing parameters will be outlined in Sections 2.2 and 2.3 and will be revisited for more complex settings in Section 4.2.2.
- In an empirical Bayes (EB) approach, σ^2 and τ^2 are considered as unknown constants which are estimated by some frequentist estimation procedure, in particular by (restricted) maximum likelihood estimation based on the *marginal* posterior of σ^2 and τ^2. Mixed model methodology (see Chapter 3 and Section 4.2.1) offers a useful framework for this EB approach.

Other model components

The simple trend model of Section 2.1.1 can be generalized in a flexible way to include a seasonal component $\boldsymbol{\xi} = (\xi_1, \dots, \xi_T)'$ that reflects regular seasonal variation in the data, such as business cycles. This leads to the additive trend-seasonal model

$$y_t = \gamma_t + \xi_t + \varepsilon_t, \qquad t = 1, \dots, T,$$

where the flexible seasonal component can be defined via a more general autoregressive smoothness prior

$$\xi_t + \xi_{t-1} + \ldots + \xi_{t-per+1} = v_t,$$

with i.i.d. error terms $v_t \sim \mathrm{N}(0, \tau^2_{season})$ and seasonal period per. For example, for monthly data and a yearly seasonal pattern, the seasonal period is $per = 12$, while for quarterly data it would be $per = 4$. The autoregressive specification of the seasonal component is motivated by the fact that parametric seasonal effects would assign one separate regression coefficient to each season under a sum to zero constraint. In contrast, the autoregressive specification allows for temporal variation in the seasonal pattern where the variance τ^2_{season} determines the magnitude of deviations from a global seasonal pattern.

The seasonal Bayesian smoothness prior can also be cast in the general form of a (partially improper) Gaussian prior

$$p(\boldsymbol{\xi}) \propto \exp\left(-\frac{1}{2\tau^2_{season}}\boldsymbol{\xi}'\boldsymbol{K}_{season}\boldsymbol{\xi}\right)$$

with an appropriately defined penalty matrix $\boldsymbol{K}_{season}$. The correspondence between Bayesian smoothing based on the mode or mean of the posterior $p(\boldsymbol{\gamma}, \boldsymbol{\xi}|\boldsymbol{y})$, and PLS smoothing based on a correspondingly extended PLS criterion, still holds.

More generally, the influence of covariates can be taken into account by extending the additive predictor $\gamma_t + \xi_t$ to

$$y_t = \gamma_t + \xi_t + \boldsymbol{x}_t'\boldsymbol{\beta} + \boldsymbol{z}_t'\boldsymbol{\alpha}_t + \varepsilon_t$$

with time-constant effects $\boldsymbol{\beta}$ for $\boldsymbol{x}_t$ and time-varying effects $\boldsymbol{\alpha}_t$ for $\boldsymbol{z}_t$.

Models of this type, including alternative priors for trend and seasonal component, for example, local linear trend models for $\boldsymbol{\gamma}$ and stochastic trigonometric seasonal models, can be cast into the form of Gaussian linear state space models, see for example Durbin and Koopman (2001), Fahrmeir and Tutz (2001, Ch. 8.1), West and Harrison (1997) or Harvey (1989). The well known Kalman filter and smoother can be used to efficiently compute posterior means, and it can also be used as an efficient algorithmic tool for minimizing corresponding PLS criteria.

Correlated errors

For time-series observations $y_t, t = 1, \ldots, T$, the assumption of i.i.d. errors ε_t in the observation model seems to be questionable. In principle, it is possible to assume that the errors follow some (stationary) stochastic process $\boldsymbol{\varepsilon} = (\varepsilon_1, \ldots, \varepsilon_T)'$. The simplest example is a stationary autoregressive process of first order,

$$\varepsilon_t = \phi\varepsilon_{t-1} + v_t$$

with $|\phi| < 1$ and i.i.d. errors $v_t \sim \mathrm{N}(0, \sigma^2)$. In the limiting case $\phi \to 1$, this AR(1)-model becomes a RW(1)-model. If we specify a RW(1)-model for the trend function, then we have two stochastic process terms of exactly the same type in the basic model

$$y_t = \gamma_t + \varepsilon_t.$$

While the RW(1)-model for $\boldsymbol{\gamma}$ aims at modelling the trend, the second component is introduced to model correlation. However, for ϕ close to 1, both components are close to each other, leading to the question of how well trend and correlation can be separated, see Section 4.3 and Fahrmeir and Kneib (2008) for a detailed discussion.

Locally adaptive smoothing

Gaussian random-walk smoothness priors with a global smoothing variance τ^2 (or quadratic penalty functions with a global smoothing parameter λ) are suitable for nonparametric estimation of smooth time trends. They are less suited, however, to recovering unsmooth trends such as trends with edges, abrupt changes in level or slope or highly fluctuating behaviour. In such situations, Gaussian random-walk priors have a tendency towards oversmoothing or blurring unsmooth parts of the underlying trend function and, on the other side, of producing too wiggly estimates of smooth parts of the trend.

There are several strategies to develop smoothers with improved local adaptivity or better 'edge preserving' properties. One way is to assume non-Gaussian error distributions in the random-walk models, for example error distributions which are scale mixtures of normals. Mixture variables may be discrete or continuous. A popular choice are χ^2- or gamma-mixture variables, leading to t-distributions for the errors. Then the RW(2)-model for the trend is

$$\gamma_t = 2\gamma_{t-1} + \gamma_{t-2} + u_t, \qquad u_t | \omega_t \sim \mathrm{N}(0, \tau^2/\omega_t).$$

Assuming the ω_t to be independently χ^2-distributed with ν degrees of freedom, the errors are marginally $t(\nu)$-distributed, leading to improved local adaptivity due to the heavier tails of the random-walk model.

A second possibility is to assume that smoothing variances of random-walk models are no longer time-constant but are time-varying, that is,

$$u_t \sim \mathrm{N}(0, \tau_t^2)$$

and that, after reparametrizing, $\delta_t = \log(\tau_t^2)$ follows a random-walk prior in a second stage of the hierarchy. This approach has been suggested more recently in a state space context (Lang, Fronk and Fahrmeir, 2002) and has also been considered for penalized spline models (Ruppert and Carroll, 2000; Krivobokova, Crainiceanu and Kauermann, 2008) and in a related context by Scheipl and Kneib (2009). Both ideas reappear in nonparametric regression based on penalized splines and will be considered in later chapters, in particular in Section 4.4.3.

Unequally spaced time-series observations

Up until now, we have tacitly assumed that the time series is equally spaced in time. Extensions to unequally spaced data are possible, either by using modified random-walk priors or by using continuous-time models. For a first-order random-walk, it seems plausible to adapt the variability of the error terms to the length of the intervals, leading to

$$\gamma_t = \gamma_{t-1} + u_t, \quad u_T \sim \mathrm{N}(0, \tau^2 \delta_t),$$

where δ_t denotes the time difference between the t-th and the $(t-1)$-th measurement. Still, the joint distribution of $\boldsymbol{\gamma}$ will be multivariate Gaussian but now the precision matrix is given by $\boldsymbol{K} = \boldsymbol{D}'\boldsymbol{\Delta}\boldsymbol{D}$, where $\boldsymbol{\Delta} = \mathrm{diag}(\delta_2, \ldots, \delta_T)$. In fact, the weighted first-order random-walk also appears naturally via the discretization of the time-continuous Brownian motion that enduces exactly the same structure for the increments in a given time interval. Similar extensions can be considered for second order random walks, see Fahrmeir and Lang (2001b) and Rue and Held (2005, Ch. 3.5) for a thorough discussion. Note that unequally spaced data are easily handled by assuming (penalized) spline models for the trend function, see Section 2.2.

2.1.3 Non-Gaussian observation models

This section extends the basic ideas for smoothing (approximately) Gaussian time series to fundamentally non-Gaussian time series. We focus on discrete-valued time series, where observations y_t may be binary, binomial or counts, but extensions to other types of non-Gaussian observations are conceptually straightforward. Categorical data will be discussed in the context of penalized spline smoothing in Section 2.2.

As an example, we first consider binary or binomial observations. Instead of a Gaussian observation model for y_t (given the trend γ_t), $t = 1, \ldots, T$, responses are now assumed to be binomial,

$$y_t \sim \mathrm{B}(n_t, \pi_t), \qquad t = 1, \ldots, T, \tag{2.20}$$

where n_t is known, and usually small or even $n_t = 1$ for binary time series. Instead of estimating the trend γ_t directly for the response variables, the aim is to estimate the probabilities $\boldsymbol{\pi} = (\pi_1, \ldots, \pi_T)'$ as a smooth curve. In complete analogy to generalized linear models we relate π_t to γ_t by a logit link, that is,

$$\gamma_t = \log\left(\frac{\pi_t}{1-\pi_t}\right) \quad \Leftrightarrow \quad \pi_t = \pi(\gamma_t) = \frac{\exp(\gamma_t)}{1+\exp(\gamma_t)}.$$

Of course, other link functions, such as the probit or complementary log-log link, are alternative possibilities for reparametrizing from $\boldsymbol{\pi}$ to $\boldsymbol{\gamma}$.

A *roughness penalty approach* replaces the least-squares criterion in (2.2), which is essentially the log-likelihood of the Gaussian observation model, through the log-likelihood corresponding to the binomial observation model. This leads to the *penalized log-likelihood* criterion

$$\begin{aligned} l_{\text{pen}}(\boldsymbol{\gamma}) &= \sum_{t=1}^{T} \{y_t \log \pi(\gamma_t) + (n_t - y_t) \log(1 - \pi(\gamma_t))\} - \frac{\lambda}{2} \sum_{t=d+1}^{T} (\Delta^d \gamma_t)^2 \\ &= l(\boldsymbol{\gamma}) - \frac{\lambda}{2} \boldsymbol{\gamma}' \boldsymbol{K} \boldsymbol{\gamma}, \end{aligned}$$

where $l(\boldsymbol{\gamma})$ is, up to constants, the log-likelihood of the (conditionally) independent binomial observations, $\Delta^d \gamma_t$ are differences of order d and $\boldsymbol{K}$ is the corresponding penalty matrix (2.5) as in Section 2.1.1. The additional factor 1/2 multiplied to the smoothing parameter is introduced for convenience to recover the PLS criterion (2.2) in case of Gaussian responses. Maximization of the penalized log-likelihood gives smooth estimates

$$\hat{\boldsymbol{\gamma}}_{\text{pen}} = \arg\max_{\boldsymbol{\gamma}} l_{\text{pen}}(\boldsymbol{\gamma})$$

and $\hat{\boldsymbol{\pi}} = \boldsymbol{\pi}(\hat{\boldsymbol{\gamma}}_{\text{pen}})$ of the trend. The solution $\hat{\boldsymbol{\gamma}}_{\text{pen}}$ has no explicit analytical form, however, and has to be computed numerically by iterative algorithms. We will introduce the corresponding concepts in Section 2.2 in the context of more general approaches based on smoothness priors.

Example 2.3: Human sleep
As an example for a non-Gaussian time-series smoothing problem, we consider a part of the human sleep data on $n = 70$ study participants introduced in Example 1.7. In this case, the time axis is defined by the 30 s epochs that form the basis for the assignment of the participants to the sleep stages and there is a total of $T = 940$ epochs available.

More specifically, we are interested in the probability of transitions from Non-REM to REM, i.e.

$$\mathbb{P}(\text{individual } i \text{ in state REM at time } t \mid \text{individual } i \text{ in state Non-REM at } t-1).$$

To implement estimation of this transition probability, we first delete all observations from the data that are not under risk for the transition of interest, that is, the database considered in this example only consists of observations that are currently in Non-REM sleep state. Then we can define the binary indicator

$$y_{it} = \begin{cases} 1 & \text{if individual } i \text{ moves from Non-REM to REM at time } t \\ 0 & \text{otherwise,} \end{cases},$$

$i = 1, \ldots, n$, $t = 1, \ldots, T$, and use a random-walk model for the logit of the transition probability

$$\text{logit}\left(\frac{\pi_{it}}{1-\pi_{it}}\right) = \gamma_t.$$

Due to the restricted database, the probability naturally turns into a conditional probability for the transition of interest. The upper row of Fig. 2.3 shows the number of observations under risk as a function of time together with the actually observed number of transitions superimposed in grey. The second row displays the resulting relative frequency of transition from Non-REM to REM in the data set.

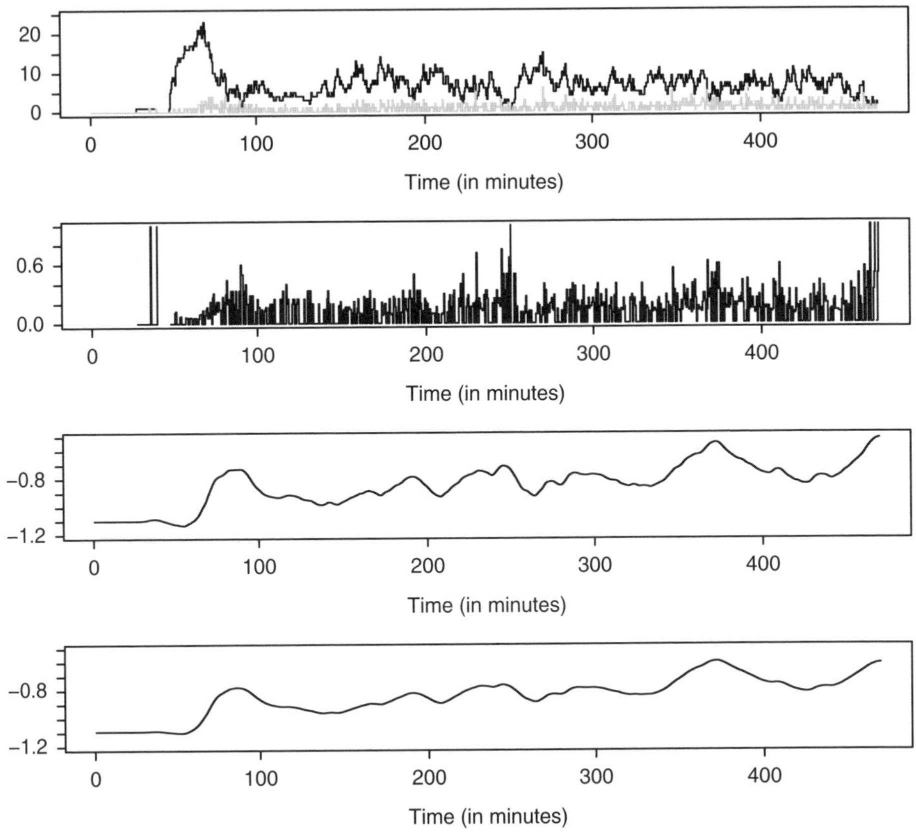

Fig. 2.3 Human sleep: The top panel shows the number of individuals under risk for the transition from Non-REM to REM (black line) and the number of actual transitions (gray line). The second panel displays the corresponding relative frequency of the transition. The third and fourth panel visualize first-order random-walk fits $\hat{\gamma}$ based on a lower ($\lambda = 200$) and a higher ($\lambda = 500$) value for the smoothing parameter.

Both figures indicate that early in the night, immediately after sleep onset, the probability for the transition is low, resulting simply from the fact that no individuals are under risk. A first peak is observed after about 90 minutes, followed by more or less unstructured fluctuation about a relatively stable level. Applying the first-order random-walk model with either lower or higher smoothing parameter to the data, yields the fitted coefficient curves $\hat{\gamma}_t$ displayed in the lower two panels of Fig. 2.3. Both provide a smoothed version of the raw relative frequencies indicating some local maxima for the probability of moving from Non-REM to REM. However, overall the changes in the probability are surprisingly small.

Note that the current example is only a very preliminary analysis to demonstrate the application of time-series smoothing for non-Gaussian observations. We will re-consider the human sleep data later on in more detail and with more complex models that allow one to analyse several transitions simultaneously and to account for additional covariate effects, see Chapter 6.5. △

For time series of counts y_t, we can proceed in the same way, assuming for example a log-linear Poisson model

$$y_t|\lambda_t \sim \text{Po}(\lambda_t), \qquad \lambda_t = \exp(\gamma_t),$$

and replacing the log-likelihood of binomial observations by the Poisson log-likelihood

$$l(\boldsymbol{\gamma}) = \sum_{t=1}^{T}(y_t \log \lambda_t - \lambda_t) = \sum_{t=1}^{T}(y_t\gamma_t - \exp(\gamma_t)).$$

Both time-series models for binomial data and Poisson counts are members of the class of *dynamic generalized linear models* (Fahrmeir and Tutz, 2001, Ch. 8). The general observation model is constructed as follows: The (conditional) density of y_t, given the (reparametrized) trend $\boldsymbol{\gamma}$, is of the simple exponential family type, and has the conditional expectation

$$\mathbb{E}(y_t|\gamma_t) = \mu_t = h(\gamma_t),$$

where the mean trend $\boldsymbol{\mu} = (\mu_1, \ldots, \mu_T)'$ is related to the reparametrized trend $\boldsymbol{\gamma}$ by a suitable response function h. More generally, we may consider non-exponential family densities, for example replacing Gaussian error distributions in (2.1) by more heavy-tailed distributions such as scale mixtures of normals. In any case, the general concept is the same: The observation model defines an appropriate log-likelihood $l(\boldsymbol{\gamma})$. For the roughness penalty approach, $l(\boldsymbol{\gamma})$ is not even restricted to be a genuine log-likelihood but can rather be a (log-)quasi-likelihood.

For a *Bayesian-smoothness prior* version of the roughness penalty approach, we proceed in complete analogy to the Gaussian case: The observation model for

$\boldsymbol{y}$ given $\boldsymbol{\gamma}$ defines the likelihood $p(\boldsymbol{y}|\boldsymbol{\gamma})$, and we again assume Gaussian random-walk smoothness priors for the trend, that is

$$p(\boldsymbol{\gamma}) \propto \exp\left(-\frac{1}{2\tau^2}\boldsymbol{\gamma}'\boldsymbol{K}\boldsymbol{\gamma}\right)$$

with the same penalty matrices $\boldsymbol{K}$ as in Section 2.1.1.

In contrast to what happens in Gaussian models, the posterior

$$p(\boldsymbol{\gamma}|\boldsymbol{y}) = \frac{p(\boldsymbol{y}|\boldsymbol{\gamma})p(\boldsymbol{\gamma})}{p(\boldsymbol{y})} \propto p(\boldsymbol{y}|\boldsymbol{\gamma})p(\boldsymbol{\gamma})$$

is now non-normal and in general not symmetric. Thus, posterior expectations and posterior modes are no longer equivalent. With a diffuse prior for initial values of the random-walk models, the posterior mode is the maximizer of $p(\boldsymbol{y}|\boldsymbol{\gamma})p(\boldsymbol{\gamma})$, that is, taking logarithms and omitting superfluous constants, of

$$l_{\text{pen}}(\boldsymbol{\gamma}) = l(\boldsymbol{\gamma}) - \frac{\lambda}{2}\boldsymbol{\gamma}'\boldsymbol{K}\boldsymbol{\gamma} = \log p(\boldsymbol{y}|\boldsymbol{\gamma}) - \frac{1}{2\tau^2}\boldsymbol{\gamma}'\boldsymbol{K}\boldsymbol{\gamma},$$

where the inverse smoothing variance $1/\tau^2$ corresponds to the smoothing parameter λ. This shows that for given smoothing parameter, the penalized log-likelihood smoother $\hat{\boldsymbol{\gamma}}_{\text{pen}}$ and the posterior mode coincide again numerically as point estimates. Since the estimate $\hat{\boldsymbol{\gamma}}_{\text{pen}}$ is not available in closed form and has to be computed numerically, the exact form of covariance matrices as well as the exact sampling distribution are no longer known. As an alternative, posterior means and covariance matrices can be computed via Markov chain Monte Carlo (MCMC), without relying on any asymptotic approximations, see Chapter 2.2.

All comments on modifications and extensions of the basic Gaussian model in Section 2.1.2 generalize almost immediately to non-Gaussian models considered in this section (at least conceptually).

2.2 Semiparametric regression based on penalized splines

2.2.1 Gaussian observation models

Random-walk models have an immediate Bayesian appeal since they involve the specification of a conditional dependency structure leading to interpretable prior assumptions. However, they also have a number of potential drawbacks: If the time series consists of a lot of data points (as in the global warming and the human sleep examples), large matrices have to be stored and processed to derive the estimates $\hat{\boldsymbol{\gamma}}$. Although sparse matrix computations can be employed to reduce the computational burden, the model still seems overly complex, in particular when trying to estimate several effects simultaneously. In empirical studies, this is reflected by relatively wiggly estimates and the inherent non-differentiability of the fitted piecewise constant curves.

A simple remedy that is frequently used in practice to circumvent the problem of high dimensionality is to group adjacent time points and to assign a constant function value to each of the groups. In this case, the random-walk prior is no longer defined for the original observations in time but for the groups. Taking this approach to a higher level leads to the consideration of polynomial splines that no longer assume a constant function within a group but allow for a lower order polynomial instead. To meet the assumption of a smooth curve, one has to impose the constraint that the polynomial pieces fuse smoothly enough at the grouping boundaries.

Polynomial splines

To introduce polynomial splines more rigorously than in the time-series context, we first write the trend γ_t as a function of t (i.e. $\gamma_t = f(t)$) and then replace the trend model

$$y_t = f(t) + \varepsilon_t,\ t = 1, \ldots, T,$$

with the more general nonparametric smoothing problem

$$y_i = f(x_i) + \varepsilon_i,\ i = 1, \ldots, n, \tag{2.21}$$

where x_i is a continuous covariate and $f(x)$ is a smooth function that describes the functional relationship between x and y. This situation is also well-known as scatterplot smoothing, since the data (x_i, y_i), $i = 1, \ldots, n$, can be inspected visually in a scatterplot.

Although random-walk models can still be applied in principle, typically the covariate x_i will be measured on a non-equidistant scale which would require weighted random-walk specifications. Instead of assigning a random-walk to $f(x)$, we approximate the function by a polynomial spline. Such a polynomial spline function (see Dierckx, 1993 or de Boor, 2001, for mathematical details) is defined based on a set of $M + 1$ (not necessarily equally spaced) knots $x_{\min} = \kappa_0 < \kappa_1 < \cdots < \kappa_{M-1} < \kappa_M = x_{\max}$ within the domain of covariate x. The function f is then called a polynomial spline of degree D, $D \in \mathbb{N}_0$ if it satisfies the following conditions:

- $f(x)$ is $(D - 1)$ times continuously differentiable and
- $f(x)$ is a polynomial of degree D (or lower) for $x \in [\kappa_m, \kappa_{m+1})$, $m = 0, \ldots, M - 1$.

The second condition specifies f as a piecewise polynomial of degree D on the intervals defined by the knots, whereas the first condition ensures that the function is qualitatively smooth with an increasing amount of smoothness for larger degrees D. For $D = 1$, $f(x)$ is a linear spline, which is continuous, but not differentiable at the knots. For $D = 0$, $f(x)$ is a step function with jumps at the knots.

The space of all polynomial splines for a fixed sequence of knots and fixed degree can be shown to be an $(M + D)$-dimensional subspace of the space of

$(D-1)$ times continuously differentiable functions. Therefore, assuming that $f(x)$ can be approximated by a polynomial spline leads to a representation in terms of a linear combination of $K = M + D$ basis functions B_k, that is,

$$f(x) = \sum_{k=1}^{K} \gamma_k B_k(x).$$

Different sets of basis functions have been considered in the literature and we will describe the two most popular versions in the following paragraph.

Truncated power series and B-splines

The basis with perhaps the most immediate appeal is the truncated power series (TPS) basis consisting of the basis functions

$$\begin{aligned} B_1(x) = 1, \quad B_2(x) = x, \quad &\ldots, \quad B_{D+1}(x) = x^D, \\ B_{D+2}(x) = (x-\kappa_1)_+^D, \quad &\ldots, \quad B_K(x) = (x-\kappa_{M-1})_+^D, \end{aligned}$$

where

$$(x-\kappa_k)_+^D = \begin{cases} (x-\kappa_k)^D & x \geq \kappa_k, \\ 0 & \text{otherwise.} \end{cases}$$

The basis consists of two components: a global polynomial defined by the first $D+1$ basis functions and truncated polynomials that allow the modification of the polynomial coefficients at each knot while ensuring the smoothness condition at the same time.

A numerically more stable basis is obtained by taking appropriate differences of the truncated polynomials. For degree $D = 0$ this yields the piecewise constant basis functions

$$B_k^0(x) = \mathbb{1}_{[\kappa_k, \kappa_{k+1})}(x) = \begin{cases} 1 & \kappa_k \leq x < \kappa_{k+1}, \\ 0 & \text{otherwise,} \end{cases} \qquad k = 0, \ldots, K-1,$$

that, combined with an appropriate penalty, effectively coincide with the grouped random-walk approach discussed at the beginning of this section. For higher degrees, B-spline basis functions are obtained recursively via

$$B_k^D(x) = \frac{x-\kappa_k}{\kappa_{k+D}-\kappa_k} B_k^{D-1}(x) + \frac{\kappa_{k+D+1}-x}{\kappa_{k+D+1}-\kappa_{k+1}} B_{k+1}^{D-1}(x).$$

Examples of single B-spline basis functions and of complete B-spline bases of varying degree defined on either equidistant or non-equidistant knots are shown in Figs. 2.4 and 2.5. Note that for the determination of all B-spline basis functions, the original knot sequence has to be extended at the boundaries to contain D additional knots both on the left and on the right of the domain. While these additional knots are easily defined when using equidistant knots by adding further knots with the same distance as for the interior knots, some additional rule has to be defined for non-equidistant knots, such as carrying forward the distance at the boundaries or using an average distance.

From Figs. 2.4 and 2.5, several useful properties of B-splines can be derived:

- *B-splines (in contrast to the TPS basis) form a local basis.* Each basis function is non-negative on an interval formed by $D+2$ adjacent knots. Correspondingly, for each fixed x in the domain, at most $D+1$ basis functions are positive.
- *Unity decomposition.* For each fixed x in the domain, the basis function evaluations sum to one, that is,

$$\sum_{k=1}^{K} B_k(x) = 1, \qquad x_{\min} \leq x \leq x_{\max}.$$

  This also shows that the B-spline basis implicitly contains an intercept which has been included explicitly in the truncated power series basis.
- *Bounded range.* The range of function evaluations is bounded from above (at least as long as all knots are equidistant). This explains why the B-spline basis has preferential numerical properties as compared to truncated polynomials. While the latter are unbounded and show a significant amount of collinearity, both problems are avoided by the B-spline basis.

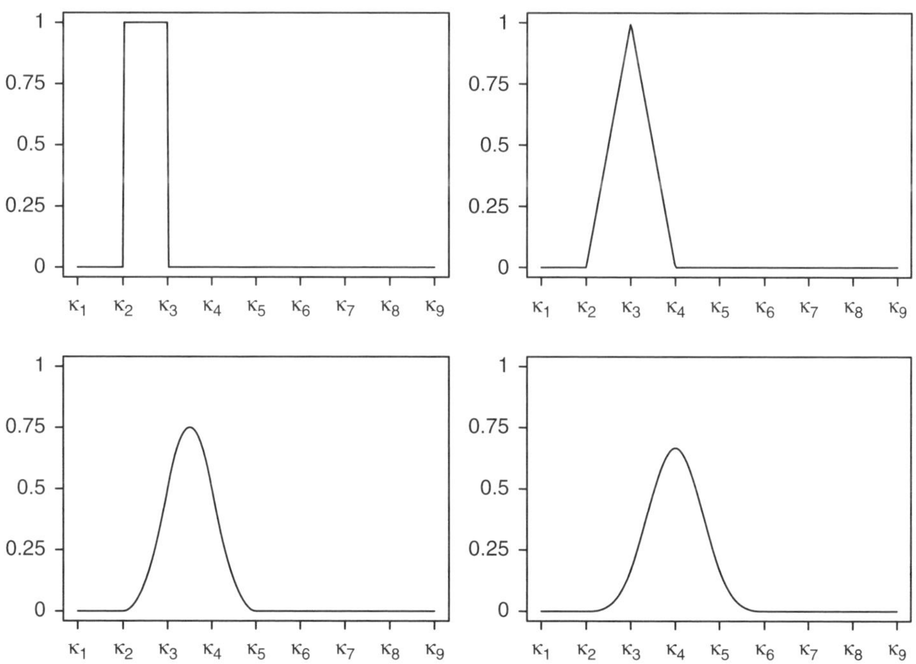

Fig. 2.4 Single B-spline basis functions of degrees $D = 0, 1, 2, 3$ based on equidistant knots.

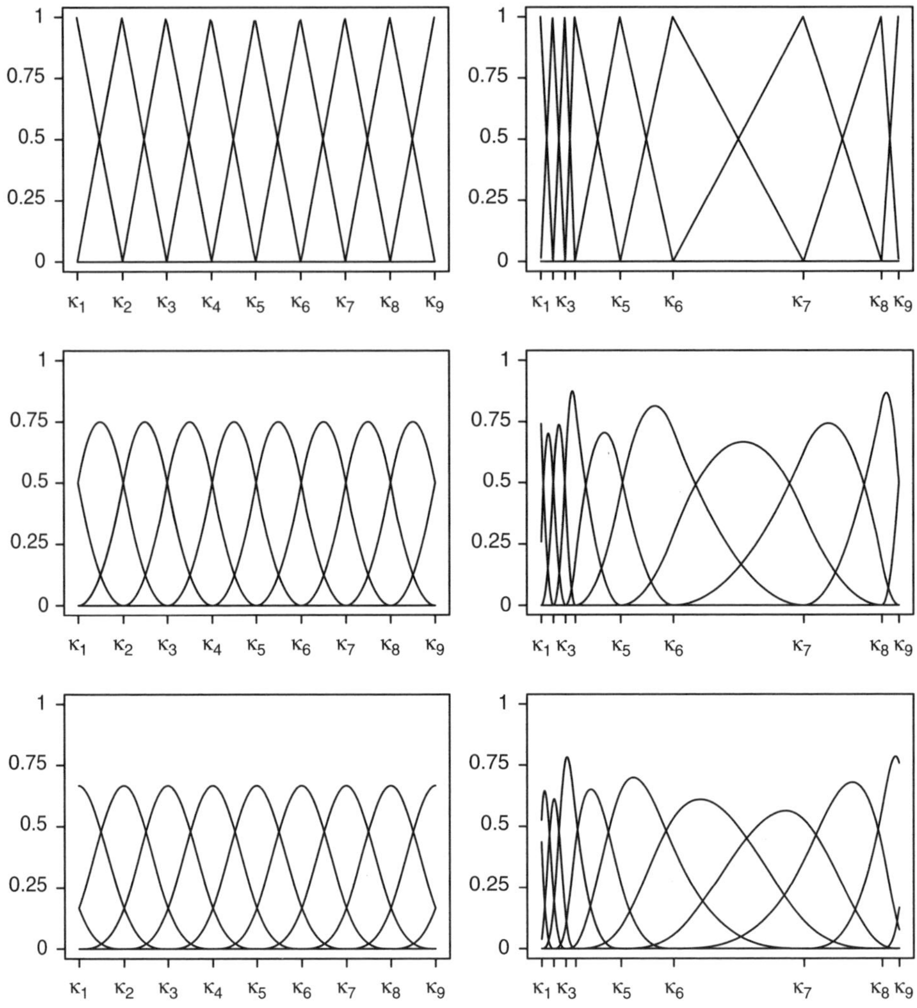

Fig. 2.5 Complete B-spline bases of degree $D = 1, 2, 3$ for equidistant (left column) and non-equidistant knots (right column).

Despite their different numerical properties, both the TPS basis and the B-spline basis define exactly the same function space (for a given knot sequence). More precisely, both are equivalent bases and, in particular, all TPS basis functions can be represented in terms of B-splines and vice versa. This is most easily seen for splines of degree $D = 0$. In this case, a specific B-spline basis function $B_k^B(x)$ is given by the difference of the two adjacent TPS basis functions $B_k^{TPS}(x)$ and $B_{k+1}^{TPS}(x)$, i.e.

$$B_k^B(x) = B_k^{TPS}(x) - B_{k+1}^{TPS}(x).$$

Reversing this relation yields an expression for TPS basis functions in terms of B-splines via

$$B_k^{TPS}(x) = \sum_{m=k}^{K} B_k^B(x).$$

For higher degrees, similar expressions can be derived. However, formulae become more complicated and do not provide additional insights and, hence, have been omitted but may be found in de Boor (2001), Chapters 8 and 9.

Nonparametric regression based on polynomial splines

In any basis representation, the scatterplot smoothing problem turns into a large linear model

$$\boldsymbol{y} = \boldsymbol{X}\boldsymbol{\gamma} + \boldsymbol{\varepsilon} \tag{2.22}$$

where

$$\boldsymbol{X} = \begin{pmatrix} B_1(x_1) & \dots & B_K(x_1) \\ \vdots & & \vdots \\ B_1(x_n) & \dots & B_K(x_n) \end{pmatrix} \tag{2.23}$$

contains the basis functions evaluated at the observed covariates whereas $\boldsymbol{y}$, $\boldsymbol{\gamma}$ and $\boldsymbol{\varepsilon}$ are stacked vectors of responses, regression coefficients and error terms. In the linear model (2.22) the usual estimate for the regression coefficients $\boldsymbol{\gamma}$ is the least squares estimate

$$\hat{\boldsymbol{\gamma}} = (\boldsymbol{X}'\boldsymbol{X})^{-1}\boldsymbol{X}'\boldsymbol{y}$$

which, under a noninformative prior for $\boldsymbol{\gamma}$, for Gaussian errors and for fixed error variance, also coincides with the posterior mode, median and mean. The estimate for the regression coefficients can be transformed into an estimate for the vector of function evaluations $\boldsymbol{f} = (f(x_1), \dots, f(x_n))'$ via

$$\hat{\boldsymbol{f}} = \boldsymbol{X}\hat{\boldsymbol{\gamma}} = \boldsymbol{X}(\boldsymbol{X}'\boldsymbol{X})^{-1}\boldsymbol{X}'\boldsymbol{y}.$$

Note that $\boldsymbol{X}'\boldsymbol{X}$ is a full matrix when using the TPS basis while the locality of the B-spline basis leads to a design matrix with many structural zeros, yielding a band structure in the matrix $\boldsymbol{X}'\boldsymbol{X}$. To be more specific, $\boldsymbol{X}'\boldsymbol{X}$ has bandwidth D. Using special algorithms for sparse matrices therefore allows us to compute $\hat{\boldsymbol{\gamma}}$ fairly efficiently by solving a system of equations with a sparse coefficient matrix. While the gain in computation time may often be negligible when computing the polynomial spline only once, it will become extremely important in later sections when considering additive models and simulation-based estimation of smoothing parameters.

Example 2.4: Global warming

A schematic representation of a polynomial spline fit for the global warming data based on B-splines can be found in Fig. 2.6. In a first step, a sequence

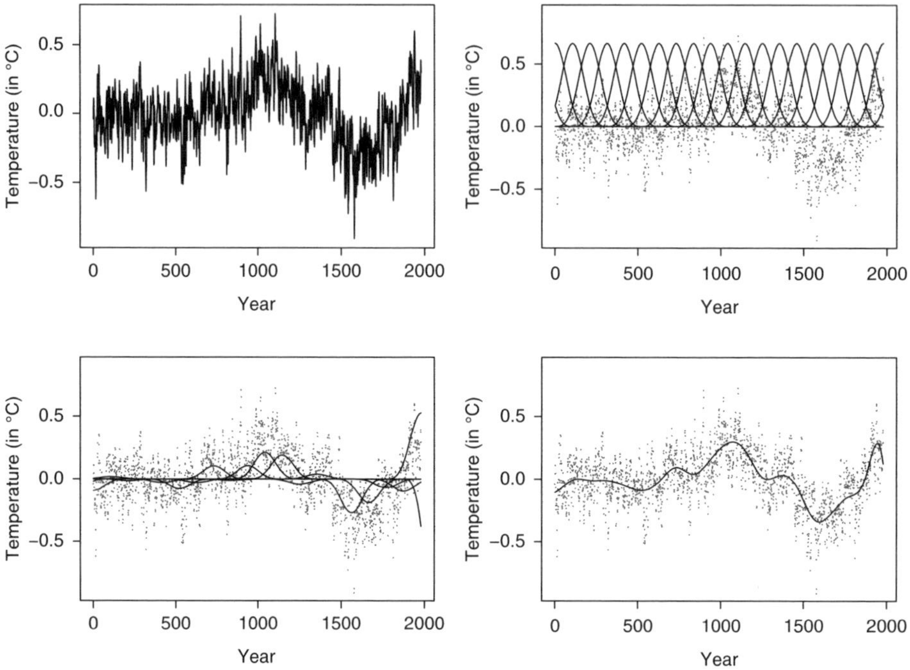

Fig. 2.6 Global warming: Schematic representation of a B-spline fit. The upper left figure displays the time series of reconstructed yearly temperature averages in the Northern hemisphere. The upper right figure overlays the data with a B-spline basis of degree $D = 3$ with 20 equidistant inner knots. In the lower left figure, the basis functions are scaled according to the data and the lower right figure shows the final fit obtained when summing up the scaled basis functions. For all figures, the data have been centred about zero to improve visibility.

of equidistant knots was defined and the corresponding basis functions were constructed. In a second step, the least squares estimate for the regression coefficients is determined and the basis functions are weighted accordingly. Summing up the weighted basis functions finally yields the B-spline reconstruction of the underlying function showing strong variation and long-time trends of the average temperature over the past 2,000 years.

The results qualitatively resemble those from the random-walk models described in Example 2.2. However, replacing random-walk smoothing by spline smoothing considerably reduces the computational burden since now only a very moderate number of regression parameters is involved. Correspondingly, computation times are dramatically reduced, which will become increasingly relevant when considering additive models with several nonparametric functions and automatic determination of the smoothing parameter. △

Characteristics of a spline fit

The exact functional form estimated with a polynomial spline depends crucially on two properties: The degree of the spline function and the number of knots employed in the basis definition.

Figure 2.7 shows various polynomial spline reconstructions for the global warming example with different degrees but a fixed set of 20 inner knots. Obviously, the degree basically determines the differentiability of the function estimate whereas the overall amount of smoothness remains largely unchanged. This observation clearly corresponds to the fact that the spline degree defines the local polynomial fitted in each of the intervals and the differentiability condition at the interval boundaries. For $D = 0$ we obtain a piecewise constant, discontinuous and hence non-differentiable function consisting of 20 steps. Increasing the degree successively leads to a continuous but non-differentiable ($D = 1$), once continuously differentiable ($D = 2$) and a twice continuously differentiable function ($D = 3$). In many applications, cubic polynomial splines (i.e. $D = 3$) will be an appropriate choice resulting in visually pleasant, smooth function estimates.

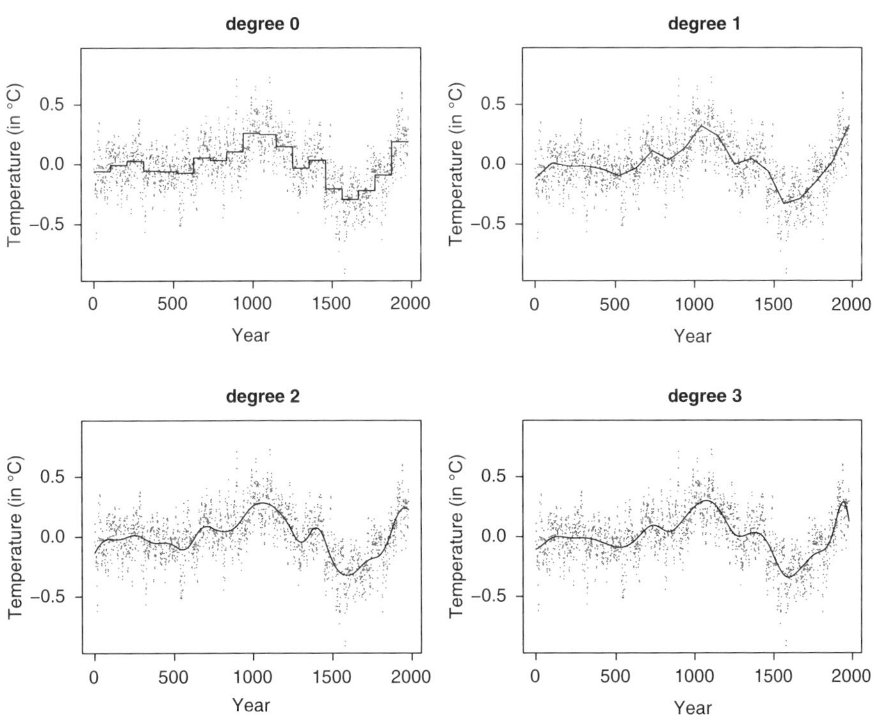

Fig. 2.7 Global warming: Polynomial spline fits based on bases with 20 inner knots and varying degrees.

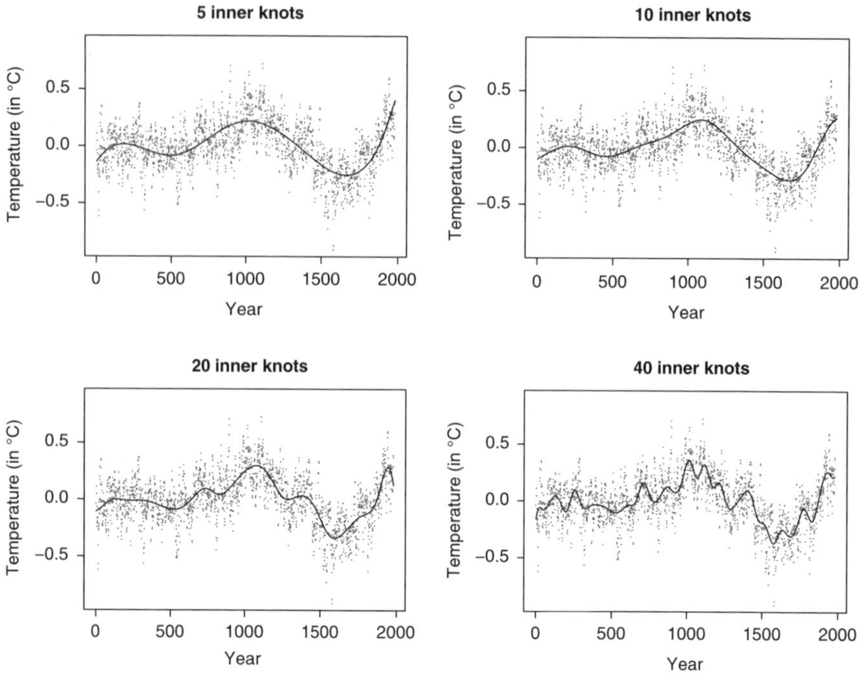

Fig. 2.8 Global warming: Polynomial spline fits based on bases of degree $D = 3$ with varying numbers of knots.

Similarly, Fig. 2.8 investigates the impact of the number of (equidistant) knots on the spline fit when using cubic splines. The number of knots significantly influences the results, with small knot sets corresponding to relatively smooth functions and large knot sets leading to wiggly curves. While the degree can typically be chosen according to subject matter considerations about the assumed differentiability of the functional relationship, such rules are hard to obtain for the number of knots. In addition, concurrent possibilities exist for positioning of the knots. Adaptive knot selection strategies based on frequentist or Bayesian model-selection strategies have been developed for a variety of situations (Friedman, 1991; Stone, Hansen, Kooperberg and Truong, 1997; Biller, 2000; Yau, Kohn and Wood, 2003, Denison, Holmes, Mallick and Smith, 2002, DiMatteo, Genovese and Kass, 2001). However, a simpler alternative working well in many practical circumstances is to utilize the penalization ideas from Section 2.1 for the polynomial spline coefficients.

P-splines

The idea of penalized splines or simply P-splines is most easily understood in the context of the truncated power series basis (Ruppert, Wand and Carroll, 2003). A very smooth curve is described by the global polynomial term consisting of the first $D + 1$ basis functions, whereas considerable additional variability is introduced for large regression coefficients of the truncated polynomials. To avoid too wiggly estimates, a penalty term can be added to the least-squares criterion that penalizes large values of these coefficients. The penalty that is simplest to work with is a quadratic penalty term, that is,

$$\text{pen}(\boldsymbol{\gamma}) = \sum_{k=D+2}^{K} \gamma_k^2 = \boldsymbol{\gamma}'\boldsymbol{K}\boldsymbol{\gamma}$$

where $\boldsymbol{K} = \text{diag}(\boldsymbol{0}_{D+1}, \boldsymbol{I}_{M-1})$. This penalty term has the advantage that it is differentiable with respect to $\boldsymbol{\gamma}$ (since it is a quadratic form) and that it corresponds to either a random effects distribution or an i.i.d. Gaussian prior, depending on the perspective. Both possibilities will be used later on in this book to derive inferential concepts for complex regression models involving several nonparametric functions and therefore several penalty terms (see Section 4.2). The quadratic penalty also has a close connection to ridge regression, as will become clear in the following.

Similarly to our considerations in Section 2.1, the penalty term is not simply added to the least-squares criterion, but multiplied with an additional smoothing parameter $\lambda \geq 0$ that determines the influence of the penalty as compared to the likelihood, that is,

$$\text{PLS}(\lambda) = \sum_{i=1}^{n} \left(y_i - \sum_{k=1}^{K} \gamma_k B_k(x_i) \right)^2 + \lambda \sum_{k=D+2}^{K} \gamma_k^2.$$

In matrix notation, we obtain

$$\text{PLS}(\lambda) = (\boldsymbol{y} - \boldsymbol{X}\boldsymbol{\gamma})'(\boldsymbol{y} - \boldsymbol{X}\boldsymbol{\gamma}) + \lambda\boldsymbol{\gamma}'\boldsymbol{K}\boldsymbol{\gamma},$$

where $\boldsymbol{X}$ is defined in (2.23).

In the limiting cases $\lambda \to 0$ and $\lambda \to \infty$, minimizing the penalized least-squares criterion $\text{PLS}(\lambda)$ yields the least squares fit for the polynomial spline without penalty and a global polynomial, respectively. Varying the smoothing parameter therefore allows to select continuously between wiggly and smooth function estimates, whereas the influence of the number and the placement of knots becomes more or less negligible. To be more precise, a large number of knots should be used, to enable enough flexibility, but the exact number is practically irrelevant with 20 to 40 knots as a suitable default choice. Note that the penalty even allows us to smooth over regions where no data is observed between two adjacent knots. In this case the penalty leads to a stable curve reproduction that smoothly connects the two parts of the function identified by the data.

The main difficulty now is to select a sensible value for the smoothing parameter. Although in the simple scatterplot smoothing model several strategies can be applied, such as a line search for minimizing AIC or a cross validation criterion, the task becomes increasingly difficult in more complex models with multiple smoothing parameters. We will outline MCMC-based full Bayes estimation later on in this section and will discuss further possibilities in more detail in Chapter 4. For the moment, we will simply assume that the smoothing parameter is given and can be treated as a fixed constant.

For B-splines, deriving a suitable penalty is a bit more complicated since no explicit differentiation between a global polynomial and local deviations from such a global function is available. Instead, we use similar penalization ideas as in the simple time-series models at the beginning of this chapter. However, instead of assigning a difference-based roughness penalty to the function evaluations, we shift the penalty to the level of the regression coefficients (Eilers and Marx, 1996). This yields the penalized least-squares criterion

$$\mathrm{PLS}(\lambda) = \sum_{i=1}^{n}\left(y_i - \sum_{k=1}^{K}\gamma_k B_k(x_i)\right)^2 + \lambda \sum_{k=d+1}^{K}(\Delta^d \gamma_k)^2,$$

where for example

$$\begin{aligned}\Delta^1\gamma_k &= \gamma_k - \gamma_{k-1}\\ \Delta^2\gamma_k &= \Delta^1\gamma_k - \Delta^1\gamma_{k-1} = \gamma_k - 2\gamma_{k-1} + \gamma_{k-2}\end{aligned}$$

are differences in the parameter sequence.

Again, the penalty term can be expressed as a quadratic form. Considering difference matrices $\boldsymbol{D}_d$ as well as penalty matrices $\boldsymbol{K}$ defined in (2.3), (2.4) and (2.5) (page 21), leads to penalties

$$\mathrm{pen}(\boldsymbol{\gamma}) = \sum_{k=d+1}^{K}(\Delta^d\gamma_k)^2 = \boldsymbol{\gamma}'\boldsymbol{D}'\boldsymbol{D}\boldsymbol{\gamma} = \boldsymbol{\gamma}'\boldsymbol{K}\boldsymbol{\gamma}$$

and the corresponding penalized least-squares criterion can be written in matrix notation as

$$\mathrm{PLS}(\lambda) = (\boldsymbol{y} - \boldsymbol{X}\boldsymbol{\gamma})'(\boldsymbol{y} - \boldsymbol{X}\boldsymbol{\gamma}) + \lambda\boldsymbol{\gamma}'\boldsymbol{K}\boldsymbol{\gamma}. \tag{2.24}$$

Note that $\mathrm{PLS}(\lambda)$ has exactly the same form for both the approach based on truncated polynomials and the B-spline approach. We will see in the remainder of the book that several modelling possibilities in nonparametric regression models and also in the context of spatial models can be represented in terms of quadratic penalties allowing us to derive quite general and generic inferential procedures.

Differentiating the PLS criterion with respect to $\boldsymbol{\gamma}$ yields

$$-2\boldsymbol{X}'\boldsymbol{y} + 2(\boldsymbol{X}'\boldsymbol{X} + \lambda\boldsymbol{K})\boldsymbol{\gamma}$$

and setting this equation equal to $\mathbf{0}$ leads to the PLS estimate

$$\hat{\gamma} = (\boldsymbol{X}'\boldsymbol{X} + \lambda\boldsymbol{K})^{-1}\boldsymbol{X}'\boldsymbol{y}.$$

The vector of function evaluations is then given by

$$\hat{\boldsymbol{f}} = \boldsymbol{X}\hat{\gamma} = \boldsymbol{X}(\boldsymbol{X}'\boldsymbol{X} + \lambda\boldsymbol{K})^{-1}\boldsymbol{X}'\boldsymbol{y},$$

which corresponds to ridge regression type estimates.

Similarly to the time-series models, the straightforward formulation of the difference penalty above only makes sense for equidistant knots. Otherwise weighted differences have to be introduced similar to those discussed for random walks on page 30. In both cases, the general structure of the quadratic penalty $\boldsymbol{\gamma}'\boldsymbol{K}\boldsymbol{\gamma}$ remains but $\boldsymbol{K}$ is of a more general form. For example, for first order differences the penalty matrix is constructed as $\boldsymbol{K} = \boldsymbol{D}'\boldsymbol{\Delta}\boldsymbol{D}$, where $\boldsymbol{\Delta} = \text{diag}(\delta_2, \dots, \delta_K)$ is a diagonal matrix containing the distances $\delta_k = \kappa_k - \kappa_{k-1}$ between adjacent knots (see Fahrmeir and Lang, 2001b; Rue and Held, 2005, Sec. 3.5; and Lindgren and Rue, 2008, for more general discussions of weighted differences). For penalized splines based on truncated polynomials, no modification is necessary in case of non-equidistant knots since the simple ridge-type penalty still enforces smooth estimates. In our experience, the consideration of non-equidistant knots is often not necessary, at least if the data are not distributed too unevenly across the domain of the covariate. In other situations, a quantile-based placement may be advantageous to place more knots where more information is available.

Similarly to the unpenalized polynomial splines, certain equivalences between penalized splines based on the TPS basis and the B-spline basis can be shown. However, results are less general and specific conditions on smoothing parameters, spline degrees and difference orders have to be fulfilled. To be more specific, for equidistant knots a penalized spline fit based on B-spline basis functions of degree D, difference order d and smoothing parameter λ is equivalent to a penalized spline fit based on the TPS-basis with degree D and smoothing parameter $\tilde{\lambda}$ if (and only if) $d = D + 1$ and $\tilde{\lambda} = (D!\,h^d)\lambda$, where $h = \kappa_{j+1} - \kappa_j$ is the distance between two adjacent knots. Note that, in particular, the difference order and the spline degree have to be linked in a very specific way to transform a B-spline-based penalized spline problem to the TPS basis. We therefore consider the B-spline basis to be the more appropriate choice since it allows for additional flexibility in formulating models by decoupling the spline degree D from the difference order d. Note also that we assume fixed smoothing parameters in the derivation of the equivalence relationships. Once the smoothing parameter is estimated, there may (and typically will) be differences in the fits obtained with the TPS basis and the B-spline basis even if the degrees and the order of penalization are matched as given above.

Figure 2.9 displays penalized spline fits for the global warming data obtained with B-splines and difference penalties of first and second order. The smoothing parameter obviously has the desired impact on the function estimates with

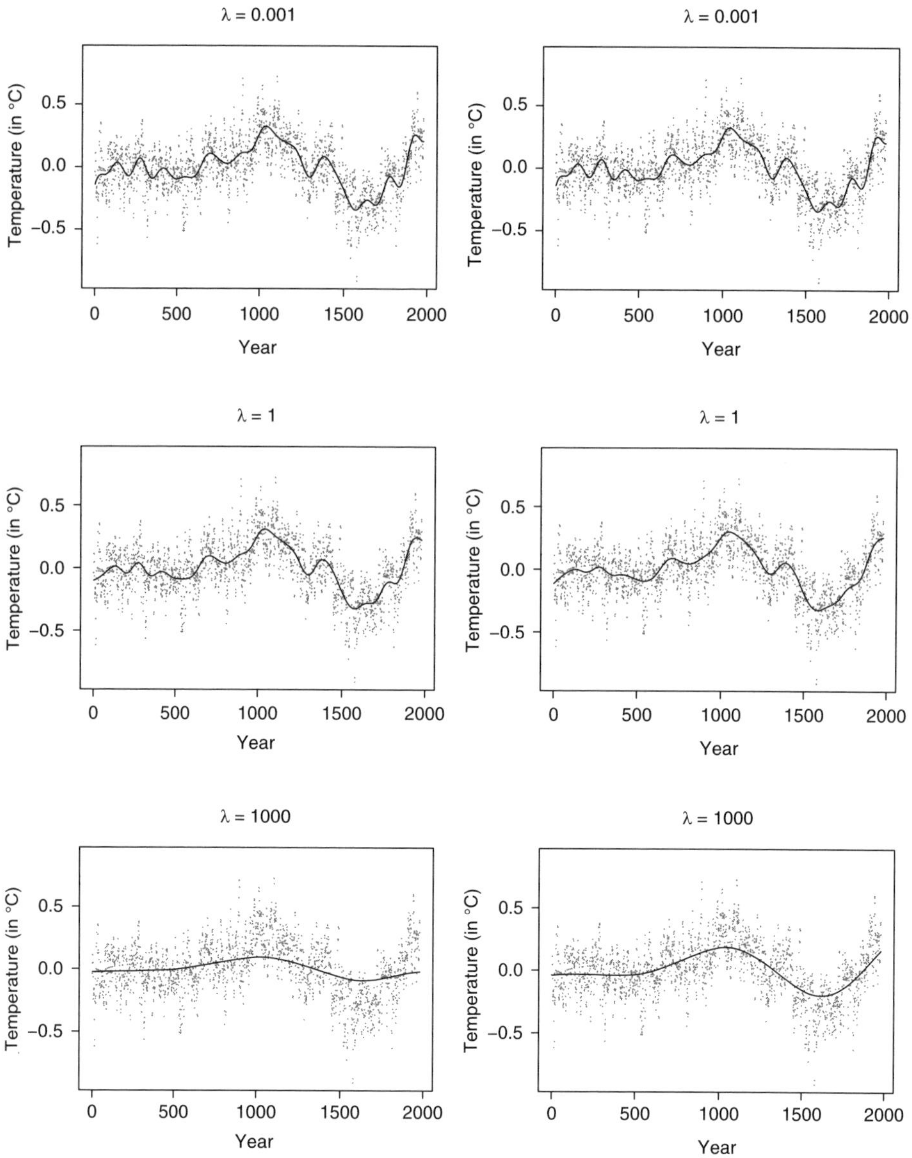

Fig. 2.9 Global warming: P-spline fits obtained with B-spline bases of degree $D = 3$, 20 inner knots and first order difference penalty (left column) and second order difference penalty (right column) for varying values of the smoothing parameter λ.

smaller values corresponding to wiggly functions and large values leading to smooth estimates. As compared to the results obtained with random walks in Section 2.1.1, the results are qualitatively similar but smoother. This is due to the assumption of piecewise cubic polynomials instead of a piecewise constant function.

In the limiting case $\lambda \to \infty$, the function estimate approaches a $(d-1)$-th order polynomial in x, provided that the degree of the spline is larger or equal to the order of the difference penalty. This property can be derived most easily from considering the derivative of a polynomial spline in the B-spline representation (compare de Boor, 2001, pp. 115). For a single basis function, the first derivative is given by

$$\frac{\partial}{\partial x} B_k^D(x) = D \cdot \left(\frac{1}{\kappa_{k+D} - \kappa_k} B_k^{D-1}(x) - \frac{1}{\kappa_{k+D+1} - \kappa_{k+1}} B_{k+1}^{D-1}(x) \right).$$

Adding the derivatives for the whole polynomial spline together yields

$$\frac{\partial}{\partial x} \sum_k \gamma_k B_k^D(x) = D \cdot \sum_k \frac{\gamma_k - \gamma_{k-1}}{\kappa_{k+D} - \kappa_k} B_k^{D-1}(x) \tag{2.25}$$

and the derivative essentially depends on differences in adjacent regression coefficients. This is particularly true for B-splines based on equidistant knots and similar expressions depending on higher order differences can be derived for higher order derivatives. Therefore, the difference penalty can, on the one hand, be interpreted as an approximation to a roughness penalty based on squared derivatives and, on the other hand, it becomes obvious that a $(d-1)$-th order polynomial remains unpenalized. If $\lambda \to \infty$, the influence of the penalty overwhelms the influence of the likelihood and we end up with a function minimizing the penalty term. This, in turn, will be a polynomial of the order $d-1$.

Customized penalties

It is also possible to derive penalties that leave different, non-polynomial function types unpenalized. For example, Eilers and Marx (1996) propose modifying the second order difference penalty to

$$\sum_{k=3}^{K} (\gamma_k - 2c\gamma_{k-1} + \gamma_{k-2})^2$$

where the only additional quantity is the constant $c = \cos(2\pi\delta_\kappa/p)$ determined by the distance δ_κ between two knots and the parameter p. In this special case, the limiting case for $\lambda \to \infty$ gives

$$\gamma_k = \gamma_0 \cos(2\pi k \delta_\kappa / p + \phi)$$

i.e. the limit is a cosine function with period p. The amplitude γ_0 and the phase ϕ are determined by the data (similar to the coefficients of the global polynomial when employing the conventional difference penalty). Penalties of this type may, for example, be useful when fitting data that approximately follow a cyclic pattern and where a global polynomial is not a suitable description for a simple model fitting the data. Other types of non-standard penalties have been derived in the context of L-splines in Welham *et al.* (2006), leaving, for example, an exponential growth curve unpenalized.

Bayesian P-splines

From a Bayesian perspective, all types of penalties considered in the previous paragraphs correspond to specific prior distributions. The ridge-type penalty in case of the *TPS basis* is equivalent to noninformative priors

$$p(\gamma_k) \propto \text{const}, \quad k = 1, \dots, D+1,$$

on the global polynomial coefficients and an i.i.d. Gaussian prior

$$\gamma_k \sim \mathrm{N}(0, \tau^2), \quad k = D+2, \dots, K,$$

on the truncated polynomials. Assuming σ^2 and τ^2 to be known or given, the posterior distribution is

$$\begin{aligned} p(\gamma|\boldsymbol{y}) &\propto p(\boldsymbol{y}|\gamma)p(\gamma) \\ &\propto \exp\left(-\frac{1}{2\sigma^2}(\boldsymbol{y} - \boldsymbol{X}\gamma)'(\boldsymbol{y} - \boldsymbol{X}\gamma)\right) \exp\left(-\frac{1}{2\tau^2}\gamma'\boldsymbol{K}\gamma\right), \end{aligned} \tag{2.26}$$

where the penalty matrix $\boldsymbol{K} = \mathrm{diag}(\boldsymbol{0}_{D+1}, \boldsymbol{I}_{M-1})$ corresponds to the precision matrix of the joint distribution of the vector $\boldsymbol{\gamma}$. Since all parameters are independent *a priori*, this joint distribution is determined as the product of $D+1$ noninformative, flat priors and $M-1$ Gaussian priors $\mathrm{N}(0, \tau^2)$. Note that the joint prior is partially improper due to the impropriety of the priors for the first $D+1$ coefficients. Therefore the matrix $\boldsymbol{K}$ does not have full rank and the corresponding covariance matrix does not exist.

Taking logarithms of the posterior and multiplying with $-2\sigma^2$ yields the penalized least-squares criterion (2.24) with smoothing parameter $\lambda = \sigma^2/\tau^2$ so that the posterior mode and the penalized least-squares estimate coincide (for fixed variances). Moreover, the posterior is Gaussian and therefore the posterior mode also equals the posterior mean (as in the random-walk models considered in Section 2.1.1). Completing squares in the quadratic form of (2.26), we obtain the posterior explicitly, and in complete analogy to (2.16) in Section 2.1, as a multivariate Gaussian

$$\gamma|\boldsymbol{y} \sim \mathrm{N}(\boldsymbol{\mu}, \boldsymbol{\Sigma}), \tag{2.27}$$

with posterior covariance matrix $\boldsymbol{\Sigma} = \mathrm{Cov}(\boldsymbol{\gamma}|\boldsymbol{y})$ and posterior mean $\boldsymbol{\mu} = \mathrm{E}(\boldsymbol{\gamma}|\boldsymbol{y})$ given by

$$\boldsymbol{\Sigma} = \left(\frac{1}{\sigma^2}\boldsymbol{X}'\boldsymbol{X} + \frac{1}{\tau^2}\boldsymbol{K}\right)^{-1} = \sigma^2\left(\boldsymbol{X}'\boldsymbol{X} + \lambda\boldsymbol{K}\right)^{-1},$$

$$\boldsymbol{\mu} = \frac{1}{\sigma^2}\boldsymbol{\Sigma}\boldsymbol{X}'\boldsymbol{y} = (\boldsymbol{X}'\boldsymbol{X} + \lambda\boldsymbol{K})^{-1}\boldsymbol{X}'\boldsymbol{y}.$$

The prior distribution that is equivalent to the difference penalty for the *B-spline basis* is a random-walk prior, for example a first-order random-walk ($d = 1$)

$$\gamma_k = \gamma_{k-1} + u_k, \quad u_k \sim \mathrm{N}(0, \tau^2), \quad k = 2, \ldots, K,$$

with noninformative prior choices $p(\gamma_1) \propto$ const for the first coefficient or a second-order random-walk ($d = 2$)

$$\gamma_k = 2\gamma_{k-1} - \gamma_{k-2} + u_k, \quad u_k \sim \mathrm{N}(0, \tau^2), \quad k = 3, \ldots, K,$$

and $p(\gamma_1, \gamma_2) \propto$ const for the starting values.

As for the random-walk models in Section 2.1, the joint prior for $\boldsymbol{\gamma}$ is a *partially improper* Gaussian smoothness prior

$$p(\boldsymbol{\gamma}) \propto \left(\frac{1}{\tau^2}\right)^{\mathrm{rank}(\boldsymbol{K})/2} \exp\left(-\frac{1}{2\tau^2}\boldsymbol{\gamma}'\boldsymbol{K}\boldsymbol{\gamma}\right), \tag{2.28}$$

with penalty matrices $\boldsymbol{K}$ as in (2.5) and $\mathrm{rank}(\boldsymbol{K}) = K - d$ for first ($d = 1$) and second ($d = 2$) order random-walk priors. Similar considerations as for the truncated power series yield that, again, the posterior mode and the posterior mean estimates equal the penalized least-squares estimate. The Bayesian formulation also allows for further interpretation of the smoothing parameter in terms of the relation of error variance σ^2 and prior variance τ^2. This reflects the following intuition: If the prior knowledge about $\boldsymbol{\gamma}$ is very precise and therefore τ^2 small relative to the error variance, this suggests a polynomial of order $d - 1$ enforced by a large value of the smoothing parameter. This property can also be derived from considering conditional distributions induced by the random-walk prior.

The (sequential) conditional distribution following from the assumption of a first-order random-walk prior is given by

$$\gamma_k|\gamma_{k-1}, \ldots, \gamma_1 \sim \mathrm{N}(\gamma_{k-1}, \tau^2),$$

that is, in expectation the sequence of regression coefficients remains constant, with the magnitude of the deviations from the expectation governed by the smoothness variance τ^2 (see Fig. 2.10 for a visual representation). As a consequence, the limiting case $\tau^2 \to 0$ (implying $\lambda \to \infty$), results in a constant sequence of regression coefficients and therefore a constant function estimate.

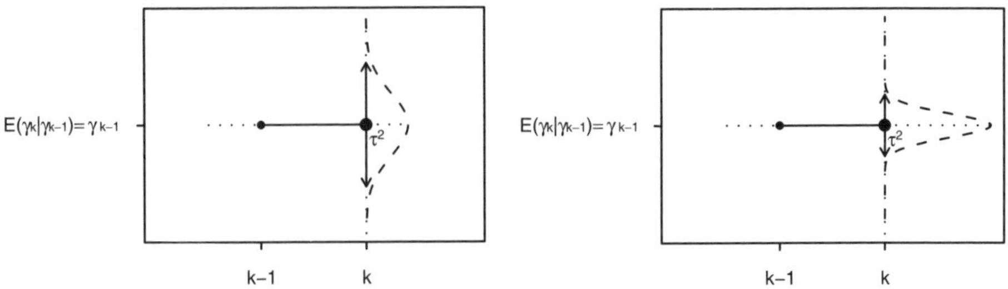

Fig. 2.10 Conditional prior distribution induced by a first-order random-walk prior for $\boldsymbol{\gamma}$. The left figure shows a prior distribution with large smoothing variance τ^2, leading to a wiggly fit. In the right figure, the smoothing variance is smaller leading to a more precise 'prediction' of γ_k based on its predecessor and therefore a smoother function estimate.

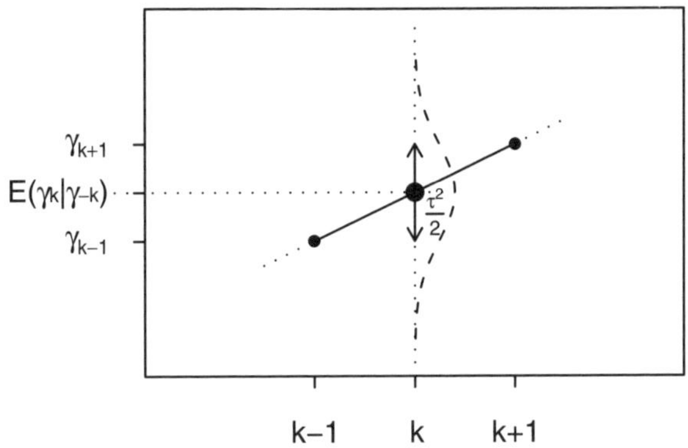

Fig. 2.11 Locally linear fit induced by a first-order random-walk prior for $\boldsymbol{\gamma}$.

Instead of considering the sequential conditional distribution, we can also calculate the conditional distribution given all other regression coefficients and not only the preceding ones. Except for the boundary coefficients, this results in

$$\gamma_k|\boldsymbol{\gamma}_{-k} \sim \mathrm{N}\left(\frac{1}{2}(\gamma_{k-1}+\gamma_{k+1}), \frac{\tau^2}{2}\right), \quad k=2,\ldots,K-1.$$

where $\boldsymbol{\gamma}_{-k}$ denotes all regression coefficients except γ_k, that is, in expectation we obtain the average of the two adjacent parameters (compare Fig. 2.11). Again, ample deviations from this locally linear fit are allowed for large values of τ^2 while small values lead to estimates that are close to the local linear fit.

Similar considerations can be conducted for the second-order random-walk prior, where the sequential conditional distribution yields a linear extrapolation based on the two preceding coefficients (Fig. 2.12) corresponding to a linear

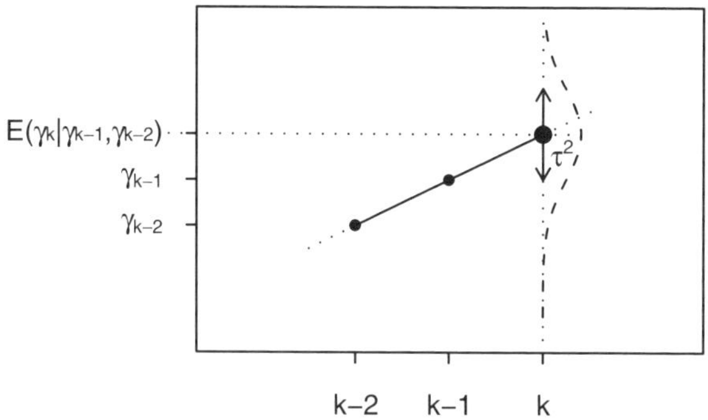

Fig. 2.12 Conditional distribution induced by a second-order random-walk prior for γ.

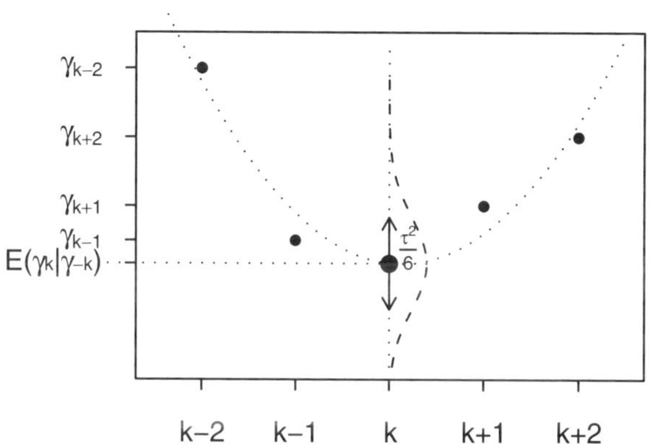

Fig. 2.13 Locally quadratic fit induced by a second-order random-walk prior for γ.

function estimate. The conditional distribution given all remaining coefficients leads to a locally quadratic fit (Fig. 2.13) based on the four nearest neighbours. This observation also explains why typically results obtained from second-order random priors are visually smoother than the ones from first-order random walks: The prior distribution pools information from a larger neighbourhood and therefore induces a larger amount of local smoothness.

Note that the random-walk models considered in Section 2.1 can actually be interpreted as special cases of penalized splines. To see this, consider splines of degree $D = 0$ leading to a piecewise constant fit. Placing a knot at each distinct observed value of the covariate x then defines a piecewise constant function with jumps at all observation points as for the random-walk model. The equivalence holds exactly when the distinct observed values are unique and equidistant (as in the time series examples considered at the beginning of the chapter) while weighted versions have to be considered in the more general case. In any case, the general structure of quadratic penalties remains and, as a consequence, estimation schemes and other considerations for penalized splines derived in the subsequent sections and chapters can straightforwardly be adapted to random-walk models.

Bayesian inference

When σ^2 and τ^2 are known or given, the posterior for $\boldsymbol{\gamma}$ is Gaussian as in (2.28), but with design matrix $\boldsymbol{X}$ in (2.23) defined through evaluations of B-spline basis functions and precision matrix $\boldsymbol{K}$ defined in terms of difference matrices as in (2.5). In practice, the observation variance σ^2 and the variance parameter τ^2, controlling the amount of smoothing, are unknown and have to be estimated from the data.

In this book, we will focus on two different concepts for estimating regression or basis function coefficients: empirical Bayes (EB) inference and full Bayes (FB) inference. *Empirical Bayes* inference considers τ^2 and σ^2 (and more generally hyperparameters in the regression model) as unknown constants which have to be estimated from their marginal posterior. We will make use of the close connection between mixed models and penalized smoothing for deriving inferential procedures in the EB framework (see Section 4.2.1).

Full Bayes inference treats the hyperparameters σ^2 and τ^2 as additional unknowns that are also assumed to be random variables, supplemented with appropriate priors. Inference is then typically based on Markov chain Monte Carlo (MCMC) simulation techniques. In the context of Bayesian P-splines, inverse gamma priors $\mathrm{IG}(a_\sigma, b_\sigma)$ and $\mathrm{IG}(a, b)$, with

$$\begin{aligned} p(\sigma^2) &\propto \frac{1}{(\sigma^2)^{a_\sigma+1}} \exp\left(-\frac{b_\sigma}{\sigma^2}\right), \\ p(\tau^2) &\propto \frac{1}{(\tau^2)^{a+1}} \exp\left(-\frac{b}{\tau^2}\right), \end{aligned} \tag{2.29}$$

which are proper for $a_\sigma > 0$, $b_\sigma > 0$ and $a > 0$, $b > 0$, are a popular choice. The main reasons are the resulting convenient computations arising from the conditional conjugacy of the Gaussian prior or the Gaussian observation model and the inverse gamma distribution for the corresponding variance parameters. More specifically, the resulting full conditionals for the variance parameters are again of the inverse gamma type, yielding a Gibbs sampler drawing directly from the full conditionals $p(\boldsymbol{\gamma}|\cdot) = p(\boldsymbol{\gamma}|\sigma^2, \tau^2, \boldsymbol{y})$, $p(\sigma^2|\cdot) = p(\sigma^2|\boldsymbol{\gamma}, \tau^2, \boldsymbol{y})$ and $p(\tau^2|\cdot) = p(\tau^2|\boldsymbol{\gamma}, \sigma^2, \boldsymbol{y})$.

The full conditional $p(\boldsymbol{\gamma}|\cdot)$, given σ^2 and τ^2, is the Gaussian posterior (2.27), that is,

$$\boldsymbol{\gamma}|\cdot \sim \mathrm{N}(\boldsymbol{\mu}, \boldsymbol{\Sigma}) \tag{2.30}$$

with

$$\boldsymbol{\Sigma} = \left(\frac{1}{\sigma^2}\boldsymbol{X}'\boldsymbol{X} + \frac{1}{\tau^2}\boldsymbol{K}\right)^{-1} = \sigma^2\left(\boldsymbol{X}'\boldsymbol{X} + \lambda\boldsymbol{K}\right)^{-1},$$

$$\boldsymbol{\mu} = \frac{1}{\sigma^2}\boldsymbol{\Sigma}\boldsymbol{X}'\boldsymbol{y} = (\boldsymbol{X}'\boldsymbol{X} + \lambda\boldsymbol{K})^{-1}\boldsymbol{X}'\boldsymbol{y}.$$

The full conditionals for σ^2 and τ^2 are derived from

$$p(\sigma^2|\cdot) \propto p(\boldsymbol{y}|\boldsymbol{\gamma}, \sigma^2)p(\sigma^2),$$
$$p(\tau^2|\cdot) \propto p(\boldsymbol{\gamma}|\tau^2)p(\tau^2)$$

with the Gaussian observation model for $\boldsymbol{y}|\boldsymbol{\gamma}, \sigma^2$ and the prior (2.28) for $\boldsymbol{\gamma}$, given τ^2. Combining likelihood and prior information, it is seen that both full conditional distributions are again inverse gamma distributions

$$\sigma^2|\cdot \sim \mathrm{IG}(\tilde{a}_\sigma, \tilde{b}_\sigma),$$
$$\tau^2|\cdot \sim \mathrm{IG}(\tilde{a}, \tilde{b})$$

but with updated parameters

$$\tilde{a}_\sigma = a_\sigma + \frac{n}{2},$$
$$\tilde{b}_\sigma = b_\sigma + \frac{1}{2}(\boldsymbol{y} - \boldsymbol{X}\boldsymbol{\gamma})'(\boldsymbol{y} - \boldsymbol{X}\boldsymbol{\gamma}),$$

and

$$\tilde{a} = a + \frac{\mathrm{rank}(\boldsymbol{K})}{2},$$
$$\tilde{b} = b + \frac{1}{2}\boldsymbol{\gamma}'\boldsymbol{K}\boldsymbol{\gamma}. \tag{2.31}$$

To make the inverse gamma priors ‘diffuse’, it is often recommended to choose small values $\varepsilon > 0$ for the hyperparameters, for example, $a = b = 0.1$, 0.01 or 0.001. This choice is motivated by the fact that the limiting case $a \to 0$, $b \to 0$ is Jeffreys’ noninformative but improper prior $p(\tau^2) \propto \tau^{-2}$. However, its use can lead to improper posteriors, with MCMC samplers unable to detect this. As a result, ‘diffuse’ IG priors with small a, b, for example, $a = b = 0.1, 0.01$ or 0.001 have been suggested. Recent research has shown, however, that these diffuse $\mathrm{IG}(\varepsilon, \varepsilon)$-priors can actually be quite informative in Bayesian variance component models with a small number of groups or units (Natarajan and Kass; 2000, Gelman, 2006). In our experience, $\mathrm{IG}(\varepsilon, \varepsilon)$-priors are a comparably robust choice for the error variance σ^2 in Gaussian models and for the variances (or inverse smoothing parameters) in Gaussian smoothness priors for nonparametric functions (unless ε is chosen to be extremely small). Nevertheless, it is good practice—at least in sparse data situations—to check sensitivity by carrying out analyses with different values of the hyperparameters a, b.

Alternative priors, suggested in Gelman (2006), are uniform priors $p(\tau) \propto$ const for standard deviations τ or the folded-t class of priors for τ^2, in particular the folded-Cauchy density

$$p(\tau^2) \propto (\tau^2 + s^2)^{-1},$$

with s^2 as a hyperparameter. The uniform prior for the standard deviation can also be expressed in form of an improper IG-type prior with hyperparameters $a = -0.5$, $b = 0$. Sun *et al.* (2001) and Fahrmeir and Kneib (2009) provide sufficient conditions for propriety of posteriors in this case. Another 'uniform shrinkage' prior has been developed in Li *et al.* (2010). Their analyses as well as those in Zhao *et al.* (2006) using folded-Cauchy priors confirm our own experience that IG$(\varepsilon, \varepsilon)$-priors are rather robust and only weakly informative for posterior estimation of functions. Frühwirth-Schnatter and Wagner (2010) suggest a normal prior on the standard deviation, that is, $\tau \sim \mathrm{N}(0, B_0)$ which marginally leads to a Ga$(1/2, 1/B_0)$ prior for τ^2 and has the advantage of improving MCMC if τ^2 is small.

Since the full conditionals are all known distributions, we obtain:

Algorithm 2.1: Gibbs sampler for Bayesian P-splines

i. Update the current value $\boldsymbol{\gamma}^c$ of the Markov chain by drawing random numbers from the multivariate Gaussian distribution (2.30), inserting current values $(\sigma^2)^c$ and $(\tau^2)^c$ in $\boldsymbol{\mu}$ and $\boldsymbol{\Sigma}$. For B-spline bases, the covariance matrix $\boldsymbol{\Sigma}$ has band structure allowing computationally efficient drawings based on Cholesky decompositions for sparse matrices (see Lang and Brezger, 2004).
ii. Update τ^2 by drawing from the inverse gamma distribution IG$(\tilde{a}, \tilde{b})$, inserting the (new) current value for $\boldsymbol{\gamma}^c$ obtained in step i.
iii. Update σ^2 by drawing from the inverse gamma distribution IG$(\tilde{a}_\sigma, \tilde{b}_\sigma)$, inserting the (new) current value for $\boldsymbol{\gamma}^c$ obtained in step i.

Example 2.5: Global warming

We applied the Gibbs sampler for Bayesian P-splines to the global warming data and are now able to obtain a data-driven estimate of the smoothing parameter too. To assess the sensitivity of the Bayesian estimates with respect to the prior assumptions, we considered four different scenarios for the inverse gamma priors of the variances. Two settings correspond to vague priors with both inverse gamma parameters small. More precisely, we chose $a = b = 0.001$ and $a = b = 0.00001$. The two further settings correspond to flat priors for the variance $(a = -1, b = 0)$ and the standard deviation $(a = -0.5, b = 0)$. In all cases, the

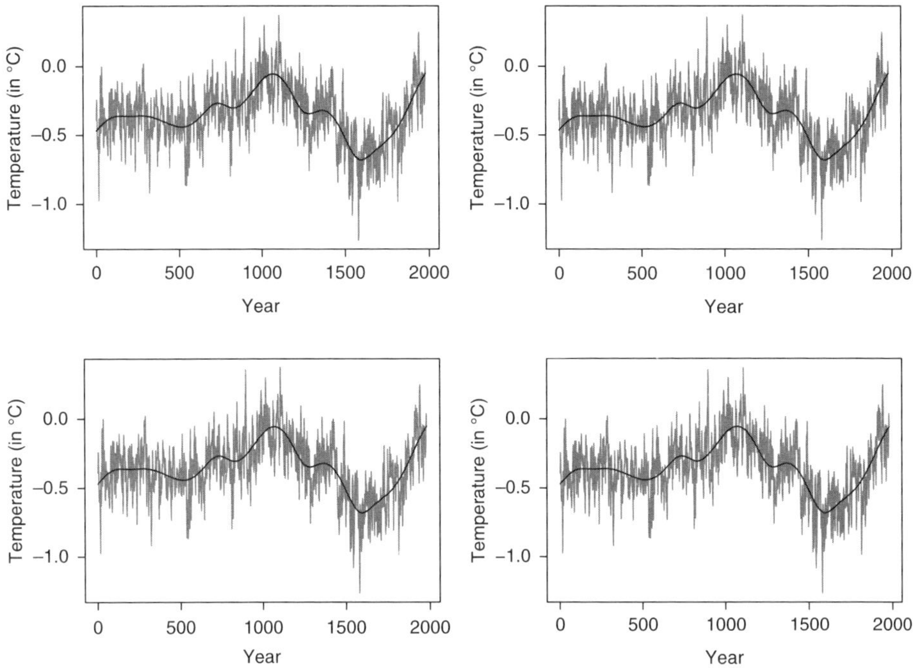

Fig. 2.14 Global warming: Bayesian P-spline fits under different prior settings (top left: $a = b = 0.001$, top right: $a = b = 0.00001$, bottom left: $a = -1, b = 0$, bottom right: $a = -0.5, b = 0$).

same prior choices were applied for the error variance σ^2. Figure 2.14 shows the resulting estimates with only very limited indication of any prior sensitivity. The latter is not too surprising in this example since we have a quite large set of data points (roughly 2,000) to identify a single function described by 22 parameters. All estimated functions indicate a strong and rapid increase of global average temperatures during the twentieth century but also identify local maxima and minima in former times. △

Degrees of freedom of a P-spline

The complexity of a penalized spline fit is characterized by the smoothing parameter λ (or equivalently by the smoothing variance τ^2 and the error variance σ^2). However, we have already seen that the same value of λ may induce different amounts of smoothness when comparing splines of different degree or with different penalty structure and also highly depends on the variability of the response. It is therefore difficult to interpret the absolute value of λ without referring explicitly to a data set. Hence, it would be of interest to obtain a general and intuitive measure of complexity for a spline fit and possibly even more generally for different types of smoothing approaches. This leads to the notion of degrees of freedom that represent a kind of parameter count, taking the effective dimension reduction induced by the penalty into account.

To make the concept of degrees of freedom more generally accessible, we base our considerations on general linear smoothers, where the estimated function $\hat{f}(x)$ in the simple smoothing problem (2.21) can be written as a weighted sum of the observed response values, that is,

$$\hat{f}(x) = \sum_{i=1}^{n} s(x, x_i) y_i = \boldsymbol{s}(x)' \boldsymbol{y},$$

with $\boldsymbol{s}(x) = (s(x, x_1), \ldots, s(x, x_n))'$. The weights are allowed to depend both on the observed covariate values x_i, $i = 1, \ldots, n$, and the point of interest x. In the case of penalized basis function approaches, the vector of weights is simply given by

$$\boldsymbol{s}(x)' = \boldsymbol{x}'(\boldsymbol{X}'\boldsymbol{X} + \lambda \boldsymbol{K})^{-1} \boldsymbol{X}',$$

where $\boldsymbol{x} = (B_1(x), \ldots, B_K(x))'$ contains the basis function evaluations at the point of interest x and $\boldsymbol{X}$ is the design matrix constructed based on the original observations $x_1, \ldots, x_n$. The hat matrix (or smoother matrix) projecting observed responses $\boldsymbol{y}$ on the predictions $\hat{\boldsymbol{y}}$ is then available by combining the weight vectors $\boldsymbol{s}(x_i)$, $i = 1, \ldots, n$, in the matrix $\boldsymbol{S}$, that is, for penalized splines

$$\boldsymbol{S} = \boldsymbol{X}(\boldsymbol{X}'\boldsymbol{X} + \lambda \boldsymbol{K})^{-1} \boldsymbol{X}'.$$

In the case of a linear model without penalization, the number of regression coefficients can be recovered as the trace of the hat matrix. In generalization, Hastie and Tibshirani (1990) define the equivalent degrees of freedom

$$\mathrm{df}(\boldsymbol{S}) = \mathrm{tr}(\boldsymbol{S}) \tag{2.32}$$

as a general complexity measure for linear smoothers. For penalized basis-function approaches, $\mathrm{df}(\boldsymbol{S})$ varies smoothly between the extremal points obtained for $\lambda \to 0$ and $\lambda \to \infty$. In the former case, the penalty effect vanishes and the complexity finally reaches the number of basis functions, that is, $\mathrm{df}(\boldsymbol{S}) = K$. The value for $\mathrm{df}(\boldsymbol{S})$ in the other extreme depends on the structure, or more precisely on the dimension of the null space of the penalty matrix $\boldsymbol{K}$. For P-splines with a B-spline basis we derived a polynomial of order $d - 1$ as the limiting case for $\lambda \to \infty$ and consequently the null space of $\boldsymbol{K}$ is d-dimensional. Hence the limiting value for the degrees of freedom is given by $\mathrm{df}(\boldsymbol{S}) \to d$ if $\lambda \to \infty$.

Note that the explicit computation of the $n \times n$ projection matrix $\boldsymbol{S}$ can (and should) be avoided in practical implementations. Using standard properties of the trace yields

$$\mathrm{tr}(\boldsymbol{S}) = \mathrm{tr}(\boldsymbol{X}(\boldsymbol{X}'\boldsymbol{X} + \lambda \boldsymbol{K})^{-1} \boldsymbol{X}') = \mathrm{tr}((\boldsymbol{X}'\boldsymbol{X} + \lambda \boldsymbol{K})^{-1} \boldsymbol{X}'\boldsymbol{X}),$$

where the last expression only depends on the $K \times K$ dimensional matrices $(\boldsymbol{X}'\boldsymbol{X}+\lambda\boldsymbol{K})^{-1}$ and $\boldsymbol{X}'\boldsymbol{X}$. These are typically required in the estimation process anyway and can therefore be obtained without any computational overhead. Moreover, it turns out that both matrices have a band matrix structure for penalized splines, allowing for additional potential numerical savings (Hutchinson and de Hoog, 1985).

The rearranged expression also reveals a further interpretation of the degrees of freedom. The Fisher information matrix for the spline coefficients is given by $\boldsymbol{F} = \boldsymbol{X}'\boldsymbol{X}/\sigma^2$ while the corresponding penalized version is given by $\boldsymbol{F}_{\text{pen}} = (\boldsymbol{X}'\boldsymbol{X} + \lambda\boldsymbol{K})/\sigma^2$. Therefore the degrees of freedom can also be written as

$$\text{df} = \text{tr}(\boldsymbol{F}\boldsymbol{F}_{\text{pen}}^{-1}),$$

i.e. as the trace of the product of unpenalized Fisher information and inverse penalized Fisher information. This allows us to also define the degrees of freedom in situations where no explicit hat matrix is available. We will, for example, use this extended version in the context of survival models in Chapter 6.

Of course, the complexity definition $\text{df}(\boldsymbol{S})$ also makes sense in a Bayesian context. For simulation-based estimation, however, there will also be a sample of values for the degrees of freedom obtained from the samples for the smoothing variance τ^2 and the error variance σ^2. The mean of these samples will then provide a natural point estimate for the degrees of freedom. Since the dependence of $\text{df}(\boldsymbol{S})$ on λ is nonlinear, one will generally obtain a different result when simply inserting the posterior mean estimate of λ into the definition of $\text{df}(\boldsymbol{S})$.

In the global warming Example 2.5, the posterior mean estimate of the degrees of freedom is in the range of 16.5 and 16.8 for the four different prior specifications. Since we used 20 inner knots and degree 3, the total number of spline parameters is given by 22. The effective degrees of freedom are somewhat lower but the relatively large values reflect that the estimated curve shows a significant amount of nonlinearity that requires enough flexibility.

Several other definitions of effective degrees of freedom have been derived in the literature by generalizing other quantities than the hat matrix from linear models to nonparametric smoothing approaches. One example is the estimated error variance, where the number of parameters appears when constructing an unbiased variance estimate. Hastie and Tibshirani (1990) provide an overview on alternative degrees of freedom definitions but definition (2.32) is still the one most frequently used in practice.

2.2.2 Univariate non-Gaussian observation models

Extending nonparametric regression models from Gaussian responses to more general response types is conceptually easy. Whenever a regression model is defined based on a linear combination of covariate effects, this linear predictor can be replaced by a nonparametric function in combination with augmenting the appropriate penalty term to the likelihood. In particular, in GLM-type exponential family regression, the (conditional) expectation of the response variable can be related to a semiparametric predictor via

$$\mathbb{E}(y|x) = \mu = h(\eta), \quad \eta = f(x) \tag{2.33}$$

where h is a known response function and $f(x)$ represents a nonparametric effect of a continuous covariate x. Expanding $f(x)$ in basis functions yields $f(x) = \boldsymbol{x}'\boldsymbol{\gamma}$, where $\boldsymbol{x} = (B_1(x), \ldots, B_K(x))'$ contains the basis functions evaluated at x and $\boldsymbol{\gamma}$ contains the corresponding regression coefficients. In analogy, the vector of function evaluations $\boldsymbol{f} = (f(x_1), \ldots, f(x_n))'$ can be represented in matrix notation as $\boldsymbol{f} = \boldsymbol{X}\boldsymbol{\gamma}$.

Penalized likelihood estimation

In case of conditionally independent observations, the likelihood for $\boldsymbol{\gamma}$ factors to the product

$$L(\boldsymbol{\gamma}) = p(\boldsymbol{y}|\boldsymbol{\gamma}) = \prod_{i=1}^{n} p(y_i|\boldsymbol{\gamma})$$

where $p(y_i|\boldsymbol{\gamma})$ denotes the likelihood contribution of the i-th observation that is determined by the exponential family type considered in the regression model. In the following, we will ignore the possible presence of an unknown dispersion parameter in the likelihood expressions. The corresponding log-likelihood is given by

$$l(\boldsymbol{\gamma}) = \log(L(\boldsymbol{\gamma})) = \sum_{i=1}^{n} l_i(\boldsymbol{\gamma}),$$

where $l_i(\boldsymbol{\gamma}) = \log p(y_i|\boldsymbol{\gamma})$. Augmenting a quadratic penalty term to the log-likelihood finally yields the penalized log-likelihood criterion

$$l_{\text{pen}}(\boldsymbol{\gamma}) = \sum_{i=1}^{n} l_i(\boldsymbol{\gamma}) - \frac{\lambda}{2}\boldsymbol{\gamma}'\boldsymbol{K}\boldsymbol{\gamma}. \tag{2.34}$$

For Gaussian responses, this penalized log-likelihood criterion effectively coincides with the penalized least-squares criterion considered in the previous section (apart from additive constants and multiplication with non-zero quantities). In this particular case, a closed form solution for $\boldsymbol{\gamma}$ was available, but this is no longer the case for general non-Gaussian response types. We will therefore introduce an iterative estimation procedure based on penalized Fisher scoring that—after suitable modifications—will also prove to be a useful tool in later chapters both in connection with empirical Bayes inference in generalized additive mixed models and for the construction of suitable proposal densities in Markov chain Monte Carlo simulation approaches.

Therefore, we first derive the score function and the Fisher information associated with criterion (2.34). Taking the first derivative yields

$$\boldsymbol{s}_{\text{pen}}(\boldsymbol{\gamma}) = \frac{\partial l_{\text{pen}}(\boldsymbol{\gamma})}{\partial \boldsymbol{\gamma}} = \boldsymbol{s}(\boldsymbol{\gamma}) - \lambda \boldsymbol{K} \boldsymbol{\gamma},$$

where $\boldsymbol{s}(\boldsymbol{\gamma})$ is the score function from an unpenalized GLM with predictor $\boldsymbol{X}\boldsymbol{\gamma}$, that is,

$$\boldsymbol{s}(\boldsymbol{\gamma}) = \sum_{i=1}^{n} \boldsymbol{s}_i(\boldsymbol{\gamma}) = \sum_{i=1}^{n} \boldsymbol{x}_i \frac{d_i}{v_i^2}(y_i - \mu_i),$$

$\boldsymbol{x}_i$ is the i-th row in the design matrix $\boldsymbol{X}$, $v_i^2 = \text{Var}(y_i)$ is the (conditional) variance of the response derived from the exponential family distribution, and

$$d_i = \frac{\partial h(\eta_i)}{\partial \eta}$$

is the derivative of the response function. Combining the respective quantities in matrix-vector notation yields the expression

$$\boldsymbol{s}_{\text{pen}}(\boldsymbol{\gamma}) = \boldsymbol{X}'\boldsymbol{D}\boldsymbol{V}^{-1}(\boldsymbol{y} - \boldsymbol{\mu}) - \lambda \boldsymbol{K}\boldsymbol{\gamma},$$

where

$$\boldsymbol{D} = \text{diag}(d_1, \ldots, d_n), \quad \boldsymbol{V} = \text{diag}(v_1^2, \ldots, v_n^2).$$

From the penalized score function, we derive the penalized Fisher information matrix as the negative derivative, that is,

$$\boldsymbol{F}_{\text{pen}}(\boldsymbol{\gamma}) = -\frac{\partial \boldsymbol{s}_{\text{pen}}(\boldsymbol{\gamma})}{\partial \boldsymbol{\gamma}} = \boldsymbol{F}(\boldsymbol{\gamma}) + \lambda \boldsymbol{K},$$

where

$$\boldsymbol{F}(\boldsymbol{\gamma}) = \sum_{i=1}^{n} \boldsymbol{x}_i w_i \boldsymbol{x}_i' = \boldsymbol{X}'\boldsymbol{W}\boldsymbol{X}$$

is the expected Fisher information from a standard GLM with predictor $\boldsymbol{\eta} = \boldsymbol{X}\boldsymbol{\gamma}$ and weights

$$w_i = \frac{d_i^2}{v_i^2}, \quad \boldsymbol{W} = \text{diag}(w_1, \ldots, w_n).$$

Based on the score function and the Fisher information matrix, an iterative Fisher scoring algorithm that determines the penalized maximum likelihood estimate $\hat{\boldsymbol{\gamma}}$ is given as follows: Based on some starting value $\hat{\boldsymbol{\gamma}}^{(0)}$ (e.g. the penalized least-squares estimate), iteratively update

$$\hat{\boldsymbol{\gamma}}^{(k+1)} = \hat{\boldsymbol{\gamma}}^{(k)} + \boldsymbol{F}_{\text{pen}}(\hat{\boldsymbol{\gamma}}^{(k)})^{-1} \boldsymbol{s}_{\text{pen}}(\hat{\boldsymbol{\gamma}}^{(k)}) \tag{2.35}$$

until the algorithm converges.

The Fisher scoring algorithm can also be rewritten in a form that emphasizes its close connection to penalized weighted least-squares estimation. Defining the working observations

$$\tilde{\boldsymbol{y}} = \boldsymbol{X}\boldsymbol{\gamma} + \boldsymbol{D}^{-1}(\boldsymbol{y} - \boldsymbol{\mu})$$

yields an alternative representation of iteration (2.35) as

$$\hat{\boldsymbol{\gamma}}^{(k+1)} = (\boldsymbol{X}'\boldsymbol{W}^{(k)}\boldsymbol{X} + \lambda\boldsymbol{K})^{-1}\boldsymbol{X}'\boldsymbol{W}^{(k)}\tilde{\boldsymbol{y}}^{(k)}$$

where $\boldsymbol{W}^{(k)}$ and $\tilde{\boldsymbol{y}}^{(k)}$ are used as shorthand notation for the corresponding quantities evaluated at the current estimate $\hat{\boldsymbol{\gamma}}^{(k)}$. In this form, the penalized Fisher scoring algorithm is also called iteratively weighted least-squares (IWLS) estimation since it relies on repeated (penalized) least-squares fits with updated weights and working responses. Quantities such as the effective degrees of freedom are then frequently defined based on the working response and working weights in the final estimation iteration. This effectively coincides with the assumption of a working model

$$\tilde{\boldsymbol{y}} \sim \mathrm{N}(\boldsymbol{X}\boldsymbol{\gamma}, \boldsymbol{W}^{-1}).$$

Bayesian inference

From a Bayesian perspective, a nonparametric regression model is defined hierarchically by the observation model

$$p(\boldsymbol{y}|\boldsymbol{\gamma}) = \prod_{i=1}^{n} p(y_i|\boldsymbol{\gamma}),$$

determined through the choice of the specific exponential family and response function, and the smoothness prior

$$p(\boldsymbol{\gamma}|\tau^2) \propto \left(\frac{1}{\tau^2}\right)^{\mathrm{rank}(\boldsymbol{K})/2} \exp\left(-\frac{1}{2\tau^2}\boldsymbol{\gamma}'\boldsymbol{K}\boldsymbol{\gamma}\right),$$

with variance (or inverse smoothing) parameter $\tau^2 = 1/\lambda$. Note that we are treating the dispersion parameter as fixed (at $\phi = 1$) so that the smoothing parameter is now given by $\lambda = 1/\tau^2$ instead of $\lambda = \sigma^2/\tau^2$ as in the Gaussian case. Assuming for the moment τ^2 to be known or given, the penalized likelihood estimate obtained via penalized Fisher scoring coincides with the posterior mode, that is, the maximizer of the posterior density

$$p(\boldsymbol{\gamma}|\boldsymbol{y}) \propto p(\boldsymbol{y}|\boldsymbol{\gamma})p(\boldsymbol{\gamma}|\tau^2).$$

However, in contrast to the Gaussian setup, the posterior is not available in closed analytical form and the posterior mode does in general no longer coincide with the posterior mean since the likelihood is not symmetric in $\boldsymbol{\gamma}$.

To estimate τ^2 jointly with $\boldsymbol{\gamma}$, we will again focus on the two concepts empirical Bayes (EB) and full Bayes (FB) inference. *Empirical Bayes* inference is described in more detail in subsequent chapters and in particular in Section 4.2.1. For *full Bayes* inference, a prior has to be specified for τ^2. As in the Gaussian case, we focus on IG(a, b) priors, with density $p(\tau^2)$ as in (2.29), page 52. For MCMC sampling from the posterior

$$p(\boldsymbol{\gamma}, \tau^2|\boldsymbol{y}) \propto p(\boldsymbol{y}|\boldsymbol{\gamma})p(\boldsymbol{\gamma}|\tau^2)p(\tau^2)$$

we consider the full conditionals

$$\begin{aligned} p(\boldsymbol{\gamma}|\tau^2, \boldsymbol{y}) &\propto p(\boldsymbol{y}|\boldsymbol{\gamma})p(\boldsymbol{\gamma}|\tau^2), \\ p(\tau^2|\boldsymbol{\gamma}, \boldsymbol{y}) &\propto p(\boldsymbol{\gamma}|\tau^2)p(\tau^2) \end{aligned}$$

for $\boldsymbol{\gamma}$ and τ^2. The full conditional $p(\tau^2|\cdot)$ is the same as in the Gaussian case, that is,

$$\tau^2|\cdot \sim \text{IG}(\tilde{a}, \tilde{b})$$

with updated hyperparameters $\tilde{a}$ and $\tilde{b}$ as in (2.31), page 53. For non-Gaussian responses, the full conditional $p(\boldsymbol{\gamma}|\cdot)$ is no longer available in closed form so that a Metropolis–Hastings (MH) algorithm is needed. Among several alternatives, we focus on an MH algorithm with *iteratively weighted least-squares* (IWLS) proposals introduced by Gamerman (1997) for generalized linear mixed models. Similar proposals have been made by Hastie and Tibshirani (2000), Rue (2001) or Brezger and Lang (2006). MH algorithms with IWLS proposals allow for automatic block-updating of the possibly high-dimensional parameter vector $\boldsymbol{\gamma}$ without tuning problems and can be used as a building block for more complex models considered in later chapters. The basic idea is to approximate the full conditional $p(\boldsymbol{\gamma}|\cdot)$ by a Gaussian distribution with a mean and precision matrix corresponding to (an approximation of) the mode and penalized Fisher information. This approximate normal distribution is then used as the proposal density in the MH algorithm. Roughly, this can be summarized as a Gaussian approximation that matches the mode of the full conditional and the curvature at the mode.

More precisely, let $\varphi(\cdot; \boldsymbol{\mu}, \boldsymbol{\Sigma})$ denote a multivariate Gaussian density with covariance matrix and mean

$$\boldsymbol{\Sigma} = \left(\boldsymbol{X}'\boldsymbol{W}\boldsymbol{X} + \frac{1}{\tau^2}\boldsymbol{K}\right)^{-1}, \qquad \boldsymbol{\mu} = \boldsymbol{\Sigma}\boldsymbol{X}'\boldsymbol{W}\tilde{\boldsymbol{y}}$$

with GLM weights $\boldsymbol{W} = \boldsymbol{W}(\boldsymbol{\gamma})$ and working observations $\tilde{\boldsymbol{y}} = \tilde{\boldsymbol{y}}(\boldsymbol{\gamma})$. Note that the covariance matrix matches the inverse penalized Fisher information utilized in penalized likelihood estimation. Given the current values $\boldsymbol{\gamma}^c$ and $(\tau^2)^c$ of the Markov chain, a proposed new value $\boldsymbol{\gamma}^p$ is drawn from $\varphi(\cdot; \boldsymbol{\mu}^c, \boldsymbol{\Sigma}^c)$, where $\boldsymbol{W}$,

$\boldsymbol{\gamma}$, $\boldsymbol{\mu}$ and $\boldsymbol{\Sigma}$ are evaluated at the current mode approximation $\hat{\boldsymbol{\gamma}}^c$. It is accepted with probability

$$\alpha(\boldsymbol{\gamma}^c, \boldsymbol{\gamma}^p) = \frac{p(\boldsymbol{y}|\boldsymbol{\gamma}^p)p(\boldsymbol{\gamma}^p|(\tau^2)^c)\varphi(\boldsymbol{\gamma}^c; \boldsymbol{\mu}^c, \boldsymbol{\Sigma}^c)}{p(\boldsymbol{y}|\boldsymbol{\gamma}^c)p(\boldsymbol{\gamma}^c|(\tau^2)^c)\varphi(\boldsymbol{\gamma}^p; \boldsymbol{\mu}^c, \boldsymbol{\Sigma}^c)}. \tag{2.36}$$

As an alternative, $\boldsymbol{\mu}$ and $\boldsymbol{\Sigma}$ in the proposal density may be evaluated at the current value $\boldsymbol{\gamma}^c$ of the chain. Then $\boldsymbol{\mu}^c$, $\boldsymbol{\Sigma}^c$ in the denominator have to be replaced by $\boldsymbol{\mu}^p$, $\boldsymbol{\Sigma}^p$, that is, $\boldsymbol{\mu}$ and $\boldsymbol{\Sigma}$ evaluated at the proposed value $\boldsymbol{\gamma}^p$.

To improve convergence to the stationary distribution, the Markov chain is initialized with posterior mode estimates obtained from penalized Fisher scoring with fixed and usually large values for the variance parameters. The resulting algorithm is a special case of the sampling scheme 1 in Brezger and Lang (2006) and will be discussed in further detail and for an expanded model class in Section 4.2.2. A brief summary is given in

Algorithm 2.2: MH algorithm with IWLS proposals

i. *Initialization*: Compute the posterior mode $\hat{\boldsymbol{\gamma}}^{(0)}$ given a fixed variance parameter τ^2 (such as $\tau^2 = 10$). Use the posterior mode as the current state $\boldsymbol{\gamma}^c$ of the chain.
ii. *Update* $\boldsymbol{\gamma}$:
   - Draw a proposal $\boldsymbol{\gamma}^p$ from the Gaussian proposal density $\varphi(\cdot; \boldsymbol{\mu}^c, \boldsymbol{\Sigma}^c)$.
   - Compute the acceptance probability (2.36) and accept $\boldsymbol{\gamma}^p$ as the new current state $\boldsymbol{\gamma}^c$ with this probability.
   - Compute the new (current) $\boldsymbol{\mu}^c$ and $\boldsymbol{\Sigma}^c$.
iii. *Update* τ^2:
   Draw the new current $(\tau^2)^c$ from $\mathrm{IG}(\tilde{a}, \tilde{b})$, with updated parameters $\tilde{a}$ and $\tilde{b}$ as in (2.31).

Usually, convergence and mixing properties of the resulting Markov chain are very satisfying. In some cases, joint updating of $\boldsymbol{\gamma}$ and τ^2 as suggested in Knorr-Held and Rue (2002) may be a useful alternative, however, at the expense of additional computing efforts, see sampling scheme 2 in Brezger and Lang (2006).

Example 2.6: Forest health
To illustrate the application of penalized splines for smoothing non-Gaussian data, we consider the forest health data from Example 1.2 with binary response. We estimated a binary logit model and obtained both posterior mode estimates (with the amount of smoothness estimated based on the empirical Bayes procedure presented in Section 4.2.1) and posterior mean estimates. The left panel of Fig. 2.15 shows the corresponding estimates. Both estimation approaches are in very close agreement and reflect the pattern found in the raw temporal trends in Fig. 1.3 with a peak of damaged trees at the end of the 1980s. The similarity of

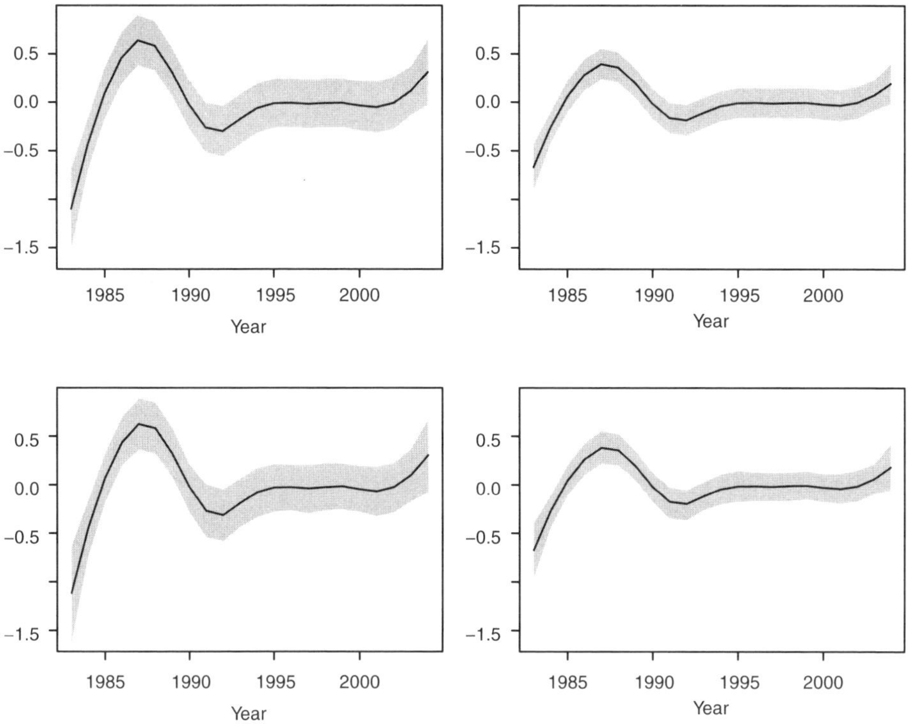

Fig. 2.15 Forest health: Point estimates and 95% pointwise credible bands for the temporal trend in the forest health example for a binary logit (left) and a binary probit model (right). The top row shows posterior mode estimates while the lower row shows corresponding posterior mean estimates.

results from empirical and full Bayes estimates does not only relate to point estimates but also carries over to uncertainty measures such as the credible bands included in Fig. 2.15. This behaviour is actually quite common, at least for larger data sets, since asymptotically we obtain the same approximate normal distribution for the posterior mode and the posterior mean (under suitable regularity conditions). △

Latent variable representations

While IWLS proposal densities typically yield good mixing performance and acceptance rates, they are still computationally more demanding than corresponding Gibbs steps since they require the computation of acceptance rates and moments of the (possibly high-dimensional) proposal density. In some special cases, these difficulties can be overcome by considering latent or auxiliary variables that translate the observation model into a Gaussian model. We will now discuss this possibility for binary response models, where an exact solution is available, but will also sketch an approximate version for Poisson responses.

The *binary probit model* relates the success probability $\pi_i = \mathbb{P}(y_i = 1)$ of a Bernoulli distributed response y_i to the nonparametric regression model $\eta_i = f(x_i)$ via the probit link, that is,

$$\pi_i = \mathbb{P}(y_i = 1) = \Phi(\eta_i),$$

where $\Phi(\cdot)$ denotes the standard normal cumulative distribution function. An equivalent model formulation is obtained by considering latent Gaussian variables

$$l_i = \eta_i + \varepsilon_i, \quad \varepsilon_i \sim \mathrm{N}(0, 1) \tag{2.37}$$

and assuming

$$y_i = 1 \iff l_i \geq 0. \tag{2.38}$$

It turns out that in this case

$$\mathbb{P}(y_i = 1) = \mathbb{P}(l_i \geq 0) = \mathbb{P}(-\varepsilon_i \leq \eta_i) = \mathbb{P}(\varepsilon_i \leq \eta_i) = \Phi(\eta_i)$$

so that the hierarchical formulation consisting of (2.37) and (2.38) is indeed equivalent to the original probit model.

Introducing the latent variables l_i has two main advantages. The first is related to the interpretation of regression effects. Often, the latent variables can be considered as an underlying, continuous variable that drives the binary decision indicator. For example, in the forest health data, the latent variable may be interpreted as a score for forest health that is dichotomized in the visual assessment of the health status. In econometric examples, the latent variable is sometimes interpreted as a utility associated with a specific decision. Since the regression effects impact the latent variable directly, it may be easier to interpret them on the latent scale. In contrast, the impact on the success probability is nonlinear due to the additional transformation with the response function and therefore more difficult to interpret.

The second advantage is related to Bayesian inference based on Markov chain Monte Carlo simulations. Imputing the latent variables as additional 'parameters' in the MCMC scheme yields a hierarchical model formulation with a Gaussian level separating the binary responses from the regression model (2.37). The posterior of the model now depends on the extra parameters l_i, and additional sampling steps for updating them are required. It is easily seen that the full conditionals for the latent variables are truncated normals, that is,

$$l_i | \cdot \sim \begin{cases} \mathrm{TN}_{[0,\infty)}(\eta_i, 1) & \text{if } y_i = 1, \\ \mathrm{TN}_{(-\infty,0)}(\eta_i, 1) & \text{if } y_i = 0 \end{cases} \tag{2.39}$$

where $\mathrm{TN}_{[a,b]}(\eta_i, \sigma^2)$ denotes a normal distribution with expectation η_i and variance σ^2 truncated to the interval $[a, b]$. Efficient algorithms for drawing from truncated normals can be found in Geweke (1991) or Robert (1995); see also Albert and Chib (1993). The resulting Gibbs sampler for probit models with Bayesian P-splines is as follows:

Algorithm 2.3: Gibbs sampler for probit models

i. *Latent variables* l_i: Draw latent variables l_i from the truncated normals in (2.39).

ii. *Update* $\boldsymbol{\gamma}$: Replace y_i through l_i in step i. of the Gibbs sampler for Gaussian responses and draw $\boldsymbol{\gamma}$ from the resulting multivariate Gaussian distribution (2.27) (page 48).

iii. *Update* τ^2: Identical to step ii. of the Gibbs sampler for Gaussian responses (again with l_i replacing y_i).

Example 2.7: Forest health

To supplement Example 2.6, we re-estimated the temporal trend with a binary probit model. For the posterior mode estimates, we again used the same iterative estimation scheme, replacing only the link function, while an auxiliary variable Gibbs sampler was used to determine the posterior mean. The main advantage is of computational nature, since the computation time is reduced to a half compared to the binary logit model, although n latent variables have to be imputed in each iteration. When comparing logit and probit estimates shown in Fig. 2.15, the qualitative functional form is almost the same. The different scaling of the y-axis is simply due to a different variability imposed by the response function (see Fahrmeir and Tutz, 2001, for details) and could be rectified by multiplying the estimates of the probit model with a factor of approximately 1.8. △

The binary probit model has an immediate relation to the Gaussian distribution via the response function Φ. However, latent Gaussian representations can be derived more generally by introducing further levels in the latent hierarchy. For the *binary logit model*, such a latent hierarchy is obtained as follows (Holmes and Held, 2006): The latent model (2.37) is replaced by

$$l_i = \eta_i + \varepsilon_i, \quad \varepsilon_i|\lambda_i \sim \mathrm{N}(0, \lambda_i),$$

that is, an additional variance parameter λ_i is introduced for the latent Gaussian error. Setting $\lambda_i = 4\psi_i^2$ where ψ_i follows a Kolmogorov–Smirnow distribution assigns a scale mixture of normal representation to the latent errors ε_i that (marginally) yields a standard logistic distribution for the errors. Again, imputing the latent variables l_i in each iteration facilitates Gibbs sampling steps in many regression situations. The full conditionals for the l_i are still truncated Gaussian conditional on the scale parameters λ_i, which have to be imputed in addition in each iteration.

For both the logit and the probit model the latent variable representation is exact in the sense that the latent formulation defines exactly the same probabilistic model after marginalization. Such exact representations are not always available and also not always necessary. In a number of recent papers (Frühwirth-Schnatter and Frühwirth 2010; Frühwirth-Schnatter *et al.* 2009; Frühwirth-Schnatter and Frühwirth, 2007; Frühwirth-Schnatter and Wagner, 2006), Frühwirth-Schnatter and co-authors advocate approximate latent variable representations as a general strategy to transform non-Gaussian to latent Gaussian models. We will discuss the *log-linear Poisson model* as one particular example to introduce the general ideas.

The first step in representing the log-linear Poisson model

$$y_i \sim \text{Po}(\lambda_i), \qquad \mathbb{E}(y_i) = \lambda_i = \exp(\eta_i)$$

in terms of latent variables is to employ basic properties of the inter-arrival times of the Poisson process. In this formulation, the count response y_i is interpreted as the number of jumps of a (homogeneous) Poisson process with rate λ_i in the time interval $[0, 1]$. The inter-arrival times τ_{ij}, $j = 1, \ldots, y_i + 1$, of this process are defined as the times between the observation of the $(j-1)$-th and the j-th event. Note that, for example, the response $y_i = 0$ corresponds to one single inter-arrival time $\tau_{i1} > 1$ so that there is always one more inter-arrival time as compared to the response y_i. In addition, it is always required that $\tau_{y_i+1} > 1$.

From the theory of Poisson processes, it is well-known that the inter-arrival times are i.i.d. exponentially distributed with rate λ_i, that is, $\tau_{ij} \sim \text{Exp}(\lambda_i)$. The family of exponential distributions is closed under scaling and we can equivalently assume

$$\tau_{ij} \sim \frac{1}{\lambda_i}\,\text{Exp}(1),$$

that is, τ_{ij} follows a scaled standard exponential distribution. Taking logarithms yields the latent variable representation

$$-\log(\tau_{ij}) = \eta_i + \varepsilon_{ij} \tag{2.40}$$

where $\varepsilon_{ij} \sim \log\text{Exp}(1)$ (independent of λ_i). Equation (2.40) already resembles equation (2.37) with the difference that the error term is not Gaussian. In contrast to the logit model, there is also no hierarchical scale mixture representation that turns ε exactly into a conditional Gaussian distribution. Instead the $\log\text{Exp}(1)$ distribution is approximated by a mixture of Gaussian distributions, that is, its density is approximated via

$$p(\varepsilon) \approx \sum_{r=1}^{R} \omega_r \varphi(\varepsilon; \mu_r, \sigma_r^2)$$

where $\varphi(\cdot; \mu, \sigma^2)$ denotes a Gaussian density with mean μ and variance σ^2, and ω_r are the mixture weights. Choosing an appropriate number R of mixture components (along with the corresponding means, variances and mixture weights) allows us to approximate the standard log-exponential distribution as accurately as required. Frühwirth-Schnatter and Wagner (2006) consider ten mixture components and obtain the means and variances of the mixture components as well as the mixture weights by minimizing the Kullback–Leibler distance between the mixture and the target distribution. Note again that the target distribution is the same for all observations and all inter-arrival times since the ε_{ij} are i. i. d. $\log \mathrm{Exp}(1)$. Therefore, the mixture parameters have to be determined only once in a pre-processing step for the desired number of mixture components.

The final step to obtain a latent Gaussian representation of the Poisson model is to introduce latent mixture indicators that assign the inter-arrival times to one of the Gaussian mixture components. Conditional on the indicators, we then have

$$-\log(\tau_{ij}) = \eta_i + \varepsilon_{ij}, \quad \varepsilon_{ij}|r \sim \mathrm{N}(\mu_r, \sigma_r^2). \tag{2.41}$$

In an MCMC sampling algorithm, the relevant step will be the imputation of the inter-arrival times (which can be obtained as the order statistic of y_i i. i. d. uniformly distributed random variables) and the imputation of latent mixture indicators, see Frühwirth-Schnatter and Wagner (2006) and Frühwirth-Schnatter and Frühwirth (2007) for details. Equation (2.41) then defines a Gaussian regression model with the same predictor as the original Poisson model. Again, MCMC sampling steps are considerably facilitated, since Gibbs steps become available.

The major drawback of any latent variable approach is the numerical effort associated with imputing additional variables. In the case of the Poisson model as formulated above, this burden can become high if the number of counts is large since $y_i + 1$ inter-arrival times have to be imputed for each original observation. Frühwirth-Schnatter, Frühwirth, Held and Rue (2009) have therefore proposed an improved latent variable scheme for Poisson responses taking advantage of the fact that the inter-arrival times of one observation are i. i. d. exponentially distributed and their sum therefore follows a gamma distribution. The imputation of the first y_i inter-arrival times can then be replaced by imputing one single random variable representing the sum. Additional effort is required to compute the mixture approximations for gamma distributions with different parameters (depending on the value y_i) but again these computations have to be performed only once in a setup step, see Frühwirth-Schnatter, Frühwirth, Held and Rue (2009) for details.

2.2.3 Categorical observation models

Extending the ideas presented in the previous section, nonparametric smoothing approaches can also be derived for (multi-)categorical response variables, where $Y_i \in \{1, \dots, C\}$. This relates to a multivariate regression problem, where the categorical response Y_i is represented in terms of C indicator variables $y_i^{(c)}$, $c = 1, \dots, C$, with

$$y_i^{(c)} = \begin{cases} 1 & \text{if } Y_i = c, \\ 0 & \text{otherwise.} \end{cases} \tag{2.42}$$

Regression models for categorical responses are then formulated based on this multivariate representation of the data. Expressing categorical data in this form is particularly suited for considering covariates that change over the categories $\{1, \ldots, C\}$ or when modelling effects that affect the choice probabilities in a category-specific way.

It is worth noting already here that one of the indicator variables is redundant since its value can be inferred from the remaining $C-1$ indicators. Without loss of generality, we therefore consider category C as the reference category that is not taken further into account in the formulation of categorical regression models. All remaining indicators can be collected in the vector $\boldsymbol{y}_i = (y_i^{(1)}, \ldots, y_i^{(C-1)})'$ following a multinomial distribution $\boldsymbol{y}_i \sim \text{Mu}(1, \boldsymbol{\pi}_i)$ where $\boldsymbol{\pi}_i = (\pi_i^{(1)}, \ldots, \pi_i^{(C-1)})'$ is the vector of multinomial probabilities

$$\pi_i^{(c)} = \mathbb{P}(Y_i = c) = \mathbb{P}(y_i^{(c)} = 1), \quad c = 1, \ldots, C-1.$$

These probabilities are related to a $(C-1)$-dimensional vector of predictors $\boldsymbol{\eta}_i = (\eta_i^{(1)}, \ldots, \eta_i^{(C-1)})'$ via a vector-valued response function $h : \mathbb{R}^{C-1} \to \mathbb{R}^{C-1}$, i.e.

$$\boldsymbol{\pi}_i = h(\boldsymbol{\eta}_i). \tag{2.43}$$

This embeds categorical regression models into the general framework of multivariate generalized linear models, see Fahrmeir and Tutz (2001, Ch. 3) or Agresti (2002). We will now discuss special model types for nominal and ordinal responses in more detail, specifying the structure of predictors and alternatives for the response function.

Nominal response models

The most well-known example of a regression model for categorical responses with unordered categories is the multinomial logit model that extends the binary logit model to

$$\pi_i^{(c)} = h^{(c)}\left(\eta_i^{(1)}, \ldots, \eta_i^{(C-1)}\right) = \frac{\exp(\eta_i^{(c)})}{1 + \sum_{s=1}^{C-1} \exp(\eta_i^{(s)})}, \quad c = 1, \ldots, C-1, \tag{2.44}$$

and

$$\pi_i^{(C)} = \frac{1}{1 + \sum_{s=1}^{C-1} \exp(\eta_i^{(s)})}$$

for the reference category.

To get a better understanding of the multinomial logit model and the possible interpretation of covariate effects, it is useful to consider a representation in terms of latent variables, also termed utilities in the context of multinomial models. These latent variables can again be used to facilitate MCMC inference, similarly to that discussed in the previous section for univariate responses. In the context of multinomial models, each of the binary responses $y_i^{(c)}$ is associated with a separate latent variable

$$l_i^{(c)} = \tilde{\eta}_i^{(c)} + \varepsilon_i^{(c)}, \quad c = 1, \ldots, C.$$

where $\varepsilon_i^{(c)}$ are i. i. d. random variables with continuous cumulative distribution function F, and $\tilde{\eta}_i^{(c)}$ are predictors related closely to the predictor of the regression model. The exact relation will become clear through the link between latent variables and the observed categorical responses in the following.

When multinomial regression models were first introduced as a statistical tool (McFadden, 1973), they were motivated as models for describing the choice of individuals between a discrete set of alternatives represented by the categories of the response. In this context, the latent variables $l_i^{(c)}$ were interpreted as the utility associated with the choice of a specific alternative. The utility is determined by covariate effects summarized in the predictor $\tilde{\eta}_i^{(c)}$ on the one hand and random fluctuations captured by the error term $\varepsilon_i^{(c)}$ on the other hand. Assuming rational behaviour, an individual should choose the alternative that maximises its utility, that is,

$$Y_i = c \quad \Leftrightarrow \quad l_i^{(c)} = \max_{s=1,\ldots,C} l_i^{(s)}. \tag{2.45}$$

Already at this stage it becomes obvious that not all predictors $\tilde{\eta}_i^{(c)}$, $c = 1, \ldots, C$ are identifiable since arbitrary shifts can be added or substracted from the predictors. Hence, we have to impose some restriction which essentially reflects the identifiability restriction discussed above for the binary indicators. The usual solution is to set the predictor in the reference category to zero, that is, $\tilde{\eta}_i^{(C)} \equiv 0$. Equivalently, we can relate the predictors of the latent model to the predictors in (2.44) via

$$\eta_i^{(c)} = \tilde{\eta}_i^{(c)} - \tilde{\eta}_i^{(C)}, \quad c = 1, \ldots, C.$$

This representation will also allow us to derive restrictions on regression coefficients and nonparametric functions when coming to the exact specification of multinomial regression models.

Assuming specific types of distributions F for the error terms $\varepsilon_i^{(c)}$ finally yields different types of multinomial regression models. For example, the standard extreme value distribution with cumulative distribution function

$$F(\varepsilon) = \exp(-\exp(-\varepsilon))$$

and density

$$f(\varepsilon) = \exp(-\exp(-\varepsilon))\exp(-\varepsilon)$$

yields

$$\mathbb{P}(Y_i = c) = \frac{\exp(\tilde{\eta}_i^{(c)})}{\sum_{s=1}^{C}\exp(\tilde{\eta}_i^{(s)})}$$

which is exactly the *multinomial logit model* when we take the identifiability restriction into account. The *multinomial (independent) probit model* is obtained when assuming i.i.d. Gaussian error terms in the latent model equation. For Gaussian errors it is also possible to consider extended models where the error terms are correlated, leading to the *multivariate probit model.*

When specifying nonparametric versions of the multinomial logit model, we have to distinguish between different types of covariates. Covariates can either be category-specific or global, that is, fixed across all categories. This leads to two types of nonparametric multinomial regression models defined as

$$\begin{aligned} l_i^{(c)} &= f^{(c)}(x_i) + \varepsilon_i^{(c)}, \\ l_i^{(c)} &= f(x_i^{(c)}) + \varepsilon_i^{(c)}. \end{aligned}$$

In both cases, specific identifiability restrictions arise, which are, however, easier to discuss after introducing the vector-matrix representation of multinomial models.

Recall from (2.43) that a categorical regression model is defined by a vector of predictors $\boldsymbol{\eta}_i$. Representing the nonparametric effects in terms of basis functions allows us to write the predictor as

$$\boldsymbol{\eta}_i = \boldsymbol{X}_i \boldsymbol{\gamma} \tag{2.46}$$

with the special choice of the design matrix $\boldsymbol{X}_i$ (with $C-1$ rows) and the regression coefficients $\boldsymbol{\gamma}$ depending on the specific model structure. When $\boldsymbol{x}_i = (B_1(x_i), \ldots, B_K(x_i))'$ denotes the design vector, the function evaluations can be written as $f^{(c)}(x_i) = \boldsymbol{x}_i'\boldsymbol{\gamma}^{(c)}$ with category-specific regression coefficients $\boldsymbol{\gamma}^{(c)}$. The identifiability constraint discussed above imposes the restriction $f^{(C)}(x) \equiv 0$ which translates to $\boldsymbol{\gamma}^{(C)} = \mathbf{0}$. The design matrix and regression coefficients in (2.46) are then given by

$$\boldsymbol{X}_i = \begin{pmatrix} \boldsymbol{x}_i' & & \\ & \ddots & \\ & & \boldsymbol{x}_i' \end{pmatrix}$$

and

$$\boldsymbol{\gamma} = (\boldsymbol{\gamma}^{(1)\prime}, \ldots, \boldsymbol{\gamma}^{(C-1)\prime})'.$$

For category-specific covariates, the basis function representation yields design vectors $\boldsymbol{x}_i^{(c)} = (B_1(x_i^{(c)}), \ldots, B_K(x_i^{(c)}))'$. In this case, the identifiability restriction yields the redefinition of all functions as $f(x^{(c)}) - f(x^{(C)})$, $c = 1, \ldots, C$, which also ensures that $f(x^{(C)}) = 0$. In matrix notation, this leads to the following expression for the function evaluations:

$$f(x_i^{(c)}) = (\boldsymbol{x}_i^{(c)} - \boldsymbol{x}_i^{(C)})'\boldsymbol{\gamma}.$$

Summarizing these expressions finally yields the following choice for the design matrix in (2.46):

$$\boldsymbol{X}_i = \begin{pmatrix} (\boldsymbol{x}_i^{(1)} - \boldsymbol{x}_i^{(C)})' \\ \vdots \\ (\boldsymbol{x}_i^{(C-1)} - \boldsymbol{x}_i^{(C)})' \end{pmatrix}.$$

Of course, both category-specific and global covariates can also be combined in a regression model and parametric effects can be added as well. This can simply be achieved by augmenting the appropriate design matrices and vectors of regression coefficients defined above to yield a large model. We will consider such extended multinomial models in Chapter 4 in the context of additive semiparametric mixed models.

Example 2.8: Brand choice

One of the main questions in marketing is the determination of variables influencing the brand choice behaviour of consumers (see, for example, Rossi, Allenby and McCulloch, 2005). In this example, we consider data on coffee purchases for the five largest coffee brands in Germany collected by the Gesellschaft für Konsumforschung, Nuremberg. The data set consists of 49,083 purchase acts of 6,407 households over a time span of 52 weeks. Available covariate information includes, for example, the price of a product or the presence of special promotional activity or adverts. In addition, the temporal dynamics underlying the market choice decisions may themselves be of interest to optimize storage capacities and adequate supply. The brand choice data have previously been analysed in Kneib, Baumgartner and Steiner (2007) and Baumgartner, Steiner and Kneib (2009).

The response variable is given by the five coffee brands considered in the data set. In a first step, we will estimate a model with calendar time as the covariate of interest, that is,

$$\eta_i^{(c)} = f^{(c)}(t_i),$$

and t_i is the time of the i-th purchase. Each category is assigned a separate function that reflects the temporal development of the purchase probability for the corresponding coffee brand (with respect to the reference brand). We considered posterior mode estimation with separate smoothing parameters for each of the temporal functions. The results are visualized in Fig. 2.16. At first sight, the estimated effects are rather wiggly and may be deemed to be over-fitting the data. However, brand choice actually shows rapid changes, so that the trend estimates may still reflect changes that are relevant. This is further supported by the observation that some of the local peaks can be attributed to certain events in time. In Fig. 2.16, vertical lines indicate the weeks of Easter and Christmas where all four curves simultaneously show a local minimum and therefore attribute a very high purchase probability to the reference brand. Since the reference brand is comparably expensive, the local minima can be attributed to a change in the

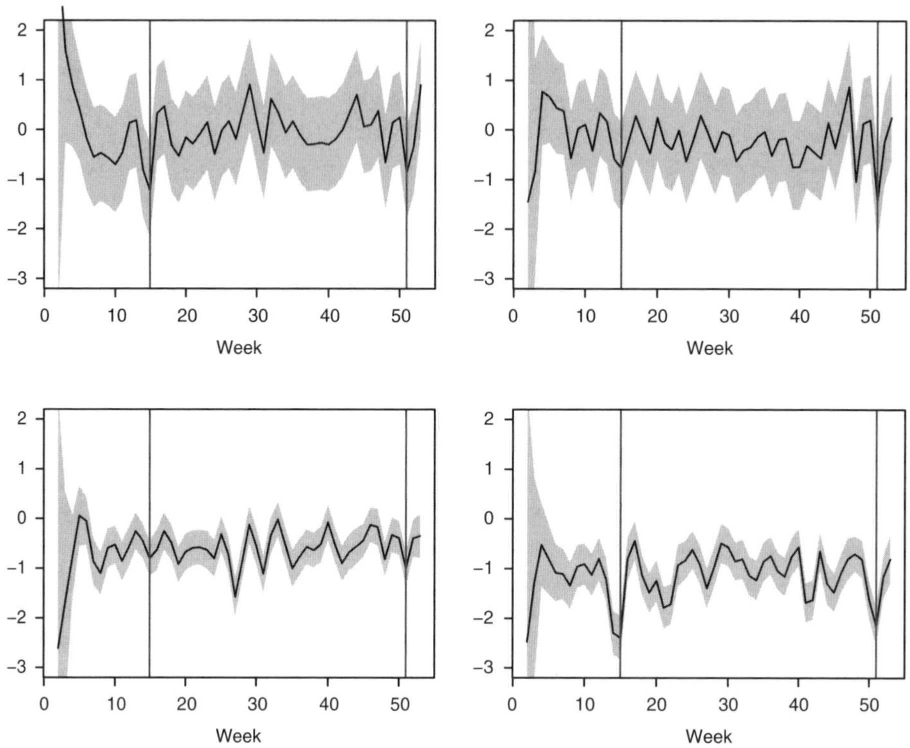

Fig. 2.16 Brand choice: Estimated temporal trends together with pointwise 95% credible intervals. The vertical lines indicate the weeks of Easter and Christmas.

purchase behaviour in the run-up to special holidays where consumers tend to buy more expensive products than usual. Note also that the overall negative effect for the third and the fourth brand (lower panel of Fig. 2.16) indicates a general preference for the reference brand as compared to the third and the fourth brand.

As a second example, we consider a regression model where a brand-specific covariate should be modelled nonparametrically, i.e. a model with predictor

$$\eta^{(c)} = f(x^{(c)}).$$

For illustration purposes, we choose loyalty as the covariate of interest. Loyalty measures the customer retention towards a specific brand and should capture whether a consumer tends to buy the same product at repeated occasions. Since it is obviously difficult to measure loyalty, it is estimated from previous purchase acts using an exponentially weighted average, i.e.

$$loyalty_{ij}^{(c)} = \varphi_L \, loyalty_{i,j-1}^{(c)} + (1 - \varphi_L) \, \mathbb{1}(Y_{i,j-1} = c).$$

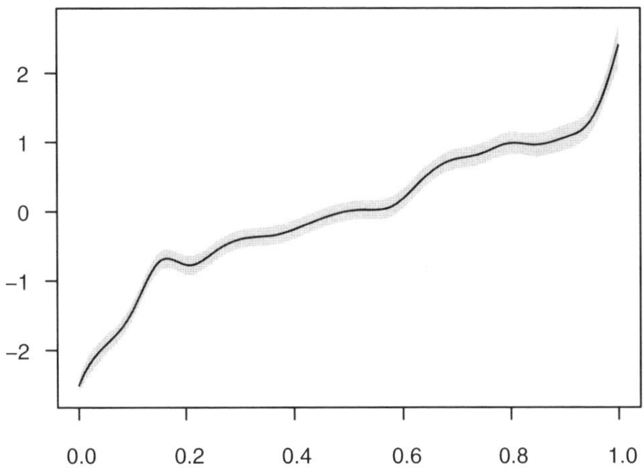

Fig. 2.17 Brand choice: Estimated loyalty effect together with pointwise 95% credible intervals.

where $loyalty_{ij}^{(c)}$ refers to the loyalty of consumer i to brand c at the j-th buying occasion, $Y_{i,j}$ is the brand chosen at this occasion and φ_L is a constant determining the persistence of loyalty. This constant was estimated as $\varphi_L = 0.75$ for the coffee data using a grid search.

Figure 2.17 shows the estimated posterior mode effect of loyalty. Note that there is only one curve available since the covariate itself is brand-specific. The effect curve has therefore to be interpreted as follows: Consider the loyalty towards the reference category is given by $x^{(C)}$ and the loyalty in the category of interest is $x^{(c)}$. In this case, the choice probability for category c as compared to the reference category is shifted by $f(x^{(c)}) - f(x^{(C)})$ on the logit scale. The estimated effect shows the expected behaviour that a high loyalty towards the category of interest increases the choice probability. The effect is not too far from linear but shows some local deviations from a purely linear curve. In particular, the increase is much steeper than predicted by a linear model for extremely small or large loyalty values. △

Cumulative models for ordinal responses

Cumulative regression models have been developed to deal with categorical regression models where the response has an ordinal scale, i.e. the categories can be ordered. In this case, only one single latent variable

$$l_i = \tilde{\eta}_i + \varepsilon_i$$

is considered. It is connected to the categorical response based on an ordered set of thresholds $-\infty = \theta^{(0)} < \theta^{(1)} < \ldots < \theta^{(C)} = \infty$ via

$$Y_i = c \quad \Leftrightarrow \quad \theta^{(c-1)} < l_i \leq \theta^{(c)}. \tag{2.47}$$

Again, the latent variable is composed of a predictor $\tilde{\eta}_i$ and an error part ε_i with cumulative distribution function F and corresponding density $f = F'$. The mechanism (2.47) can then be interpreted as follows: The thresholds slice the density of the error term at specific points whereas the predictor $\tilde{\eta}_i$ defines a shift of the location of the latent variable. Hence, increasing values of the predictor translate to higher probabilities for larger response categories and vice versa (see Fig. 2.18 for a graphical representation).

Model (2.47) is called a cumulative regression model since it defines the cumulative distribution function of the categorical response via

$$\mathbb{P}(Y_i \leq c) = F(\theta^{(c)} - \tilde{\eta}_i). \tag{2.48}$$

This also enables a second interpretation of regression coefficients since these directly impact the cumulative distribution function of Y_i. Choosing F as the logistic distribution function leads to the *cumulative logit model*, and the *cumulative probit model* is obtained for $F = \Phi$, the standard normal distribution function. From this equation it also becomes obvious that the response $\tilde{\eta}_i$ is not allowed to contain an intercept in order to obtain an identifiable model. For nonparametric functions, this relates to a centering restriction that will re-appear in additive models in Section 2.3. An alternative solution, sometimes employed in the literature, is to fix one of the thresholds at a pre-specified value such as $\theta^{(1)} = 0$.

In extended cumulative models, parts of the predictor $\tilde{\eta}_i$ can be defined to be category-specific yielding, for example, the nonparametric model $\tilde{\eta}_i^{(c)} = f^{(c)}(x_i)$. More generally, category-specific effects lead to the latent model

$$l_i = \tilde{\eta}_i^{(c)} + \varepsilon_i.$$

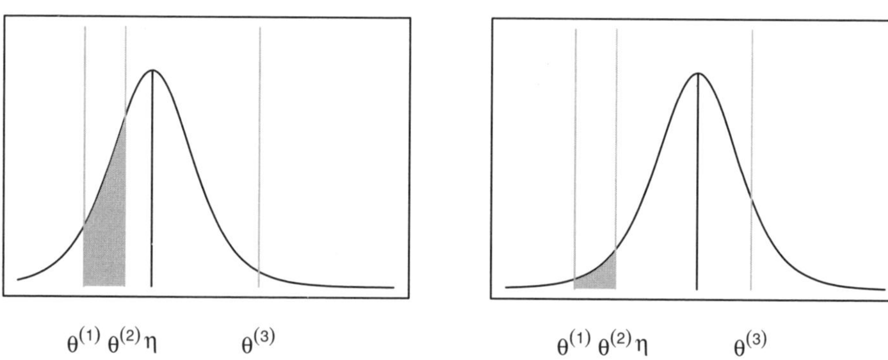

Fig. 2.18 Interpretation of cumulative regression models: The linear predictor $\tilde{\eta}_i$ shifts the density of l_i. The shaded region denotes the probability $\mathbb{P}(Y_i = 2)$.

However, such extensions have to be used with caution in practice, since in this case the ordering restrictions

$$\theta^{(1)} - \tilde{\eta}_i^{(1)} < \ldots < \theta^{(C-1)} - \tilde{\eta}_i^{(C-1)}$$

have to hold for all values of the predictors, leading to complicated restrictions on the functions $f^{(c)}(x_i)$. Hence, the usual strategy is to keep most of the regression effects constant across categories and to assign category-specific effects only to a very small subset of effects.

To cast cumulative models into the general framework of categorical regression models, we first derive the response function from (2.48). From the cumulative distribution function, the category probabilities are easily obtained as

$$\begin{aligned}
\mathbb{P}(Y_i = 1) &= \pi^{(1)} = h^{(1)}\left(\eta_i^{(1)}, \ldots, \eta_i^{(C-1)}\right) = F(\eta_i^{(1)}) \\
\mathbb{P}(Y_i = c) &= \pi^{(c)} = h^{(c)}\left(\eta_i^{(1)}, \ldots, \eta_i^{(C-1)}\right) = F(\eta_i^{(c)}) - F(\eta_i^{(c-1)}), \quad c > 1,
\end{aligned}$$

with predictor

$$\eta_i^{(c)} = \theta^{(c)} - \tilde{\eta}_i$$

in case of only one single predictor $\tilde{\eta}_i$ and

$$\eta_i^{(c)} = \theta^{(c)} - \tilde{\eta}_i^{(c)}$$

in case of category-specific predictors $\tilde{\eta}_i^{(c)}$.

For the special case of a nonparametric model $\tilde{\eta}_i = f(x_i) = \boldsymbol{x}_i'\tilde{\boldsymbol{\gamma}}$, the design matrix and the regression coefficients are given by

$$\boldsymbol{X}_i = \begin{pmatrix} 1 & & & & -\boldsymbol{x}_i' \\ & 1 & & & -\boldsymbol{x}_i' \\ & & \ddots & & \vdots \\ & & & 1 & -\boldsymbol{x}_i' \end{pmatrix}$$

and

$$\boldsymbol{\gamma} = (\theta^{(1)}, \ldots, \theta^{(C-1)}, \tilde{\boldsymbol{\gamma}}')'$$

such that $\boldsymbol{\eta}_i = \boldsymbol{X}_i\boldsymbol{\gamma}$. For category-specific nonparametric effects $\tilde{\eta}_i^{(c)} = f^{(c)}(x_i) = \boldsymbol{x}_i'\tilde{\boldsymbol{\gamma}}^{(c)}$, these expressions have to be altered to

$$\boldsymbol{X}_i = \begin{pmatrix} 1 & -\boldsymbol{x}_i' & & & & & \\ & & 1 & -\boldsymbol{x}_i' & & & \\ & & & & \ddots & & \\ & & & & & 1 & -\boldsymbol{x}_i' \end{pmatrix}$$

and

$$\gamma = (\theta^{(1)}, \tilde{\gamma}^{(1)\prime}, \ldots, \theta^{(C-1)}, \tilde{\gamma}^{(C-1)\prime})'.$$

Example 2.9: Forest health

Cumulative models allow us to take the ordinal structure of the defoliation indicator into account that serves as the response when analysing the forest health data described in Example 1.2. To be more specific, we now consider the three-categorical response reflecting no, medium and severe defoliation. Our current focus is on estimating the temporal trend and in a first step we specify a cumulative logit model with predictor

$$\eta_{it}^{(c)} = \theta^{(c)} - f(t).$$

that consists only of a global trend function and the category-specific thresholds. Note that we are indexing the data both in time (t) and observation plots (i) to stick to the longitudinal structure of the data. However, this longitudinal structure is completely neglected when estimating the parameters. We will come back to the forest health data in later chapters when extended methodology allows us to analyse the longitudinal structure and temporal trends simultaneously.

The resulting posterior mode estimate is shown in Fig. 2.19 and generally resembles the estimate for the simplified binary model shown in Fig. 2.15. However, when comparing the global trend with the empirical trends shown in Fig. 1.3, it seems at least questionable whether the estimated effect adequately separates between severe defoliation and at most medium defoliation. More specifically, the nonparametric estimate seems to separate well between no and moderate defoliation since it recovers the two corresponding empirical trend lines, but the the slight increase in the trend separating medium and severe defoliation is not recovered. This is simply due to the fact that the model only contains one trend function $f(t)$ and this function is determined by the larger fraction of the data in the categories corresponding to no and medium defoliation.

To circumvent this difficulty, we extended the regression model to contain a category-specific trend function $f^{(c)}(t)$, that is,

$$\eta_{it}^{(c)} = \theta^{(c)} - f^{(c)}(t).$$

Now, we obtain the two trend functions shown in Fig. 2.20 that correspond more closely to the empirical time trends reflecting both the long-term increasing trend for severe defoliation and the peak at the end of the 1980s for no versus at least medium defoliation. More specifically, $f^{(1)}(t)$ separates between no and at least medium defoliation while $f^{(2)}(t)$ separates at most medium and severe defoliation. Alternatively, the impact of the two functions on the cumulative probabilities $\mathbb{P}(Y_{it} \leq c)$ can be interpreted. △

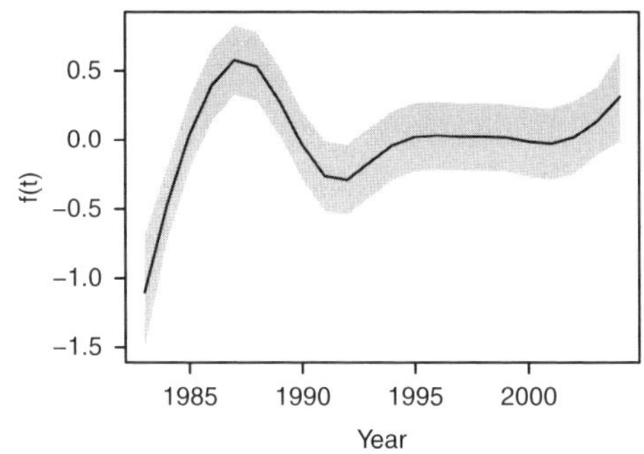

Fig. 2.19 Forest health: Estimated global trend function.

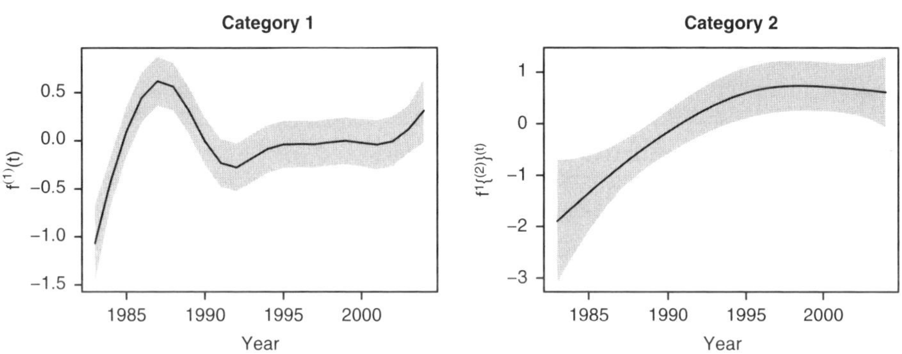

Fig. 2.20 Forest health: Estimated category-specific trend function.

The curvature of the first function $f^{(1)}(t)$ shown in the left panel of Fig. 2.20 indicates a higher frequency of larger defoliation degrees at the end of the 1980s. Note that the function does not simply reflect the probability for category 2 but in fact separates the healthy trees from trees with either moderate or severe defoliation. Correspondingly, the second function $f^{(2)}(t)$ shown in the right panel reflects the increasing share of trees with severe defoliation contrasted to the joint frequency of no and medium defoliation.

Sequential models for ordinal responses

A second possibility for defining categorical regression models with ordered response categories is the sequential model. It is the appropriate choice if the response is the result of a sequential mechanism. This is, for example, the case if the response categories can only be achieved successively, such as in discrete time survival models or when analysing stages of increasing severity of a disease. Correspondingly, the sequential model consists of a sequence of decisions on whether a higher category is achieved or whether an observation remains in the current response category.

The sequential model is based on latent variables

$$l_i^{(c)} = \tilde{\eta}_i + \varepsilon_i^{(c)}$$

with i. i. d. random errors $\varepsilon_i^{(c)}$ and predictor $\tilde{\eta}_i$. The probability of achieving category c is then modelled conditionally on achieving the previous category $c-1$ based on a set of thresholds $\theta^{(1)}, \dots, \theta^{(C-1)}$ via

$$Y_i = c \,|\, Y_i \geq c \quad \Leftrightarrow \quad l_i^{(c)} \leq \theta^{(c)}.$$

In contrast to the cumulative model, the thresholds do not have to fulfill any ordering restriction since only one of them is associated with a specific transition from category $c-1$ to category c.

Inserting the cumulative distribution function F of the random errors leads to the conditional probabilities

$$\mathbb{P}(Y_i = c \,|\, Y_i \geq c) = F(\theta^{(c)} - \tilde{\eta}_i). \tag{2.49}$$

Note that the left-hand side defines a discrete-time analogon to the hazard rate, forming the basis for most of the regression models for survival times considered in Chapter 6. Due to this analogy, the sequential model is often considered in discrete-time survival analyses.

From the conditional probabilities (2.49), the unconditional probabilities and therefore the response function of the sequential model can be derived as

$$\mathbb{P}(Y_i = c) = \pi_i^{(c)} = h^{(c)}(\eta_i^{(1)}, \dots, \eta_i^{(C-1)}) = F(\eta_i^{(c)}) \prod_{s=1}^{c-1} [1 - F(\eta_i^{(s)})]$$

with linear predictors

$$\eta_i^{(c)} = \theta^{(c)} - \tilde{\eta}_i.$$

Different choices for the cumulative distribution function F result in different models, for example sequential logit and sequential probit models with the logistic or the standard normal distribution, respectively.

In complete analogy to the cumulative model, extended sequential models with category-specific effects can be defined. In this case, there are also no difficulties with identifying the model since the thresholds of the sequential model are not ordered. The design matrices and vectors of regression coefficients take exactly the same form as for the cumulative model and are therefore omitted.

Penalised likelihood inference

Estimation of semiparametric categorical regression models can be based on the same principles of penalized likelihood as for univariate non-Gaussian responses. The general structure of the penalized likelihood remains, that is,

$$l_{\text{pen}}(\boldsymbol{\gamma}) = l(\boldsymbol{\gamma}) - \text{pen}(\boldsymbol{\gamma})$$

where $l(\boldsymbol{\gamma})$ denotes the unpenalized likelihood and $\text{pen}(\boldsymbol{\gamma})$ is a penalty term (with the smoothing parameter absorbed into the penalty). The unpenalized likelihood is simply obtained from the multinomial distribution of the individual response vectors, that is, from the assumption $\boldsymbol{y}_i \sim \text{Mu}(1, \boldsymbol{\pi}_i)$, where $\boldsymbol{\pi}_i$ is derived from the corresponding model by applying the response function to the predictors $\boldsymbol{\eta}_i = \boldsymbol{X}_i\boldsymbol{\gamma}$.

When considering penalized splines for the nonparametric effects, the penalty term will be quadratic, that is

$$\text{pen}(\boldsymbol{\gamma}) = \boldsymbol{\gamma}'\boldsymbol{K}\boldsymbol{\gamma}$$

where we have absorbed the smoothing parameters in the penalty matrix $\boldsymbol{K}$. The specific choice of $\boldsymbol{K}$ depends on the model under consideration. For example, in case of a multinomial model with predictor $\eta^{(c)} = f^{(c)}(x) = \boldsymbol{x}'\boldsymbol{\gamma}^{(c)}$, the penalty matrix is given by

$$\boldsymbol{K} = \text{blockdiag}\left(\frac{1}{\tau_1^2}\boldsymbol{K}^{(1)}, \dots, \frac{1}{\tau_{C-1}^2}\boldsymbol{K}^{(C-1)}\right)$$

where $\boldsymbol{K}^{(c)}$ are penalty matrices associated with the functions $f^{(c)}(x)$. For a cumulative or sequential model with predictor

$$\eta^{(c)} = \theta^{(c)} - f(x) = \theta^{(c)} - \boldsymbol{x}'\tilde{\boldsymbol{\gamma}}$$

the penalty matrix is given by

$$\boldsymbol{K} = \text{blockdiag}\left(\boldsymbol{0}_{C-1}, \frac{1}{\tau^2}\tilde{\boldsymbol{K}}\right)$$

where the $(C-1)\times(C-1)$ matrix of zeros $\boldsymbol{0}_{C-1}$ corresponds to the unpenalized estimation of the thresholds and $\tilde{\boldsymbol{K}}$ is the penalty matrix associated with $f(x)$.

In any case, extending the IWLS algorithm for univariate responses yields the update step

$$\hat{\boldsymbol{\gamma}}^{(k+1)} = (\boldsymbol{X}'\boldsymbol{W}^{(k)}\boldsymbol{X} + \boldsymbol{K})^{-1}\boldsymbol{X}'\boldsymbol{W}^{(k)}\tilde{\boldsymbol{y}}^{(k)}.$$

that has to be performed iteratively to obtain the penalized likelihood estimate $\hat{\boldsymbol{\gamma}}$. The overall design matrix $\boldsymbol{X}$ is obtained by stacking the individual design matrices $\boldsymbol{X}_i$. The matrix of working weights $\boldsymbol{W} = \boldsymbol{D}\boldsymbol{V}^{-1}\boldsymbol{D}'$

has block-diagonal structure established by the block-diagonal matrices $\boldsymbol{D} = \text{blockdiag}(\boldsymbol{D}_1, \ldots, \boldsymbol{D}_n)$ and $\boldsymbol{V} = \text{blockdiag}(\boldsymbol{V}_1, \ldots, \boldsymbol{V}_n)$, with $(C-1) \times (C-1)$ matrices $\boldsymbol{D}_i$ and $\boldsymbol{V}_i$ given by

$$\boldsymbol{D}_i = \frac{\partial h(\boldsymbol{\eta}_i)}{\partial \boldsymbol{\eta}} = \begin{pmatrix} \dfrac{\partial h^{(1)}(\boldsymbol{\eta}_i)}{\partial \eta^{(1)}} & \cdots & \dfrac{\partial h^{(C-1)}(\boldsymbol{\eta}_i)}{\partial \eta^{(1)}} \\ \vdots & \ddots & \vdots \\ \dfrac{\partial h^{(1)}(\boldsymbol{\eta}_i)}{\partial \eta^{(C-1)}} & \cdots & \dfrac{\partial h^{(C-1)}(\boldsymbol{\eta}_i)}{\partial \eta^{(C-1)}} \end{pmatrix}$$

and $\boldsymbol{V}_i = \text{Cov}(\boldsymbol{y}_i)$, that is,

$$\boldsymbol{V}_i = \begin{pmatrix} \pi_i^{(1)}(1-\pi_i^{(1)}) & -\pi_i^{(1)}\pi_i^{(2)} & \cdots & -\pi_i^{(1)}\pi_i^{(C-1)} \\ -\pi_i^{(1)}\pi_i^{(2)} & \ddots & & \vdots \\ \vdots & & \ddots & -\pi_i^{(C-2)}\pi_i^{(C-1)} \\ -\pi_i^{(1)}\pi_i^{(C-1)} & \cdots & -\pi_i^{(C-2)}\pi_i^{(C-1)} & \pi_i^{(C-1)}(1-\pi_i^{(C-1)}) \end{pmatrix}.$$

The working observations $\tilde{\boldsymbol{y}}$ are defined by

$$\tilde{\boldsymbol{y}} = \boldsymbol{\eta} + (\boldsymbol{D}^{-1})'(\boldsymbol{y} - \boldsymbol{\pi}),$$

where $\tilde{\boldsymbol{y}}$, $\boldsymbol{\eta}$, $\boldsymbol{y}$ and $\boldsymbol{\pi}$ are stacked versions of the corresponding individual quantities.

The derivatives of the response function with respect to the predictors have to be derived for each of the model types individually. For the multinomial logit model, these derivatives can be shown to be given by

$$\frac{\partial h^{(c)}(\boldsymbol{\eta})}{\partial \eta^{(j)}} = \begin{cases} \dfrac{\exp(\eta^{(c)})\left(1 + \sum_{s=1,s\neq c}^{C-1} \exp(\eta^{(s)})\right)}{\left(1 + \sum_{s=1}^{C-1} \exp(\eta^{(s)})\right)^2} = \pi^{(c)}(1-\pi^{(c)}) & c = j \\ -\dfrac{\exp(\eta^{(c)})\exp(\eta^{(j)})}{\left(1 + \sum_{s=1}^{C-1} \exp(\eta^{(s)})\right)^2} = -\pi^{(c)}\pi^{(j)} & c \neq j. \end{cases}$$

Therefore, we have $\boldsymbol{D}_i = \boldsymbol{D}_i' = \boldsymbol{V}_i$ leading to the simplified working weights $\boldsymbol{W}_i = \boldsymbol{D}_i = \boldsymbol{V}_i$.

In case of cumulative and sequential models, the derivatives of h depend on the density f of the latent variables. For cumulative models, we obtain

$$\frac{\partial h^{(c)}(\boldsymbol{\eta})}{\partial \eta^{(j)}} = \begin{cases} f(\eta^{(j)}) & j = c, \\ -f(\eta^{(j)}) & j = c-1, \\ 0 & \text{otherwise,} \end{cases}$$

while for sequential models the derivatives are given by

$$\frac{\partial h^{(c)}(\boldsymbol{\eta})}{\partial \eta^{(j)}} = \begin{cases} f(\eta^{(c)}) \prod_{s=1}^{c-1} (1 - F(\eta^{(s)})) & j = c, \\ -F(\eta^{(c)}) f(\eta^{(j)}) \prod_{s=1, s \neq j}^{c-1} (1 - F(\eta^{(s)})) & j < c, \\ 0 & j > c. \end{cases}$$

Bayesian inference

Empirical Bayes inference with Gaussian smoothness priors for basis coefficients γ or $\gamma^{(c)}$ and variance parameters $\tau^2 = 1/\lambda$ or $\tau_c^2 = 1/\lambda_c$, respectively, considered as unknown constants is as closely related to penalized likelihood inference as for univariate non-Gaussian responses in Section 2.2.2. A mixed model approach for estimating variance parameters jointly with basis coefficients is described in detail in Section 4.2.1.

Full Bayes inference may be based on extensions of the MH algorithm with IWLS proposals, for example for multinomial logit models, or on latent variable representations of categorical regression models. In particular, in combination with Gaussian or conditionally Gaussian latent error terms, the second strategy leads to considerable simplification and often allows us to construct Gibbs samplers.

We exemplify this for the cumulative probit model (2.48) with $F = \Phi$. Then, the latent variables l_i follow a latent Gaussian nonparametric regression model

$$l_i = f(\boldsymbol{x}_i) + \varepsilon_i = \boldsymbol{x}_i'\tilde{\boldsymbol{\gamma}} + \varepsilon_i,\ i = 1, \ldots, n,$$

with i. i. d. errors $\varepsilon_i \sim \mathrm{N}(0, 1)$. Each observable response Y_i is related to l_i through the thresholding mechanism (2.47). In addition to $\tilde{\boldsymbol{\gamma}}$ and τ^2, we have to estimate the threshold parameters $\boldsymbol{\theta} = (\theta^{(1)}, \ldots, \theta^{(C-1)})'$ as well as the latent variables. Assuming a diffuse prior $p(\boldsymbol{\theta}) \propto$ const and proceeding conceptually as for binary probit models, we obtain

Algorithm 2.4: Gibbs sampler for cumulative probit models

i. *Latent variables* l_i: If $Y_i = c$, then l_i is generated from $\mathrm{TN}_{(\theta^{(c-1)}, \theta^{(c)}]}(\boldsymbol{x}_i'\tilde{\boldsymbol{\gamma}}, 1)$ with $\tilde{\boldsymbol{\gamma}}$ evaluated at its current value. Drawing from this truncated normal can be performed as described in Geweke (1991) and Robert (1995).

ii. *Thresholds* $\theta^{(c)}$: The full conditional for $\theta^{(c)}$, $c = 1, \ldots, C-1$, is uniform on the interval

$$[\max\{l_i : Y_i = c\}, \min\{l_i : Y_i = c + 1\}],$$

see Albert and Chib (1993). Posterior samples may exhibit bad mixing, however, because the intervals can become quite small so that the chain moves slowly. Chen and Dey (2000) suggested other parametrizations as possible alternatives. For the special but quite common case of three categories $C = 3$, such problems can be avoided as follows: First include an intercept term $\tilde{\gamma}_0$ with diffuse prior to set $\theta^{(1)} = 0$. Second, assume that errors ε_i are i. i. d. $\mathrm{N}(0, \sigma^2)$, with unknown σ^2, instead of $\sigma^2 = 1$. This rescaling allows to set $\theta^{(2)} = 1$. With an IG prior for σ^2, the full conditionals are again IG densities with updated parameters as for Gaussian responses, replacing y_i through l_i.

iii. *Update* $\tilde{\gamma}$ and τ^2: Identical to the Gibbs sampler for Gaussian responses, replacing y_i through l_i, see page 54.

Gibbs samplers for multinomial probit models can be based on latent variables and derived from the principle of maximal utility (2.45). If each category c is associated with a separate Gaussian model

$$l_i^{(c)} = f^c(\boldsymbol{x}_i) + \varepsilon_i^{(c)} = \boldsymbol{x}_i'\boldsymbol{\gamma}^{(c)} + \varepsilon_i^{(c)}, \; c = 1, \ldots, C-1$$

for the latent variables and $l_i^{(C)} \equiv 0$ for identifiability reasons, we obtain

Algorithm 2.5: Gibbs sampler for the multinomial probit model

i. *Latent variables* $l_i^{(c)}$, $c = 1, \ldots, C-1$:
   - If $Y_i = c$, $c \neq C$, then $l_i^{(c)}$ is generated from $\mathrm{N}(\boldsymbol{x}_i'\boldsymbol{\gamma}^{(c)}, 1)$ subject to the constraints $l_i^{(c)} > l_i^{(h)}$, $h \neq c$, and $l_i^{(c)} > 0 \; (\equiv l_i^{(C)})$. Next $l_i^{(h)}$ for $h \neq c$ is generated from $\mathrm{N}(\boldsymbol{x}_i'\boldsymbol{\gamma}^{(h)}, 1)$ subject to $l_i^{(h)} < l_i^{(c)}$ generated just as before.
   - If $Y_i = C$, then $l_i^{(h)}$, $h = 1, \ldots, C-1$, are generated from $\mathrm{N}(\boldsymbol{x}_i'\boldsymbol{\gamma}^{(h)}, 1)$ subject to $l_i^{(h)} < 0 \; (\equiv l_i^{(C)})$.

ii. *Updates* $\boldsymbol{\gamma}^{(c)}$ and τ_c^2: Identical to separate Gibbs samplers for Gaussian responses, replacing y_i through $l_i^{(c)}$, $c = 1, \ldots, C-1$, see page 54.

Variants and improvements of this basic algorithm are, for example, discussed in McCulloch and Rossi (1994), Nobile (1998), van Dyk and Meng (2001) and Imai and van Dyk (2005).

2.2.4 Related smoothing approaches

Integral penalties

A general principle for constructing penalties in basis-function approaches (not necessarily based on polynomial splines) is to define a roughness penalty based on squared derivatives. For a B-spline basis, we saw that this is essentially equivalent to the difference penalty proposed before if equidistant knots are employed. However, this connection no longer holds for other types of basis functions where the difference penalty would have only a heuristic justification.

The most popular variant of derivative-based penalties is the squared second order derivative

$$\text{pen}(f) = \int (f''(x))^2 dx$$

since it enforces visually smooth function estimates (but of course requires basis functions that are twice continuously differentiable). It also corresponds to the square of the most popular norm in the space of twice continuously differentiable functions. This property will be used in the context of reproducing kernel Hilbert space representations discussed later in this section.

In principle, the integral penalty does not require a basis function representation but is defined on the space of all twice differentiable functions, see the following paragraph on smoothing splines. If, however, a basis function approach is considered, the penalty immediately reduces to a quadratic form in the basis function coefficients since

$$\int (f''(x))^2 dx = \sum_{k_1=1}^{K} \sum_{k_2=1}^{K} \gamma_{k_1}\gamma_{k_2} \int B''_{k_1}(x)B''_{k_2}(x)dx = \boldsymbol{\gamma}'\boldsymbol{K}\boldsymbol{\gamma}. \tag{2.50}$$

The entries in the penalty matrix $\boldsymbol{K}$ depend on the specific form of the basis functions. In particular, if an orthonormal basis is employed, the penalty matrix reduces to an identity matrix. When embedding the basis-function approach into a reproducing kernel Hilbert space (RKHS) framework, the penalty matrix is also particularly easy to construct since it only contains evaluations of the basis functions (see the paragraph on RKHS below).

Note that the quadratic form of the penalty also implies that an equivalent Bayesian formulation can be considered, where a Gaussian prior with precision matrix $\boldsymbol{K}$ is assigned to $\boldsymbol{\gamma}$. This distribution will always be partially improper since linear functions remain unpenalized by the second derivative integral penalty. This results in a non-trivial null-space of dimension two for the penalty matrix.

Smoothing splines

The most well-known smoothing approach that is introduced in terms of an integral penalty are smoothing splines. We will briefly review the frequentist perspective on smoothing splines before considering a Bayesian smoothing spline variant. To introduce smoothing splines, we do not start with a basis function representation for $f(x)$ but seek the solution of a specific optimization problem. This optimality property partly explains the popularity of smoothing splines although they have a number of practical disadvantages that, in our opinion, make them less suitable for complex models as compared to penalized polynomial splines.

Consider the following optimization problem: We are searching for the minimizing function of the penalized least-squares criterion

$$\sum_{i=1}^{n} (y_i - f(x_i))^2 + \lambda \int (f''(x))^2 dx \tag{2.51}$$

in the class of all twice continuously differentiable functions. It turns out that, without further assumptions about $f(x)$, the solution is a natural cubic spline, with knots at all distinct observed values for the covariate x (Reinsch, 1967). A natural cubic spline differs from a cubic polynomial spline only in one further assumption: At the boundaries (beyond the extremal inner knots), the spline function is restricted to be linear. This corresponds to the behaviour of a curve template (the origin of the name spline) beyond its fixed points.

Hence, despite starting from a nonparametric smoothing problem in the function space of twice continuously differentiable functions, we end up with a solution that belongs to the very specific function space of natural cubic splines with finite-dimensional basis. It is therefore representable in terms of basis functions via

$$f(x) = \sum_{k=1}^{K} \gamma_k B_k(x)$$

where K denotes the number of distinct covariate values. In the interior of the domain, these basis functions coincide with the basis of a polynomial spline with the same knot set, but at the boundaries appropriate modifications have to be employed to ensure the linearity condition. There are also other basis function representations for smoothing splines that allow us to embed them into the framework of reproducing kernel Hilbert spaces, see the paragraph below.

Estimation of a smoothing spline, or more precisely, the corresponding basis coefficients, can now be based on the penalized least-squares criterion

$$(\boldsymbol{y} - \boldsymbol{X}\boldsymbol{\gamma})'(\boldsymbol{y} - \boldsymbol{X}\boldsymbol{\gamma}) + \lambda\boldsymbol{\gamma}'\boldsymbol{K}\boldsymbol{\gamma},$$

where $\boldsymbol{X}$ is composed of the basis function evaluations and the penalty matrix $\boldsymbol{K}$ is obtained as specified in (2.50). Again, we end up with a penalized least-squares estimate

$$\hat{\boldsymbol{\gamma}} = (\boldsymbol{X}'\boldsymbol{X} + \lambda\boldsymbol{K})^{-1}\boldsymbol{X}'\boldsymbol{y}$$

and the corresponding function estimate is given by $\hat{\boldsymbol{f}} = \boldsymbol{X}\hat{\boldsymbol{\gamma}}$. Other representations of $\hat{\boldsymbol{f}}$ work directly with the vector of function evaluations and avoid the specification in terms of basis functions and regression coefficients, see for example Fahrmeir and Tutz (2001, Ch. 5).

Note that in contrast to P-splines, smoothing splines typically require a much larger basis and are therefore numerically much more intense. To be more specific, the dimension of the system of equations that has to be solved to obtain the PLS estimate $\hat{\boldsymbol{\gamma}}$ is given by the number of distinct values observed for the covariate, that is, the number of knots for the natural cubic spline. In contrast to the usual dimension of a cubic polynomial spline (number of knots plus two), the dimension is reduced due to the linearity conditions at the boundaries. If smoothing splines are employed as a building block in more complex model classes, computations easily become numerically intractable. In practice, the large basis is therefore often reduced in an adhoc manner to a natural cubic spline with a reduced number of knots. The resulting function estimate will usually still be close to the estimate from the full spline basis but no longer has the optimality properties that originally motivated the use of smoothing splines. In this case, it might be preferable to employ the simpler P-spline idea right from start instead of relying on smoothing splines.

Hastie (1996) proposes a dimension reduction technique for smoothing splines that is based on a spectral decomposition and a sparse representation in terms of the eigenvectors associated with the largest eigenvalues. Thereby this approach also avoids an ad hoc selection of knots. The resulting smoothing spline approximation is called a low-rank smoother since its basis is of only moderate size. In this sense, penalized splines also have the low-rank property but might be preferred due to their simpler construction and more intuitive appeal.

Bayesian interpretation of smoothing splines

For a given data configuration, the prior corresponding to the quadratic penalty of smoothing splines is simply given by the improper normal distribution

$$p(\boldsymbol{\gamma}) \propto \exp\left(-\frac{1}{2\tau^2}\boldsymbol{\gamma}'\boldsymbol{K}\boldsymbol{\gamma}\right).$$

However, it is also possible to derive a generic Bayesian interpretation of smoothing splines that is based on the original optimization problem in the class of twice continuously differentiable functions. Following the presentation in Wahba (1978), we consider the somewhat more general optimization problem

$$\sum_{i=1}^{n}(y_i - f(x_i))^2 + \lambda \int (f^{(m)}(x))^2 dx \to \min_f \tag{2.52}$$

within the space of all m times continuously differentiable functions. For $m = 2$ we obtain the smoothing spline problem considered above as a special case, whereas in general the solution is a natural spline of degree $2m-1$. In particular, for $m = 2$, the result is a natural spline of degree 3, that is, a natural cubic spline. Note that only splines of odd degree can be obtained as optimal solutions to a criterion of the form (2.52). The notion of reproducing kernel Hilbert spaces will provide an explanation for this fact.

Without assuming a basis function representation, a continuous stochastic process prior has to be assigned to $f(x)$. Wahba (1978) assumes the prior

$$f(x) = \sum_{j=1}^{m}\beta_j \frac{x^{j-1}}{(j-1)!} + \tau Z(x)$$

where the polynomial coefficients $\boldsymbol{\beta} = (\beta_1, \dots, \beta_m)'$ are independent and identically Gaussian distributed, that is, $\boldsymbol{\beta} \sim \mathrm{N}(\mathbf{0}, c\boldsymbol{I}_m)$, and $Z(x)$ equals the m-fold integrated Wiener process, that is,

$$Z(x) = \int_0^x \frac{(x-u)^{m-1}}{(m-1)!} dW(u)$$

and $W(u)$ is a standard Wiener process with $\mathbb{E}(W(u)) = 0$ and $\text{Var}(W(u)) = 1$. The integral with respect to the Wiener process can be re-written in a more standard form. For example, in the special case $m = 2$ we obtain

$$Z(x) = \int_0^x (x-u) dW(u) = \int_0^x W(u) du.$$

In general, the integral reduces to

$$Z(x) = \int_0^x \frac{(x-u)^{m-1}}{(m-1)!} dW(u) = \int_0^x \frac{(x-u)^{m-2}}{(m-2)!} W(u) du.$$

The natural spline solution $\hat{f}(x)$ for the optimization problem (2.52) is obtained as the posterior expectation

$$\hat{f}(x) = \lim_{c \to \infty} \mathbb{E}(f(x)|\boldsymbol{y}), \tag{2.53}$$

that is, under diffuse priors for the polynomial coefficients $\boldsymbol{\beta}$, the estimate $\hat{f}(x)$ results as a Bayesian estimate without referring to a basis function representation. This Bayesian perspective also yields an alternative understanding of the optimality criterion itself. Since the posterior mean minimizes the squared error loss, it can be understood as a Bayesian analogue to the Gauß–Markov property of the least-squares estimate. In that sense, the posterior mean estimate (2.53) provides a natural extension of Gauß–Markov estimation to nonparametric regression.

The Bayesian perspective on smoothing splines can also be employed to reveal another connection between smoothing by spline functions and random-walk approaches. Kohn and Ansley (1988) derive an alternative characterization of the continuous process prior (2.52) for $f(x)$ from the stochastic differential equation

$$\frac{\partial^m}{\partial x^m} f(x) = \tau \frac{\partial W(x)}{\partial x}.$$

For the special case of first order derivatives ($m = 1$), an approximation in terms of difference quotients yields

$$\frac{f(x+\Delta x) - f(x)}{|\Delta x|} \approx \tau \frac{W(x+\Delta x) - W(x)}{|\Delta x|}$$

or equivalently

$$f(x+\Delta x) - f(x) \approx \tau[W(x+\Delta x) - W(x)] \sim \text{N}(0, |\Delta x|\tau^2).$$

Hence, the process prior for $f(x)$ implies that the increments follow the same distribution as the increments of a Wiener process with variance τ^2 which, in turn, form a white noise sequence for non-overlapping, equidistant intervals. Comparing this result with the random-walk priors for time series in Section 2.1.1 reveals that a first-order random-walk prior is in fact a discrete approximation to the continuous stochastic process prior leading to a natural spline of degree 1. Similarly, it can be shown that higher order difference quotients yield discrete approximations to natural spline priors with higher degree. For example, second-order differences correspond to the natural cubic spline employed in our original formulation of the smoothing spline approach, see Biller and Fahrmeir (1997).

Reproducing kernel Hilbert spaces

A unifying representation of several smoothing approaches in terms of general function spaces and penalties can be derived from the notion of reproducing kernel Hilbert spaces. The underlying theory allows us to reduce everything we have considered so far to the core of smoothing by dealing with general function spaces as well as general optimality criteria. Following the introductions in Evgeniou, Pontil and Poggio (2000) and Hastie, Tibshirani and Friedman (2003), we consider the optimization problem

$$\sum_{i=1}^{n} l(y_i, f(x_i)) - \frac{\lambda}{2}||f||_{\mathcal{F}} \rightarrow \max_{f \in \mathcal{F}} \tag{2.54}$$

where l is some concave utility function describing the estimation problem (such as the log-likelihood), $\mathcal{F}$ is a suitable space of functions characterizing the admissible solutions and $||\cdot||_{\mathcal{F}}$ is a corresponding norm. In the following, $\mathcal{F}$ will be assumed to be a reproducing kernel Hilbert space (RKHS).

A Hilbert space $\mathcal{F}$ is a complete vector space with an associated scalar product $\langle f, g \rangle$, $f, g \in \mathcal{F}$, that defines the norm $||f|| = \sqrt{\langle f, f \rangle}$. Note that we denote the Hilbert space and its elements as $\mathcal{F}$ and f to emphasize the relation to function spaces, although Hilbert spaces can of course be considered more generally. A RKHS is obtained as a special Hilbert space by assuming in addition that all elements of $\mathcal{F}$ are bounded linear functionals.

It can be shown that a RKHS can be uniquely characterized in terms of a positive definite function $k(x, y)$ (the reproducing kernel) that fulfils the reproducing property

$$f(x) = \langle f(y), k(y, x) \rangle, \quad f \in \mathcal{F}.$$

Under mild regularity conditions on the utility function $l(\cdot)$ it then follows that the maximizer of (2.54) has a finite dimensional representation as

$$\hat{f}(x) = \sum_{j=1}^{n} \hat{\gamma}_j k(x, x_j) = \boldsymbol{x}'\hat{\boldsymbol{\gamma}}$$

where $\boldsymbol{x} = (k(x, x_1), \ldots, k(x, x_n))'$.

If the utility function is the likelihood of a Gaussian model, (2.54) is equivalent to

$$(\boldsymbol{y} - \boldsymbol{X}\boldsymbol{\gamma})'(\boldsymbol{y} - \boldsymbol{X}\boldsymbol{\gamma}) + \lambda||f||_{\mathcal{F}} \to \min_{\boldsymbol{\gamma}} \tag{2.55}$$

where $\boldsymbol{X}$ is obtained by defining basis functions $B_j(x) = k(x, x_j)$ and $\boldsymbol{X}[i,j] = k(x_i, x_j)$. In this representation, x_j can be considered a knot while the kernel is interpreted as a function of the first argument only. Similarly, the penalty term can be written as a quadratic form

$$\text{pen}(f) \propto ||f||_{\mathcal{F}} = \sum_{i=1}^{n}\sum_{j=1}^{n} k(x_i, x_j)\gamma_i\gamma_j = \boldsymbol{\gamma}'\boldsymbol{K}\boldsymbol{\gamma}$$

where $\boldsymbol{K}[i,j] = k(x_i, x_j)$. As a consequence, $\boldsymbol{X} = \boldsymbol{K}$ and the minimizer of (2.55) is given by

$$\hat{\boldsymbol{\gamma}} = (\boldsymbol{X}'\boldsymbol{X} + \lambda\boldsymbol{K})^{-1}\boldsymbol{X}'\boldsymbol{y} = (\boldsymbol{K} + \lambda\boldsymbol{I})^{-1}\boldsymbol{y}$$

while

$$\hat{\boldsymbol{f}} = \boldsymbol{X}\hat{\boldsymbol{\gamma}} = (\boldsymbol{I} + \lambda\boldsymbol{K}^{-1})^{-1}\boldsymbol{y}.$$

More generally, the estimation problem (2.54) can be recast as

$$\sum_{i=1}^{n} l(y_i, \boldsymbol{x}_i'\boldsymbol{\gamma}) - \frac{\lambda}{2}\boldsymbol{\gamma}'\boldsymbol{K}\boldsymbol{\gamma},$$

immediately revealing the close connection to the penalization approaches considered in the previous part of this chapter. However, the structure of the estimation problem is considerably simpler in RKHS representation. Both the design vectors and the penalty simply contain evaluations of the reproducing kernel. Polynomial splines and smoothing splines, as well as several further statistical learning approaches, can actually be recast in terms of RKHS.

Since the representing kernel is a positive definite function, the relation to Bayesian smoothing approaches in function spaces is also obvious: The kernel $k(x, y)$ defines the covariance function of the zero mean Gaussian-process prior for $f(x)$. Therefore, the optimization criterion in (2.54) can be considered the posterior in a Bayesian formulation of the estimation problem.

To obtain a deeper understanding of RKHS and their properties, an alternative characterization of the reproducing kernel is valuable. Suppose that linearly independent functions $\phi_1(x), \phi_2(x), \ldots$ and positive numbers $\lambda_i \in \mathbb{R}$, $i = 1, 2, \ldots$ are given such that the reproducing kernel can be represented as

$$k(x, y) = \sum_{i=1}^{\infty} \lambda_i \phi(x)\phi(y).$$

Then it can be shown that the functions $\phi_1(x), \phi_2(x), \ldots$ form a basis of the Hilbert space $\mathcal{F}$, that is,

$$\mathcal{F} = \{f : f(x) = \sum_{i=1}^{\infty} \alpha_i \phi_i(x),\ \alpha_i \in \mathbb{R},\ i = 1, 2, \dots\}$$

subject to $||f|| < \infty$. The scalar product in $\mathcal{F}$ can now be represented as

$$\langle f, g \rangle = \left\langle \sum_{i=1}^{\infty} \alpha_i \phi_i(x), \sum_{i=1}^{\infty} \beta_i \phi_i(x) \right\rangle_{\mathcal{F}} = \sum_{i=1}^{\infty} \frac{\alpha_i \beta_i}{\lambda_i}$$

and the associated norm is defined by

$$||f||^2_{\mathcal{F}} = \langle f, f \rangle = \sum_{i=1}^{\infty} \frac{\alpha_i^2}{\lambda_i}.$$

In the above presentation, we have tacitly assumed that all elements of $\mathcal{F}$ are subject to penalization. If special functions, such as linear ones in the case of smoothing splines, shall remain unpenalized, the function space of interest is defined as the orthogonal sum $\mathcal{F} = \mathcal{F}_0 \oplus \mathcal{F}_1$ where $\mathcal{F}_0$ defines the null space of the penalty and $\mathcal{F}_1$ defines the orthogonal deviation from the null space. In most cases, $\mathcal{F}_0$ will consist of simple smooth functions such as low-order polynomials. In this case, the penalty $\text{pen}(f)$ is replaced by the penalty of the orthogonal projection of f on $\mathcal{F}_1$ and each function in $\mathcal{F}$ can be represented as

$$f(x) = \sum_{j=1}^{p} \beta_j h_j(x) + \sum_{i=1}^{n} \gamma_i k(x, x_i)$$

where $h_j(x)$, $j = 1, \dots, p$ is a basis of $\mathcal{F}_0$ that is orthogonal to the kernels $k(\cdot, x_i)$.

Wood (2006, p. 126) presents a RKHS basis for the space of cubic polynomial splines. Minor modifications of his basis allow us to represent smoothing splines in the context of RKHS (Wahba, 1990) as well as several further statistical learning approaches (Evgeniou, Pontil and Poggio, 2000). However, note that only splines of odd degree can be represented in terms of RKHS and therefore the generality of the approach does not come without restrictions. It also does not allow us to consider the class of penalized splines with difference penalties that we will use throughout this book.

However, RKHS allows us to consider non-standard data situations. For example, Vishwanathan, Schraudolph, Kondor and Borgwardt (2009) define kernels on graphs, Karatzoglou and Feinerer (2007) approach the problem of text mining based on string kernels, and Smola, Gretton, Song and Schölkopf (2007) embed probability distributions in RKHS.

Other types of basis functions

Besides polynomial spline and natural spline bases, other types of basis functions have been employed in smoothing approaches for different contexts. For example, the popular multivariate adaptive regression spline (MARS) approach by Friedman (1991) adds two-sided truncated polynomials

$$B_\kappa(x) = (\kappa - x)_+^l + (x - \kappa)_+^l$$

to the basis function catalogue. In most MARS implementations, linear versions with $l = 1$ have been employed.

Basis functions motivated from spatial smoothing approaches are derived from correlation functions such as the exponential correlation function

$$B_\kappa(x) = \exp(-|x - \kappa|/\phi)$$

or the Gaussian correlation function

$$B_\kappa(x) = \exp(-|x - \kappa|^2/\phi).$$

In both cases, the knots κ are typically given by the set of the observed covariate values, making the basis adaptive to the distribution of the covariate. The parameter ϕ is called the range parameter since it determines the scaling or effective range of the correlation function. We will consider bivariate extensions and further examples of correlation functions in the context of spatial smoothing in Section 5.4. In these extensions, the distance between a point $\boldsymbol{x} \in \mathbb{R}^2$ and a bivariate knot $\boldsymbol{\kappa} \in \mathbb{R}^2$ is determined by a suitable norm instead of the absolute value. Using the above formulae then yields radial basis functions since the contour lines of the basis functions are circles.

Note that smoothing approaches based on correlation functions have an immediate relation to reproducing kernel Hilbert spaces, where the kernel is defined as $k(\kappa, x) = B_\kappa(x)$. By definition, the correlation function defining the basis functions has to be positive definite and therefore there exists a corresponding RKHS. The associated penalty can then be written as

$$\text{pen}(f) = \boldsymbol{\gamma}'\boldsymbol{K}\boldsymbol{\gamma},$$

where $\boldsymbol{K}[i, j] = B_{x_i}(x_j)$.

2.3 Generalized additive models

2.3.1 Gaussian additive models

Practical situations where smoothing with respect to one single covariate is sufficient are rather rare, since typically several covariates of interest $x_1, \ldots, x_p$ are to be considered simultaneously. In other situations, the effect of confounding variables has to be taken into account, leading also to the necessity of combining several effects. A direct generalization of the scatterplot smoothing approach would try to estimate the functional relationship

$$y = f(x_1, \ldots, x_p) + \varepsilon$$

from given data. However, in most situations the sample size will be too small to estimate a completely flexible p-dimensional function. The underlying reason is that the data become increasingly sparse in the high-dimensional covariate space, a phenomenon referred to as the curse of dimensionality (see Fig. 5.27 on page 366 and Hastie and Tibshirani (1990, p. 83–84) for a graphical explanation). As a remedy, additional structure has to be imposed on the nonlinear relationship between response and covariates. A significant reduction in complexity is assumed in additive models where each effect is allowed to be a flexible function but the overall model is composed additively, i.e.

$$f(x_1, \ldots, x_p) = f_1(x_1) + \ldots + f_p(x_p).$$

Note that this assumption may be quite restrictive since it does not allow for interactions between covariate effects. We will therefore introduce extended models, where at least two-way interactions $f(x_1, x_2)$ can be included, in Chapter 5. This is particularly important in spatial regression models, where x_1 and x_2 may represent spatial coordinate information and a purely additive effect will rarely be observed in practice.

Instead of allowing for nonlinear effects for all covariates, it may also be necessary to restrict some covariate effects to be linear. This will for example be the case for categorical covariates or if some effects turn out to be linear in preliminary analyses. We will therefore focus on the semiparametric model

$$y = \boldsymbol{u}'\boldsymbol{\beta} + f_1(x_1) + \ldots + f_p(x_p) + \varepsilon \tag{2.56}$$

in the following.

Before actually discussing estimation of the semiparametric model, we have to take care of a specific identifiability constraint. Noting that

$$f_1(x_1) + f_2(x_2) = f_1(x_1) + c + f_2(x_2) - c$$

shows us that the predictor in the semiparametric model remains unchanged when replacing $f_1(x_1)$ with $\tilde{f}_1(x_1) = f_1(x_1) + c$ and $f_2(x_2)$ with $\tilde{f}_2(x_2) = f_2(x_2) - c$. As a consequence, the levels of the functions are not identified, since they can be shifted arbitrarily between the functions. To overcome this difficulty, we always include an intercept β_0 in the model that describes the overall level and assume that all nonparametric functions have a fixed level. The latter can for example be achieved by 'centring' the functions via

$$\sum_{i=1}^{n} f_1(x_{i1}) = \ldots = \sum_{i=1}^{n} f_p(x_{ip}) = 0.$$

Other centring strategies can be applied for example based on the integral of the functions.

Each of the nonparametric components in the semiparametric model (2.56) can be represented in terms of a polynomial basis, that is,

$$f_l(x_l) = \sum_{k=1}^{K_l} \gamma_{lk} B_k(x_l).$$

Combining the function evaluations in the vectors $\boldsymbol{f}_l = (f_l(x_{l1}), \ldots, f_l(x_{ln}))'$ yields

$$\boldsymbol{f}_l = \boldsymbol{X}_l \boldsymbol{\gamma}_l$$

where $\boldsymbol{X}_l$ is a design matrix containing the basis function evaluations. Applying this matrix notation to all model terms and introducing the design matrix $\boldsymbol{U}$ for the covariates $\boldsymbol{u}$ with linear effects, leads to the model equation

$$\boldsymbol{y} = \boldsymbol{U}\boldsymbol{\beta} + \boldsymbol{X}_1\boldsymbol{\gamma}_1 + \ldots + \boldsymbol{X}_p\boldsymbol{\gamma}_p + \boldsymbol{\varepsilon}. \tag{2.57}$$

As in the univariate scatterplot smoothing context, basis function-approaches suffer from the difficulty of selecting an appropriate number of basis functions, which can be avoided by introducing penalty terms $\lambda_l \boldsymbol{\gamma}_l' \boldsymbol{K}_l \boldsymbol{\gamma}_l$ for each of the nonparametric components.

Simultaneous penalized least-squares (PLS) smoothing

Estimation of all regression effects is based on the extended PLS criterion

$$\text{PLS}(\boldsymbol{\gamma}_1, \ldots, \boldsymbol{\gamma}_p, \boldsymbol{\beta}) = (\boldsymbol{y} - \boldsymbol{\eta})' (\boldsymbol{y} - \boldsymbol{\eta}) + \sum_{l=1}^{p} \lambda_l \boldsymbol{\gamma}_l' \boldsymbol{K}_l \boldsymbol{\gamma}_l \tag{2.58}$$

where $\boldsymbol{\eta} = \boldsymbol{U}\boldsymbol{\beta} + \boldsymbol{X}_1\boldsymbol{\gamma}_1 + \ldots + \boldsymbol{X}_p\boldsymbol{\gamma}_p$. The PLS criterion combines the sum of squared residuals as a model fit criterion with the penalty terms for the different functions. Note that each function is assigned a separate smoothing parameter λ_l, making their appropriate choice more important but also more difficult. Strategies for estimating the smoothing parameters within a penalized likelihood will be discussed in Chapter 4, while a fully Bayesian solution is described later in this section.

In the context of *low-dimensional* basis-function approaches, such as P-splines with moderate numbers of knots, the PLS criterion (2.58) can be simultaneously minimized with respect to all regression coefficients even with a larger number of nonparametric functions. Setting the first derivative of the PLS criterion to zero, yields the system of equations

$$\begin{pmatrix} \boldsymbol{X}_1'\boldsymbol{X}_1 + \lambda_1\boldsymbol{K}_1 & \cdots & \boldsymbol{X}_1'\boldsymbol{X}_p & \boldsymbol{X}_1'\boldsymbol{U} \\ \vdots & \ddots & & \\ \boldsymbol{X}_p'\boldsymbol{X}_1 & & \boldsymbol{X}_p'\boldsymbol{X}_p + \lambda_p\boldsymbol{K}_p & \boldsymbol{X}_p'\boldsymbol{U} \\ \boldsymbol{U}'\boldsymbol{X}_1 & \cdots & \boldsymbol{U}'\boldsymbol{X}_p & \boldsymbol{U}'\boldsymbol{U} \end{pmatrix} \begin{pmatrix} \boldsymbol{\gamma}_1 \\ \vdots \\ \boldsymbol{\gamma}_p \\ \boldsymbol{\beta} \end{pmatrix} = \begin{pmatrix} \boldsymbol{X}_1'\boldsymbol{y} \\ \vdots \\ \boldsymbol{X}_p'\boldsymbol{y} \\ \boldsymbol{U}'\boldsymbol{y} \end{pmatrix} \tag{2.59}$$

which resembles the structure from the univariate case. The simultaneous estimation of all effects also has the advantage that the smoother matrix $\boldsymbol{S}$ for the full model (projecting the observed responses $\boldsymbol{y}$ to the predicted values $\hat{\boldsymbol{y}}$) is available as

$$\boldsymbol{S} = (\boldsymbol{X}_1 \dots \boldsymbol{X}_p \boldsymbol{U}) \begin{pmatrix} \boldsymbol{X}_1'\boldsymbol{X}_1 + \lambda_1 \boldsymbol{K}_1 & \cdots & \boldsymbol{X}_1'\boldsymbol{X}_p & \boldsymbol{X}_1'\boldsymbol{U} \\ \vdots & \ddots & & \\ \boldsymbol{X}_p'\boldsymbol{X}_1 & & \boldsymbol{X}_p'\boldsymbol{X}_p + \lambda_p \boldsymbol{K}_p & \boldsymbol{X}_p'\boldsymbol{U} \\ \boldsymbol{U}'\boldsymbol{X}_1 & \cdots & \boldsymbol{U}'\boldsymbol{X}_p & \boldsymbol{U}'\boldsymbol{U} \end{pmatrix}^{-1} \begin{pmatrix} \boldsymbol{X}_1' \\ \vdots \\ \boldsymbol{X}_p' \\ \boldsymbol{U}' \end{pmatrix}$$

and the overall model complexity can be derived by taking the trace. Taking traces only of diagonal blocks corresponding to specific functions yields the degrees of freedom associated with the respective function. In addition, confidence intervals not only for the function estimates themselves but also for contrasts of functions or other linear combinations of parameters can be determined based on information contained in the smoother matrix $\boldsymbol{S}$, see Section 4.2.2. This is not possible with results from iterative procedures such as the backfitting algorithm (discussed in the following paragraph) that typically only provide point estimates for the regression coefficients.

Backfitting

For *high-dimensional* basis-function approaches, such as smoothing splines, minimization of the PLS criterion (2.58) is traditionally achieved with the backfitting algorithm. The basic idea is that, after neglecting the error term ε, a function $\boldsymbol{f}_l$ can be approximated as

$$\boldsymbol{f}_l \approx \boldsymbol{y} - \boldsymbol{f}_1 - \dots - \boldsymbol{f}_{l-1} - \boldsymbol{f}_{l+1} - \dots - \boldsymbol{f}_p - \boldsymbol{U}\boldsymbol{\beta}.$$

Of course, in practice we do not know the remaining functions but given current estimates we can derive the partial working residuals

$$\boldsymbol{y} - \sum_{k \neq l} \hat{\boldsymbol{f}}_k - \boldsymbol{U}\hat{\boldsymbol{\beta}}$$

and obtain a working estimate for $\boldsymbol{f}_l$ by applying the corresponding smoother to these residuals, that is,

$$\hat{\boldsymbol{f}}_l = \boldsymbol{S}_l(\boldsymbol{y} - \sum_{k \neq l} \hat{\boldsymbol{f}}_k - \boldsymbol{U}\hat{\boldsymbol{\beta}}).$$

The backfitting algorithm proceeds iteratively through the sequence of functions and updates each function estimate until no further changes are observed. This can be summarized in the following algorithm:

Algorithm 2.6: Backfitting

i. Initialize $\hat{\boldsymbol{f}}_1 = \ldots = \hat{\boldsymbol{f}}_p = \mathbf{0}$ and $\hat{\boldsymbol{\beta}} = \mathbf{0}$.

ii. Obtain updated estimates $\hat{\boldsymbol{f}}_l, l = 1, \ldots, p$, as

$$\hat{\boldsymbol{f}}_l = \boldsymbol{S}_l(\boldsymbol{y} - \sum_{k \neq l} \hat{\boldsymbol{f}}_k - \boldsymbol{U}\hat{\boldsymbol{\beta}}).$$

iii. Obtain an updated estimate $\hat{\boldsymbol{\beta}}$ as

$$\hat{\boldsymbol{\beta}} = (\boldsymbol{U}'\boldsymbol{U})^{-1}\boldsymbol{U}'(\boldsymbol{y} - \sum_{k} \hat{\boldsymbol{f}}_k).$$

iv. Iterate steps ii. and iii. until the estimates no longer change.

As discussed in detail in Buja *et al.* (1989), the backfitting algorithm belongs to the class of Gauß–Seidel algorithms for iteratively solving high-dimensional systems of linear equations.

Bayesian backfitting: the Gibbs sampler

In a Bayesian formulation, the penalties are equivalent to assigning (mutually independent) Gaussian prior distributions

$$p(\boldsymbol{\gamma}_l \,|\, \tau_l^2) \propto \left(\frac{1}{\tau_l^2}\right)^{\operatorname{rank}(\boldsymbol{K}_l)/2} \exp\left(-\frac{1}{2\tau_l^2}\boldsymbol{\gamma}_l'\boldsymbol{K}_l\boldsymbol{\gamma}_l\right)$$

to the basis function coefficients $\boldsymbol{\gamma}_l$. For the regression coefficients $\boldsymbol{\beta}$ of the linear part we assume a flat noninformative prior $p(\boldsymbol{\beta}) \propto$ const, corresponding to unpenalized parameters in the frequentist approach, or a weakly-informative prior $\boldsymbol{\beta} \sim \mathrm{N}(\boldsymbol{c}, \boldsymbol{C})$ with density

$$p(\boldsymbol{\beta}) \propto \exp\left(-\frac{1}{2}(\boldsymbol{\beta} - \boldsymbol{c})'\boldsymbol{C}^{-1}(\boldsymbol{\beta} - \boldsymbol{c})\right),$$

and 'large' covariance matrix $\boldsymbol{C}$, reducing to a flat prior in the limiting case $\boldsymbol{C}^{-1} = \mathbf{0}$.

As in Section 2.2.1, we assume inverse gamma priors

$$\sigma^2 \sim \mathrm{IG}(a_\sigma, b_\sigma),$$
$$\tau_l^2 \sim \mathrm{IG}(a_l, b_l), \quad l = 1, \ldots, p$$

as a standard choice for the observation variance σ^2 and for the variance of the inverse smoothing parameters τ_l^2.

The posterior of all parameters is given by

$$p(\boldsymbol{\beta}, \boldsymbol{\gamma}_1, \ldots, \boldsymbol{\gamma}_p, \tau_1^2, \ldots, \tau_p^2, \sigma^2 | \boldsymbol{y}) \propto p(\boldsymbol{y}|\boldsymbol{\eta}) p(\boldsymbol{\beta}) \cdot \prod_{l=1}^{p} \left(p(\boldsymbol{\gamma}_l | \tau_l^2) p(\tau_l^2) \right) p(\sigma^2)$$

$$\propto \left(\frac{1}{\sigma^2}\right)^{n/2} \exp\left(-\frac{1}{2\sigma^2}(\boldsymbol{y} - \boldsymbol{\eta})'(\boldsymbol{y} - \boldsymbol{\eta})\right) \cdot \exp\left(-\frac{1}{2}(\boldsymbol{\beta} - \boldsymbol{c})'\boldsymbol{C}^{-1}(\boldsymbol{\beta} - \boldsymbol{c})\right)$$

$$\cdot \prod_{l=1}^{p} \left(\frac{1}{\tau_l^2}\right)^{\operatorname{rank}(\boldsymbol{K}_l)/2} \exp\left(-\frac{1}{2\tau_l^2} \boldsymbol{\gamma}_l' \boldsymbol{K}_l \boldsymbol{\gamma}_l\right)$$

$$\cdot \prod_{l=1}^{p} (\tau_l^2)^{-a_l - 1} \exp\left(-\frac{b_l}{\tau_l^2}\right) \cdot (\sigma^2)^{-a_\sigma - 1} \exp\left(-\frac{b_\sigma}{\sigma^2}\right).$$

Straightforward calculations show that the full conditionals for regression coefficients are Gaussian:

$$\boldsymbol{\gamma}_l | \cdot \sim \mathrm{N}(\boldsymbol{\mu}_l, \boldsymbol{\Sigma}_l),\ l = 1, \ldots, p \tag{2.60}$$

with covariance matrix and mean

$$\boldsymbol{\Sigma}_l = \left(\frac{1}{\sigma^2} \boldsymbol{X}_l' \boldsymbol{X}_l + \frac{1}{\tau_l^2} \boldsymbol{K}_l\right)^{-1},$$

$$\boldsymbol{\mu}_l = \frac{1}{\sigma^2} \boldsymbol{\Sigma}_l \boldsymbol{X}_l' (\boldsymbol{y} - \boldsymbol{\eta}_{-l}),$$

where

$$\boldsymbol{\eta}_{-l} = \boldsymbol{\eta} - \boldsymbol{X}_l \boldsymbol{\gamma}_l = \boldsymbol{\eta} - \boldsymbol{f}_l = \boldsymbol{U}\boldsymbol{\beta} + \sum_{k \neq l} \boldsymbol{f}_k$$

is the remaining part of the predictor after deleting $\boldsymbol{f}_l = \boldsymbol{X}_l \boldsymbol{\gamma}_l$. Similarly

$$\boldsymbol{\beta} | \cdot \sim \mathrm{N}(\boldsymbol{\mu}_\beta, \boldsymbol{\Sigma}_\beta), \tag{2.61}$$

with

$$\boldsymbol{\Sigma}_\beta = \sigma^2 (\boldsymbol{U}'\boldsymbol{U} + \sigma^2 \boldsymbol{C}^{-1})^{-1},$$

$$\boldsymbol{\mu}_\beta = \frac{1}{\sigma^2} \boldsymbol{\Sigma}_\beta (\boldsymbol{U}'(\boldsymbol{y} - \boldsymbol{\eta}_{-\beta}) + \sigma^2 \boldsymbol{C}^{-1} \boldsymbol{c}),$$

where $\boldsymbol{\eta}_{-\beta} = \boldsymbol{\eta} - \boldsymbol{U}\boldsymbol{\beta}$.

Full conditionals for σ^2 and τ_l^2 are inverse Gamma $\mathrm{IG}(\tilde{a}_\sigma, \tilde{b}_\sigma)$ and $\mathrm{IG}(\tilde{a}_l, \tilde{b}_l)$, respectively, with updated parameters as in Section 2.2.1:

$$\tilde{a}_\sigma = a_\sigma + \frac{n}{2}, \quad \tilde{b}_\sigma = b_\sigma + \frac{\boldsymbol{\varepsilon}'\boldsymbol{\varepsilon}}{2},$$

$$\tilde{a}_l = a_l + \frac{\operatorname{rank}(\boldsymbol{K}_l)}{2}, \quad \tilde{b}_l = b_l + \frac{\boldsymbol{\gamma}_l' \boldsymbol{K}_l \boldsymbol{\gamma}_l}{2}.$$

The resulting *Gibbs sampler* can be summarized as follows.

Algorithm 2.7: Gibbs sampler for additive models

i. Update γ_l, $l = 1, \ldots, p$ by drawing from the Gaussian distributions $\mathrm{N}(\boldsymbol{\mu}_l, \boldsymbol{\Sigma}_l)$, inserting current values $(\sigma^2)^c$ and $(\tau^2)^c$ in $\boldsymbol{\mu}_l$ and $\boldsymbol{\Sigma}_l$. Take advantage of band matrix structures in $\boldsymbol{P}_l = \boldsymbol{\Sigma}_l^{-1}$.
ii. Update $\boldsymbol{\beta}$ by drawing from $\mathrm{N}(\boldsymbol{\mu}_\beta, \boldsymbol{\Sigma}_\beta)$, inserting the current value $(\sigma^2)^c$ in $\boldsymbol{\mu}_\beta$ and $\boldsymbol{\Sigma}_\beta$.
iii. Update τ_l^2, $l = 1, \ldots, p$, by drawing from $\mathrm{IG}(\tilde{a}_l, \tilde{b}_l)$, inserting the (new) current value γ_l^c obtained in step i.
iv. Update σ^2 by drawing from $\mathrm{IG}(\tilde{a}_\sigma, \tilde{b}_\sigma)$, inserting the new current values γ_l^c and $\boldsymbol{\beta}^c$ obtained in steps i. and ii.

For given σ^2 and $\tau_1^2 = \sigma^2/\lambda_1, \ldots, \tau_p^2 = \sigma^2/\lambda_p$, steps i. and ii. constitute the Bayesian analogue of the backfitting algorithm. Therefore, the Gibbs sampler has been termed 'Bayesian backfitting' by Hastie and Tibshirani (2000) in the context of smoothing splines. For Bayesian P-splines, this is the Gibbs sampler in Lang and Brezger (2004).

Usual standard choices for the hyperparameters a_l, b_l, $l = 1, \ldots, p$ and a_σ, b_σ are small values such as $a_l = b_l = 0.001$. As already discussed in Section 2.2.1, other choices have been suggested in the literature. A simple diffuse choice are improper $\mathrm{IG}(a_l, b_l)$ priors obtained with $a_l = -0.5$, $b_l = 0$, equivalent to diffuse priors $p(\tau_l) \propto \mathrm{const}$ for standard deviations. The full conditionals $\mathrm{IG}(\tilde{a}_l, \tilde{b}_l)$ are proper as well as (under mild restrictions) the entire posterior, see Fahrmeir and Kneib (2009).

Example 2.10: Childhood malnutrition in India

To illustrate the application of semiparametric additive models employing either penalized least-squares estimation (EB inference) or Markov chain Monte Carlo simulations (FB inference), we analyse the data set on childhood malnutrition in India, presented in Example 1.3. We include all continuous covariates as non-linear effects using cubic penalized splines with 20 inner knots and second-order random-walk penalties. Subsuming categorical covariates in the vector $\boldsymbol{u}$ leads to the model equation

$$\eta_i = \boldsymbol{u}_i'\boldsymbol{\beta} + f_1(cage_i) + f_2(cfeed_i) + f_3(mbmi_i) + f_4(mage_i) + f_5(medu_i) + f_5(edup_i).$$

Note that we are neglecting the spatial effect for the moment, although this may lead to smaller standard deviations due to spatial correlations that have not been included in the model. We will deal with this issue in Chapter 5, where we discuss prior distributions for spatial effects based on regional data.

For full Bayes inference, the optimal amount of smoothing was determined by assigning inverse gamma priors to the smoothing variances where all hyperparameters in the inverse gamma priors were fixed at 0.001. For empirical Bayes inference, the smoothing variance was estimated using mixed model methodology as introduced in Section 4.2.1.

Figures 2.21 and 2.22 visualize the resulting posterior mean and posterior mode estimates for the nonparametric effects. For most of the effects, the point estimates are in very good agreement, though MCMC estimation shows a tendency towards somewhat more wiggly effects. In particular, some effects are estimated to be close to linear with EB inference while FB inference detects some local variation. This is even more pronounced in the credible bands, where MCMC estimation shows more sensitivity with respect to areas with sparse data, such as very high values for the mother's BMI or education of the partner. Here, the credible intervals for FB inference are considerably wider than for EB inference. This is a common phenomenon that reflects the additional uncertainty due to the estimation of the smoothing parameter. While this uncertainty is carried forward to the credible intervals in the FB approach, it is not accounted for with EB inference.

The estimated nonparametric effects lead to the following conclusions: Most children start in good or at least satisfactory nutritional conditions when they are born, but a steady decline begins immediately after birth. This decline continues up to an age of about two years followed by an unexpected increase. The latter can, however, be explained by an artefact in the data: At the age of two years, the reference population that is considered in the construction of the Z-score for malnutrition changes. This results in a general increase of response values that actually does not relate to a better nutritional status. The effects of duration of breastfeeding seems to be surprising at first sight since breastfeeding is expected to have a protective effect with respect to malnutrition. The estimated effect, however, seems to suggest the reverse, where longer breastfeeding leads to smaller values of the Z-score. An explanation is that long durations of breastfeeding may relate to the absence of alternative food sources. Both the body mass index of the mother and the education of the mother show the expected effects with a protective effect of better nutritional status of the mother and better education. In contrast, the education of the partner does not seem to have a very pronounced effect on childhood malnutrition.

Estimated parametric effects are summarized in Tables 2.1 and 2.2. In this case, the agreement between FB and EB inference is even closer, both for the point estimates and the posterior standard deviation. We only interpret some selected effects. Twin children have a significantly reduced Z-score which is intuitively plausible since resources have to be distributed across two children. Moreover, twin children will typically start already with a lower Z-score. Being the first child in the birth order results in a significant plus in the Z-score, possibly reflecting the parents preference for first born children. Most asset variables have positive effect on the Z-score with bicycle as a sole exception. However, the effect of all asset variables is not too large in magnitude. △

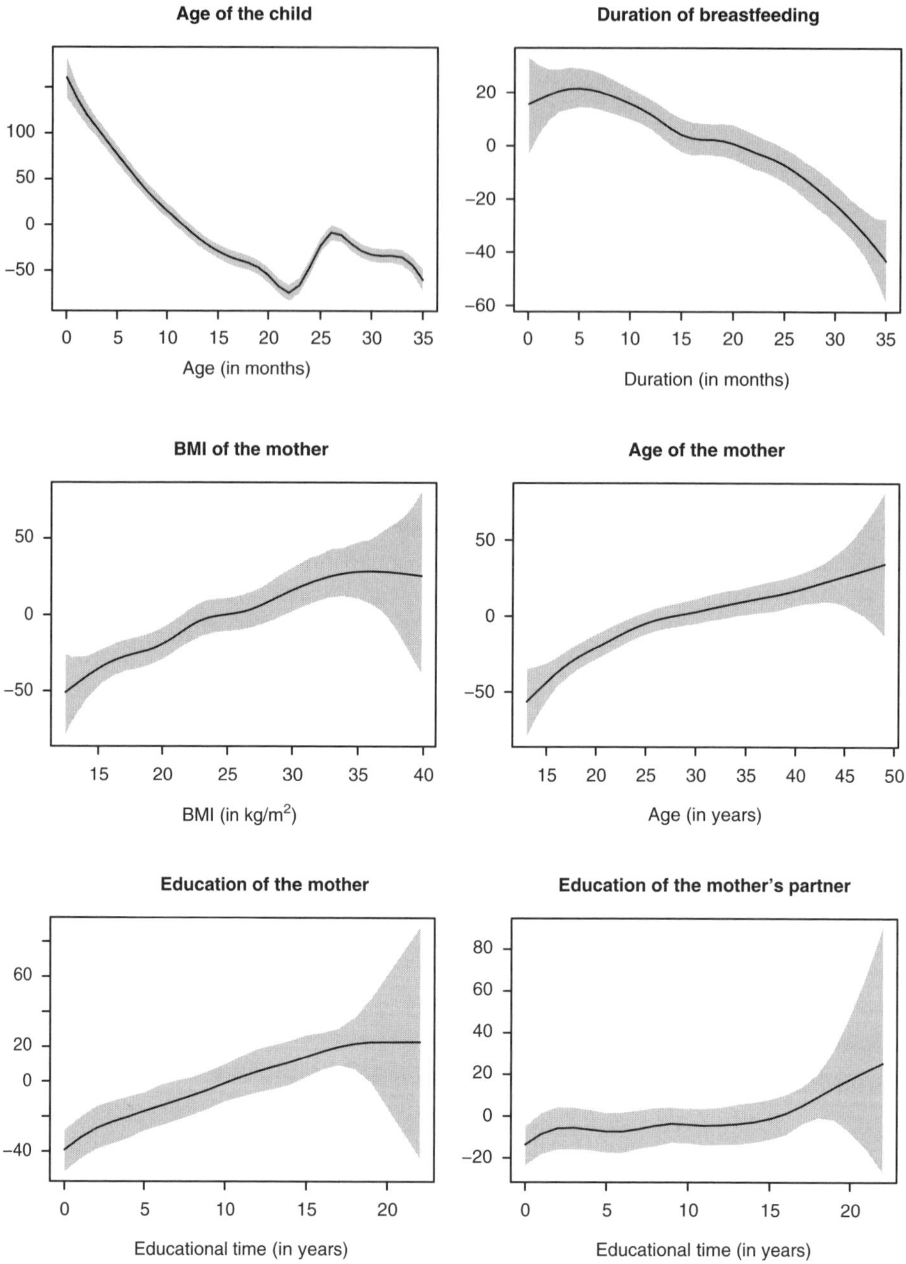

Fig. 2.21 Childhood malnutrition in India: Posterior mean estimates for nonparametric effects using Markov chain Monte Carlo simulation techniques (solid lines) and 95% pointwise credible bands.

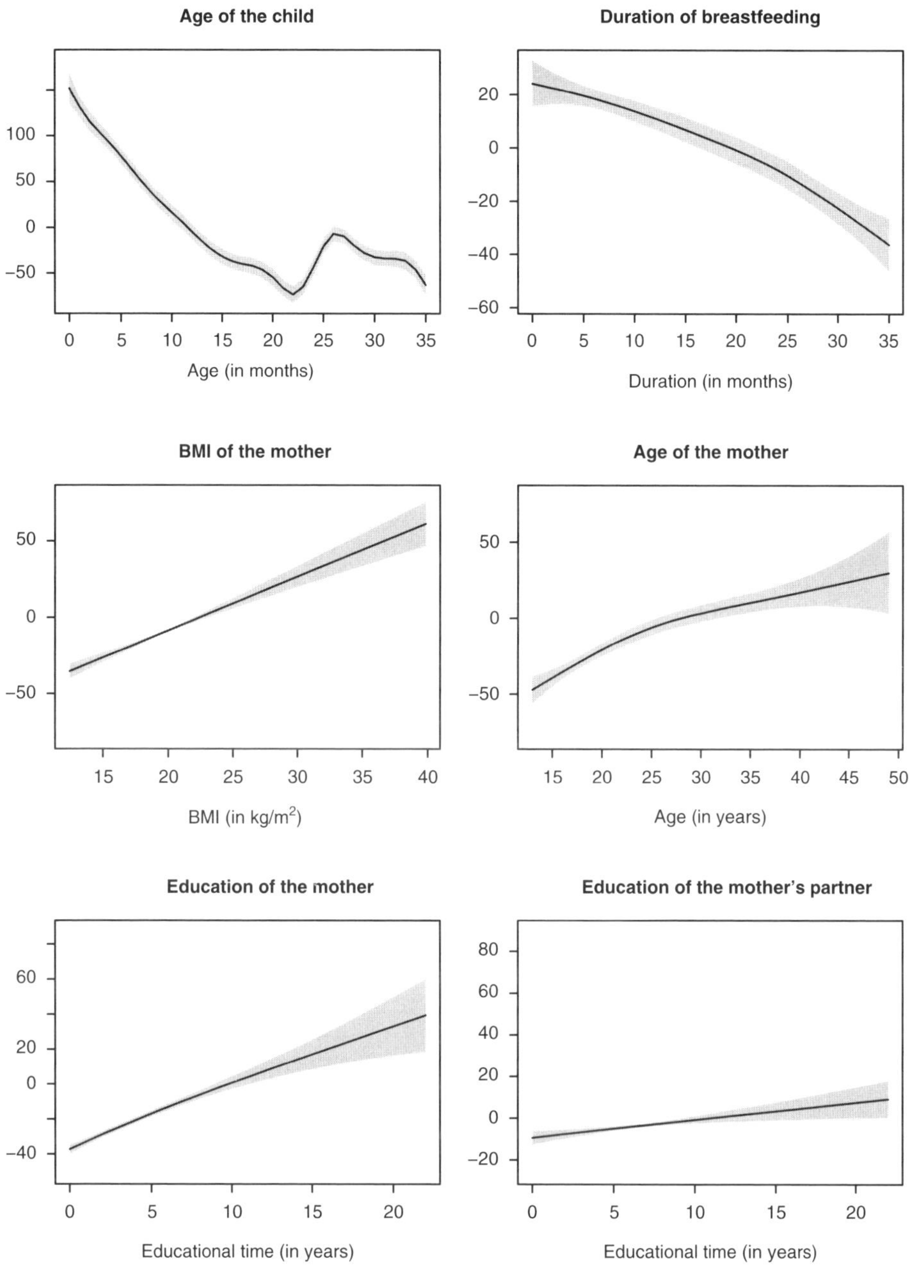

Fig. 2.22 Childhood malnutrition in India: Posterior mode estimates for nonparametric effects using empirical Bayes inference (solid lines) and 95% pointwise credible bands.

Table 2.1 Childhood malnutrition in India: Posterior means, standard deviations and 95% credible intervals for parametric effects.

variable	estimate	sd	95% credible	interval
intercept	−119.837	9.228	−138.184	−101.917
csex	−2.260	1.938	−6.113	1.546
ctwin	−66.027	9.605	−86.091	−47.560
munemployed	0.924	2.226	−3.543	5.249
cbirthorder (5+)	−41.336	4.705	−50.832	−32.260
cbirthorder (4)	−19.390	4.226	−27.324	−10.935
cbirthorder (3)	−21.185	3.363	−27.586	−14.430
cbirthorder (2)	−10.931	2.678	−16.037	−6.000
cbirthorder(1)	92.842			
deadchildren (3+)	−19.893	7.290	−34.734	−6.486
deadchildren (2)	−6.467	5.271	−16.616	4.222
deadchildren (1)	−5.811	3.075	−11.518	0.495
deadchildren (0)	32.171			
mreligion (Other)	35.338	6.023	23.957	47.143
mreligion (Sikh)	1.199	6.713	−11.236	13.517
mreligion (Christian)	35.345	3.910	27.785	43.491
mreligion (Muslim)	2.432	2.864	−3.103	8.014
mreligion (Hindu)	−74.314			
mresidence	−2.345	2.465	−7.365	2.313
electricity	10.543	2.378	5.698	15.048
radio	9.292	2.164	5.199	13.692
television	13.926	2.650	8.741	18.929
refrigerator	6.780	3.994	−1.755	14.776
bicycle	−9.331	2.027	−13.232	−5.573
motorcycle	7.916	3.444	1.237	15.086
car	2.784	7.558	−11.523	17.292

2.3.2 Non-Gaussian additive models

Having the quantities for a univariate model with only one influential variable available, it is easy to extend estimation to semiparametric additive non-Gaussian models. When using low-dimensional basis-function representations, such as penalized splines, all regression coefficients can be estimated simultaneously in one large Fisher scoring algorithm. Therefore we first write the model in matrix notation as

$$\mathbb{E}(\boldsymbol{y}) = \boldsymbol{\mu} = h(\boldsymbol{\eta}), \quad \boldsymbol{\eta} = \boldsymbol{U}\boldsymbol{\beta} + \boldsymbol{X}_1\boldsymbol{\gamma}_1 + \ldots + \boldsymbol{X}_p\boldsymbol{\gamma}_p, \tag{2.62}$$

where $\boldsymbol{y}$, $\boldsymbol{\mu}$, and $\boldsymbol{\eta}$ are stacked versions of the univariate quantities and $\boldsymbol{U}$, $\boldsymbol{X}_1, \ldots, \boldsymbol{X}_p$ are appropriately defined design matrices as in (2.57). Extending the Fisher scoring algorithm and the equivalent IWLS representation to the additive modelling setting yields the system of equations

Table 2.2 Childhood malnutrition in India: Posterior modes, standard deviations, 95% credible intervals and p-values for parametric effects.

variable	estimate	sd	95% credible	interval	p-value
intercept	−133.023	4.737	−142.309	−123.738	0.000
csex	−2.176	1.883	−5.867	1.515	0.248
ctwin	−66.315	9.768	−85.465	−47.166	0.000
munemployed	0.731	2.162	−3.508	4.970	0.735
cbirthorder (5+)	−41.043	4.661	−50.180	−31.906	0.000
cbirthorder (4)	−18.826	4.151	−26.963	−10.688	0.000
cbirthorder (3)	−20.524	3.306	−27.005	−14.044	0.000
cbirthorder (2)	−10.385	2.668	−15.615	−5.156	0.000
cbirthorder(1)	90.778				
deadchildren (3+)	−20.257	7.533	−35.024	−5.490	0.008
deadchildren (2)	−6.665	5.382	−17.217	3.886	0.216
deadchildren (1)	−6.186	3.052	−12.169	−0.203	0.042
deadchildren (0)	33.108				
mreligion (Other)	35.951	6.113	23.967	47.936	0.000
mreligion (Sikh)	0.926	6.680	−12.169	14.022	0.890
mreligion (Christian)	36.180	3.898	28.538	43.822	0.000
mreligion (Muslim)	2.551	2.838	−3.013	8.114	0.369
mreligion (Hindu)	−75.608				
mresidence	−2.333	2.472	−7.179	2.512	0.346
electricity	11.073	2.355	6.457	15.689	0.000
radio	9.328	2.138	5.136	13.521	0.000
television	13.742	2.646	8.554	18.929	0.000
refrigerator	6.013	3.890	−1.613	13.640	0.122
bicycle	−9.259	2.010	−13.199	−5.319	0.000
motorcycle	7.635	3.489	0.796	14.474	0.028
car	2.208	7.496	−12.488	16.904	0.768

$$\begin{pmatrix} \boldsymbol{X}_1'\boldsymbol{W}^{(k)}\boldsymbol{X}_1 + \lambda_1\boldsymbol{K}_1 & \cdots & \boldsymbol{X}_1'\boldsymbol{W}^{(k)}\boldsymbol{U} \\ \vdots & \ddots & \\ \boldsymbol{U}'\boldsymbol{W}^{(k)}\boldsymbol{X}_1 & \cdots & \boldsymbol{U}'\boldsymbol{W}^{(k)}\boldsymbol{U} \end{pmatrix} \begin{pmatrix} \boldsymbol{\gamma}_1 \\ \vdots \\ \boldsymbol{\beta} \end{pmatrix} = \begin{pmatrix} \boldsymbol{X}_1'\boldsymbol{W}^{(k)}\tilde{\boldsymbol{y}}^{(k)} \\ \vdots \\ \boldsymbol{U}'\boldsymbol{W}^{(k)}\tilde{\boldsymbol{y}}^{(k)} \end{pmatrix} \tag{2.63}$$

that has to be updated iteratively until no further changes are observed in the regression coefficients.

Based on equation (2.63), we can also derive analogies to some of the quantities discussed for Gaussian responses. The inverse of the left-hand side of (2.63) in the final Fisher scoring iteration provides us with the Fisher information matrix for the complete vector of regression coefficients. Assuming (approximate or asymptotic) normality of the corresponding estimates, confidence intervals and confidence bands for linear combinations of the regression coefficients can be deduced. The most important example will be confidence intervals for the function evaluations $\hat{\boldsymbol{f}}_l = \boldsymbol{X}_l\hat{\boldsymbol{\gamma}}_l$. From the IWLS algorithm, we can also derive the smoother matrix in the working model in the final iteration. The trace of this matrix provides the degrees of freedom for the complete model, that is, a measure of model complexity.

When employing high-dimensional bases, as with smoothing splines, it is common to base estimation on a combination of backfitting and Fisher scoring. Since Fisher scoring can be interpreted as an iteratively reweighted least squares fit, a natural extension is to replace the penalized least-squares fit by an iteratively reweighted backfitting step. However, similar problems as in the Gaussian setting are observed in the more general setting, that is, no confidence bands are available, no degrees of freedom can be computed and additional iterations are required to perform the internal backfitting loop. Therefore, we do not provide further details on the backfitting algorithm for non-Gaussian responses.

For *full Bayesian* inference, the MH algorithm with IWLS proposals can be extended in a conceptually straightforward way to generalized additive models. We assume for simplicity that the dispersion parameter of the exponential family (corresponding to σ^2 in the Gaussian case) is known. Priors for $\boldsymbol{\beta}, \gamma_1, \ldots, \gamma_p$ and $\tau_1^2, \ldots, \tau_p^2$ are the same as in the Gaussian case. The basic idea is the same as in Section 2.2.2: The full non-Gaussian conditionals $p(\gamma_l|\cdot)$ and $p(\boldsymbol{\beta}|\cdot)$ are approximated by Gaussian distributions with means and precision matrices evaluated at the current mode approximations obtained from one (or more) Fisher scoring steps within backfitting, that is, IWLS applied to current residuals. More precisely, we define $\varphi(\gamma_l; \boldsymbol{\mu}_l^c, \boldsymbol{\Sigma}_l^c)$ as a multivariate Gaussian density with covariance matrix and mean

$$
\begin{aligned}
\boldsymbol{\Sigma}_l^c &= \left(\boldsymbol{X}_l' \boldsymbol{W}^c \boldsymbol{X}_l + \frac{1}{\tau_l^2} \boldsymbol{K}_l \right)^{-1}, \\
\boldsymbol{\mu}_l^c &= \boldsymbol{\Sigma}_l^c \boldsymbol{X}_l' \boldsymbol{W}^c (\tilde{\boldsymbol{y}}^c - \boldsymbol{\eta}_{-l}^c)
\end{aligned}
$$

where $\boldsymbol{\eta}_{-l}^c$ is the (current) predictor after deleting $\boldsymbol{X}_l \gamma_l^c$, while $\boldsymbol{W}^c$ and $\tilde{\boldsymbol{y}}^c$ are the weight matrix and the working observation, evaluated at the current predictor

$$
\boldsymbol{\eta}^c = \boldsymbol{U}\boldsymbol{\beta}^c + \boldsymbol{X}_1 \gamma_1^c + \ldots + \boldsymbol{X}_l \gamma_l^c + \ldots + \boldsymbol{X}_p \gamma_p^c.
$$

As an alternative, the current value γ_l^c of the Markov chain may be replaced through a current mode approximation $\hat{\gamma}_l^c$.

Similarly, $\varphi(\boldsymbol{\beta}; \boldsymbol{\mu}_\beta^c, \boldsymbol{\Sigma}_\beta^c)$ is the Gaussian density with covariance matrix and mean

$$
\begin{aligned}
\boldsymbol{\Sigma}_\beta^c &= (\boldsymbol{U}' \boldsymbol{W}^c \boldsymbol{U} + \boldsymbol{C}^{-1})^{-1}, \\
\boldsymbol{\mu}_\beta^c &= \boldsymbol{\Sigma}_\beta^c (\boldsymbol{U}' \boldsymbol{W}^c (\tilde{\boldsymbol{y}}^c - \boldsymbol{\eta}_{-\beta}^c) + \boldsymbol{C}^{-1} \boldsymbol{c}).
\end{aligned}
$$

Let the acceptance probability for a new proposal γ_l^p—and analogously for $\boldsymbol{\beta}^p$—be defined as

$$
\alpha(\gamma_l^c, \gamma_l^p) = \frac{p(\boldsymbol{y}|\boldsymbol{\eta}^p) p(\gamma_l^p|(\tau_l^2)^c) \varphi(\gamma_l^c; \boldsymbol{\mu}_l^c, \boldsymbol{\Sigma}_l^c)}{p(\boldsymbol{y}|\boldsymbol{\eta}^c) p(\gamma_l^c|(\tau_l^2)^c) \varphi(\gamma_l^p; \boldsymbol{\mu}_l^p, \boldsymbol{\Sigma}_l^p)}
$$

where $\boldsymbol{\eta}^c$ is the current predictor, and $\boldsymbol{\mu}_l^p$ and $\boldsymbol{\Sigma}_l^p$ are obtained by evaluating $\boldsymbol{\mu}_l$ and $\boldsymbol{\Sigma}_l$ for the proposed predictor $\boldsymbol{\eta}^p = \boldsymbol{U\beta}^c + \boldsymbol{X}_1\boldsymbol{\gamma}_1^p + \ldots + \boldsymbol{X}_l\boldsymbol{\gamma}_l^p + \ldots + \boldsymbol{X}_p\boldsymbol{\gamma}_p^c$.

Then the MH algorithm with IWLS proposals is extended to GAMs as follows:

Algorithm 2.8: Metropolis Hastings sampler for generalized additive models

i. Initialization: Compute the posterior modes $\boldsymbol{\mu}_l^{(0)}$, $l = 1, \ldots, p$, and $\boldsymbol{\beta}^{(0)}$ via (backfitting within) Fisher scoring, given fixed variance parameters $\tau_l^2 = c_l$ (e.g. $c_l = 10$).

ii. For $l = 1, \ldots, p$ update $\boldsymbol{\gamma}_l^c$ by drawing a proposal $\boldsymbol{\gamma}_l^p$ from $\varphi(\boldsymbol{\gamma}_l; \boldsymbol{\mu}_l^c, \boldsymbol{\Sigma}_l^c)$ and accept it with probability $\alpha(\boldsymbol{\gamma}_l^c, \boldsymbol{\gamma}_l^p)$ as the new current $\boldsymbol{\gamma}_l^c$.

iii. Update $\boldsymbol{\beta}$ in complete analogy with $\boldsymbol{\beta}^p$ drawn from $\varphi(\boldsymbol{\beta}; \boldsymbol{\mu}_\beta^c, \boldsymbol{\Sigma}_\beta^c)$.

iv. Updates for $\tau_1^2, \ldots \tau_p^2$ are obtained as for Gaussian additive models, see page 95.

As an alternative, Gibbs samplers may be constructed based on latent variable representations, where latent variables follow (conditionally) Gaussian models with additive predictor $\boldsymbol{\eta}$. The main difference to models with only one function and predictor $\boldsymbol{\eta} = \boldsymbol{X\gamma}$ is to replace updating of $\boldsymbol{\gamma}$ given the latent variables by steps i. and ii. of the Gibbs sampler for Gaussian additive models, again replacing observable variables y_i through latent variables l_i, drawn from truncated normals.

Example 2.11: Forest health

The specification of a semiparametric additive cumulative model allows us to analyse the forest health data in a more realistic setup. Therefore, we specify a cumulative logit model with predictor

$$\eta_{it}^{(c)} = \theta^{(c)} - f^{(c)}(t) - f(age_{it}) - \boldsymbol{u}_{it}'\boldsymbol{\beta}.$$

containing category-specific trend functions $f^{(c)}(t)$, a nonparametric effect of age and further categorical and continuous covariates in the vector $\boldsymbol{u}_{it}$. Note that the model still ignores spatial correlations and also the longitudinal structure of the data and therefore the obtained results, and in particular the associated uncertainty measures, should be interpreted with care.

Figure 2.23 visualizes the three estimated nonparametric effects. Both temporal effects closely follow the pattern detected already in the simpler model without any further covariates and are smoothed versions of the raw empirical trends visualized in Fig. 1.3. For the age effect, we obtain a nearly u-shaped function with less damage being associated with very old and very young trees.

Table 2.3 shows the estimated thresholds and parametric effects together with 95% credible intervals and p-values. There are several covariates that appear to have highly significant effects and we will check the validity of this conclusion

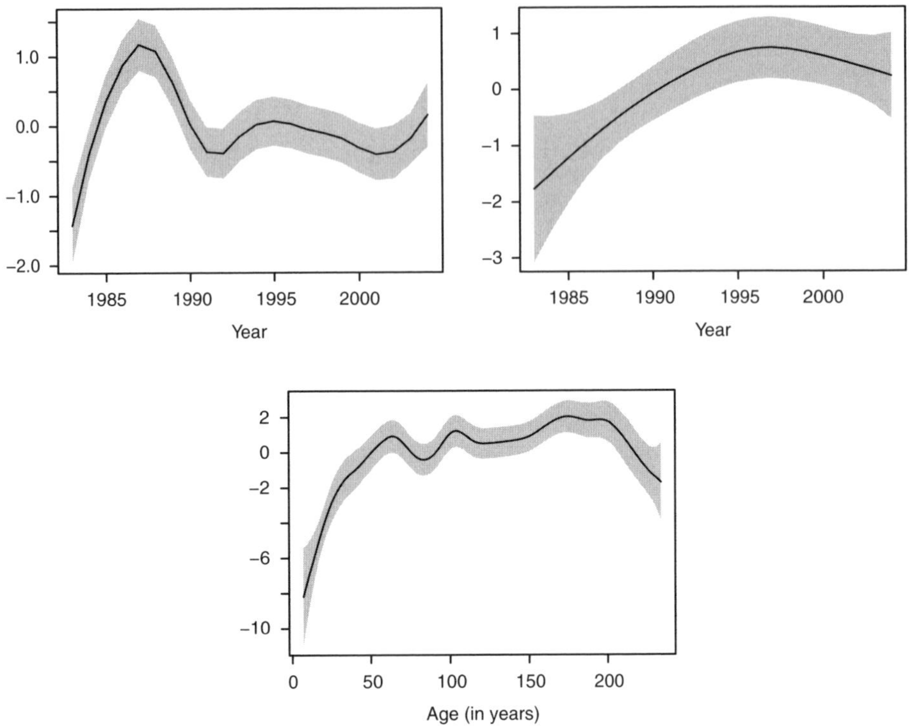

Fig. 2.23 Forest health: Estimated nonparametric effects in an additive model with category-specific trends. The two top-level functions correspond to the estimated category-specific time trends while the lower panel visualizes the estimated age effect.

in later chapters where the longitudinal and the spatial structure are considered appropriately in the analyses. △

2.4 Notes and further reading

Throughout this book, concepts for smoothing nonlinear functional effects will be based on basis functions expansions in combination with smoothness priors or, in frequentist terms, roughness penalties for basis-function coefficients. More specifically, we focus on (Bayesian) P-splines. As pointed out in this chapter, Bayesian inference, in particular empirical Bayes inference, and penalized likelihood inference are then closely related. Therefore, any literature relying on the latter concept can be of interest for further reading. Wood (2006) gives a recent and thorough account of generalized additive models using P-splines and penalized likelihood inference, with smoothing parameters estimated through generalized cross validation (GCV) or a version of Akaike's

Table 2.3 Forest health: Estimated parametric effects and thresholds.

variable	estimate	sd	95% credible	interval	p-value
θ_1	−4.990	1.447	−7.827	−2.153	0.001
θ_2	−0.347	1.442	−3.174	2.480	0.810
elevation	0.001	0.002	−0.002	0.004	0.450
inclination	0.007	0.007	−0.008	0.021	0.366
soil	−0.023	0.008	−0.038	−0.008	0.003
ph	−0.695	0.269	−1.222	−0.168	0.010
canopy	−4.100	0.351	−4.788	−3.413	0.000
standcat	0.446	0.083	0.284	0.608	0.000
fertilization	−0.268	0.136	−0.534	−0.001	0.049
humus0	−0.244	0.141	−0.520	0.032	0.082
humus1	−0.109				
humus2	0.323	0.122	0.084	0.561	0.008
humus3	0.156	0.141	−0.122	0.433	0.271
humus4	−0.126	0.187	−0.493	0.242	0.502
moisture1	−0.732	0.162	−1.050	−0.415	0.000
moisture2	0.17				
moisture3	0.562	0.115	0.337	0.787	0.000
alkali1	0.431	0.171	0.095	0.767	0.012
alkali2	−0.395				
alkali3	−0.676	0.176	−1.020	−0.332	0.000
alkali4	0.640	0.235	0.180	1.100	0.007

Information Criterion (AIC). The book by Ruppert, Wand and Carroll (2003) has brought the mixed model approach, closely related to empirical Bayes inference, to the attention of a broad audience. Ruppert, Wand and Carroll (2009) provide an overview on developments in semiparametric regression during 2003–2007. Tutz (2003) considers P-spline modelling for a variety of categorical regression models.

Because of our focus on basis-function expansions and smoothness priors, we have not included semi- and nonparametric regression based on the principle of localized estimation, such as kernel and local regression. Hastie and Loader (1993) give a good introduction to local regression, highlighting advantages over classical kernel regression. A comprehensive presentation of local polynomial regression is Fan and Gijbels (1996), while Loader (1999) gives an extensive treatment of local regression and likelihood. A Bayesian approach comparably close to kernel regression are relevance vector machines, using Gaussian kernels as basis functions (Tipping, 2001).

Smoothness of P-splines, as described in this chapter, is controlled through a global smoothing variance τ^2. For functions with areas of high curvature, such as distinct peaks, jumps or oscillations, this will lead to oversmoothing or blurring of such local features while the estimate will tend to be too wiggly in areas where the function is locally very smooth or close to a straight line. Locally adaptive variants of splines, which are less sensitive to such local features, can be derived by appropriately extending the hyperprior specification for τ^2. Most proposals either assume that τ^2 is replaced by a sequence τ_k^2 of independent prior variances (see e.g. Lang and Brezger, 2004), or the variance is itself a smooth function $\tau^2(x)$ (see e.g. Baladandayuthapani, Mallick and Carroll, 2005). Section 4.4.2 provides more detailed information about these and related approaches.

An alternative strategy to improve local adaptivity is adaptive knot or basis coefficient selection. Adaptive Bayesian knot selection using reversible jump MCMC is suggested in Biller (2000), Biller and Fahrmeir (2001), Denison, Mallick and Smith (1998), and Denison, Holmes, Mallick and Smith (2002) for spline bases, and in Rummel (2006) for relevance vector machines. Basis coefficient selection transfers to Bayesian variable selection with spike and slab priors to include or exclude basis functions from a high-dimensional basis-function representation, see for example Smith and Kohn (1996) and Section 4.5.1 for more details and additional references.

Wavelet regression is another alternative method to effectively model spatially heterogeneous regression functions and irregular signals. In principle, wavelets are used as an orthonormal basis for representing a function, see for example Ogden (1997) and Gençay, Selçuk and Whitcher (2002) for introductions to wavelets, and Donoho and Johnstone (1994, 1995) for popular frequentist wavelet shrinkage estimators. Bayesian wavelet regression is mostly based on spike and slab priors for wavelet coefficients, see for example Abramovich, Sapatinas and Silverman (1998), Clyde and George (2000) or Clyde, Parmigiani and Vidakovic (1998).

A different Bayesian approach to flexible regression modelling is based on Dirichlet process priors, as a generalization of finite mixture models, and related stochastic process priors, see the survey in Müller and Quintana (2004) and some contributions in Dey, Müller and Sinha (1998). In this book, Dirichlet priors are used as flexible nonparametric random effects priors in mixed models (Chapters 3 and 4).

The examples in this chapter and many others provide evidence that penalized likelihood inference, empirical Bayes inference and full Bayes inference often yield comparable results, although standard deviations, confidence intervals, etc. of the former rely on approximate normality of estimators. Recent research supports this empirical evidence through asymptotic theory (Li and Ruppert, 2008; Kauermann, Krivobokova and Fahrmeir 2009; Claeskens, Krivobokova and Opsomer, 2009).

In this book, full Bayes inference relies on MCMC techniques. Recently, Rue, Martino and Chopin (2009) have proposed a computationally attractive, numerical alternative using integrated nested Laplace approximation (INLA). We outline this method in some more detail in Section 4.6.2.

3 Generalized Linear Mixed Models

This chapter gives an introduction to linear and generalized linear mixed models. Our primary goal is to describe concepts for statistical modelling and inference in this class of models that are needed as basic building blocks or inferential tools in the following chapters. More detailed expositions (focusing on mixed models for longitudinal and cluster data) can be found, for example, in McCulloch and Searle (2001), Diggle *et al.* (2002), Verbeke and Molenberghs (2000), Fitzmaurice, Laird and Ware (2004) and Molenberghs and Verbeke (2005). Gelman and Hill (2006) also include multilevel models, with an emphasis on social science applications. State of the art surveys on various aspects of longitudinal data analysis, including generalized linear mixed models, can be found in the handbook edited by Fitzmaurice *et al.* (2009).

Most commonly, mixed models are considered for the analysis of clustered data. Clustering may be due to subsampling in primary sampling units in cross-sectional studies, due to repeated measurements over time in longitudinal studies, due to spatially aligned measurements, or due to any other data generating mechanism that induces a cluster structure. In any of these cases, cluster-specific random effects are added to the usual predictors of regression models both to account for correlations between observations within clusters and to introduce (random) cluster-specific effects. We will discuss the resulting model structure and motivations for treating the cluster-specific effects as random in detail in Section 3.1.1 after introducing the relevant notation.

For clustered observations, the data consist of n_i repeated observations $(y_{ij}, \boldsymbol{x}_{ij})$, $j = 1, \ldots, n_i$, for each individual or cluster $i = 1, \ldots, n$, where y denotes a response variable, $\boldsymbol{x}$ is a vector of covariates and n_i may denote the the number of repeated measurements in time for longitudinal data or the number of elements in a primary sampling unit such as a hospital or a geographical region. Collecting all individual- or cluster-specific responses in the vector

$$\boldsymbol{y}_i = (y_{i1}, \ldots, y_{in_i})'$$

and all covariate vectors in the matrix

$$\boldsymbol{X}_i = (\boldsymbol{x}'_{i1}, \ldots, \boldsymbol{x}'_{in_i})'$$

leads to the complete data representation $(\boldsymbol{y}_i, \boldsymbol{X}_i)$ for individual or cluster i.

In the following, we first describe linear mixed models (LMM) for longitudinal data with (conditionally) Gaussian responses y_{ij}, making the conventional assumption that the random effects are i. i. d. Gaussian variables. We then extend LMMs by allowing correlated Gaussian random effects. This leads to a very broad class of models that are appropriate for analysing spatial and spatio-temporal data and for Bayesian approaches to semiparametric smoothing. Section 3.2 introduces LMMs with flexible non-Gaussian priors for random effects. In particular, we describe nonparametric modelling of random effects distributions through Dirichlet process-based priors. Section 3.3 provides extensions of LMMs to generalized linear mixed models (GLMM) for non-Gaussian and categorical responses.

3.1 Linear mixed models with Gaussian random effects

3.1.1 Linear mixed models for longitudinal data

We first define the general model and exemplify through familiar special cases subsequently. Let

$$(y_{ij}, \boldsymbol{x}_{ij}, j = 1, \ldots, n_i)$$

denote the values of a response variable y and a covariate vector $\boldsymbol{x}$ observed at times $t_{i1} < \ldots < t_{ij} < \ldots < t_{in_i}$ for individuals $i = 1, \ldots, n$ in a longitudinal study. A LMM as considered by Laird and Ware (1982) is defined hierarchically in the following way:

At the *first stage*, responses y_{ij} are assumed to depend linearly on unknown population-specific fixed effects $\boldsymbol{\beta}$ and on unknown individual- or unit-specific random effects $\boldsymbol{\alpha}_i$ through the *measurement model*

$$y_{ij} = \boldsymbol{u}'_{ij}\boldsymbol{\beta} + \boldsymbol{z}'_{ij}\boldsymbol{\alpha}_i + \varepsilon_{ij}. \tag{3.1}$$

In (3.1), $\boldsymbol{u}_{ij}$ and $\boldsymbol{z}_{ij}$ represent design vectors and the error terms ε_{ij} are i.i.d. $\mathrm{N}(0, \sigma^2)$ random variables. The design vectors $\boldsymbol{u}_{ij}$ and $\boldsymbol{z}_{ij}$ are often constructed from original covariates $\boldsymbol{x}_{ij}$, but we may also admit that they depend on past responses $y_{il}, l < j$, as in transitional models for longitudinal data. In most cases, $\boldsymbol{z}_{ij}$ is a subvector of $\boldsymbol{u}_{ij}$. At the *second stage*, the model is supplemented through a distributional assumption for the random effects $\boldsymbol{\alpha}_i$, reflecting the idea that the data are generated by drawing individuals from a population. Therefore the effects $\boldsymbol{\alpha}_i$ are assumed to vary independently from one individual to another according to a *mixing* or—from a Bayesian perspective—*prior distribution*. In this section we make the standard parametric assumption of a Gaussian random effects distribution

$$\boldsymbol{\alpha}_i \stackrel{\text{i.i.d.}}{\sim} \mathrm{N}(\boldsymbol{0}, \boldsymbol{Q}) \tag{3.2}$$

with unknown covariance matrix $\boldsymbol{Q}$. In addition, error terms and random effects are assumed to be mutually independent.

The most prominent special case is a *random intercept model*, where $z_{ij} \equiv 1$ and $\alpha_i \sim \mathrm{N}(0, \delta^2)$ is a random intercept with variance $\delta^2 > 0$ yielding the model equation

$$y_{ij} = \boldsymbol{u}'_{ij}\boldsymbol{\beta} + \alpha_i + \varepsilon_{ij}. \tag{3.3}$$

If the covariate vector $\boldsymbol{u}$ contains an intercept, the corresponding regression coefficient (β_0, say) can be interpreted as the population intercept whereas α_i accounts for individual-specific deviations from this population average. The random effects $\beta_0 + \alpha_i \sim \mathrm{N}(\beta_0, \delta^2)$ may be interpreted as effects of omitted covariates and account for unobserved heterogeneity, see the following paragraph on advantages of mixed models for a detailed discussion.

Random intercept models are still based on the assumption that all covariate effects (or slopes) are equal for each individual. This may be too restrictive when intercepts and some slope coefficients are specific for each cluster. For example, in case of longitudinal data, a model with individual-specific linear trends is given through

$$\begin{aligned} y_{ij} &= \beta_0 + \beta_1 t_{ij} + \alpha_{0i} + \alpha_{1i} t_{ij} + \varepsilon_{ij} \\ &= (\beta_0 + \alpha_{0i}) + (\beta_1 + \alpha_{1i}) t_{ij} + \varepsilon_{ij}, \end{aligned} \tag{3.4}$$

where further covariates have been omitted to ease notation. In (3.4), $\beta_0 + \beta_1 t$ corresponds to the linear population trend whereas $\alpha_{0i} + \alpha_{1i} t$ is the individual deviation from the population average. Obviously, more general individual-specific trend functions, such as (piecewise) polynomials, or models with random effects of other covariates than time can be formulated in complete analogy. Models with any kind of random covariate effects (besides a random intercept) are commonly referred to as *random slope models*.

The following examples illustrate some of the ideas underlying random effects models.

Example 3.1: Height of schoolgirls

This longitudinal data on growth curves of 20 preadolescent schoolgirls, with yearly measurements from age 6 to 10, has first been reported in Goldstein (1979, Table 4.3, p. 101). The girls are divided into three groups according to the height of their mothers (small mothers with height < 155 cm, medium mothers with height between 155 cm and 164 cm, tall mothers with height > 164 cm). The individual data are shown separately for each group in Fig. 3.1. The original data were slightly modified by Verbeke and Molenberghs (2000), replacing the second obviously incorrect measurement 112 of girl 5 by 122. These data are also available in the DPpackage of Jara (2007). We will analyse the data with a simple linear random slope model

$$y_{ij} = \beta_0 + \beta_1 t_{ij} + \alpha_{i0} + \alpha_{i1} t_{ij} + \beta_2 \mathrm{group2} + \beta_3 \mathrm{group3} + \varepsilon_{ij}, \quad \varepsilon_{ij} \overset{\text{i.i.d.}}{\sim} \mathrm{N}(0, \sigma^2),$$

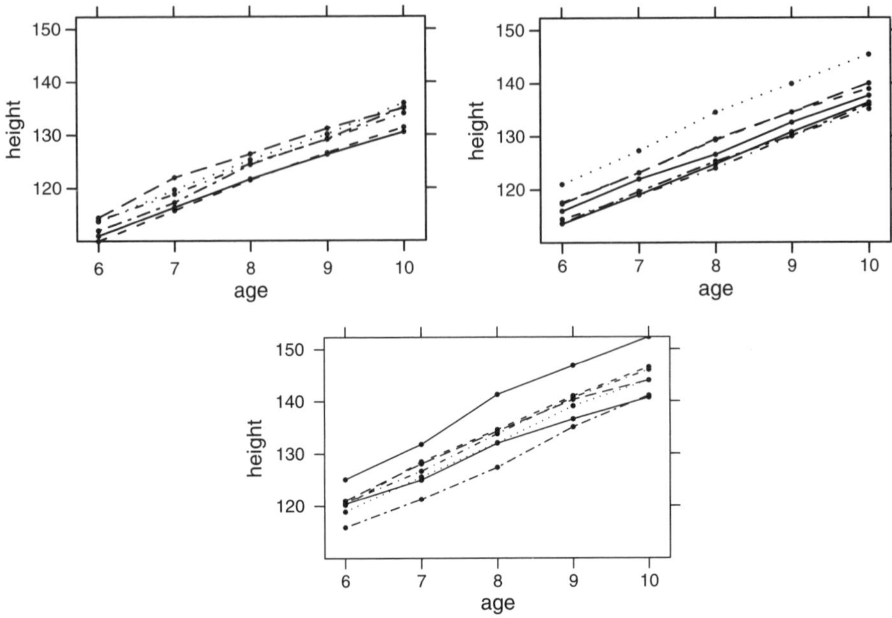

Fig. 3.1 Height of schoolgirls: Individual growth trajectories in the three groups corresponding to small (top left), medium (top right) and tall (bottom) mothers.

$i = 1, \ldots, 20$, $j = 1, \ldots, 4$, with linear population trend $\beta_0 + \beta_1 t$ for age t (in years), individual-specific linear growth curves $(\beta_0 + \alpha_{i0}) + (\beta_1 + \alpha_{i1})t$, and fixed effects for the dummy-coded group variable (with group1 'small mothers' as reference group). We will compare results based on Gaussian random effects, using empirical and full Bayes inference, and with more flexible Dirichlet process priors for random effects in Section 3.2.3. △

Example 3.2: Childhood obesity

As already discussed in Example 1.1 and illustrated in Figs. 1.1 and 1.2, the influence of age on a child's BMI is nonlinear. Assuming a linear population trend would clearly result in a misspecified model. Taking age of children as the basic time scale t, we therefore consider a linear mixed model of the form

$$y_{ij} = \boldsymbol{u}_{ij}'\boldsymbol{\beta} + f(t_{ij}) + \boldsymbol{z}_{ij}'\boldsymbol{\alpha}_i + \varepsilon_{ij}, \quad \varepsilon_{ij} \overset{\text{i.i.d.}}{\sim} \mathrm{N}(0, \sigma^2), \tag{3.5}$$

to assess the influence of a child's age and risk factors on its BMI. The vector $\boldsymbol{u}$ consists of the binary covariates sex, breastfeeding, maternal smoking and area, and of the two continuous (centred) covariates maternal BMI at pregnancy beginning (mBMI) and maternal BMI gain during pregnancy (mDiffBMI), see

Tables 1.1 and 1.2 in Example 1.1. To apply methodology for linear mixed models, the nonlinear effect $f(t)$ has to be expressed in appropriate parametric form, but linear in the unknown parameters. While global polynomials of low degree are not flexible enough, simple piecewise polynomials, such as a hockey stick function $\beta_0 + \beta_1 t + \beta_2(t-1)_+$ or splines with a few knots, might be considered. Justified by the model assessment in Chapter 4, we prefer the nonlinear parametric function

$$f(t) = \beta_0 + \beta_1 t + \beta_2 \frac{\log(t+1)}{(t+1)^2}. \tag{3.6}$$

The random effects part $z'_{ij}\alpha_i$ models individual-specific deviations from the population trend $f(t)$ in (3.6). We compare a linear random slope specification $\alpha_{i0} + \alpha_{i1}t$ with the nonlinear random function specification

$$g_i(t) = \alpha_{i0} + \alpha_{i1} t + \alpha_{i2} \frac{\log(t+1)}{(t+1)^2}, \tag{3.7}$$

where $g_i(t)$ has the same functional form as $f(t)$. Figure 3.6 (page 137) shows longitudinal measurements together with the population trend $f(t)$, (estimated) individual curves without random effects $u_i'\beta + f(t)$ and individual curves with random effects $u_i'\beta + f(t) + g_i(t)$, using full Bayesian estimation with the Gibbs sampler 3.1, see Example 3.4 for more details. In Chapter 4, we will apply an additive mixed model, where $f(t)$ is specified through a P-spline, and compare with the parametric specification considered in this chapter. △

Advantages of mixed models

The examples already illustrate some advantages of analysing longitudinal data with linear mixed models. In particular, this class of models enables the incorporation of individual-specific parameters allowing us to model individual deviations from population effects and trends. Such individual-specific effects may be considered as a surrogate for effects of covariates that have not been collected or that have been omitted in the analyses. If the omitted covariates are related to the grouping structure underlying the data, for example if the covariates characterize properties of the individuals in longitudinal data, ignoring their effects makes observations within groups more similar to each other than randomly selected observations. This problem is commonly referred to as *unobserved heterogeneity* since effects explaining the heterogeneity of observations have been missed in the model formulation. Obviously, ignoring such missing sources of heterogeneity induces the risk of biased and inefficient estimates for the remaining covariate effects. In particular, the random intercept is then often considered as a surrogate for missing covariates that are approximately constant across observations within a single cluster.

In addition to accounting for unobserved heterogeneity, random effects automatically allow us to take correlations of observations within subjects into account. In particular, any random effects model can be considered equivalent to a specific model with correlated error terms in marginal formulation (see the next paragraph for details). For illustration purposes, consider again the random intercept model (3.3). In this model, the correlations between observations belonging to the same individual are given by

$$\mathrm{Cor}(y_{ij}, y_{ik}) = \frac{\mathrm{Cov}(y_{ij}, y_{ik})}{\sqrt{\mathrm{Var}(y_{ij})\,\mathrm{Var}(y_{ik})}} = \frac{\delta^2}{\sigma^2 + \delta^2} \tag{3.8}$$

for $j \neq k$ where $\delta^2 = \mathrm{Var}(\alpha_i)$ is the variance of the random intercept. As a consequence, all observations within a subject are equally correlated while observations of different subjects remain independent. More complex correlation structures arise when considering random slopes or mixed models with nested random effect structures for more than one grouping level. Note that the correlations between observations of the same subject arise from a marginal perspective on the model, while in a conditional interpretation (for given random effects) the observations will again be independent since (by assumption)

$$\mathrm{Cor}(y_{ij}, y_{ik} | \alpha_i) = 0.$$

From the marginal view on the model it becomes obvious that efficiency is lost in the estimation of the fixed effects when treating the observations as independent. This will also have the consequence that standard errors of the corresponding estimates will be wrong, leading to potentially false conclusions when performing tests on the model. Typically, standard errors will be too small when ignoring random effects, although the reverse behaviour may also be observed in practice. The latter behaviour may result when, for example, the overall mean level of individual curves is well identified but there is a lot of variability due to the difference in clusters. In this case, the variability of the intercept may be overstated in a model without random effects structures since parts of the individual-specific variability are falsely attributed to the pure sampling variability. When introducing a random effect, this allows to separate the two sources of variability and may therefore yield more precise estimates for the global level. Similar arguments hold for random slopes, too.

In addition, the introduction of individual-specific effects as fixed effects without a random effects distribution assumption would induce computational difficulties. The number of parameters is large compared to the number of individuals in the sample and also potentially large compared to the total number of observations. Traditional parametric least-squares or likelihood inference, treating individual-specific parameters as if they were fixed and deterministic, would become extremely unstable or even impossible. Treating these effects as random variables, drawn from a common mixing distribution, automatically regularizes the inferential problem, as will be seen in the next subsections: Instead of relying on the classical principle of maximum likelihood, estimation of fixed and random parameters is based on a penalized likelihood (or least-squares) criterion, where the penalty is defined through the log-density of the mixing distribution. From a Bayesian perspective, penalized likelihood inference may also be interpreted as an empirical Bayes approach, where the mixing distribution defines a prior for the random effects, while fixed effects follow a flat noninformative or a weakly-informative prior.

The random effects distribution is also sometimes motivated from the assumption that the individuals in the data set form a random sample from the population of individuals. Hence, the random effects distribution describes the distribution of individual-specific effects in the population. This has sometimes led to a debate about the appropriateness of random effects models when there is no underlying population available. For example, random effects are also used in spatial regression models where spatial regions define the clustering structure. In this case, there are typically observations available on all regions, i.e. the complete 'population' of regions has been observed. Still, mixed models are often employed in this situation due to their built in regularization properties. While the assumption of a random effect in such settings may be debatable from a frequentist perspective, there are no conceptual difficulties in a Bayesian interpretation of mixed models. From a Bayesian perspective, the random effects distribution corresponds to a prior assumption for the individual-specific effects. Since prior beliefs can be expressed without referring to a hypothetical population of individuals, one is free in setting up appropriate random effects distributions to achieve the desired regularization properties.

In summary, random effects models have the following advantages:

- *Individual-specific effects* can be introduced to account for individual departures from the population behaviour.
- They allow us to correct for *unobserved heterogeneity* induced by omitted covariates.
- *Correlations* between observations of the same subject can be taken into account (at least to some extent).
- *Estimation is stabilized* by assuming a common mixing distribution for the random effects as compared to classical least-squares or likelihood inference.

In particular the last point naturally leads to the consideration of a full Bayesian framework, where all parameters are regarded as random variables with suitable priors. In matrix notation the measurement model (3.1) takes the form

$$\boldsymbol{y}_i = \boldsymbol{U}_i\boldsymbol{\beta} + \boldsymbol{Z}_i\boldsymbol{\alpha}_i + \boldsymbol{\varepsilon}_i, \tag{3.9}$$

where $\boldsymbol{y}_i = (y_{i1}, \ldots, y_{in_i})'$ is the response vector, $\boldsymbol{U}_i = (\boldsymbol{u}'_{i1}, \ldots, \boldsymbol{u}'_{in_i})'$ and $\boldsymbol{Z}_i = (\boldsymbol{z}'_{i1}, \ldots, \boldsymbol{z}'_{in_i})'$ are design matrices, and $\boldsymbol{\varepsilon}_i = (\varepsilon_{i1}, \ldots, \varepsilon_{in_i})'$ is the vector of errors for unit i.

The assumption of uncorrelated errors, that is, $\boldsymbol{\varepsilon}_i \sim \mathrm{N}(0, \sigma^2 \boldsymbol{I})$, may be relaxed by assuming $\boldsymbol{\varepsilon}_i \sim \mathrm{N}(0, \boldsymbol{\Sigma}_i)$, where the covariance matrix $\boldsymbol{\Sigma}_i$ is specified through certain types of error processes, for example a first-order autoregression or a stationary process with some autocorrelation function. In the following we will mostly consider the uncorrelated case $\boldsymbol{\Sigma}_i = \sigma^2 \boldsymbol{I}$ for simplicity.

Marginal and conditional formulation

The measurement model (3.9) implies the hierarchical, or more precisely, *conditional Gaussian model*

$$\boldsymbol{y}_i | \boldsymbol{\alpha}_i \sim \mathrm{N}(\boldsymbol{U}_i \boldsymbol{\beta} + \boldsymbol{Z}_i \boldsymbol{\alpha}_i, \sigma^2 \boldsymbol{I}) \tag{3.10}$$

for the response vector $\boldsymbol{y}_i$, given the random effect $\boldsymbol{\alpha}_i$. From the conditional perspective, the individual-specific effects $\boldsymbol{\alpha}_i$ are interpreted analogously to the usual population effects $\boldsymbol{\beta}$ with the only difference that they only apply to one single subject.

Together with the assumption of Gaussian random effects, the corresponding *marginal Gaussian model* is given by

$$\boldsymbol{y}_i \sim \mathrm{N}(\boldsymbol{U}_i \boldsymbol{\beta}, \sigma^2 \boldsymbol{I} + \boldsymbol{Z}_i \boldsymbol{Q} \boldsymbol{Z}_i'). \tag{3.11}$$

This is in fact a linear Gaussian model

$$\boldsymbol{y}_i = \boldsymbol{U}_i \boldsymbol{\beta} + \boldsymbol{\varepsilon}_i^*$$

with heteroscedastic errors

$$\boldsymbol{\varepsilon}_i^* \sim \mathrm{N}(\boldsymbol{0}, \boldsymbol{V}_i), \quad \boldsymbol{V}_i = \sigma^2 \boldsymbol{I} + \boldsymbol{Z}_i \boldsymbol{Q} \boldsymbol{Z}_i',$$

where the fixed regression parameters are equal to the population parameters $\boldsymbol{\beta}$ in the (conditional) mixed model. In other words, interpretation of population parameters $\boldsymbol{\beta}$ in the conditional model and in the marginal model is the same. This useful property of linear mixed models is lost for generalized linear mixed models with non-normal responses in Section 3.3.

The non-diagonal covariance matrix $\boldsymbol{V}_i$ induces a correlation structure between repeated measurements y_{ij} of individual i in the marginal model, even if the errors ε_{ij} in the conditional model are uncorrelated. For the special case of a random intercept model we obtain

$$\boldsymbol{Z}_i \boldsymbol{Q} \boldsymbol{Z}_i' = \delta^2 \begin{pmatrix} 1 & \dots & 1 \\ \vdots & & \vdots \\ 1 & \dots & 1 \end{pmatrix},$$

resulting in the *equicorrelation model* (3.8). More flexible correlation structures can be specified if the errors $\boldsymbol{\varepsilon}_i$ in the conditional model are correlated with $\mathrm{Cov}(\boldsymbol{\varepsilon}_i) = \boldsymbol{\Sigma}_i$, for example due to an autoregressive error structure. Then the covariance matrix in the marginal model generalizes to $\boldsymbol{V}_i = \boldsymbol{\Sigma}_i + \boldsymbol{Z}_i \boldsymbol{Q} \boldsymbol{Z}_i'$.

Multilevel models

Multilevel models extend the ideas of random effects models to data structures with more than one grouping level. For longitudinal data, the only grouping level is given by the subjects, while in multilevel data each observation may belong to several, possibly nested, groups. The classical example for this type of data structure are results on school tests where pupils are grouped within classes and classes are grouped in schools. If we assume nested data, that is, one pupil is member of exactly one class and each class belongs to exactly one school, then this leads to three-level models such as

$$y_{ijk} = \boldsymbol{u}'_{ijk}\boldsymbol{\beta} + \alpha_i + \alpha_{ij} + \varepsilon_{ijk}$$

where only random intercepts are considered, i indexes schools, j indexes classes within schools and k indexes individual pupils within classes. Of course, random slopes can also be included at any stage of the model and it is of particular interest to compare effects of the same type of covariate information on different levels of the model. For example, the covariate 'maths score' can be constructed both on the class level as an average score and on the individual level. To separate class level from individual-level effects, it is common practice to include both the average score and the individual deviation from the average score as covariates.

The nested three-level model allows for a variance decomposition on three levels, similar to the longitudinal mixed model with only two levels. In the latter, the intra-group correlation coefficient of the equicorrelation random intercept model provides an intuitive measure for the percentage of variability that can be attributed to the grouping of the data (after correcting for covariates). In the three level model, similar correlation coefficients can be derived for each of the levels. Of course, the three-level model can be extended to further hierarchical stages in complete analogy. However, additional grouping comes at the cost of extremely increased computational needs, especially if the random effects of different stages are allowed to be correlated.

It is also possible to consider non-nested multilevel models, where individuals are members of two types of groups indexed as $i = 1, \dots, I$ and $j = 1, \dots, J$. In this case, the two levels represented by these indices are no longer hierarchically ordered but are considered to represent different groupings on the same stage. For example, pupils might be members of different types of classes (related to a specific course subject). This could, for example, be accounted for by introducing random effects for both groups such as in the random intercept model

$$y_{ijk} = \boldsymbol{u}'_{ijk}\boldsymbol{\beta} + \alpha_i + \alpha_j + \varepsilon_{ijk},$$

where i and j index two different types of courses while k indexes pupils. Note that we kept the triple index for convenience, although not all combinations will actually be observed in the data.

Multilevel models will not be considered in detail in this book, although they have become a standard tool for applied regression analyses in several areas of statistical research. Examples include meta analyses in clinical research or models for learning achievements in sociology. In-depth introductions are provided for example in Bryk and Raudenbush (1992), Goldstein (2002), Gelman and Hill (2006), Skrondal and Rabe-Hesketh (2004) or de Leeuw and Meijer (2007).

3.1.2 General linear mixed models

The LMM $\boldsymbol{y}_i = \boldsymbol{U}_i\boldsymbol{\beta} + \boldsymbol{Z}_i\boldsymbol{\alpha}_i + \boldsymbol{\varepsilon}_i, i = 1, \ldots, n$, for longitudinal data, and also the multilevel extensions and several further types of mixed models considered later on, can be formulated in rather compact form by defining the vectors

$$\boldsymbol{y} = (\boldsymbol{y}_1', \ldots, \boldsymbol{y}_n')', \quad \boldsymbol{\varepsilon} = (\boldsymbol{\varepsilon}_1', \ldots, \boldsymbol{\varepsilon}_n')', \quad \boldsymbol{\alpha} = (\boldsymbol{\alpha}_1', \ldots, \boldsymbol{\alpha}_n')'$$

of all responses, errors and random effects as well as the design matrices

$$\boldsymbol{U} = \begin{pmatrix} \boldsymbol{U}_1 \\ \vdots \\ \boldsymbol{U}_n \end{pmatrix} \quad \text{and} \quad \boldsymbol{Z} = \begin{pmatrix} \boldsymbol{Z}_1 & \cdots & 0 \\ \vdots & \ddots & \vdots \\ 0 & \cdots & \boldsymbol{Z}_n \end{pmatrix}.$$

Then the LMM can be written as

$$\boldsymbol{y} = \boldsymbol{U}\boldsymbol{\beta} + \boldsymbol{Z}\boldsymbol{\alpha} + \boldsymbol{\varepsilon} \tag{3.12}$$

with distributional assumptions

$$\begin{pmatrix} \boldsymbol{\varepsilon} \\ \boldsymbol{\alpha} \end{pmatrix} \sim \mathrm{N}\left(\begin{pmatrix} \boldsymbol{0} \\ \boldsymbol{0} \end{pmatrix}, \begin{pmatrix} \boldsymbol{\Sigma} & \boldsymbol{0} \\ \boldsymbol{0} & \boldsymbol{G} \end{pmatrix} \right) \tag{3.13}$$

and block-diagonal covariance matrices

$$\boldsymbol{\Sigma} = \text{blockdiag}(\boldsymbol{\Sigma}_1, \ldots, \boldsymbol{\Sigma}_n), \quad \boldsymbol{G} = \text{blockdiag}(\boldsymbol{Q}, \ldots, \boldsymbol{Q}),$$

where $\boldsymbol{\Sigma} = \sigma^2\boldsymbol{I}$ for the case of uncorrelated, homoscedastic errors ε_{ij}.

A *general linear mixed model* is defined through the general measurement model (3.12), with arbitrary design matrices $\boldsymbol{U}$ and $\boldsymbol{Z}$, and the general Gaussian assumption (3.13) for errors and random effects with positive definite covariance matrices $\boldsymbol{\Sigma}$ and $\boldsymbol{G}$. Note that it is still assumed that $\boldsymbol{\varepsilon}$ and $\boldsymbol{\alpha}$ are independent. However, we allow for *correlated* random effects $\boldsymbol{\alpha}$, because we do not necessarily restrict $\boldsymbol{G}$ to be (block-)diagonal. Correlated random effects with specific forms for $\boldsymbol{G}$ arise quite naturally if we want to incorporate nonparametric, temporal or spatial effects in LMMs within a Bayesian framework as in Chapters 4, 5 and 6. The observation model (3.12) implies the *conditional model*

$$\boldsymbol{y}|\boldsymbol{\alpha} \sim \mathrm{N}(\boldsymbol{U}\boldsymbol{\beta} + \boldsymbol{Z}\boldsymbol{\alpha}, \boldsymbol{\Sigma})$$

and the *marginal model*

$$\boldsymbol{y} \sim \mathrm{N}(\boldsymbol{U}\boldsymbol{\beta}, \boldsymbol{\Sigma} + \boldsymbol{Z}\boldsymbol{G}\boldsymbol{Z}')$$

for responses in a general LMM.

3.1.3 Bayesian linear mixed models

In Bayesian linear mixed models, both 'fixed effects' and 'random effects' are random variables and must be supplemented by appropriate prior assumptions. Compared to the common frequentist framework, fixed and random effects are treated more symmetrically, and are distinguished only through different priors. The mixing distribution for random effects is simply interpreted as a prior distribution. Thus, for the general LMM the prior corresponding to (3.13) is Gaussian

$$p(\boldsymbol{\alpha}) \propto \exp\left(-\frac{1}{2}\boldsymbol{\alpha}'\boldsymbol{G}^{-1}\boldsymbol{\alpha}\right), \tag{3.14}$$

neglecting the normalizing constant. Note that the Bayesian formulation does not require the prior precision $\boldsymbol{G}^{-1}$ to be of full rank but permits partially improper Gaussian priors. These have already been relevant in Chapter 2 for prior distributions in the context of penalized spline smoothing and will reappear in the up-coming chapters. In contrast, most of the frequentist inferential procedures will require positive definite precision or covariance matrices. We will tackle this problem by a reparametrization strategy in Chapter 4.2.1 that allows us to utilize mixed model methodology for empirical Bayes inference also with rank-deficient precision matrices.

The frequentist assumption that fixed effects $\boldsymbol{\beta}$ are 'unknown constants' corresponds to the flat, noninformative prior

$$p(\boldsymbol{\beta}) \propto \text{const}\,. \tag{3.15}$$

A Gaussian prior $\boldsymbol{\beta} \sim \mathrm{N}(\boldsymbol{c}, \boldsymbol{C})$

$$p(\boldsymbol{\beta}) \propto \exp\left(-\frac{1}{2}(\boldsymbol{\beta}-\boldsymbol{c})'\boldsymbol{C}^{-1}(\boldsymbol{\beta}-\boldsymbol{c})\right) \tag{3.16}$$

is the standard choice for an informative prior. Weakly-informative choices are obtained when considering 'large' positive definite covariance matrices $\boldsymbol{C}$, for example $\boldsymbol{C} = k\boldsymbol{I}$, k large. The noninformative flat prior $p(\boldsymbol{\beta}) \propto$ const is obtained as a limiting case if the precision matrix $\boldsymbol{C}^{-1}$ tends to $\boldsymbol{0}$. The assignment of an informative prior for $\boldsymbol{\beta}$ may for example be useful when multicollinearity problems appear in the design matrix $\boldsymbol{U}$ or if high-dimensional vectors $\boldsymbol{\beta}$ are considered. It is commonly assumed that $\boldsymbol{\beta}$ and $\boldsymbol{\alpha}$ are independent.

Bayesian inference is based on the posterior of parameters given the data. Regarding for the moment variances or association parameters in $\boldsymbol{\Sigma}$ and $\boldsymbol{G}$ as known or given, the posterior

$$p(\boldsymbol{\beta}, \boldsymbol{\alpha}|\boldsymbol{y}) \propto p(\boldsymbol{y}|\boldsymbol{\alpha}, \boldsymbol{\beta})\, p(\boldsymbol{\beta})\, p(\boldsymbol{\alpha})$$

for $\boldsymbol{\beta}$ and $\boldsymbol{\alpha}$ is Gaussian. The posterior mean estimates for fixed and random effects can therefore be obtained simultaneously by maximizing the posterior. Taking logarithms, this maximization can be shown to be equivalent to minimizing the penalized least-squares criterion

$$(\boldsymbol{y} - \boldsymbol{U\beta} - \boldsymbol{Z\alpha})'\boldsymbol{\Sigma}^{-1}(\boldsymbol{y} - \boldsymbol{U\beta} - \boldsymbol{Z\alpha}) + (\boldsymbol{\beta} - \boldsymbol{c})'\boldsymbol{C}^{-1}(\boldsymbol{\beta} - \boldsymbol{c}) + \boldsymbol{\alpha}'\boldsymbol{G\alpha}.$$

For a flat prior $p(\boldsymbol{\beta}) \propto$ const, the first penalty term vanishes. Comparing with similar expressions in Section 2.1 (Bayesian smoothing) and Section 2.2 (Bayesian penalized splines), we see that informative priors induce analogous Bayesian regularization, yielding ridge-type shrinkage estimators for both the fixed effects and the random effects. For a flat prior, estimation of $\boldsymbol{\beta}$ is unpenalized while random effects $\boldsymbol{\alpha}$ are still obtained as shrinkage estimators.

In practice, of course, $\boldsymbol{\Sigma}$ and $\boldsymbol{G}$ will contain unknown parameters which have to be estimated jointly with $\boldsymbol{\beta}$ and $\boldsymbol{\alpha}$. We will again distinguish between *empirical Bayes* (*EB*) and *full Bayes* (*FB*) approaches. For FB inference, unknown variance or association parameters are considered as random variables with suitably specified hyperpriors and can be estimated jointly with fixed and random effects using MCMC techniques. The form of hyperpriors depends on the structure of $\boldsymbol{\Sigma}$ and $\boldsymbol{G}$. In Bayesian LMMs (3.9) for longitudinal data, with $\text{Cov}(\boldsymbol{\varepsilon}_i) = \sigma^2\boldsymbol{I}$ and $\boldsymbol{\alpha}_i \sim \text{N}(\boldsymbol{0}, \boldsymbol{Q})$, the hyperpriors are often specified as follows: In the absence of prior knowledge, Jeffreys' noninformative, but improper priors

$$p(\sigma^2) \propto \sigma^{-2}, \quad p(\boldsymbol{Q}) \propto |\boldsymbol{Q}|^{-\frac{q+1}{2}},$$

with $q = \dim(\boldsymbol{\alpha}_i)$, seem to be a reasonable choice. These priors can also be regarded as the limiting case for inverse gamma and Wishart priors, respectively, the conjugate priors for variances and covariance matrices in Gaussian models. However, as has been shown by Hobert and Casella (1996), noninformative priors can lead to improper posteriors. Therefore, proper but highly dispersed priors are sometimes considered as alternatives, see for example Besag, Green, Higdon and Mengersen (1995). For σ^2, an example of such a weakly informative prior is an inverse gamma distribution

$$\sigma^2 \sim \text{IG}(\epsilon, \epsilon)$$

with small ϵ, for example $\epsilon = 0.001$. Assuming, for simplicity, $\boldsymbol{Q}$ to be diagonal, similar weakly-informative inverse Gamma priors are often suggested for the variances $\delta_j^2 = \text{Var}(\alpha_{ij})$ for the components $\alpha_{ij}, j = 1, \dots, q$, of the i.i.d. random effects $\boldsymbol{\alpha}_i$. For non-diagonal $\boldsymbol{Q}$, the hyperparameters a and $\boldsymbol{B}$ in an inverted Wishart prior $\boldsymbol{Q} \sim \text{IW}(a, \boldsymbol{B})$ have to be carefully selected. In some data situations (e.g. for small sample sizes), estimates may react sensitively to the particular choice of the hyperparameters ϵ, a and $\boldsymbol{B}$. It is therefore good practice to estimate models using a (small) number of different choices for the hyperparameters to assess the dependence of results on minor changes in the prior assumptions. In addition, the choice of non-diagonal $\boldsymbol{Q}$ and therefore dependent random effects can be computationally disadvantageous, in particular for a larger number of random effects q. Numerical difficulties due to singular or close to singular inverse Wishart full conditionals are then often encountered in the MCMC simulation process. Generally, specification of hyperpriors in mixed models is still an area of methodological research, see Gelman (2006) as well as Chapter 4.

As a particular advantage, Bayesian mixed models can be extended in a comparably straightforward manner to parametric *non-Gaussian error distributions* and *non-Gaussian priors* for fixed and random effects with improved robustness and shrinkage properties. This can be achieved by incorporating (finite or continuous) mixtures of normals, since they can be defined hierarchically as well.

For illustration, we focus here on Student distributions, where the degree of freedom ν characterizes the amount of nonnormality. The specification of the random effects prior is extended by introducing additional hyperparameters $\lambda_1, \ldots, \lambda_n$:

$$\boldsymbol{\alpha}_i | \boldsymbol{Q}, \lambda_i \sim \mathrm{N}(\mathbf{0}, \boldsymbol{Q}/\lambda_i),$$

with hyperparameters following gamma priors

$$\lambda_i \sim \mathrm{Ga}(\nu/2, \nu/2)$$

in a further stage of the hierarchy. It is well known that the resulting marginal distribution for the random effects is a multivariate t-distribution, that is,

$$\boldsymbol{\alpha}_i \sim \mathrm{t}_\nu(\mathbf{0}, \boldsymbol{Q}),$$

with location parameter $\mathbf{0}$, scale parameter $\boldsymbol{Q}$ and ν degrees of freedom. The Cauchy distribution results for $\nu = 1$ and distributions with $\nu > 30$ are considered to be approximately normal. A prior $p(\nu)$, for example for a finite set of ν-values, may be introduced in a next stage of such a hierarchical mixed model.

Note that the hierarchical prior specification corresponds to the assumption of a scale mixture of normals prior for the random effects, that is the prior is Gaussian conditional on some hyperparameters influencing the prior variance. Specifying appropriate hyperpriors (or mixing distributions) for these hyperparameters allows us to generate a large class of flexible random effects distributions. Such flexible parametric random effects priors as well semi- or nonparametric priors, in particular mixtures of normals and Dirichlet process priors, are considered in more detail in Section 3.2.

3.1.4 Likelihood-based inference

This section reviews the basic concepts of likelihood-based estimation (or prediction) and hypothesis testing for fixed effects, random effects, and variance components, that is, unknown parameters in the covariance matrices. The following section will then introduce the corresponding Bayesian estimation approaches and will relate them to the likelihood-based principles.

Estimation and prediction

Traditional fitting of mixed models, in particular of mixed models for longitudinal data, is often primarily concerned with estimation of fixed effects $\boldsymbol{\beta}$ and unknown parameters $\boldsymbol{\vartheta}$ in the covariance matrices $\boldsymbol{\Sigma} = \mathrm{Cov}(\boldsymbol{y}|\boldsymbol{\alpha}) = \mathrm{Cov}(\boldsymbol{\varepsilon})$ and

$\boldsymbol{G} = \text{Cov}(\boldsymbol{\alpha})$. Estimation of random effects $\boldsymbol{\alpha}$ is often termed prediction in this context since, from a frequentist perspective, there is no such thing as a 'true' random effect that could be estimated. Instead, $\boldsymbol{\alpha}$ is considered a random quantity and therefore it has to be predicted based on the observed data. Moreover, prediction of random effects is often considered to be of minor importance in itself since the random effects are treated as a convenient tool to construct marginal models with a sensible correlation structure. Correspondingly, inference for $\boldsymbol{\beta}$ and $\boldsymbol{\vartheta}$ is then typically based on the marginal model

$$\boldsymbol{y} \sim \text{N}(\boldsymbol{U\beta}, \boldsymbol{V}),$$

where $\boldsymbol{V} = \boldsymbol{V}(\boldsymbol{\vartheta}) = \boldsymbol{\Sigma} + \boldsymbol{ZGZ}'$ is the marginal covariance matrix of the response depending on the vector of all unknown variance components $\boldsymbol{\vartheta}$. Given the variance parameters, the weighted least-squares estimate for $\boldsymbol{\beta}$ is given by

$$\hat{\boldsymbol{\beta}} = (\boldsymbol{U}'\boldsymbol{V}^{-1}\boldsymbol{U})^{-1}\boldsymbol{U}'\boldsymbol{V}^{-1}\boldsymbol{y}.$$

Plugging this estimate back into the likelihood of the marginal model yields the profile likelihood for $\boldsymbol{\vartheta}$ that can be maximized to obtain the corresponding maximum likelihood estimates.

Throughout the rest of this book, joint determination of the fixed and random effects will be essential since random effects are both of interest in longitudinal models, where they provide information about individual-specific effects, and in more general Bayesian smoothing approaches, where the random effects carry important information about smooth function estimates, see Section 4.2.1. We also consider the joint estimation framework to provide richer information for the analyst anyway, and therefore focus on joint estimation of fixed and random effects in the following. In the Bayesian paradigm, there is no necessity to distinguish between estimation and prediction since all unknown parameters are treated as random variables anyway. Hence, joint determination of all unknowns in one estimation loop is the usual standard in Bayesian inferential procedures.

When inference is concerned not only with fixed but also with random effects, the conditional model is typically employed as the basis. Conceptually, likelihood-based estimation is then carried out iteratively in the following two steps (outlined in more detail in the following paragraphs):

1. Obtain updated estimates $\hat{\boldsymbol{\beta}}$ and $\hat{\boldsymbol{\alpha}}$, given the current variance components, as the maximizers of the joint likelihood $L(\boldsymbol{\beta}, \boldsymbol{\alpha}) = p(\boldsymbol{y}|\boldsymbol{\beta}, \boldsymbol{\alpha})p(\boldsymbol{\alpha})$. This step will be accomplished by penalized least-squares estimation.
2. Update estimates $\hat{\boldsymbol{\vartheta}}$ for the variance components by (restricted) maximum likelihood estimation, given the current estimates $\hat{\boldsymbol{\beta}}$ for fixed effects and $\hat{\boldsymbol{\alpha}}$ for random effects. Maximization can be carried out through Fisher scoring iterations or EM-type algorithms as described for example in Harville (1977).

The two estimation steps are iterated until convergence.

Estimation of regression coefficients for given variance components

Simultaneous estimation of fixed and random effects can be based on the joint likelihood

$$L(\boldsymbol{\beta}, \boldsymbol{\alpha}) = p(\boldsymbol{y}|\boldsymbol{\beta}, \boldsymbol{\alpha})p(\boldsymbol{\alpha})$$

of observations and random effects. Taking logarithms and omitting constants, maximization of $L(\boldsymbol{\beta}, \boldsymbol{\alpha})$ is equivalent to minimization of the least squares criterion

$$(\boldsymbol{y} - \boldsymbol{U\beta} - \boldsymbol{Z\alpha})'\boldsymbol{\Sigma}^{-1}(\boldsymbol{y} - \boldsymbol{U\beta} - \boldsymbol{Z\alpha}) + \boldsymbol{\alpha}'\boldsymbol{G}^{-1}\boldsymbol{\alpha} \to \min_{\boldsymbol{\beta},\boldsymbol{\alpha}}. \tag{3.17}$$

Note that the structure of (3.17) exactly coincides with the structure of PLS criteria considered in Chapter 2 except that we allow for a more general covariance structure $\boldsymbol{\Sigma}$ of the error terms. Not very surprisingly, the resulting estimates are therefore also determined as the solutions of the following systems of equations:

$$\begin{pmatrix} \boldsymbol{U}'\boldsymbol{\Sigma}^{-1}\boldsymbol{U} & \boldsymbol{U}'\boldsymbol{\Sigma}^{-1}\boldsymbol{Z} \\ \boldsymbol{Z}'\boldsymbol{\Sigma}^{-1}\boldsymbol{U} & \boldsymbol{Z}'\boldsymbol{\Sigma}^{-1}\boldsymbol{Z} + \boldsymbol{G}^{-1} \end{pmatrix} \begin{pmatrix} \beta \\ \alpha \end{pmatrix} = \begin{pmatrix} \boldsymbol{U}'\boldsymbol{\Sigma}^{-1}\boldsymbol{y} \\ \boldsymbol{Z}'\boldsymbol{\Sigma}^{-1}\boldsymbol{y} \end{pmatrix}. \tag{3.18}$$

Alternative, equivalent formulae derived in Harville (1977) are

$$\hat{\beta} = (\boldsymbol{U}'\boldsymbol{V}^{-1}\boldsymbol{U})^{-1}\boldsymbol{U}'\boldsymbol{V}^{-1}\boldsymbol{y} \tag{3.19}$$

which coincides with the weighted least-squares estimate obtained from maximizing the marginal likelihood $L(\boldsymbol{\beta})$, and

$$\begin{aligned} \hat{\alpha} &= \left(\boldsymbol{Z}'\boldsymbol{\Sigma}^{-1}\boldsymbol{Z} + \boldsymbol{G}^{-1}\right)^{-1} \boldsymbol{Z}'\boldsymbol{\Sigma}^{-1}(\boldsymbol{y} - \boldsymbol{U}\hat{\beta}) \\ &= \boldsymbol{G}\boldsymbol{Z}'\boldsymbol{V}^{-1}(\boldsymbol{y} - \boldsymbol{U}\hat{\beta}) \\ &= \boldsymbol{G}\boldsymbol{Z}'\boldsymbol{P}\boldsymbol{y} \end{aligned} \tag{3.20}$$

where

$$\begin{aligned} \boldsymbol{P} &= \boldsymbol{V}^{-1} - \boldsymbol{V}^{-1}\boldsymbol{U}(\boldsymbol{U}'\boldsymbol{V}^{-1}\boldsymbol{U})^{-1}\boldsymbol{U}'\boldsymbol{V}^{-1} \\ &= \boldsymbol{\Sigma}^{-1} - \boldsymbol{\Sigma}^{-1}(\boldsymbol{U}, \boldsymbol{Z})\boldsymbol{H}^{-1}(\boldsymbol{U}, \boldsymbol{Z})'\boldsymbol{\Sigma}^{-1} \end{aligned} \tag{3.21}$$

and

$$\boldsymbol{H} = \begin{pmatrix} \boldsymbol{U}'\boldsymbol{\Sigma}^{-1}\boldsymbol{U} & \boldsymbol{U}'\boldsymbol{\Sigma}^{-1}\boldsymbol{Z} \\ \boldsymbol{Z}'\boldsymbol{\Sigma}^{-1}\boldsymbol{U} & \boldsymbol{Z}'\boldsymbol{\Sigma}^{-1}\boldsymbol{Z} + \boldsymbol{G}^{-1} \end{pmatrix}.$$

Note that equation (3.20) provides an interpretation of $\hat{\boldsymbol{\alpha}}$ as a shrinkage estimate based on the residuals obtained from the preceding estimation of the fixed effects. Omitting the normality assumption on the errors $\boldsymbol{\varepsilon}$, $\hat{\boldsymbol{\beta}}$ and $\hat{\boldsymbol{\alpha}}$ can also be derived as best linear unbiased estimators or predictors, respectively (abbreviated BLUE and BLUP), see Harville (1976).

Maximum likelihood (ML) estimation of variance components

Maximum likelihood estimation of $\boldsymbol{\vartheta}$ would be based on the log-likelihood of the marginal model, that is,

$$l(\boldsymbol{\beta}, \boldsymbol{\vartheta}) = -\frac{1}{2}\log(|\boldsymbol{V}|) - \frac{1}{2}(\boldsymbol{y} - \boldsymbol{U}\boldsymbol{\beta})'\boldsymbol{V}^{-1}(\boldsymbol{y} - \boldsymbol{U}\boldsymbol{\beta}). \tag{3.22}$$

Plugging in the ML estimate $\hat{\boldsymbol{\beta}} = \hat{\boldsymbol{\beta}}(\boldsymbol{\vartheta})$ as given in (3.19), the resulting profile log-likelihood $l_p(\boldsymbol{\vartheta})$ would then be maximized to obtain the ML estimate $\hat{\boldsymbol{\vartheta}}$. However, ML estimation for variances is well-known to be biased in this context, a property that can be most easily exemplified in the classical linear model where the ML estimate for the error variance

$$\hat{\sigma}^2_{ML} = \frac{1}{n}(\boldsymbol{y} - \boldsymbol{U}\hat{\boldsymbol{\beta}})'(\boldsymbol{y} - \boldsymbol{U}\hat{\boldsymbol{\beta}}),$$

is biased towards zero whereas the unbiased estimate is given by

$$\hat{\sigma}^2 = \frac{1}{n-m}(\boldsymbol{y} - \boldsymbol{U}\hat{\boldsymbol{\beta}})'(\boldsymbol{y} - \boldsymbol{U}\hat{\boldsymbol{\beta}}),$$

where $m = \dim(\boldsymbol{\beta})$ denotes the number of regression coefficients. The underlying reason for the bias of ML estimates is that they do not take into account the loss in degrees of freedom caused by the preceding estimation of $\boldsymbol{\beta}$. It is therefore of interest to develop a general principle for deriving unbiased or less biased estimates such as the usual estimate for the error variance in the linear model. For mixed models, this general principle is given by restricted maximum likelihood (REML) estimation, developed by Patterson and Thompson (1971) in the context of blocked experiments where closed form solutions for the estimates were still available. Later, the ideas of Patterson and Thompson (1971) were extended to more general linear mixed models as outlined in the following paragraph.

REML estimation for the variance parameters

The principal idea of REML estimation is to base inference for the variance components not on the original likelihood $l(\boldsymbol{\beta}, \boldsymbol{\vartheta})$ but on a modified likelihood that only depends on the variance components. This likelihood is derived from a number of error contrasts $\boldsymbol{a}'\boldsymbol{y}$ (i.e. linear combinations of response variables) with $\mathbb{E}(\boldsymbol{a}'\boldsymbol{y}) = 0$. The distribution of an error contrast is given by $\mathrm{N}(0, \boldsymbol{a}'\boldsymbol{V}\boldsymbol{a})$. Hence, since $\boldsymbol{y}$ depends on $\boldsymbol{\beta}$ only through its expectation, the distribution is independent from $\boldsymbol{\beta}$. Still, the distribution depends on the same variance parameters $\boldsymbol{\vartheta}$, enabling inference about the latter based on a set of error contrasts.

Typical examples for error contrasts are the estimated residuals

$$\hat{\varepsilon}_l = y_l - \boldsymbol{u}_l'\hat{\boldsymbol{\beta}}$$

where the vector $\boldsymbol{a}$ equals the l-th row of the residual matrix $\boldsymbol{R} = \boldsymbol{I}_N - \boldsymbol{U}(\boldsymbol{U}'\boldsymbol{U})^{-1}\boldsymbol{U}'$ and $N = \sum_{i=1}^{n} n_i$ denotes the total sample size. Deriving the distribution of the complete vector of residuals yields

$$\hat{\varepsilon} \sim \mathrm{N}(\boldsymbol{0}, \boldsymbol{R}\boldsymbol{V}\boldsymbol{R})$$

and the corresponding likelihood could be used to estimate the variance parameters. However, the joint distribution of all residuals is partially improper since the matrix $\boldsymbol{R}$ does not have full rank. To be more specific, we have $\mathrm{rank}(\boldsymbol{R}) = N - m$ and the rank-deficiency simply reflects the fact that $m = \dim(\boldsymbol{\beta})$ parameters have already been estimated, causing a corresponding loss in degrees of freedom in the likelihood.

To avoid impropriety, one could select $N - m$ linear independent rows from the residual matrix $\boldsymbol{R}$. However, it is more convenient to start with an arbitrary $N \times (N - m)$ dimensional matrix $\boldsymbol{A}$ defining error contrasts $\boldsymbol{e} = \boldsymbol{A}'\boldsymbol{y}$. Such a matrix can, for example, be derived from the residual matrix $\boldsymbol{R}$ based on the spectral decomposition

$$\boldsymbol{A}\boldsymbol{A}' = \boldsymbol{I}_N - \boldsymbol{U}(\boldsymbol{U}'\boldsymbol{U})^{-1}\boldsymbol{U}' \text{ subject to } \boldsymbol{A}'\boldsymbol{A} = \boldsymbol{I}_{N-m}.$$

The resulting density of the error contrasts is then given by

$$\begin{aligned} p(\boldsymbol{e}) = \left(\frac{1}{2\pi}\right)^{\frac{N-m}{2}} & |\boldsymbol{U}'\boldsymbol{U}|^{\frac{1}{2}} |\boldsymbol{V}|^{-\frac{1}{2}} |\boldsymbol{U}'\boldsymbol{V}^{-1}\boldsymbol{U}|^{-\frac{1}{2}} \\ & \cdot \exp\left[-\frac{1}{2}(\boldsymbol{y} - \boldsymbol{U}\hat{\boldsymbol{\beta}})'\boldsymbol{V}^{-1}(\boldsymbol{y} - \boldsymbol{U}\hat{\boldsymbol{\beta}})\right], \end{aligned} \tag{3.23}$$

independent of the specific choice of $\boldsymbol{A}$ (as long as linearly independent error contrasts are used, see Verbeke and Molenberghs, 2000). Note that, despite reducing the number of free observations by m, there is no loss in information about $\boldsymbol{\vartheta}$ when replacing $\boldsymbol{y}$ by $\boldsymbol{e}$ since the error contrasts are marginally sufficient for the variance parameters in the sense of Sprott (1975). The right-hand side of (3.23) also shows that the density of the error contrasts is actually a function of the original observations only, and therefore the error contrasts are not required explicitly in the REML estimation but only serve as a motivating construct.

As a consequence, the restricted log-likelihood can be defined as

$$\begin{aligned} l^*(\boldsymbol{\vartheta}) &= -\frac{1}{2}\log(|\boldsymbol{V}|) - \frac{1}{2}\log(|\boldsymbol{U}'\boldsymbol{V}^{-1}\boldsymbol{U}|) - \frac{1}{2}(\boldsymbol{y} - \boldsymbol{U}\hat{\boldsymbol{\beta}})'\boldsymbol{V}^{-1}(\boldsymbol{y} - \boldsymbol{U}\hat{\boldsymbol{\beta}}) \\ &= l(\boldsymbol{\vartheta}) - \frac{1}{2}\log(|\boldsymbol{U}'\boldsymbol{V}^{-1}\boldsymbol{U}|) \end{aligned}$$

without referring explicitly to the error contrasts. Surprisingly, the restricted log-likelihood differs from the profile log-likelihood considered in ML estimation only by the additional quantity $-1/2\log(|\boldsymbol{U}'\boldsymbol{V}^{-1}\boldsymbol{U}|)$.

Maximization of $l^*(\boldsymbol{\vartheta})$ yields the restricted maximum likelihood estimate for $\boldsymbol{\vartheta}$. Since, however, the restricted log-likelihood $l^*(\boldsymbol{\vartheta})$ depends on the variance parameters in a complicated, nonlinear way, no explicit solutions for the REML estimates (and similarly for the ML estimates) are available. Instead, the restricted log-likelihood has to be maximized numerically, for example using a Newton–Raphson or Fisher scoring procedure as outlined in the following paragraph describing details of REML estimation.

Details on REML estimation for variance parameters

To actually compute the REML estimates $\hat{\boldsymbol{\vartheta}}$, we consider Newton–Raphson and Fisher scoring algorithms that rely on first and second derivatives of the restricted log-likelihood. The key ingredient for deriving these derivatives are rules for matrix differentiation as provided for example in Harville (1997).

The score function is generally given by

$$\boldsymbol{s}^*(\boldsymbol{\vartheta}) = \frac{\partial l^*(\boldsymbol{\vartheta})}{\partial \boldsymbol{\vartheta}} = (s_j^*(\boldsymbol{\vartheta})),$$

$j = 1, \ldots, \dim(\boldsymbol{\vartheta})$, and contains the elements

$$\begin{aligned} s_j^*(\boldsymbol{\vartheta}) &= \frac{\partial l^*(\boldsymbol{\vartheta})}{\partial \vartheta_j} \\ &= -\frac{1}{2} \operatorname{tr}\left(\boldsymbol{P}\frac{\partial \boldsymbol{V}}{\partial \vartheta_j}\right) + \frac{1}{2}(\boldsymbol{y} - \boldsymbol{U}\hat{\boldsymbol{\beta}})'\boldsymbol{V}^{-1}\frac{\partial \boldsymbol{V}}{\partial \vartheta_j}\boldsymbol{V}^{-1}(\boldsymbol{y} - \boldsymbol{U}\hat{\boldsymbol{\beta}}) \\ &= -\frac{1}{2} \operatorname{tr}\left(\boldsymbol{P}\frac{\partial \boldsymbol{V}}{\partial \vartheta_j}\right) + \frac{1}{2}(\boldsymbol{y} - \boldsymbol{U}\hat{\boldsymbol{\beta}} - \boldsymbol{Z}\hat{\boldsymbol{\alpha}})'\boldsymbol{\Sigma}^{-1}\frac{\partial \boldsymbol{V}}{\partial \vartheta_j}\boldsymbol{\Sigma}^{-1}(\boldsymbol{y} - \boldsymbol{U}\hat{\boldsymbol{\beta}} - \boldsymbol{Z}\hat{\boldsymbol{\alpha}}) \end{aligned}$$

where $\boldsymbol{P}$ is defined in (3.21). Taking second derivatives yields the observed Fisher information matrix

$$\boldsymbol{F}^*_{\text{obs}}(\boldsymbol{\vartheta}) = -\frac{\partial \boldsymbol{s}^*(\boldsymbol{\vartheta})}{\partial \boldsymbol{\vartheta}} = -\frac{\partial^2 l^*(\boldsymbol{\vartheta})}{\partial \boldsymbol{\vartheta}\partial \boldsymbol{\vartheta}'} = (\boldsymbol{F}^*_{\text{obs},jh}(\boldsymbol{\vartheta}))$$

with entries

$$\begin{aligned} F^*_{\text{obs},jh}(\boldsymbol{\vartheta}) &= -\frac{\partial^2 l^*(\boldsymbol{\vartheta})}{\partial \vartheta_j d\vartheta_h} \\ &= \frac{1}{2} \operatorname{tr}\left(\boldsymbol{P}\frac{\partial^2 \boldsymbol{V}}{\partial \vartheta_j \partial \vartheta_h} - \boldsymbol{P}\frac{\partial \boldsymbol{V}}{\partial \vartheta_j}\boldsymbol{P}\frac{\partial \boldsymbol{V}}{\partial \vartheta_h}\right) \\ &\quad - \frac{1}{2}(\boldsymbol{y} - \boldsymbol{U}\hat{\boldsymbol{\beta}})'\boldsymbol{V}^{-1}\left(\frac{\partial^2 \boldsymbol{V}}{\partial \vartheta_j \partial \vartheta_h} - 2\frac{\partial \boldsymbol{V}}{\partial \vartheta_j}\boldsymbol{P}\frac{\partial \boldsymbol{V}}{\partial \vartheta_h}\right)\boldsymbol{V}^{-1}(\boldsymbol{y} - \boldsymbol{U}\hat{\boldsymbol{\beta}}). \end{aligned}$$

The structure of the Fisher information considerably simplifies when switching to the expected Fisher information

$$\boldsymbol{F}^*(\boldsymbol{\vartheta}) = \mathbb{E}(\boldsymbol{F}^*_{\text{obs}}(\boldsymbol{\vartheta})) = (F^*_{jh}(\boldsymbol{\vartheta})).$$

In this case, the individual entries are given by

$$F^*_{jh}(\boldsymbol{\vartheta}) = \mathbb{E}(F^*_{\text{obs},jh}(\boldsymbol{\vartheta}))$$
$$= \frac{1}{2}\,\text{tr}\left(\boldsymbol{P}\frac{\partial \boldsymbol{V}}{\vartheta_j}\boldsymbol{P}\frac{\partial \boldsymbol{V}}{\vartheta_h}\right).$$

From either the observed or the expected Fisher information matrix, we can now determine the usual updates in Newton–Raphson and Fisher scoring algorithms. In the latter case, for example, this update is given by

$$\hat{\boldsymbol{\vartheta}}^{(k+1)} = (\boldsymbol{F}^*(\hat{\boldsymbol{\vartheta}}^{(k)}))^{-1}\boldsymbol{s}^*(\hat{\boldsymbol{\vartheta}}^{(k)})).$$

The exact structure of the Fisher information and the score function employed in such updates depends on the model structure via the derivatives of the covariance matrix $\boldsymbol{V}$ with respect to the elements of $\boldsymbol{\vartheta}$. In case of the error variance σ^2 and the simple model $\boldsymbol{\Sigma} = \sigma^2 \boldsymbol{I}_N$, this derivative is given by

$$\frac{\partial \boldsymbol{V}}{\partial \sigma^2} = \boldsymbol{I}_N.$$

For variance components models with $\boldsymbol{Q} = \text{blockdiag}(\delta_1^2\boldsymbol{I}, \dots, \delta_q^2\boldsymbol{I})$ and $\boldsymbol{Z} = (\boldsymbol{Z}_1, \dots, \boldsymbol{Z}_q)$ the derivative simplifies to

$$\frac{\partial \boldsymbol{V}}{\partial \delta_j^2} = \boldsymbol{Z}_j\boldsymbol{Z}_j'.$$

In this case, the expressions can also be simplified further, and in particular the computation, storage and inversion of $N \times N$ matrices can be completely avoided (see Fahrmeir, Kneib and Lang, 2004, for details).

An alternative to iterative estimation based on Newton–Raphson or Fisher scoring is given by expectation-maximization (EM) algorithms. Since the latter do not provide the information needed to construct credible intervals or tests, we will not discuss them in detail here but refer the reader to Fahrmeir and Tutz (2001, Ch. 7).

Bayesian interpretation of ML and REML estimation

ML and REML estimation can also be compared in a Bayesian formulation of mixed models, leading to additional insights about their properties as statistical estimates. Therefore we consider a fully Bayesian model, where $\boldsymbol{\beta}$, $\boldsymbol{\alpha}$ and $\boldsymbol{\vartheta}$ are all treated as random variables. We further assume noninformative flat priors $p(\boldsymbol{\beta}) \propto$ const and $p(\boldsymbol{\vartheta}) \propto$ const for fixed effects and variance components, respectively, whereas the informative prior for the random effects is given by the random effects distribution $\boldsymbol{\alpha} \sim \text{N}(\boldsymbol{0}, \boldsymbol{Q})$. In this case, maximum likelihood and restricted maximum likelihood estimate can be characterized as either the variance parameter component of the mode of the joint posterior $p(\boldsymbol{\beta}, \boldsymbol{\vartheta}|\boldsymbol{y})$ (ML) or as the mode of the marginal posterior $p(\boldsymbol{\vartheta}|\boldsymbol{y})$ (REML).

This can be seen as follows: With flat priors for $\boldsymbol{\beta}$ and $\boldsymbol{\vartheta}$, the marginal posterior for $\boldsymbol{\vartheta}$ is proportional to

$$\begin{aligned} p(\boldsymbol{\vartheta}|\boldsymbol{y}) &= \int p(\boldsymbol{\beta}, \boldsymbol{\vartheta}|\boldsymbol{y}) d\boldsymbol{\beta} \\ &\propto \int p(\boldsymbol{y}|\boldsymbol{\beta}, \boldsymbol{\vartheta}) p(\boldsymbol{\beta}) p(\boldsymbol{\vartheta}) d\boldsymbol{\beta} \\ &= \int p(\boldsymbol{y}|\boldsymbol{\beta}, \boldsymbol{\vartheta}) d\boldsymbol{\beta}. \end{aligned}$$

On the other hand, the full conditional distribution of $\boldsymbol{\beta}$ is given by

$$p(\boldsymbol{\beta}|\boldsymbol{y}, \boldsymbol{\vartheta}) = \frac{p(\boldsymbol{y}|\boldsymbol{\beta}, \boldsymbol{\vartheta}) p(\boldsymbol{\beta})}{\int p(\boldsymbol{y}|\boldsymbol{\beta}, \boldsymbol{\vartheta}) p(\boldsymbol{\beta}) d\boldsymbol{\beta}} = \frac{p(\boldsymbol{y}|\boldsymbol{\beta}, \boldsymbol{\vartheta})}{\int p(\boldsymbol{y}|\boldsymbol{\beta}, \boldsymbol{\vartheta}) d\boldsymbol{\beta}}.$$

Since this full conditional is the Gaussian distribution of the weighted least-squares estimate, that is,

$$\boldsymbol{\beta}|\boldsymbol{y}, \boldsymbol{\vartheta} \sim \mathrm{N}(\hat{\boldsymbol{\beta}}, (\boldsymbol{U}'\boldsymbol{V}^{-1}\boldsymbol{U})^{-1})$$

we finally obtain

$$p(\boldsymbol{\vartheta}|\boldsymbol{y}) \propto \int p(\boldsymbol{y}|\boldsymbol{\beta}, \boldsymbol{\vartheta}) d\boldsymbol{\beta} = \frac{p(\boldsymbol{y}|\boldsymbol{\beta}, \boldsymbol{\vartheta})}{p(\boldsymbol{\beta}|\boldsymbol{y}, \boldsymbol{\vartheta})} \propto p(\boldsymbol{e}).$$

Hence, maximizing the restricted likelihood with respect to $\boldsymbol{\vartheta}$ is equivalent to determining the marginal posterior mode estimate for the variance parameters.

Note that using the mode of the marginal posterior for the estimation of hyperparameters is the recommended strategy in empirical Bayes analyses in general. Hence the Bayesian interpretation reveals a second argument in favour of the REML estimate as compared to ML estimation.

Testing hypotheses

To test linear hypotheses

$$H_0 : \boldsymbol{C}\boldsymbol{\beta} = \boldsymbol{d} \quad \text{versus} \quad H_1 : \boldsymbol{C}\boldsymbol{\beta} \neq \boldsymbol{d}$$

for *fixed effects*, one may consider the *Wald statistic*

$$W = (\boldsymbol{C}\hat{\boldsymbol{\beta}} - \boldsymbol{d})'[\boldsymbol{C}(\boldsymbol{X}'\hat{\boldsymbol{V}}^{-1}\boldsymbol{X})^{-1}\boldsymbol{C}']^{-1}(\boldsymbol{C}\hat{\boldsymbol{\beta}} - \boldsymbol{d})$$

or the *likelihood ratio statistic*

$$\mathrm{LRT} = 2\{l(\hat{\boldsymbol{\beta}}, \hat{\boldsymbol{\vartheta}}) - l(\tilde{\boldsymbol{\beta}}, \tilde{\boldsymbol{\vartheta}})\}$$

where $l(\boldsymbol{\beta}, \boldsymbol{\vartheta})$ is the log-likelihood of the marginal model, and $(\hat{\boldsymbol{\beta}}, \hat{\boldsymbol{\vartheta}})$ and $(\tilde{\boldsymbol{\beta}}, \tilde{\boldsymbol{\vartheta}})$ are ML estimates under H_1 and H_0, respectively. However, $(\boldsymbol{X}'\hat{\boldsymbol{V}}^{-1}\boldsymbol{X})^{-1}$ might be a poor estimate of $\mathrm{Cov}(\hat{\boldsymbol{\beta}})$ for small samples, and the asymptotic χ^2-approximation for W is lacking a rigorous and general foundation because of correlated observations in $\boldsymbol{y}$. Therefore, Wald tests, or F-tests, with $F = W/\operatorname{rank}(\boldsymbol{C})$ as implemented in SAS or R, have to used with caution. Similar cautionary remarks have to be made for the LRT statistic, see for example Crainiceanu and Ruppert (2004b). Results on testing hypothesis about *random effects* or *random effects variances* are even scarcer. Suppose there is only one scalar random effect, for example a random intercept. Testing if this random effect α_i is zero, that is,

$$H_0 : \alpha_i = 0 \text{ for all } i \quad \text{versus} \quad H_1 : \alpha_i \neq 0 \text{ for at least one } i$$

is equivalent to testing

$$H_0 : \delta^2 = \mathrm{Var}(\alpha_i) = 0 \quad \text{versus} \quad H_1 : \delta^2 > 0.$$

Standard asymptotic theory is violated because $\delta^2 = 0$ is on the boundary of the parameter space. Self and Liang (1987) and Stram and Lee (1994) show that the asymptotic distribution of the LRT is a mixture of chi-square distributions (with a point mass in zero) under H_0 if the response vector can be decomposed into i.i.d. subvectors. This is, for example, the case in balanced ANOVA models with random intercepts, but already precludes situations where the individual numbers of replications n_i are different. More general problems are still an active area of research, see for example Greven (2007), Scheipl, Greven and Küchenhoff (2008), and further details presented in Chapter 4.2.1.

3.1.5 Bayesian inference

Empirical Bayes inference

The closest relation between maximum likelihood and Bayesian inference is obtained in an empirical Bayes (EB) approach where the unknown variance components are not considered as a part of the Bayesian estimation scheme but have to be determined in advance from their marginal posterior. This also provides a Bayesian justification why REML estimates are to be preferred over ML estimates, since the former exactly correspond to the modes of the marginal posterior for $\boldsymbol{\vartheta}$. Given the estimates of the variance parameters and assuming a flat prior $p(\boldsymbol{\beta}) \propto \text{const}$ for fixed effects, the posterior for the regression coefficients is given by

$$p(\boldsymbol{\beta}, \boldsymbol{\alpha}|\boldsymbol{y}) \propto p(\boldsymbol{y}|\boldsymbol{\alpha}, \boldsymbol{\beta})p(\boldsymbol{\alpha}),$$

and the corresponding log-posterior is

$$-\frac{1}{2}(\boldsymbol{y}-\boldsymbol{U}\boldsymbol{\beta}-\boldsymbol{Z}\boldsymbol{\alpha})'\boldsymbol{\Sigma}^{-1}(\boldsymbol{y}-\boldsymbol{U}\boldsymbol{\beta}-\boldsymbol{Z}\boldsymbol{\alpha})-\frac{1}{2}\boldsymbol{\alpha}\boldsymbol{G}^{-1}\boldsymbol{\alpha}$$

up to an additive constant. Maximizing the (log-) posterior gives the posterior mode and—since the posterior is Gaussian—the posterior mean $\mathbb{E}(\boldsymbol{\beta},\boldsymbol{\alpha}|\boldsymbol{y})$. Comparing with the previous section, we see immediately that the posterior mode and mean coincides with the penalized MLE $(\hat{\boldsymbol{\beta}},\hat{\boldsymbol{\alpha}})$, given $\boldsymbol{\vartheta}$. The final EB estimate is obtained by replacing $\boldsymbol{\Sigma}$ and $\boldsymbol{G}$ through $\hat{\boldsymbol{\Sigma}}=\boldsymbol{\Sigma}(\hat{\boldsymbol{\vartheta}})$, $\hat{\boldsymbol{G}}=\boldsymbol{G}(\hat{\boldsymbol{\vartheta}})$, with the REML estimate $\hat{\boldsymbol{\vartheta}}$.

Full Bayes inference

For full Bayes (FB) inference we also have to specify a prior $p(\boldsymbol{\vartheta})$ in addition to the Gaussian priors $\boldsymbol{\beta}\sim\mathrm{N}(\boldsymbol{c},\boldsymbol{C})$ (with $p(\boldsymbol{\beta})\propto$ const for $\boldsymbol{C}^{-1}=\boldsymbol{0}$ contained as a special case) and $\boldsymbol{\alpha}\sim\mathrm{N}(\boldsymbol{0},\boldsymbol{G})$. Assuming prior independence for $\boldsymbol{\beta},\boldsymbol{\alpha}$ and $\boldsymbol{\vartheta}$, the full posterior is

$$p(\boldsymbol{\beta},\boldsymbol{\alpha},\boldsymbol{\vartheta}|\boldsymbol{y})\propto p(\boldsymbol{y}|\boldsymbol{\beta},\boldsymbol{\alpha},\boldsymbol{\vartheta})p(\boldsymbol{\beta})p(\boldsymbol{\alpha})p(\boldsymbol{\vartheta}).$$

For valid posterior inference, the posterior has to be proper, that is,

$$p(\boldsymbol{y})=\int p(\boldsymbol{y}|\boldsymbol{\beta},\boldsymbol{\alpha},\boldsymbol{\vartheta})\,p(\boldsymbol{\beta})p(\boldsymbol{\alpha})p(\boldsymbol{\vartheta})\,d\boldsymbol{\beta}d\boldsymbol{\alpha}d\boldsymbol{\vartheta}<\infty.$$

This holds for proper priors $p(\boldsymbol{\vartheta})$, but not necessarily for improper variants, see Hobert and Casella (1996) or Fahrmeir and Kneib (2009) for characterizations of admissible prior specifications.

In general, $p(\boldsymbol{y})$ and the posterior $p(\boldsymbol{\beta},\boldsymbol{\alpha},\boldsymbol{\vartheta}|\boldsymbol{y})$ are not available in closed analytical form. Therefore FB inference typically relies on numerical integration (for small dimension of unknown parameter vectors) or on MCMC techniques for models with moderate to large numbers of unknown parameters. Approximate solutions also receive increasing attention and we will sketch one of the very recent approaches based on integrated nested Laplace approximations in Section 4.6.

For MCMC simulation, the structure of full conditionals is as follows:

$$\begin{aligned}p(\boldsymbol{\beta}|\cdot)&\propto p(\boldsymbol{y}\,|\,\boldsymbol{\beta},\boldsymbol{\alpha},\boldsymbol{\vartheta})p(\boldsymbol{\beta})\\&\propto\exp\left\{-\frac{1}{2}(\boldsymbol{y}-\boldsymbol{U}\boldsymbol{\beta}-\boldsymbol{Z}\boldsymbol{\alpha})'\boldsymbol{\Sigma}^{-1}(\boldsymbol{y}-\boldsymbol{U}\boldsymbol{\beta}-\boldsymbol{Z}\boldsymbol{\alpha})-\frac{1}{2}(\boldsymbol{\beta}-\boldsymbol{c})'\boldsymbol{C}^{-1}(\boldsymbol{\beta}-\boldsymbol{c})\right\},\end{aligned}$$

$$\begin{aligned}p(\boldsymbol{\alpha}|\cdot)&\propto p(\boldsymbol{y}\,|\,\boldsymbol{\beta},\boldsymbol{\alpha},\boldsymbol{\vartheta})p(\boldsymbol{\alpha}|\boldsymbol{\vartheta})\\&\propto\exp\left\{-\frac{1}{2}(\boldsymbol{y}-\boldsymbol{U}\boldsymbol{\beta}-\boldsymbol{Z}\boldsymbol{\alpha})'\boldsymbol{\Sigma}^{-1}(\boldsymbol{y}-\boldsymbol{U}\boldsymbol{\beta}-\boldsymbol{Z}\boldsymbol{\alpha})-\frac{1}{2}\boldsymbol{\alpha}'\boldsymbol{G}^{-1}\boldsymbol{\alpha}\right\},\end{aligned}$$

$$\begin{aligned} p(\boldsymbol{\vartheta}|\cdot) &\propto p(\boldsymbol{y}\,|\,\boldsymbol{\beta},\boldsymbol{\alpha},\boldsymbol{\vartheta})p(\boldsymbol{\alpha}|\boldsymbol{\vartheta})p(\boldsymbol{\vartheta}) \\ &\propto \exp\left\{-\frac{1}{2}(\boldsymbol{y}-\boldsymbol{U\beta}-\boldsymbol{Z\alpha})'\boldsymbol{\Sigma}^{-1}(\boldsymbol{y}-\boldsymbol{U\beta}-\boldsymbol{Z\alpha})-\frac{1}{2}\boldsymbol{\alpha}'\boldsymbol{G}^{-1}\boldsymbol{\alpha}\right\}p(\boldsymbol{\vartheta}). \end{aligned}$$

Simple algebraic manipulations show that the full conditionals $p(\boldsymbol{\beta}|\cdot)$ and $p(\boldsymbol{\alpha}|\cdot)$ are Gaussian:

$$p(\boldsymbol{\beta}\,|\,\cdot) \propto \exp\left\{-\frac{1}{2}(\boldsymbol{\beta}-\boldsymbol{\mu}_{\boldsymbol{\beta}})'\boldsymbol{\Sigma}_{\boldsymbol{\beta}}^{-1}(\boldsymbol{\beta}-\boldsymbol{\mu}_{\boldsymbol{\beta}})\right\},$$

with

$$\begin{aligned} \boldsymbol{\mu}_{\boldsymbol{\beta}} &= (\boldsymbol{U}'\boldsymbol{\Sigma}^{-1}\boldsymbol{U}+\boldsymbol{C}^{-1})^{-1}(\boldsymbol{U}'\boldsymbol{\Sigma}^{-1}(\boldsymbol{y}-\boldsymbol{Z\alpha})+\boldsymbol{C}^{-1}\boldsymbol{c}) \\ \boldsymbol{\Sigma}_{\boldsymbol{\beta}} &= (\boldsymbol{U}'\boldsymbol{\Sigma}^{-1}\boldsymbol{U}+\boldsymbol{C}^{-1})^{-1}, \end{aligned}$$

and

$$p(\boldsymbol{\alpha}\,|\,\cdot) \propto \exp\left\{-\frac{1}{2}(\boldsymbol{\alpha}-\boldsymbol{\mu}_{\boldsymbol{\alpha}})'\boldsymbol{\Sigma}_{\boldsymbol{\alpha}}^{-1}(\boldsymbol{\alpha}-\boldsymbol{\mu}_{\boldsymbol{\alpha}})\right\}$$

with

$$\begin{aligned} \boldsymbol{\mu}_{\boldsymbol{\alpha}} &= (\boldsymbol{Z}'\boldsymbol{\Sigma}^{-1}\boldsymbol{Z}+\boldsymbol{G}^{-1})^{-1}\boldsymbol{Z}'\boldsymbol{\Sigma}^{-1}(\boldsymbol{y}-\boldsymbol{U\beta}) \\ \boldsymbol{\Sigma}_{\boldsymbol{\alpha}} &= (\boldsymbol{Z}'\boldsymbol{\Sigma}^{-1}\boldsymbol{Z}+\boldsymbol{G}^{-1})^{-1}. \end{aligned}$$

The expectations in these full conditionals show a close connection to (penalized) least-squares estimation applied to residuals with respect to the other regression effects. This also reveals a close connection to the backfitting algorithm discussed in Chapter 2, where penalized least-squares estimates were fitted to iteratively updated residuals.

The specific structure of the full conditionals for variance parameters depend more strongly on the exact model specification and in particular on the prior assumptions for the variances. An example is provided in the following paragraph for the special case of longitudinal data with uncorrelated errors.

Full Bayes inference for longitudinal data

In the case of longitudinal data with uncorrelated errors, that is, $\boldsymbol{\Sigma} = \sigma^2\boldsymbol{I}$, a Gaussian prior $\boldsymbol{\beta} \sim \mathrm{N}(\boldsymbol{c}, \boldsymbol{C}^{-1})$ or, with $\boldsymbol{C}^{-1} = \boldsymbol{0}$, a flat prior $p(\boldsymbol{\beta}) \propto \text{const}$ for fixed effects, i. i. d. Gaussian random effects $\boldsymbol{\alpha}_i \sim \mathrm{N}(\boldsymbol{0}, \boldsymbol{Q})$, an inverse Wishart prior $\boldsymbol{Q} \sim \mathrm{IW}(a, \boldsymbol{B})$ for the random effects covariance matrix, and an inverse gamma prior $\sigma^2 \sim \mathrm{IG}(a_\sigma, b_\sigma)$ for the error variance, the full conditionals for the fixed effects reduce to

$$\boldsymbol{\beta}|\cdot \sim \mathrm{N}(\boldsymbol{\mu}_{\boldsymbol{\beta}}, \boldsymbol{\Sigma}_{\boldsymbol{\beta}})$$

with $\boldsymbol{\mu}_{\boldsymbol{\beta}} = (\boldsymbol{U}'\boldsymbol{U}+\sigma^2\boldsymbol{C}^{-1})^{-1}(\boldsymbol{U}'(\boldsymbol{y}-\boldsymbol{Z\alpha})+\sigma^2\boldsymbol{C}^{-1}\boldsymbol{c})$ and $\boldsymbol{\Sigma}_{\boldsymbol{\beta}} = \sigma^2(\boldsymbol{U}'\boldsymbol{U}+\sigma^2\boldsymbol{C}^{-1})^{-1}$, while correspondingly the random effects full conditional is given by

$$\boldsymbol{\alpha}_i|\cdot \sim \mathrm{N}(\boldsymbol{\mu}_{\boldsymbol{\alpha}_i}, \boldsymbol{\Sigma}_{\boldsymbol{\alpha}_i}), \quad i = 1, \ldots, n,$$

with $\boldsymbol{\mu}_{\boldsymbol{\alpha}_i} = (\boldsymbol{Z}_i'\boldsymbol{Z}_i + \sigma^2\boldsymbol{Q}^{-1})^{-1}\boldsymbol{Z}_i'(\boldsymbol{y}_i - \boldsymbol{U}_i\boldsymbol{\beta})$ and $\boldsymbol{\Sigma}_{\boldsymbol{\alpha}_i} = \sigma^2(\boldsymbol{Z}_i'\boldsymbol{Z}_i + \sigma^2\boldsymbol{Q}^{-1})^{-1}$. The full conditional for the error variance is again inverse gamma but with updated parameters, yielding

$$\sigma^2|\cdot \sim \mathrm{IG}(\tilde{a}_\sigma, \tilde{b}_\sigma)$$

with

$$\tilde{a}_\sigma = a_\sigma + \frac{1}{2}\sum_{i=1}^{n} n_i \text{ and } \tilde{b}_\sigma = b_\sigma + \frac{1}{2}\sum_{i=1}^{n}(\boldsymbol{y}_i - \boldsymbol{U}_i\boldsymbol{\beta} - \boldsymbol{Z}_i\boldsymbol{\alpha}_i)'(\boldsymbol{y}_i - \boldsymbol{U}_i\boldsymbol{\beta} - \boldsymbol{Z}_i\boldsymbol{\alpha}_i).$$

Analogously, the full conditional of the covariance matrix is given by an updated inverse Wishart

$$\boldsymbol{Q}|\cdot \sim \mathrm{IW}(\tilde{a}, \tilde{\boldsymbol{B}})$$

where

$$\tilde{a} = a + \frac{n}{2} \quad \text{and} \quad \tilde{\boldsymbol{B}} = \boldsymbol{B} + \frac{1}{2}\sum_{i=1}^{n}\boldsymbol{\alpha}_i\boldsymbol{\alpha}_i'.$$

In case of *a priori* independent components α_{il}, $l = 1, \ldots, q$, of the random effects vector, i.e. $\boldsymbol{Q} = \mathrm{diag}(\delta_1^2, \ldots, \delta_q^2)$, the inverse Wishart prior resolves to independent inverse gamma priors

$$\delta_l^2 = \mathrm{Var}(\alpha_{il}) \sim \mathrm{IG}(a_l, b_l), \quad l = 1, \ldots, q,$$

and the corresponding full conditionals are given by

$$\delta_l^2|\cdot \sim \mathrm{IG}(\tilde{a}_l, \tilde{b}_l), \quad l = 1, \ldots, q,$$

with

$$\tilde{a}_l = a_l + \frac{n}{2} \quad \text{and} \quad \tilde{b}_l = b_l + \frac{1}{2}\sum_{i=1}^{n}\alpha_{il}^2.$$

All full conditionals are well-known distributions with standard samplers being available for drawing random samples. Hence, we obtain:

Algorithm 3.1: Gibbs sampler for longitudinal LMMs
Initialize all parameters at suitable starting values and draw samples for

i. $\boldsymbol{\beta}$ from $\mathrm{N}(\boldsymbol{\mu}_{\boldsymbol{\beta}}, \boldsymbol{\Sigma}_{\boldsymbol{\beta}})$, inserting current values $\boldsymbol{\alpha}_i^c$ and $(\sigma^2)^c$.
ii. $\boldsymbol{\alpha}_i$, $i = 1, \ldots, n$, from $\mathrm{N}(\boldsymbol{\mu}_{\boldsymbol{\alpha}_i}, \boldsymbol{\Sigma}_{\boldsymbol{\alpha}_i})$, inserting current values $\boldsymbol{\beta}^c$, $(\sigma^2)^c$ and $\boldsymbol{Q}^c$.
iii. σ^2 from $\mathrm{IG}(\tilde{a}_\sigma, \tilde{b}_\sigma)$, inserting current values $\boldsymbol{\beta}^c$ and $\boldsymbol{\alpha}_i^c$.

iv. $\boldsymbol{Q}$ from $\text{IW}(\tilde{a}, \tilde{\boldsymbol{B}})$, inserting current values $\boldsymbol{\alpha}_i^c$ or, if $\boldsymbol{Q} = \text{diag}(\delta_1^2, \ldots, \delta_q^2)$, δ_l^2 from $\text{IG}(\tilde{a}_l, \tilde{b}_l)$, inserting current values α_{il}^c, $l = 1, \ldots, q$.

MCMC sampling schemes encompassing parameter expansions to improve mixing and convergence are discussed, for example, in van Dyk and Meng (2001) and Kinney and Dunson (2007).

One of the main advantages of inference based on MCMC is its modularity that allows to modify parts of the model specification while leaving all other levels of the hierarchy untouched. For example, the Gibbs sampler outlined for the longitudinal LMM can be easily extended to Bayesian LMMs with non-normal errors or random effects modelled through scale mixtures of normals as indicated in the following Section 3.2.

Example 3.3: Height of schoolgirls

We analyse the longitudinal data on height of schoolgirls from Example 3.1 with the linear mixed model

$$y_{ij} = \beta_0 + \beta_1 t_{ij} + \alpha_{0i} + \alpha_{1i} t_{ij} + \beta_2 \text{group2} + \beta_3 \text{group3} + \varepsilon_{ij}, \quad \varepsilon_{ij} \overset{\text{i.i.d.}}{\sim} \text{N}(0, \sigma^2),$$

using EB (REML) and FB (Gibbs sampling) inference. In both cases we assume flat priors for fixed effects and Gaussian random effects $(\alpha_{i0}, \alpha_{i1})' \overset{\text{i.i.d.}}{\sim} \text{N}(\boldsymbol{0}, \boldsymbol{Q})$, $\boldsymbol{Q} = \text{diag}(\delta_0^2, \delta_1^2)$ as *a priori* independent. For FB inference, hyperpriors are

$$\sigma^2 \sim \text{IG}(a_\sigma, b_\sigma), \quad \delta_0^2 \sim \text{IG}(a_0, b_0), \quad \delta_1^2 \sim \text{IG}(a_1, b_1)$$

with $a_\sigma = b_\sigma = a_0 = b_0 = a_1 = b_1 = 0.005$.

Estimates for the fixed effects, the error variance and the random effects variances, together with 95% credibility intervals, are given in Table 3.1. Comparing FB and EB estimation results, we see that the point estimates for the population intercept and trend are in close agreement. FB credibility intervals are a bit wider than EB confidence intervals: this reflects the fact that FB inference automatically accounts for uncertainty in estimating variances while EB inference relies on plug in estimates of variances. The point estimates of group effects are similar but differ a bit more. This is also reflected in the comparably large credibility and confidence intervals: The small sample size does not provide enough information for assessing statistically significant group effects. Estimates of linear population trends for the three groups are visualized in Fig. 3.2.

The point estimates for random effects variances in Table 3.1 are still similar but seem to differ more from each other. As an advantage of FB inference, credibility intervals are available. They are rather wide, reflecting uncertainty in estimating individual-specific random intercepts and slopes based on only a few observations.

Figure 3.3 displays kernel density estimates for random intercepts and slopes, confirming that FB and EB inference lead to similar conclusions: While the

Table 3.1 Height of schoolgirls: Posterior means (FB inference) and posterior modes (EB inference) together with 95% credible intervals for the fixed effects and variance components.

	FB inference			EB inference		
	mean	2.5%	97.5%	mode	2.5%	97.5%
β_0	80.77	78.18	83.51	80.68	78.28	83.08
β_1	5.71	5.44	6.00	5.72	5.48	5.95
β_2	1.99	−1.65	5.43	2.09	−1.15	5.32
β_3	2.96	−0.59	6.48	3.18	−0.06	6.41
	mean	2.5%	97.5%	mode	2.5%	97.5%
σ^2	0.54	0.37	0.79	0.50	–	–
δ_0^2	6.80	1.40	16.68	6.05	–	–
δ_1^2	0.33	0.15	0.67	0.24	–	–

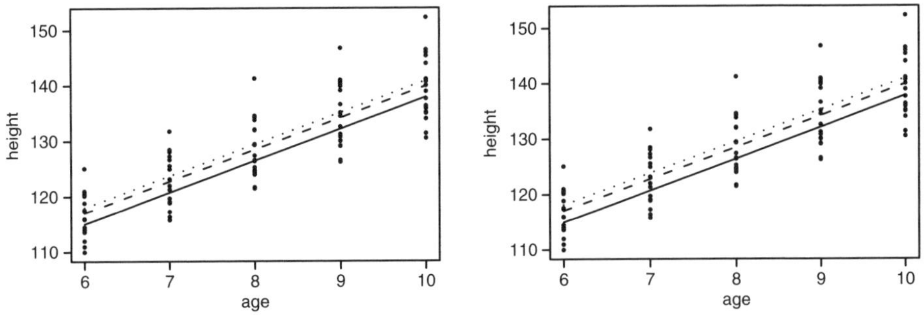

Fig. 3.2 Height of schoolgirls: Estimated population trends for the three groups defined by small (solid line), medium (dashed line) and tall mothers (dotted line).

empirical distribution of estimated intercepts resembles a Gaussian density, estimated slopes seem to follow a non-Gaussian density, perhaps representable through a mixture of Gaussian densities. We explore this more formally in Example 3.5 on page 153 based on Dirichlet process mixture priors. Figure 3.4 shows population trends and individual-specific trends for selected girls in each of the three groups in case of FB inference. △

Example 3.4: Childhood obesity

For analysing the childhood obesity data from the LISA study (Example 1.1), we apply FB (Gibbs sampling) and EB (REML) inference to the linear mixed model

$$y_{ij} = \boldsymbol{u}_{ij}'\boldsymbol{\beta} + f(t_{ij}) + \boldsymbol{z}_{ij}'\boldsymbol{\alpha}_i + \varepsilon_{ij}, \quad \varepsilon_{ij} \overset{\text{i.i.d.}}{\sim} \mathrm{N}(0, \sigma^2),$$

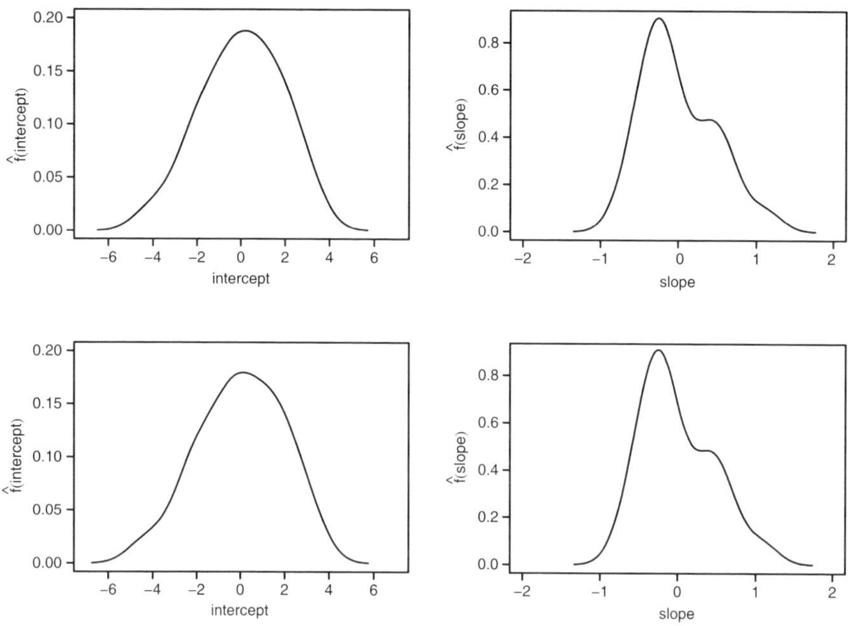

Fig. 3.3 Height of schoolgirls: Kernel density estimates for the estimated random intercepts (left) and random slopes (right) in case of FB inference (top) and EB inference (bottom).

with the nonlinear function

$$f(t) = \beta_0 + \beta_1 t + \beta_2 \frac{\log(t+1)}{(t+1)^2},$$

and the covariate vector $\boldsymbol{u}$ described in Example 3.2 (page 110). Individual-specific deviations from the population trend are modelled either as a random line $\alpha_{i0} + \alpha_{i1} t_{ij}$ or as the nonlinear random function $g_i(t)$ in (3.7). We assume flat priors for fixed effects and Gaussian random effects $\alpha_{il} \sim \mathrm{N}(0, \delta_l^2)$, $l = 0, 1, 2$, as *a priori* independent. For FB inference, hyperpriors are

$$\sigma^2 \sim \mathrm{IG}(a_\sigma, b_\sigma), \quad \delta_l^2 \sim \mathrm{IG}(a_l, b_l)$$

with $a_\sigma = b_\sigma = a_l = b_l = 0.001$.

Estimates for the fixed effects, the error variance and the random effects variances, together with 95% credibility intervals, are given in Table 3.2 for the model with the random line specification and in Table 3.3 for the nonlinear random curve specification (3.7). Comparing FB and EB estimation results in Table 3.2, we see that the estimated effects are in good or reasonable agreement. This is in particular true for the significant effects of sex, mBMI (maternal

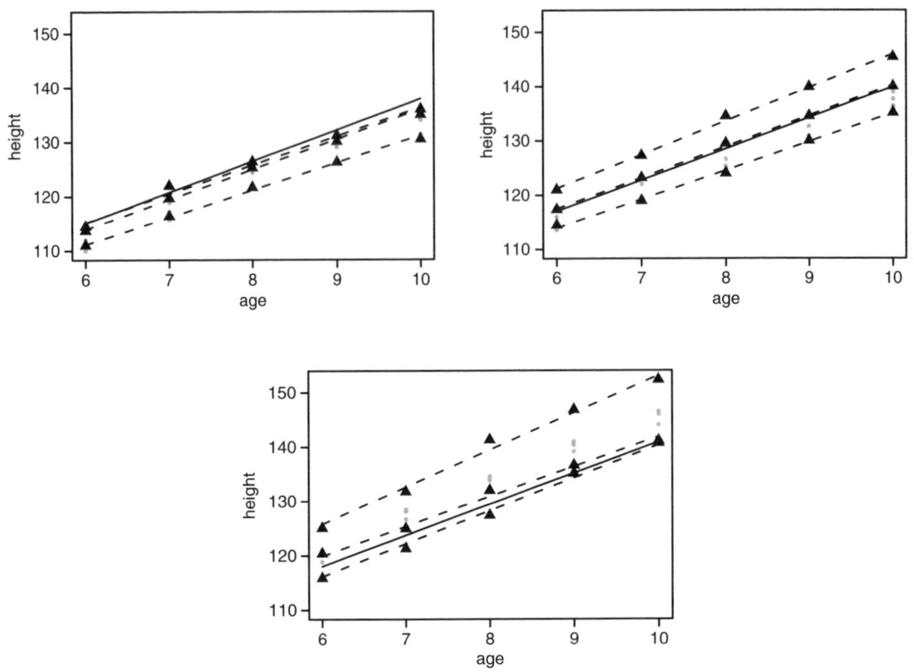

Fig. 3.4 Height of schoolgirls: Estimated population trends (solid line) together with selected individual-specific growth trajectories (dashed lines) for three individuals from each of the groups with small (top left), medium (top right) and tall (bottom) mothers obtained with FB inference. The individual observations are visualized as triangles.

BMI at pregnancy begin), mDiffBMI (maternal BMI gain during pregnancy) and the coefficients (const, age, functionAge) of the nonlinear population trend $f(t)$. As one might have expected, the effects of being a male child, mBMI and mDiffBMI significantly increase a child's BMI on average. On the other hand, breastfeeding, mSmoke (maternal smoking) and the type of area (rural or urban) have no significant effect. This seems to be surprising, in particular for breastfeeding, since some experts suppose that exclusive breastfeeding causes lower BMI values.

Compared with estimation results for the model with the random curve specification in Table 3.3, we see that the effect of breastfeeding becomes significant with FB inference while it remains nonsignificant with EB inference. (Note, however, that the credibility interval for FB inference almost includes zero.) The general conclusion for other effects does not change dramatically, although the positive effect of sex becomes smaller. The intercept and the coefficients of age and functionAge are in very close agreement with those of Table 3.2, indicating that estimation of the population trend $f(t)$ is quite robust, regardless of the

Table 3.2 Childhood obesity: Posterior means (FB inference) and posterior modes (EB inference) together with 95% credible intervals for the fixed effects and variance components for the model with random lines. mBMI and mDiffBMI have been centred for the analyses.

	FB inference			EB inference		
	mean	2.5%	97.5%	mode	2.5%	97.5%
const	12.56	12.44	12.68	12.60	12.49	12.71
sex	0.3081	0.2263	0.3909	0.3001	0.2212	0.3789
breast	0.0466	−0.0336	0.1267	0.0225	−0.0600	0.1051
mSmoke	0.0578	−0.0641	0.1775	0.0803	−0.0360	0.1966
area	0.0221	−0.0794	0.1190	−0.0124	−0.1091	0.0843
mBMI	0.0450	0.0332	0.0569	0.0493	0.0386	0.0600
mDiffBMI	0.0829	0.0571	0.1094	0.0733	0.0489	0.0977
age	0.3048	0.2929	0.3170	0.3038	0.2914	0.3162
functionAge	21.40	21.18	21.65	21.40	21.17	21.64
	mean	2.5%	97.5%	mode	2.5%	97.5%
σ^2	1.0005	0.9776	1.0248	0.9858	–	–
δ_0^2	0.7120	0.6603	0.7676	0.8012	–	–
δ_1^2	0.0336	0.0291	0.0383	0.0411	–	–

Table 3.3 Childhood obesity: Posterior means (FB inference) and posterior modes (EB inference) together with 95% credible intervals for the fixed effects and variance components for the model with random curves. mBMI and mDiffBMI have been centred for the analyses.

	FB inference			EB inference		
	mean	2.5%	97.5%	mode	2.5%	97.5%
const	12.61	12.48	12.72	12.64	12.52	12.75
sex	0.1530	0.0771	0.2305	0.2245	0.1483	0.3008
breast	0.1318	0.0440	0.2195	0.0548	−0.0251	0.1346
mSmoke	−0.0317	−0.1592	0.0949	0.0492	−0.0633	0.1617
area	0.0157	−0.0801	0.1233	−0.0267	−0.1202	0.0667
mBMI	0.0506	0.0390	0.0623	0.0531	0.0427	0.0634
mDiffBMI	0.1193	0.0937	0.1436	0.0881	0.0645	0.1117
age	0.3051	0.2934	0.3164	0.3037	0.2910	0.3165
functionAge	21.43	21.09	21.76	21.43	21.06	21.80
σ^2	0.7633	0.7437	0.7839	0.7155	–	–
δ_0^2	0.6984	0.6392	0.7603	1.1013	–	–
δ_1^2	0.0429	0.0385	0.0478	0.0561	–	–
δ_2^2	36.88	33.64	40.68	50.70	–	–

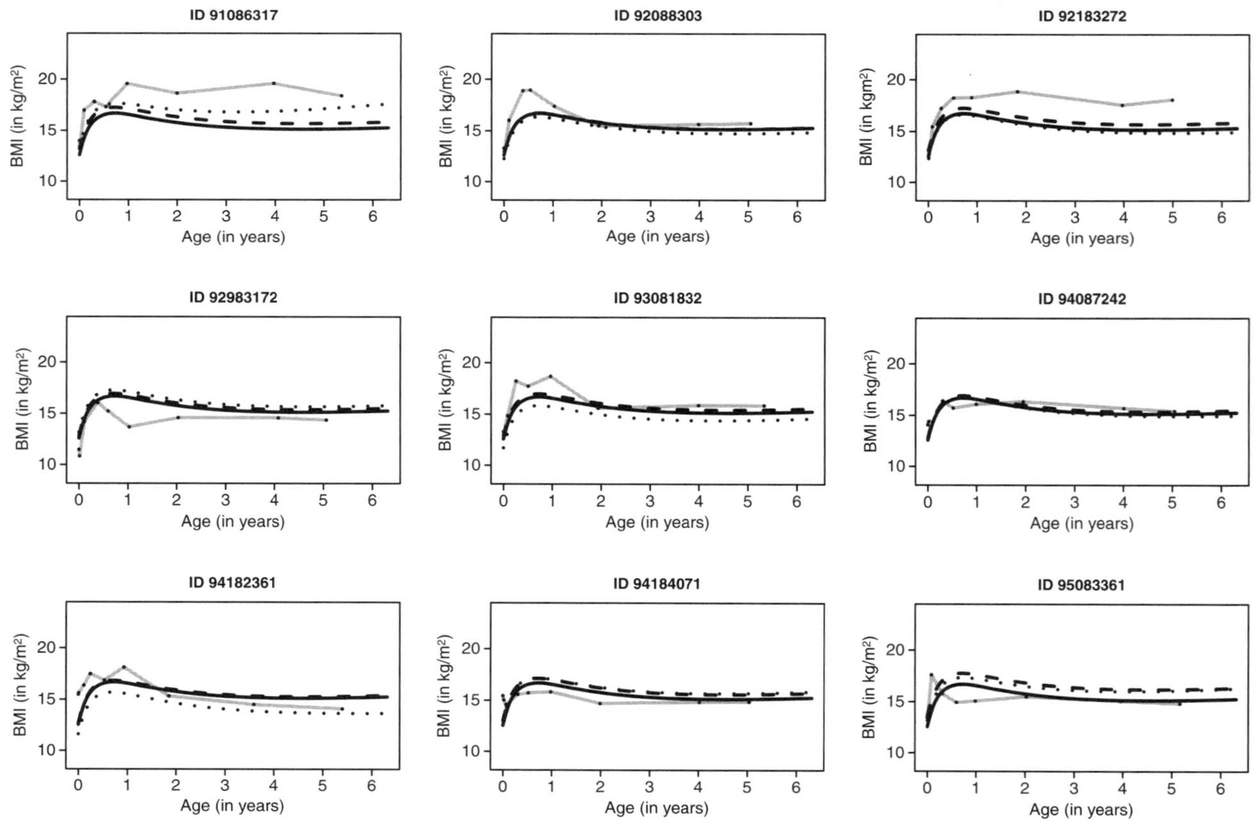

Fig. 3.5 Childhood obesity: Observed BMI curves for 9 randomly selected individuals (grey lines) in comparison to the estimated population trend (solid lines), the individual trend without random effects (dashed lines) and the individual trend with random effects (dotted lines) for the model with random lines.

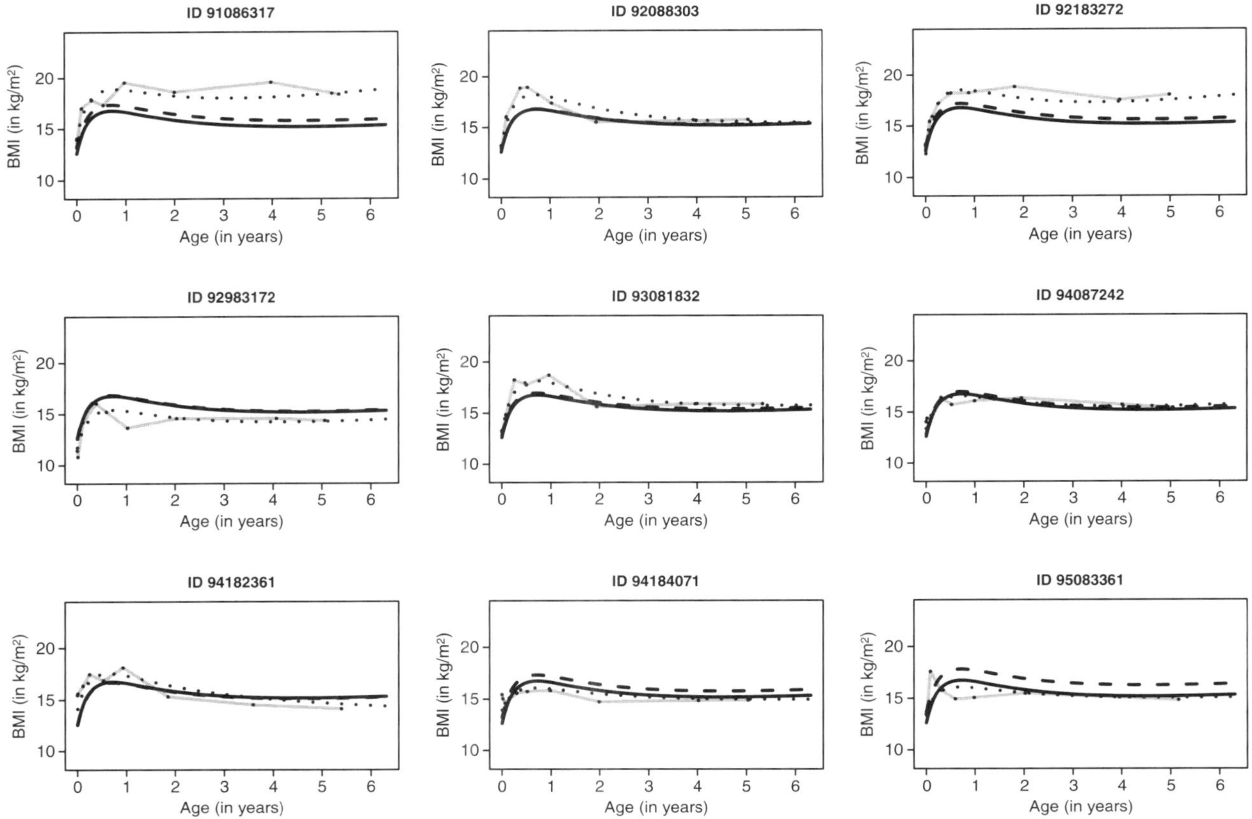

Fig. 3.6 Childhood obesity: Observed BMI curves for 9 randomly selected individuals (grey lines) in comparison to the estimated population trend (solid lines), the individual trend without random effects (dashed lines) and the individual trend with random effects (dotted lines) for the model with random curves.

linear or nonlinear random effects specification. Figures 3.5 and 3.6 display the (estimated) population trend $f(t)$ and the individual trends $\boldsymbol{u}_i'\boldsymbol{\beta} + f(t)$ without random effects, together with observed BMI values and individual-specific curves $\boldsymbol{u}_i'\boldsymbol{\beta} + f(t) + g_i(t)$ for nine randomly selected children. The estimated population trends for both models are almost indistinguishable and show the typical pattern of a strongly increasing BMI during the first months after birth, a peak at about 10 to 12 months, and a moderate decrease thereafter. The individual curves, of course, differ between the two models because of the different random deviation specifications: For some individuals, however, the difference is negligible, while for others there seems to be an improvement in terms of goodness-of-fit. △

3.2 Linear mixed models with flexible random effects priors

Conceptually, LMM for longitudinal data are very attractive in allowing subject-specific random variability and, at the same time, borrowing information across subjects. The standard assumption of i.i.d. Gaussian random effects, often made for mathematical convenience, may however be questionable, mainly for two reasons: First, how sensitive reacts estimates of fixed effects if the true random effets distribution is non-Gaussian? Second, how substantial is the impact of such a misspecification on the estimation of the random effects? The Gaussian distribution has light tails, it is symmetric and it has only one mode. Therefore, effects of outlying subjects may be over-shrunk towards the population mean, possible skewness may be blurred, and clusters of subjects with similar effects, corresponding to a multimodal random effects distribution, may remain undetected. In the following, we will discuss flexible extensions of Gaussian random effects models that allow to study such effects by allowing flexible random effects distributions.

Some limitations of Gaussian random effects may be overcome with heavy-tailed random effects distributions, for example Student distributions with small degrees of freedom or the Laplace distribution. Student distributions as well as the Laplace distribution can be represented hierarchically through scale mixtures of normals, which is very convenient for full Bayesian inference. In the example of a Student prior $\boldsymbol{\alpha}_i|\boldsymbol{Q} \sim t_\nu(\boldsymbol{0}, \boldsymbol{Q})$ with ν degrees of freedom, location parameter $\boldsymbol{0}$ and scale parameter $\boldsymbol{Q}$, such a scale mixture representation is obtained as

$$\boldsymbol{\alpha}_i|\boldsymbol{Q}, \lambda_i \sim \mathrm{N}(\boldsymbol{0}, \boldsymbol{Q}/\lambda_i), \quad \boldsymbol{Q} \sim \mathrm{IW}(a, \boldsymbol{B}), \quad \lambda_i \sim \mathrm{Ga}(\nu/2, \nu/2).$$

When marginalizing out the parameters λ_i, we obtain $\boldsymbol{\alpha}_i|\boldsymbol{Q} \sim \mathrm{t}_\nu(\boldsymbol{0}, \boldsymbol{Q})$. Gibbs steps can then be used to update the hyperparameters $\lambda_1, \ldots, \lambda_n$ and $\boldsymbol{Q}$ employing the full conditionals

$$\lambda_i|\cdot \sim \text{Ga}\left(\frac{\nu}{2}+\frac{q}{2}, \frac{\nu}{2}+\frac{1}{2}\boldsymbol{\alpha}_i'\boldsymbol{Q}^{-1}\boldsymbol{\alpha}_i\right)$$

$$\boldsymbol{Q}|\cdot \sim \text{IW}\left(a+\frac{n}{2}, \boldsymbol{B}+\frac{1}{2}\sum_{i=1}^{n}\lambda_i\boldsymbol{\alpha}_i\boldsymbol{\alpha}_i'\right),$$

where $q = \dim(\boldsymbol{\alpha}_i)$.

Skewness in the random effects distribution may be introduced into any unimodal density $f(\cdot)$ symmetric about 0 using a scale factor on each side of 0 (Fernández and Steel, 1998). This gives the prior

$$p(\alpha_i|\lambda) = \frac{2}{\lambda+1/\lambda}\left[f\left(\frac{\alpha_i}{\lambda}\right)\mathbb{1}_{[0,\infty)}(\alpha_i) + f(\lambda\alpha_i)\mathbb{1}_{(-\infty,0)}(\alpha_i)\right],$$

where the scale factor λ controls skewness. For $\lambda = 1$ this reduces to $f(\alpha_i)$, and $0 < \lambda < 1$ and $\lambda > 1$ correspond to negative and positive skewing. Lee and Thompson (2008) apply this idea to linear mixed models.

Parametric unimodal priors, such as extensions of the Student or Laplace distribution that allow heavy-tails and skewness, are still too restrictive to deal with the issue of clustered random effects, which may arise due to unobserved covariates or latent sub-populations. Finite mixture models, considered in Section 3.2.1, define flexible yet parametric priors which allow multimodality. More flexibility can be achieved with Bayesian nonparametric models. Sections 3.2.2 and 3.2.3 focus on Dirichlet process (DP) and Dirichlet process mixture (DPM) priors as the most prominent type of nonparametric approaches for modelling random effects distributions. These priors may be interpreted as a natural nonparametric extension of finite mixture models.

3.2.1 Finite mixture models

Parametric discrete or continuous multimodal random effects priors can be obtained through finite mixture priors

$$p(\boldsymbol{\alpha}_i|\boldsymbol{\pi},\boldsymbol{\phi}) = \sum_{k=1}^{K}\pi_k p(\boldsymbol{\alpha}_i|\boldsymbol{\phi}_k), \tag{3.24}$$

where $\boldsymbol{\pi} = (\pi_1, \ldots, \pi_k)'$, with $\pi_1 + \ldots + \pi_K = 1$, is the vector of mixture weights, and $\boldsymbol{\phi} = (\boldsymbol{\phi}_1', \ldots, \boldsymbol{\phi}_K')'$ are the parameters of the parametric families $p(\boldsymbol{\alpha}_i|\boldsymbol{\phi}_k), k = 1, \ldots, K$. If $\boldsymbol{\mu}_k = \int \boldsymbol{\alpha}_i p(\boldsymbol{\alpha}_i|\boldsymbol{\phi}_k)d\boldsymbol{\alpha}_i$ denotes the mean in cluster k, then

$$\mathbb{E}(\boldsymbol{\alpha}_i) = \sum_{k=1}^{K}\pi_k\boldsymbol{\mu}_k,$$

and $\mathbb{E}(\boldsymbol{\alpha}_i) \neq \mathbf{0}$, in general. To avoid centring restrictions on $\boldsymbol{\pi}$ and $\boldsymbol{\phi}$ that guarantee $\mathbb{E}(\boldsymbol{\alpha}_i) = \mathbf{0}$, a common strategy is to redefine the random effects as

$$\boldsymbol{\alpha}_i = \tilde{\boldsymbol{\alpha}} + \tilde{\boldsymbol{\alpha}}_i$$

where $\tilde{\boldsymbol{\alpha}} = \mathbb{E}(\boldsymbol{\alpha}_i)$ corresponds to fixed effects of the covariate vectors $\boldsymbol{z}_{ij}$, and $\tilde{\boldsymbol{\alpha}}_i$, with $\mathbb{E}(\tilde{\boldsymbol{\alpha}}_i) = \mathbf{0}$ as before, is the 'random' effects part. However, this requires that the random effects covariate vector $\boldsymbol{z}_{ij}$ is a subset of the fixed effect covariate vector $\boldsymbol{u}_{ij}$. With an analogous redefinition of $\boldsymbol{U}_i$ as the design matrix constructed from covariates with fixed effects $\boldsymbol{\beta}$ only, the observation model of a LMM is reformulated as

$$\boldsymbol{y}_i = \boldsymbol{U}_i\boldsymbol{\beta} + \boldsymbol{Z}_i\boldsymbol{\alpha}_i + \boldsymbol{\varepsilon}_i, \quad \boldsymbol{\varepsilon}_i \sim \mathrm{N}(\mathbf{0}, \sigma^2\boldsymbol{I}), \tag{3.25}$$

The random effects $\boldsymbol{\alpha}_i, i = 1, \ldots, n$, are assumed to be generated i. i. d. from the finite mixture (3.24). Implicitly, finite mixture models make the assumption that each random effect comes from one of the mixture components $p(\boldsymbol{\alpha}_i|\boldsymbol{\phi}_k)$ with probability π_k. This can be expressed explicitly via a latent classification variable $c_i \in \{1, \ldots, K\}$, associating $\boldsymbol{\alpha}_i$ with the cluster it belongs to:

$$p(\boldsymbol{\alpha}_i|\boldsymbol{\pi}, \boldsymbol{\phi}) = \sum_{k=1}^{K} p(c_i = k|\boldsymbol{\pi})p(\boldsymbol{\alpha}_i|c_i = k, \boldsymbol{\phi}) = \sum_{k=1}^{K} \pi_k p(\boldsymbol{\alpha}_i|\boldsymbol{\phi}_k),$$

and

$$\boldsymbol{\alpha}_i|c_i \sim p(\boldsymbol{\alpha}_i|\boldsymbol{\phi}_{c_i}).$$

Finite mixture of normals prior: the heterogeneity model

Specifying the mixture components as multivariate Gaussian densities leads to a mixture of normals prior

$$\boldsymbol{\alpha}_i|\boldsymbol{\pi}, \boldsymbol{\mu}, \boldsymbol{Q} \sim \sum_{k=1}^{K} \pi_k \, \mathrm{N}(\boldsymbol{\mu}_k, \boldsymbol{Q}_k),$$

with unknown component means $\boldsymbol{\mu} = (\boldsymbol{\mu}_1, \ldots, \boldsymbol{\mu}_k)$, unknown component covariance matrices $\boldsymbol{Q} = (\boldsymbol{Q}_1, \ldots, \boldsymbol{Q}_K)$ and mixture weights. Verbeke and Molenberghs (2000) call LMMs with this random effects prior the *heterogeneity model.* A special version with homogeneous covariance matrices $\boldsymbol{Q}_1 = \ldots = \boldsymbol{Q}_K$ has been suggested by Verbeke and Lesaffre (1996). The models have been considered and applied in diverse fields, from social science to biostatistics, see Frühwirth-Schnatter, Tüchler and Otter (2004), Rossi, Allenby and McCulloch (2005, Ch. 5) and Frühwirth-Schnatter (2006) for further references. For $K = 1$, the heterogeneity model reduces to the conventional LMM with a normal random effects prior. Conditional on the latent classification variables $c_i, i = 1 \ldots, n$, we have

$$\boldsymbol{y}_i = \boldsymbol{U}_i\boldsymbol{\beta} + \boldsymbol{Z}_i\boldsymbol{\alpha}_i + \boldsymbol{\varepsilon}_i, \quad \boldsymbol{\varepsilon}_i \sim \mathrm{N}(0, \sigma^2\boldsymbol{I}),$$

with

$$\boldsymbol{\alpha}_i|c_i \sim \mathrm{N}(\boldsymbol{\mu}_{c_i}, \boldsymbol{Q}_{c_i}).$$

The marginal model, after integrating out the random effects but still conditioning on c_i, is

$$\boldsymbol{y}_i = \boldsymbol{U}_i\boldsymbol{\beta} + \boldsymbol{Z}_i\boldsymbol{\mu}_i + \tilde{\varepsilon}_i, \quad \tilde{\boldsymbol{\varepsilon}}_i \sim \mathrm{N}(0, \boldsymbol{V}_i),$$

with

$$\boldsymbol{V}_i = \boldsymbol{Z}_i\boldsymbol{Q}_{c_i}\boldsymbol{Z}_i' + \sigma^2\boldsymbol{I}.$$

Therefore, the heterogeneity model can also be regarded as a finite mixture of linear mixed models.

Formally, the heterogeneity model also encompasses *latent class mixed models* if each covariance matrix is the null matrix, that is, $\boldsymbol{Q}_1 = \ldots = \boldsymbol{Q}_K = \boldsymbol{0}$. Then $p(\boldsymbol{\alpha}_i|\boldsymbol{\pi}, \boldsymbol{\mu})$ is a discrete distribution with K unknown support points $\boldsymbol{\mu}_1, \ldots, \boldsymbol{\mu}_K$ and $\mathbb{P}(\boldsymbol{\alpha}_i = \boldsymbol{\mu}_k) = \pi_k$. The marginal distribution then is

$$p(\boldsymbol{y}_i|\boldsymbol{\pi}, \boldsymbol{\mu}) = \sum_{k=1}^{K} \pi_k \varphi(\boldsymbol{y}_i|\boldsymbol{U}_i\boldsymbol{\beta} + \boldsymbol{Z}_i\boldsymbol{\mu}_k; \sigma^2\boldsymbol{I}),$$

i.e. we obtain a finite mixture of linear regression models.

MCMC inference for finite mixture models

For full Bayesian inference we need to place priors on the parameters $\boldsymbol{\pi} = (\pi_1, \ldots, \pi_K)'$ and $\boldsymbol{\phi} = (\boldsymbol{\phi}_1', \ldots, \boldsymbol{\phi}_K')'$. The prior for $\boldsymbol{\phi}$ is specified through

$$\boldsymbol{\phi}_k \overset{\text{i.i.d.}}{\sim} G_0,$$

where the choice of the prior distribution G_0 is model-specific. For example, conjugate Gaussian/inverse Wishart priors are a popular choice for the Gaussian heterogeneity model. The prior distribution G_0 may depend on hyperparameters and we might place hyperpriors on them in a further stage of the hierarchy.

For the mixing weights $\boldsymbol{\pi}$, a symmetric Dirichlet prior

$$\boldsymbol{\pi} \sim \mathrm{Dir}(\nu_0/K, \ldots, \nu_0/K)$$

is a natural choice: The symmetry reflects the usual assumption that the labels of the mixture components could be exchanged without changing the model. Due to the properties of the Dirichlet distribution, the prior mean of π_k is equal to $1/K$, and the prior variance is $(1 - 1/K)/K(\nu_0 + 1)$, so that ν_0 is a precision parameter.

A variety of MCMC algorithms can be designed to draw samples from the posterior for the fixed effects $\boldsymbol{\beta}$, random effects $\boldsymbol{\alpha} = (\boldsymbol{\alpha}_1', \ldots, \boldsymbol{\alpha}_n')'$, the mixture parameters $\boldsymbol{\phi}, \boldsymbol{\pi}$ and/or the classification variables $\boldsymbol{c} = (c_1, \ldots, c_n)'$. Augmenting the model with the latent cluster indicators $\boldsymbol{c}$, Gibbs samplers can be constructed. Given $\boldsymbol{c}$, we know which random effects belong to which cluster. Then the general structure of Gibbs samplers is defined by the following full conditionals:

- $p(\boldsymbol{\pi}|\boldsymbol{c}, \boldsymbol{\phi}, \boldsymbol{y}) = p(\boldsymbol{\pi}|\boldsymbol{c})$: use standard multinominal-Dirichlet conjugacy;
- $p(\boldsymbol{\beta}|\boldsymbol{c}, \boldsymbol{\phi}, \boldsymbol{\alpha}, \sigma^2, \boldsymbol{y})$: update as in linear models;
- $p(\boldsymbol{\phi}|\boldsymbol{c}, \boldsymbol{\beta}, \boldsymbol{\alpha}, \sigma^2, \boldsymbol{y})$: separate updates for each $\boldsymbol{\phi}_k$ (using conjugacy);
- $p(\boldsymbol{\alpha}_i|\boldsymbol{\beta}, \boldsymbol{\phi}_{\boldsymbol{c}_i}, \sigma^2, \boldsymbol{y}_i)$: separate updates for each $\boldsymbol{\alpha}_i$ (using conjugacy);
- $p(\sigma^2|\boldsymbol{\beta}, \boldsymbol{\alpha}, \boldsymbol{y})$: inverse Gamma with updated parameters.

Special cases result under specific assumptions for the mixture model. For example, for the heterogeneity model priors can be specified as follows: A (weakly) informative Gaussian prior

$$\boldsymbol{\beta} \sim \mathrm{N}(\boldsymbol{c}, \boldsymbol{C})$$

for fixed effects, conjugate Gaussian/inverse Wishart priors

$$\boldsymbol{\mu}_k \sim \mathrm{N}(\boldsymbol{m}, \boldsymbol{M}), \quad \boldsymbol{Q}_k \sim \mathrm{IW}(a, \boldsymbol{B})$$

for G_0, an inverse gamma prior

$$\sigma^2 \sim \mathrm{IG}(a_\sigma, b_\sigma)$$

for the error variance, and the symmetric Dirichlet prior with parameter ν_0/K for the weights. A detailed Gibbs sampling algorithm for this setting is described in (Frühwirth-Schnatter, 2006, Ch. 8.5).

Penalized mixture of normals priors

In practice, the dimension of $\boldsymbol{\alpha}$ as well as the number K of mixture components in (3.24) is low for computational reasons. Moreover, the number K of mixture components was tacitly assumed to be known or given. A data driven choice of K is not as straightforward as it seems to be. Various generic model selection procedures, from crossvalidation to DIC, may be considered, or one could place a parametric prior, for example a Poisson prior, on K. Dirichlet process (DP)-based priors provide a nonparametric alternative, see the following sections.

Another idea is based on penalization, conceptually close to Bayesian P-spline smoothing (Section 2.2). However, when modelling a density, it seems more natural to use parametric densities, like Gaussian densities, as 'basis functions'. Komárek and Lesaffre (2008) suggest such penalized Gaussian mixtures as a random effects distribution in the context of generalized linear mixed models. To simplify presentation of this approach, we consider only scalar random effects α_i, that is, random intercepts. Assume a random effect prior

$$\alpha_i \overset{\text{i.i.d.}}{\sim} g(\alpha_i) = \frac{1}{\delta} g^*(\alpha_i^*),$$

where δ is an unknown scale parameter and $\alpha_i^* = \alpha_i/\delta$ are standardized random effects. Then the basic idea is to approximate the density g^* of α_i^* through

$$g^*(\alpha_i^*) = \frac{1}{\sigma_0} \sum_{k=1}^{K} \pi_k \varphi\left(\frac{\alpha_i^* - \mu_k}{\sigma_0}\right), \tag{3.26}$$

where $\pi = (\pi_1, \ldots, \pi_K)$ are unknown mixture weights, while the modes μ_k of the normal densities $\varphi(\cdot)$ are equidistant knots, defining a fine grid on the α-axis, that is, K is comparably large as in the P-spline approach for estimating unknown regression functions. Komárek and Lesaffre (2008) suggest choosing the value of σ_0 as $\sigma_0 = 2\Delta/3$, where Δ is the distance between two successive knots of the grid. Then (3.26) appears as a basis function approach for approximating the random effects prior, with basis functions $\varphi((\alpha_i^* - \mu_k)/\sigma_0)$ and unknown coefficients $\pi_1, \ldots, \pi_K$. Following the concept of P-splines, differences between successive weights should be penalized to achieve smoothness of the (estimated) prior. Because weights are restricted to $0 < \pi_k < 1$ and $\pi_1 + \ldots + \pi_K = 1$, they are reparametrized to unrestricted coefficients $\gamma_1, \ldots, \gamma_K$ according to the multinomial logit transformation

$$\pi_j = \frac{\exp(\gamma_j)}{\sum_{k=1}^{K} \exp(\gamma_k)}.$$

Because the probabilities sum to one, a restriction can be imposed on one of the resulting parameters and, without loss of generality, we can assume $\gamma_K = 0$ for convenience. Then a random-walk prior is assumed for the sequence $\gamma_1, \ldots, \gamma_{K-1}$, as for Bayesian P-splines. Further priors and details are described in Komárek and Lesaffre (2008) in the more general context of generalized linear mixed models, with an implementation for scalar or bivariate random effects provided by the `R` contributed package `glmmAK`.

3.2.2 Dirichlet processes

In Section 3.2.3, we pursue a different route for nonparametric modelling of random effects priors, based on Dirichlet process (DP) priors. To make the similarities between finite mixture priors and DP-based priors more transparent, we look at finite mixture modelling from another perspective. Let

$$G(\cdot) = \sum_{k=1}^{K} \pi_k \delta_{\phi_k}(\cdot)$$

define a discrete distribution, where $\delta_{\phi_k}(\cdot)$ is an atom at ϕ_k, that is, it has probability mass 1 at ϕ_k and 0 otherwise. We can also consider such a discrete distribution as a special case of a mixture distribution where the realizations from G correspond to the expectations of the random effects. Due to the discrete nature of G, there is only a fixed collection of expectations such that the random effects will be grouped according to their expectation.

In a further stage of the hierarchy, the atoms of the discrete distribution G are also assigned a prior distribution G_0. More specifically, the generation of random effects can be represented in the following hierarchy:

$$\begin{aligned}
\boldsymbol{\phi}_k &\overset{\text{i.i.d.}}{\sim} G_0 \\
\boldsymbol{\pi} &\sim \text{Dir}(\nu_0/K, \ldots, \nu_0/K) \\
G(\cdot) &= \sum_{k=1}^{K} \pi_k \delta_{\boldsymbol{\phi}_k}(\cdot) \\
\theta_i &\overset{\text{i.i.d.}}{\sim} G \\
\boldsymbol{\alpha}_i &\overset{\text{ind.}}{\sim} p(\boldsymbol{\alpha}_i|\theta_i).
\end{aligned} \tag{3.27}$$

In a first step, the atoms ϕ_k are generated as i.i.d. draws from the base distribution G_0. In combination with the mixture weights $\boldsymbol{\pi}$ obtained from a Dirichlet distribution, these atoms are then used to define the discrete mixture G. Hyperparameters of the random effects distribution are then generated as i.i.d. draws from this discrete mixture. Because $\boldsymbol{\phi} = (\boldsymbol{\phi}_1', \ldots, \boldsymbol{\phi}_K')'$ and $\boldsymbol{\pi} = (\pi_1, \ldots, \pi_K)'$ are random, G is a random probability measure. Since G is discrete, each θ_i is equal to one of the underlying atoms $\boldsymbol{\phi}_k$ generated i. i. d. from G_0 and there will typically be a number of observations with the same hyperparameter values. The subset of θ_i values that is equal to $\boldsymbol{\phi}_k$ belongs to cluster k.

Instead of introducing the latent parameters θ_i, we may equivalently rewrite a finite mixture model in terms of latent classification variables $c_i \in \{1, \ldots, K\}$, with $c_i = k \Leftrightarrow \theta_i = \boldsymbol{\phi}_k$. Then the finite mixture model is defined by

$$\begin{aligned}
\boldsymbol{\phi}_k &\overset{\text{i.i.d.}}{\sim} G_0, \boldsymbol{\pi} \sim \text{Dir}(\nu_0/K, \ldots, \nu_0/K), \\
c_i|\boldsymbol{\pi} &\overset{\text{ind.}}{\sim} \mathbb{P}(c_i = k|\boldsymbol{\pi}) = \pi_k, \quad k = 1, \ldots, K, \\
\boldsymbol{\alpha}_i &\overset{\text{ind.}}{\sim} p(\boldsymbol{\alpha}_i|\boldsymbol{\phi}_{c_i}).
\end{aligned}$$

In the previous considerations, we have always assumed finite mixtures with fixed K. DP-based priors for $\boldsymbol{\alpha}_i$ extend the discrete mixture hierarchy to $K \to \infty$ (in theory) and K large (in practice).

A nonparametric Bayesian approach for the unknown random effects density is to define a prior probability model on some infinite-dimensional space of densities or probability distributions. This requires the definition of probability measures on a set of distributions, with densities $p(\cdot)$ as realizations. The random measure $G(\cdot)$ in (3.27) is a simple example. Generally, such probability measures are referred to as *random probability measures* (RPM). Dirichlet processes (DP) and related RPMs based on DPs, such as DP mixtures, are a popular class of priors in nonparametric Bayesian data analysis, from density estimation to nonparametric regression, survival analysis and more complex models; see Müller and Quintana (2004) and Dunson (2007) for reviews, including references to generalizations, for example Polya trees.

Dirichlet process: descriptive definition

Ferguson (1973) gave the following non-constructive definition of a DP, characterizing it through marginal Dirichlet distributions. Let $\Theta \subseteq \mathbb{R}^d, d \geq 1$, denote some scalar or vector-valued sample space or, as in our context, a parameter space. Then let G_0 be a probability distribution on Θ and $\nu_0 > 0$ a scalar parameter.

Per definition, a random probability measure G is generated by a DP if for any finite (measurable) partition $A_1 \cup \ldots \cup A_r = \Theta$ the vector $(G(A_1), \ldots, G(A_r))'$ of probabilities is Dirichlet distributed with

$$(G(A_1), \ldots, G(A_r)) \sim \text{Dir}(\nu_0 G_0(A_1), \ldots, \nu_0 G_0(A_r)).$$

We write $G \sim \text{DP}(\nu_0, G_0)$, where G_0 is the *base distribution* and ν_0 the *concentration* (or *strength*) parameter.

The interpretation of G_0 and ν_0 follows directly from the properties of the Dirichlet distribution:

$$\mathbb{E}(G(A)) = G_0(A)$$

$$\text{Var}(G(A)) = \frac{G_0(A)(1 - G_0(A))}{\nu_0 + 1}$$

for any (measurable) $A \subseteq \Theta$.

Thus, $G_0(A)$ is basically the mean of the DP, and the concentration parameter ν_0 can be understood as an inverse variance parameter. The larger ν_0 is, the smaller the variance, and the DP will concentrate more of its mass around G_0. Indeed, $G(A) \to G_0(A)$ for $\nu_0 \to \infty$, but this does not imply $G \to G_0$. On the contrary, realisations from a DP are *discrete distributions* with probability one, even if G_0 is continuous. This becomes immediately evident by the following constructive definition of a DP.

Stick breaking representation

Sethuraman (1994) showed that a DP can be constructed through a *stick breaking process*. Any $G \sim \text{DP}(\nu_0, G_0)$ can be represented as

$$G(\cdot) = \sum_{k=1}^{\infty} \pi_k \delta_{\phi_k}(\cdot),$$

where the (random) locations ϕ_k of the point masses $\delta_{\phi_k}(\cdot)$ are generated from the base distribution G_0, that is,

$$\phi_k \overset{\text{i.i.d.}}{\sim} G_0,$$

independently from the (random) weights π_k. The weights are generated through a *stick breaking process*:

$$\pi_1 = v_1, \quad \pi_k = v_k \left(1 - \sum_{j=1}^{k-1}(1-\pi_j)\right) = v_k \prod_{j=1}^{k-1}(1-v_j), \quad k = 2, 3, \ldots,$$

with

$$v_k \overset{\text{i.i.d.}}{\sim} \text{Be}(1, \nu_0).$$

Thus, realizations G of the DP can be obtained as infinite mixtures of point masses. The locations $\boldsymbol{\phi}_k$ are an i. i. d. sample from G_0, and the weights π_k are generated through a 'stick breaking procedure', first breaking the part $\pi_1 = v_1$ from a stick of the length $1 = \pi_1 + \pi_2 + \ldots$, then breaking off the part $\pi_2 = v_2(1-\pi_1)$ from the remaining stick of length $1-\pi_1$, and so on. The stick breaking procedure guarantees that $\sum_{k=1}^{\infty} \pi_k = 1$ (almost surely).

The stick breaking presentation immediately shows that a DP is (almost surely) a discrete RPM, and how to generate realizations G from $\text{DP}(\nu_0, G_0)$. The concentration parameter ν_0 controls the amount of discreteness: For small ν_0, realizations of G look quite different from G_0, while for ν_0 large, realizations of G appear like a discrete approximation to the (continuous) base distribution G_0 (see Fig. 3.7 for an illustration with base measure $G_0 = \text{N}(0,1)$).

This property of a DP is also reflected by the fact that (Ishwaran and James, 2001)

$$\mathbb{E}(1 - (\pi_1 + \ldots + \pi_K)) = \mathbb{E}\left(\sum_{k=K+1}^{\infty} \pi_k\right) = \left(\frac{\nu_0}{\nu_0 + 1}\right)^K \to 0$$

for the sum of weights of point masses beyond the point mass in $\boldsymbol{\phi}_K$: Convergence to 0 is fast for ν_0 small, and it is comparably slow for ν_0 large.

This also motivates the definition of *truncated DPs*, defined by truncation of the stick breaking process after $\boldsymbol{\phi}_K$ by setting $v_K = 1$:

$$G_K(\cdot) \sim \sum_{k=1}^{K} \pi_k \delta_{\boldsymbol{\phi}_k}(\cdot),$$

where $\boldsymbol{\phi}_k \overset{\text{i.i.d.}}{\sim} G_0$, and $\pi_k, k = 1, \ldots, K-1$, are given by the truncated stick breaking procedure.

The stick breaking construction of a DP and its truncated version are quite useful in constructing MCMC algorithms. Comparing with the finite random measure representation (3.27) of finite mixture models in the previous section, we recognize the similarity between $G \sim \text{DP}(\nu_0, G_0)$ and the random measure introduced there.

Another important implication of the stick breaking representation is the clustering property of DP processes: If we draw a sample $\theta_i, i = 1, \ldots, n$, from (a realization) of G, this creates ties among the θ_i, with clusters of θ_i sharing the same value $\boldsymbol{\phi}_k$, say. This clustering property becomes even more evident in the following.

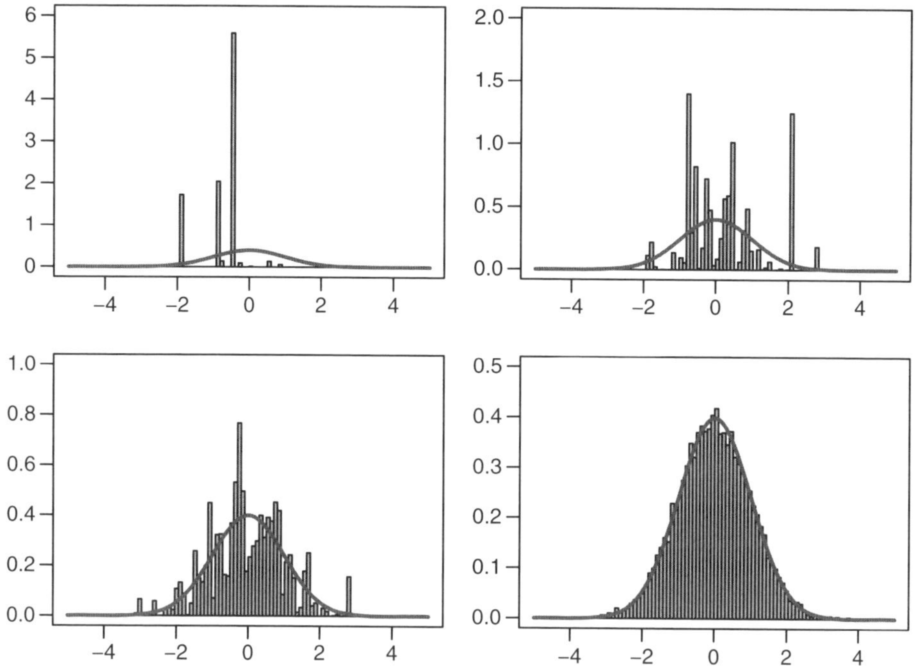

Fig. 3.7 Influence of the concentration parameter ν_0 on realizations of a Dirichlet process with base measure $G_0 = \mathrm{N}(0, 1)$. The chosen values for ν_0 are $\nu_0 = 1$ (top left), $\nu_0 = 10$ (top right), $\nu_0 = 100$ (bottom left) and $\nu_0 = 10000$ (bottom right).

Posterior Dirichlet process

Suppose that we sample G from a DP, and then draw an i. i. d. sample $\theta_1, \dots, \theta_n$, with $\theta_i \in \Theta$, from G itself. What is the posterior distribution of G given the observed values $\theta_1, \dots, \theta_n$? Let $A_1, \dots, A_r$ be a finite (measurable) partition of Θ and let $n_1, \dots, n_r$ be the number of observed values in the partitioning sets. By the conjugacy between the Dirichlet and the multinomial distribution we obtain

$$(G(A_1), \dots, G(A_r))|\theta_1, \dots, \theta_n \sim \mathrm{Dir}(\nu_0 G_0(A_1) + n_1, \dots, \nu_0 G_0(A_r) + n_r).$$

This shows that the posterior must be a DP again. Rewriting $n_k = \sum_{i=1}^n \delta_{\theta_i}(A_k)$, where $\delta_{\theta_i}(\cdot)$ is a point mass located at θ_i, it is easily seen that the posterior is a DP with updated parameters:

$$G|\theta_1, \dots, \theta_n \sim \mathrm{DP}\left(\nu_0 + n, \frac{\nu_0}{\nu_0 + n} G_0 + \frac{n}{\nu_0 + n} \frac{1}{n} \sum_{i=1}^n \delta_{\theta_i}\right) \sim \mathrm{DP}(\nu_0^*, G_0^*)$$

The posterior base distribution $G_0^*(\cdot)$ is a weighted average of the (prior) base distribution $G_0(\cdot)$ and the empirical distribution $(\sum_{i=1}^n \delta_{\theta_i}(\cdot))/n$. The weights are controlled by the concentration parameter ν_0: The larger ν_0, the larger the weight for G_0 in comparsion to the weight for the empirical distribution and vice versa. For $n \gg \nu_0$, the empirical distribution will dominate. For $n \to \infty$, this gives the consistency property of a DP: the posterior DP converges to the true underlying distribution over Θ.

Predictive distribution and clustering property

Consider now the predictive distribution for $\theta_{n+1}|\theta_1, \dots, \theta_n$, with G marginalized out. We have

$$\begin{aligned} \mathbb{P}(\theta_{n+1} \in A|\theta_1, \dots, \theta_n) &= \mathbb{E}(G(A)|\theta_1, \dots, \theta_n) \\ &= G_0^*(A) \\ &= \frac{\nu_0}{\nu_0 + n} G_0(A) + \frac{n}{\nu_0 + n} \frac{1}{n} \sum_{i=1}^{n} \delta_{\theta_i}(A). \end{aligned}$$

Therefore, the predictive distribution $p(\theta_{n+1}|\theta_1, \dots, \theta_n)$ is the posterior base distribution $G_0^*(A)$. Sequentially drawing $\theta_1, \dots, \theta_n, \theta_{n+1}$ can be interpreted as an urn (the Blackwell–MacQueen) scheme: In the beginning we draw $\theta_1 \sim G_0$, label a ball with 1 and drop it into an urn. In step $n + 1$, we either draw θ_{n+1} with probability $\nu_0/(\nu_0 + n)$ from G_0, label a new ball with $n + 1$, and drop it into the urn, or, with probability $n/(\nu_0 + n)$ randomly pick out a ball from the urn (i.e. draw θ_{n+1} from the empirical distribution), label a new ball with the label of the ball drawn, say k, and drop balls back into the urn. This obviously reflects the clustering property of a DP, which is due to the discreteness of the DP and not to G_0, which will usually be assumed to be continuous.

Formally, the clustering property can also be seen as follows. Let $\theta_1^*, \dots, \theta_m^*$, $m \leq n$, be the unique values (the labels) among $\theta_1, \dots, \theta_n$, and n_k the number of repeats of θ_k^*. In the stick breaking representation, $\theta_1^*, \dots, \theta_m^*$ correspond to m different values ϕ_k, drawn from G_0, where ϕ_k with large π_k have better chances to be among the $\theta_1^*, \dots, \theta_m^*$. The predictive distribution can then equivalently be expressed as

$$\theta_{n+1}|\theta_1, \dots, \theta_n \sim \frac{1}{\nu_0 + n} \left(\nu_0 G_0(\cdot) + \sum_{k=1}^{m} n_k \delta_{\theta_k^*}(\cdot) \right).$$

An existing value θ_k^* will be drawn for θ_{n+1} with probability proportional to n_k, while a new value will be drawn from G_0^* with probability proportional to ν_0. This motivates the interpretation as a *Chinese restaurant process*: Guests arrive at a Chinese restaurant. Guest 1 sits down at table 1 ($\theta_1 = \theta_1^*$). When guest $n + 1$ arrives, m tables (labelled $\theta_k^*, k = 1, \dots, m$) are occupied, each with n_k clients. Guest $n + 1$ chooses one of these tables with probability proportional to n_k, or sits down at a new table, labelled with $\theta_{n+1} \sim G_0$, with probability proportional to ν_0.

Dirichlet process mixtures

Due to the discreteness and the clustering property of a DP, choosing a DP directly (i.e. with $\boldsymbol{\alpha}_i = \theta_i$) as a prior for the unknown random effects distribution $p(\cdot)$ in mixed models will automatically create ties among the $\boldsymbol{\alpha}_i, i = 1, \ldots, n$, and will form clusters of subjects having the same random effect $\boldsymbol{\alpha}_k^*$, say. This leads to a nonparametric Bayesian version of latent class mixed models, and may be a useful approximation in some applications. In other situations, it will be more realistic to allow clusters with similar subjects as in finite Gaussian mixture models, where each subject still has its own unique random effect value. This can be accomplished with Dirichlet process mixtures (DPM). The simple idea is: Consider $\theta_1, \ldots, \theta_n$ generated from a DP prior $G \sim \text{DP}$ as latent parameters of a continuous random effects prior $p(\boldsymbol{\alpha}_i|\theta_i)$, and, given θ_i, draw $\boldsymbol{\alpha}_i$ from $p(\boldsymbol{\alpha}_i|\theta_i)$. In hierarchical form we have

$$\begin{aligned} G &\sim \text{DP}(\nu_0, G_0) \\ \theta_i|G &\sim G \\ \boldsymbol{\alpha}_i|\theta_i &\overset{\text{ind.}}{\sim} p(\boldsymbol{\alpha}_i|\theta_i). \end{aligned}$$

Additionally, hyperpriors may be specified for ν_0 and hyperparameters in G_0, such as $\boldsymbol{\mu}_0$ and $\boldsymbol{Q}_0$ if the base distribution is Gaussian. With the stick breaking construction

$$G(\cdot) = \sum_{k=1}^{\infty} \pi_k \delta_{\boldsymbol{\phi}_k}(\cdot)$$

a DPM can be represented as an infinite mixture model

$$p(\boldsymbol{\alpha}_i|\boldsymbol{\pi}, \boldsymbol{\phi}) \overset{\text{i.i.d.}}{\sim} \sum_{k=1}^{\infty} \pi_k p(\boldsymbol{\alpha}_i|\boldsymbol{\phi}_k),$$

reducing to a finite mixture model with K large for a truncated DP. Now the close relationship to the representation (3.28) of finite mixture models in Section 3.2.1 becomes apparent. As in that section, we may also introduce latent class indicators c_i, with $c_i = k \Leftrightarrow \theta_i = \boldsymbol{\phi}_k$, or shortly $\theta_i = \boldsymbol{\phi}_{c_i}$. Then the DPM, with DP truncated at K, is defined by the hierarchy

$$\begin{aligned} c_i|\boldsymbol{\pi} &\overset{\text{i.i.d.}}{\sim} \sum_{k=1}^{K} \pi_k \boldsymbol{\phi}_k, \\ \boldsymbol{\alpha}_i &\overset{\text{ind.}}{\sim} p(\boldsymbol{\alpha}_i|\boldsymbol{\phi}_{c_i}), \end{aligned}$$

where $\boldsymbol{\pi}$ and $\boldsymbol{\phi}$ are generated as in the (truncated) stick breaking representation.

3.2.3 LMM with DP-based random effects priors

We consider LMMs for longitudinal data in the reparametrized form (3.25)

$$\boldsymbol{y}_i = \boldsymbol{U}_i\boldsymbol{\beta}_i + \boldsymbol{Z}_i(\boldsymbol{\alpha} + \tilde{\boldsymbol{\alpha}}_i) + \boldsymbol{\varepsilon}_i = \boldsymbol{U}_i\boldsymbol{\beta} + \boldsymbol{Z}_i\boldsymbol{\alpha}_i + \boldsymbol{\varepsilon}_i, \quad \boldsymbol{\varepsilon}_i \sim \mathrm{N}(\boldsymbol{0}, \sigma^2\boldsymbol{I}), \tag{3.28}$$

$i = 1, \ldots, n$, where $\boldsymbol{\beta}$ are the fixed effects of the covariates $\boldsymbol{u}_{ij}$ without random effects, $\boldsymbol{\alpha}$ are the fixed (or population) effects of covariates $\boldsymbol{z}_{ij}$ which have additional subject-specific random deviations $\tilde{\boldsymbol{\alpha}}_i$ from $\boldsymbol{\alpha}$, with $\mathbb{E}(\tilde{\boldsymbol{\alpha}}_i) = 0$, and $\boldsymbol{\alpha}_i = \boldsymbol{\alpha} + \tilde{\boldsymbol{\alpha}}_i$ are the uncentred random effects.

We distinguish two types of nonparametric prior distributions for random effects: First, *DP priors* implying discrete random effects distributions with clusters of subjects sharing the same random effects vector. Second, *DPM priors* for continuous random effects distributions, with the potential of clustering subjects with similar but not identical random effects. LMMs with DP priors can be considered as semiparametric extensions of latent class mixed models, and LMMs with DPM priors as extensions of the finite mixture heterogeneity model in Section 3.2.1. In both cases, it seems natural to assume a Gaussian base distribution G_0 for DPs.

Longitudinal data LMMs with DP random effects priors

LMMs with DP priors are defined hierarchically through the observation model (3.28) and the following prior specifications for fixed and random effects:

$$\begin{aligned} \boldsymbol{\beta}|\boldsymbol{c}, \boldsymbol{C} &\sim \mathrm{N}(\boldsymbol{c}, \boldsymbol{C}) \\ \boldsymbol{\alpha}_i|G &\sim G \\ G|\nu_0, G_0 &\sim \mathrm{DP}(\nu_0, G_0) \\ G_0 &\sim \mathrm{N}(\boldsymbol{\mu}_0, \boldsymbol{Q}_0) \end{aligned} \tag{3.29}$$

Setting $\boldsymbol{c} = \boldsymbol{0}$ and $\boldsymbol{C}$ large gives a weakly informative prior for $\boldsymbol{\beta}$, with a flat prior $p(\boldsymbol{\beta}) \propto \text{const}$ as the limiting case $\boldsymbol{C}^{-1} \to \boldsymbol{0}$. To increase flexibility, we may assume the fixed effects hyperpriors

$$\boldsymbol{c} \sim \mathrm{N}(\boldsymbol{m_c}, \boldsymbol{S_c}), \quad \boldsymbol{S_c} = \mathrm{diag}(s^2_{c_1}, \ldots, s^2_{c_m}) \tag{3.30}$$

with default values such as $\boldsymbol{m_c} = \boldsymbol{0}$, $\boldsymbol{S_c} = 1000\boldsymbol{I}$, and

$$\boldsymbol{C} = \mathrm{diag}(\sigma^2_{\beta_1}, \ldots, \sigma^2_{\beta_m}), \quad \sigma^2_{\beta_l} \overset{\text{i.i.d.}}{\sim} \mathrm{IG}(a_{\boldsymbol{\beta}}, b_{\boldsymbol{\beta}}).$$

For the hyperparameters $\boldsymbol{\mu}_0$ and $\boldsymbol{Q}_0$ of the DP prior base distribution G_0 we assume

$$\boldsymbol{\mu}_0 \sim \mathrm{N}(\boldsymbol{m}_0, \boldsymbol{S}_0), \quad \boldsymbol{S}_0 = \mathrm{diag}(s^2_{01}, \ldots, s^2_{0q}) \tag{3.31}$$

with default values $\boldsymbol{m}_0 = \boldsymbol{0}$, $\boldsymbol{S}_0 = 100\boldsymbol{I}$, and

$$\boldsymbol{Q}_0 = \text{diag}(\delta_{01}^2, \ldots, \delta_{0q}^2), \quad \delta_{0l}^2 \overset{\text{i.i.d.}}{\sim} \text{IG}(a_0, b_0), \quad l = 1, \ldots, q.$$

As an alternative, an inverse Wishart prior $\boldsymbol{Q}_0 \sim \text{IW}(a_0, \boldsymbol{B}_0 = b_0\boldsymbol{I})$ is sometimes specified, for example in the R-package `DPlmm`. For the concentration parameter ν_0, a Gamma prior $\nu_0 \sim \text{Ga}(a_\nu, b_\nu)$ is often suggested. To avoid difficulties with too small values for ν_0 in Gibbs sampling based on the stick breaking presentation, we consider a discrete prior

$$\nu_0 \sim \sum_{\omega \in \Omega} \mathbb{P}(\nu_0 = \omega)\delta_\omega \tag{3.32}$$

with support $\Omega = \{0.5, 0.6, \ldots, 100\}$ and probabilities which resemble a Gamma distribution. For the error variance we choose

$$\sigma^2 \sim \text{IG}(a_\sigma, b_\sigma).$$

The stick breaking representation

$$G(\cdot) = \sum_{k=1}^{\infty} \pi_k \delta_{\boldsymbol{\phi}_k}(\cdot)$$

or its truncated approximation makes the discreteness of the random effects prior and the clustering property evident.

A Bayesian LMM with a similar hierarchical structure was first proposed by Kleinman and Ibrahim (1998a), see also Ibrahim and Kleinman (1998). They assumed a LMM in original parametrization, corresponding to a normal random effects prior with $\mathbb{E}(\boldsymbol{\alpha}_i) = \boldsymbol{0}$, and specified a centred Gaussian base distribution $G_0 \sim \text{N}(\boldsymbol{0}, \boldsymbol{Q}_0)$ instead of an uncentred G_0 in (3.29). However, constraining the base distribution to have zero mean will not ensure that G is centred about zero *a posteriori*. The `DPpackage` in `R` (Jara, 2007) is based on the reparametrized observation model (3.28) together with prior specifications as in (3.29). Two recent alternatives are proposed in Li, Lin and Müller (2010), using post-processing of the sampled parameters to adjust for remaining bias, and in Dunson, Yang and Baird (2007), using a centred DP prior.

Posterior inference can be based on the marginal approach, integrating out G as in the Blackwell–MacQueen urn scheme, or on the conditional approach using stick breaking priors. The marginal approach is faster, but the conditional approach provides richer MCMC output. In any case, the full conditionals for $\boldsymbol{\beta}$ and σ^2 are

$$\boldsymbol{\beta}|\boldsymbol{\alpha}, \sigma^2, \boldsymbol{y} \sim \text{N}(\boldsymbol{\mu}_{\boldsymbol{\beta}}, \boldsymbol{\Sigma}_{\boldsymbol{\beta}})$$

with

$$\boldsymbol{\Sigma}_{\boldsymbol{\beta}} = \left(\frac{1}{\sigma^2} \sum_{i=1}^{n} \boldsymbol{U}_i' \boldsymbol{U}_i + \boldsymbol{C}^{-1} \right)^{-1},$$

$$\boldsymbol{\mu}_{\boldsymbol{\beta}} = \frac{1}{\sigma^2} \boldsymbol{\Sigma}_{\boldsymbol{\beta}} \left(\sum_{i=1}^{n} \boldsymbol{U}_i' (\boldsymbol{y}_i - \boldsymbol{Z}_i \boldsymbol{\alpha}_i) + \sigma^2 \boldsymbol{C}^{-1} \boldsymbol{c} \right),$$

and

$$\sigma^2 | \boldsymbol{\beta}, \boldsymbol{\alpha}, \boldsymbol{y} \sim \mathrm{IG}(\tilde{a}_\sigma, \tilde{b}_\sigma)$$

with

$$\tilde{a}_\sigma = a_\sigma + \frac{1}{2} \sum_{i=1}^{n} n_i, \quad \tilde{b}_\sigma = b_\sigma + \frac{1}{2} \sum_{i=1}^{n} (\boldsymbol{y}_i - \boldsymbol{U}_i \boldsymbol{\beta} - \boldsymbol{Z}_i \boldsymbol{\alpha}_i)' (\boldsymbol{y}_i - \boldsymbol{U}_i \boldsymbol{\beta} - \boldsymbol{Z}_i \boldsymbol{\alpha}_i).$$

For the marginal approach, as used in Escobar and West (1995), Kleinman and Ibrahim (1998b), and in the `DPpackage`, we obtain

$$p(\boldsymbol{\alpha}_i | \boldsymbol{\beta}, \sigma^2, \boldsymbol{\alpha}_{-i}, \boldsymbol{y}) \propto \sum_{j \neq i} \varphi(\boldsymbol{y}_i | \boldsymbol{U}_j \boldsymbol{\beta} + \boldsymbol{Z}_j \boldsymbol{\alpha}_j, \sigma^2 \boldsymbol{I}) \delta_{\boldsymbol{\alpha}_j}$$
$$+ \ \nu_0 H_i \varphi(\boldsymbol{\alpha}_i | \boldsymbol{\mu}_0, \boldsymbol{Q}_0) \varphi(\boldsymbol{y}_i | \boldsymbol{U}_i \boldsymbol{\beta} + \boldsymbol{Z}_i \boldsymbol{\alpha}_i, \sigma^2 \boldsymbol{I}),$$

where $\boldsymbol{\alpha}_{-i}$ are the random effects for the subjects excluding subject i, and

$$H_i = \int \varphi(\boldsymbol{y}_i | \boldsymbol{U}_i \boldsymbol{\beta} + \boldsymbol{Z}_i \boldsymbol{\alpha}_i, \sigma^2 \boldsymbol{I}) \varphi(\boldsymbol{\alpha}_i | \boldsymbol{\mu}_0, \boldsymbol{Q}_0) d\boldsymbol{\alpha}_i.$$

Thus, with probability proportional to $\varphi(\boldsymbol{y}_i | \boldsymbol{U}_i \boldsymbol{\beta} + \boldsymbol{Z}_i \boldsymbol{\alpha}_i; \sigma^2 \boldsymbol{I})$ we set $\boldsymbol{\alpha}_i = \boldsymbol{\alpha}_j$ and with probability proportional to $\nu_0 H_i$ we sample from the Gaussian full conditional

$$\boldsymbol{\alpha}_i | \boldsymbol{\beta}, \sigma^2, \boldsymbol{y}_i \propto \varphi(\boldsymbol{\alpha}_i | \boldsymbol{\mu}_0, \boldsymbol{Q}_0) \varphi(\boldsymbol{y}_i | \boldsymbol{U}_i \boldsymbol{\beta} + \boldsymbol{Z}_i \boldsymbol{\alpha}_i, \sigma^2 \boldsymbol{I}).$$

Note that the integral for H_i can be evaluated analytically since it only involves Gaussian densities.

In the full conditionals for $\boldsymbol{\beta}, \sigma^2$ and $\boldsymbol{\alpha}_i, i = 1, \ldots, n$, we have (tacitly) assumed that $\nu_0, \boldsymbol{\mu}_0$ and $\boldsymbol{Q}_0$ are given. Additionally, steps for drawing from corresponding full conditionals for these hyperparameters can be based on Escobar and West (1995) for ν_0, and on Kleinman and Ibrahim (1998a). The implementation of the DPlmm model in the `DPpackage` (Jara, 2007) follows these lines.

As an alternative, a conditional sampler based on a (truncated) stick breaking prior and latent class indicators can be derived. It is a special case of Algorithm 4.6 for semiparametric mixed models in Section 4.4.1 (page 238).

Longitudinal data LMMs with DPM priors

For LMMs with DPM priors for continuous random effects distributions, the hierarchical specification of priors can be extended as follows:

$$\begin{aligned} \boldsymbol{\alpha}_i|\boldsymbol{\mu}_i &\overset{\text{ind.}}{\sim} \mathrm{N}(\boldsymbol{\mu}_i, \boldsymbol{Q}) \\ \boldsymbol{\mu}_i|G &\sim G \\ G|\nu_0, G_0 &\sim \mathrm{DP}(\nu_0, G_0) \\ G_0 &\sim \mathrm{N}(\boldsymbol{\mu}_0, \boldsymbol{Q}_0). \end{aligned} \tag{3.33}$$

We assume either $\boldsymbol{Q} = \mathrm{diag}(\delta_1^2, \ldots, \delta_q^2)$, with $\delta_l^2 \overset{\text{i.i.d.}}{\sim} \mathrm{IG}(a_{\boldsymbol{\alpha}}, b_{\boldsymbol{\alpha}})$, or again an inverse Wishart prior for non-diagonal $\boldsymbol{Q}$ as in the R-package `DPMlmm`. The remaining priors are the same as for LMMs with DP priors. The model DPMlmm in the `DPpackage` (Jara, 2007) and a corresponding MCMC algorithm marginalising out G are based on this prior specification. Assuming the hyperpriors (3.30), (3.31), (3.32), a conditional block Gibbs sampler is obtained as a special case of Algorithm 4.6 for semiparametric mixed models in Section 4.4.1 (page 238), omitting step i to update basis coefficients of nonlinear functions in the predictor.

Example 3.5: Height of schoolgirls

In Example 3.3 we assumed a LMM with a Gaussian prior for random intercepts and slopes. To check this prior assumption and to see how sensitive results depend on it, we replace it with a flexible DPM prior as specified in (3.33). We reanalyse the data with the R-package `DPMlmm` and, for comparison, with the conditional block Gibbs sampler (block DPM) of Section 4.4.1 using the stick breaking presentation truncated at $K = 200$ for the underlying DP prior. In both cases, we assume a hyperprior $\sigma^2 \sim \mathrm{IG}(0.005, 0.005)$ for the error variance. For the `DPMlmm`-package, inverse Wishart priors are assumed for $\boldsymbol{Q}$ and $\boldsymbol{Q}_0$, with hyperparameters $a = a_0 = 4.01$ and $\boldsymbol{B} = \boldsymbol{B}_0 = \mathrm{diag}(10, 10)$ (which is the default option in DPMlmm). For the conditional block Gibbs sampler, we specify diagonal covariance matrices $\boldsymbol{Q}$ and $\boldsymbol{Q}_0$, with inverse gamma priors for the variances and all hyperparameters set to 0.005 (again a default option). For the concentration parameter ν_0, a gamma prior is assumed in the `DPMlmm`-package, while we assume a discrete prior, approximating a gamma prior, for the block DPM sampler. Due to the different prior settings as well as sampling schemes, results from both DPM models differ in some details, but are generally close. In particular, estimated population trends and selected individual-specific trends in Fig. 3.8 are in close agreement with Example 3.3. Also, kernel density estimates of random intercepts and slopes in Fig. 3.9 are in close agreement with those obtained with Gaussian random effects priors in Example 3.3. Thus, at least in this example, results for fixed and random effects are not very sensitive with regard to the prior assumption on the random effects distribution. △

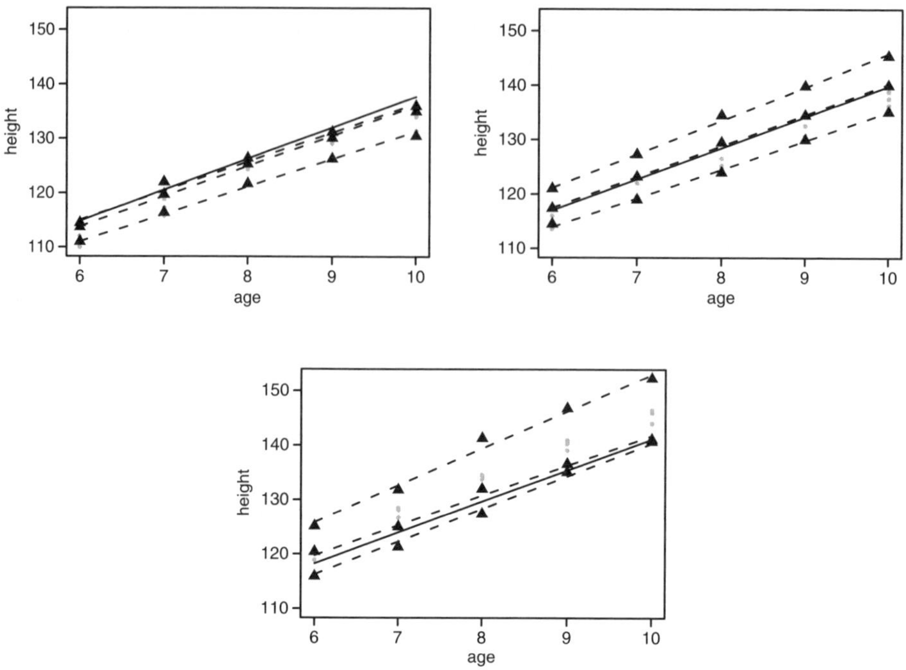

Fig. 3.8 Height of schoolgirls: Estimated population trends (solid line) together with selected individual-specific growth trajectories (dashed lines) for three individuals from each of the groups with small (top left), medium (top right) and tall (bottom) mothers obtained with block DPM. The individual observations are visualized as triangles.

Example 3.6: Childhood obesity

We reanalyse the obesity data, replacing the Gaussian random effects prior in Example 3.4 through a DPM prior and assuming nonlinear individual-specific random curves $g_i(t)$ defined in (3.7). With a DPM prior, however, the random coefficients of these curves are no longer centred around zero, compare the reparametrized form (3.28) of LMMs. To avoid identifiability problems, we therefore have to delete the population trend $f(t)$ from the predictor. The remaining covariates with fixed effects are the same as in Example 3.4. The results in Table 3.4 and Figs. 3.10 and 3.11 have been obtained with the conditional block Gibbs sampler, that is, Algorithm 4.6 (after deleting step i) described in full detail in Section 4.4.1. The estimates for the fixed effects in Table 3.4 differ to some extent from corresponding FB estimates in Table 3.2. All estimates lie within the credibility intervals of Table 3.2, however. Astonishingly, agreement with the EB estimates is much better. Note also that credibility intervals in Table 3.4 are partly considerably smaller. The population curve (solid line) in Fig. 3.4 has been obtained by averaging over all individual curves and comes

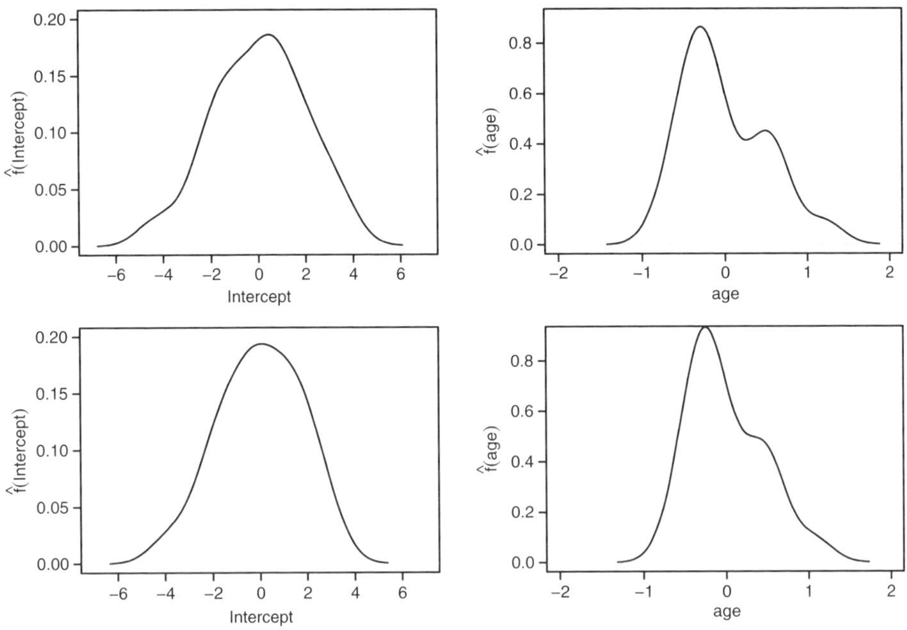

Fig. 3.9 Height of schoolgirls: Kernel density estimates for the estimated random intercepts (left) and random slopes (right) obtained with DPMlmm (top) and block DPM (bottom).

close to the population trend in Example 3.4. Selected individual curves differ a bit more. In particular, there is one individual curve that further increases after the usual peak at the end of the first year, instead of decreasing. It turns out that this individual belongs to a small cluster of individuals, represented by the non-decreasing cluster-curve in Fig. 3.11. All other cluster-curves have shapes similar to the population curve. We discuss this cluster detecting feature of DPM modelling, which is not possible with Gaussian priors, in more detail in Example 4.8 (page 242), where the population trend will be modelled through a Bayesian P-spline. It will also be seen that the results there are in good agreement with the ones just discussed. △

3.3 Generalized linear mixed models

Generalized linear mixed models (GLMM) extend regression models for non-normal responses, such as generalized linear models and categorical regression models, by incorporating random effects in the linear predictor in complete analogy as for LMMs. Apart from special cases, however, the close and simple relation between conditional and marginal models is lost, implying differences in interpretation and additional methodological problems for statistical inference.

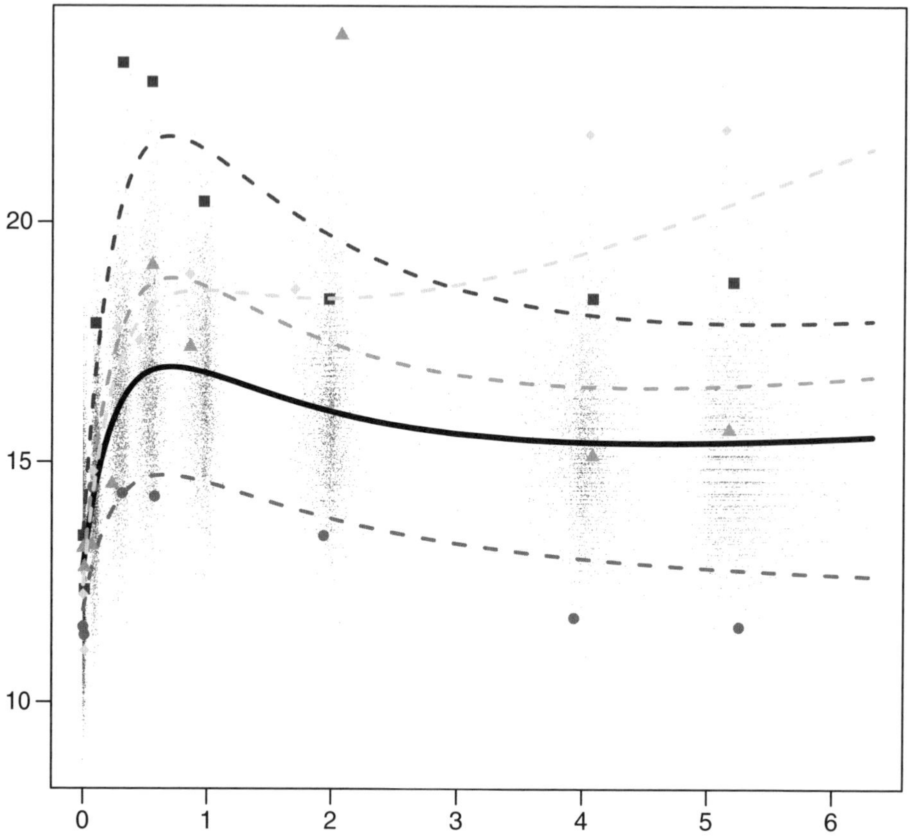

Fig. 3.10 Childhood obesity: DPM mixed model with random slopes.

Table 3.4 Childhood obesity: Estimates for fixed effects and the error variance in the DPM model. mBMI and mDiffBMI have been centred for the analyses.

	mean	median	sd.	2.5%	97.5%
sex	0.2080	0.2087	0.0379	0.1811	0.2332
breast	0.0617	0.0605	0.0385	0.0350	0.0882
mSmoke	0.0011	0.0024	0.0557	−0.0404	0.0409
area	−0.0356	−0.0348	0.0472	−0.0669	−0.0062
mBMI	0.0497	0.0496	0.0044	0.0468	0.0528
mDiffBMI	0.0846	0.0846	0.0114	0.0770	0.0922
σ^2	0.7180	0.7182	0.0099	0.7113	0.7240

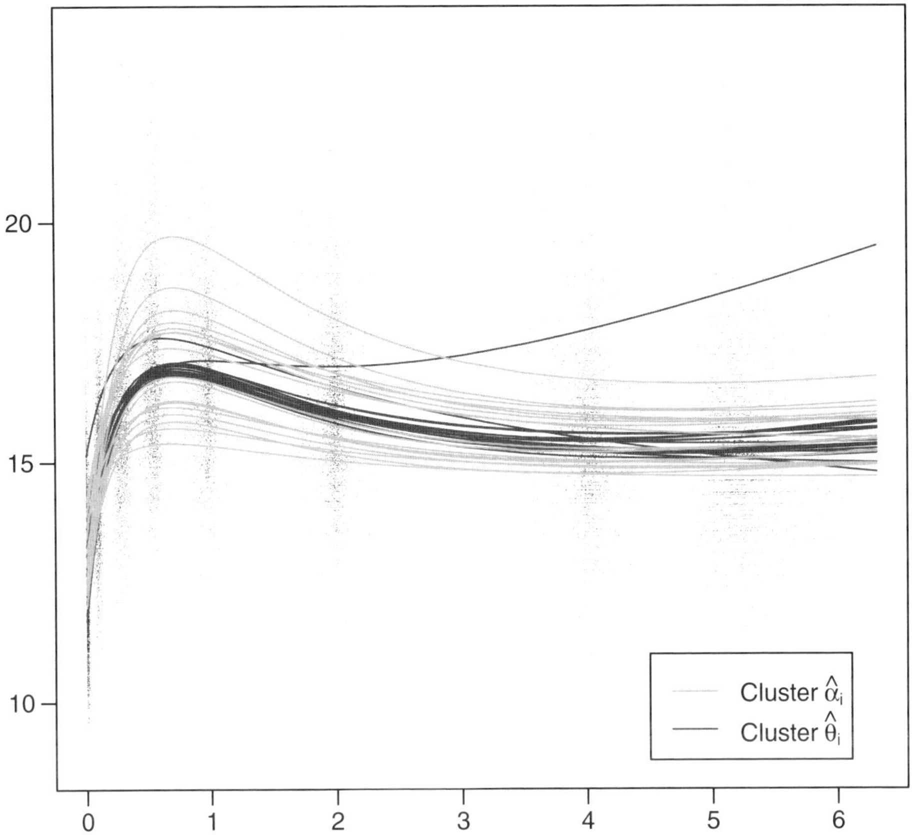

Fig. 3.11 Childhood obesity: Clustering of $\hat{\boldsymbol{\theta}}_i$ and $\hat{\boldsymbol{\alpha}}_i$ for the DPM model.

3.3.1 Generalized linear mixed models for longitudinal data

GLMMs for univariate responses

For univariate non-normal responses $y_{ij}, i = 1, \ldots, n, j = 1, \ldots, n_i$, GLMMs are defined hierarchically in two stages, combining the specification of GLMs and LMMs as follows:

At the *first stage* it is assumed that, given the random effects $\boldsymbol{\alpha}_i$ (as well as covariates), the responses y_{ij} are conditionally independent with exponential family densities

$$f(y_{ij}|\boldsymbol{\alpha}_i) = \exp\left(\frac{y_{ij}\theta_{ij} - b(\theta_{ij})}{\phi} + c(y_{ij}, \phi)\right) \tag{3.34}$$

with natural parameter θ_{ij} and scale (or dispersion) parameter ϕ. The conditional mean $\mu_{ij} = \mathbb{E}(y_{ij}|\boldsymbol{\alpha}_i)$ is linked to the linear predictor

$$\eta_{ij} = \boldsymbol{u}'_{ij}\boldsymbol{\beta} + \boldsymbol{z}'_{ij}\boldsymbol{\alpha}_i$$

through

$$\mu_{ij} = \mathbb{E}(y_{ij}|\boldsymbol{\alpha}_i) = h(\eta_{ij}),$$

where h is a known response function as in GLMs. Hence, conditional on the random effects, we end up with a GLM that contains additional individual-specific effects while the structure relating the predictor and the expectation remains unchanged between GLMs and GLMMS.

At the *second stage*, the observation model is supplemented by a distributional assumption for the i. i. d. random effects $\boldsymbol{\alpha}_i$. The standard specification assumes i. i. d. Gaussian random effects

$$\boldsymbol{\alpha}_i|\boldsymbol{Q} \overset{\text{i.i.d.}}{\sim} \text{N}(\boldsymbol{0}, \boldsymbol{Q}),$$

as for LMMs. This assumption again enforces regularized estimation of the individual-specific effects and can alternatively be considered as a penalty or as prior information depending on the context.

Special GLMMs are obtained by specific assumptions on the distribution of response variables, the response function, and the linear predictor.

Binary mixed models are defined through

$$\pi_{ij} = \mathbb{P}(y_{ij} = 1|\eta_{ij}) = h(\boldsymbol{u}'_{ij}\boldsymbol{\beta} + \boldsymbol{z}'_{ij}\boldsymbol{\alpha}_i),$$

with the *logit mixed model* resulting from the logistic distribution function $h(\eta) = \exp(\eta)/(1 + \exp(\eta))$, and the *probit mixed model* resulting from the standard normal distribution function $h(\eta) = \Phi(\eta)$. As for binary regression models with nonparametric effects (Section 2.2.2), both models can be derived from a latent LMM

$$l_{ij} = \boldsymbol{u}'_{ij}\boldsymbol{\beta} + \boldsymbol{z}'_{ij}\boldsymbol{\alpha}_i + \varepsilon_{ij}$$

in combination with a threshold mechanism

$$y_{ij} = 1 \quad \Leftrightarrow \quad l_{ij} \geq 0,$$

linking the unobservable l_{ij} to the observable response y_{ij}. The logit and the probit mixed model are obtained by assuming a logistic distribution or a standard normal distribution for the errors ε_{ij}, respectively. In combination with Gaussian random effects, the models are also called *logit-normal* or *probit-normal* models.

For count variables, a *loglinear Poisson mixed model* $y_{ij}|\boldsymbol{\alpha}_i \sim \text{Po}(\lambda_{ij})$ with rate

$$\lambda_{ij} = \exp(\boldsymbol{u}'_{ij}\boldsymbol{\beta} + \boldsymbol{z}'_{ij}\boldsymbol{\alpha}_i)$$

is the benchmark model. In the presence of remaining overdispersion, not explained through covariates and random effects, a *loglinear negative binomial model* might be considered, alternatively. Note that the loglinear negative binomial model may actually be equivalently represented as a loglinear Poisson model with individual-specific i. i. d. gamma random effects and therefore can be seen as a combination of different types of random effects, see for example Fahrmeir and Osuna (2006).

The term GLMM is often restricted to the class of mixed models with exponential family densities (3.34) and the Gaussian specification for the random effects. More generally, GLMMs may also be defined for response distributions outside of exponential family models, and other specifications for random effects densities $p(\boldsymbol{\alpha}_i)$ may be useful. A typical example are random intercept models based on conjugate distributions for the random effects as in Gamma-Poisson and Beta-Binomial models (see e.g. the h-likelihood approach by Lee and Nelder, 1996, and Lee, Nelder and Pawitan, 2006) or non-Gaussian priors considered in Sections 3.3.5 and 4.4. In this section, we will focus on Gaussian random effects, however.

Bayesian GLMMs are obtained in complete analogy to LMMs: The random effects distribution is interpreted as a prior $p(\boldsymbol{\alpha}|\boldsymbol{Q}) = \prod_{i=1}^{n} p(\boldsymbol{\alpha}_i|\boldsymbol{Q})$ for $\boldsymbol{\alpha} = (\boldsymbol{\alpha}_1', \ldots, \boldsymbol{\alpha}_n')'$, and a flat prior $p(\boldsymbol{\beta}) \propto$ const or a (weakly) informative Gaussian prior $\boldsymbol{\beta} \sim \mathrm{N}(\boldsymbol{c}, \boldsymbol{C})$ is specified for the 'fixed effects' $\boldsymbol{\beta}$. For full Bayesian inference, priors also have to be specified for unknown parameters or elements in $\boldsymbol{Q}$ and, if necessary, for the scale parameter ϕ.

Marginal and conditional models

Conceptually, the marginal density $f(y_{ij})$ can be determined from the conditional density by integrating out random effects, that is,

$$f(y_{ij}) = \int f(y_{ij}|\boldsymbol{\alpha}_i)p(\boldsymbol{\alpha}_i)d\boldsymbol{\alpha}_i,$$

where $p(\boldsymbol{\alpha}_i)$ is the random effects density. Apart from Gaussian LMMs and rather special models using conjugate random effects distributions, this integration cannot be carried out analytically, and the marginal density is not available in closed form. This is one of the main reasons why inference in GLMMs gets more complicated compared to LMMs.

Marginal expectations

$$\mathbb{E}(y_{ij}) = \mathbb{E}(\mathbb{E}(y_{ij}|\boldsymbol{\alpha}_i)) = \mathbb{E}(h(\boldsymbol{u}_{ij}'\boldsymbol{\beta} + \boldsymbol{z}_{ij}'\boldsymbol{\alpha}_i))$$

as well as marginal variances and covariances are, in general, also not available in closed form. Moreover, because

$$\mathbb{E}(y_{ij}) = \mathbb{E}(h(\boldsymbol{u}_{ij}'\boldsymbol{\beta} + \boldsymbol{z}_{ij}'\boldsymbol{\alpha}_i)) \neq h(\boldsymbol{u}_{ij}'\boldsymbol{\beta}),$$

the fixed effects $\boldsymbol{\beta}$ of the conditional models cannot be interpreted as the population effects in a corresponding marginal model with the same response function and with fixed effects only, as it was the case in LMMs. Still, the relationship between conditional fixed effects $\boldsymbol{\beta}$ in a random intercept model and population effects $\tilde{\boldsymbol{\beta}}$ in a corresponding marginal model can be clarified to a certain extent for binary responses.

For a *probit-normal random intercept model*

$$\mathbb{E}(y_{ij}|\alpha_i) = \mathbb{P}(y_{ij} = 1|\alpha_i) = \Phi(\boldsymbol{u}'_{ij}\boldsymbol{\beta} + \alpha_i), \quad \alpha_i \sim \mathrm{N}(0, \delta^2)$$

it can be shown that

$$\mathbb{E}(y_{ij}) = \mathbb{P}(y_{ij} = 1) = \Phi(\boldsymbol{u}'_{ij}\tilde{\boldsymbol{\beta}}),$$

where $\tilde{\boldsymbol{\beta}} = \boldsymbol{\beta}/\sqrt{\delta^2 + 1}$. Thus, the marginal model is again a probit model, but with population effects $\tilde{\boldsymbol{\beta}}$ shrunk towards zero by the factor $1/\sqrt{\delta^2 + 1}$ in comparison to the fixed effects $\boldsymbol{\beta}$ in the conditional probit-normal intercept model. The amount of shrinkage increases with the variance δ^2 of the random intercept, only for $\delta^2 = 0$ we have $\boldsymbol{\beta} = \tilde{\boldsymbol{\beta}}$.

For the *logit random intercept model* there is only an approximate relation between conditional and marginal parameters with

$$\tilde{\boldsymbol{\beta}} \approx \boldsymbol{\beta}/\sqrt{c^2\delta + 1}, \quad c = 16\sqrt{3}/(15\pi) \approx 0.588$$

(Diggle *et al.*, 2002, Sec. 7.4). For the *loglinear Poisson-normal random intercept model* we have

$$\mathbb{E}(y_{ij}|\alpha_i) = \exp(\boldsymbol{u}'_{ij}\boldsymbol{\beta} + \alpha_i) = \nu_i \exp(\boldsymbol{u}'_{ij}\boldsymbol{\beta})$$

where $\nu_i = \exp(\alpha_i)$ is a log-normal random effect. Therefore,

$$\begin{aligned}\mathbb{E}(y_{ij}) &= \mathbb{E}(\nu_i \exp(\boldsymbol{u}'_{ij}\boldsymbol{\beta})) = \exp(\delta^2/2)\exp(\boldsymbol{u}'_{ij}\boldsymbol{\beta}) \\ &= \exp(\delta^2/2 + \boldsymbol{u}'_{ij}\boldsymbol{\beta}),\end{aligned}$$

so that fixed effects $\boldsymbol{\beta}$ and marginal effects are the same, with the exception of the intercept term.

Generally, however, the relationship between conditional and marginal models cannot be clarified analytically.

Interpretation of regression parameters

As shown by these examples, the impact of fixed effects $\boldsymbol{\beta}$ on the expectation of the response in GLMMs has to be interpreted carefully and differently compared to LMMs: Apart from special cases, they are *not* the population parameters in a corresponding marginal model with the same link function. They still have, however, a subject-specific interpretation in the conditional model and can also be interpreted on the level of the predictor.

Consider, for example, a logit random slope model

$$\mathbb{P}(y_{ij} = 1|\boldsymbol{\alpha}_i) = \exp(\eta_{ij})/(1 + \exp(\eta_{ij})),$$

with predictor

$$\eta_{ij} = \beta_0 + \beta_1 t_{ij} + \alpha_{0i} + \alpha_{1i} t_{ij} + \boldsymbol{u}'_{ij}\boldsymbol{\beta}_{\boldsymbol{u}}.$$

Given an individual trend

$$(\beta_0 + \alpha_{0i}) + (\beta_1 + \alpha_{1i})t_{ij},$$

the parameters $\boldsymbol{\beta}_{\boldsymbol{u}}$ have the same interpretation as covariate effects in a usual logit model when considering them on the logit scale

$$\log\left(\frac{\mathbb{P}(y_{ij} = 1|\boldsymbol{\alpha}_i)}{\mathbb{P}(y_{ij} = 0|\boldsymbol{\alpha}_i)}\right) = \eta_{ij}.$$

On this scale,

$$\mathbb{E}(\eta_{ij}) = \beta_0 + \beta_1 t_{ij} + \boldsymbol{u}'_{ij}\boldsymbol{\beta}_{\boldsymbol{u}},$$

so that fixed effects may be still interpreted as population effects, however on the level of linear predictors and not for the expectation of the response. Similarly, we can interpret the fixed effects as population effects on the loglinear predictor scale of Poisson models. This interpretation also makes particular sense in combination with latent variable representations. In this case, the mixed model is specified for the latent (Gaussian) variables and both fixed and random effects can be interpreted on the latent scale as in LMMs.

GLMMs for categorical responses

For categorical responses $Y_{ij} \in \{1, \ldots, C\}$, where C is the number of ordered or unordered categories, mixed models can be defined in analogy to GLMMs for scalar responses through appropriate extension of linear predictors in categorical response models with fixed effects only.

A *multinomial mixed logit model* for unordered categories $\{1, \ldots, C\}$ with category-specific fixed and random effects is given by

$$\mathbb{P}(Y_{ij} = c|\boldsymbol{\alpha}_i^{(1)}, \ldots, \boldsymbol{\alpha}_i^{(C-1)}) = \frac{\exp(\eta_{ij}^{(c)})}{1 + \sum_{s=1}^{C-1} \exp(\eta_{ij}^{(s)})}, \quad c = 1, \ldots, C-1,$$

with category-specific linear mixed predictors

$$\eta_{ij}^{(c)} = \boldsymbol{u}'_{ij}\boldsymbol{\beta}^{(c)} + \boldsymbol{z}'_{ij}\boldsymbol{\alpha}_i^{(c)},$$

and Gaussian random effects $\boldsymbol{\alpha}_i^{(c)} \overset{\text{i.i.d.}}{\sim} \mathrm{N}(\boldsymbol{0}, \boldsymbol{Q}_c)$ that are mutually independent.

Correspondingly a *cumulative mixed logit model* for ordered categories is defined by

$$\mathbb{P}(Y_{ij} \le c|\boldsymbol{\alpha}_i) = \frac{\exp(\theta^{(c)} - \boldsymbol{u}'_{ij}\boldsymbol{\beta} - \boldsymbol{z}'_{ij}\boldsymbol{\alpha}_i)}{1 + \exp(\theta^{(c)} - \boldsymbol{u}'_{ij}\boldsymbol{\beta} - \boldsymbol{z}'_{ij}\boldsymbol{\alpha}_i)}$$

or, equivalently,

$$\log\left(\frac{\mathbb{P}(Y_{ij} \le c|\boldsymbol{\alpha}_i)}{\mathbb{P}(Y_{ij} > c|\boldsymbol{\alpha}_i)}\right) = \eta_{ij}^{(c)} = \theta^{(c)} - \boldsymbol{u}'_{ij}\boldsymbol{\beta} - \boldsymbol{z}'_{ij}\boldsymbol{\alpha}_i$$

with ordered thresholds

$$-\infty = \theta^{(0)} < \theta^{(1)} < \ldots < \theta^{(C-1)} < \theta^{(C)} = \infty.$$

The extension from fixed to individual-specific random thresholds $\theta_i^{(c)}$ is not completely straightforward: Assuming $\boldsymbol{\theta}_i = (\theta_i^{(1)}, \ldots, \theta_i^{(C-1)})' \sim \mathrm{N}(\mathbf{0}, \boldsymbol{Q}_\theta)$ would in general violate the order restriction, so that either a suitable reparametrization (e.g. Fahrmeir and Tutz, 2001, Ch. 7.2) or non-Gaussian mixing distributions are needed.

Following Section 2.2.3, categorical mixed models can also be derived from latent variable representations. A multinomial mixed probit model for example, is obtained by assuming separate latent Gaussian LMMs

$$l_{ij}^{(c)} = \eta_{ij}^{(c)} + \varepsilon_{ij}^{(c)} = \boldsymbol{u}'_{ij}\boldsymbol{\beta}^{(c)} + \boldsymbol{z}'_{ij}\boldsymbol{\alpha}_i^{(c)} + \varepsilon_{ij}^{(c)}, \quad c = 1, \ldots, C$$

(with $\eta_{ij}^{(C)} \equiv 0$ for identifiability reasons). Relating them to the observed responses through the principle of maximal utility, that is,

$$Y_{ij} = c \quad \Leftrightarrow \quad l_{ij}^{(c)} = \max_{s=1,\ldots,C} l_{ij}^{(s)},$$

leads to a mixed probit model with category-specific linear mixed predictors. Cumulative mixed probit models can be derived in analogy to Section 2.2.3.

Example 3.7: Habitat selection

As an example for longitudinal data with unordered categorical responses, we analyse a data set on habitat selection considered in Kneib, Knauer and Küchenhoff (2010a). The major aim of habitat selection studies is to detect preferred and avoided habitat types and to study the impact of covariates on the habitat selection mechanism. The data are often collected by radio-tracking or visual observation studies, where repeated observations are made on a population of animals. The response variable corresponds to the different available habitat types.

In this example, we are considering data on habitat selection by a population of 22 brown bears in Slovenia collected between 1993 and 1998. The number of individual observations varies quite a bit for the different individuals and yields a total number of 1560 observations. The principal interest in this study was to differentiate between the preference for open areas and forested areas. In a more detailed account, the habitat type 'forested areas' is split further into three mutually exclusive habitat types, namely 'paved road' (forested areas within 100 m distance to a paved road), 'forest road' (forested areas outside 'paved road' but within 100 m distance to forest roads), 'highway' (other forested areas within 1,000 m distance to highways) and 'remote forest' (the remaining part of the habitat type 'forested areas'). Hence, the response variable of interest is unordered categorical with five categories. 'Remote forest' was chosen as the reference category.

When analysing habitat selection, it is important to take different availability into account. Therefore we first defined home ranges for the brown bears, obtained from a minimum convex polygon containing all observations of an individual animal, buffered by an additional range of 2,500 m which is the average distance between two daily observations. The choice probabilities then have to be corrected for the percentage of a particular habitat type within the home range (the areas that are in principle available for habitat selection). This is achieved by introducing a category-specific offset term, leading to the model equation

$$\pi_{ij}^{(c)} = \mathbb{P}(Y_{ij} = c) \propto a_{ij}^{(c)} \exp\left(\boldsymbol{u}_{ij1}'\boldsymbol{\beta}_1^{(c)} + \boldsymbol{u}_{ij2}^{(c)\prime}\boldsymbol{\beta}_2 + \alpha_i^{(c)}\right)$$

where the offset $a_{ij}^{(c)}$ is proportional to the available area of habitat type c within the home range and can simply be set to the percentage of habitat type c.

The model contains both category-specific covariates $\boldsymbol{u}_{ij2}^{(c)}$ with regression parameters $\boldsymbol{\beta}_2$ and global covariates $\boldsymbol{u}_{ij1}$ with effects $\boldsymbol{\beta}_1^{(c)}$:

- Bear-specific discrete covariates were included as three indicator variables: age class with categories yearling, subadult and adult (two indicators), and dispersal status with categories resident and dispersing (one indicator).
- Habitat-specific continuous covariates are elevation, slope, aspect (circular data, represented as the sum of a sine and a cosine transform in the analyses), resulting in a total of four habitat-specific covariates.

For the habitat-specific covariates, average values defined on a grid of random points (1 point/ha) were used in the following analyses. The category-specific random intercept $\alpha_i^{(c)}$ accounts for correlations due to the longitudinal data collection and also introduces animal-specific deviations from the global preference parameters contained in the vector $\boldsymbol{\beta}_1^{(c)}$. △

3.3.2 General mixed models for non-Gaussian responses

To make GLMMs useful beyond the context of longitudinal data, we will now introduce a general formulation of GLMMs that comprises all special cases considered in this book and in particular enables mixed model-based smoothing in empirical Bayes approaches (see Section 4.2.1). First, we write the stacked vector $\boldsymbol{\eta} = (\eta_{ij}, i = 1, \ldots, n, j = 1, \ldots, n_i)'$ of all predictors in compact matrix notation as

$$\boldsymbol{\eta} = \boldsymbol{U\beta} + \boldsymbol{Z\alpha}.$$

Then the (conditional) mean vector $\mathbb{E}(\boldsymbol{y}|\boldsymbol{\alpha})$ of all observations $\boldsymbol{y}$ is linked to $\boldsymbol{\eta}$ through

$$\mathbb{E}(\boldsymbol{y}|\boldsymbol{\alpha}) = h(\boldsymbol{\eta}) = h(\boldsymbol{U\beta} + \boldsymbol{Z\alpha}),$$

where the mapping h from $\boldsymbol{\eta} = \boldsymbol{U\beta} + \boldsymbol{Z\alpha}$ to $\boldsymbol{y}$ is to be understood elementwise, that is, for a typical element y_l of $\boldsymbol{y}$ we have

$$\mathbb{E}(y_l|\boldsymbol{\alpha}) = h(\eta_l) = h(\boldsymbol{u}_l'\boldsymbol{\beta} + \boldsymbol{z}_l'\boldsymbol{\alpha}),$$

where $\boldsymbol{u}_l'$ and $\boldsymbol{z}_l'$ are the corresponding l-th rows of $\boldsymbol{U}$ and $\boldsymbol{Z}$), and h is the usual scalar response function.

As for LMMs, a Gaussian mixing distribution $\boldsymbol{\alpha} \sim \mathrm{N}(\boldsymbol{0}, \boldsymbol{G})$ is the standard choice. For full Bayesian modelling, the same priors can be chosen for $\boldsymbol{\beta}$ and unknown variance components $\boldsymbol{\vartheta}$ as in Section 3.1.

To summarize, a GLMM in general form is defined by the following assumptions:

1. The conditional density $f(y_l|\boldsymbol{\alpha})$ of the components y_l of $\boldsymbol{y}$ is of the exponential family type (3.34), for example, binomial, Poisson, etc.
2. The components of $\boldsymbol{y}$ are conditionally independent.
3. The conditional expectation $\boldsymbol{\mu} = \mathbb{E}(\boldsymbol{y}|\boldsymbol{\alpha})$ is linked to the predictor $\boldsymbol{\eta} = \boldsymbol{U\beta} + \boldsymbol{Z\alpha}$ through
$$\boldsymbol{\mu} = h(\boldsymbol{\eta}).$$
4. The mixing distribution for the random effects is Gaussian $\boldsymbol{\alpha} \sim \mathrm{N}(\boldsymbol{0}, \boldsymbol{G})$.

For full Bayesian GLMMs, $f(y_l|\boldsymbol{\alpha})$ is to be understood conditionally on $\boldsymbol{\alpha}$ as well as $\boldsymbol{\beta}$ and unknown variance components $\boldsymbol{\vartheta}$. Also, $\boldsymbol{\alpha}, \boldsymbol{\beta}$ and $\boldsymbol{\vartheta}$ are assumed to be *a priori* independent. This definition can be extended to other non-normal responses, in particular categorical responses, by admitting other types of conditional densities. Another extension are non-normal mixing distributions, such as scale mixtures of normals.

3.3.3 Likelihood-based and empirical Bayes inference

For likelihood-based inference, fixed effects $\boldsymbol{\beta}$ and parameters $\boldsymbol{\vartheta}$ in $\boldsymbol{G} = \boldsymbol{G}(\boldsymbol{\vartheta})$ are considered as unknown but nonrandom quantities.Conceptually the *marginal* likelihood $L(\boldsymbol{\beta}, \boldsymbol{\vartheta})$ is obtained by integrating out random effects $\boldsymbol{\alpha}$ in the conditional density,

$$L(\boldsymbol{\beta}, \boldsymbol{\vartheta}) = \int p(\boldsymbol{y}|\boldsymbol{\beta}, \boldsymbol{\alpha}) p(\boldsymbol{\alpha}|\boldsymbol{\vartheta}) d\boldsymbol{\alpha},$$

where $p(\boldsymbol{y}|\boldsymbol{\beta}, \boldsymbol{\alpha})$ is the product of the conditional densities of the components of $\boldsymbol{y}$, and $p(\boldsymbol{\alpha}|\boldsymbol{\vartheta})$ is a (possibly) high-dimensional (Gaussian) prior. The integral can be solved analytically only in special cases. Numerical or Monte Carlo integration can be used for low-dimensional $\boldsymbol{\alpha}$ (Fahrmeir and Tutz, 2001, Ch. 7.3; Molenberghs and Verbeke, 2005). However, in the following chapters $\boldsymbol{\alpha}$ will often represent random effects corresponding to penalized spline smoothers or spatial effects and will therefore be high-dimensional, and we will prefer a Laplace approximation of the integral that provides a simple yet accurate solution. This approach leads to a penalized likelihood criterion, generalizing the corresponding penalized least-squares criterion considered in Section 3.1.4: For fixed or given $\boldsymbol{\vartheta}$ we maximize the joint likelihood

$$L(\boldsymbol{\beta}, \boldsymbol{\alpha}; \boldsymbol{\vartheta}) = p(\boldsymbol{y}|\boldsymbol{\beta}, \boldsymbol{\alpha}) p(\boldsymbol{\alpha}|\boldsymbol{\vartheta})$$

with respect to $\boldsymbol{\beta}$ and $\boldsymbol{\alpha}$. Taking logarithms, this is equivalent to maximizing the *penalized log-likelihood*

$$l_{\text{pen}}(\boldsymbol{\beta}, \boldsymbol{\alpha}; \boldsymbol{\vartheta}) = l(\boldsymbol{\beta}, \boldsymbol{\alpha}) - \frac{1}{2}\boldsymbol{\alpha}'\boldsymbol{G}\boldsymbol{\alpha},$$

where the conditional likelihood $l(\boldsymbol{\beta}, \boldsymbol{\alpha})$ is defined as for a usual GLM with predictor $\boldsymbol{\eta} = \boldsymbol{U}\boldsymbol{\beta} + \boldsymbol{Z}\boldsymbol{\alpha}$. In an EB approach with a flat prior $p(\boldsymbol{\beta}) \propto \text{const}$ for fixed effects, the posterior is

$$p(\boldsymbol{\beta}, \boldsymbol{\alpha}|\boldsymbol{y}) \propto p(\boldsymbol{y}|\boldsymbol{\beta}, \boldsymbol{\alpha}) p(\boldsymbol{\alpha}|\boldsymbol{\vartheta}),$$

so that the penalized likelihood estimate $(\hat{\boldsymbol{\beta}}, \hat{\boldsymbol{\alpha}})$ is equivalent to the posterior mode estimate. Imputing an estimate $\hat{\boldsymbol{\vartheta}}$ for $\boldsymbol{\vartheta}$ gives the EB estimates for $\boldsymbol{\beta}$ and $\boldsymbol{\alpha}$.

Penalized likelihood estimates $\hat{\boldsymbol{\beta}}, \hat{\boldsymbol{\alpha}}$ are computed numerically through Fisher scoring iterations. Score functions are

$$s_{\boldsymbol{\beta}}(\boldsymbol{\beta}, \boldsymbol{\alpha}) = \frac{\partial l_{\text{pen}}(\boldsymbol{\beta}, \boldsymbol{\alpha}; \boldsymbol{\vartheta})}{\partial \boldsymbol{\beta}} = \boldsymbol{U}'\boldsymbol{D}\boldsymbol{S}^{-1}(\boldsymbol{y} - \boldsymbol{\mu}),$$

$$s_{\boldsymbol{\alpha}}(\boldsymbol{\beta}, \boldsymbol{\alpha}) = \frac{\partial l_{\text{pen}}(\boldsymbol{\beta}, \boldsymbol{\alpha}; \boldsymbol{\vartheta})}{\partial \boldsymbol{\alpha}} = \boldsymbol{Z}'\boldsymbol{D}\boldsymbol{S}^{-1}(\boldsymbol{y} - \boldsymbol{\mu}) - \boldsymbol{G}^{-1}\boldsymbol{\alpha},$$

with GLM weights $\boldsymbol{\mu} = \mathbb{E}(\boldsymbol{y}|\boldsymbol{\alpha}), \boldsymbol{D} = \text{diag}(\partial h(\boldsymbol{\eta}_l)/\partial \boldsymbol{\eta}), \boldsymbol{S} = \text{diag}(\text{Var}(\boldsymbol{y}_l|\boldsymbol{\alpha}))$ depending on $\boldsymbol{\beta}$ and $\boldsymbol{\alpha}$ (and possibly on the scale parameter ϕ).

The correspondingly partitioned (expected) information matrix is

$$F(\boldsymbol{\beta}, \boldsymbol{\alpha}) = \begin{pmatrix} \boldsymbol{U}'\boldsymbol{W}\boldsymbol{U} & \boldsymbol{U}'\boldsymbol{W}\boldsymbol{Z} \\ \boldsymbol{Z}'\boldsymbol{W}\boldsymbol{U} & \boldsymbol{Z}'\boldsymbol{W}\boldsymbol{Z} + \boldsymbol{G}^{-1} \end{pmatrix},$$

with $\boldsymbol{W} = \boldsymbol{D}\boldsymbol{S}^{-1}\boldsymbol{D}$ depending on $\boldsymbol{\beta}$ and $\boldsymbol{\alpha}$. Defining the working observations

$$\tilde{\boldsymbol{y}} = \boldsymbol{U}\boldsymbol{\beta} + \boldsymbol{Z}\boldsymbol{\alpha} + \boldsymbol{D}^{-1}(\boldsymbol{y} - \boldsymbol{\mu}),$$

Fisher scoring iterations can be written as iteratively weighted (penalized) least-squares iterations

$$\begin{pmatrix} \boldsymbol{U}'\boldsymbol{W}^{(k)}\boldsymbol{U} & \boldsymbol{U}'\boldsymbol{W}^{(k)}\boldsymbol{Z} \\ \boldsymbol{Z}'\boldsymbol{W}^{(k)}\boldsymbol{U} & \boldsymbol{Z}'\boldsymbol{W}^{(k)}\boldsymbol{Z} + \boldsymbol{G}^{-1} \end{pmatrix} \begin{pmatrix} \hat{\boldsymbol{\beta}}^{(k+1)} \\ \hat{\boldsymbol{\alpha}}^{(k+1)} \end{pmatrix} = \begin{pmatrix} \boldsymbol{U}'\boldsymbol{W}^{(k)}\tilde{\boldsymbol{y}}^{(k)} \\ \boldsymbol{Z}'\boldsymbol{W}^{(k)}\tilde{\boldsymbol{y}}^{(k)} \end{pmatrix}, \tag{3.35}$$

$k = 1, 2, \ldots$, where $\tilde{\boldsymbol{y}}$ and $\boldsymbol{W}$ are evaluated at the current iterates $\hat{\boldsymbol{\beta}}^{(k)}, \tilde{\boldsymbol{y}}^{(k)}$. Note that the score function, the Fisher information and the complete estimation algorithm resemble exactly the structure that we found in Section 2.3 in the context of penalized generalized additive models when the penalty matrix is replaced with the inverse covariance matrix of the random effects. This observation in combination with the estimation schemes for variance parameters discussed below will also provide a possibility to determine the smoothing parameters in semiparametric regression models, see Section 4.2.1.

Variance components $\boldsymbol{\vartheta}$ can be estimated by maximizing an (approximative) marginal restricted log-likelihood: A Laplace argument as in Breslow and Clayton (1993) shows that the conditional log-likelihood can be approximated through the log-likelihood of a LMM for the working response, that is,

$$\tilde{\boldsymbol{y}}|\boldsymbol{\beta}, \boldsymbol{\alpha} \sim \mathrm{N}(\boldsymbol{U}\boldsymbol{\beta} + \boldsymbol{Z}\boldsymbol{\alpha}, \boldsymbol{W}^{-1}).$$

Hence we obtain a working LMM with marginal covariance matrix

$$\boldsymbol{V} = \boldsymbol{W}^{-1} + \boldsymbol{Z}'\boldsymbol{G}\boldsymbol{Z}$$

and can now apply methodology developed in Section 3.1 for Gaussian mixed models. In particular, we obtain the (approximate) *marginal restricted log-likelihood*

$$l^*(\boldsymbol{\vartheta}) = -\frac{1}{2}\log(|\boldsymbol{V}|) - \frac{1}{2}\log(|\boldsymbol{U}'\boldsymbol{V}^{-1}\boldsymbol{U}|) - \frac{1}{2}(\hat{\boldsymbol{y}} - \boldsymbol{U}\hat{\boldsymbol{\beta}})'\boldsymbol{V}^{-1}(\hat{\boldsymbol{y}} - \boldsymbol{U}\hat{\boldsymbol{\beta}})$$

where $\boldsymbol{V}$ and $\tilde{\boldsymbol{y}}$ depend on $\boldsymbol{\vartheta}$, $\hat{\boldsymbol{\beta}}$ and $\hat{\boldsymbol{\alpha}}$.

Switching between IWLS iterations for $\boldsymbol{\beta}$ and $\boldsymbol{\alpha}$, given a current estimate $\hat{\boldsymbol{\vartheta}}^{(k)}$ of $\boldsymbol{\vartheta}$, and iteration steps for maximizing $l^*(\boldsymbol{\vartheta})$, given current estimates of $\boldsymbol{\beta}$ and $\boldsymbol{\alpha}$, leads to the following algorithm:

Algorithm 3.2: Penalized likelihood/empirical Bayes estimation in mixed models

i. Choose initial values $\hat{\boldsymbol{\beta}}^{(0)}, \hat{\boldsymbol{\alpha}}^{(0)}, \hat{\boldsymbol{\vartheta}}^{(0)}$.

For $k = 1, 2, \ldots,$

ii. Compute the working observations

$$\tilde{\boldsymbol{y}}^{(k)} = \tilde{\boldsymbol{y}}^{(k)} = \boldsymbol{U}\hat{\boldsymbol{\beta}}^{(k)} + \boldsymbol{Z}\hat{\boldsymbol{\alpha}}^{(k)} + (\boldsymbol{D}^{(k)})^{-1}(\boldsymbol{y} - \boldsymbol{\mu}^{(k)}),$$

and working weights $\boldsymbol{W}^{(k)}$. Solve the equations (3.35) for $\hat{\boldsymbol{\beta}}^{(k+1)}, \hat{\boldsymbol{\alpha}}^{(k+1)}$.

iii. Compute $\hat{\boldsymbol{\vartheta}}^{(k+1)}$ by carrying out one step of an iterative numerical algorithm, for example a Newton–Raphson step, for maximizing $l^*(\boldsymbol{\vartheta})$ as in Section 3.1.4.

Approximate inference relies on the assumption that the sampling distribution is approximately Gaussian,

$$\begin{pmatrix} \hat{\boldsymbol{\beta}} \\ \hat{\boldsymbol{\alpha}} \end{pmatrix} \overset{\text{a}}{\sim} \text{N}\left(\begin{pmatrix} \hat{\boldsymbol{\beta}} \\ \hat{\boldsymbol{\alpha}} \end{pmatrix}, \hat{\boldsymbol{F}}^{-1} \right),$$

where $\hat{\boldsymbol{F}}$ is the information matrix evaluated at $\hat{\boldsymbol{\beta}}, \hat{\boldsymbol{\alpha}}$. The quality of this normal approximation depends on the specific mixed model and on the number of unknown parameters in relation to the sample size. If this approximation is questionable, full Bayesian inference is recommended.

Example 3.8: Habitat selection

We will now apply the empirical Bayes procedure to the habitat selection data from Example 3.7. To assess the impact of the random effects on the estimation results, we will also consider a model without the random intercept for comparison.

Figure 3.12 visualizes the estimated random effects, that is, the individual-specific deviations from global preferences. Obviously, there is quite large variation with respect to the preference of habitat type 'highway' while variation for 'forest road' is rather low. Parametric fixed effects are collected in Tables 3.5 (model without random effects) and 3.6 (model with random effects). When considering the intercept estimates first, the general ordering of preferences is the same for the random intercept and the purely parametric model. However, the estimated standard deviations and the corresponding p-values are quite different with overly optimistic results obtained when treating the observations as independent. While three out of four intercepts are considered to be different from zero in this case, the p-values are much closer to or even above the 5% level when including random effects. Note that the intercepts are in fact interesting in our model since they reflect the overall population preference for a specific habitat type. The surprising result that highways (category 3) are actually preferred over remote forest can be explained as follows: highways are

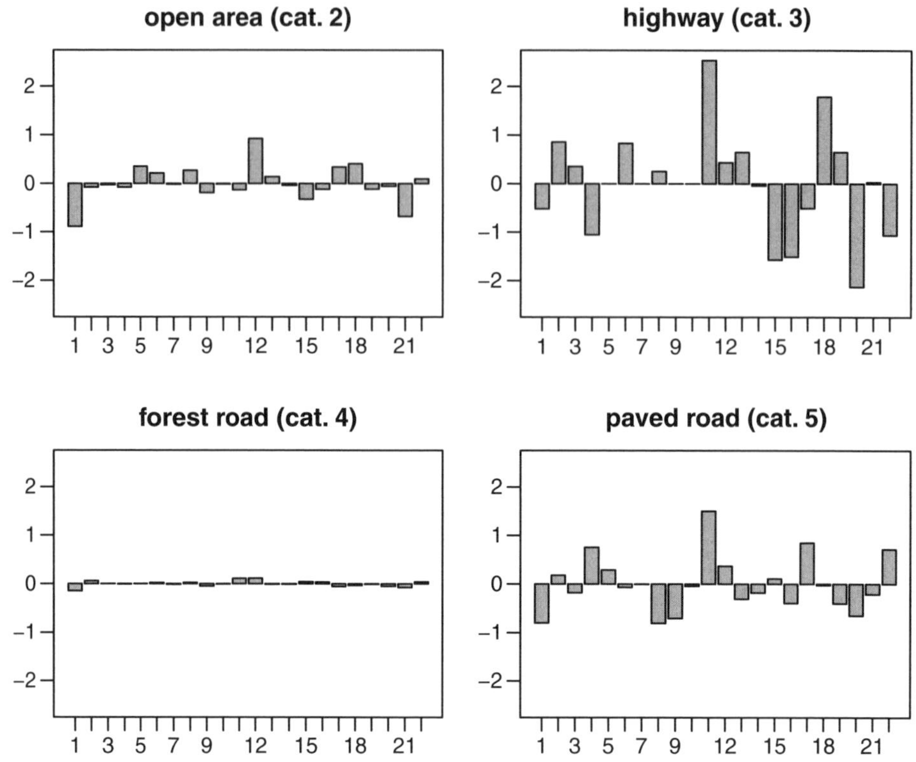

Fig. 3.12 Habitat selection: Estimated random effects representing individual-specific deviations from the overall habitat preferences.

not only characterized by the presence of noise but also by an almost total absence of human intervention. More generally, the point estimates for the regression coefficients are typically not too far from each other or at least have the same sign while the assessment of their significance changes with increased standard deviations in the random intercept model. △

Example 3.9: Forest health

As a second example, we reconsider the forest health data (Example 1.2) and introduce a plot-specific random intercept that accounts for both unobserved spatially varying influential factors and the longitudinal structure of the repeated observations. To be more specific, we specify a cumulative logit model with predictor

$$\eta_{ij}^{(c)} = \theta^{(c)} - \boldsymbol{u}_{ij}'\boldsymbol{\beta}^{(c)} - \alpha_i$$

Table 3.5 Habitat selection: Posterior modes, standard deviations, 95% credible intervals and p-values in a model without random effects.

variable	estimate	sd	95% credible	interval	p-value
intercept (cat. 2)	−1.739	0.402	−2.527	−0.950	0.000
intercept (cat. 3)	1.886	0.253	1.391	2.381	0.000
intercept (cat. 4)	−0.170	0.145	−0.454	0.115	0.244
intercept (cat. 5)	1.721	0.311	1.112	2.329	0.000
subadult (cat. 2)	0.480	0.456	−0.415	1.375	0.293
subadult (cat. 3)	−0.088	0.258	−0.593	0.418	0.733
subadult (cat. 4)	0.219	0.218	−0.208	0.645	0.316
subadult (cat. 5)	−1.204	0.311	−1.814	−0.594	0.000
yearling (cat. 2)	0.831	0.523	−0.193	1.856	0.111
yearling (cat. 3)	−0.897	0.289	−1.464	−0.331	0.002
yearling (cat. 4)	0.244	0.219	−0.185	0.673	0.265
yearling (cat. 5)	0.386	0.254	−0.113	0.885	0.129
disperser (cat. 2)	0.166	0.549	−0.911	1.242	0.763
disperser (cat. 3)	0.860	0.520	−0.160	1.880	0.098
disperser (cat. 4)	0.097	0.241	−0.377	0.570	0.688
disperser (cat. 5)	0.051	0.318	−0.573	0.675	0.874
elevation	0.016	0.002	0.012	0.019	0.000
slope	−0.121	0.046	−0.211	−0.031	0.009
sin(aspect)	−0.694	0.149	−0.987	−0.402	0.000
cos(aspect)	0.278	0.180	−0.075	0.632	0.122

comprising ordered thresholds $\theta^{(1)}$ and $\theta^{(2)}$, a plot-specific random intercept α_i and further covariate effects in the linear predictor $\boldsymbol{u}_{ij}'\boldsymbol{\beta}^{(c)}$. Apart from the linear time trend, all effects are considered as fixed across covariates, so that in fact

$$\boldsymbol{u}_{ij}'\boldsymbol{\beta}^{(c)} = t_{ij}\beta_t^{(c)} + \tilde{\boldsymbol{u}}_{ij}'\tilde{\boldsymbol{\beta}}.$$

Again, inference is based on the penalized likelihood or empirical Bayes approach. Figure 3.13 visualizes the estimated random effects that indicate strong spatial variation in the data. However, the spatial structure does not seem to be too smooth, which might suggest that correlations are mostly induced by the longitudinal structure and only to a smaller extent by the spatial structure. A possible explanation is that several spatial covariates are already included such that spatial correlations are effectively dealt with by inclusion of these covariates.

Tables 3.7 and 3.8 show the estimated fixed effects for a model without random intercept and including the random intercept. We observe a qualitatively similar behaviour as in the habitat selection example with mostly comparable point estimates but increased standard errors when including the random intercepts. We do not give a detailed interpretation of the estimated covariate effects here since several important aspects have not been taken into account in this analysis. In particular, non-linearity of some of the effects is not investigated and

Table 3.6 Habitat selection: Posterior modes, standard deviations, 95% credible intervals and p-values in a model including random effects.

variable	estimate	sd	95% credible	interval	p-value
intercept (cat. 2)	−1.858	0.809	−3.444	−0.273	0.022
intercept (cat. 3)	1.203	0.801	−0.367	2.773	0.133
intercept (cat. 4)	−0.296	0.237	−0.760	0.168	0.211
intercept (cat. 5)	1.263	0.666	−0.042	2.568	0.057
subadult (cat. 2)	0.282	0.725	−1.139	1.702	0.697
subadult (cat. 3)	−1.942	1.048	−3.997	0.112	0.063
subadult (cat. 4)	0.206	0.278	−0.340	0.751	0.460
subadult (cat. 5)	−1.364	0.614	−2.568	−0.160	0.026
yearling (cat. 2)	0.919	0.709	−0.471	2.308	0.195
yearling (cat. 3)	−0.116	0.910	−1.900	1.667	0.899
yearling (cat. 4)	0.329	0.263	−0.185	0.844	0.210
yearling (cat. 5)	0.683	0.542	−0.379	1.745	0.208
disperser (cat. 2)	0.037	0.764	−1.461	1.535	0.962
disperser (cat. 3)	1.313	1.055	−0.755	3.381	0.214
disperser (cat. 4)	0.030	0.274	−0.507	0.567	0.914
disperser (cat. 5)	0.461	0.512	−0.544	1.465	0.369
elevation	0.014	0.003	0.007	0.020	0.000
slope	−0.119	0.091	−0.297	0.059	0.191
sinasp	−0.693	0.246	−1.174	−0.211	0.005
cosasp	0.512	0.296	−0.069	1.093	0.084

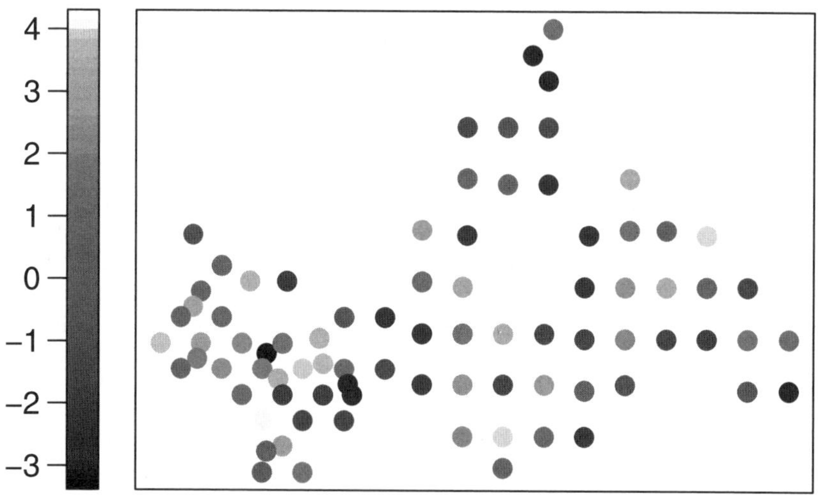

Fig. 3.13 Forest health: Estimated random effects.

Table 3.7 Forest health: Posterior modes, standard deviations, 95% credible intervals and p-values in a model without random effects.

variable	estimate	sd	95% credible	interval	p-value
$\theta^{(1)}$	−4.369	1.263	−6.845	−1.892	0.001
$\theta^{(2)}$	0.646	1.305	−1.913	3.204	0.620
time (cat. 1)	−0.001	0.010	−0.021	0.019	0.914
time (cat. 2)	0.072	0.025	0.022	0.121	0.005
age	0.020	0.002	0.017	0.023	0.000
elevation	0.001	0.001	−0.002	0.004	0.481
inclination	0.000	0.007	−0.013	0.013	0.989
soil	−0.021	0.007	−0.035	−0.007	0.003
ph	−1.223	0.243	−1.700	−0.746	0.000
canopy	−2.476	0.256	−2.978	−1.975	0.000
standcat	0.339	0.073	0.195	0.482	0.000
fertilization	−0.548	0.118	−0.779	−0.318	0.000
humus0	−0.103	0.127	−0.351	0.145	0.416
humus1	−0.025				
humus2	0.269	0.114	0.045	0.492	0.019
humus3	0.096	0.133	−0.165	0.356	0.471
humus4	−0.237	0.175	−0.579	0.106	0.175
moisture1	−0.630	0.150	−0.925	−0.335	0.000
moisture2	0.009				
moisture3	0.621	0.105	0.416	0.826	0.000
alkali1	0.276	0.150	−0.018	0.570	0.065
alkali2	−0.770				
alkali3	−0.462	0.150	−0.756	−0.167	0.003
alkali4	0.956	0.212	0.541	1.372	0.000

there is no structured spatial effect contained in the model. We will investigate the forest health data in more detail in Chapters 4 and 5. △

3.3.4 Full Bayesian inference for longitudinal data

We consider a GLMM for longitudinal data with

$$\mathbb{E}(y_{ij}|\boldsymbol{\alpha}_i) = \mu_{ij} = h(\boldsymbol{u}'_{ij}\boldsymbol{\beta} + \boldsymbol{z}'_{ij}\boldsymbol{\alpha}_i), \quad i = 1, \dots, n, j = 1, \dots, n_i,$$

a Gaussian prior $\mathrm{N}(\boldsymbol{c}, \boldsymbol{C})$ or a flat prior for $\boldsymbol{\beta}$, a Gaussian prior

$$\boldsymbol{\alpha}_i|\boldsymbol{Q} \sim \mathrm{N}(\boldsymbol{0}, \boldsymbol{Q})$$

for i. i. d. random effects $\boldsymbol{\alpha}_i$, and an inverse Wishart prior

$$\boldsymbol{Q} \sim \mathrm{IW}(a, \boldsymbol{B})$$

Table 3.8 Forest health: Posterior modes, standard deviations, 95% credible intervals and p-values in a model including random effects.

variable	estimate	sd	95% credible	interval	p-value
$\theta^{(1)}$	−5.112	2.730	−10.463	0.239	0.061
$\theta^{(2)}$	2.545	2.784	−2.913	8.003	0.361
time (cat. 1)	−0.020	0.013	−0.046	0.007	0.141
time (cat. 2)	0.109	0.035	0.040	0.178	0.002
age	0.034	0.005	0.025	0.043	0.000
elevation	0.001	0.005	−0.009	0.010	0.914
inclination	−0.003	0.026	−0.053	0.047	0.900
soil	−0.015	0.025	−0.064	0.035	0.561
ph	−1.522	0.320	−2.149	−0.895	0.000
canopy	−3.796	0.560	−4.894	−2.697	0.000
standcat	0.415	0.237	−0.050	0.880	0.080
fertilization	−0.906	0.428	−1.745	−0.068	0.034
humus0	−0.103	0.163	−0.423	0.218	0.530
humus1	−0.120				
humus2	0.270	0.136	0.003	0.538	0.047
humus3	0.164	0.166	−0.161	0.490	0.323
humus4	−0.211	0.235	−0.672	0.250	0.370
moisture1	−0.946	0.565	−2.053	0.161	0.093
moisture2	0.110				
moisture3	0.836	0.393	0.065	1.607	0.033
alkali1	0.548	0.553	−0.536	1.632	0.322
alkali2	−1.166				
alkali3	−0.465	0.569	−1.582	0.651	0.414
alkali4	1.083	0.726	−0.341	2.506	0.136

for $\boldsymbol{Q}$. The full conditionals are

$$p(\boldsymbol{\beta}|\cdot) \propto \prod_{i=1}^{n}\prod_{j=1}^{n_i} p(y_{ij}|\boldsymbol{\beta},\boldsymbol{\alpha}_i)p(\boldsymbol{\beta})$$

$$p(\boldsymbol{\alpha}_i|\cdot) \propto \prod_{i=1}^{n}\prod_{j=1}^{n_i} p(y_{ij}|\boldsymbol{\beta},\boldsymbol{\alpha}_i)p(\boldsymbol{\alpha}_i|\boldsymbol{Q}), \quad i = 1,\dots,n,$$

$$p(\boldsymbol{Q}|\cdot) \propto \prod_{i=1}^{n} p(\boldsymbol{\alpha}_i|\boldsymbol{Q})p(\boldsymbol{Q}).$$

Due to the conjugate setting in the model definition for $p(\boldsymbol{Q})$, its full conditional is still inverted Wishart with parameters

$$\tilde{a} = a + \frac{n}{2}, \quad \tilde{\boldsymbol{B}} = \boldsymbol{B} + \frac{1}{2}\sum_{i=1}^{n} \boldsymbol{\alpha}_i\boldsymbol{\alpha}_i'.$$

For the special case $\boldsymbol{Q} = \text{diag}(\delta_1^2, \ldots, \delta_q^2)$ and independent inverse gamma priors $\delta_l^2 \sim \text{IG}(a_l, b_l)$, the full conditionals $p(\delta_l^2|\cdot)$ are inverse gamma with updated parameters

$$\tilde{a}_l = a_l + \frac{n}{2}, \quad \tilde{b}_l = b_l + \frac{1}{2}\sum_{i=1}^{n} \alpha_{il}^2.$$

The full conditionals for $\boldsymbol{\beta}$ and $\boldsymbol{\alpha}_i, i = 1, \ldots, n$, are generally not available in closed form. Therefore, MH-sampling instead of Gibbs sampling has to be applied. Our preferred choice is the MH algorithm with IWLS proposals suggested by Gamerman (1997), modifying the algorithms in Sections 2.2.2 and 2.3.2 as follows.

For fixed effects $\boldsymbol{\beta}$, let $\varphi(\boldsymbol{\beta}; \boldsymbol{\mu}_{\boldsymbol{\beta}}, \boldsymbol{\Sigma}_{\boldsymbol{\beta}})$ denote the multivariate Gaussian density with covariance matrix and mean

$$\boldsymbol{\Sigma}_{\boldsymbol{\beta}} = (\boldsymbol{U}'\boldsymbol{W}\boldsymbol{U} + \boldsymbol{C}^{-1})^{-1}, \quad \boldsymbol{\mu}_{\boldsymbol{\beta}} = \boldsymbol{\Sigma}_{\boldsymbol{\beta}}(\boldsymbol{U}'\boldsymbol{W}(\tilde{\boldsymbol{y}} - \boldsymbol{Z}\boldsymbol{\alpha}) + \boldsymbol{C}^{-1}\boldsymbol{c}), \tag{3.36}$$

with weight matrix $\boldsymbol{W}$ and working observations $\tilde{\boldsymbol{y}}$ corresponding to all observations $\boldsymbol{y} = (\boldsymbol{y}_1', \ldots, \boldsymbol{y}_n')'$. For random effects, let $\varphi(\boldsymbol{\alpha}_i; \boldsymbol{\mu}_{\boldsymbol{\alpha}_i}, \boldsymbol{\Sigma}_{\boldsymbol{\alpha}_i})$ denote the multivariate Gaussian density with

$$\boldsymbol{\Sigma}_{\boldsymbol{\alpha}_i} = (\boldsymbol{Z}_i'\boldsymbol{W}_i\boldsymbol{Z}_i + \boldsymbol{Q}^{-1})^{-1}, \quad \boldsymbol{\mu}_{\boldsymbol{\alpha}_i} = \boldsymbol{\Sigma}_{\boldsymbol{\alpha}_i}\boldsymbol{Z}_i'\boldsymbol{W}_i(\tilde{\boldsymbol{y}}_i - \boldsymbol{U}_i\boldsymbol{\beta}), \tag{3.37}$$

with weight matrix $\boldsymbol{W}_i$, working observations $\tilde{\boldsymbol{y}}_i$ and design matrix $\boldsymbol{U}_i$ corresponding to observations $\boldsymbol{y}_i = (y_{i1}, \ldots, y_{in_i})'$.

Given current values $\boldsymbol{\beta}^c, \boldsymbol{\alpha}_1^c, \ldots, \boldsymbol{\alpha}_n^c$ (and $\boldsymbol{Q} = \boldsymbol{Q}^c$), a proposal $\boldsymbol{\beta}^p$ is drawn from $\varphi(\boldsymbol{\beta}; \boldsymbol{\mu}_{\boldsymbol{\beta}}^c, \boldsymbol{\Sigma}_{\boldsymbol{\beta}}^c)$, with $\boldsymbol{\mu}_{\boldsymbol{\beta}}$ and $\boldsymbol{\Sigma}_{\boldsymbol{\beta}}$ evaluated for the current values $\boldsymbol{\beta}^c$, $\boldsymbol{\alpha}^c$. It is accepted with probability

$$\alpha(\boldsymbol{\beta}^c, \boldsymbol{\beta}^p) = \frac{p(\boldsymbol{y}|\boldsymbol{\beta}^p, \boldsymbol{\alpha}^c)p(\boldsymbol{\beta}^p)\varphi(\boldsymbol{\beta}^c; \boldsymbol{\mu}_{\boldsymbol{\beta}}^p, \boldsymbol{\Sigma}_{\boldsymbol{\beta}}^p)}{p(\boldsymbol{y}|\boldsymbol{\beta}^c, \boldsymbol{\alpha}^c)p(\boldsymbol{\beta}^c)\varphi(\boldsymbol{\beta}^p; \boldsymbol{\mu}_{\boldsymbol{\beta}}^c, \boldsymbol{\Sigma}_{\boldsymbol{\beta}}^c)}.$$

In the nominator, the mean and covariance matrix of the Gaussian density are evaluated at $\boldsymbol{\beta}^c$ and $\boldsymbol{\alpha}^c$.

Similarly, given current values, a proposal $\boldsymbol{\alpha}_i^p$, $i = 1, \ldots, n$, is drawn from $\varphi(\boldsymbol{\alpha}_i; \boldsymbol{\mu}_{\boldsymbol{\alpha}_i}^c, \boldsymbol{\Sigma}_{\boldsymbol{\alpha}_i}^c)$ and accepted with probability

$$\alpha(\boldsymbol{\alpha}_i^c, \boldsymbol{\alpha}_i^p) = \frac{p(\boldsymbol{y}|\boldsymbol{\beta}^c, \boldsymbol{\alpha}_1^c, \ldots, \boldsymbol{\alpha}_i^p, \ldots, \boldsymbol{\alpha}_n^c)p(\boldsymbol{\alpha}_i^p|\boldsymbol{Q}^c)\varphi(\boldsymbol{\alpha}_i^c; \boldsymbol{\mu}_{\boldsymbol{\alpha}_i}^p, \boldsymbol{\Sigma}_{\boldsymbol{\alpha}_i}^p)}{p(\boldsymbol{y}|\boldsymbol{\beta}^c, \boldsymbol{\alpha}_1^c, \ldots, \boldsymbol{\alpha}_i^c, \ldots, \boldsymbol{\alpha}_n^c)p(\boldsymbol{\alpha}_i^c|\boldsymbol{Q}^c)\varphi(\boldsymbol{\alpha}_i^p; \boldsymbol{\mu}_{\boldsymbol{\alpha}_i}^c, \boldsymbol{\Sigma}_{\boldsymbol{\alpha}_i}^c)}$$

with $\boldsymbol{\mu}_{\boldsymbol{\alpha}_i}$ and $\boldsymbol{\Sigma}_{\boldsymbol{\alpha}_i}$ in the nominator evaluated at $\boldsymbol{\beta}^c, \boldsymbol{\alpha}_1^c, \ldots, \boldsymbol{\alpha}_i^p, \ldots, \boldsymbol{\alpha}_n^c$.

Combining these MH steps for fixed and random effects with Gibbs steps for drawing from IW or IG full conditionals for $\boldsymbol{Q}$ or $\delta_1^2, \ldots, \delta_q^2$, respectively, we obtain a MH algorithm with IWLS proposal for GLMMs.

This general strategy can be easily extended to categorical mixed models. As an alternative to MH algorithms, Gibbs samplers can be developed by introducing latent auxiliary variables as described in Chapter 2.2.

Algorithm 3.3: MH algorithm with IWLS proposal for GLMMs

i. *Initialization*: Compute the posterior modes $\hat{\boldsymbol{\beta}}^{(0)}$ and $\hat{\boldsymbol{\alpha}}_i^{(0)}$ given a fixed covariance matrix $\boldsymbol{Q}$. Use the posterior modes as the current states $\boldsymbol{\beta}^c$ and $\boldsymbol{\alpha}_i^c$ of the chain.
ii. Update $\boldsymbol{\beta}$ by drawing a proposal $\boldsymbol{\beta}^p$ from the Gaussian distribution with moments given in (3.36) and accept it with probability $\alpha(\boldsymbol{\beta}^c, \boldsymbol{\beta}^p)$.
iii. For $i = 1, \ldots, n$ draw proposals $\boldsymbol{\alpha}_i^p$ from the Gaussian distribution with moments given in (3.37) and accept with probability $\alpha(\boldsymbol{\alpha}_i^c, \boldsymbol{\alpha}_i^p)$.
iv. *Update* $\boldsymbol{Q}$: The new current $\boldsymbol{Q}$ can be simulated as for Gaussian LMMs in Algorithm 3.1 on page 130.

3.3.5 Longitudinal data GLMMs with flexible random effects priors

We consider GLMMs for longitudinal data where the linear predictor is modelled in reparametrized form as in (3.25) for LMMs:

$$\boldsymbol{\eta}_i = \boldsymbol{U}_i\boldsymbol{\beta} + \boldsymbol{Z}_i\boldsymbol{\alpha}_i, \quad i = 1, \ldots, n \tag{3.38}$$

where $\boldsymbol{\alpha}_i = \boldsymbol{\alpha} + \tilde{\boldsymbol{\alpha}}_i$ is the sum of the 'fixed' effect vector $\boldsymbol{\alpha} = \mathbb{E}(\boldsymbol{\alpha}_i)$ of the covariate vectors $\boldsymbol{z}_{ij}$ and $\tilde{\boldsymbol{\alpha}}_i$, with $\mathbb{E}(\tilde{\boldsymbol{\alpha}}_i) = \boldsymbol{0}$, is the 'random' effects part. Flexible random effects priors can be specified in complete analogy to Section 3.2. For example, Lenk and DeSarbo (2000) consider finite mixture models for GLMMs. Such conventional finite mixture models can be extended to models with penalized Gaussian mixture priors, see Komárek and Lesaffre (2008) for concepts and computational details. GLMMs with DP and DPM random effects priors can be defined as follows.

GLMMs with DP random effects priors

Priors for the observation model (3.38) can be specified through the same hierarchy (3.29) as in Section 3.2.3 for LMMs. The priors for the DPglmm model in the `DPpackage` (Jara, 2007) are

$$\begin{aligned}
\boldsymbol{\alpha}_i | G &\sim G \\
G | \nu_0, G_0 &\sim \mathrm{DP}(\nu_0, G_0) \\
G_0 &\sim \mathrm{N}(\boldsymbol{\mu}_0, \boldsymbol{Q}_0) \\
\boldsymbol{\beta} &\sim \mathrm{N}(\boldsymbol{c}, \boldsymbol{C}),
\end{aligned}$$

with hyperpriors

$$\begin{aligned}\nu_0 &\sim \text{Ga}(a_\nu, b_\nu)\\ \boldsymbol{\mu}_0 &\sim \text{N}(\boldsymbol{m}_0, \boldsymbol{S}_0)\\ \boldsymbol{Q}_0 &\sim \text{IW}(a_0, \boldsymbol{B}_0 = b_0\boldsymbol{I}).\end{aligned}$$

Similar priors were suggested in Kleinman and Ibrahim (1998b), without taking care of the centring problem, and other authors. A tutorial, based on a WinBUGS implementation, is presented in Ohlssen, Sharples and Spiegelhalter (2007).

The computational implementation in the `DPpackage` is based on marginalization of the DP. For binomial, Poisson and Gamma models, methods for nonconjugate priors (MacEachern and Müller, 1998; Neal, 2000) are applied, in particular Algorithm 8 of Neal (2000), in combination with IWLS proposals for MH steps to draw from full conditionals for fixed and random effects.

For logit and probit Bernoulli models, MCMC is based on auxiliary variables l_{ij}, related to observed binary responses y_{ij} through the threshold mechanism

$$Y_{ij} = 1 \quad \Leftrightarrow \quad l_{ij} = \boldsymbol{u}'_{ij}\boldsymbol{\beta} + \boldsymbol{z}'_{ij}\boldsymbol{\alpha}_i + \varepsilon_{ij} \geq 0, \tag{3.39}$$

with Gaussian i. i. d. errors $\varepsilon_{ij} \sim \text{N}(0, \sigma^2)$.
In the probit case $\sigma^2 = 1$, while

$$\sigma^2_{ij} = 4\psi^2_{ij}, \quad \psi_{ij} \sim \text{KS},$$

where KS is the Kolmogorov–Smirnov distribution, for the logit model (see Section 2.2.2 and Holmes and Held, 2006). The introduction of auxiliary Gaussian mixed models (3.39) allows us to apply sampling methods for conjugate priors (Escobar and West, 1998). Extending the threshold mechanism (3.39) from binary to multicategorical ordered responses Y_{ij}, ordinal mixed probit models with DP priors for random effects are obtained, see the DPolmm model of the `DPpackage`.

GLMMs with DPM random effects priors

As for LMMs, nonparametric continuous random effects priors are specified through DPM priors, adding an additional layer to the hierarchical structure of the DP model:

$$\begin{aligned}\boldsymbol{\alpha}_i|\boldsymbol{\mu}_i &\overset{\text{ind.}}{\sim} \text{N}(\boldsymbol{\mu}_i, \boldsymbol{Q})\\ \boldsymbol{\mu}_i|G &\sim G, \quad \boldsymbol{Q} \sim \text{IW}(a, \boldsymbol{B} = b\boldsymbol{I})\\ G|\nu_0, G_0 &\sim \text{DP}(\nu_0, G_0)\\ G_0 &\sim \text{N}(\boldsymbol{\mu}_0, \boldsymbol{Q}_0)\\ \boldsymbol{\mu}_0 &\sim \text{N}(\boldsymbol{m}_0, \boldsymbol{S}_0)\\ \boldsymbol{Q}_0 &\sim \text{IW}(a_0, \boldsymbol{B}_0 = b_0\boldsymbol{I})\end{aligned}$$

with the same priors for $\boldsymbol{\beta}$ and ν_0 as in the DP model. MCMC algorithms based on marginalization can be implemented by adding a further full conditional to algorithms for DP priors. Inference for the models DPMglmm and DPMolmm (for ordinal responses) in the `DPpackage` is implemented in this way.

3.4 Notes and further reading

The contents of this chapter have been primarily selected to provide mixed model concepts and methodology as a basis for semiparametric longitudinal, spatial and event history regression models considered in the following chapters. Therefore, several aspects have been omitted or are only alluded to. Methods for selecting and assessing models are presented in Section 4.5. We have also deliberately neglected the important issue of incomplete data. An extensive account of methods for incomplete and missing data is given in Verbeke and Molenberghs (2000), Molenberghs and Verbeke (2001) for linear mixed models, Molenberghs and Verbeke (2005) for discrete longitudinal data, in Part V of the handbook edited by Fitzmaurice *et al.* (2009) and with a focus on Bayesian modelling in Daniels and Hogan (2008). Other books including (chapters on) mixed models are Diggle, Heagerty, Liang and Zeger (2002), Fitzmaurice, Laird and Ware (2004), McCulloch and Searle (2001), Demidenko (2004) and Fitzmaurice *et al.* (2009).

With the exception of the Gaussian heterogeneity model in Verbeke and Molenberghs (2000), these texts mainly deal with Gaussian i. i. d. random effects. Lee and Nelder (1996) and Lee, Nelder and Pawitan (2006) define hierarchical generalized linear models (HGLMs) that include linear mixed models with Gaussian random effects priors as a special case. For non-Gaussian exponential family responses, the random effects α_i are some monotone function $\alpha_i = \alpha(v_i)$ of random components v_i, which follow a distribution conjugate to a GLM family of distributions. It is not necessary for the distribution of v_i to be the conjugate family of that for y_i given v_i. If it is, the model is a conjugate HGLM. This class of models includes LMMs with conditionally Gaussian responses and Gaussian random effects as a special case, but it also comprises mixed models with non-Gaussian responses and random effects. For example, suppose that $y_i|\mu_i$ is Poisson with $\mu_i = \exp(\boldsymbol{u}_i'\boldsymbol{\beta}+\alpha_i)$ and $\alpha_i = \log(v_i)$. If the distribution of v_i is gamma, then α_i is log-gamma distributed and the model is a Poisson-gamma HGLM. A corresponding Poisson GLMM assumes a normal distribution for α_i so that v_i is log-normal and it would be called a Poisson-log-normal HGLM. Inference in HGLMs is based on a so-called h-likelihood, a generalization of the joint likelihood of observations and random effects in Gaussian LMMs.

Multilevel models have only been mentioned in Section 3.1. In principle our formulation of general mixed models in Sections 3.1.2 and 3.3.2 comprises models with nested random effects through appropriate definition of design matrices and correlation structures, but we have omitted any further details on modelling and inference. General treatments of multilevel models are given for example in

Goldstein (2002), Skrondal and Rabe-Hesketh (2004) and Gelman and Hill (2006). Due to the hierarchical structure of multilevel modelling, Bayesian MCMC inference is quite attractive for multilevel models, see for example Browne and Draper (2000), Browne, Draper, Goldstein and Rasbash (2002), and Browne and Draper (2006) for a comparison of Bayesian and likelihood-based inference.

4 Semiparametric Mixed Models for Longitudinal Data

In this chapter, we consider Bayesian inference in semiparametric mixed models (SPMMs) for longitudinal data. This class of models extends conventional parametric linear mixed models (LMMs) and generalized linear mixed models (GLMMs) in several ways. An important subclass of SPMMs is obtained by including smooth functions of time or of continuous covariates as additional nonparametric effects in the predictor while keeping the assumption of a Gaussian random effects prior. For example, if it is difficult to assume a simple parametric time trend *a priori*, a LMM could be extended to

$$y_{ij} = \boldsymbol{u}_{ij}'\boldsymbol{\beta} + f(t_{ij}) + \boldsymbol{z}_{ij}'\boldsymbol{\alpha}_i + \varepsilon_{ij},$$

where $f(t)$ is a smooth nonparametric trend function, modelled through a smoothing spline or a P-spline. In Examples 1.1, 3.2 and 3.4 on childhood obesity, $f(t)$ corresponds to the population trend of the effect of a child's age. Because this nonlinear trend is difficult to specify parametrically, it is reasonable to model it nonparametrically, at least for exploratory purposes. In many applications, there are further continuous covariates with possibly nonlinear effects. A typical example for binary and categorical responses are the longitudinal data on forest damage, described in Example 1.2, and analysed in Examples 2.6, 2.7, 2.9, 2.11 and 3.9. For each tree i in this study, the degree of defoliation serves as an indicator for its damage state. Defining the binary response $y_{it} = 1$ (tree i is damaged in year t, $t = 1983, \ldots, 2004$) and $y_{it} = 0$ (no damage) a possible mixed logit model is

$$\log \frac{\mathbb{P}(y_{it} = 1)}{\mathbb{P}(y_{it} = 0)} = \boldsymbol{u}_{it}'\boldsymbol{\beta} + f_1(t) + f_2(age_{it}) + \boldsymbol{z}_{it}'\boldsymbol{\alpha}_i,$$

where $\boldsymbol{\beta}$ are usual linear fixed effects, $f_1(t)$ is a smooth time trend, $f_2(age)$ is a nonparametric effect of the time-varying covariate *age*, and $\boldsymbol{\alpha}_i$ are tree-specific Gaussian random effects, for example random intercepts (if $z_{it} \equiv 1$). One particular advantage of this model is that it allows for proper significance statements for the nonlinear effects accounting for the correlations induced by the longitudinal data collection. Note, however, that we are still ignoring the spatial alignment of the data that may attribute to further correlation in the data. We will tackle this issue with spatial smoothing approaches in Chapter 5.

Such additive extensions of LMMs and GLMMs are often termed additive mixed models (AMMs) and generalized additive mixed models (GAMMs), respectively. More general models are obtained if we admit time-varying effects or effects varying over the domain of another continuous covariate, leading to varying coefficient mixed models (VCMMs), or if unit-specific 'random' time trends $h_i(t)$ are of interest in addition to the population trend $f(t)$. We consider GAMMs, VCMMs and further extensions including nonparametric functions as subclasses of SPMMs.

For Bayesian inference, nonparametric smooth functions are modelled by specifying appropriate smoothness priors. In Section 4.1, we assume Gaussian smoothness priors, focusing on Bayesian P-splines in combination with Gaussian priors for random effects, and outline various model specifications that are included as special cases in SPMMs. Section 4.2 describes inferential techniques, detailing both empirical Bayes (EB) estimation based on mixed model technology and full Bayes (FB) techniques. Section 4.3 discusses the relation between Bayesian smoothing and correlation. Section 4.4 considers some additional or alternative semiparametric extensions of GLMMs: First, as in Section 3.2, the assumption of Gaussian random effects can be removed by allowing nonparametric DP or DPM priors in combination with Gaussian smoothness priors for functional effects. Second, local adaptivity of functional effects can be improved by scale mixtures of Gaussian smoothness priors, with variance parameters following stochastic process priors in another hierarchical stage. Third, we also consider the case of high-dimensional fixed effects $\boldsymbol{\beta}$, with Bayesian shrinkage priors regularizing the resulting ill-posed inferential problem. Shrinkage priors can also be used for model choice and variable selection. The final Section 4.5 describes strategies for model choice and model checking in SPMMs.

4.1 Semiparametric mixed models based on Gaussian priors

4.1.1 Observation models for univariate responses from exponential families

As in Chapter 3, suppose longitudinal data consisting of n subjects, with subject i having n_i repeated observations at times $t_{ij}, j = 1, \ldots, n_i$. Let y_{ij} be the response for subject i at time t_{ij}, and $x_{ij1}, \ldots, x_{ijp}$ the observed values of continuous covariates $x_1, \ldots, x_p$.

Generalized additive mixed models

Additive mixed models (AMMs) combine the LMM (3.1) of Section 3.1.1 with the additive model (2.56) of Section 2.3.1 to

$$y_{ij} = \boldsymbol{u}'_{ij}\boldsymbol{\beta} + f_1(x_{ij1}) + \ldots + f_p(x_{ijp}) + \boldsymbol{z}'_{ij}\boldsymbol{\alpha}_i + \varepsilon_{ij}. \tag{4.1}$$

As for LMMs, $\boldsymbol{\beta}$ is a vector of fixed effects interpreted as population effects, $\boldsymbol{\alpha}_i$ are subject-specific random effects, $\boldsymbol{u}_{ij}$ and $\boldsymbol{z}_{ij}$ denote the corresponding design vectors, and ε_{ij} are i. i. d. $\mathrm{N}(0, \sigma^2)$ errors. The functions $f_1, \ldots, f_p$ of the continuous covariates $x_1, \ldots, x_p$ are nonlinear population effects and will be modelled by (Bayesian) P-splines or some other basis function representation as introduced in Section 2.2. Thus, each function $f_l, l = 1, \ldots, p$, is approximated by a linear combination

$$f_l(x_l) = \sum_{k=1}^{K} \gamma_{kl} B_k(x_l)$$

of appropriate basis functions. Note that we have notationally suppressed the possibility that the number of knots as well as the degree and the type of basis functions employed for a specific covariate may be different. An important special case of model (4.1) is obtained when a continuous covariate, say x_1, is chosen as time t. Then, the nonparametric function $f_1(x_1) = f_{time}(t)$ represents the population time trend.

Generalized additive mixed models (GAMMs) for responses y_{ij} with (conditional) exponential family densities (3.34) as in Section 3.3.1 extend the linear mixed predictor of GLMMs to the additive mixed predictor

$$\begin{aligned}\eta_{ij}^{add} &= \boldsymbol{u}_{ij}'\boldsymbol{\beta} + f_1(x_{ij1}) + \ldots + f_p(x_{ijp}) + \boldsymbol{z}_{ij}'\boldsymbol{\alpha}_i \\ &= \eta_{ij}^{lin} + f_1(x_{ij1}) + \ldots + f_p(x_{ijp}).\end{aligned} \tag{4.2}$$

To obtain a compact representation of GAMMs, we can represent all the nonparametric effects in vector-matrix notation. Arranging predictors in vectors $\boldsymbol{\eta}_i = (\eta_{i1}, \ldots, \eta_{in_i})$ and defining appropriate design matrices as for GLMMs and GAMs, predictors for SPMMs can always be expressed as

$$\boldsymbol{\eta}_i = \boldsymbol{U}_i\boldsymbol{\beta} + \boldsymbol{X}_{i1}\boldsymbol{\gamma}_1 + \ldots + \boldsymbol{X}_{ip}\boldsymbol{\gamma}_p + \boldsymbol{Z}_i\boldsymbol{\alpha}_i, \quad i = 1, \ldots, n. \tag{4.3}$$

Let $\boldsymbol{f}_{il} = (f(x_{i1l}), \ldots, f(x_{in_il}))'$ denote the vector of function values of a typical function $f_l(x_l)$ evaluated at observed covariate values $x_{ijl}, j = 1, \ldots, n_i$. From the basis function representation we obtain

$$\boldsymbol{f}_{il} = \boldsymbol{X}_{il}\boldsymbol{\gamma}_l,$$

where $\boldsymbol{\gamma}_l = (\gamma_{l1}, \ldots, \gamma_{lK})'$ denotes the regression coefficients, and $\boldsymbol{X}_{il}$ is the unit-specific $(n_i \times K)$-design matrix

$$\boldsymbol{X}_{il} = \begin{pmatrix} B_1(x_{i1l}) & \cdots & B_K(x_{i1l}) \\ \vdots & \ddots & \vdots \\ B_1(x_{in_il}) & \cdots & B_K(x_{in_il}) \end{pmatrix}.$$

Stacking individual predictors $\boldsymbol{\eta}_i$ to the 'grand' predictor vector $\boldsymbol{\eta} = (\boldsymbol{\eta}_1', \ldots, \boldsymbol{\eta}_n')'$ and defining 'grand' design matrices $\boldsymbol{U}$, $\boldsymbol{X}_1, \ldots, \boldsymbol{X}_p$, $\boldsymbol{Z}$ by

$$\boldsymbol{U} = \begin{pmatrix} \boldsymbol{U}_1 \\ \vdots \\ \boldsymbol{U}_n \end{pmatrix}, \quad \boldsymbol{X}_l = \begin{pmatrix} \boldsymbol{X}_{1l} \\ \vdots \\ \boldsymbol{X}_{nl} \end{pmatrix}, \quad \boldsymbol{Z} = \begin{pmatrix} \boldsymbol{Z}_1 & \cdots & 0 \\ \vdots & \ddots & \vdots \\ 0 & \cdots & \boldsymbol{Z}_n \end{pmatrix}$$

finally yields the model

$$\boldsymbol{\eta} = \boldsymbol{U\beta} + \boldsymbol{X}_1\boldsymbol{\gamma}_1 + \ldots + \boldsymbol{X}_p\boldsymbol{\gamma}_p + \boldsymbol{Z\alpha} \tag{4.4}$$

in complete analogy to the general form of GLMMs in Section 3.3.

GAMMs (and the special case of AMMs) are main effect models with respect to the effects of the continuous covariates: While usual parametric interactions may be included through appropriate specification of the design vectors $\boldsymbol{u}_{ij}$ and $\boldsymbol{z}_{ij}$, they do not allow for nonparametric interactions.

Varying coefficient mixed models

A special form of interaction is possible in *varying coefficient mixed models* (VCMMs): They extend the additive predictor (4.2) to

$$\eta_{ij}^{vc} = \eta_{ij}^{add} + g_1(x_{ij1})v_{ij1} + \ldots + g_m(x_{ijm})v_{ijm}, \tag{4.5}$$

where the functions g_l, $l = 1, \ldots, m$, can be interpreted as nonlinear effects of binary or continuous covariates $v_1, \ldots, v_m$ (the so-called interaction variables), varying over the domain of the covariates $x_1, \ldots, x_m$ (therefore called the effect modifiers). For the special cases $g_l(x_l) \equiv \beta_l$ and $g_l(x_l) = x_l\beta_l$ we reobtain a parametric effect of v_l and a parametric interaction between x_l and v_l, respectively.

These special cases also reveal that varying coefficient models impose additional identifiability constraints supplementing the centring condition discussed in Section 2.3.1 for additive models. For example, a model of the form

$$\eta_{ij} = \ldots + v_{ij}\beta + v_{ij}g(x_{ij}) + \ldots$$

is not identifiable since the varying coefficient term $vg(x)$ includes the parametric effect $v\beta$ as a special case when the function $g(x)$ is constant. Imposing the same centring restriction on the functions $g_l(x_l)$ as in the additive model in Section 2.3.1 allows us to overcome this type of identifiability constraint. In fact, the centring condition in additive models appears as a special case with $v \equiv 1$. In addition, when centring the functions $g_l(x_l)$, it is of vital importance to include the main effect $v_l\beta_l$ separately in the model. An advantage of the centred model is that it permits inclusion of an interaction variable v with different effect modifiers, for example

$$\eta_{ij} = \ldots + v_{ij}\beta + v_{ij}g_1(x_{ij1}) + v_{ij}g_2(x_{ij2}) + \ldots$$

Without the centring condition, the two varying coefficient terms would be collinear since they both contain the parametric effect $v\beta$.

An important example of varying coefficient models are models with time-varying effects. In this case, time takes the role as the effect modifier, leading to

$x_1 = t$ and $g_1(t_{ij})$ as time-varying effect of v_1. Such a specification for example permits us to assign different time trends to subgroups coded through binary indicators v defining the groups, or to investigate time-stability of covariate effects.

Applying the same vector-matrix notation as for GAMMs, VCMMs can be cast into a similar framework. In a first step, all effect modifiers are represented in terms of basis functions, leading to a representation for the vectors of function evaluations $\boldsymbol{g}_{il}^* = (g_h(x_{i1l}), \ldots, g_h(x_{in_il}))'$ as

$$\boldsymbol{g}_{il}^* = \boldsymbol{X}_{il}^* \boldsymbol{\gamma}_l,$$

where $\boldsymbol{X}_{il}^*$ is defined in complete analogy to the GAMM case by applying the basis functions representation and $\boldsymbol{\gamma}_l$ represents the corresponding regression coefficients. To account for the additional multiplication with the interaction variable, the design matrix $\boldsymbol{X}_{il}^*$ then has to be multiplied row-wise with the corresponding values of the interaction variable. In matrix notation this corresponds to a pre-multiplication with a diagonal matrix, that is,

$$\boldsymbol{g}_{il} = \text{diag}(v_{i1l}, \ldots, v_{in_il}) \boldsymbol{X}_{il}^* \boldsymbol{\gamma}_l = \boldsymbol{X}_{il} \boldsymbol{\gamma}_l$$

where $\boldsymbol{g}_{il} = (v_{i1l} g_l(x_{i1l}), \ldots, v_{in_il} g_l(x_{in_il}))'$. The complete model is then again obtained by stacking all individual-specific matrices yielding a similar expression as (4.4).

ANOVA type interactions

Obviously, the terms $v_l g_l(x_l)$ may be interpreted as a special form of interaction. However, this specific form may be too restrictive for interactions between two continuous covariates x_1 and x_2, say, if the effect varies with *both* x_1 and x_2. In this case, the consideration of interaction surfaces will be more appropriate and we will introduce so-called *ANOVA type interaction models* in Chapter 5 in connection with spatial models. The predictor of such an ANOVA model is of the form

$$\eta_{ij}^{ANOVA} = \eta_{ij}^{add} + f_{1|2}(x_{ij1}, x_{ij2}),$$

where $f_{1|2}(x_{ij1}, x_{ij2})$ is a two-dimensional surface.

Generic representation

As we have seen in the examples discussed so far, any type of effect in an SPMM can be represented in terms of a (possibly high-dimensional) design matrix and a corresponding vector of regression coefficients. To ease notation and to facilitate description of inferential procedures, we will therefore consider the following generic notation: A SPMM is given by the individual-specific predictor

$$\boldsymbol{\eta}_i = \boldsymbol{U}_i \boldsymbol{\beta} + \boldsymbol{X}_{i1} \boldsymbol{\gamma}_1 + \ldots + \boldsymbol{X}_{ir} \boldsymbol{\gamma}_r + \boldsymbol{Z}_i \boldsymbol{\alpha}_i \tag{4.6}$$

where r denotes the number of various semiparametric effects in the model. These semiparametric effects may comprise arbitrary combinations of nonlinear effects and varying coefficient terms and will be further extended to include interaction surfaces and spatial effects in Chapter 5.

4.1.2 Observation models for categorical responses

In the case of *nominal responses* Y_{ij} with unordered categories $c = 1, \ldots, C$, a *semiparametric mixed multinomial logit model* with C as reference category is specified by the probabilities for categories $c = 1, \ldots, C-1$ as

$$\mathbb{P}(Y_{ij} = c) = \frac{\exp(\eta_{ij}^{(c)})}{1 + \sum_{s=1}^{C-1} \exp(\eta_{ij}^{(s)})},$$

where $\eta_{ij}^{(c)}$ is a category-specific predictor. For example, $\eta_{ij}^{(c)}$ may be specified as a GAMM-type predictor

$$\eta_{ij}^{(c)} = \boldsymbol{u}_{ij}'\boldsymbol{\beta}^{(c)} + f_1^{(c)}(x_{ij1}) + \ldots + f_m^{(c)}(x_{ijm}) + \boldsymbol{z}_{ij}'\boldsymbol{\alpha}_i^{(c)} \tag{4.7}$$

with *category-specific* fixed effects $\boldsymbol{\beta}^{(c)}, f_1^{(c)}, \ldots, f_m^{(c)}$ and random effects $\boldsymbol{\alpha}_i^{(c)}$, and with *global* design vectors and covariates, not depending on the specific category.

If *category-specific* covariates $\boldsymbol{u}^{(c)}$, $\boldsymbol{x}^{(c)}$ or $\boldsymbol{z}^{(c)}$ are available, GAMM-type predictors can be extended with terms such as

$$\boldsymbol{w}_{ij}^{(c)\prime}\boldsymbol{\beta}, \quad f_l(x_{ijl}^{(c)}),\ l = 1, \ldots, m, \quad \text{or} \quad \boldsymbol{z}_{ij}^{(c)\prime}\boldsymbol{\alpha}_i$$

where the fixed effect vector $\boldsymbol{\beta}$, the functions f_l, $l = 1, \ldots, m$, and the random effect vectors $\boldsymbol{\alpha}_i$ are postulated to be *global*, that is, identical for all categories. The common, global random effect $\boldsymbol{\alpha}_i$ introduces correlation between categories to overcome the 'independence from irrelevant alternatives' property inherent in traditional parametric multinomial models (see for example McFadden, 1973). As before, various other extensions such as to VCMMs can be derived.

Bayesian inference is considerably facilitated by introducing *auxiliary, latent variables* (or 'utilities') $l^{(c)}$ associated with category c as considered in Section 2.2.3. In the case of longitudinal data assume that $l_{ij}^{(c)}$ follows a latent SPMM

$$l_{ij}^{(c)} = \eta_{ij}^{(c)} + \varepsilon_{ij}^{(c)}, \quad c = 1, \ldots, C-1,$$

with a semiparametric, for example GAMM-type, predictor $\eta_{ij}^{(c)}$.

For ordered responses we assume that the response Y_{ij} is a categorized version of the latent variable

$$l_{ij} = \tilde{\eta}_{ij} + \varepsilon_{ij},$$

defined through the threshold mechanism

$$Y_{ij} = c \quad \Leftrightarrow \quad \theta^{(c-1)} < l_{ij} \leq \theta^{(c)}, \quad c = 1, \ldots, C,$$

with ordered thresholds $-\infty = \theta^{(0)} < \theta^{(1)} < \ldots < \theta^{(C)} = \infty$. With a semiparametric mixed predictor $\tilde{\eta}_{ij}$, for example a GAMM-type predictor, one gets *semiparametric cumulative mixed models*. Recall that for identifiability reasons the fixed effects term $\boldsymbol{u}_{ij}'\boldsymbol{\beta}$ must not contain an intercept term β_0. Otherwise one of the thresholds, for example $\theta^{(1)}$, has to be set to zero.

In the basic cumulative models defined so far, covariates have no impact on the values of the thresholds. However, in some applications it is likely that the thresholds themselves are covariate-dependent or, equivalently, that some of the fixed or random effects are category-specific. This leads to extended semiparametric cumulative mixed models where the predictor is

$$\eta_{ij}^{(c)} = \theta^{(c)} - \boldsymbol{u}_{ij}'\boldsymbol{\beta} - \boldsymbol{w}_{ij}'\boldsymbol{\beta}^{(c)} - f_1^{(c)}(x_{ij1}) - \ldots - \boldsymbol{z}_{ij}'\boldsymbol{\alpha}_i^{(c)},$$

for example. In such a model, the covariate-dependent thresholds are given by

$$\tilde{\theta}_{ij}^{(c)} = \theta^{(c)} - \boldsymbol{w}_{ij}'\boldsymbol{\beta}^{(c)} - f_1^{(c)}(x_{ij1}) - \ldots - \boldsymbol{z}_{ij}'\boldsymbol{\alpha}_i^{(c)}.$$

Although easily defined, such extended cumulative models are of considerably increased complexity, since the ordering restrictions

$$\tilde{\theta}_{ij}^{(1)} < \ldots < \tilde{\theta}_{ij}^{(C-1)}$$

have to be fulfilled for all possible values of covariates and their effects. This may cause severe problems in estimating such models and category-specific effects should therefore only be used with care.

Semiparametric extensions of sequential regression models can be derived in complete analogy to the cumulative case.

Example 4.1: Forest health
Cumulative additive mixed models allow for a more detailed analysis of the forest health data as compared to a binary regression model. We may now consider the predictor structure

$$\eta_{it}^{(c)} = \theta^{(c)} - \boldsymbol{u}_{it}'\boldsymbol{\beta} - f^{(c)}(t) - f(age_{it}) - \alpha_i$$

for a cumulative logit or probit model with more than two categories. Both the thresholds $\theta^{(c)}$ and the category-specific trends $f^{(c)}(t)$ allow us to separate the categories from each other while the remaining effects are assumed to be constant across categories to ensure identifiability of the model. △

4.1.3 Gaussian priors for regression parameters and functions

Throughout this section, we take *fixed effect priors* to be uniform and noninformative,

$$p(\boldsymbol{\beta}) \propto \text{const}$$

or to be Gaussian priors

$$\boldsymbol{\beta} \sim \text{N}(\boldsymbol{c}, \boldsymbol{C})$$

as for GLMMs in Chapter 3. Uniform priors correspond to 'fixed' effects in frequentist approaches. They are improper, but with appropriate choice of other priors, such as those for variance parameters, posteriors are usually proper if the number of observations is large enough compared to the (fixed) dimension of $\boldsymbol{\beta}$. This is the case in most standard situations for longitudinal data. Theoretical results in Sun, Tsutakawa and He (2001) for hierarchical linear mixed models and Fahrmeir and Kneib (2009) for generalized SPMMs provide sufficient conditions for propriety of posteriors. Loosely speaking, such conditions resemble conditions for the existence of MLEs in GLMs for a finite sample: They are close to necessity and comparably easy to check for Gaussian responses, but generally much more difficult to check and farther away from necessity for non-normal responses.

In more sparse data situations, a (weakly) informative Gaussian prior $\boldsymbol{\beta} \sim \text{N}(\boldsymbol{c}, \boldsymbol{C})$ leads to Bayesian ridge-type shrinkage and regularization. In practice it is common to take $\boldsymbol{C}$ to be diagonal and to set $\boldsymbol{c} = \boldsymbol{0}$. The diagonal elements control the amount of shrinkage towards zero, with large diagonal elements corresponding to weakly informative or approximately flat priors. With appropriate choice of priors for the variance components, such a strategy ensures that posteriors will be proper.

We also assume that *random effect priors* are (conditionally) Gaussian,

$$\boldsymbol{\alpha}_i|\boldsymbol{Q} \overset{\text{i.i.d.}}{\sim} \text{N}(\boldsymbol{0}, \boldsymbol{Q}),$$

or, equivalently, with $\boldsymbol{\alpha} = (\boldsymbol{\alpha}_1', \ldots, \boldsymbol{\alpha}_n')'$

$$\boldsymbol{\alpha}|\boldsymbol{Q} \sim \text{N}(\boldsymbol{0}, \boldsymbol{G}),$$

$\boldsymbol{G} = \text{blockdiag}(\boldsymbol{Q}, \ldots, \boldsymbol{Q})$. More flexible priors, in particular DP(M) priors, will be discussed in Section 4.4.

Priors for functions f_l are defined through *Gaussian smoothness priors* for the corresponding vector $\boldsymbol{\gamma}_l$ in the chosen basis function representation. We prefer Bayesian P-splines with a B-spline basis as described in Section 2.2. Then the Gaussian smoothness prior is specified by

$$p(\boldsymbol{\gamma}_l|\tau_l^2) \propto \frac{1}{(\tau_l^2)^{\text{rank}(\boldsymbol{K}_l)/2}} \exp\left(-\frac{1}{2\tau_l^2}\boldsymbol{\gamma}_l'\boldsymbol{K}_l\boldsymbol{\gamma}_l\right), \tag{4.8}$$

where $\boldsymbol{K}_l = \boldsymbol{D}'\boldsymbol{D}$ is the penalty matrix and $\boldsymbol{D}$ the matrix of first or second order differences of basis coefficients. As explained in Section 2.2, $\boldsymbol{K}_l$ enforces smoothness and the variance component τ_l^2 corresponds to the (inverse) smoothing parameter. This Gaussian smoothness prior is partially improper, because $\text{rank}(\boldsymbol{K}_l) = \dim(\boldsymbol{\gamma}_l) - d$, where $d = 1$ or $d = 2$ for first or second order differences, respectively. Therefore, in theory, the question of posterior propriety arises again. Theoretical results are given in Fahrmeir and Kneib (2009). In practice, problems with improper posteriors are quite unusual.

With a truncated polynomial basis (as discussed in Section 2.2.1), and similarly with radial basis functions, the problem can be avoided by incorporating the coefficient of the global polynomial term into the vector $\boldsymbol{\beta}$ of fixed effects, and penalizing only the coefficient vector $\boldsymbol{\gamma}_l$ corresponding to the truncated power terms. Then the prior is

$$p(\boldsymbol{\gamma}_l|\tau_l^2) \propto \frac{1}{(\tau_l^2)^{\dim(\boldsymbol{\gamma}_l)/2}} \exp\left(-\frac{1}{2\tau_l^2}\boldsymbol{\gamma}_l'\boldsymbol{\gamma}_l\right),$$

because $\boldsymbol{K}_l = \boldsymbol{I}$ and $\text{rank}(\boldsymbol{K}_l) = \dim(\boldsymbol{\gamma}_l)$.

Similar strategies have been suggested for P-splines in Fahrmeir, Kneib and Lang (2004) by reparametrization and deletion of unpenalized parts, see Section 4.2.1 of this book, as well as in Chib and Jeliazkov (2006) and Panagiotelis and Smith (2008) by placing additional restrictions on the priors.

Another choice, which leads to computationally more demanding but possibly less stable algorithms, are Bayesian smoothing splines, see Hastie and Tibshirani (2000) for GAMs, Lin and Zhang (1999) for GAMMs and Li, Lin and Müller (2010) in the context of SPMMs with Gaussian and Dirichlet process priors for random effects.

For empirical Bayes (EB) inference, variance components are treated as unknown constants and are estimated through (marginal or restricted) ML methods. For full Bayes (FB) inference, elicitation of *variance components priors* is still an area of recent research, see, for example Gelman (2006). Conditionally conjugate priors have the computational advantage that Gibbs sampling is possible for drawing from posteriors. Inverse Gamma priors $\text{IG}(a, b)$,

$$p(\tau^2) \propto \frac{1}{(\tau^2)^{a+1}} \exp\left(-\frac{b}{\tau^2}\right)$$

which are proper for $a > 0$, $b > 0$, are a popular choice for a typical variance component τ^2. For a general variance-covariance matrix, the corresponding extension is given by an inverse Wishart prior. The limiting case $a \to 0$, $b \to 0$ is Jeffreys' noninformative but improper prior $p(\tau^2) \propto \tau^{-2}$. However, its use can lead to improper posteriors, with MCMC samplers unable to detect this. As a result, 'diffuse' IG priors with small a, b, e.g. $a = b = 0.1, 0.01$ or 0.001 have been suggested. Recent research has shown, however, that these 'diffuse' $\text{IG}(\varepsilon, \varepsilon)$-priors can actually be quite informative in Bayesian variance component models with a small number of groups or units (Natarajan and Kass, 2000, Gelman, 2006). In our experience, $\text{IG}(\varepsilon, \varepsilon)$-priors are a comparably robust choice for the error variance σ^2 in Gaussian models and for the variances (or inverse smoothing parameters) in Gaussian smoothness priors for nonparametric functions (unless ε is chosen extremely small). Nevertheless, it is good practice—at least in sparse data situations—to check sensitivity by carrying out analyses with different values of the hyperparameters a, b.

Alternative priors, suggested in Gelman (2006), are uniform priors $p(\tau) \propto$ const for standard deviations τ or the folded-t class of priors for τ^2, in particular the folded-Cauchy density

$$p(\tau^2) \propto (\tau^2 + s^2)^{-1},$$

with s as a hyperparameter. The uniform prior for the standard deviation can also be expressed in form of an improper IG-type prior with hyperparameters $a = -0.5$, $b = 0$. Sun, Tsutakawa and He (2001) and Fahrmeir and Kneib (2009) provide sufficient conditions for propriety of posteriors in this case. Another 'uniform shrinkage' prior has been developed in Li, Lin and Müller (2010). Their analyses as well as those in Zhao *et al.* (2006) using folded-Cauchy priors confirm our own experience that IG$(\varepsilon, \varepsilon)$-priors are rather robust and only weakly informative for posterior estimation of functions.

4.2 Estimation

4.2.1 Empirical Bayes inference

Empirical Bayes inference in SPMMs relies on the close connection between mixed models and penalized estimation of regression coefficients. It is therefore also closely related to penalized maximum likelihood estimation and can be interpreted in a frequentist context. The essential tool will be a mixed model representation of regression terms with partially improper priors and, hence, we will start with a reconsideration of this type of priors. Afterwards we will show how SPMMs can be recast as mixed models, making inferential techniques introduced in Chapter 3 available. Finally, the construction of pointwise and simultaneous credible regions and tests on the functional form of nonlinear effects will be discussed.

In most parts of this section, we will deal with the simple problem of univariate nonparametric smoothing, that is, a regression model with predictor

$$\eta_i = f(x_i), \quad i = 1, \ldots, n, \tag{4.9}$$

for illustration. The corresponding considerations will then form the building block for estimation in general SPMMs with longitudinal structure.

Intuitive example

To get an intuitive understanding of the relationship between smoothing and mixed models, we consider the very basic example of penalized splines in truncated power series representation in combination with a ridge penalty. In this case, a nonparametric effect is expanded in the basis

$$\begin{aligned} B_1(x) &= 1, \quad B_2(x) = x, \quad \dots, \quad B_{D+1}(x) = x^D, \\ B_{D+2}(x) &= (x - \kappa_1)_+^D, \quad \dots, \quad B_K(x) = (x - \kappa_{M-1})_+^D, \end{aligned}$$

consisting of a polynomial part and the truncated polynomials. This allows us to represent the smoothing problem (4.9) in matrix notation as

$$\boldsymbol{\eta} = \boldsymbol{X}\boldsymbol{\gamma}$$

where $\boldsymbol{X}$ contains the evaluations of the basis functions and $\boldsymbol{\gamma}$ is the vector of corresponding regression coefficients. As described in Section 2.2.1, a suitable penalty for splines in truncated power series representation is a ridge penalty for the coefficients of the truncated polynomials. In a Bayesian formulation, this is equivalent to assuming noninformative prior distributions

$$p(\gamma_k) \propto \text{const}, \quad k = 1, \dots, D+1$$

for the polynomial basis and i. i. d. Gaussian priors

$$\gamma_k \sim \mathrm{N}(0, \tau^2), \quad k = D+2, \dots, K$$

for the truncated polynomials.

It can already be noted at this point that this is exactly the Bayesian formulation of a mixed model, where flat priors are assigned to the fixed effects whereas i. i. d. Gaussian priors are assigned to the random effects. This can be made even more obvious by defining the design matrices

$$\boldsymbol{U} = \begin{pmatrix} 1 & x_1 & x_1^2 & \dots & x_1^D \\ \vdots & \vdots & \vdots & & \vdots \\ 1 & x_n & x_n^2 & \dots & x_n^D \end{pmatrix} \qquad \boldsymbol{Z} = \begin{pmatrix} (x_1 - \kappa_1)_+^D & \dots & (x_1 - \kappa_{M-1})_+^D \\ \vdots & & \vdots \\ (x_n - \kappa_1)_+^D & \dots & (x_n - \kappa_{M-1})_+^D \end{pmatrix}$$

and the corresponding subvectors of coefficients

$$\boldsymbol{\beta} = (\gamma_1, \dots, \gamma_{D+1})' \qquad \boldsymbol{\alpha} = (\gamma_{D+2}, \dots, \gamma_K)'.$$

Decomposing the model in this way yields the model equation

$$\boldsymbol{\eta} = \boldsymbol{U}\boldsymbol{\beta} + \boldsymbol{Z}\boldsymbol{\alpha}$$

supplemented with the distributional assumptions

$$p(\boldsymbol{\beta}) \propto \text{const} \qquad \boldsymbol{\alpha} \sim \mathrm{N}(\boldsymbol{0}, \boldsymbol{G})$$

where $\boldsymbol{G} = \tau^2 \boldsymbol{I}_{M-1}$. This is exactly the structure of a general mixed model as introduced in Sections 3.1.2 and 3.3.2 (with the restriction $\boldsymbol{c} = \boldsymbol{0}$, $\boldsymbol{C}^{-1} = \boldsymbol{0}$ to obtain a noninformative prior for the fixed effects). Note that, in particular, there is no grouping involved in this mixed model and that the random effects assumption is not motivated by a hypothetical population individuals are drawn from. In contrast, the same random effects apply to all observations and the random effects assumption either represents prior knowledge or can be interpreted as an attempt to regularize the estimation of basis coefficients.

The main advantage of establishing the connection between mixed models and penalized smoothing is the availability of automatic procedures for determining the random effects variance in mixed models such as restricted maximum likelihood estimation. Such procedures can now be turned into a strategy for estimating the smoothing parameter, providing a fully automatic estimation routine for the nonparametric smoothing problem. We will also discuss some further advantages and properties of the mixed model representation later on in this section.

It is now tempting to extend the mixed model representation concept to general nonparametric effects $\boldsymbol{\eta} = \boldsymbol{X}\boldsymbol{\gamma}$ with quadratic penalty

$$\text{pen}(\boldsymbol{\gamma}) = \boldsymbol{\gamma}'\boldsymbol{K}\boldsymbol{\gamma}$$

or multivariate Gaussian prior

$$p(\boldsymbol{\gamma}) \propto \left(\frac{1}{\tau^2}\right)^{\frac{\text{rank}(\boldsymbol{K})}{2}} \exp\left(-\frac{1}{2\tau^2}\boldsymbol{\gamma}'\boldsymbol{K}\boldsymbol{\gamma}\right) \tag{4.10}$$

by interpreting $p(\boldsymbol{\gamma})$ as a random effects distribution. When comparing the random effects prior $\boldsymbol{\alpha} \sim \text{N}(\boldsymbol{0}, \boldsymbol{G})$ with density

$$p(\boldsymbol{\alpha}) \propto \left(\frac{1}{|\boldsymbol{G}|}\right)^{\frac{1}{2}} \exp\left(-\frac{1}{2}\boldsymbol{\alpha}'\boldsymbol{G}\boldsymbol{\alpha}\right)$$

with the density of the smoothness prior (4.10), both are of exactly the same form since they correspond to multivariate Gaussian priors and would coincide if

$$\frac{1}{\tau^2}\boldsymbol{K} = \boldsymbol{G}^{-1}.$$

In this special case, both $\boldsymbol{G}$ and $\boldsymbol{K}$ would have full rank and we would have

$$|\boldsymbol{G}| = (\tau^2)^{\text{rank}(\boldsymbol{K})}|\boldsymbol{K}^{-1}| = (\tau^2)^{\dim(\boldsymbol{\gamma})}|\boldsymbol{K}^{-1}| \propto (\tau^2)^{\dim(\boldsymbol{\gamma})}.$$

However, in many cases of practical interest, the penalty matrix $\boldsymbol{K}$ is rank-deficient, leading to a non-trivial null space of the matrix $\boldsymbol{K}$. This is in contradiction to the standard assumptions in mixed models. In particular, the Fisher scoring estimation scheme for determining REML estimates depends on the availability of derivatives of the covariance matrix $\tau^2 \boldsymbol{K}^{-1}$ which are not available unless $\boldsymbol{K}$ is of full rank.

Often the null space of $\boldsymbol{K}$ is not only a mathematical artefact but corresponds to interpretable properties of the penalized smoother. For example, for penalized splines with d-th order differences, the null space corresponds to the polynomial of order $d-1$ that remains unpenalized by the difference penalty. From a Bayesian perspective, this means that a flat prior is assigned to a part of the function whereas another part is assigned an informative prior to enforce smoothness. For truncated power series splines, these two parts are easily accessible since they simply correspond to specific basis functions and coefficients. In general, however, there is no immediate separation between these two parts that would facilitate the decomposition into a 'fixed effects' and a 'random effects' part. Still, it will turn out in the following paragraph that a linear reparametrization of $\boldsymbol{\gamma}$ provides exactly such a decomposition.

We will heuristically introduce this reparametrization for penalized splines based on B-splines first and will then discuss general strategies applicable for any type of penalized smoother with partially improper Gaussian prior. In the latter case, the key to the reparametrization will be to study a spectral decomposition of the penalty matrix $\boldsymbol{K}$ that will provide us with suitable bases for both the random and the fixed part of $\boldsymbol{\gamma}$.

Mixed model representation for P-splines

We will now consider nonparametric smoothing where $f(x)$ is represented in terms of a K-dimensional B-spline basis with *equidistant* knots, i.e.

$$f(x) = \sum_{k=1}^{K} \gamma_k B_k(x).$$

Again this allows to rewrite the model in matrix notation as

$$\boldsymbol{\eta} = \boldsymbol{X}\boldsymbol{\gamma},$$

and a d-th order random-walk prior with density (4.10) is assigned to $\boldsymbol{\gamma}$. In this special case, we have $\boldsymbol{K} = \boldsymbol{D}'\boldsymbol{D}$, where $\boldsymbol{D}$ is a d-th order difference matrix, and $R = \text{rank}(\boldsymbol{K}) = K - d$, with $K = \dim(\boldsymbol{\gamma})$. As a consequence, the null space of $\boldsymbol{K}$ is d-dimensional, corresponding to the d regression coefficients describing the polynomial of order $d-1$ that remains unpenalized by $\boldsymbol{K}$.

To separate the penalized and the unpenalized part of $\boldsymbol{\gamma}$, we have to find suitable bases for corresponding subspaces of $\mathbb{R}^K$. For the polynomial space, such a basis can be defined by the columns of the $(K \times d)$ matrix

$$\tilde{\boldsymbol{U}} = \begin{pmatrix} 1 & 1 & 1^2 & \dots & 1^{d-1} \\ 1 & 2 & 2^2 & \dots & 2^{d-1} \\ \vdots & \vdots & \vdots & & \vdots \\ 1 & K & K^2 & \dots & K^{d-1} \end{pmatrix}. \tag{4.11}$$

Note that this choice relies on the fact that the knots are equidistant and any polynomial basis in a sequence of K equidistant values will be suitable. For non-equidistant knots (in combination with an accordingly modified difference penalty), a matrix of basis vectors for the polynomial part is given by

$$\tilde{\boldsymbol{U}} = \begin{pmatrix} 1 & \kappa_1 & \kappa_1^2 & \dots & \kappa_1^{d-1} \\ 1 & \kappa_2 & \kappa_2^2 & \dots & \kappa_2^{d-1} \\ \vdots & \vdots & \vdots & & \vdots \\ 1 & \kappa_K & \kappa_K^2 & \dots & \kappa_K^{d-1} \end{pmatrix}$$

where $\kappa_1, \dots, \kappa_K$ denote the sequence of knots used to construct the spline.

To capture the deviation from the unpenalized polynomial, consider the $(K \times R)$ dimensional matrix

$$\tilde{\boldsymbol{Z}} = \boldsymbol{D}'(\boldsymbol{D}\boldsymbol{D}')^{-1} \tag{4.12}$$

where $R = \text{rank}(\boldsymbol{K})$ denotes the rank of the penalty matrix. We will see in the following considerations that this matrix contains a basis of the orthogonal complement of the polynomial part and also translates the penalty for the corresponding part of γ into a particularly simple form. To see this, we rewrite γ as

$$\gamma = \tilde{\boldsymbol{U}}\boldsymbol{\beta} + \tilde{\boldsymbol{Z}}\boldsymbol{\alpha}$$

providing a one-to-one transformation of γ into the vectors $\boldsymbol{\beta}$ and $\boldsymbol{\alpha}$ of dimensions $(d \times 1)$ and $(R \times 1)$, respectively, since the composed matrix $(\tilde{\boldsymbol{U}}, \tilde{\boldsymbol{Z}})$ has full rank.

Inserting the reparametrization in the quadratic form defining the prior distribution yields

$$\begin{aligned} \gamma'\boldsymbol{K}\gamma &= (\tilde{\boldsymbol{U}}\boldsymbol{\beta} + \tilde{\boldsymbol{Z}}\boldsymbol{\alpha})'\boldsymbol{K}(\tilde{\boldsymbol{U}}\boldsymbol{\beta} + \tilde{\boldsymbol{Z}}\boldsymbol{\alpha}) \\ &= \boldsymbol{\beta}'\tilde{\boldsymbol{U}}'\boldsymbol{K}\tilde{\boldsymbol{U}}\boldsymbol{\beta} + 2\boldsymbol{\beta}'\tilde{\boldsymbol{U}}'\boldsymbol{K}\tilde{\boldsymbol{Z}}\boldsymbol{\alpha} + \boldsymbol{\alpha}'\tilde{\boldsymbol{Z}}'\boldsymbol{K}\tilde{\boldsymbol{Z}}\boldsymbol{\alpha} \\ &= \boldsymbol{\alpha}(\boldsymbol{D}\boldsymbol{D}')^{-1}\boldsymbol{D}\boldsymbol{D}'\boldsymbol{D}\boldsymbol{D}'(\boldsymbol{D}\boldsymbol{D}')^{-1}\boldsymbol{\alpha} \\ &= \boldsymbol{\alpha}'\boldsymbol{\alpha}. \end{aligned}$$

Obviously, the parameters corresponding to the polynomial basis completely drop out of the penalty term, corresponding to the fact that they are not penalized. The penalty for $\boldsymbol{\alpha}$ transforms into a simple ridge penalty. Hence, from a Bayesian perspective, replacing $\boldsymbol{\gamma}$ with $\boldsymbol{\beta}$ and $\boldsymbol{\alpha}$ results in the prior distributions

$$p(\boldsymbol{\beta}) \propto \text{const}$$

and

$$\boldsymbol{\alpha} \sim \mathrm{N}(\mathbf{0}, \tau^2 \boldsymbol{I}_R)$$

providing a clear separation of unpenalized or 'fixed' effects in $\boldsymbol{\beta}$ and penalized or 'random' effects in $\boldsymbol{\alpha}$.

Similarly, inserting the reparametrization into the model equation yields a decomposition of the vector of function evaluations into fixed and random effects, i.e.

$$\begin{aligned}\eta &= \boldsymbol{X}\gamma \\ &= \boldsymbol{X}(\tilde{\boldsymbol{U}}\boldsymbol{\beta} + \tilde{\boldsymbol{Z}}\boldsymbol{\alpha}) \\ &= \boldsymbol{U}\boldsymbol{\beta} + \boldsymbol{Z}\boldsymbol{\alpha}\end{aligned}$$

where $\boldsymbol{U} = \boldsymbol{X}\tilde{\boldsymbol{U}}$ and $\boldsymbol{Z} = \boldsymbol{X}\tilde{\boldsymbol{Z}}$. As a consequence, we can interpret the penalized spline as a mixed model with fixed effects $\boldsymbol{\beta}$ and i. i. d. Gaussian random effects $\boldsymbol{\alpha}$ in the reparametrized version, and can employ mixed model methodology for estimation. In particular, ML or REML estimates are available for the smoothing variance, enabling a completely data-driven fit without the need for manual tuning of the smoothing parameter.

Note that the polynomial of order $d-1$ in an arbitrary equidistant sequence of values contained in $\tilde{\boldsymbol{U}}$ actually turns into a polynomial of order $d-1$ in x when premultiplying the basis matrix $\tilde{\boldsymbol{U}}$ with the B-spline matrix $\boldsymbol{X}$ (see Greven, 2007, for a proof).

Mixed model representation for general penalized smoothers

The previous paragraph discussed a mixed model representation for the special case of B-splines with difference penalties. We will now introduce a general mixed model representaton for penalized smoothers with rank-deficient penalty matrix and discuss some corresponding theoretical results. Therefore we consider the regression model

$$\eta = \boldsymbol{X}\gamma, \qquad p(\gamma) \propto \exp\left(-\frac{1}{2\tau^2}\gamma'\boldsymbol{K}\gamma\right).$$

where $K = \dim(\gamma)$ and $\operatorname{rank}(\boldsymbol{K}) = R < K$. For $K = R$, there is no need to reparametrize γ since its distribution is already proper and can readily be interpreted as a random effects distribution with covariance matrix $\boldsymbol{G} = \tau^2\boldsymbol{K}^{-1}$.

While the basis of the unpenalized part of γ was easy to derive from the properties of the random-walk prior, we want to restrict assumptions about $\boldsymbol{K}$ to positive semi-definiteness in the following. This property simply ensures that $\gamma'\boldsymbol{K}\gamma \geq 0$ for all $\gamma \in \mathbb{R}^K$, which is essential to let $\boldsymbol{K}$ define a penalty. The unpenalized part of $\boldsymbol{K}$ then corresponds to those γ for which $\gamma'\boldsymbol{K}\gamma = 0$, i.e. to the parameter vectors γ contained in the null space of $\boldsymbol{K}$. A basis for this null space can be derived from the spectral decomposition of the penalty matrix

$$\boldsymbol{K} = \boldsymbol{\Gamma}\boldsymbol{\Omega}\boldsymbol{\Gamma}', \qquad \boldsymbol{\Omega} = \text{diag}(\omega_1, \ldots, \omega_K)$$

where $\omega_1 \leq \omega_2 \leq \ldots \leq \omega_K$ are the eigenvalues (in ascending order) and $\boldsymbol{\Gamma}$ contains the corresponding orthonormal eigenvectors, that is, $\boldsymbol{\Gamma}'\boldsymbol{\Gamma} = \boldsymbol{I}$.

Since $\boldsymbol{K}$ is positive semidefinite, all eigenvalues are non-negative, that is, $\omega_k \geq 0$. If the rank of $\boldsymbol{K}$ is given by $\text{rank}(\boldsymbol{K}) = R$, then the null space has dimension $K - R$ and correspondingly there are exactly $K - R$ zero eigenvalues $\omega_1 = \ldots = \omega_{K-R} = 0$. We then proceed by splitting the spectral decomposition defined by $\boldsymbol{\Gamma}$ and $\boldsymbol{\Omega}$ according to null space and orthogonal deviation as $\boldsymbol{\Gamma} = (\boldsymbol{\Gamma}_1, \boldsymbol{\Gamma}_2)$ where $\boldsymbol{\Gamma}_1$ and $\boldsymbol{\Gamma}_2$ are of dimensions $K \times (K - R)$ and $K \times R$, respectively. Correspondingly, we decompose the matrix of eigenvalues as $\boldsymbol{\Omega} = \text{blockdiag}(\boldsymbol{\Omega}_1, \boldsymbol{\Omega}_2)$ where $\boldsymbol{\Omega}_1 = \text{diag}(\omega_1, \ldots, \omega_{K-R})$ and $\boldsymbol{\Omega}_2 = \text{diag}(\omega_{K-R+1}, \ldots, \omega_K)$.

In analogy to the previous section, we then consider the reparametrization

$$\boldsymbol{\gamma} = \tilde{\boldsymbol{U}}\boldsymbol{\beta} + \tilde{\boldsymbol{Z}}\boldsymbol{\alpha} \tag{4.13}$$

where $\tilde{\boldsymbol{U}} = \boldsymbol{\Gamma}_1$ and $\tilde{\boldsymbol{Z}} = \boldsymbol{\Gamma}_2\boldsymbol{\Omega}_2^{-1/2}$. Then the following properties hold for this decomposition:

1. The composed $(K \times K)$ matrix $(\tilde{\boldsymbol{U}}, \tilde{\boldsymbol{Z}})$ has full rank and therefore $\boldsymbol{\beta}$ and $\boldsymbol{\alpha}$ are a one-to-one transformation of the regression coefficients $\boldsymbol{\gamma}$. We can also invert the expression utilizing

$$(\tilde{\boldsymbol{U}}, \tilde{\boldsymbol{Z}})^{-1} = \begin{pmatrix} \boldsymbol{\Gamma}_1' \\ \boldsymbol{\Omega}_2^{1/2}\boldsymbol{\Gamma}_2' \end{pmatrix}$$

   to yield expressions for $\boldsymbol{\beta}$ and $\boldsymbol{\alpha}$ in terms of $\boldsymbol{\gamma}$, i.e.

$$\begin{aligned} \boldsymbol{\beta} &= \boldsymbol{\Gamma}_1'\boldsymbol{\gamma} \\ \boldsymbol{\alpha} &= \boldsymbol{\Omega}_2^{1/2}\boldsymbol{\Gamma}_2'\boldsymbol{\gamma}. \end{aligned}$$

2. The matrices $\tilde{\boldsymbol{U}}$ and $\tilde{\boldsymbol{Z}}$ are orthogonal since

$$\tilde{\boldsymbol{U}}'\tilde{\boldsymbol{Z}} = \boldsymbol{\Gamma}_1'\boldsymbol{\Gamma}_2\boldsymbol{\Omega}_2^{-1/2} = \boldsymbol{0}.$$

   Hence, $\boldsymbol{\beta}$ and $\boldsymbol{\alpha}$ are *a priori* independent.
3. The prior for $\boldsymbol{\beta}$ is $p(\boldsymbol{\beta}) \propto \text{const}$ since

$$\tilde{\boldsymbol{U}}'\boldsymbol{K}\tilde{\boldsymbol{U}} = \boldsymbol{\Gamma}_1'\boldsymbol{K}\boldsymbol{\Gamma}_1 = \boldsymbol{0}$$

4. The prior for $\boldsymbol{\alpha}$ is $\boldsymbol{\alpha} \sim \text{N}(\boldsymbol{0}, \tau^2\boldsymbol{I}_R)$ since

$$\begin{aligned} \tilde{\boldsymbol{Z}}'\boldsymbol{K}\tilde{\boldsymbol{Z}} &= \boldsymbol{\Omega}_2^{-1/2}\boldsymbol{\Gamma}_2'\boldsymbol{K}\boldsymbol{\Gamma}_2\boldsymbol{\Omega}_2^{-1/2} \\ &= \boldsymbol{\Omega}_2^{-1/2}\boldsymbol{\Omega}_2\boldsymbol{\Omega}_2^{-1/2} \\ &= \boldsymbol{I}_R. \end{aligned}$$

Inserting reparametrization (4.13) into the model equation leads to

$$\boldsymbol{\eta} = \boldsymbol{U}\boldsymbol{\beta} + \boldsymbol{Z}\boldsymbol{\alpha}$$

with $\boldsymbol{U} = \boldsymbol{X}\tilde{\boldsymbol{U}}$, $\boldsymbol{Z} = \boldsymbol{X}\tilde{\boldsymbol{Z}}$,

$$p(\boldsymbol{\beta}) \propto \text{const} \qquad \text{and} \qquad \boldsymbol{\alpha} \sim \text{N}(\boldsymbol{0}, \tau^2 \boldsymbol{I}_R),$$

that is, a mixed model with fixed effects $\boldsymbol{\beta}$ and i. i. d. random effects $\boldsymbol{\alpha}$ where the random effects variance τ^2 corresponds to the smoothing variance. As a consequence, any type of penalized smoother with quadratic penalty can be expressed in terms of an equivalent mixed model and REML estimation can be applied to determine the smoothing parameter.

When comparing the general reparametrization based on the spectral decomposition with the special reparametrization introduced for penalized splines in the previous paragraph, it turns out that both decompositions share the properties 1 to 4. In fact, we can use any transformation defined by matrices $\tilde{\boldsymbol{U}}$ and $\tilde{\boldsymbol{Z}}$ that fulfil these four requirements and obtain a valid mixed model representation. The decomposition based on the spectral decomposition has the advantage of general applicability and should also lead to numerically stable estimation schemes, while other choices may be more convenient and easier to understand.

As an illustration, we consider the mixed model representation of a cubic penalized B-spline $f(x)$ with second-order random-walk penalty and 18 inner knots between $[-3, 3]$. Figure 4.1 shows the columns of the matrix $\boldsymbol{U} = \boldsymbol{X}\tilde{\boldsymbol{U}}$ obtained from the spectral decomposition of the penalty matrix. At first sight, it seems surprising that the columns should actually span the space of linear functions in x. However, it turns out numerically that this is actually the case. Moreover, it should be noted that the second basis function starts at zero and therefore has an anchored level that in some way represents the intercept of the linear term. By construction, the first basis function is orthogonal to the second one and therefore represents the deviation from the first regression line.

Figure 4.2 depicts the first eight columns of the design matrix $\boldsymbol{Z} = \boldsymbol{X}\tilde{\boldsymbol{Z}}$ corresponding to the penalized deviation from the linear part of the function.

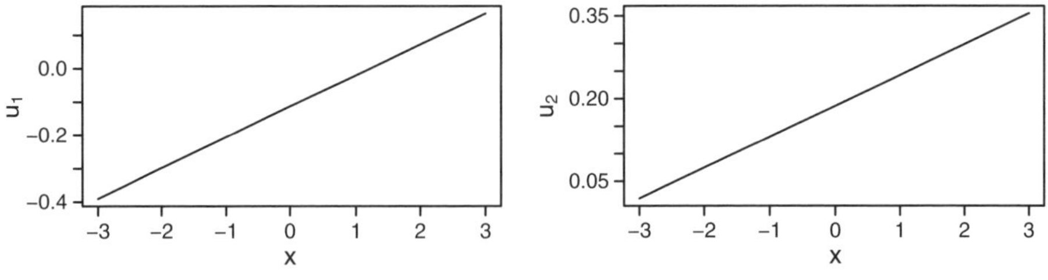

Fig. 4.1 Basis of the fixed effects part for a cubic penalized spline with second order random-walk penalty. Shown are the two columns of $\boldsymbol{U}$.

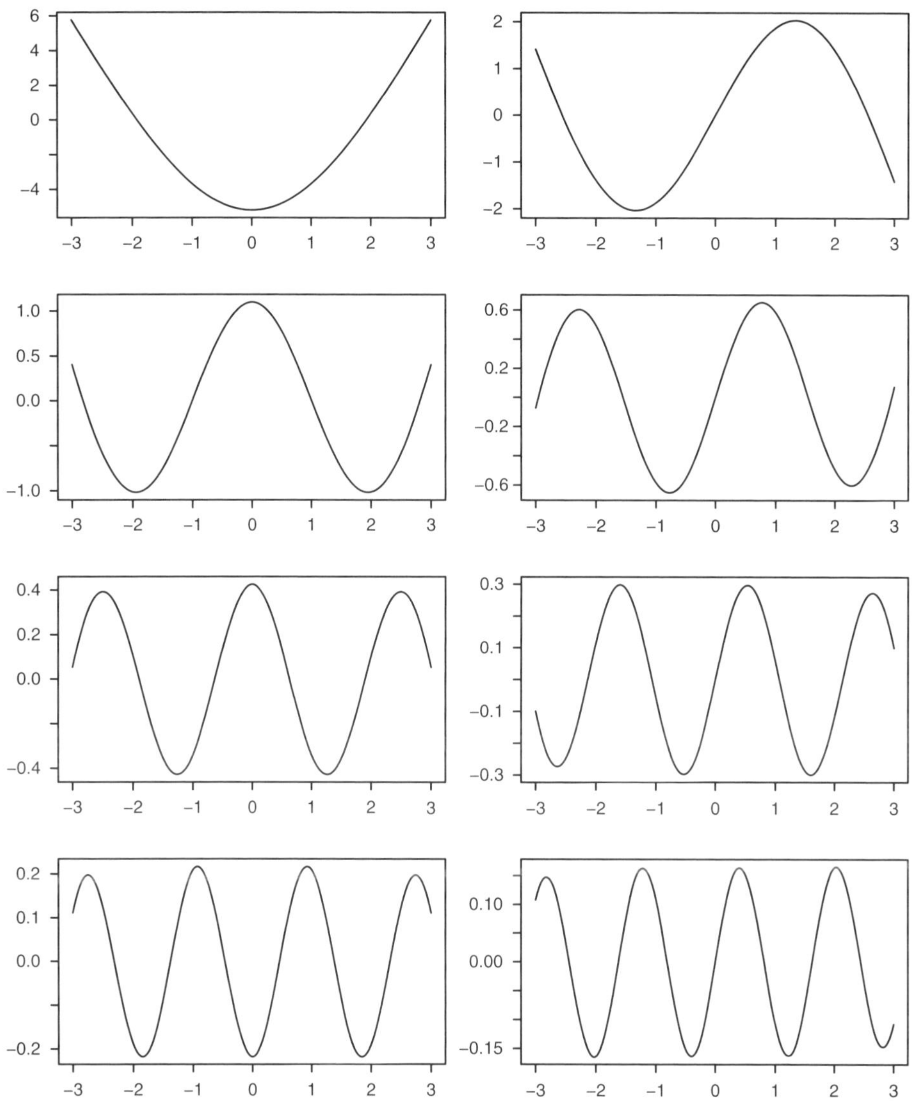

Fig. 4.2 First eight basis functions of the random effects part for a cubic penalized spline with second-order random-walk penalty.

Note that the columns are ordered according to ascending eigenvalues and therefore represent increasingly flexible and nonlinear deviations from the linear effect. In fact, the first column corresponds to a quadratic effect while the following columns represent basis functions of sinusoidal form with increasing frequency.

As a second example, we consider one of the customised penalties introduced in Section 2.2.1, page 47 for penalizing deviations from a cosine function. Therefore the quadratic penalty had to be modified to

$$\sum_{k=3}^{K}(\gamma_k - 2c\gamma_{k-1} + \gamma_{j-2})^2$$

where the constant $c = \cos(2\pi\delta_\kappa/p)$ is obtained from the distance δ_κ between two knots and the additional parameter p. Constructing again a cubic penalized spline with 18 inner knots, Figures 4.3 and 4.4 show columns of the fixed effects and random effects design matrices. The fixed effects basis now consists of two orthogonal cosine functions determined by the parameter p. The first random effects design vectors are quite different from those obtained with the simple second-order random-walk penalty while columns corresponding to larger eigenvalues again are high-frequency sinusoidal curves.

Mixed model-based estimation of SPMMs

The results derived for one single function can of course easily be extended to the general SPMM (4.6) with predictor

$$\boldsymbol{\eta} = \boldsymbol{U}\boldsymbol{\beta} + \boldsymbol{X}_1\boldsymbol{\gamma}_1 + \ldots + \boldsymbol{X}_r\boldsymbol{\gamma}_r + \boldsymbol{Z}\boldsymbol{\alpha}.$$

In this case, the model equation already contains a set of fixed effects $\boldsymbol{\beta}$ and random effects $\boldsymbol{\alpha} \sim \mathrm{N}(\boldsymbol{0}, \boldsymbol{G})$ in its original formulation. Applying the mixed model reparametrization to each of the nonparametric effects $\boldsymbol{X}_l\boldsymbol{\gamma}_l$ adds further fixed and random effects. More specifically, we obtain

$$\boldsymbol{X}_l\boldsymbol{\gamma}_l = \boldsymbol{U}_l\boldsymbol{\beta}_l + \boldsymbol{Z}_l\boldsymbol{\alpha}_l$$

with $p(\boldsymbol{\beta}_l) \propto \text{const}$ and $\boldsymbol{\alpha}_l \sim \mathrm{N}(\boldsymbol{0}, \tau_l^2\boldsymbol{I})$. Inserting these expressions in the predictor yields

$$\begin{aligned}\boldsymbol{\eta} &= \boldsymbol{U}\boldsymbol{\beta} + \boldsymbol{U}_1\boldsymbol{\beta}_1 + \ldots + \boldsymbol{U}_r\boldsymbol{\beta}_r + \boldsymbol{Z}_1\boldsymbol{\alpha}_1 + \ldots + \boldsymbol{Z}_r\boldsymbol{\alpha}_r + \boldsymbol{Z}\boldsymbol{\alpha} \\ &= \boldsymbol{U}_{\text{fix}}\boldsymbol{\beta}_{\text{fix}} + \boldsymbol{Z}_{\text{ran}}\boldsymbol{\alpha}_{\text{ran}}\end{aligned}$$

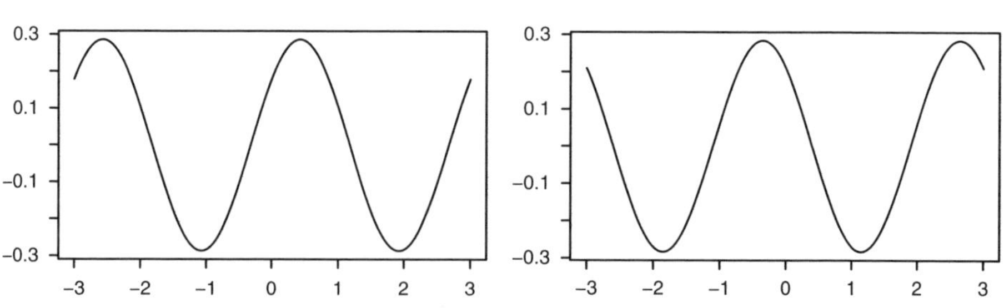

Fig. 4.3 Basis of the fixed effects part for a cubic penalized spline with cosine penalty.

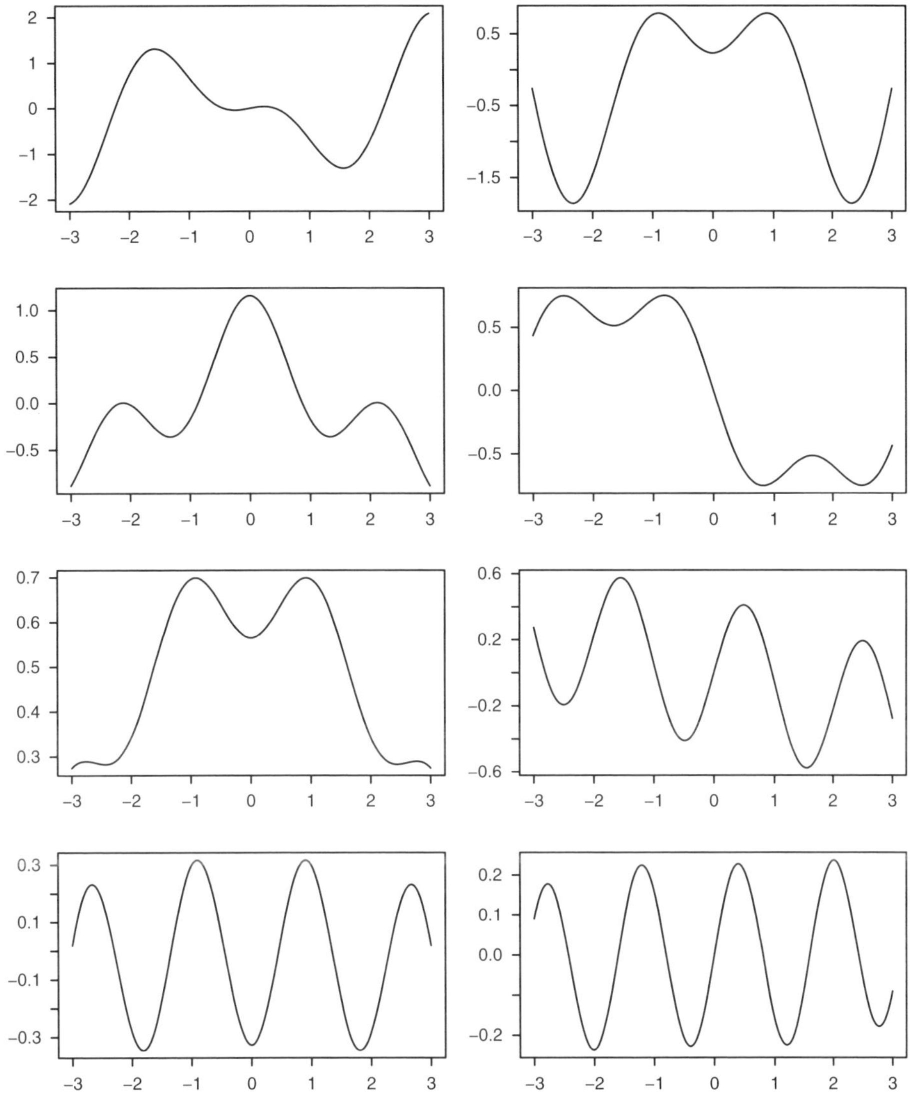

Fig. 4.4 First eight basis functions of the random effects part for a cubic penalized spline with cosine penalty.

where

$$\boldsymbol{U}_{\text{fix}} = (\boldsymbol{U}, \boldsymbol{U}_1, \ldots, \boldsymbol{U}_r) \qquad \boldsymbol{Z}_{\text{ran}} = (\boldsymbol{Z}_1, \ldots, \boldsymbol{Z}_r, \boldsymbol{Z})$$
$$\boldsymbol{\beta}_{\text{fix}} = (\boldsymbol{\beta}', \boldsymbol{\beta}_1', \ldots, \boldsymbol{\beta}_r')' \qquad \boldsymbol{\alpha}_{\text{ran}} = (\boldsymbol{\alpha}_1', \ldots, \boldsymbol{\alpha}_r', \boldsymbol{\alpha}')'$$

and

$$p(\boldsymbol{\beta}_{\text{fix}}) \propto \text{const} \qquad \boldsymbol{\alpha}_{\text{ran}} \sim \text{N}(\mathbf{0}, \boldsymbol{G}_{\text{ran}})$$

with

$$G_{\text{ran}} = \text{blockdiag}(\tau_1^2 I, \dots, \tau_r^2 I, G).$$

Since the partially improper prior structures for Bayesian penalized splines leave a constant function estimate unpenalized, each of the fixed effect design matrices $U_1, \dots, U_r$ corresponding to such a spline will contain an intercept column such that the combined model would not be identifiable due to multicollinearities in the matrix U_{fix}. A simple workaround is to delete superfluous intercept columns from the matrices $U_1, \dots, U_r$ and to include an intercept term in U. Similarly, identifiability problems may occur when including varying coefficient terms and we will discuss this issue in more detail after summarizing mixed model-based inference and providing a first example.

The complete mixed model-based estimation scheme can now be compactly summarized in the following algorithm extending Algorithm 3.2 (page 167):

Algorithm 4.1: Mixed model-based estimation of SPMMs

i. Initialize all parameters with suitable starting values $\hat{\beta}_{\text{fix}}^{(0)}$, $\hat{\alpha}_{\text{ran}}^{(0)}$ and $\hat{G}_{\text{ran}}^{(0)}$.

ii. Update fixed and random regression coefficients based on one step of a penalized Fisher scoring algorithm based on the current values for the variance parameters, that is, solve

$$\begin{pmatrix} U_{\text{fix}}' W U_{\text{fix}} & U_{\text{fix}}' W Z_{\text{ran}} \\ Z_{\text{ran}}' W U_{\text{fix}} & Z_{\text{ran}}' W Z_{\text{ran}} + G_{\text{ran}}^{-1} \end{pmatrix} \begin{pmatrix} \beta_{\text{fix}} \\ \alpha_{\text{ran}} \end{pmatrix} = \begin{pmatrix} U_{\text{fix}}' W \tilde{y} \\ Z_{\text{ran}}' W \tilde{y} \end{pmatrix} \tag{4.14}$$

for β_{fix} and α_{ran} (with working weights and working responses as in Section 2.3).

iii. Update the variance parameters for the random effects based on one REML Fisher scoring step as in Section 3.1.4, page 124.

iv. Assess whether there are still significant changes in either the regression coefficients or the variance parameters. If both types of parameter estimates converged, terminate the algorithm, otherwise go back to step ii.

Algorithm 4.1 also covers the case of Gaussian SPMMs after minor modifications. The weight matrix is then $W = I/\sigma^2$ and the working observations $\tilde{y}$ reduce to the actual observations y. In step iii, the usual estimate for σ^2, given the current predictor, has to be included.

Based on the estimates $\hat{\beta}$ and $\hat{\alpha}$, both the original model parameters and the function estimates can be recovered as

$$\hat{\gamma}_l = \tilde{X}_l \hat{\beta}_l + \tilde{Z}_l \hat{\alpha}_l$$

and

$$\hat{f}_l = X_l \hat{\beta}_l + Z_l \hat{\alpha}_l.$$

Note that for mixed model-based estimates of SPMMs the equivalence between penalized splines in B-spline representation and truncated power series representation does no longer hold in general even if the degree and order of penalization are chosen according to the discussion in Section 2.2.1. The estimates are equivalent when utilizing REML estimation but they differ when employing ML estimation for the determination of the smoothness variances. The underlying reason is that REML estimation corrects for the difference in the fixed effects design matrix while differences are pertinent for ML estimation (see Greven, 2007, for details).

Example 4.2: Childhood obesity
In Examples 3.2 and 3.4, the influence of a child's age t and of other covariates on its BMI was analysed with the mixed model specification (3.5)

$$y_{ij} = \boldsymbol{u}_{ij}'\boldsymbol{\beta} + f(t_{ij}) + \boldsymbol{z}_{ij}'\boldsymbol{\alpha}_i + \varepsilon_{ij}, \quad \varepsilon_{ij} \overset{\text{i.i.d.}}{\sim} \mathrm{N}(0, \sigma^2), \tag{4.15}$$

assuming the parametric form (3.6) for the population trend $f(t)$. Keeping all other covariates in the model, we now replace this parametric function through a cubic P-spline with a second-order random-walk prior and 12 (quantile based) knots. As in Example 3.4 the random effects part is specified through a random line or the nonlinear random function (3.7). EB inference has been carried out with the R-package `amer` (Scheipl, 2010a). The results in Table 4.1 show that estimates of fixed covariate effects are very close to corresponding estimates in

Table 4.1 Childhood obesity: Posterior modes (EB inference) together with 95% credible intervals for the fixed effects and variance components for the AMM with random lines and the AMM with random curves. mBMI and mDiffBMI have been centred for the analyses.

	Random lines			Random curves		
	mode	2.5%	97.5%	mode	2.5%	97.5%
const	−103.71	−116.08	−91.33	−105.05	−115.86	−94.2362
sex	0.3008	0.2219	0.3797	0.2257	0.1495	0.3020
breast	0.0260	−0.0567	0.1086	0.0604	−0.0195	0.1403
mSmoke	0.0772	−0.0392	0.1936	0.0451	−0.0674	0.1576
area	−0.0110	−0.1079	0.0860	−0.0278	−0.1214	0.0658
mBMI	0.0492	0.0384	0.0599	0.0530	0.0426	0.0633
mDiffBMI	0.0728	0.0484	0.0972	0.0877	0.0641	0.1113
σ^2	0.9185	–	–	0.6379	–	–
δ_0^2	0.8168	–	–	1.1323	–	–
δ_1^2	0.0435	–	–	0.0590	–	–
δ_2^2	–	–	–	52.65	–	–

Tables 3.2 and 3.3. (Note that intercepts are not directly comparable because of the different parametrization of $f(t)$ in Example 3.4). Figures 4.5 (model with random lines) and 4.6 (model with random curves) display the estimated population trend as well as the individual trend with and without random effects for the same selected individuals as in Examples 3.2 and 3.4. We see that the overall shape of the nonparametric population trend closely resembles the parametric trend in these examples. A more formal assessment of this claim will be carried out in Section 4.5.2, see Example 4.12. △

Example 4.3: Forest health
As an illustration for mixed model-based estimation in categorical SPMMs, we determined the posterior mode estimates for the model from Example 4.1 (page 184). Note that this is basically an extension of the model in Example 2.11 (page 103) that includes plot-specific random effects in addition. To shorten the discussion, we will focus on nonparametric and random effects in the following.

Figure 4.7 shows the estimated nonlinear time trends (top row) $f^{(1)}(t)$ (left) and $f^{(2)}(t)$ (right) together with 95% pointwise credible intervals (we will describe on page 206 how these confidence bands can be obtained). Both are very close to the estimates shown in Figure 2.23. While the second trend function $f^{(2)}(t)$ exhibits somewhat more variability reflected in the wider credible intervals, the credible intervals for $f^{(1)}(t)$ are even more narrow when including plot-specific random effects as compared to the independence model from Example 2.11. This may be caused by the absorption of some plot-specific features in the trend function.

Larger changes are observed for the age effect. While a decreasing effect for very old plots had been observed in the independence model, the effect is now mostly increasing over the whole age-domain with a peak for plots of about 50 years of age as a sole exception. The estimated curve also appears to be somewhat smoother, indicating again that parts of the plot-specific effects may have been captured in the age effect in the independence model.

Finally, Fig. 4.8 visualizes the estimated plot-specific random effects. In fact, the inter-plot variability is quite expressed with not too much indication of a spatially smooth effect. This will also be confirmed in later analyses comprising both random and spatial effects where a dominance of the random effect is observed. △

Identifiability

The mixed model representation is also useful when considering the identifiability restrictions arising in SPMMs. We already discussed an example for such identifiability constraints in the special case of additive models in Section 2.3 where the different levels of the nonparametric functions are not identified. The

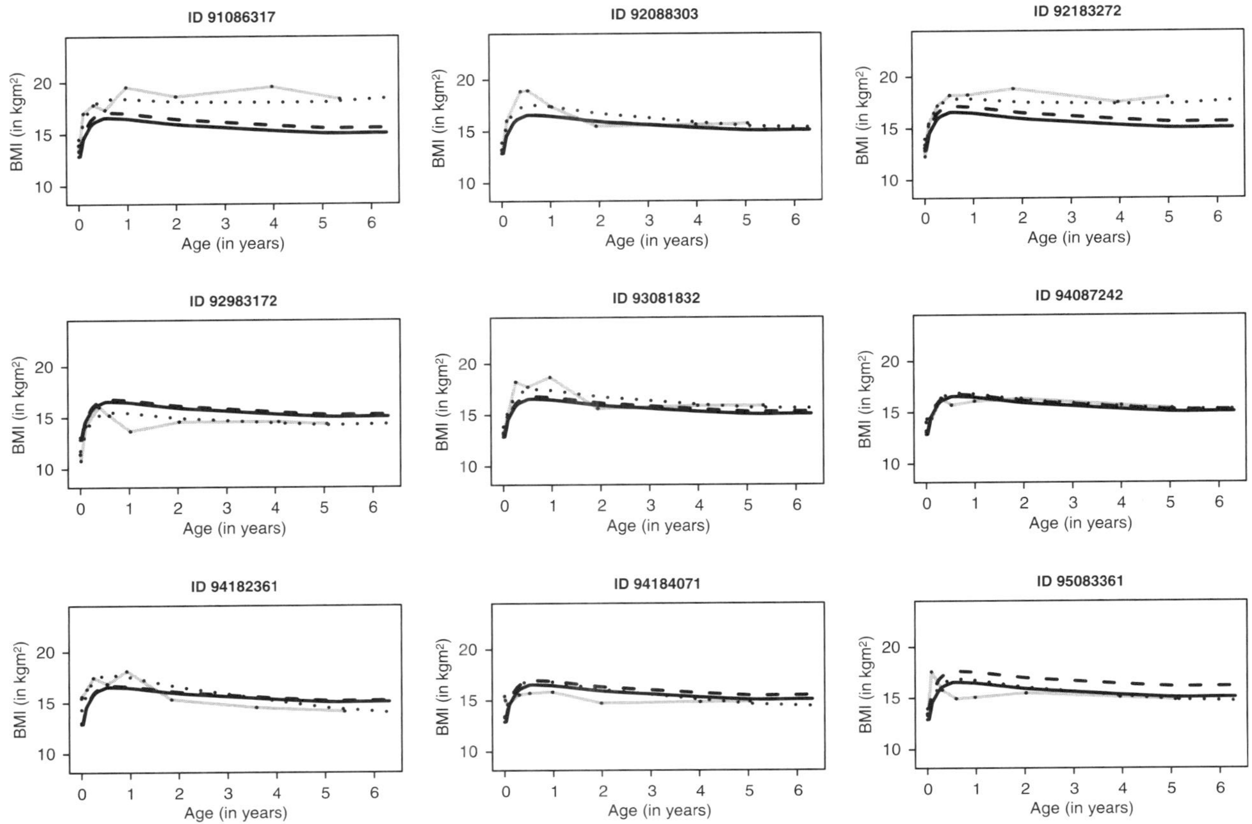

Fig. 4.5 Childhood obesity: Observed BMI curves for 9 randomly selected individuals (grey lines) in comparison to the estimated population trend (solid lines), the individual trend without random effects (dashed lines) and the individual trend with random effects (dotted lines) for the EB AMM with random lines.

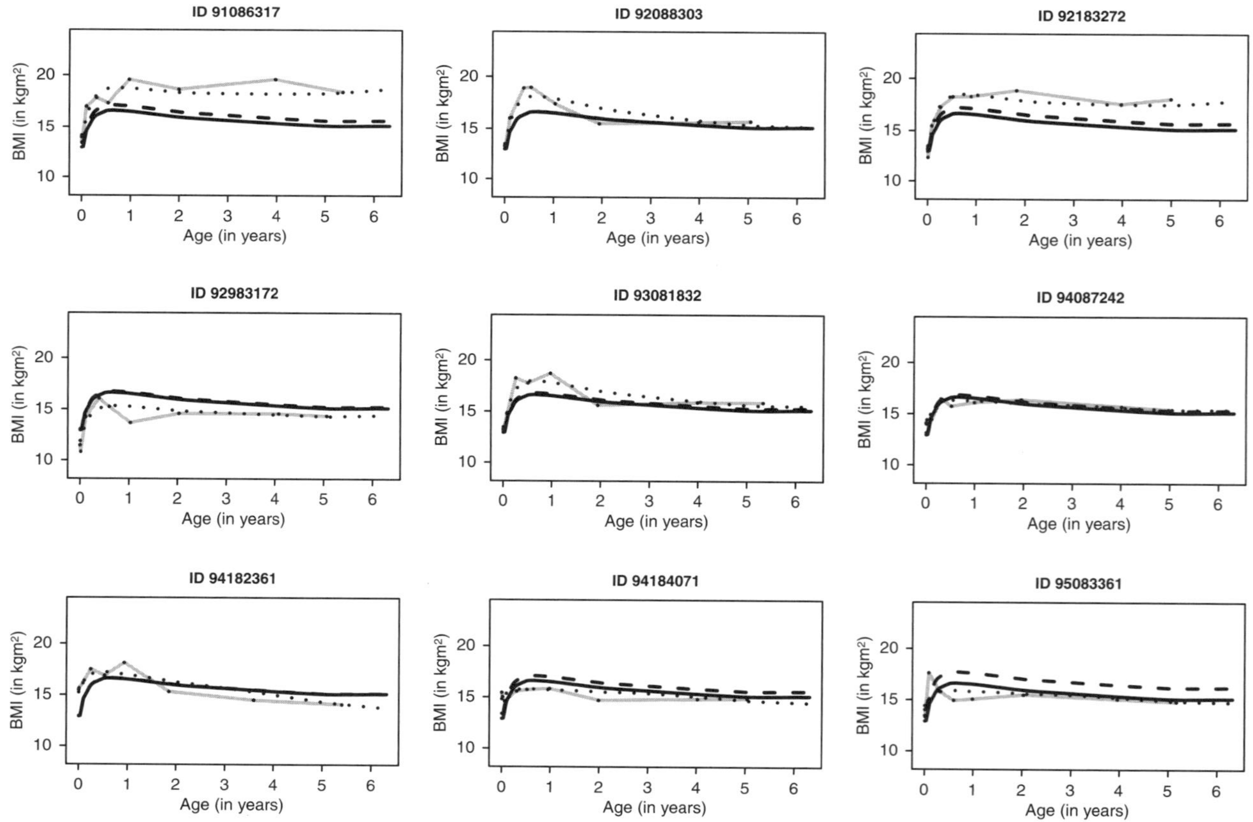

Fig. 4.6 Childhood obesity: Observed BMI curves for 9 randomly selected individuals (grey lines) in comparison to the estimated population trend (solid lines), the individual trend without random effects (dashed lines) and the individual trend with random effects (dotted lines) for the EB AMM with random curves.

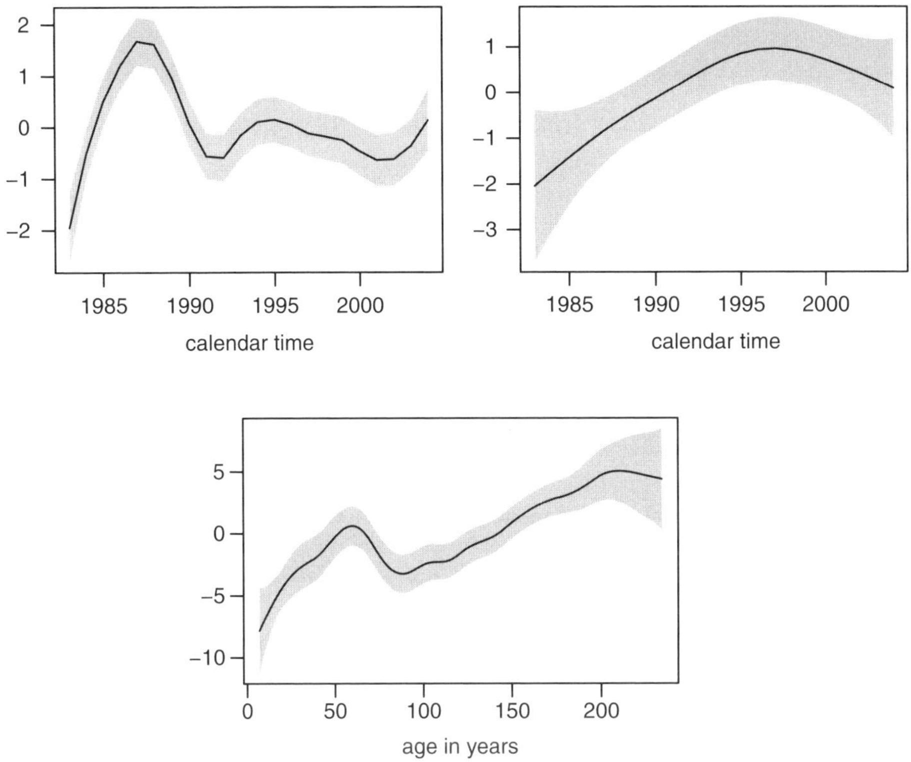

Fig. 4.7 Forest health: Estimated nonparametric effects together with pointwise 95% credible intervals.

mixed model representation now allows a closer look on this issue: If all nonparametric effects in an additive model are represented as penalized splines based on B-spline bases and difference penalties, the null spaces of the corresponding penalty matrices all contain the constant effect that is not penalized. Correspondingly, the space spanned by the columns of $\boldsymbol{U}_l$ contains a constant effect. Combining several fixed effects design matrices containing the constant leads to an unidentifiable model where the system of equations (4.14) can no longer be solved. Hence, identifiability of a model can now be checked by testing whether $\boldsymbol{U}_{\text{fix}}$ has full column rank. Moreover, identifiability can be achieved by reducing the basis of fixed effects, that is, by centring the basis around the constant. This effectively reduces the dimension of the null space to $R - K - 1$.

The deletion of the constant effect can be most easily achieved when considering a representation of the fixed effects part of a function that explicitly contains a constant. While this is generally not the case for the mixed model representation obtained from the spectral decomposition, the variant based on (4.11) and (4.12) has this property. Figure 4.9 shows the corresponding basis

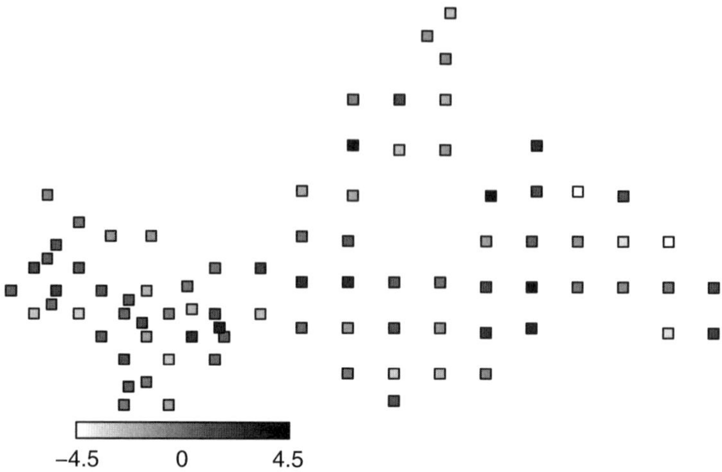

Fig. 4.8 Forest health: Posterior mode estimates for the plot-specific random effects.

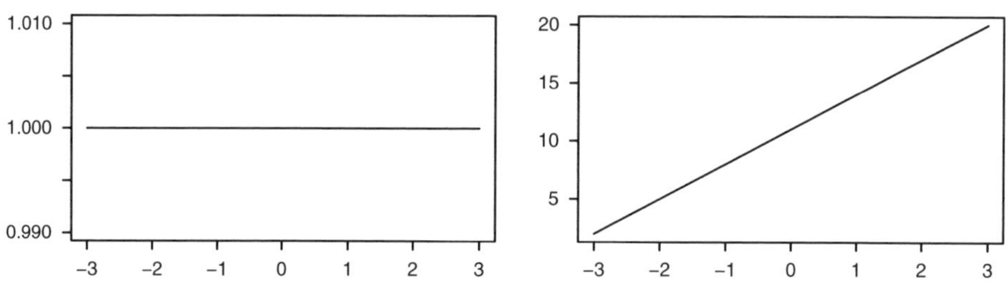

Fig. 4.9 Basis of the fixed effects part constructed as in (4.11) for a cubic penalized spline with second-order random-walk penalty.

resulting for the fixed effects part that exactly reflects the constant and the linear effect forming the columns in $\boldsymbol{U}_l$. In this case, identifiability can be achieved by simply deleting the columns of ones from all fixed effects design matrices $\boldsymbol{U}_l$.

The unidentified level of functional effects is not the only identifiability restriction that might arise in SPMMs. For example, including both a 'fixed', linear time trend βt and a flexible time trend $f(t)$ represented as a P-spline with second-order random-walk prior in the model

$$\eta_{it} = \ldots + t\beta + f(t) + \ldots$$

leads to a non-identifiable model.

Further examples arise in the context of varying coefficient models. Consider, for example, that a time-varying effect of a covariate u, say, shall be estimated and a model of the form

$$\eta_{it} = \ldots + u_{it}\beta + u_{it}f(t) + \ldots$$

would be specified. In this case, the linear effect β is not separable from the level of the time-varying effect $f(t)$ and therefore the model is not identifiable. This becomes particularly clear in the mixed model representation where the fixed effects matrix representing $f(t)$ will contain a constant vector (at least in an appropriate parametrization). Multiplying this vector with u results in a column of fixed effects that is exactly collinear to the original u included in the model specification. A solution to this problem is to either exclude separate linear effects for covariates that should be considered as interaction variables in varying coefficient models or to centre the effect $f(t)$. The latter also has the advantage that several effect modifiers can appear in multiple varying coefficient terms, such as in

$$\eta_{it} = \ldots + u_{it}\beta + u_{it}f_1(x_1) + u_{it}f_2(x_2) + \ldots$$

Note that it is crucial in this case to include a separate effect of u since both f_1 and f_2 are assumed to be centred.

Constrained smoothness priors

While the major aim of the mixed model representation is to facilitate empirical Bayes inference utilizing mixed model methodology, full Bayes analyses can also benefit from the analytic considerations. In particular, the mixed model representation allows us to investigate the partial impropriety of the multivariate Gaussian prior and allows us to assess the qualitative nature of the improper part of the prior.

Fahrmeir and Kneib (2009) utilized this to derive conditions for the propriety of posteriors in SPMMs based on results for mixed models. Panagiotelis and Smith (2008) employ the mixed model representation to turn the partially improper prior into a proper prior, enabling, for example, the computation of Bayes factors and Bayesian variable selection. They replace the prior distribution

$$p(\boldsymbol{\gamma}) \propto \exp\left(-\frac{1}{2\tau^2}\boldsymbol{\gamma}'\boldsymbol{K}\boldsymbol{\gamma}\right)$$

with

$$p(\boldsymbol{\gamma}) \propto \exp\left(-\frac{1}{2\tau^2}\boldsymbol{\gamma}'\boldsymbol{K}\boldsymbol{\gamma}\right) \mathbb{1}(\boldsymbol{\Gamma}_1'\boldsymbol{\gamma} = \mathbf{0})$$

where $\mathbb{1}(\boldsymbol{\Gamma}_1'\boldsymbol{\gamma} = 0)$ is an indicator function that restricts the prior to be positive only for those values of $\boldsymbol{\gamma}$ that fulfil the linear constraint $\boldsymbol{\Gamma}_1'\boldsymbol{\gamma} = \mathbf{0}$. This constraint effectively removes the impropriety from the prior but of course also deletes the corresponding 'unpenalized' part of the effect from the model.

For example, for a penalized spline with second-order random-walk prior, the matrix $\boldsymbol{\Gamma}_1$ contains a basis of the null space of $\boldsymbol{K}$, which is given by a linear trend in this case (see equation (4.13) and the corresponding definition of $\boldsymbol{\Gamma}_1$). As a consequence, the constraint $\boldsymbol{\Gamma}_1'\boldsymbol{\gamma} = \mathbf{0}$ removes the linear portion of the nonparametric effect from the model, and if the original model should be recovered, an additional linear parameter (and an intercept) have to be included in the model separately. Still, the constraint prior allows us to combine a proper prior for the nonlinear effect with a proper prior for the linear effect fairly easily, yielding only proper priors at the stage of the regression coefficients.

Sampling from multivariate Gaussian distributions subject to linear constraints can easily be implemented and allows to construct MCMC simulation schemes for models comprising constraint effects (see Rue and Held, 2005, for details).

Credible intervals and bands

In the mixed model representation, construction of pointwise and simultaneous credible bands is based on the assumption that the posterior of the estimated regression coefficients is approximately Gaussian with covariance matrix

$$\mathrm{Cov}\begin{pmatrix}\hat{\boldsymbol{\beta}}_{\text{fix}} - \boldsymbol{\beta}_{\text{fix}} \\ \hat{\boldsymbol{\alpha}}_{\text{ran}} - \boldsymbol{\alpha}_{\text{ran}}\end{pmatrix} = \begin{pmatrix}\boldsymbol{U}_{\text{fix}}'\boldsymbol{W}\boldsymbol{U}_{\text{fix}} & \boldsymbol{U}_{\text{fix}}'\boldsymbol{W}\boldsymbol{Z}_{\text{ran}} \\ \boldsymbol{Z}_{\text{ran}}'\boldsymbol{W}\boldsymbol{U}_{\text{fix}} & \boldsymbol{Z}_{\text{ran}}'\boldsymbol{W}\boldsymbol{Z}_{\text{ran}} + \hat{\boldsymbol{G}}_{\text{ran}}^{-1}\end{pmatrix}^{-1} = \boldsymbol{H}$$

where $\boldsymbol{W}$ denotes the working weights in the final iteration of (4.14). From the covariance matrix of the regression coefficients, the approximate covariance matrix of a functional effect $\hat{\boldsymbol{f}}_l = \boldsymbol{U}_l\hat{\boldsymbol{\beta}}_l + \boldsymbol{Z}_l\hat{\boldsymbol{\alpha}}_l$ is obtained as

$$\mathrm{Cov}(\hat{\boldsymbol{f}}_l - \boldsymbol{f}_l) = (\boldsymbol{U}_l \boldsymbol{Z}_l)\begin{pmatrix}\boldsymbol{H}_{\hat{\boldsymbol{\beta}}_l,\hat{\boldsymbol{\beta}}_l} & \boldsymbol{H}_{\hat{\boldsymbol{\beta}}_l,\hat{\boldsymbol{\alpha}}_l} \\ \boldsymbol{H}_{\hat{\boldsymbol{\alpha}}_l,\hat{\boldsymbol{\beta}}_l} & \boldsymbol{H}_{\hat{\boldsymbol{\alpha}}_l,\hat{\boldsymbol{\alpha}}_l}\end{pmatrix}\begin{pmatrix}\boldsymbol{U}_l' \\ \boldsymbol{Z}_l'\end{pmatrix} = \boldsymbol{H}_{\boldsymbol{f}_l} \tag{4.16}$$

with appropriate submatrices of $\boldsymbol{H}$ inserted. The diagonal elements of this covariance matrix provide the variances required for the construction of pointwise credible intervals of level α via

$$\hat{\boldsymbol{f}}_l \pm z_{1-\alpha/2}\sqrt{\mathrm{diag}(\boldsymbol{H}_{\boldsymbol{f}_l})}$$

where $z_{1-\alpha/2}$ denotes the $1-\alpha/2$ quantile of the standard normal distribution and the diag operator extracts the diagonal from the covariance matrix $\boldsymbol{H}_{\boldsymbol{f}_l}$.

Of course, the same formulae can also be employed to construct credible intervals for new observations $\boldsymbol{y}^*$ by replacing the design matrices $\boldsymbol{U}_l$ and $\boldsymbol{Z}_l$ in (4.16) with the corresponding design matrices $\boldsymbol{U}_l^*$ and $\boldsymbol{Z}_l^*$ for the new observations. Similarly, prediction intervals can be constructed for new observations.

If simultaneous credible bands for the vector $\hat{\boldsymbol{f}}_l$ are desired, these can be obtained by generalizing the considerations for pointwise credible intervals. Based on the assumption

$$\hat{\boldsymbol{f}}_l - \boldsymbol{f}_l \overset{\text{a}}{\sim} \mathrm{N}(\mathbf{0}, \boldsymbol{H}_{\boldsymbol{f}_l}) \tag{4.17}$$

a simultaneous credible band is then given by

$$\hat{\boldsymbol{f}}_l + m_{1-\alpha}\sqrt{\text{diag}(\boldsymbol{H}_{\boldsymbol{f}_l})}$$

where $m_{1-\alpha}$ is the $(1-\alpha)$ quantile of

$$\max_x \left| \frac{\hat{f}_l(x) - f_l(x)}{\sqrt{\text{Var}(\hat{f}_l(x))}} \right| \tag{4.18}$$

(compare Ruppert, Wand and Carroll, 2003). The distribution of the maximal standardized distance between the estimated and the true curve (4.18) and in particular its quantiles are hard to evaluate analytically. Therefore, the latter are typically approximated by simulation based on (4.17). Approaches that circumvent the necessity for simulation-based determination of the quantile $m_{1-\alpha}$ but instead work with Sun and Loader (1994)'s volume of tubes formula are presented in Krivobokova, Kneib and Claeskens (2010).

Although we have only discussed credible intervals and bands for the nonparametric components of SPMMs, exactly the same approaches can of course also be considered for random effects and parametric effects.

We close our considerations on credible bands with a word of caution on the interpretation of pointwise and simultaneous credible bands. Quite often, credible intervals and bands are used to 'test' whether a nonparametric function estimate can be safely reduced to a parametric effect such as a linear curve using the following procedure: If a line is completely contained in the credible intervals or bands, this is seen as an indication of linearity (or more precisely as lack of evidence for a nonlinear effect). However, this neglects the fact that the credibility region depicted in a graph is only a projection of an originally high-dimensional credibility region (see also the discussions in Hastie and Tibshirani, 1990, Sec. 3.8.2, and Ruppert, Wand and Carroll, 2003, Sec. 6.5). For example, if the true effect is quadratic and all uncertainty in the data is associated with the level of the function, this may in extreme cases lead to credible intervals that are wide enough to contain a simple linear effect although the original credibility region had a complex shape reflecting the sole uncertainty in the level. Hence, formal significance tests should be used when trying to differentiate between linear and nonlinear curves.

Tests on the functional form

Utilizing the mixed model representation, tests for $\boldsymbol{\alpha}_l = \mathbf{0}$ provide a possibility to determine whether a function f_l can be restricted to the null space of the penalty. For B-spline based penalized splines with d-th order difference penalty, this corresponds to testing a polynomial of order $d-1$ against a more general, nonparametric alternative. The special case of most practical interest relates to testing a linear versus a nonlinear functional relationship, that is, to a second-order difference penalty. Similarly, tests can be introduced in the truncated power series representation, where now the null hypothesis is defined by the degree of the spline. In principle, any of the classical testing approaches, that is, likelihood ratio tests, Wald tests and score tests could now be considered, but we will concentrate on likelihood ratio tests in the following and will comment on alternatives later on.

Tests on the functional form are still an area of active research, in particular for additive and generalized additive models. We will therefore restrict our attention to Gaussian responses and mixed models with only one variance component to ease presentation. The principal difficulties present in this case carry over to more complicated situations. Hence, we consider the model

$$\boldsymbol{y} = \boldsymbol{U}\boldsymbol{\beta} + \boldsymbol{Z}\boldsymbol{\alpha} + \boldsymbol{\varepsilon}$$

with distributional assumptions

$$\boldsymbol{\alpha} \sim \mathrm{N}(\boldsymbol{0}, \tau^2 \boldsymbol{I}) \quad \text{and} \quad \boldsymbol{\varepsilon} \sim \mathrm{N}(\boldsymbol{0}, \sigma^2, \boldsymbol{I}).$$

This comprises penalized splines in their mixed model representation (which will be the focus in the following) but also a random intercept model. We are interested in testing

$$H_0 : \boldsymbol{\alpha} = \boldsymbol{0} \quad \text{vs.} \quad H_1 : \boldsymbol{\alpha} \neq \boldsymbol{0}.$$

However, since $\boldsymbol{\alpha}$ is a random effect, we can equivalently express this hypothesis based on the variance of the random effect via

$$H_0 : \tau^2 = 0 \quad \text{vs.} \quad H_1 : \tau^2 > 0$$

since $\boldsymbol{\alpha} = \boldsymbol{0}$ if and only if $\tau^2 = 0$. It will turn out in the following that it is advantageous to further rewrite the testing problem in terms of the inverse smoothing parameter $\tilde{\lambda} = \tau^2/\sigma^2$ as

$$H_0 : \tilde{\lambda} = 0 \quad \text{vs.} \quad H_1 : \tilde{\lambda} > 0.$$

The (finite sample) likelihood ratio test statistic for this testing problem is then defined as

$$\mathrm{LRT}_N = 2 \sup_{H_1} l(\boldsymbol{\beta}, \tilde{\lambda}, \sigma^2) - 2 \sup_{H_0} l(\boldsymbol{\beta}, \tilde{\lambda}, \sigma^2)$$

where $N = \dim(\boldsymbol{y})$ denotes the total number of observations and the marginal log-likelihood

$$l(\boldsymbol{\beta}, \tilde{\lambda}, \sigma^2) = -\frac{N}{2} \log(\sigma^2) - \frac{1}{2} \log(|\boldsymbol{V}_{\tilde{\lambda}}|) - \frac{1}{2\sigma^2} (\boldsymbol{y} - \boldsymbol{U}\boldsymbol{\beta})' \boldsymbol{V}_{\tilde{\lambda}}^{-1} (\boldsymbol{y} - \boldsymbol{U}\boldsymbol{\beta})$$

is obtained from the marginal model $\boldsymbol{y} \sim \mathrm{N}(\boldsymbol{U}\boldsymbol{\beta}, \sigma^2 \boldsymbol{V}_{\tilde{\lambda}})$ where

$$\mathrm{Cov}(\boldsymbol{y}) = \sigma^2 \boldsymbol{V}_{\tilde{\lambda}} = \sigma^2 \boldsymbol{I} + \tau^2 \boldsymbol{Z}\boldsymbol{Z}' = \sigma^2(\boldsymbol{I} + \tilde{\lambda}\boldsymbol{Z}\boldsymbol{Z}')$$

and $\boldsymbol{V}_{\tilde{\lambda}} = \boldsymbol{I} + \tilde{\lambda}\boldsymbol{Z}\boldsymbol{Z}'$.

Since estimation in mixed models is mostly based on restricted maximum likelihood instead of maximum likelihood principles, an alternative attempt for the testing procedure can be based on the restricted likelihood, leading to the restricted likelihood ratio test statistic

$$\mathrm{RLRT}_N = 2 \sup_{H_1} l^*(\boldsymbol{\beta}, \tilde{\lambda}, \sigma^2) - 2 \sup_{H_0} l^*(\boldsymbol{\beta}, \tilde{\lambda}, \sigma^2)$$

where

$$\begin{aligned} l^*(\boldsymbol{\beta}, \tilde{\lambda}, \sigma^2) = & -\frac{N-m}{2}\log(\sigma^2) - \frac{1}{2}\log(|\boldsymbol{V}_{\tilde{\lambda}}|) - \frac{1}{2}\log(|\boldsymbol{U}'\boldsymbol{V}_{\tilde{\lambda}}^{-1}\boldsymbol{U}|) \\ & - \frac{1}{2\sigma^2}(\boldsymbol{y} - \boldsymbol{U}\boldsymbol{\beta})'\boldsymbol{V}_{\tilde{\lambda}}^{-1}(\boldsymbol{y} - \boldsymbol{U}\boldsymbol{\beta}) \end{aligned}$$

is the restricted likelihood (considered as a function of fixed effects and variance components) and $m = \dim(\boldsymbol{\beta})$ is the dimension of fixed effects.

For both test statistics, classical likelihood theory would yield an asymptotic χ^2-distribution with one degree of freedom for our testing problem. However, the parameter space corresponding to our testing problems is $\tilde{\lambda} \in [0, \infty)$ and therefore contradicts the usual regularity condition that the parameter space should be an open interval. Note also that we cannot transform the parameter space to an open interval using for example the log-transform, since the boundary of the parameter space is actually the value to be tested. In fact, testing problems on the boundary of parameter spaces has already gained considerable attention in the statistical literature, dating back at least to Chernoff (1954), Moran (1971) and in particular Self and Liang (1987). Stram and Lee (1994) applied the results of Self and Liang (1987) to the context of Gaussian mixed models and showed that under certain regularity conditions the asymptotic distribution of the (restricted) likelihood ratio test statistic under the null hypothesis is a $0.5\chi_0^2 : 0.5\chi_1^2$ mixture, where the χ_0^2 distribution corresponds to a point mass in zero. The latter has to be included since with a chance of (asymptotically) 50% the random effects variance is estimated to be zero under the null hypothesis. Since the parameter space of the variance parameter is no longer an open interval but includes zero as a boundary point, the maximum of the likelihood can be achieved on the boundary, yielding a positive probability for estimating $\hat{\tau}^2 = 0$.

Stram and Lee (1994) assume that the response vector $\boldsymbol{y}$ can be decomposed into subvectors $\boldsymbol{y}_i$, $i = 1, \ldots, n$, which are independent under both the null hypothesis and the alternative. In our formulation, this corresponds to the assumption that the marginal covariance matrix has a block-diagonal structure. This results, in particular if the design matrix of random effects has a block-diagonal structure $\boldsymbol{Z} = \text{blockdiag}(\boldsymbol{Z}_1, \ldots, \boldsymbol{Z}_n)$ such that

$$\boldsymbol{V}_{\tilde{\lambda}} = \text{blockdiag}(\boldsymbol{V}_i) = \text{blockdiag}(\boldsymbol{I}_{n_i} + \tilde{\lambda}\boldsymbol{Z}_i\boldsymbol{Z}_i')$$

where $n_i = \dim(\boldsymbol{y}_i)$ denotes the number of observations in subvector $\boldsymbol{y}_i$. Such a decomposition is, for example, possible in the random intercepts model for longitudinal data

$$y_{ij} = \boldsymbol{u}_{ij}'\boldsymbol{\beta} + \alpha_i + \varepsilon_{ij}, \quad \alpha_i \sim \text{N}(0, \tau^2), \quad i = 1, \ldots, n, j = 1, \ldots, n_i.$$

Collecting all observations for one individual i leads to

$$\boldsymbol{y}_i = \boldsymbol{X}_i\boldsymbol{\beta} + \boldsymbol{Z}_i\alpha_i + \varepsilon_i$$

where $\boldsymbol{Z}_i = (1, \ldots, 1)'$.

Unfortunately, the regularity conditions assumed in Stram and Lee (1994) do not hold in general mixed models with one variance component and in particular when testing the functional form of a nonparametric effect. Even in cases where the vector of responses can be partitioned according to the longitudinal structure of the data, the conditions for the asymptotic χ^2-mixture have to be carefully evaluated since, for example, a requirement is that the number of clusters grows to infinity. In particular, it is not sufficient that the total number of observations

$$N = \sum_{i=1}^{n} n_i$$

tends to infinity, which could also be achieved by growing numbers of observations within the clusters. Moreover, Crainiceanu and Ruppert (2004a) and Crainiceanu *et al.* (2005) have investigated the finite sample and asymptotic distribution of the likelihood ratio test statistic under various conditions and found that even in situations where the assumptions underlying Stram and Lee (1994) are fulfilled, the finite sample deviations from the $0.5\chi_0^2 : 0.5\chi_1^2$ mixture may be considerable.

However, Crainiceanu and Ruppert (2004a) also present representations of the finite sample and the asymptotic distribution of likelihood ratio and restricted likelihood ratio tests that allow us to construct tests for zero variances in general Gaussian mixed models with one variance component. The basis for these tests is a spectral representation of the profile log-likelihood. The profile likelihood estimates for $\boldsymbol{\beta}$ and σ^2 are given by

$$\hat{\boldsymbol{\beta}} = (\boldsymbol{X}'\boldsymbol{V}_{\tilde{\lambda}}^{-1}\boldsymbol{X})^{-1}\boldsymbol{X}'\boldsymbol{V}_{\tilde{\lambda}}^{-1}\boldsymbol{y}, \quad \hat{\sigma}^2 = \frac{1}{N}(\boldsymbol{y} - \boldsymbol{X}\hat{\boldsymbol{\beta}})'\boldsymbol{V}_{\tilde{\lambda}}^{-1}(\boldsymbol{y} - \boldsymbol{X}\hat{\boldsymbol{\beta}}).$$

Plugging these expressions back into the log-likelihood yields the profile log-likelihood for $\tilde{\lambda}$:

$$2l(\tilde{\lambda}) = -\log(|\boldsymbol{V}_{\tilde{\lambda}}|) - N\log(\boldsymbol{y}'\boldsymbol{P}'_{\tilde{\lambda}}\boldsymbol{V}_{\tilde{\lambda}}^{-1}\boldsymbol{P}_{\tilde{\lambda}}\boldsymbol{y})$$

where

$$\boldsymbol{P}_{\tilde{\lambda}} = \boldsymbol{I} - \boldsymbol{U}(\boldsymbol{U}'\boldsymbol{V}_{\tilde{\lambda}}^{-1}\boldsymbol{U})^{-1}\boldsymbol{U}'\boldsymbol{V}_{\tilde{\lambda}}^{-1} \quad \text{and} \quad \boldsymbol{P}_0 = \boldsymbol{I} - \boldsymbol{U}(\boldsymbol{U}'\boldsymbol{U})^{-1}\boldsymbol{U}'$$

Crainiceanu and Ruppert (2004a) then show that using the spectral decomposition of the profile log-likelihood the likelihood ratio test statistic can be written as

$$\text{LRT}_N \stackrel{\mathcal{D}}{=} \sup_{\tilde{\lambda}\geq 0}\left(N\log\left(1+\frac{D_1(\tilde{\lambda})}{D_2(\tilde{\lambda})}\right) - \sum_{s=1}^{K}\log(1+\tilde{\lambda}\xi_{s,N})\right)$$

where $\stackrel{\mathcal{D}}{=}$ denotes equality in distribution,

$$D_1(\tilde{\lambda}) = \sum_{s=1}^{K}\frac{\tilde{\lambda}\mu_{s,N}}{1+\tilde{\lambda}\mu_{s,N}}w_s^2, \quad D_2(\tilde{\lambda}) = \sum_{s=1}^{K}\frac{w_s^2}{1+\tilde{\lambda}\mu_{s,N}} + \sum_{s=K+1}^{N-m}w_s^2,$$

$\mu_{s,N}$ and $\xi_{s,N}$ are the K eigenvalues of the matrices $\boldsymbol{Z}'\boldsymbol{P}_0\boldsymbol{Z}$ and $\boldsymbol{Z}'\boldsymbol{Z}$ and w_s, $s = 1, \dots, N-m$, are i.i.d. $\mathrm{N}(0,1)$ distributed.

Analogous derivations for restricted maximum likelihood estimates yield the spectral decomposition of the restricted likelihood ratio test statistic based on the restricted profile log-likelihood

$$2l^*(\tilde{\lambda}) = -\log(|\boldsymbol{V}_{\tilde{\lambda}}|) - \log(|\boldsymbol{U}'\boldsymbol{V}_{\tilde{\lambda}}^{-1}\boldsymbol{U}|) - (N-m)\log(\boldsymbol{y}'\boldsymbol{P}'_{\tilde{\lambda}}\boldsymbol{V}_{\tilde{\lambda}}^{-1}\boldsymbol{P}_{\tilde{\lambda}}\boldsymbol{y})$$

as

$$\text{RLRT}_N \stackrel{\mathcal{D}}{=} \sup_{\tilde{\lambda}\geq 0}\left((N-m)\log\left(1+\frac{D_1(\tilde{\lambda})}{D_2(\tilde{\lambda})}\right) - \sum_{s=1}^{K}\log(1+\tilde{\lambda}\mu_{s,N})\right).$$

Although both expressions look quite complicated at first sight, they enable relatively easy access to the finite sample distributions of (restricted) likelihood ratio tests under the null hypothesis based on simulations. Crainiceanu and Ruppert (2004a) outline such a simulation scheme that replaces the supremum by the maximum over a grid of values for $\tilde{\lambda}$. Since the eigenvalues have to be computed only once, the simulation actually boils down to the simulation of i.i.d. Gaussian variables and the determination of the maximum. For moderately sized problems, the simulation of thousands of replications from the distributions is fast and allows for an accurate determination of critical values or p-values.

Possible simplifications arise when considering the asymptotic distributions instead of the finite sample versions. The asymptotic distributions essentially depend on the asymptotic behaviour of the eigenvalues $\mu_{s,N}$ and $\xi_{s,N}$.

Crainiceanu and Ruppert (2004a) derive the asymptotic distribution of the LRT and RLRT statistics under the condition that

$$\lim_{N\to\infty} \frac{\mu_{s,N}}{N^\alpha} = \mu_s, \quad \lim_{N\to\infty} \frac{\xi_{s,N}}{N^\alpha} = \xi_s,$$

for some $\alpha \geq 0$ and show that the asymptotic distribution equals the $0.5\chi_0^2 : 0.5\chi_1^2$ mixture under the regularity conditions in Stram and Lee (1994) but may be quite far from this mixture when these conditions are violated. They furthermore provide formulae for the probability that the inverse smoothing parameter is estimated to be zero. Depending on the design, this probability may well be as large as 90% for maximum likelihood estimation and 70% for restricted maximum likelihood estimation.

Greven *et al.* (2008) extend the methodology presented in Crainiceanu and Ruppert (2004a) to additive mixed models with several variance components. Since in this case the simulation of the null distribution may become prohibitively large, they suggest approximations to the null distribution that either completely avoid simulations or allow to reduce the necessary number of simulations by using a parametric approximation. Still, testing is restricted to Gaussian models and the extension to non-Gaussian response types is challenging. Moreover, the null distributions always depend on the observed data situation in a complicated manner such that it is not possible to make the distribution available in a look up table.

So far we have focused on (restricted) likelihood ratio testing. Alternatively, score and Wald tests may be considered but none of them will avoid the difficulties observed for likelihood ratio tests due to the non-standard testing situation. In particular, there will always be a positive probability of estimating $\tau^2 = 0$ and therefore $\boldsymbol{\alpha} = \mathbf{0}$. Hence, the null distribution for any test

$$H_0 : \tau^2 = 0 \quad \text{vs.} \quad H_1 : \tau^2 > 0$$

will always have a point mass in zero. Wood (2006), Cantoni and Hastie (2002), or Hastie and Tibshirani (1990) (amongst others) propose F-type tests for functional forms that are constructed in analogy to F-tests in the linear model with equivalent degrees of freedom df replacing the usual parameter count. However, none of the suggested F-tests accounts for the positive probability to observe $\hat{\tau}^2 = 0$. Scheipl, Greven and Küchenhoff (2008) compare (restricted) likelihood ratio tests and F-type tests in a comprehensive simulation study and find that ignoring this fact leads to a serious loss in power for the F-type tests. Exact restricted likelihood ratio testing is implemented in the R-package `RLRsim` (Scheipl, 2010c).

4.2.2 Full Bayes estimation with Gaussian priors for regression parameters

For full Bayes inference, we assume SPMMs with predictors in general from, that is, individual-specific and complete predictors

$$\boldsymbol{\eta}_i = \boldsymbol{U}_i\boldsymbol{\beta} + \boldsymbol{X}_{i1}\gamma_1 + \ldots + \boldsymbol{X}_{ir}\gamma_r + \boldsymbol{Z}_i\boldsymbol{\alpha}_i$$

and

$$\boldsymbol{\eta} = \boldsymbol{U}\boldsymbol{\beta} + \boldsymbol{X}_1\gamma_1 + \ldots + \boldsymbol{X}_r\gamma_r + \boldsymbol{Z}\boldsymbol{\alpha},$$

respectively. As in Section 4.1.3, priors for $\boldsymbol{\beta}$, $\gamma = (\gamma_1', \ldots, \gamma_r')$, and $\boldsymbol{\alpha} = (\boldsymbol{\alpha}_1', \ldots, \boldsymbol{\alpha}_n')$ are (conditionally) Gaussian,

$$\begin{aligned} \boldsymbol{\beta} &\sim \mathrm{N}(\boldsymbol{c}, \boldsymbol{C}) \quad \text{or} \quad p(\boldsymbol{\beta}) \propto \text{const} \;\; \text{for} \;\; \boldsymbol{C}^{-1} = \boldsymbol{0} \\ \boldsymbol{\alpha}_i|\boldsymbol{Q} &\overset{\text{i.i.d.}}{\sim} \mathrm{N}(\boldsymbol{0}, \boldsymbol{Q}) \quad \text{or} \quad \boldsymbol{\alpha}|\boldsymbol{Q} \sim \mathrm{N}(\boldsymbol{0}, \boldsymbol{G}),\ \boldsymbol{G} = \text{blockdiag}(\boldsymbol{Q}) \\ \gamma_l|\tau_l^2 &\sim \mathrm{N}(\boldsymbol{0}, \tau_l^2\boldsymbol{K}_l^-), \quad l = 1, \ldots, r, \end{aligned}$$

where $\boldsymbol{K}_l^-$ denotes a generalized inverse of the possibly rank-deficient penalty matrix $\boldsymbol{K}_l$, see (4.8). With priors $p(\tau_l^2)$ for the variance components $\boldsymbol{\tau}^2 = (\tau_1^2, \ldots, \tau_r^2)', p(\boldsymbol{Q})$ as well as $p(\sigma^2)$ for the error variance $\sigma^2 = \text{Var}(\varepsilon_{ij})$ in Gaussian observation models, and under usual (conditional) independence assumptions, the posterior is given by

$$p(\boldsymbol{\beta}, \gamma, \boldsymbol{\alpha}, \boldsymbol{\tau}^2, \boldsymbol{Q}, \sigma^2) \propto p(\boldsymbol{y}|\boldsymbol{\eta}, \sigma^2)p(\sigma^2)p(\boldsymbol{\beta})p(\boldsymbol{\alpha}|\boldsymbol{Q})p(\boldsymbol{Q})\prod_{l=1}^{r} p(\gamma_l|\tau_l^2)p(\tau_l^2)$$

$$= \left(\prod_{i=1}^{n} p(\boldsymbol{y}_i|\boldsymbol{\eta}_i, \sigma^2)p(\boldsymbol{\alpha}_i|\boldsymbol{Q})\right) p(\sigma^2)p(\boldsymbol{Q}) \left(\prod_{l=1}^{r} p(\gamma_l|\tau_l^2)p(\tau_l^2)\right) p(\boldsymbol{\beta}),$$

where $p(\boldsymbol{y}|\boldsymbol{\eta})$ is the complete data likelihood and $p(\boldsymbol{y}_i|\boldsymbol{\eta}_i)$ is the likelihood contribution of subject i. MCMC simulations are based on drawings from full conditionals of blocks of parameters given the other parameters and the data.

Gaussian SPMMs

For Gaussian SPMMs with

$$\boldsymbol{y}_i = \boldsymbol{\eta}_i + \varepsilon_i, \quad \varepsilon_i \sim \mathrm{N}(\boldsymbol{0}, \sigma^2\boldsymbol{I}), \quad i = 1, \ldots, n,$$

the full conditionals for $\boldsymbol{\beta}, \gamma$ and $\boldsymbol{\alpha}$ are multivariate Gaussian. Straightforward calculations show that the covariance matrix $\boldsymbol{\Sigma}_{\boldsymbol{\beta}}$ and the mean $\boldsymbol{\mu}_{\boldsymbol{\beta}}$ of $\boldsymbol{\beta}|\cdot \sim \mathrm{N}(\boldsymbol{\mu}_{\boldsymbol{\beta}}, \boldsymbol{\Sigma}_{\boldsymbol{\beta}})$ are given by

$$\boldsymbol{\Sigma}_{\boldsymbol{\beta}} = \sigma^2\left(\boldsymbol{U}'\boldsymbol{U} + \sigma^2\boldsymbol{C}^{-1}\right)^{-1}, \; \boldsymbol{\mu}_{\boldsymbol{\beta}} = \frac{1}{\sigma^2}\boldsymbol{\Sigma}_{\boldsymbol{\beta}}\left[\boldsymbol{U}'(\boldsymbol{y} - \boldsymbol{\eta}_{-\boldsymbol{\beta}}) + \sigma^2\boldsymbol{C}^{-1}\boldsymbol{c}\right] \tag{4.19}$$

where $\boldsymbol{C}^{-1} = \boldsymbol{0}$ for a flat prior $p(\boldsymbol{\beta}) \propto$ const, and

$$\boldsymbol{\eta}_{-\boldsymbol{\beta}} = \boldsymbol{X}_1\gamma_1 + \ldots + \boldsymbol{X}_r\gamma_r + \boldsymbol{Z}\boldsymbol{\alpha} = \boldsymbol{\eta} - \boldsymbol{U}\boldsymbol{\beta}$$

is the partial residual obtained by deleting $\boldsymbol{U\beta}$ from the predictor. Similarly, the covariance matrices $\boldsymbol{\Sigma}_l$ and means $\boldsymbol{\mu}_l$ of Gaussian full conditionals $p(\boldsymbol{\gamma}_l|\cdot)$, $l = 1, \ldots, r$, are given by

$$\boldsymbol{\Sigma}_l = \left(\frac{1}{\sigma^2}\boldsymbol{X}_l'\boldsymbol{X}_l + \frac{1}{\tau_l^2}\boldsymbol{K}_l\right)^{-1}, \quad \boldsymbol{\mu}_l = \frac{1}{\sigma^2}\boldsymbol{\Sigma}_l\boldsymbol{X}_l'(\boldsymbol{y} - \boldsymbol{\eta}_{-l}), \tag{4.20}$$

where $\boldsymbol{\eta}_{-l} = \boldsymbol{\eta} - \boldsymbol{X}_l\boldsymbol{\gamma}_l$ is the remaining predictor after deleting $\boldsymbol{X}_l\boldsymbol{\gamma}_l$. Finally, precision matrices and means of the Gaussian full conditionals $p(\boldsymbol{\alpha}_i|\cdot), i = 1, \ldots, n$, of unit-specific parameters are

$$\boldsymbol{\Sigma}_{\boldsymbol{\alpha}_i} = \left(\frac{1}{\sigma^2}\boldsymbol{Z}_i'\boldsymbol{Z}_i + \boldsymbol{Q}^{-1}\right)^{-1}, \quad \boldsymbol{\mu}_{\boldsymbol{\alpha}_i} = \frac{1}{\sigma^2}\boldsymbol{\Sigma}_{\boldsymbol{\alpha}_i}\boldsymbol{Z}_i'(\boldsymbol{y}_i - \boldsymbol{\eta}_{-i}) \tag{4.21}$$

where

$$\boldsymbol{\eta}_{-i} = \boldsymbol{U}_i\boldsymbol{\beta} + \boldsymbol{X}_{i1}\boldsymbol{\gamma}_1 + \ldots + \boldsymbol{X}_{ir}\boldsymbol{\gamma}_r = \boldsymbol{\eta}_i - \boldsymbol{Z}_i\boldsymbol{\alpha}_i$$

is the unit-specific predictor $\boldsymbol{\eta}_i$ after removing the random effects component $\boldsymbol{Z}_i\boldsymbol{\alpha}_i$.

Because the precision matrices of the high-dimensional Gaussian full-conditionals $p(\boldsymbol{\gamma}_l|\cdot)$ have band matrix structure, computationally efficient draws can be performed as follows (Rue, 2001):

1. Compute the Cholesky decomposition $\boldsymbol{P}_l = \boldsymbol{\Sigma}_l^{-1} = \boldsymbol{L}_l'\boldsymbol{L}_l$ using sparse matrix operations.
2. Solve $\boldsymbol{L}_l'\tilde{\boldsymbol{\gamma}}_l = \boldsymbol{z}$ for $\tilde{\boldsymbol{\gamma}}_l$, where $\boldsymbol{z}$ is a vector of independent standard Gaussians. It follows that $\tilde{\boldsymbol{\gamma}}_l \sim \mathrm{N}(\boldsymbol{0}, \boldsymbol{\Sigma}_l)$, with covariance matrix $\boldsymbol{\Sigma}_l = \boldsymbol{P}_l^{-1}$.
3. Compute the mean $\boldsymbol{\mu}_l$ by solving

$$\boldsymbol{P}_l\boldsymbol{\mu}_l = \boldsymbol{L}_l'\boldsymbol{L}_l\boldsymbol{\mu}_l = \frac{1}{\sigma^2}\boldsymbol{X}_l'(\boldsymbol{y} - \boldsymbol{\eta}_{-l})$$

   with forward-backward steps using the Cholesky decomposition again. Setting $\boldsymbol{\gamma}_l = \tilde{\boldsymbol{\gamma}}_l + \boldsymbol{\mu}_l$ yields $\boldsymbol{\gamma}_l \sim \mathrm{N}(\boldsymbol{\mu}_l, \boldsymbol{\Sigma}_l)$.

Full conditionals for the error variance σ^2, the smoothing variances $\tau_l^2, l = 1, \ldots, r$, and the variance-covariance matrix $\boldsymbol{Q}$ are computationally most convenient in a conditionally conjugate prior setting, where

$$\sigma^2 \sim \mathrm{IG}(a_\sigma, b_\sigma), \quad \tau_l^2 \sim \mathrm{IG}(a_l, b_l), \quad \boldsymbol{Q} \sim \mathrm{IW}(a, \boldsymbol{B}).$$

Then the full conditionals are inverse Gamma or inverse Wishart with updated parameters

$$\tilde{a}_\sigma = a_\sigma + \frac{N}{2}, \quad \tilde{b}_\sigma = b_\sigma + \frac{1}{2}\sum_{i,j}(y_{ij} - \eta_{ij})^2_{ij}, \tag{4.22}$$

$$\tilde{a}_l = a_l + \frac{\text{rank}(\boldsymbol{K}_l)}{2}, \quad \tilde{b}_l = b_l + \frac{1}{2}\gamma_l' \boldsymbol{K}_l \gamma_l, \tag{4.23}$$

$$\tilde{a} = a + \frac{n}{2}, \quad \tilde{\boldsymbol{B}} = \boldsymbol{B} + \frac{1}{2}\sum_{i=1}^{n} \boldsymbol{\alpha}_i \boldsymbol{\alpha}_i', \tag{4.24}$$

where $N = \sum_{i=1}^n n_i$ is the total number of observations.

In practice $\boldsymbol{Q}$ is often assumed to be diagonal, with $\boldsymbol{Q} = \text{diag}(\delta_1^2, \ldots, \delta_q^2)$, and, further assuming prior independence of the variances, the inverse Wishart reduces to inverse Gamma priors

$$\delta_l^2 \sim \text{IG}(c_l, d_l), \quad l = 1, \ldots, q.$$

Then $\delta_l^2 | \cdot \sim \text{IG}(\tilde{c}_l, \tilde{d}_l)$ with

$$\tilde{c}_l = \tilde{c}_l + \frac{n}{2}, \quad \tilde{d}_l = d_l + \frac{1}{2}\sum_{i=1}^{n} \alpha_{il}^2, \tag{4.25}$$

where α_{il} is the l-th component of $\boldsymbol{\alpha}_i$.

As discussed in Section 4.1.3, the choice of diffuse, but proper IG$(\varepsilon, \varepsilon)$-priors is usually not crucial for σ^2 and for smoothing variances τ_l^2. We also admit improper IG-type priors with $a = -0.5, b = 0$, corresponding to uniform priors for standard deviations as recommended for practical work in Gelman (2006). Full conditionals are proper IG-distributions, and Sun, Tsutakawa and He (2001) as well as Fahrmeir and Kneib (2009) provide conditions for propriety of posteriors. Other choices are possible but computationally more demanding, see e.g. Li, Lin and Müller (2010).

Algorithm 4.2: Gibbs sampler for Gaussian SPMMs
Initialize all parameters at suitable starting values and draw samples for

i. $\boldsymbol{\beta}$ from N$(\boldsymbol{\mu}_{\boldsymbol{\beta}}, \boldsymbol{\Sigma}_{\boldsymbol{\beta}})$ with moments defined in (4.19), inserting current values for all other parameters.
ii. γ_l from N$(\boldsymbol{\mu}_l, \boldsymbol{\Sigma}_l)$ with moments defined in (4.20) utilizing the efficient draws for multivariate Gaussian variables with band matrix structure for the precision matrix, inserting current values for all other parameters.
iii. $\boldsymbol{\alpha}_i$ from N$(\boldsymbol{\mu}_{\boldsymbol{\alpha}_i}, \boldsymbol{\Sigma}_{\boldsymbol{\alpha}_i})$ with moments defined in (4.21), inserting current values for all other parameters.

iv. σ^2 from $\text{IG}(\tilde{a}_\sigma, \tilde{b}_\sigma)$, inserting current values for all other parameters.
v. τ_l^2 from $\text{IG}(\tilde{a}_l, \tilde{b}_l)$, inserting current values for all other parameters.
vi. $\boldsymbol{Q}$ from $\text{IW}(\tilde{a}, \tilde{\boldsymbol{B}})$, inserting current values $\boldsymbol{\alpha}_i^c$ or, if $\boldsymbol{Q} = \text{diag}(\delta_1^2, \ldots, \delta_q^2)$, δ_l^2 from $\text{IG}(\tilde{a}_l, \tilde{b}_l)$, inserting current values α_{il}^c, $l = 1, \ldots, q$.

Example 4.4: Childhood obesity
FB inference with Gibbs sampling is now applied to the additive mixed model (4.15) of Example 4.2. The population trend $f(t)$ is specified through a cubic Bayesian P-spline with a second-order random-walk prior and 20 (equidistant) knots. Hyperpriors for the variance σ^2 and for the variance components are chosen as in Example 3.4. The hyperprior for the smoothing variance τ^2 of the Bayesian P-spline is $\tau^2 \sim \text{IG}(a, b)$, with $a = b = 0.001$. Table 4.2 contains the estimates for the fixed effects as well as variance components. Comparing with corresponding results in Tables 3.2 and 3.3 in Example 3.4, we see that estimation results are quite close to each other. Figures 4.10 (model with random lines) and 4.11 (model with random curves) show estimated population trends as well as individual curves with and without random effects for the same selected individuals as in the examples before. We see that the nonparametric population curve is a bit more wiggly than the estimated parametric curve in Example 3.4. Its shape, however, confirms that the choice of the parametric population trend (3.6) seems to be a good one. We will check this through model assessment techniques in Section 4.5.2, see Example 4.12. △

Table 4.2 Childhood obesity: Posterior means (FB inference) together with 95% credible intervals for the fixed effects and variance components for the AMM with random lines and the AMM with random curves. mBMI and mDiffBMI have been centred for the analyses.

	Random lines			Random curves		
	mean	2.5%	97.5%	mean	2.5%	97.5%
const	15.53	15.34	15.72	15.56	15.38	15.73
sex	0.3063	0.2238	0.3836	0.1489	0.0597	0.2358
breast	0.0433	−0.0461	0.1303	0.1316	0.0428	0.2169
mSmoke	0.0513	−0.0671	0.1699	−0.0308	−0.1556	0.1042
area	0.0196	−0.0783	0.1217	0.0174	−0.0797	0.1133
mBMI	0.0449	0.0331	0.0554	0.0506	0.0387	0.0627
mDiffBMI	0.0815	0.0574	0.1068	0.1201	0.0938	0.1480
σ^2	0.9858	0.9627	1.0090	0.7444	0.7234	0.7653
δ_0^2	0.7143	0.6643	0.7679	0.7103	0.6524	0.7728
δ_1^2	0.0338	0.0295	0.0383	0.0434	0.0391	0.0484
δ_2^2	–	–	–	37.47	33.86	41.05

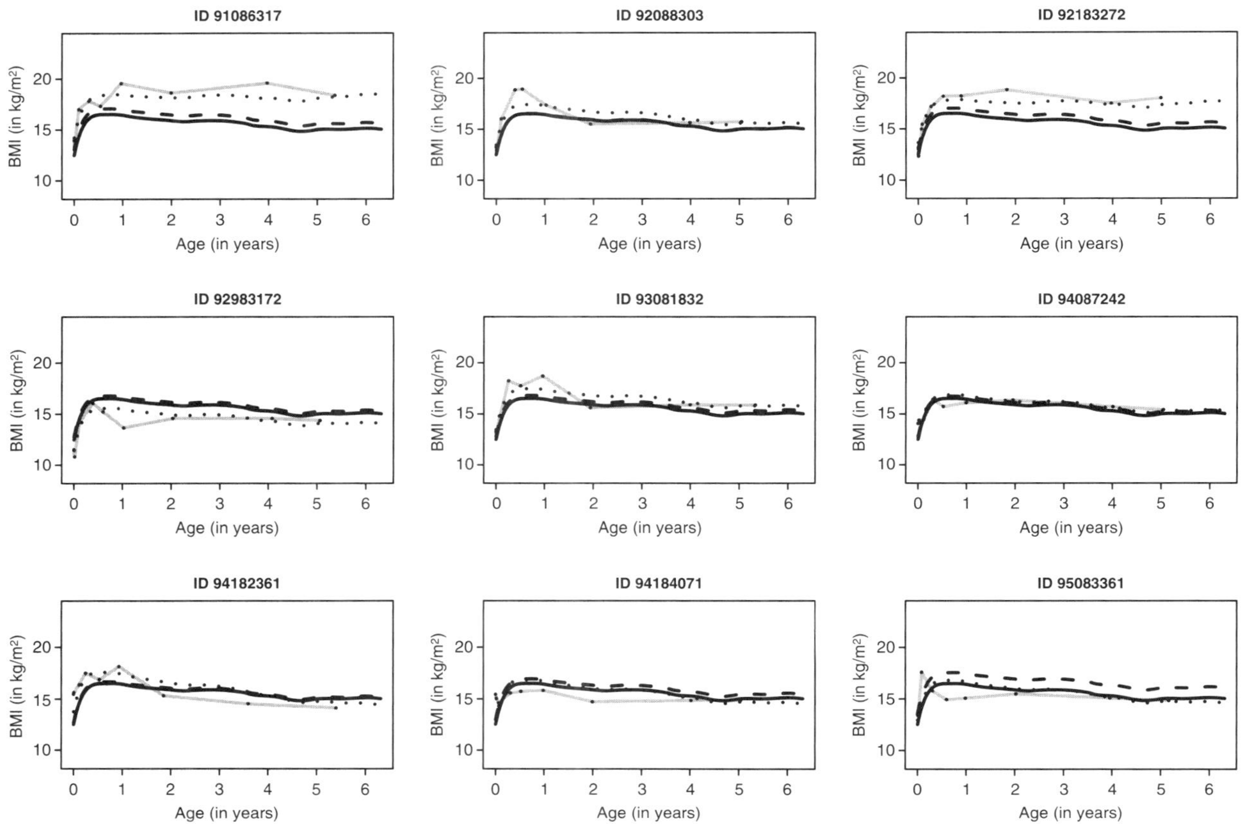

Fig. 4.10 Childhood obesity: Observed BMI curves for 9 randomly selected individuals (grey lines) in comparison to the estimated population trend (solid lines), the individual trend without random effects (dashed lines) and the individual trend with random effects (dotted lines) for the FB AMM with random lines.

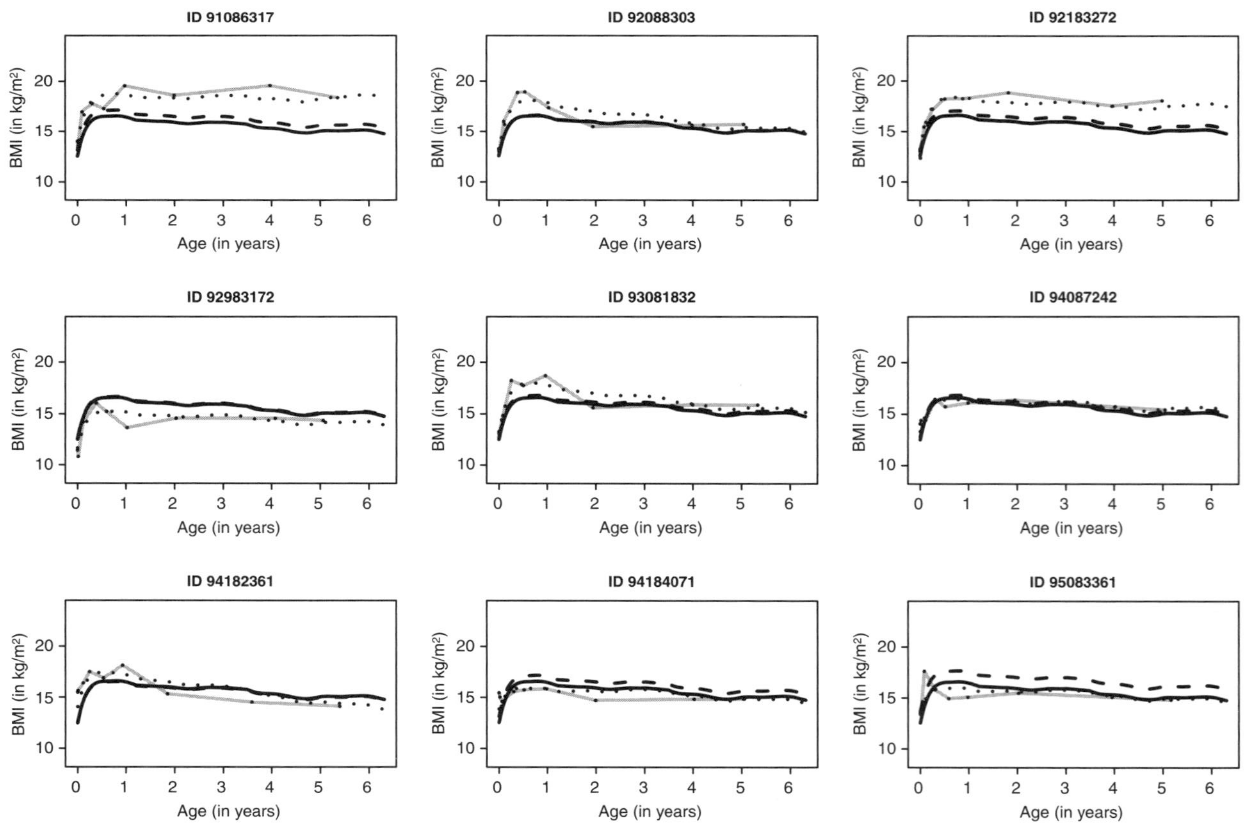

Fig. 4.11 Childhood obesity: Observed BMI curves for 9 randomly selected individuals (grey lines) in comparison to the estimated population trend (solid lines), the individual trend without random effects (dashed lines) and the individual trend with random effects (dotted lines) for the FB AMM with random curves.

Exponential family SPMMs

For non-Gaussian SPMMs with univariate responses from an exponential family, full conditionals for regression parameters are no longer Gaussian but analytically intractable, so that direct Gibbs sampling as for Gaussian SPMMs is not possible any more. As in Chapters 2 and 3, there are two strategies for designing MCMC algorithms in this situation: First, use an MH-algorithm with appropriately chosen proposal densities. Second, introduce auxiliary (latent) variables following a SPMM with Gaussian errors, or errors which follow a mixture of Gaussians. Here we describe the first strategy in some detail while the second strategy is exemplified for categorical SPMMs.

Our preferred technique is a generalization of the MH-algorithm with iteratively weighted least-squares (IWLS-) proposals, originally suggested by Gamerman (1997) for generalized linear (mixed) models and extended to SPMMs (with a random intercept only) by Brezger and Lang (2006). To simplify, we assume that the exponential family does not contain an unknown dispersion parameter ϕ, i.e. we set $\phi = 1$, which corresponds to $\sigma^2 = 1$ for Gaussian models. Then $p(y_{ij}) \propto \exp(y_{ij}\theta_{ij} + b(\theta_{ij}))$ with canonical parameter θ_{ij}.

The basic idea is to combine Fisher scoring or IWLS as outlined in Section 2.2.2 for estimating regression parameters in generalized linear models and the Metropolis–Hastings algorithm. More precisely, the goal is to approximate the full conditionals of regression parameters by Gaussian distributions with precision matrices and means as used in penalized likelihood or empirical Bayes estimation. For fixed effects $\boldsymbol{\beta}$, let $\varphi(\boldsymbol{\beta}; \boldsymbol{\mu}_{\boldsymbol{\beta}}, \boldsymbol{\Sigma}_{\boldsymbol{\beta}})$ denote the multivariate Gaussian density with covariance matrix and mean

$$\boldsymbol{\Sigma}_{\boldsymbol{\beta}} = \left(\boldsymbol{U}'\boldsymbol{W}\boldsymbol{U} + \boldsymbol{C}^{-1}\right)^{-1}, \quad \boldsymbol{\mu}_{\boldsymbol{\beta}} = \boldsymbol{\Sigma}_{\boldsymbol{\beta}}\left[\boldsymbol{U}'\boldsymbol{W}(\tilde{\boldsymbol{y}} - \boldsymbol{\eta}_{-\boldsymbol{\beta}}) + \boldsymbol{C}^{-1}\boldsymbol{c}\right], \tag{4.26}$$

where $\boldsymbol{\eta}_{-\boldsymbol{\beta}} = \boldsymbol{\eta} - \boldsymbol{U}\boldsymbol{\beta}$ is defined as in the Gaussian case. The term $\boldsymbol{U}'\boldsymbol{W}\boldsymbol{U}$ is the (expected) Fisher information matrix of GLMs, where the weight matrix $\boldsymbol{W} = \boldsymbol{W}(\boldsymbol{\eta})$ is diagonal with usual GLM weights

$$w_{ij}(\eta_{ij}) = \{b''(\theta_{ij})(g'(\mu_{ij}))^2\}^{-1},$$

and $g = h^{-1}$ as the chosen link (or inverse response) function. The elements of $\tilde{\boldsymbol{y}} = \tilde{\boldsymbol{y}}(\boldsymbol{\eta})$ are the working observations

$$\tilde{y}_{ij}(\eta_{ij}) = \eta_{ij} + g'(\mu_{ij})(y_{ij} - \mu_{ij})$$

known from IWLS. In complete analogy, we define $\varphi(\boldsymbol{\gamma}_l; \boldsymbol{\mu}_l, \boldsymbol{\Sigma}_l)$ for regression parameters $\boldsymbol{\gamma}_l$ of functions as multivariate Gaussian densities with covariance matrix and mean

$$\boldsymbol{\Sigma}_l = \left(\boldsymbol{X}_l'\boldsymbol{W}\boldsymbol{X}_l + \frac{1}{\tau_l^2}\boldsymbol{K}_l\right)^{-1}, \quad \boldsymbol{\mu}_l = \boldsymbol{\Sigma}_l\boldsymbol{X}_l'\boldsymbol{W}(\tilde{\boldsymbol{y}} - \boldsymbol{\eta}_{-l}), \tag{4.27}$$

where $\boldsymbol{\eta}_{-l} = \boldsymbol{\eta} - \boldsymbol{X}_l\boldsymbol{\gamma}_l$ is the predictor after removing $\boldsymbol{X}_l\boldsymbol{\gamma}_l$. Finally, Gaussian densities $\varphi(\boldsymbol{\alpha}_i; \boldsymbol{\mu}_{\boldsymbol{\alpha}_i}, \boldsymbol{\Sigma}_{\boldsymbol{\alpha}_i})$ for $\boldsymbol{\alpha}_i$ are defined by

$$\boldsymbol{\Sigma}_{\boldsymbol{\alpha}_i} = \left(\boldsymbol{Z}_i'\boldsymbol{W}_i\boldsymbol{Z}_i + \boldsymbol{Q}^{-1}\right)^{-1}, \quad \boldsymbol{\mu}_{\boldsymbol{\alpha}_i} = \boldsymbol{\Sigma}_{\boldsymbol{\alpha}_i}\boldsymbol{Z}_i'\boldsymbol{W}_i(\tilde{\boldsymbol{y}}_i - \boldsymbol{\eta}_{-i}), \tag{4.28}$$

with $\boldsymbol{\eta}_{-i} = \boldsymbol{\eta}_i - \boldsymbol{Z}_i\boldsymbol{\alpha}_i$, diagonal weight matrix $\boldsymbol{W}_i$ and working observation $\tilde{\boldsymbol{y}}_i$ corresponding to observation $\boldsymbol{y}_i, i = 1, \ldots, n$.

These Gaussian densities are used as proposal densities in MH steps for drawing from the full conditionals $p(\boldsymbol{\beta}|\cdot), p(\boldsymbol{\gamma}_l|\cdot)$, and $p(\boldsymbol{\alpha}_i|\cdot)$. Suppose we want to update $\boldsymbol{\gamma}_l$ with current state $\boldsymbol{\gamma}_l^c$. Then, according to IWLS, a new value $\boldsymbol{\gamma}_l^p$ is proposed by drawing a random number from the Gaussian proposal density $\varphi(\boldsymbol{\gamma}_l; \boldsymbol{\mu}_l, \boldsymbol{\Sigma}_l)$. As in the Gaussian case, computationally efficient drawing uses Cholesky decompositions of precision matrices. Then the proposed $\boldsymbol{\gamma}_l^p$ is accepted with probability

$$\alpha(\boldsymbol{\gamma}_l^c, \boldsymbol{\gamma}_l^p) = \frac{p(\boldsymbol{y}|\boldsymbol{\eta}^p)p(\boldsymbol{\gamma}_l^p|(\tau_l^2)^c)\varphi(\boldsymbol{\gamma}_l^c; \boldsymbol{\mu}_l^p, \boldsymbol{\Sigma}_l^p)}{p(\boldsymbol{y}|\boldsymbol{\eta}^c)p(\boldsymbol{\gamma}_l^c|(\tau_l^2)^c)\varphi(\boldsymbol{\gamma}_l^p; \boldsymbol{\mu}_l^c, \boldsymbol{\Sigma}_l^c)}. \tag{4.29}$$

The observation likelihood $p(\boldsymbol{y}|\boldsymbol{\eta}^c)$, the prior $p(\boldsymbol{\gamma}_l^c|(\tau_l^2)^c)$, as well as the mean $\boldsymbol{\mu}_l^c$ and covariance matrix $\boldsymbol{\Sigma}_l^c$ of the Gaussian proposal density in the denominator are evaluated at current values of parameters $\boldsymbol{\beta}^c, \ldots, \boldsymbol{\gamma}_l^c, \boldsymbol{\alpha}^c, \ldots, (\tau_l^2)^c, \ldots$, of the predictor $\boldsymbol{\eta}^c$ and weight matrix $\boldsymbol{W}^c = \boldsymbol{W}(\boldsymbol{\eta}^c)$. In the nominator $\boldsymbol{\gamma}_l^c$ is replaced through the proposed value $\boldsymbol{\gamma}_l^p$, holding the remaining parameters at their current values, so that

$$\boldsymbol{\eta}^p = \boldsymbol{U}\boldsymbol{\beta}^c + \ldots + \boldsymbol{X}_l\boldsymbol{\gamma}_l^p + \ldots + \boldsymbol{Z}\boldsymbol{\alpha}^c,$$

and $\boldsymbol{\mu}_l^p, \boldsymbol{\Sigma}_l^p$ are evaluated at $\boldsymbol{W}^p = \boldsymbol{W}(\boldsymbol{\eta}^p), \tilde{\boldsymbol{y}}^p = \tilde{\boldsymbol{y}}(\boldsymbol{\eta}^p)$ etc. (instead of $\boldsymbol{W} = \boldsymbol{W}(\boldsymbol{\eta}^c), \tilde{\boldsymbol{y}}^c = \tilde{\boldsymbol{y}}(\boldsymbol{\eta}^c)$ etc.). If the proposal is accepted, we set $\boldsymbol{\gamma}_l^c = \boldsymbol{\gamma}_l^p$, otherwise we keep the current $\boldsymbol{\gamma}_l^c$. A slightly different sampling scheme uses the current posterior mode approximation $\boldsymbol{\mu}_l^c$ rather than $\boldsymbol{\gamma}_l^c$ in the predictor $\boldsymbol{\eta}^c$ for computing the weight matrix $\boldsymbol{W}^c = \boldsymbol{W}(\boldsymbol{\eta}^c)$ and the working observation $\tilde{\boldsymbol{y}}^c = \tilde{\boldsymbol{y}}(\boldsymbol{\eta}^c)$. It turns out that acceptance rates can be considerably higher compared to the sampling scheme based on the current $\boldsymbol{\gamma}_l^c$. MH-steps with IWLS-proposals for $\boldsymbol{\beta}$ and $\boldsymbol{\alpha}_i$ are performed in complete analogy, with proposal densities $\varphi(\boldsymbol{\beta}; \boldsymbol{\mu}_\beta, \boldsymbol{\Sigma}_\beta)$ and $\varphi(\boldsymbol{\alpha}_i|\boldsymbol{\mu}_{\boldsymbol{\alpha}_i}, \boldsymbol{\Sigma}_{\boldsymbol{\alpha}_i})$, respectively. Gibbs steps for variance parameters $\tau_l^2, l = 1, \ldots, r$, and for $\boldsymbol{Q}$ or variance parameters $\delta_1^2, \ldots, \delta_q^2$ are identical to the Gaussian case.

We summarize the IWLS updating scheme based on current modes in the following, see Brezger and Lang (2006) for additional details and remarks. It extends the MH algorithm for GAMs (Algorithm 2.8 in Section 2.3.2) and GLMMs (Algorithm 3.3 in Section 3.3.4) to SPMMs. The order of evaluations is stressed because it is crucial for computational efficiency.

Algorithm 4.3: MH algorithm with IWLS proposal

i. *Initialization:* Compute the posterior modes $\boldsymbol{\mu}_l^c$, $l = 1, \ldots, r$, and $\boldsymbol{\beta}^c$ for $\boldsymbol{\gamma}_l$ and $\boldsymbol{\beta}$ given the fixed variance parameters $\tau_l^2 = c_l$ (e.g. $c_l = 10$). The mode is computed via backfitting within Fisher scoring. Use the posterior mode estimates as the current state $\gamma_l^c, \boldsymbol{\beta}^c$ of the chain. Set $(\tau_l^2)^c = c_l$. Store the current predictor in the vector $\boldsymbol{\eta}^c$.

ii. *For* $l = 1, \ldots, r$ *update* γ_l:
- Compute the likelihood $p(\boldsymbol{y}|\boldsymbol{\eta}^c)$.
- Exchange $\boldsymbol{X}_l\gamma_l^c$ in the current predictor $\boldsymbol{\eta}^c$ by $\boldsymbol{X}_l\boldsymbol{\mu}_l^c$, i.e. $\boldsymbol{\eta}^c = \boldsymbol{\eta}^c + \boldsymbol{X}_l(\boldsymbol{\mu}_l^c - \gamma_l^c)$.
- Draw a proposal γ_l^p from the Gaussian proposal density $\varphi(\gamma_l; \boldsymbol{\mu}_l^c, \boldsymbol{\Sigma}_l^c)$ and covariance matrix $\boldsymbol{\Sigma}_l^c$ given in (4.27).
- Exchange $\boldsymbol{X}_l\boldsymbol{\mu}_l^c$ in the current predictor $\boldsymbol{\eta}^c$ by $\boldsymbol{X}_l\boldsymbol{\beta}_l^p$, i.e. $\boldsymbol{\eta}^p = \boldsymbol{\eta}^c + \boldsymbol{X}_l(\gamma_l - \boldsymbol{\mu}_l^c)$.
- Compute the likelihood $p(\boldsymbol{y}|\boldsymbol{\eta}^p)$
- Compute $\varphi(\gamma_l^p; \boldsymbol{\mu}_l^c, \boldsymbol{\Sigma}_l^c)$, $\varphi(\gamma_l^c; \boldsymbol{\mu}_l^p, \boldsymbol{\Sigma}_l^p)$, $p(\gamma_l^p|(\tau_l^2)^c)$ and $p(\gamma_l^c|(\tau_l^2)^c)$.
- Accept γ_l^p as the new state of the chain with probability (4.29). If the proposal is rejected, exchange $\boldsymbol{X}_l\boldsymbol{\beta}_l^p$ in the current predictor $\boldsymbol{\eta}^c$ by $\boldsymbol{X}_l\boldsymbol{\beta}_l^c$, i.e. $\boldsymbol{\eta}^c = \boldsymbol{\eta}^c + \boldsymbol{X}_l(\gamma_l^c - \gamma_l^p)$.
- Set $\boldsymbol{\mu}_l^c = \boldsymbol{\mu}_l$.

iii. *Update fixed effects parameters* by similar steps as for updating of γ_l, compare Algorithm 3.3. page 174.

iv. *For* $i = 1, \ldots, n$ *update random effects* $\boldsymbol{\alpha}_i$ by similar steps as for updating of γ_l, compare Algorithm 3.3. page 174.

v. *For* $l = 1, \ldots, r$ *update variance parameters* τ_r^2: Variance parameters are updated by drawing from the inverse gamma full conditionals with hyperparameters given in (4.27).

vi. *Update variance parameters* $\delta_1^2, \ldots, \delta_q^2$: Draw from inverse gamma full conditionals given in (4.25).

Some properties and possible modifications of the algorithm are summarized in the following remarks:

1. For Gaussian SPMMs the MH-steps with IWLS proposals in ii., iii. and iv. simplify to Gibbs steps, drawing directly from the Gaussian full conditionals (4.19), (4.20) and (4.21). An additional Gibbs step is required for updating the error variance via the IG full conditional (4.22). Then Algorithm 4.3 basically reduces to Algorithm 4.2, see also Lang and Brezger (2004).
2. Usually, convergence and mixing of Markov chains is excellent with IWLS proposals. In some cases, for example if two-dimensional interaction effects are included (see Chapter 5), convergence problems may occur. To overcome such difficulties, joint updating of (γ_l, τ_l^2) as suggested by Knorr-Held and Rue (2002) can be useful, see Brezger and Lang (2006) for details.
3. A natural source for saving computing time is to avoid the (re)computation of the IWLS weight matrix $\boldsymbol{W}$ (and thereby the matrix $\boldsymbol{X}'\boldsymbol{W}\boldsymbol{X}$) in every iteration of the sampler. A possible strategy is to recompute the weights only

every t-th iteration. It is even possible to keep the weights fixed after the burn in period. Experience suggests that for most distributions and models the acceptance rates and the mixing of the chains is almost unaffected by keeping the weights fixed.

4. Another variant is to replace the (current) posterior mode approximation by the current value γ_l^c.

Example 4.5: Forest health

To illustrate the estimation of exponential family SPMMs based on MCMC simulation, we consider a binary logit model for the forest health data with nonlinear trend and age effect in combination with plot-specific random effects and some further covariates with parametric effects. The resulting estimates are visualized in Figs. 4.12 and 4.13 and basically confirm our conclusions from previous analyses. Note that, as compared to Example 4.3, we now have only one estimated trend function available since we are dealing with a binary regression model. Moreover, the scale of the effect estimates is different since we are considering a logit instead of a probit model. △

Categorical SPMMs

MH algorithms with IWLS proposals (or other variants) can be extended to categorical SPMMs by embedding categorical regression into multivariate GLMs or GAMs and generalizing the MH algorithm with IWLS proposals accordingly. Here we outline an alternative strategy based on auxiliary latent models, already described for Bayesian P-Splines in Section 2.2.3.

We focus on probit models, augmenting categorical outcomes Y_{ij} by auxiliary latent variables $l_{ij}^{(c)} = \eta_{ij}^{(c)} + \varepsilon_{ij}^{(c)}, c = 1, \ldots, C-1$, $l_{ij}^{(C)} \equiv 0$, for unordered and $l_{ij} = \eta_{ij} + \varepsilon_{ij}$, for ordered outcomes. Posterior analyses are then based on

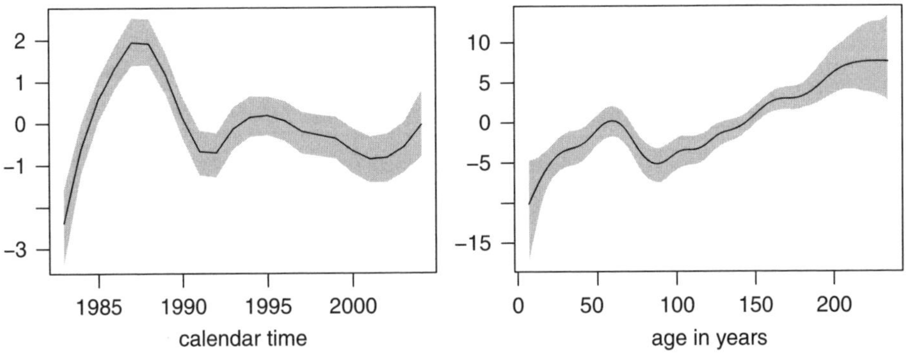

Fig. 4.12 Forest health: Estimated nonparametric effects together with pointwise 95% credible intervals.

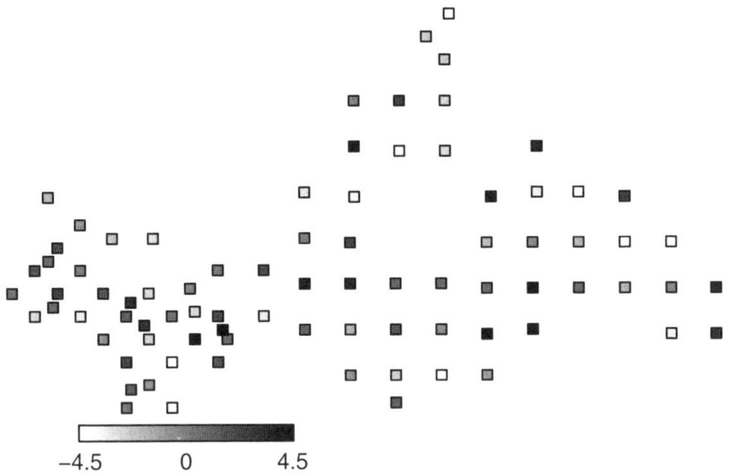

Fig. 4.13 Forest health: Posterior mean estimates for the random effect.

$$p(\boldsymbol{\beta}, \boldsymbol{\gamma}, \boldsymbol{\alpha}, \boldsymbol{\tau}, \boldsymbol{Q}, \boldsymbol{l}|\boldsymbol{y}) \propto p(\boldsymbol{y}|\boldsymbol{l})p(\boldsymbol{l}|\boldsymbol{\beta}, \boldsymbol{\gamma}, \boldsymbol{\alpha})p(\boldsymbol{\beta})p(\boldsymbol{\gamma}|\boldsymbol{\tau})p(\boldsymbol{\tau})p(\boldsymbol{\alpha}|\boldsymbol{Q})p(\boldsymbol{Q}),$$

with $p(\boldsymbol{y}|\boldsymbol{l}) = \prod_{i,j} p(Y_{ij}|l_{ij})$, where $l_{ij} = (l_{ij}^{(1)}, \dots, l_{ij}^{(C)})$ for nominal responses, $p(\boldsymbol{\gamma}|\boldsymbol{\tau}) = \prod_l p(\gamma_l|\tau_l^2)p(\tau_l^2), p(\boldsymbol{\alpha}|\boldsymbol{Q}) = \prod_{i=1}^n p(\boldsymbol{\alpha}_i|\boldsymbol{Q})$, etc. The conditional likelihood $p(Y_{ij}|l_{ij})$ is determined by the mechanisms relating observed Y_{ij}'s to latent l_{ij}'s. For a nominal response, we have

$$p(Y_{ij}|l_{ij}) = \sum_{c=1}^{C} \mathbb{1}(\max(l_{ij}^{(1)}, \dots, l_{ij}^{(C)}) = l_{ij}^{(Y_{ij})})\mathbb{1}(Y_{ij} = c), \tag{4.30}$$

and for a cumulative model, we get

$$p(Y_{ij}|l_{ij}) = \sum_{c=1}^{C} \mathbb{1}(\theta^{(c-1)} < l_{ij} \leq \theta^{(c)})\mathbb{1}(Y_{ij} = c), \tag{4.31}$$

due to the fact that $p(Y_{ij}|l_{ij})$ is one if l_{ij} obeys the constraint imposed by the observed value of Y_{ij}. Compared to MH-sampling, additional drawings from the full conditionals for the latent variables l_{ij} are necessary. As an advantage, full conditionals for regression parameters become Gaussian, allowing computationally efficient Gibbs sampling. The full conditionals for l_{ij} are

$$p(l_{ij}|\boldsymbol{\beta}, \boldsymbol{\gamma}, \boldsymbol{\alpha}, Y_{ij}) \propto p(Y_{ij}|l_{ij})p(l_{ij}|\boldsymbol{\beta}, \boldsymbol{\gamma}, \boldsymbol{\alpha}).$$

Since latent variables l_{ij} have (conditional) Gaussian distributions with means η_{ij} and unit variances, their full conditionals are truncated normals, with truncation points determined by the restrictions in (4.30) and (4.31), compare also Section 2.2.3.

In the following we summarize the resulting Gibbs sampling schemes.

Algorithm 4.4: Sampling scheme for cumulative probit SPMMs

i. *Latent variables* l_{ij}: If $Y_{ij} = c$, then l_{ij} is generated from $\mathrm{TN}_{(\theta^{(c-1)},\theta^{(c)}]}(\eta_{ij}, 1)$ with η_{ij} evaluated at the current values of all regression coefficients. Drawing from this truncated normal can be performed as described in Geweke (1991) and Robert (1995).

ii. Following Albert and Chib (1993), the full conditional for threshold $\theta^{(c)}$, $c = 1, \dots, C-1$ is uniform on the interval

$$[\max\{l_{ij} : Y_{ij} = c\}, \min\{l_{ij} : Y_{ij} = c+1\}].$$

Posterior samples from these uniform distributions may exhibit bad mixing. A reason is that intervals can become quite small, and as a consequence, the chain moves slowly. In such a case, other parametrizations as suggested for example those in Chen and Dey (2000) are a valuable alternative. For $C = 3$ such a reparametrization becomes particularly convenient: First, include a constant β_0 in the vector $\boldsymbol{\beta}$ of fixed effects to set $\theta^{(1)} = 0$. Second, because parameters in the predictor of the latent Gaussian model are only identifiable up to a multiplicative factor, we assume that errors ε_{ijt} are $\mathrm{N}(0, \sigma^2)$ distributed with unknown variance σ^2. This allows us to set $\theta^{(2)} = 1$. For σ^2 we specify an inverse gamma prior, leading to posterior samples from an inverse gamma full conditional.

iii. Regression parameters $\boldsymbol{\beta}, \boldsymbol{\gamma}$ and $\boldsymbol{\alpha}$ are generated from Gaussian full conditionals $p(\boldsymbol{\beta}|\cdot), p(\boldsymbol{\gamma}_l|\cdot)$ and $p(\boldsymbol{\alpha}_i|\cdot)$ given in (4.19), (4.20) and (4.21), with $\sigma^2 = 1$, replacing y_{ij} with l_{ij}.

iv. Samples for variance components $\tau_1^2, \dots, \tau_r^2, \delta_1^2, \dots, \delta_q^2$ (or $\boldsymbol{Q}$) are obtained from inverse gamma (or Wishart) distributions (4.23), (4.25) or (4.24).

Algorithm 4.5: Sampling scheme for multinomial probit SPMMs

Since only differences of latent variables can be identified, we set $l_{ij}^{(C)} = 0$ for the reference category C.

i. *Latent variables:* $l_{ij}^{(c)}$, $c = 1, \dots, C-1$:
   - If $Y_{ij} = c$, $c \neq C$, then $l_{ij}^{(c)}$ is generated from $\mathrm{N}(\eta_{ij}^{(c)}, 1)$ subject to the constraints $l_{ij}^{(c)} > l_{ij}^{(h)}$, $h \neq c$, and $l_{ij}^{(c)} > 0\ (\equiv l_{ij}^{(C)})$. Next $l_{ij}^{(h)}$ for $h \neq c$ is generated from $\mathrm{N}(\eta_{ij}^{(h)}, 1)$ subject to $l_{ij}^{(h)} < l_{ij}^{(c)}$ generated just before.
   - If $Y_{ij} = C$, then $l_{ij}^{(h)}$, $h = 1, \dots, C-1$, are generated from $\mathrm{N}(\eta_{ij}^{(h)}, 1)$ subject to $l_{ij}^{(h)} < 0\ (\equiv l_{ij}^{(C)})$.

ii. *Updates:* Posterior samples for regression parameters and variance components are generated as in the steps iii. and iv. of the sampling scheme for ordered responses.

Example 4.6: Forest health

Extending Example 4.5, we now estimate a categorical SPMM for the forest health data based on MCMC simulations. To enable the application of a latent Gaussian formulation, we also replace the logit model with a probit model, since this considerably speeds up computations. Note that no category-specific trend functions are available in the full Bayesian model since category-specific effects apart from the thresholds prevent the re-formulation of the three-categorical cumulative probit model as a simple linear model.

Figures 4.14 and 4.15 visualize the estimated nonlinear and random effects obtained from the MCMC simulations. All effects are in good agreement with the effects found in the binary logit model from Example 4.5 apart from the different scaling resulting from the change in the link function. △

Credible intervals and bands

Fully Bayesian credible intervals and simultaneous credible bands attempt to find credible regions for either a single function evaluation $f_l(x_l)$ or for a vector of function evaluations $\boldsymbol{f}_l$. To ease notation, we will drop the index l in the following discussion. In principle, the goal is to find a credible region $\mathcal{C}$ such that the posterior probability

$$\mathbb{P}_{\boldsymbol{f}|\boldsymbol{y}}(\boldsymbol{f} \in \mathcal{C}) = \int_{\mathcal{C}} p(\boldsymbol{f}|\boldsymbol{y}) d\boldsymbol{f} \tag{4.32}$$

equals the desired coverage probability α. Note that in contrast to the interpretation of classical confidence bands, the probability statement (4.32) relates to the posterior distribution and not to independent replications of the data generating mechanism as in the frequentist setting. Still, Bayesian credible bands are

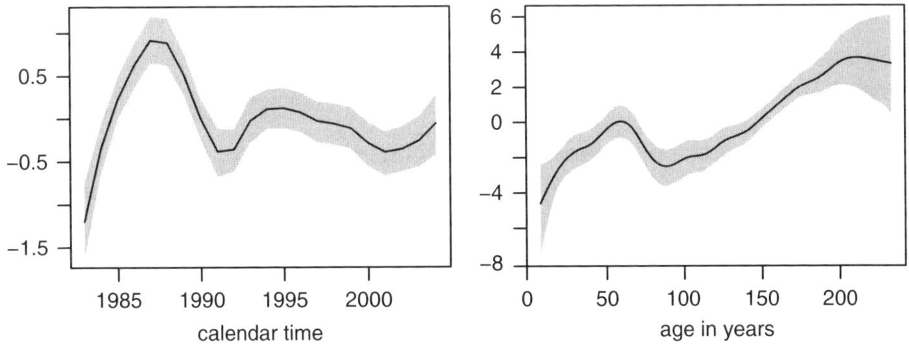

Fig. 4.14 Forest health: Estimated nonparametric effects together with pointwise 95% credible intervals.

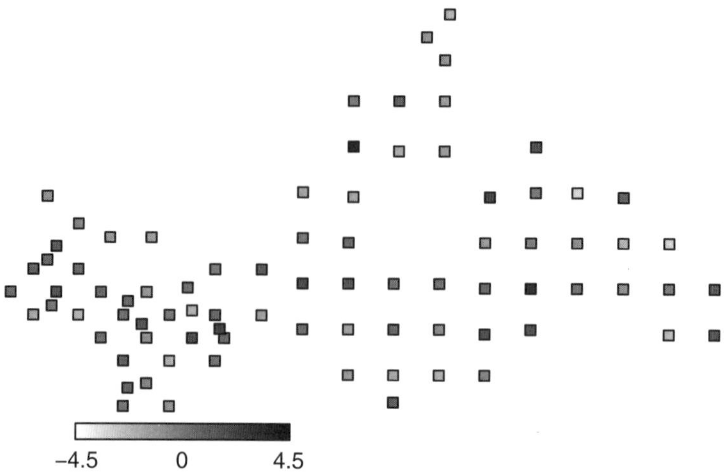

Fig. 4.15 Forest health: Posterior mean estimates for the random effect.

often interpreted analogously to frequentist credible bands and have also shown good coverage properties in simulation studies when being used in this way.

From a decision theoretic perspective, highest posterior density credible regions would be optimal, where $\mathcal{C}$ is chosen according to a threshold for the posterior density and the threshold is determined such that the coverage probability is achieved. However, such credible bands require the analytic availability of the posterior, which is seldom the case. In contrast, fully Bayesian credible bands have to be derived based on samples from the posterior that are available from the MCMC simulations. Although large samples in principle allow us to estimate the complete posterior, it is common to base credible intervals on empirical quantiles instead. In particular, if a sample for $\hat{f}(x)$ is available, a simple credible interval of level α can be estimated by the $\alpha/2$ and the $(1-\alpha/2)$ quantile of the samples. Provided that the sample is long enough, this will result in an interval covering a fraction α of the marginal posterior for $\hat{f}(x)$ but will in general not correspond to a highest posterior density interval unless the posterior is symmetric and unimodal.

Simultaneous credible regions for a vector $\hat{\boldsymbol{f}} = (\hat{f}_1, \ldots, \hat{f}_r)' = (\hat{f}(x_1), \ldots, \hat{f}(x_r))'$ are more difficult to derive from samples, since no simultaneous ordering of samples of a vector can be accomplished. In the literature, only a very limited number of approaches has been discussed, and there is still no consensus on which strategy is preferable.

Besag *et al.* (1995) propose a simultaneous credible region based on order statistics of the samples $f_1^{(t)}, \ldots, f_r^{(t)}$, where $t = 1, \ldots, T$ is the MCMC iteration index. If $f_j^{[t]}$ denotes the order statistic and $r_j^{(t)}$ the rank associated with $f_j^{(t)}$, determine the smallest integer t^* such that the hyperrectangular

$$\left[f_j^{[T-t^*+1]}, f_j^{[t^*]}\right], \quad j = 1, \dots, r$$

contains $\alpha \cdot 100\%$ of the sampled vectors $\boldsymbol{f}_r^{(1)}, \dots, \boldsymbol{f}_r^{(T)}$. As Besag *et al.* (1995) indicate, the index t^* can be derived as the $(\alpha \cdot T)$-th order statistic of the set

$$\left\{ \max_{t=1,\dots,T} \left\{ T - \min_{j=1,\dots,r} r_j^{(t)} + 1, \max_{j=1,\dots,r} r_j^{(t)} \right\} \right\}.$$

In the univariate case, that is for $r = 1$ the simultaneous credible band reduces to the usual symmetric credible interval derived from the $\alpha/2$ and $(1 - \alpha/2)$ sample quantiles.

An alternative simultaneous credible region for $\hat{\boldsymbol{f}}_r$ is proposed in Crainiceanu *et al.* (2007). Suppose that $\hat{\boldsymbol{f}}_r$ is the posterior mean estimate and that the posterior standard deviation for each point contained in $\hat{\boldsymbol{f}}_r$ has been computed as $\hat{s}_j$, $j = 1, \dots, r$. Assuming approximate posterior normality and deriving the $(1 - \alpha)$ sample quantile $z_{1-\alpha}$ of

$$\max_{j=1,\dots,r} \left| \frac{f_j^{(t)} - \hat{f}_j}{\hat{s}_j} \right|, \tag{4.33}$$

a simultaneous credible region is given by the hyperrectangular

$$\left[\hat{f}_j - \hat{s}_j, \hat{f}_j + \hat{s}_j\right], j = 1, \dots, r.$$

Note the similarity to the simultaneous confidence band derived in the context of mixed model-based smoothing. In particular, (4.33) can be considered the sampling based equivalent to (4.18) (page 207). The credible band proposed by Crainiceanu *et al.* (2007) relies on posterior normality, using the posterior standard deviation as the measure of uncertainty and the posterior mean as the centre point for the credible band. Hence, it does not make use of the full posterior distribution information contained in the sample. Moreover, if the posterior is asymmetric, the posterior normality credible band may be misleading.

A third alternative can be obtained by combining the two previous approaches. The basic idea is to use the posterior mean as the centre point of a credible band and the pointwise credible intervals as local information on the variability of the estimate. The pointwise credible intervals are then scaled by a constant factor until a fraction α of all sampled curves $\boldsymbol{f}_r^{(t)}$ is contained in the credible band. In making use of the pointwise credible intervals, this third alternative avoids some of the problems associated with approximate posterior normality. To be more specific, the pointwise credible intervals can be asymmetric and can react flexibly with respect to local uncertainty in the curve estimation. Therefore, this variant may be particularly attractive in situations with non-Gaussian responses where the posterior is likely to be asymmetric. Still the credible band is centred at the posterior mean and therefore combines information from this point estimate with quantile information. This type of simultaneous credible band has first been proposed in Krivobokova, Kneib and Claeskens (2010).

When comparing the three credible bands with respect to their computational demands, the posterior normality credible band is the most attractive one since it avoids processing the complete sample of curves. In contrast, both alternative proposals rely on ordering of the samples, making their use computationally more intense in situations with large r.

Similarly to the case of mixed model-based credible bands it should be noted that all simultaneous credible bands are only projections of high-dimensional hyperrectangulars. Hence, it may be misleading to try to perform a test for simplified functional forms based on the simultaneous credible bands. Even if, for example, a line is contained in the simultaneous band this does not necessarily provide the possibility of concluding that the true effect is linear (compare the discussion of credible bands in Section 4.2.1 on page 206). Fully Bayesian counterparts of mixed model based tests on functional forms are discussed within the framework of model selection in Section 4.5.3.

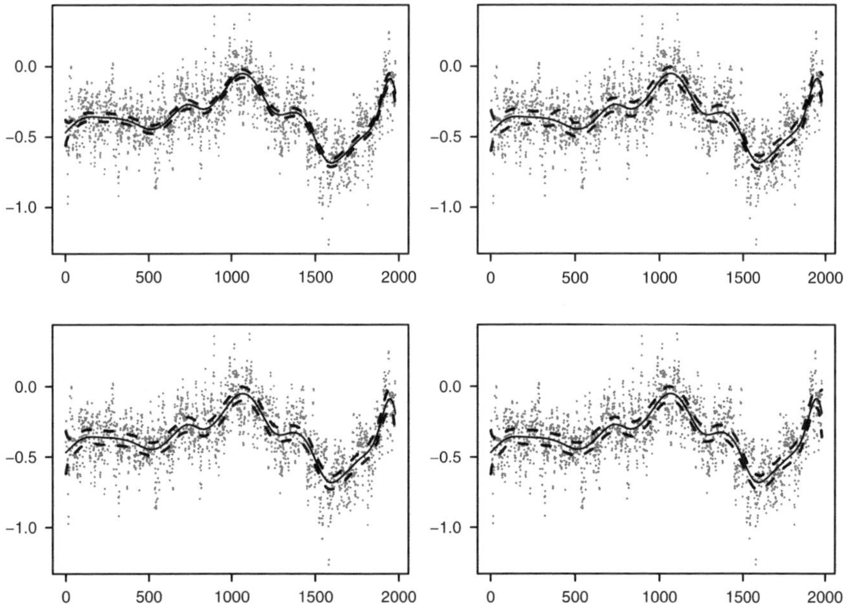

Fig. 4.16 Global warming: Posterior mean estimates for the global warming data together with 95% pointwise credible intervals (top left) and 95% simultaneous credible bands according to Besag *et al.* (top right), according to Crainiceanu, Ruppert and Carroll (bottom left) and according to Krivobokova, Kneib and Claeskens (bottom right).

Example 4.7: Global warming
As an example on pointwise and simultaneous Bayesian credible bands we reconsider the global warming data that we used in Chapter 2 to introduce basic concepts of Bayesian smoothing and semiparametric regression. We estimated a cubic penalized spline with 20 inner knots and second-order random-walk penalty based on 12,000 iterations, 2,000 burn in iterations and a final thinning step where only every 10th iteration was stored. Based on the resulting 1,000 iterations, we estimated the posterior mean, 95% pointwise credible intervals and the three 95% simultaneous credible bands.

The results are visualized in Fig. 4.16 and lead to the following conclusions: As expected, the simultaneous credible bands are wider than their pointwise counterparts. Note, however, that the simultaneous credible bands will contain 95% of the sampled curves and not of the observed data as would be the case for simultaneous 95% prediction intervals. The three simultaneous bands are very close to each other and in fact are hardly distinguishable when being plotted in one figure. Simulations on the frequentist coverage properties of the Bayesian simultaneous credible bands in Krivobokova, Kneib and Claeskens (2010) have shown that the credible bands according to Besag *et al.* (1995) tend to be somewhat to narrow, leading to undercoverage. The two remaining methods are typically quite close to each other, at least in the settings with Gaussian responses considered in Krivobokova, Kneib and Claeskens (2010). △

4.3 Smoothing and correlation

From a Bayesian perspective, the difference in modelling 'fixed' and 'random effects' is the choice of the associated prior: While flat, noninformative priors are assigned to the former, the latter are supplemented with proper distributional assumptions. In frequentist smoothing approaches, the coefficients of a penalized spline or some other penalized estimation procedure are commonly interpreted as fixed unknown parameters, while the penalty is introduced for regularization. This contradicts the distinction made before since penalties correspond to at least partially informative priors. In fact, partially improper priors as the ones we used for penalized splines seem to fall in between 'fixed' and 'random' effects. The mixed model representation introduced in Section 4.2.1 allowed for partial clarification in this regard since the proper and the improper part of the penalty could be clearly separated. In this section, we will further investigate changes in model interpretation, when the coefficients of penalized smoothers are treated as (partially) random. In particular, we will be able to show that smoothing and modelling of correlated responses actually have quite a lot in common. Moreover, the inclusion of both correlated errors and smooth model terms (in particular smooth trend functions) may lead to identifiability difficulties. We will restrict our attention to Gaussian time-series models in the following but essentially the same conclusions are valid for generalized regression models for longitudinal data.

Correlations induced by penalized smoothing approaches

For simplicity, consider a Gaussian time series

$$y_t = f(t) + \varepsilon_t,\ t = 1, \dots, T,$$

where the time trend $f(t)$ is to be recovered using a penalized spline. In matrix notation, the model can be represented as

$$\boldsymbol{y} = \boldsymbol{X}\boldsymbol{\gamma} + \boldsymbol{\varepsilon}, \quad p(\boldsymbol{\gamma}|\tau^2) \propto \exp\left(-\frac{1}{2\tau^2}\boldsymbol{\gamma}'\boldsymbol{K}\boldsymbol{\gamma}\right),$$

and the mixed model representation allows us to rewrite the specification as

$$\boldsymbol{y} = \boldsymbol{U}\boldsymbol{\beta} + \boldsymbol{Z}\boldsymbol{\alpha} + \boldsymbol{\varepsilon}, \quad p(\boldsymbol{\beta}) \propto \text{const}, \quad \boldsymbol{\alpha}|\tau^2 \sim \text{N}(\boldsymbol{0}, \tau^2\boldsymbol{I}).$$

The frequentist perspective on the penalized spline model where $\boldsymbol{\gamma}$ is treated as a fixed parameter, corresponds to the conditional perspective on the mixed model representation. In this case, the random effects $\boldsymbol{\alpha}$ enter the model as an additional vector of regression coefficients that allow for nonlinearity of the function $f(t)$ and are estimated subject to regularization. However, one can also take the marginal perspective on the mixed model, where

$$\boldsymbol{y} \sim \text{N}(\boldsymbol{U}\boldsymbol{\beta}, \sigma^2\boldsymbol{I} + \tau^2\boldsymbol{Z}\boldsymbol{Z}').$$

In particular, a nondiagonal covariance matrix $\boldsymbol{V} = \text{Cov}(\boldsymbol{y}) = \sigma^2\boldsymbol{I} + \tau^2\boldsymbol{Z}\boldsymbol{Z}'$ arises, where the structure of the random effects design matrix determines the correlations in $\boldsymbol{V}$ induced by the assumption that $\boldsymbol{\alpha}$ is random. This marginal interpretation reveals that including a nonlinear effect supplemented with a penalty, actually induces correlations in the response vector. We can also extract the exact form of this correlation structure from the marginal covariance matrix via

$$\text{Cor}(y_t, y_{t+h}) = \frac{\text{Cov}(y_t, y_{t+h})}{\sqrt{\text{Var}(y_t)\,\text{Var}(y_{t+h})}} = \frac{\boldsymbol{V}[t, t+h]}{\sqrt{\boldsymbol{V}[t,t]\boldsymbol{V}[t+h, t+h]}}$$

for selected values of $t = 1, \dots, T$, and $h = -t+1, \dots, T-t$.

To get a better understanding of the correlation induced by a penalized spline, we consider a simulated data set, where $f(t) = \sin(2\pi(t-1)/T)$, $t = 1, \dots, T = 601$ and $\varepsilon_t \sim \text{N}(0, 0.25)$. The resulting data and the true underlying trend are visualized in Fig. 4.17. Penalized splines based on either B-splines or truncated power series splines were fitted to the data. In both cases, cubic splines and twenty inner knots have been employed and the B-spline based spline has been supplemented with a second order difference penalty.

Figure 4.18 shows the estimated correlation function for P-splines based on B-spline basis functions and time points $t = 100, 200, 300$ and 400. The dominant spike that is present in all four cases corresponds to the error variance and

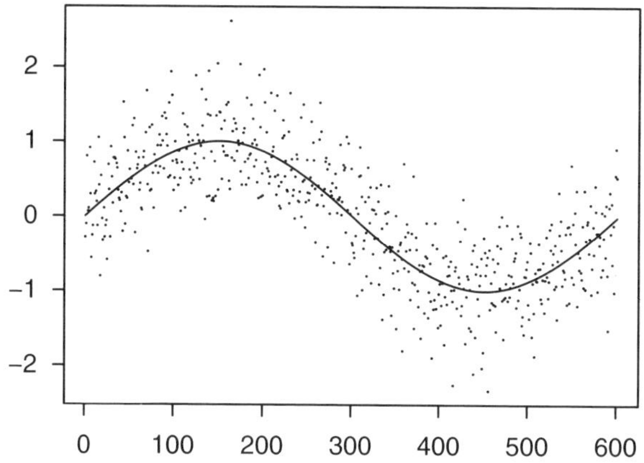

Fig. 4.17 Simulated data that are used to illustrate the marginal correlation structured induced by a penalized spline.

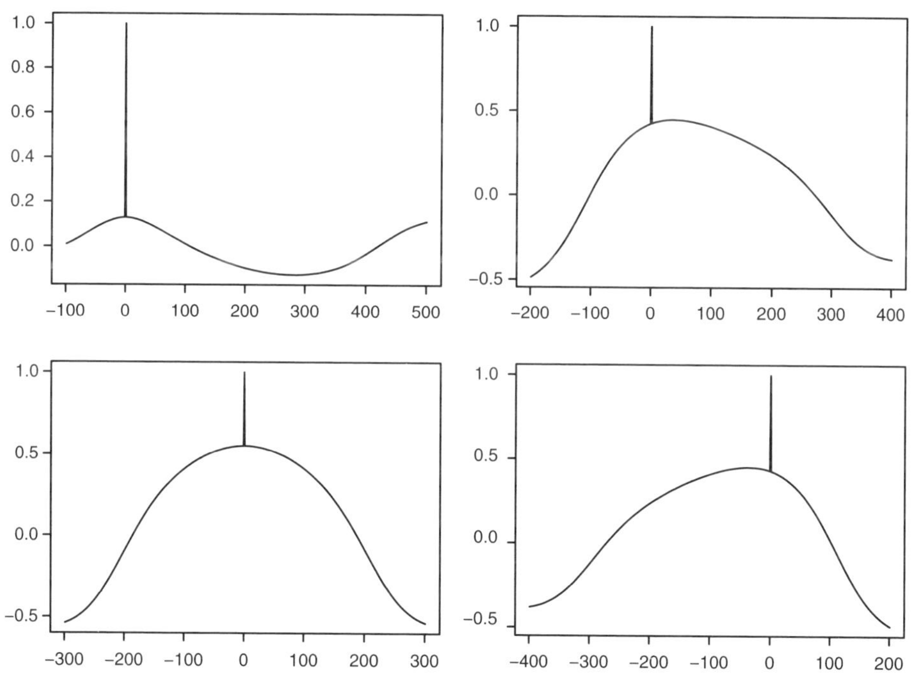

Fig. 4.18 Induced correlation functions for B-spline-based penalized splines at time points $t = 100, 200, 300, 400$.

accounts for the fact that two realizations y_t and y_{t+h} tend to be similar if h is close but they will not be the same even for very small h since independent noise ε is added to the trend function. Another important observation is that the correlation functions are quite different from each other and therefore the correlation induced by the spline is not of a stationary form. Still, the general structure of the correlation functions confirms the intuition that two observations that are close to each other in time will be correlated since they share comparable values of the trend function. The correlation is then decaying with increasing distance and approaches zero for large distances at least in the three cases $t = 200, 300$ and 400. In these cases, the correlation functions are also not too far from being symmetric. For $t = 100$, we observe a qualitatively different behaviour. The correlation function is first decreasing but rises again for large values of h. This somewhat counterintuitive behaviour is caused by the special form of the trend function that has been chosen to have sinusoidal form. Since the sine curve follows a cyclic trend, distances in the range of 2π, 4π, etc. will always yield positive correlations.

Figure 4.19 presents similar results for the penalized spline based on a truncated power series basis. Here, the correlation functions have a completely

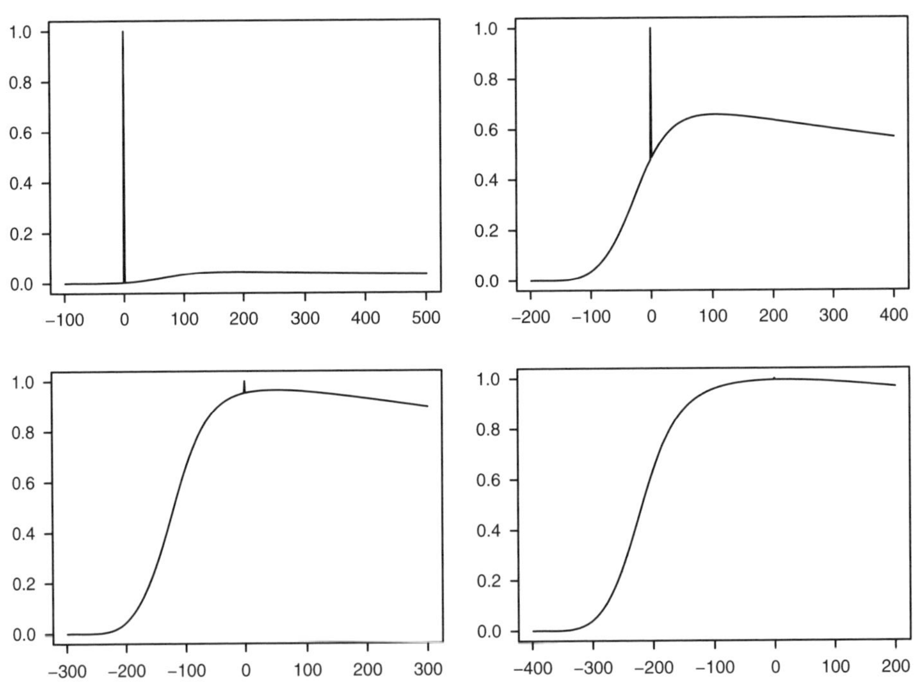

Fig. 4.19 Induced correlation functions for truncated power series-based penalized splines at time points $t = 100, 200, 300, 400$.

different form and are extremely asymmetric. In particular, the correlation is mostly rising, apart from the peak induced by the error component. This surprising behaviour reflects the special structure of the truncated power series basis, where basis functions are not defined locally but remain positive starting from a specific knot. The extremely high correlations that are observed for $t = 300$ and $t = 400$ also reflect numerical problems that are associated with the truncated power series basis, where the design matrix may be numerically almost collinear.

Identifiability problems

In the previous section, we have dealt with the estimation of a trend function but kept the assumption of independent error terms. However, in the context of time-series smoothing it is common practice to include temporally correlated error terms. Since we have seen that penalized spline smoothers also induce temporal correlations, identification problems may be expected when including both sources of correlation.

For illustration purposes, we again consider the model

$$y_t = f(t) + \varepsilon_t,\ t = 1, \dots, T,$$

but now assume a random-walk prior for the trend function and an autoregressive structure for the error terms. More specifically, a first-order random-walk prior is assigned to $f(t)$, where

$$f(t) = f(t-1) + u_t, \quad u_t \sim \mathrm{N}(0, \tau^2)$$

and the error term is assumed to follow a first order autoregressive process with

$$\varepsilon_t = \alpha\varepsilon_{t-1} + v_t, \quad v_t \sim \mathrm{N}(0, \sigma^2)$$

and $|\alpha| < 1$. This leads to a weighted penalized least-squares criterion

$$(\boldsymbol{y} - \boldsymbol{f})\boldsymbol{K}_\alpha(\boldsymbol{y} - \boldsymbol{f}) + \lambda\boldsymbol{f}'\boldsymbol{K}_1\boldsymbol{f} \tag{4.34}$$

for the estimation of $(\boldsymbol{f} = f(1), \dots, f(T))'$, where $\lambda = \sigma^2/\tau^2$ denotes the smoothing parameter and

$$\boldsymbol{K}_\alpha = \begin{pmatrix} 1 & -\alpha & & & \\ -\alpha & 1+\alpha^2 & -\alpha & & \\ & \ddots & \ddots & \ddots & \\ & & -\alpha & 1+\alpha^2 & -\alpha \\ & & & -\alpha & 1 \end{pmatrix}$$

is the inverse of a first order autoregressive correlation matrix that accounts for correlations induced by the autoregressive process assumption for the error term. Note that $\boldsymbol{K}_1$, the first order difference penalty, actually results as a special case of $\boldsymbol{K}_\alpha$ when inserting $\alpha = 1$.

Maximizing (4.34) with respect to $\boldsymbol{f}$ yields the penalized least-squares estimate

$$\hat{\boldsymbol{f}}_\alpha = (\boldsymbol{K}_\alpha + \lambda \boldsymbol{K}_1)^{-1} \boldsymbol{K}_\alpha \boldsymbol{y}$$

that is also the posterior mode and posterior mean (for given variance parameters). To compute $\hat{\boldsymbol{f}}_\alpha$, the matrix

$$\boldsymbol{P}_\alpha = \boldsymbol{K}_\alpha + \lambda \boldsymbol{K}_1$$

has to be inverted and numerical difficulties are to be expected when errors with autoregressive parameter close to one are used. To be more specific, for $\alpha \to 1$ we find that

$$\boldsymbol{P}_\alpha \to (1 + \lambda)\boldsymbol{K}_1$$

which is singular. This simply reflects the fact that the stochastic assumptions for the trend function $\boldsymbol{f}$ and the error term $\boldsymbol{\varepsilon}$ tend to be very similar for α close to one and can therefore no longer be separated. The numerical difficulties can be quantified in terms of the condition index

$$\kappa_\alpha = \frac{\omega_{\max}(\boldsymbol{P}_\alpha)}{\omega_{\min}(\boldsymbol{P}_\alpha)}$$

where $\omega_{\max}(\boldsymbol{P}_\alpha)$ and $\omega_{\min}(\boldsymbol{P}_\alpha)$ denote the largest and the smallest eigenvalue of $\boldsymbol{P}_\alpha$, respectively.

Figure 4.20 visualizes the condition index for a time series with $T = 100$ time points and, as expected, indicates ill-conditioned inversion of $\boldsymbol{P}_\alpha$ for high values of the autoregressive parameter. Smaller values of the smoothing parameter seem to mitigate this effect to some extent but in fact the observed reduction would hardly make any practical difference. Note also that the condition index has been visualized on the log-scale so that even relatively small values still correspond to ill-conditioned matrices. Fahrmeir and Kneib (2008) further investigate properties of $\hat{\boldsymbol{f}}$ and find that it retains validity and meaningfulness as a point estimate even for large values of α but that the identifiability problems are reflected in rather large uncertainty attached to the point estimate.

Note that similar identifiability problems arise when the random-walk assumption for the trend function is replaced by the assumption of a penalized spline prior. Intuitively, the penalized spline is empirically close to a random-walk at least for a large number of knots. This intuition is confirmed when inspecting the corresponding condition number, obtained in analogy to the random-walk specification (Fig. 4.21) In addition, Fig. 4.21 leads to the conclusion that the increase in the condition number is faster for models with larger autoregressive parameter.

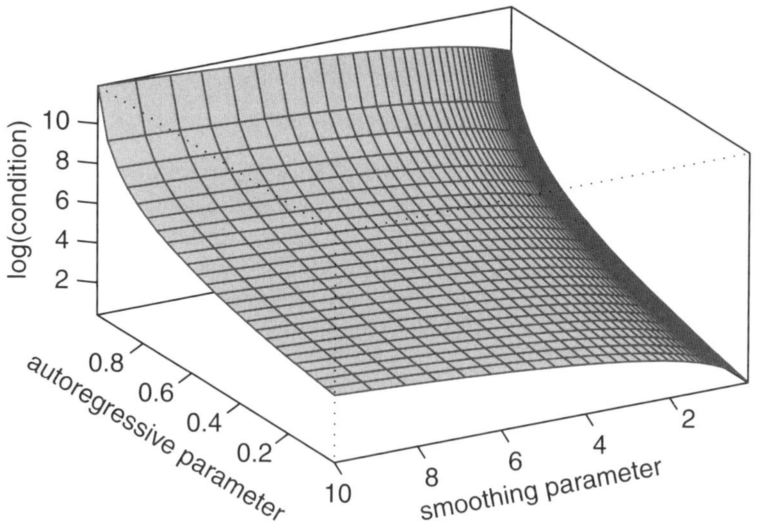

Fig. 4.20 Condition index κ_α for different combinations of smoothing parameters and autoregressive parameters.

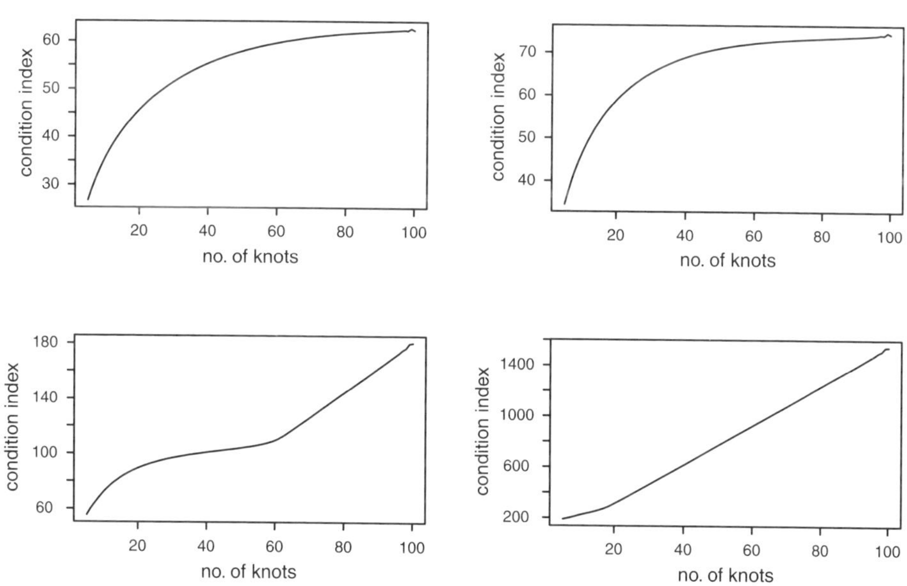

Fig. 4.21 Condition index κ_α depending on the number of knots of a cubic P-spline with second-order difference penalty and autoregressive parameter $\alpha = 0.3$ (top left), $\alpha = 0.5$ (top right), $\alpha = 0.7$ (bottom left) and $\alpha = 0.9$ (bottom right).

Fahrmeir and Kneib (2008) also investigate the separability of autoregressive errors and nonparametric trends in simulated data obtained from the extended model equation

$$y_t = f(t) + \varepsilon_t + \delta_t \tag{4.35}$$

where $f(t)$ is modelled using a cubic P-spline with second-order random-walk penalty and 20 inner knots, ε_t represents an autoregressive error component and δ_t represents additional Gaussian i.i.d. noise variables. As an illustration, Fig. 4.22 visualizes one selected simulated data set together with the point estimates for the trend and the autoregressive error component (when the true autoregressive parameter is set to $\alpha = 0.9$). Since we have rather large autocorrelations in the data, the stochastic structure of the autoregressive error is relatively close to a random-walk, resulting in close to non-stationary behaviour with longer deviations from the zero line. As expected, these longer deviations are partially captured by the trend when estimating the model. The true trend has been chosen as a sinusoidal curve which, however, in estimation is overlayed with parts of the autoregressive error. Vice versa, the estimated autoregressive component does not show enough fluctuation and in fact the estimated autoregressive parameter is much smaller than the true value of $\alpha = 0.9$. Note that the sum of trend and autoregressive components is still well recovered by the sum of the two estimates, but one should be cautious in interpreting the separation in real data.

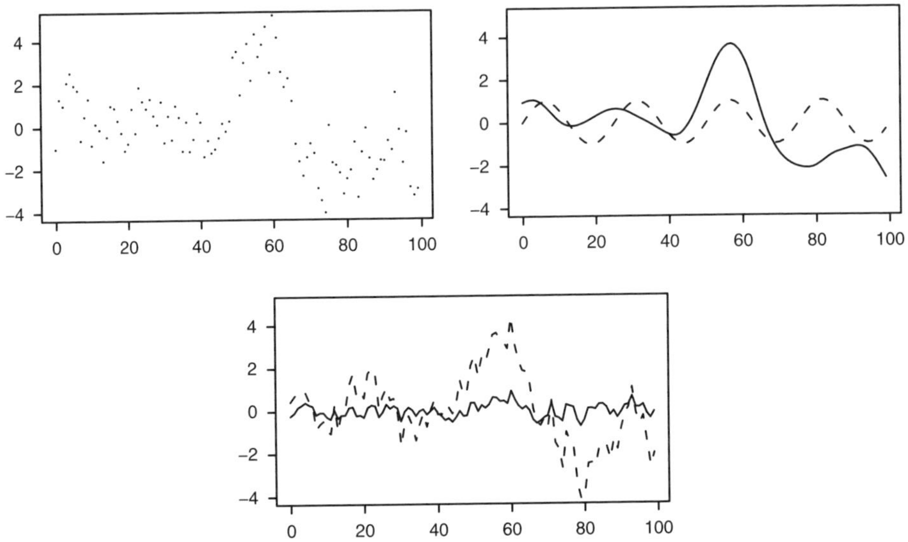

Fig. 4.22 Simulated data from model (4.35) (top left), estimated trend component (top right) and estimate autoregressive component (bottom). The true effects have been added as dashed lines.

Radial basis functions and correlated errors

The autoregressive error component considered in the previous paragraph induces an exponentially decaying correlation between the response observations. Instead of assigning an autoregressive prior to ε_t, we will now model it as a stationary Gaussian process with an exponential correlation function. To be more specific, we again consider the model

$$y_t = f(t) + \varepsilon_t + \delta_t$$

where $f(t)$ denotes the trend function, possibly estimated by a penalized spline smoother and δ_t are Gaussian i.i.d. errors. For the vector of correlated error terms $\boldsymbol{\varepsilon} = (\varepsilon_1, \ldots, \varepsilon_T)'$, we assume a Gaussian process prior

$$\boldsymbol{\varepsilon} \sim \mathrm{N}(\mathbf{0}, \sigma^2 \boldsymbol{R})$$

where the correlation matrix $\boldsymbol{R}$ is generated from a parametric correlation function $\rho(h) = \mathrm{Cor}(\varepsilon_t, \varepsilon_{t+h})$ such that

$$\boldsymbol{R}[t, t+h] = \rho(h).$$

For example, $\rho(h) = \exp(-h/\phi)$ results in the special case of an exponential correlation function with range parameter ϕ that relates to the exponential correlation function of an autoregressive error term via

$$\rho(h) = \exp(-h/\phi) = \alpha^h$$

with $\alpha = \exp(-1/\phi)$.

Representing $f(t)$ in terms of a spline basis yields the vector-matrix representation

$$\boldsymbol{y} = \boldsymbol{Z\gamma} + \boldsymbol{\varepsilon} + \boldsymbol{\delta}.$$

Reparametrizing the vector $\boldsymbol{\varepsilon}$ as $\tilde{\boldsymbol{\varepsilon}} = \boldsymbol{R}^{-1}\boldsymbol{\varepsilon}$ gives the equivalent model specification

$$\boldsymbol{y} = \boldsymbol{Z\gamma} + \boldsymbol{R}\tilde{\boldsymbol{\varepsilon}} + \boldsymbol{\delta},$$

where now $\tilde{\boldsymbol{\varepsilon}} \sim \mathrm{N}(\mathbf{0}, \sigma^2 \boldsymbol{R}^{-1})$. From a smoothing perspective, this allows to interpret $\tilde{\boldsymbol{\varepsilon}}$ as a vector of regression coefficients associated with basis function evaluations collected in the design matrix $\boldsymbol{R}$. Each of the columns of $\boldsymbol{R}$ corresponds to one such basis function, defined as

$$\boldsymbol{R}[t, s] = B_t(s) = \rho(|t - s|),$$

that is, the basis functions take the form of the correlation function and each basis function is centred at one time point t that may be interpreted as a knot.

Furthermore, the distributional assumption $\tilde{\varepsilon} \sim \mathrm{N}(\mathbf{0}, \sigma^2 \boldsymbol{R}^{-1})$ corresponds to the prior density

$$p(\tilde{\varepsilon}) \propto \exp\left(-\frac{1}{2\sigma^2}\tilde{\varepsilon}'\boldsymbol{R}\tilde{\varepsilon}\right)$$

where $\boldsymbol{R}$ can be interpreted as penalty matrix. Indeed, since $\boldsymbol{R}$ contains exponentially decaying correlations, it will force coefficients of basis functions that are close in time to be similar.

In summary, the assumption of a stationary Gaussian process for $\boldsymbol{\varepsilon}$ can be seen to be equivalent to a basis function approach with a design matrix and penalty derived from the correlation function. If a further temporal trend $f(t)$ is included in the model there will obviously be identifiability problems, at least if the basis functions describing $f(t)$ and the correlation functions share similar smoothness properties. We will further pursue this connection in Chapter 5.4 where we extend the ideas sketched in this paragraph to spatial regression models with spatial effects modelled as stationary Gaussian random fields.

Summary

In summary, the three previous sections all indicate that a clear separation between nonparametric 'fixed' effects and 'random' effects accounting for correlations is not possible. This has immediate consequences for the specification of regression models since complex models combining different types of smooth and random effects should only be used with care if both types of effects share similar smoothness properties. Still, such combinations may be required to adequately reflect uncertainty in the effect estimates, for example when both the longitudinal and the spatial structure of a spatio-temporally aligned data set should be taken into account. This is the case in the forest health example and we will use a combination of unstructured and correlated random effects in Chapter 5 to account for both temporal and spatial correlation in the data. However, the considerations from the previous sections tell us that it will be hard to separate both sources of correlations from each other and that one should be cautious in interpreting the separated effects.

4.4 Extensions based on non-Gaussian priors

4.4.1 SPMMs with DP-based random effects priors

As for LMMs and GLMMs, a natural extension of SPMMs is to replace the Gaussian assumption for random effects through more flexible random effects priors such as Dirichlet process (DP) or Dirichlet process mixture (DPM) priors. An important special case is an additive mixed model

$$y_{ij} = \boldsymbol{u}_{ij}'\boldsymbol{\beta} + f(t_{ij}) + \boldsymbol{z}_{ij}'\boldsymbol{\alpha}_i + \varepsilon_{ij}, \tag{4.36}$$

with i. i. d. errors $\varepsilon_{ij} \sim \mathrm{N}(0, \sigma^2)$. Li, Lin and Müller (2010) consider this model, assuming a Bayesian smoothing spline for the time trend $f(t)$ while Heinzl,

Kneib and Fahrmeir (2009) propose a similar model based on Bayesian P-splines. Extensions to more general observation models described in Section 4.1.1, with generic predictor

$$\boldsymbol{\eta}_i = \boldsymbol{U}_i\boldsymbol{\beta} + \boldsymbol{X}_{i1}\boldsymbol{\gamma}_1 + \ldots + \boldsymbol{X}_{ir}\boldsymbol{\gamma}_r + \boldsymbol{Z}_i\boldsymbol{\alpha}_i,$$

are possible. To avoid identifiability problems already discussed in Section 3.2, we assume that $\boldsymbol{u}_{ij}$ is the vector of covariates with *fixed effects only*. Covariates with *fixed and random effects* are part of $\boldsymbol{z}_{ij}$. For example in a model with individual-specific linear trends $\alpha_{0i} + \alpha_{1i}t$, the random intercept and the random slope are the sum $\alpha_{0i} = \tilde{\alpha}_0 + \tilde{\alpha}_{0i}$ and $\alpha_{1i} = \tilde{\alpha}_1 + \tilde{\alpha}_{1i}$ of 'fixed' effects $\tilde{\alpha}_0, \tilde{\alpha}_1$ and random deviations $\tilde{\alpha}_{0i}, \tilde{\alpha}_{1i}$ with $\mathbb{E}(\tilde{\alpha}_{0i}) = \mathbb{E}(\tilde{\alpha}_{1i}) = 0$. In the generic form of the predictor, the design matrix $\boldsymbol{U}_i$ is therefore constructed from covariates with fixed effects only as in Equation (3.25), page 140.

However, an additional problem arises with the inclusion of nonlinear covariate effects. Consider, as a typical example, the additive model (4.36) with population trend $f(t)$ modelled through a Bayesian P-spline with second-order random-walk prior or a Bayesian cubic smoothing spline. Suppose, we want to include individual-specific linear trends $\alpha_{0i} + \alpha_{1i}t$ in addition to the population trend. Then we are faced with the same problems as with a linear population trend: Without further restrictions, P-splines or smoothing splines comprise linear trends as special cases. There are (at least) two possible strategies: Either the population trend $f(t)$ models only *deviations from a linear population trend*, or random intercepts and slopes have to be centred about zero, modelling only individual linear deviations from $f(t)$. Li, Lin and Müller (2010) deal with this problem in a post-processing step while Dunson, Yang and Baird (2007) introduce a centred DP prior. Heinzl, Kneib and Fahrmeir (2009) suggest centring random effects about zero in the MCMC algorithm as a simple but effective device and we will use this strategy in the following. The specification of $f(t)$ as a nonlinear deviation from a linear population trend would require additional, comparably complicated linear constraints for B-splines with a second-order random-walk prior or for cubic smoothing splines, see Panagiotelis and Smith (2008) as well as Section 4.2.1 for such a possibility. For linear regression splines represented through a TPS basis, a simple approach is to delete the 'fixed' linear effect corresponding to the basis functions 1 and t. Obviously the same strategies for assuring identification are relevant for nonlinear functions of other continuous covariates.

Taking care of the identification problems just discussed, SPMMs with DP or DPM random effects priors are defined by adding Gaussian smoothness priors for basis function coefficients and inverse Gamma priors for variance parameters to the priors for fixed effects and random effects as specified already for LMMs and GLMMs with DP and DPM random effects priors in Sections 3.2.3 and 3.3.5. This leads to the following priors for SPMMs with DP random effects:

$$
\begin{aligned}
\boldsymbol{\beta}|\boldsymbol{c},\boldsymbol{C} &\sim \mathrm{N}(\boldsymbol{c},\boldsymbol{C})\\
p(\boldsymbol{\gamma}_l|\tau_l^2) &\propto \exp\left(-\frac{1}{2\tau_l^2}\boldsymbol{\gamma}_l'\boldsymbol{K}\boldsymbol{\gamma}_l\right), \quad l=1,\ldots,r\\
\tau_l^2 &\sim \mathrm{IG}(a_l,b_l)\\
\boldsymbol{\alpha}_i|G &\sim G\\
G|\nu_0,G_0 &\sim \mathrm{DP}(\nu_0,G_0)\\
G_0 &\sim \mathrm{N}(\boldsymbol{\mu}_0,\boldsymbol{Q}_0).
\end{aligned}
$$

Further priors are $\sigma^2 \sim \mathrm{IG}(a_\sigma, b_\sigma)$ for Gaussian SPMMs and hyperpriors (3.30), (3.31) and (3.32) for $\boldsymbol{c}$, $\boldsymbol{C}$, $\boldsymbol{\mu}_0$, $\boldsymbol{Q}_0$ and ν_0 as defined in Section 3.2.3.

For SPMMs with DPM priors, the random effects part is modified to

$$
\begin{aligned}
\boldsymbol{\alpha}_i|\boldsymbol{\mu}_i,\boldsymbol{Q} &\overset{\text{ind.}}{\sim} \mathrm{N}(\boldsymbol{\mu}_i,\boldsymbol{Q})\\
\boldsymbol{\mu}_i|G &\sim G\\
G|\nu_0,G_0 &\sim \mathrm{DP}(\nu_0,G_0),
\end{aligned}
$$

while the other priors remain the same.

A conditional MCMC approach can be based on a truncated stick breaking presentation

$$
G(\cdot) = \sum_{k=1}^{K} \pi_k \delta_{\boldsymbol{\phi}_k}(\cdot),
$$

where $\boldsymbol{\phi}_k \overset{\text{i.i.d.}}{\sim} \mathrm{N}(\boldsymbol{\mu}_0, \boldsymbol{Q}_0)$, and $\boldsymbol{\pi} = (\pi_1, \ldots, \pi_K)$ is generated through the truncated stick breaking procedure. Introducing latent class indicators c_i, with $c_i = k \Leftrightarrow \boldsymbol{\mu}_i = \boldsymbol{\phi}_k$, or in more compact notation $\boldsymbol{\mu}_i = \boldsymbol{\phi}_{c_i}$, as in Section 3.2.2, the DPM prior is defined by the hierarchy

$$
\begin{aligned}
\boldsymbol{\mu}_i &\overset{\text{ind.}}{\sim} \mathrm{N}(\boldsymbol{\phi}_{c_i},\boldsymbol{Q}), \quad i=1,\ldots,n\\
c_i|\boldsymbol{\pi} &\overset{\text{ind.}}{\sim} \mathbb{P}(c_i=k|\boldsymbol{\pi})=\pi_k, \quad k=1,\ldots,K.
\end{aligned}
$$

For Gaussian SPMMs with DPM random effects priors, the following Gibbs sampler can be derived (Heinzl, Kneib and Fahrmeir, 2009):

Algorithm 4.6: Gibbs sampler for Gaussian SPMMs with DPM priors

i. *Update P-spline basis coefficients* $\boldsymbol{\gamma}_l$, $l = 1, \ldots, r$, by drawing from

$$
\boldsymbol{\gamma}_l|\cdot \sim \mathrm{N}(\boldsymbol{\mu}_l, \boldsymbol{\Sigma}_l),
$$

with $\boldsymbol{\mu}_l, \boldsymbol{\Sigma}_l$ defined as in (4.20), page 214.

ii. *Update fixed effects* $\boldsymbol{\beta}$ by drawing from

$$\boldsymbol{\beta}|\cdot \sim \mathrm{N}(\boldsymbol{\mu}_{\boldsymbol{\beta}}, \boldsymbol{\Sigma}_{\boldsymbol{\beta}}),$$

with $\boldsymbol{\mu}_{\boldsymbol{\beta}}, \boldsymbol{\Sigma}_{\boldsymbol{\beta}}$ defined as in (4.19). If hyperpriors $\boldsymbol{c} \sim \mathrm{N}(\boldsymbol{m_c}, \boldsymbol{S_c})$, $\boldsymbol{S_c} = \mathrm{diag}(s_{c_1}^2, \ldots, s_{c_m}^2)$ and $\boldsymbol{C} = \mathrm{diag}(\sigma_{\beta_1}^2, \ldots, \sigma_{\beta_m}^2)$, $\sigma_{\beta_l}^2 \overset{\text{i.i.d.}}{\sim} \mathrm{IG}(a_{\boldsymbol{\beta}}, b_{\boldsymbol{\beta}})$, $l = 1, \ldots, m = \dim(\boldsymbol{\beta})$, are assumed, then update c_l and $\sigma_{\beta_l}^2$ by drawing from

$$c_l|\beta_l, \sigma_{\beta_l}^2 \sim \mathrm{N}\left(\left(\frac{1}{\sigma_{\beta_l}^2} + \frac{1}{s_{c_l}^2}\right)^{-1}\left(\frac{\beta_l}{\sigma_{\beta_l}^2} + \frac{m_{c_l}}{s_{c_l}^2}\right), \left(\frac{1}{\sigma_{\beta_l}^2} + \frac{1}{s_{c_l}^2}\right)^{-1}\right),$$

$$\sigma_{\beta_l}^2|\beta_l, c_l \sim \mathrm{IG}(a_{\boldsymbol{\beta}} + 0.5, b_{\boldsymbol{\beta}} + 0.5(\beta_l - c_l)^2).$$

iii. *Update random effects* $\boldsymbol{\alpha}_i$, $i = 1, \ldots, n$, through the following steps:

- Draw $\boldsymbol{\alpha}_i$ from the full conditional

$$\boldsymbol{\alpha}_i|\boldsymbol{\mu}_i, \boldsymbol{Q}, \cdot \sim \mathrm{N}(\boldsymbol{\mu}_{\boldsymbol{\alpha}_i}, \boldsymbol{\Sigma}_{\boldsymbol{\alpha}_i}),$$

  where $\boldsymbol{\Sigma}_{\boldsymbol{\alpha}_i} = (\boldsymbol{Z}_i'\boldsymbol{Z}_i/\sigma^2 + \boldsymbol{Q}^{-1})^{-1}$ is identical with (4.21), and

$$\boldsymbol{\mu}_{\boldsymbol{\alpha}_i} = \boldsymbol{\Sigma}_{\boldsymbol{\alpha}_i}\left(\frac{1}{\sigma^2}\boldsymbol{Z}_i'(\boldsymbol{y}_i - \boldsymbol{\eta}_{-i}) + \boldsymbol{Q}^{-1}\boldsymbol{\mu}_i\right).$$

  (For $\boldsymbol{\mu}_i = \boldsymbol{0}$ this also reduces to (4.21).
- *Centring of random effects*:
  Compute $\bar{\boldsymbol{\alpha}} = \sum \boldsymbol{\alpha}_i/n$. Replace $\boldsymbol{\alpha}_i$ through $\boldsymbol{\alpha}_i - \bar{\boldsymbol{\alpha}}$.
- *Update coefficients* $\boldsymbol{\phi}_k$, $k = 1, \ldots, K$, of the truncated stick-breaking representation by drawing $\boldsymbol{\phi}_k$ from

$$\boldsymbol{\phi}_k|\boldsymbol{\mu}_0, \boldsymbol{Q}_0 \sim \mathrm{N}(\boldsymbol{\mu}_0, \boldsymbol{Q}_0)$$

  if there is no $c_i = k$ (i.e. cluster k is currently empty), and from

$$\phi_{k_l}|\delta_l^2, \mu_{0l}, \delta_{0l}^2, \boldsymbol{\alpha}, \boldsymbol{c} \sim \mathrm{N}(\mu_{0l}^*, \delta_{0l}^*)$$

$$\delta_{0l}^* = \left(\frac{n_k}{\delta_l^2} + \frac{1}{\delta_{0l}^2}\right)^{-1}, \quad n_k = |\{i : c_i = k\}|$$

$$\mu_{0l}^* = \delta_{0l}^*\left(\frac{n_k}{\delta_l^2}\bar{\alpha}_{l,k} + \frac{\mu_{0l}}{\delta_{0l}^2}\right).$$

- *Update latent class indicators* c_i by drawing from

$$c_i|\cdot \sim \sum_{k=1}^{K} c^* \varphi(\boldsymbol{\alpha}_i|\boldsymbol{\phi}_k, \boldsymbol{Q})\pi_k \delta_{\boldsymbol{\phi}_k}$$

where c^* is a normalizing constant, and set

$$\boldsymbol{\mu}_i = \boldsymbol{\phi}_{c_i}.$$

- *Update* π_k, $k = 1, \ldots, K$ through the stick breaking procedure:

$$V_k|\boldsymbol{c} \sim \text{Be}\left(1 + n_k, \nu_0 + \sum_{l=k+1}^{K} n_l\right)$$
$$\pi_k = V_k \prod_{l<k}(1 - V_l).$$

- *Update* ν_0 by drawing a new value $\omega \in \Omega = \{0.5, 0.6. \ldots, 100\}$ for ν_0:

$$\nu_0|\boldsymbol{\pi} \sim \sum_{\omega\in\Omega} \exp\left((K-1)\log(\omega) + (\omega - 1)\sum_{k=1}^{K}\log(1 - V_k)\right) p_\omega \delta_\omega$$

where p_ω denotes the prior probability of ω.

- For $l = 1, \ldots, m$ draw hyperparameters from

$$\mu_{0l}|\cdot \sim \text{N}\left(\left(\frac{n}{\delta^2_{0l}} + \frac{1}{s^2_{0l}}\right)^{-1}\left(\frac{n}{\delta^2_{0l}}\bar{\mu}_l + \frac{m_{0l}}{s^2_{0l}}\right), \left(\frac{n}{\delta^2_{0l}} + \frac{1}{s^2_{0l}}\right)^{-1}\right)$$
$$\delta^2_{0l}|\cdot \sim \text{IG}\left(a_0 + 0.5n, b_0 + \sum_{i=1}^{n}(\mu_{il} - \mu_{0l})^2\right)$$
$$\delta^2_l|\cdot \sim \text{IG}\left(a_{\boldsymbol{\alpha}} + 0.5n, b_{\boldsymbol{\alpha}} + 0.5\sum_{i=1}^{n}(\alpha_{il} - \mu_{il})^2\right).$$

iv. *Update error variance* σ^2 by drawing from

$$\sigma^2|\cdot \sim \text{IG}\left(a_\sigma + 0.5N, b_\sigma \sum_{i=1}^{n}(\boldsymbol{y}_i - \boldsymbol{\eta}_i)'(\boldsymbol{y}_i - \boldsymbol{\eta}_i)\right),$$

with $N = \dim(\boldsymbol{y})$, the total number of observations.

Example 4.8: Childhood obesity

Following Heinzl, Kneib and Fahrmeir (2009), we reanalyse the Lisa data from Example 1.1 based on the additive mixed model (4.15)

$$y_{ij} = \boldsymbol{u}'_{ij}\boldsymbol{\beta} + f(t_{ij}) + \boldsymbol{z}'_{ij}\boldsymbol{\alpha}_i + \varepsilon_{ij}, \quad \varepsilon_{ij} \overset{\text{i.i.d.}}{\sim} \text{N}(0, \sigma^2),$$

of Examples 4.2 and 4.4, but replacing the Gaussian random effects prior through a DPM prior as described in this section and applying the Gibbs sampler for Gaussian SPMMs with DPM priors. As in the examples before, the random effects part is specified through individual random lines or, in the extended model, through individual random curves $g_i(t)$, defined in (3.7) (page 111).

Table 4.3 contains estimation results for fixed effects in the extended model. The results are in line with the estimates obtained in Example 4.4, assuming Gaussian random effects. Population trends and individual curves for the model with random lines and curves are shown in Fig. 4.24 for three selected children. The population trend looks quite similar to the estimated trend using Gaussian random effects priors. Comparing individual curves, we can observe a better fit with the extended model.

An interesting feature of DPM priors is the automatic clustering structure induced by the Dirichlet process. However, although we get a clustering of

Table 4.3 Childhood obesity: Estimates for fixed effects and the error variance in the extended DPM model with random curves. mBMI and mDiffBMI have been centred for the analyses.

	mean	median	sd.	2.5%	97.5%
sex	0.2010	0.2008	0.0380	0.1244	0.7217
breast	0.0528	0.0512	0.0375	−0.0232	0.1265
mSmoke	−0.0073	−0.0063	0.0529	−0.1069	0.0951
area	−0.0512	−0.0516	0.0481	−0.1460	0.0411
mBMI	0.0464	0.0465	0.0052	0.0363	0.0566
mDiffBMI	0.0804	0.0800	0.0120	0.0574	0.1037
σ^2	0.7026	0.7025	0.0093	0.6846	0.7217

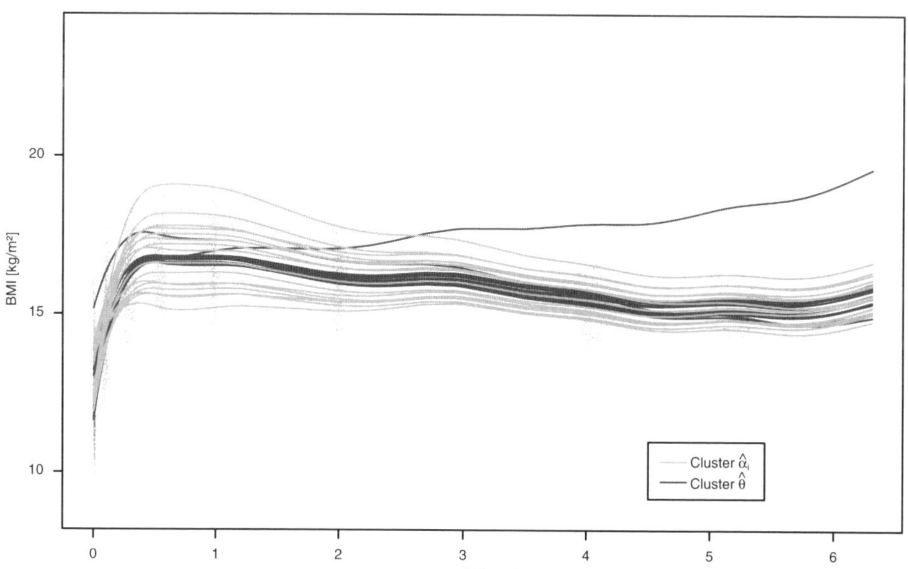

Fig. 4.23 Childhood obesity: Clustering of $\hat{\boldsymbol{\theta}}_i$ and $\hat{\boldsymbol{\alpha}}_i$ for the extended DPM model.

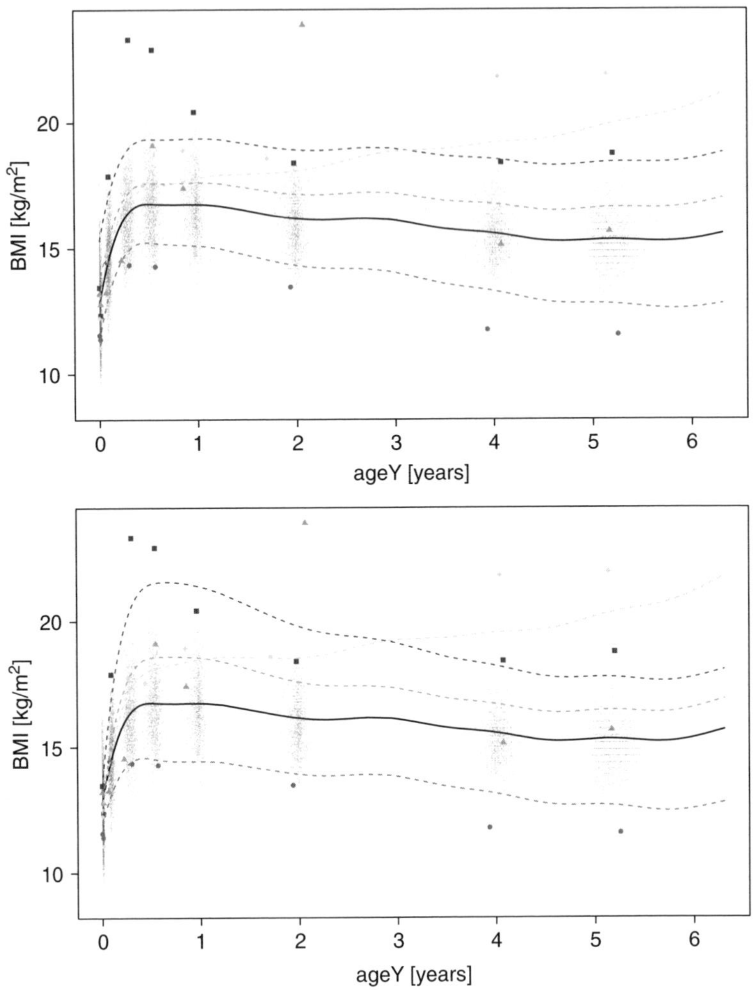

Fig. 4.24 Childhood obesity: Additive mixed DPM model with random lines (top) and random curves (bottom). The solid line corresponds to the mean curve while the dashed lines show estimated curves for three selected children.

subjects at each iteration of the Gibbs sampler, we have to construct a single overall clustering pattern for interpretation. We proceed as follows: First, we estimate the number of clusters using the median of all numbers of clusters in the MCMC iterations. Second, we allocate the $\hat{\boldsymbol{\theta}}_i$-parameters to clusters by k-means clustering. In addition, we do this for the random effects $\hat{\boldsymbol{\alpha}}_i$, $i = 1, \ldots, n$, too. In Fig. 4.23 we see that there are mainly level shifts between clusters both for $\hat{\boldsymbol{\theta}}_i$ and $\hat{\boldsymbol{\alpha}}_i$.

However, for $\hat{\boldsymbol{\theta}}_i$ one cluster attracts particular attention: For most of the children the BMI steeply increases until the age of about six months and then decreases slowly. In contrast, there is a group for which the development of BMI is very similar to the others only during the first year while a permanent ascent of the BMI is observed afterwards. It is remarkable that this cluster could not be detected for $\hat{\boldsymbol{\alpha}}_i$. More specifically, there are 30 children which belong to that special cluster. It is of interest if there are meaningful differences in the values of covariates between the subjects in the extreme cluster and the remaining observations. Indeed, especially the covariates breast and area show noticeable differences. Most of the children in the extreme cluster were bottlefed and/or breastfed until the age of four months. By contrast, across the remaining observations the majority of children were breastfed only. So in this sense breast-feeding only is essentially an indicator for a normal and lower development of the BMI like some experts expect, although there is no general negative effect of the covariate breast. Furthermore, the ratio of children living in an urban area to a rural area is nearly balanced in the extreme cluster where most of the other children live in cities. △

4.4.2 Shrinkage priors for high-dimensional regression parameters

In general terms, Bayesian prior specification resembles the frequentist notion of regularization. Regularization approaches that induce specific forms of shrinkage have gained considerable attention in recent years through the advent of high-dimensional regression models, where effects of high-dimensional covariate vectors are incorporated into regression models. This includes the possibility of considering more covariates than observations. Classical, purely likelihood-based estimation approaches do not allow for such model specifications since there is not enough data available to identify all regression coefficients, but the models can be made identifiable by imposing a penalty on the regression coefficients leading to penalized likelihood estimation

$$\hat{\boldsymbol{\beta}} = \arg\max_{\boldsymbol{\beta}} \left[l(\boldsymbol{\beta}) - \lambda \operatorname{pen}(\boldsymbol{\beta}) \right].$$

As already noted at several places throughout this book, the specification of a penalty term corresponds to assigning an informative prior distribution to parameters. In the following, we will discuss a number of penalization approaches proposed in the literature and consider their Bayesian counterparts. Several important variants can be expressed as scale mixtures of normals, enabling the construction of efficient Bayesian inferential procedures based on MCMC simulations.

Ridge prior

A ridge penalty for an m-dimensional vector of regression coefficients $\boldsymbol{\beta}$ is given by

$$\lambda \operatorname{pen}(\boldsymbol{\beta}) = \lambda \boldsymbol{\beta}'\boldsymbol{\beta} = \frac{1}{\tau_{\boldsymbol{\beta}}^2} \sum_{l=1}^{m} \beta_l^2, \quad m = \dim(\boldsymbol{\beta}),$$

that is, it relies on a quadratic penalty that shrinks all coefficients towards zero. The Bayesian analogue is given by an i.i.d. Gaussian prior

$$\boldsymbol{\beta} \sim \mathrm{N}(\mathbf{0}, \tau^2_{\boldsymbol{\beta}} \boldsymbol{I}_m),$$

where the inverse smoothing parameter appears as the prior variance, i.e. $\lambda = 1/\tau^2_{\boldsymbol{\beta}}$. This specification exactly resembles the priors we have previously considered for the specification of Bayesian random effects but follows a different rationale. While random effects are commonly introduced to model unobserved heterogeneity induced by a cluster or grouping structure in the data, no such grouping is assumed in the ridge case. Here, the regression coefficients do not correspond to group-specific effects but are common to all observations. As a consequence, the prior is imposed to enforce smoothing/shrinkage of the estimates in $\boldsymbol{\beta}$ towards zero to stabilize estimation.

While the smoothing parameter will typically be determined by a procedure like cross validation in a frequentist approach, the Bayesian interpretation immediately allows us to incorporate estimation of $\tau^2_{\boldsymbol{\beta}}$ by assigning a suitable hyperprior. The conjugate choice is an inverse gamma prior, which leads to the hierarchical prior specification

$$\boldsymbol{\beta} | \tau^2_{\boldsymbol{\beta}} \sim \mathrm{N}(\mathbf{0}, \tau^2_{\boldsymbol{\beta}} \boldsymbol{I}), \quad \tau^2_{\boldsymbol{\beta}} \sim \mathrm{IG}(a, b).$$

Marginally, each β_l will then follow a scaled t-distribution with 2a degrees of freedom and scale parameter $\sqrt{a/b}$. As a consequence, the Bayesian ridge prior has heavier tails and is more spiked then the original frequentist penalty. In fact, this is a desirable behaviour since it leads to more penalization of small regression coefficients close to zero but only moderate penalization for larger regression coefficients. A visual comparison of the conditional Gaussian prior and the marginal t-prior is given in Fig. 4.25. Note that while the Bayesian ridge prior in the conditional formulation for $p(\boldsymbol{\beta}|\tau^2_{\boldsymbol{\beta}})$ assumes *a priori* independence of the coefficients this does no longer hold in the marginal t-distributed prior $p(\boldsymbol{\beta})$.

One main advantage of the hierarchical specification of the Bayesian ridge lies in the fact that the regression coefficients separate the hyperprior for the variance $\tau^2_{\boldsymbol{\beta}}$ from the observation model. Hence, when setting up an MCMC sampler, the full conditional of the variance is always given by

$$\tau^2_{\boldsymbol{\beta}} | \cdot \sim \mathrm{IG}\left(a + \frac{m}{2}, b + \frac{1}{2}\sum_{l=1}^{m} \beta_l^2\right) \tag{4.37}$$

leading to a simple Gibbs update regardless of the type of regression model considered. For the regression coefficients, we have to differentiate between Gaussian models and other types of observation models. In the former case, the full conditional is given by $\boldsymbol{\beta}|\cdot \sim \mathrm{N}(\boldsymbol{\mu}_{\boldsymbol{\beta}}, \boldsymbol{\Sigma}_{\boldsymbol{\beta}})$ where the moments are

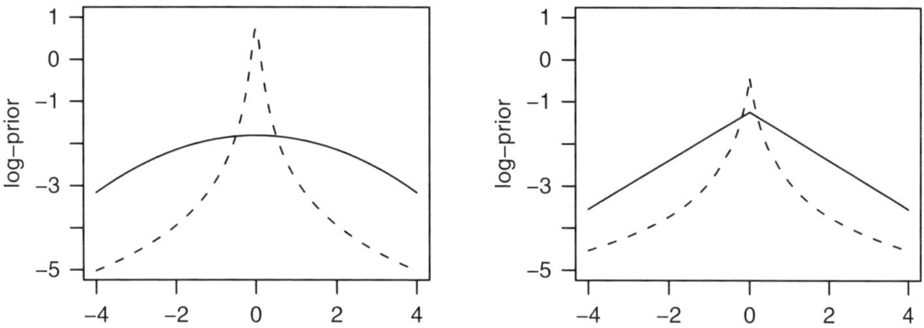

Fig. 4.25 Conditional (solid lines) and marginal (dashed lines) log-priors for the Bayesian ridge (left panel) and the Bayesian lasso (right panel).

$$\boldsymbol{\mu}_{\boldsymbol{\beta}} = \boldsymbol{\Sigma}_{\boldsymbol{\beta}} \frac{1}{\sigma^2} \boldsymbol{X}'(\boldsymbol{y} - \boldsymbol{\eta}_{-\boldsymbol{\beta}}), \quad \boldsymbol{\Sigma}_{\boldsymbol{\beta}} = \left(\frac{1}{\sigma^2} \boldsymbol{X}'\boldsymbol{X} + \frac{1}{\tau_{\boldsymbol{\beta}}^2} \boldsymbol{I}_m \right)^{-1},$$

σ^2 denotes the error variance of the Gaussian observation model and $\boldsymbol{\eta}_{-\boldsymbol{\beta}} = \boldsymbol{\eta} - \boldsymbol{X}\boldsymbol{\beta}$ denotes a current partial residual. In the latter case, IWLS proposals will be drawn from a Gaussian proposal distribution with expectation and covariance matrix

$$\boldsymbol{\mu}_{\boldsymbol{\beta}} = \boldsymbol{\Sigma}_{\boldsymbol{\beta}} \boldsymbol{X}'\boldsymbol{W}(\tilde{\boldsymbol{y}} - \boldsymbol{\eta}_{-\boldsymbol{\beta}}), \quad \boldsymbol{\Sigma}_{\boldsymbol{\beta}} = \left(\boldsymbol{X}'\boldsymbol{W}\boldsymbol{X} + \frac{1}{\tau_{\boldsymbol{\beta}}^2} \boldsymbol{I}_m \right)^{-1}$$

where $\boldsymbol{W}$ and $\tilde{\boldsymbol{y}}$ are the usual generalized linear model weights and working responses.

The full conditionals for regularized effects $\boldsymbol{\beta}$ can now simply be inserted into the general MCMC algorithms for SPMMs considered previously. For example, when including regularized effects into a longitudinal SPMM, step iii in Algorithm 4.3 from page 220 has to be replaced by the Gibbs/MH update from above and an additional update step for $\tau_{\boldsymbol{\beta}}^2$. Of course it is also possible to include both regularized and unregularized parametric effects by splitting up $\boldsymbol{\beta}$ into a part with flat, noninformative prior and a part that is assigned a Bayesian ridge prior. For example, the intercept will frequently remain unpenalized to avoid shrinkage of the overall level and may therefore be assigned a flat prior.

Lasso prior

The lasso (Tibshirani, 1996) has become one of the most popular regularization and shrinkage approaches in large m, small n settings and replaces the quadratic penalty of the ridge with an absolute value penalty, yielding

$$\lambda \operatorname{pen}(\boldsymbol{\beta}) = \lambda \sum_{l=1}^{m} |\beta_l|.$$

Due to the L_1 geometry, penalized likelihood estimates obtained with the lasso will typically contain a number of estimates that are shrunk exactly to zero. Hence, the lasso does not only provide regularized estimates but inherently leads to variable selection. In particular this property has made the lasso an extremely attractive regularization approach. Moreover, it puts more regularization weights on small regression coefficients while larger coefficients are left with a smaller amount of regularization.

However, the improved regularization properties of the lasso come at the price of increased computational complexity that stems from the fact that the absolute value penalty is not differentiable in zero. While efficient algorithms such as least angle regression (Efron *et al.*, 2004) allow for the determination of solutions for a given regularization parameter, the data-driven determination of λ often relies on computationally intense methods such as cross validation.

In a Bayesian formulation, the lasso penalty corresponds to i.i.d. Laplace priors, i.e.

$$\beta_l \overset{\text{i.i.d.}}{\sim} \text{Lap}(\lambda).$$

While this in principle enables the Bayesian treatment of lasso regularized estimation, it is hard to come up with an efficient MCMC sampler since the Laplace prior is not conjugate to any of the common observation models and the non-differentiability of the prior in zero prohibits the construction of IWLS proposals. Fortunately a hierarchical representation of the Laplace distribution as a scale mixture of normals allows us to overcome this difficulty (Park and Casella, 2008; Kneib *et al.*, 2010b). To be more specific, consider the following hierarchical prior specification:

$$\beta_l|\tau^2_{\beta,l} \sim \text{N}(0, \tau^2_{\beta,l}), \quad \tau^2_{\beta,l} \sim \text{Exp}(0.5\lambda^2).$$

Then, marginally β_l follows a Laplace distribution with parameter λ, i.e.

$$p(\beta_l) = \int_0^\infty p(\beta_l|\tau^2_{\beta,l})p(\tau^2_{\beta,l})d\tau^2_{\beta,l} \propto \exp(-\lambda|\beta_l|).$$

While on first sight the introduction of a further hierarchical stage in the prior specification seems to make the model more complicated (in particular since one additional parameter $\tau^2_{\beta,l}$ has to be introduced for each parameter β_l), it in fact greatly facilitates MCMC-based estimation since the lasso recurs to a conditionally Gaussian specification. In particular, updates for the regression coefficients can be performed exactly as for the Bayesian ridge relying on Gibbs sampling for (latent) Gaussian responses or IWLS proposals for more general observation models. The only change is in the prior structure for the regression coefficients: Since each regression coefficient is assigned a separate variance $\tau^2_{\beta,l}$, the covariance matrix has to be changed to

$$\boldsymbol{\Sigma}_{\boldsymbol{\beta}} = \left(\frac{1}{\sigma^2}\boldsymbol{X}'\boldsymbol{X} + \boldsymbol{C}^{-1}\right)^{-1}$$

where

$$\boldsymbol{C} = \text{diag}(\tau^2_{\beta,1}, \ldots, \tau^2_{\beta,m}).$$

Still, the smoothing parameter λ remains unspecified but in a Bayesian formulation we can simply assign a further hyperprior to it. In particular, a gamma distribution $\lambda^2 \sim \text{Ga}(a, b)$ yields a closed form full conditional for λ. In this case, the marginal distribution of β_l (marginalizing over both $\tau^2_{\beta,l}$ and λ) is given by

$$p(\beta_l|a, b) = \frac{a2^a}{\sqrt{\pi 2b}}\Gamma\left(a + \frac{1}{2}\right)\exp\left(\frac{\beta_l^2}{8b}\right) D_{-2a-1}\left(\sqrt{\frac{\beta_l^2}{2b}}\right)$$

where D_{-2a-1} is the parabolic cylinder function (Griffin and Brown, 2005; Griffin and Brown, 2007). Figure 4.25 compares the conditional and the marginal specification of the Bayesian lasso. Again, as for the Bayesian ridge, the conditional and the marginal variant are quite different and the marginal prior is much more peaked in zero while having heavier tails than its conditional counterpart. Moreover, the marginal versions of the Bayesian ridge and the Bayesian lasso are qualitatively very similar. Hence, we may suspect that their empirical performance is also closer to each other than based on their frequentist analogues (which correspond to the conditional priors in Fig. 4.25).

Again, the regression coefficients (conditionally) separate all hyperparameters from the observation model such that the full conditionals are always the same. When using the exponential prior for the variances $\tau^2_{\beta,l}$ and the gamma prior for the smoothing parameter λ, the following distributions result:

$$\left.\frac{1}{\tau^2_{\beta,l}}\right| \cdot \sim \text{InvGauss}\left(\frac{|\lambda|}{|\beta_l|}, \lambda^2\right), \quad \lambda^2|\cdot \sim \text{Ga}\left(a + m, b + \frac{1}{2}\sum_{l=1}^{m}\tau^2_{\beta,l}\right). \quad (4.38)$$

In summary, for inclusion of lasso priors in MCMC algorithms for AMs (Algorithm 2.7), GAMs (Algorithm 2.8), LMMs (Algorithm 3.1), GLMMs (Algorithm 3.3) and SPMMs (Algorithms 4.2 and 4.3) as well as for categorical regression models, the Gaussian prior $\boldsymbol{\beta} \sim \text{N}(\boldsymbol{c}, \boldsymbol{C})$ has to be replaced by a conditional Gaussian prior with $\boldsymbol{c} = \boldsymbol{0}$ and $\boldsymbol{C} = \text{diag}(\tau_1^2, \ldots, \tau_m^2)$ for the lasso prior. Of course, $\boldsymbol{C}$ has to be understood in the same way in corresponding full conditionals. In addition, hyperparameters have to be drawn from the full conditionals (4.38).

In particular, for inclusion in the general MCMC Algorithm 4.3 for SPMMs from page 220, step iii of the algorithm has to be replaced by the following hierarchy:

iii.a *Update $\boldsymbol{\beta}$:* Sample $\boldsymbol{\beta}$ directly from the full conditional $\boldsymbol{\beta}|\cdot \sim \text{N}(\boldsymbol{\mu}_{\boldsymbol{\beta}}, \boldsymbol{\Sigma}_{\boldsymbol{\beta}})$ with

$$\boldsymbol{\mu}_{\boldsymbol{\beta}} = \boldsymbol{\Sigma}_{\boldsymbol{\beta}}\frac{1}{\sigma^2}\boldsymbol{X}'(\boldsymbol{y} - \boldsymbol{\eta}_{-\boldsymbol{\beta}}), \quad \boldsymbol{\Sigma}_{\boldsymbol{\beta}} = \left(\frac{1}{\sigma^2}\boldsymbol{X}'\boldsymbol{X} + \boldsymbol{C}^{-1}\right)^{-1}$$

for (latent) Gaussian responses or propose $\boldsymbol{\beta}$ from a Gaussian proposal density with expectation and covariance matrix

$$\boldsymbol{\mu}_{\boldsymbol{\beta}} = \boldsymbol{\Sigma}_{\boldsymbol{\beta}} \boldsymbol{X}' \boldsymbol{W} (\tilde{\boldsymbol{y}} - \boldsymbol{\eta}_{-\boldsymbol{\beta}}), \quad \boldsymbol{\Sigma}_{\boldsymbol{\beta}} = \left(\boldsymbol{X}' \boldsymbol{W} \boldsymbol{X} + \boldsymbol{C}^{-1} \right)^{-1}.$$

iii.b *Update hyperparameters:* Sample $\tau^2_{\beta,l}$, $l = 1, \ldots, m$ and λ directly from the full conditionals given in (4.38).

We first consider a binary probit model for cross-sectional credit scoring data to illustrate the application of regularization priors in generalized additive models.

Example 4.9: Credit scoring
In this example, we analyse the defaults of 1,000 consumer credits collected by a Southern German bank following an analysis presented in Kneib, Konrath and Fahrmeir (2010b). The major aim is to predict credit-worthiness of a client based on a larger number of covariates. The response variable is the binary indicator y_i that specifies whether the credit has been paid back ($y_i = 1$, credit-worthy) or not ($y_i = 0$, not credit-worthy). In addition to the continuous covariates age of the client, amount of the credit and duration of the credit which will be included with nonlinear effects, 51 categorical covariates are available that will be assigned ridge and lasso priors for regularization. We therefore consider the following binary additive probit model

$$\mathbb{P}(y_i = 1) = \Phi(\beta_0 + \boldsymbol{x}_i' \boldsymbol{\beta} + f_1(age) + f_2(amount) + f_3(duration)).$$

where the intercept is separated from the remaining fixed effects since it will not be subject to regularization and is therefore always assigned a flat, noninformative prior. In contrast, the vector $\boldsymbol{\beta}$ is supplemented with a Bayesian ridge or Bayesian lasso prior.

To provide a benchmark for regularized estimation, we also consider an 'expert model' developed in the literature for the same data set that only contains nonlinear effects and information on three further categorical covariates (payment of previous credits with categories good and bad; intended use with categories professional and private; running account with categories no running account, normal running account and good running account). We investigate the predictive performance of our models based on 10-fold cross validation where the area under the receiver operator characteristic (ROC) curve (AUC, Hosmer and Lemeshow, 2000) serves as the performance measure. Therefore we split the data set in ten separate data sets, use 90% of the data for estimation and evaluate the AUC on the remaining out-of-bag sample, i.e. the 10% of the data left out of estimation.

Figure 4.26 shows a parallel coordinate plot of the resulting AUC that combines results for one distinct cross validation sample obtained with the different model specifications. The most obvious conclusion that can be drawn from this

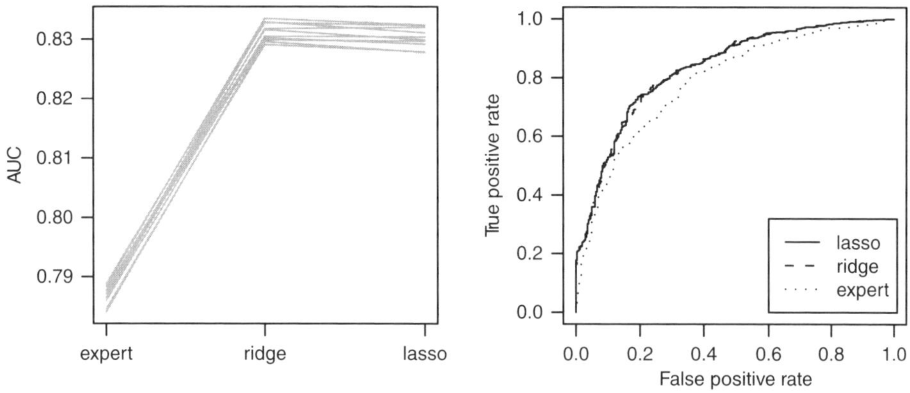

Fig. 4.26 Credit scoring: Parallel coordinate plots of the area under the curve obtained from 10-fold cross validation (left) and receiver operator characteristic curve for the complete data set (right).

figure is that any type of regularized estimation pays off in terms of predictive ability. Both lasso and ridge regularization show a comparable predictive performance as might have been expected from the comparable marginal prior structures that they induce. For illustrative purposes, Fig. 4.26 also shows the ROC curves for the three models obtained with the full data set. It confirms the improved predictive ability of the regularized estimation schemes since the ROC curves are consistently higher than those of the literature model. Also, the ROC curves for lasso and ridge are quite close over the complete domain.

Figure 4.27 summarizes the estimated nonlinear effects for the three models. All functions show only moderate deviations from linearity and only very little sensitivity with respect to the model specified for the parametric effects. The age effect is approximately constant for older credit-holders but indicates a somewhat reduced credit-worthiness for young clients up to an age of about 35 years. Both larger credits and long-term credits have a higher probability of defaulting.

Table 4.4 shows the estimated regularized effects for the subset of covariates with (estimated) posterior probability less than 0.05 in either the lasso or the ridge model. Posterior probabilities are defined here as $2\min\{\mathbb{P}(\beta_j > 0), \mathbb{P}(\beta_j < 0)\}$ and may be interpreted as a kind of Bayesian p-values. The covariates include information on several variables characterizing the running account, confirming the use of analogous variables in the literature example. In addition, some variables characterizing the intended use have a highly significant impact on the probability of credit defaults. Note that, in contrast to its frequentist counterpart, the Bayesian lasso does not set coefficients exactly to zero. This is simply due to the fact that MCMC does not determine the posterior mode but the posterior mean. Both quantities do not typically coincide and moreover sampling variability will typically lead to deviations from zero even if mean and mode would

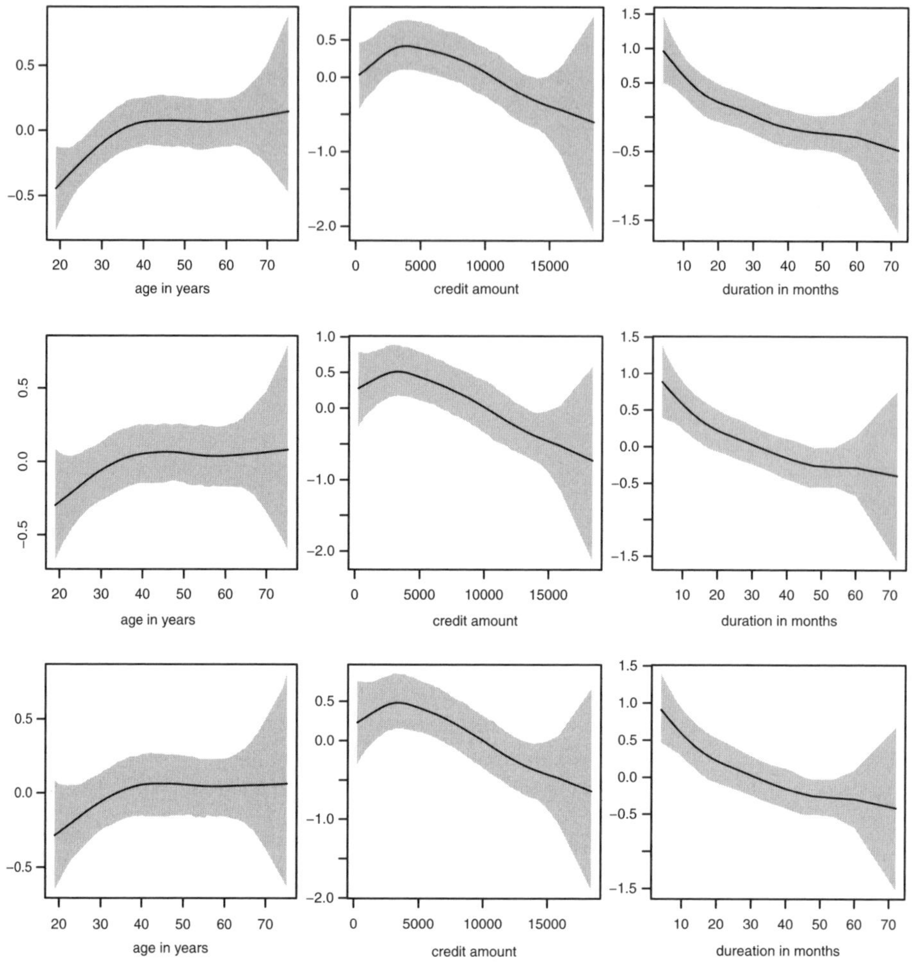

Fig. 4.27 Credit scoring: Estimated nonparametric effects and 95% pointwise credible bands in the expert model (top row), the Bayesian ridge model (middle row) and the Bayesian lasso model (bottom row).

be the same in theory. However, Bayesian regularization also shrinks small and unstable coefficients to zero such that they effectively drop out of the model. △

The second example illustrates shrinkage priors for fixed effects of longitudinal data in an additive mixed model.

Example 4.10: Childhood obesity
To reanalyse the childhood obesity data, we include the remaining covariates TV watching and computer playing (less than 1 hour as reference category, tvpc2 = 1

Table 4.4 Credit scoring: Posterior mean and posterior probabilities for regularized effects with posterior probability lower than 0.05 in either the Bayesian lasso or the Bayesian ridge model.

	lasso		ridge	
variable	mean	prob.	mean	prob.
no current account	−0.891	0.000	−0.820	0.000
no positive balance	−0.637	0.000	−0.593	0.000
no savings account	−0.390	0.004	−0.422	0.002
for new vehicle	0.491	0.012	0.445	0.010
poor paying habits	−0.509	0.018	−0.479	0.012
risky account	−0.554	0.018	−0.530	0.006
alternative purpose	−0.309	0.036	−0.353	0.008
medium term employment	0.270	0.040	0.269	0.070
no previous loans	−0.233	0.092	−0.269	0.024
no assets	0.206	0.130	0.277	0.046

up to 2 hours, tvpc3 = more than 3 hours), mother's education (no degree or CSE as reference category, mEdu2 = secondary school, mEdu3 = subject specific high school diploma, mEdu4 = high school diploma), hours spent outdoors (outdoor) and weight gain until the age of 2 years (wgain2y), described in Example 1.1, into the covariate vector $\boldsymbol{u}$ with linear population effects $\boldsymbol{\beta}$ of the additive mixed model (4.15) in Examples 4.2 and 4.4. The population trend is again specified through a cubic Bayesian P-spline and the individual-specific trends through the nonlinear random curves (3.7), already introduced in Examples 3.2, 3.4, 4.2 and 4.4. Apart from the intercept term, we impose ridge and lasso priors on the components β_j of linear population effects. Table 4.5 contains posterior mean estimates together with (estimated) posterior probabilities $2\min\{\mathbb{P}(\beta_j > 0), \mathbb{P}(\beta_j < 0)\}$ and, for comparison, posterior means and credibility intervals obtained with flat priors for all effects. Comparing posterior means, we can recognize the shrinkage effect of lasso and ridge priors. Lasso and ridge estimates are quite close to each other, with a bit more shrinkage for the lasso. Posterior probabilities and credibility intervals provide similar evidence about the effect of covariates: Among the additional covariates mother's education (more exactly mEdu3 and mEdu4 compared to the reference category) and weight gain have significant impact. Compared to previous analyses with a subset of the covariates, the effect of sex has lost influence. Breastfeeding now seems to bear the risk of increasing a child's BMI. This is at least surprising or even in conflict with prior expectations of some epidemiologists. A closer look at the covariates reveals comparably high correlations between some of them, so that our results have to be interpreted with care, motivating additional research. The population trend $f(t)$ as well as individual curves are less affected from changes in the model and estimates are therefore not shown here. △

Table 4.5 Childhood obesity: Posterior mean estimates for the fixed effects in the AMM with Gaussian prior (together with 95% credible intervals) and in the AMMs with lasso and ridge shrinkage priors (together with posterior probabilities). wgain2y, outdoor, mBMI and mDiffBMI have been centred for the analyses.

	Gaussian prior			lasso prior		ridge prior	
	mean	2.5%	97.5%	mean	prob.	mean	prob.
const	15.49	15.27	15.71	15.59	–	15.58	–
sex	0.0881	0.0028	0.1674	0.0739	0.076	0.0772	0.068
breast	0.1783	0.0917	0.2664	0.1566	0.000	0.1571	0.000
tvpc2	0.0135	−0.0798	0.1140	0.0041	0.898	0.0079	0.870
tvpc3	0.2229	−0.1179	0.5799	0.0767	0.486	0.0775	0.490
mSmoke	−0.0781	−0.2056	0.0479	−0.0601	0.234	−0.0771	0.176
mEdu2	0.1220	−0.0459	0.3017	0.0580	0.502	0.0558	0.402
mEdu3	0.3755	0.1890	0.5679	0.3062	0.000	0.2849	0.000
mEdu4	−0.1922	−0.3551	−0.0181	−0.2357	0.002	−0.2358	0.000
area	0.0545	−0.0482	0.1616	0.0313	0.482	0.0425	0.334
wgain2y	0.1271	0.0927	0.1629	0.1240	0.000	0.1262	0.000
outdoor	0.0342	−0.0011	0.0692	0.0314	0.076	0.0334	0.066
mBMI	0.0432	0.0317	0.0548	0.0423	0.000	0.0429	0.000
mDiffBMI	0.1158	0.0889	0.1395	0.1130	0.000	0.1146	0.000

L_q priors

Both the ridge and the lasso penalty are special cases of L_q penalties, where

$$\lambda \, \text{pen}(\boldsymbol{\beta}) = \lambda \sum_{l=1}^{m} |\beta_l|^q, \quad 0 < q \le 2.$$

Fu (1998) in particular considers the case of L_q priors with $1 \le q \le 2$ and calls this the bridge regression penalty because it bridges between lasso ($q = 1$) and ridge ($q = 2$). The corresponding class of prior distributions is given by the power exponential prior

$$p(\beta_l|\lambda) \propto \exp(-\lambda|\beta_l|^q), \quad 0 < q \le 2.$$

Park and Casella (2008) show that the entire class of power exponential priors can be represented as a scale mixture of normals since

$$\exp(-|\beta_l|^q) \propto \int_0^\infty \exp\left(-\frac{\beta_l^2}{2\tau_{\beta,l}^2}\right) \frac{1}{\tau_{\beta,l}^6} s_{q/2}\left(\frac{1}{2\tau_{\beta,l}^2}\right) d\tau_{\beta,l}^2$$

where $s_q(\cdot)$ is the density of the positive stable distribution with index q. However, in this case the mixing distribution is quite complicated and it seems at least difficult to exploit the scale mixture property in Bayesian inferential procedures.

Still, such a procedure would be extremely helpful, for example to approximate solutions close to L_0 regularization, where the number of non-zero regression coefficients is penalized. Of course, such penalties are extremely non-smooth, inducing multiple modes in the posterior. In principle, MCMC algorithms with mode jumping behaviour or Hamiltonian MCMC samplers designed for such situations may be valuable tools, see for example Tjelmeland and Hegstad (2001) and Chen, Qin and Lius (2001).

Further examples

Some further examples of regularization penalties proposed in the literature can also be related to Bayesian regularization priors. For example, the elastic net (Zou and Hastie, 2005) with penalty

$$\lambda \operatorname{pen}(\beta_l) = \lambda \left[\alpha|\beta_l| + (1-\alpha)\beta_l^2\right], \quad 0 \leq \alpha \leq 1,$$

is simply a mixture of a Gaussian and a Laplace distribution from a Bayesian perspective, compare Li and Lin (2010). More specifically, let I_l be a binary indicator variable with $\mathbb{P}(I_l = 1) = \alpha$. Then, the conditional prior of β_l is given by

$$p(\beta_l|I_l, \lambda, \alpha) \propto I_l \exp(-\lambda|\beta_l|) + (1 - I_l)\exp(-\lambda\beta_l^2)$$

where suitable hyperpriors have to be specified for the unknown hyperparameters λ and α. In a second step, the Laplace prior part of the elastic net could again be represented as a scale mixture of normals so that a conditionally Gaussian prior specification results.

The elastic net has been motivated by the fact that the combination of ridge and lasso penalty tends to select groups of correlated regressors instead of a single representative. A further penalty function proposed in this context is the octogonal shrinkage and clustering algorithm for regression (OSCAR, Bondell and Reich, 2008), where the joint penalty for all regression coefficients is given by

$$\lambda \operatorname{pen}(\boldsymbol{\beta}) = \lambda \left[\alpha \sum_{l=1}^{m} |\beta_l| + (1-\alpha)\sum_{l<k} \max(|\beta_l|, |\beta_k|)\right], \quad 0 \leq \alpha \leq 1.$$

Again, the penalty corresponds to a mixture of two components with mixture weights determined by the parameter α. The second term in the penalty, that is, $\max(|\beta_l|, |\beta_k|)$ corresponds to the L_∞ norm of the two coefficients involved. Whether a conditionally Gaussian formulation of this penalty exists is still an open question.

Other penalties such as the hard shrinkage penalty

$$\lambda \operatorname{pen}(\beta_l) = \lambda^2 - (|\beta_l| - \lambda)^2 \mathbb{1}(|\beta_l| < \lambda)$$

or the smoothly clipped absolute deviation (SCAD, Fan and Li, 2001) with penalty function

$$\lambda\,\mathrm{pen}(\beta_l) = \begin{cases} \lambda|\beta_l| & |\beta_l| \leq \lambda \\ -(\beta_l^2 - 2a\lambda|\beta_l| + \lambda^2)/(2a-2) & \lambda < |\beta_l| < a\lambda \\ 0.5(a+1)\lambda^2 & |\beta_l| > a\lambda \end{cases}$$

and $a > 2$ do not seem to have (proper) Bayesian prior counterparts since the penalties are bounded which turns into $p(\beta_l) \to 0$ for $|\beta_l| \to \infty$ when inverting the penalty to a prior. The SCAD penalty is derived as a quadratic spline function that defines a compromise between hard shrinkage and LASSO with knots at $-a\lambda, -\lambda, \lambda$ and $a\lambda$.

4.4.3 Locally adaptive priors for functions

The smoothing approaches discussed in Chapter 2 and considered so far in this chapter are appropriate if the function of interest is smooth and the amount of smoothness does not vary too much over the codomain of the covariate. If, however, the function shows strong local variation in the amount of smoothness or even discontinuities, locally adaptive function estimation may be required. In the following, we will discuss two groups of approaches: The first still relies on penalization or corresponding smoothness priors but adapts the structure such that local variability is supported. The second class of approaches uses knot or basis function selection strategies to adapt the function representation to local variability.

Locally adaptive penalties

Bayesian penalized B-splines assume that a function of interest $f(x)$ can be approximated in terms of B-spline basis functions such that

$$f(x) = \sum_{k=1}^{K} \gamma_k B_k(x)$$

subject to a random-walk prior of order d for the coefficients γ_k. In the standard formulation, the variance of the random-walk is assumed to be constant, yielding a constant variance for the increments of the random-walk, that is,

$$\Delta^d \gamma_k \sim \mathrm{N}(0, \tau^2).$$

Hence the most obvious way of introducing locally adaptive variability is to replace the constant τ^2 with τ_k^2. Since the knots are placed within the codomain of the covariate x, we can also consider τ_k^2 a function of the covariate and write $\tau_k^2 = \tau^2(x)$. The various approaches proposed for this setting vary mostly in the specification of $\tau^2(x)$ and the particular inferential technique applied for estimating it.

If the variability is changing very rapidly, it may be useful to completely disentangle the variances and to assume that the τ_k^2 are independent from each other. If we again assume inverse gamma distributions for the independent variances such that

$$\tau_k^2 \overset{\text{ind.}}{\sim} \text{IG}(a_k, b_k),$$

this corresponds to independent and marginally t-distributed increments in the random-walk prior (see Lang and Brezger, 2004). Note that also the increments in the original penalized spline formulation are marginally t-distributed but they are not independent since they all depend on the same variance parameter. The advantage of the independent inverse gamma assumption is that inferential procedures have to be changed only in details. In particular, the penalty matrix that was given by

$$\frac{1}{\tau^2}\boldsymbol{K} = \frac{1}{\tau^2}\boldsymbol{D}'\boldsymbol{D}$$

for the random-walk with global smoothing parameter is now replaced with

$$\boldsymbol{D}'\boldsymbol{T}^{-1}\boldsymbol{D}$$

where $\boldsymbol{T} = \text{diag}(\tau_{d+1}^2, \dots, \tau_K^2)$. Apart from this re-definition, estimation of the regression coefficients $\boldsymbol{\gamma}$ remains completely unchanged. In the case of fully Bayesian estimation based on MCMC, the update relies on either Gibbs sampling in the case of (latent) Gaussian responses or IWLS proposals for more general response types, exactly as described in Section 2.2 and used throughout this chapter. For the variance parameters, the corresponding full conditionals are again inverse gamma with updated parameters, that is,

$$\tau_k^2 | \cdot \sim \text{IG}\left(a_k + \frac{1}{2}, b_k + \frac{1}{2}(\Delta^d \gamma_k)^2\right).$$

For empirical Bayes inference, the mixed model representation remains valid although the random effects arising in the reparametrization are no longer identically distributed. Still, the random effects are normal and independent such that REML estimation can be used for both regression coefficients and variances.

While being conceptually easy to understand and to implement, the assumption of independent variance parameters significantly increases the number of parameters and does not allow us to borrow strength between neighbouring increments of the random-walk. However, in many situations such borrowing of information will be extremely useful, especially if the variability is changing smoothly over the codomain of x. Hence, we consider $\tau^2(x)$ to be itself a function of x and represent it using another polynomial spline basis as

$$\tau^2(x) = \exp\left(\sum_{l=1}^{L} \alpha_l B_l(x)\right).$$

In a further stage of the hierarchy, a random-walk prior may be placed on the sequence α_l, $l = 1, \ldots, L$ to enforce specific assumptions about the smoothness of the variance function. Baladandayuthapani, Mallick and Carroll (2005) propose a fully Bayesian implementation of this approach while Krivobokova, Crainiceanu and Kauermann (2008) consider mixed model-based estimation.

As a compromise between independent variances and smoothly varying variances, Scheipl and Kneib (2009) have proposed step functions for the variance parameter sequence τ_k^2. On the one hand, this allows for very abrupt changes in the variability at discontinuities but on the other hand enforces smoothness in areas of almost constant variability. Three types of models with increasing complexity have been considered: In the basic version, both the number and the location of the changepoints of the variance function are fixed. Of course, this approach requires some knowledge about the underlying form of the function to be estimated. To obtain a more data-driven variant, the locations of the changepoints are allowed to vary in a second step. The most complex model also allows the number of changepoints to be determined from the data. While basic MCMC algorithms can be used in the two simpler model variants, the third requires reversible jump MCMC (RJ-MCMC) steps since the dimension of the parameter space changes with the number of changepoints.

To further increase the adaptivity of the penalized spline estimate, Scheipl and Kneib (2009) also replace the inverse gamma prior for the pieces of the variance function with a prior distribution that is more peaked in zero and has heavier tails. The rationale is that small jumps should be shrunk towards zero in areas with only very limited variability of the function while large increments should be left almost unpenalized. In principle, this corresponds to ideas discussed in the previous section on shrinkage priors for high-dimensional covariate vectors. More specifically, Scheipl and Kneib (2009) borrow the normal-exponential-gamma prior proposed in Griffin and Brown (2007) and assume an exponential prior for the variances with a gamma distributed rate. The model hierarchy for fixed number of changepoints and changepoint locations can be characterized as follows:

- Regression coefficients: The parameters $\boldsymbol{\gamma}$ conditionally follow a Gaussian random-walk with

$$\boldsymbol{D}\boldsymbol{\gamma} \sim \mathrm{N}(\boldsymbol{0}, \boldsymbol{T}), \quad \boldsymbol{T} = \mathrm{blockdiag}(\tau_1^2 \boldsymbol{I}_{l_1}, \ldots, \tau_B^2 \boldsymbol{I}_{l_B})$$

  where B denotes the number of constant pieces of the variance function (i.e. there are $B - 1$ changepoints) and l_b, $b = 1, \ldots, B$, denotes the lengths of the variance pieces.
- Variances: The variances τ_b^2 are assumed to be independent and follow exponential distributions with rate λ_b:

$$\tau_b^2 \sim \mathrm{Exp}(\lambda_b).$$

- Rates: The rates of the exponential distributions follow independent gamma distributions
$$\lambda_b \sim \text{Ga}(a_\lambda, b_\lambda).$$
- Hyperprior on the gamma rate: Further flexibility is achieved by assigning a further hyperprior on the parameters of the gamma prior. Following Griffin and Brown (2007) it is sufficient to let b_λ vary while a_λ can be fixed at $a_\lambda = 0.5$. A simple prior specification for b_λ could then for example be a uniform distribution on a $\log_{10}$ equidistant grid of 550 values between 10^{-3} and 10^5.

Marginally, this hierarchy induces exactly the desired prior structure with heavy tails and a spike in zero. Due to conjugate choices, all updates can be implemented in a Gibbs sampler, see Scheipl and Kneib (2009) for details.

If the changepoints are in addition, allowed to vary the distribution of the order statistic of $B - 1$ discrete uniform distributions on the set of possible changepoints is a useful prior choice. In a RJ-MCMC extension, a Poisson distribution or a truncated Poisson distribution are the most obvious candidates for the prior on the number of changepoints. Scheipl and Kneib (2009) describe the corresponding prior setting as well as birth and death moves of the RJ-MCMC algorithm in detail.

Example 4.11: Fractionation curves
We illustrate adaptive function estimation with data from 'Specificity Assessment From Fractionation Experiments' (SAFE, Drobyshev *et al.*, 2003). The purpose of these experiments is to control the quality of cDNA microarray experiments. This is achieved by investigating the degree of undesired cross-hybridization of specific probe strands, i.e. how often cDNA sections pair with cDNA probes on the chip which have a similar, but not exactly equal, base sequence. Therefore, microarray chips are repeatedly treated with formamide solutions of increasing concentration and intensities are recorded after each washing. A fractionation curve is then obtained by plotting the intensities for each probe against the concentration of the formamide solution. Since the cohesion between cross-hybridizing cDNA strands is weaker than the cohesion for perfect matches, they will be already washed away at lower concentrations. Once a critical concentration of the formamide solution is reached, the cross-hybridized probe sequence is abruptly washed away, yielding a sudden drop in the signal intensity. This induces abrupt changes in the local variability of the fractionation curve that makes adaptive function estimation necessary.

Since the fractionation curves are expected to be close to a step function, we employ splines of degree zero with 20 basis functions and first-order random-walk priors both for the adaptive and the non-adaptive curve estimation, see Scheipl and Kneib (2009) for further details on hyperprior choices.

Figure 4.28 shows an example of a spot binding only the correct complementary cDNA. Therefore the corresponding fractionation curve shows only one

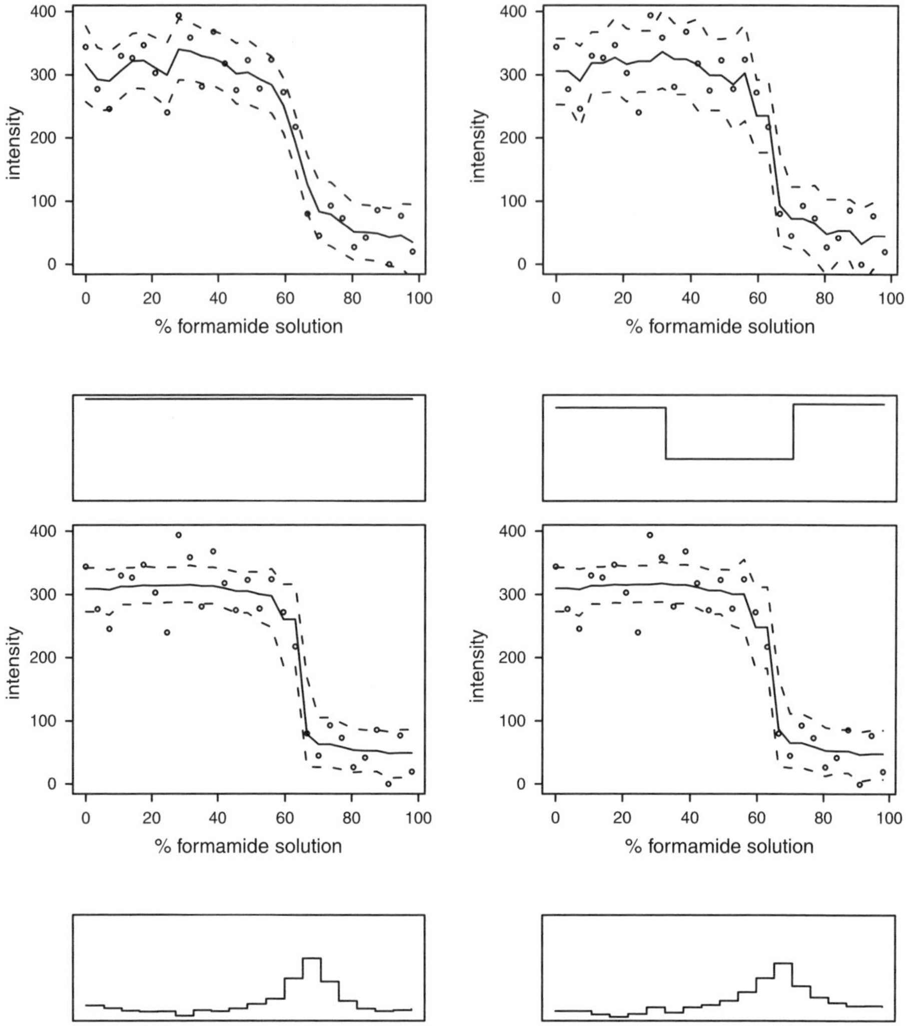

Fig. 4.28 Fractionation curves: Example of a spot binding only the correct complementary cDNA. The upper left panel shows the results of a non-adaptive fit with global smoothing variance. The remaining three panels show results obtained for adaptive splines based on step functions for the variance with fixed changepoints (upper right), varying changepoint positions (lower left) and varying number and position of changepoints (lower right). The lower parts of the figures present the square-root of the estimated variance functions on the log-scale.

step at a concentration of about 65% that corresponds to the point where the binding energy between complementary cDNA strands was no longer sufficient for cohesion. The situation is different for the spot shown in Fig. 4.29 where we observe an additional step in the fractionation curve at about 15%. For this spot, cross-hybridization was present and the corresponding strands are washed away at 15% concentration.

Even without the explicit monotonicity constraints appropriate for this data, the adaptive fits fairly well reproduce the piecewise constant and decreasing structure that is expected. Note also that despite ignoring the expected structure of a decreasing step function, the results give an indication of the number and the location of jumps in the estimated function when looking at the estimated variance function for those variants that allow for varying changepoint positions. While the variance functions in Fig. 4.28 clearly show a single jump, two distinct jump points are present in Fig. 4.29. As a consequence, the adaptive procedures would also support inference about the location of the jump points that could be extracted from the MCMC samples making it a valuable alternative to step-function based approaches where this is typically difficult.

Of course, the quality of the adaptive fits may deteriorate when fixing the locations of the changepoints at inappropriate places. For example in Fig. 4.28, both the non-adaptive fit and the adaptive fit with fixed changepoints are very wiggly in the first interval containing the first third of the data, corresponding to a large value of the variance function. This seems to be caused by an attempt to fit local outliers due to the low signal in this area. A similar behaviour is observed for the third interval, whereas the second interval, containing the jump, is assigned a small variance to obtain a smooth fit due to the high signal induced by the jump. In general, the non-adaptive fit exhibits excessive wiggliness for low concentrations in Fig. 4.28 and for intermediate concentrations in Fig. 4.29, which shows the improvement that can be gained by an adaptive fit in this context. △

Knot selection strategies

Instead of making the penalty adaptive to the variability of the function, an alternative strategy is to make the placement of the knots adaptive. In fact, this often allows us to completely drop the penalty from the model specification and to work with a flat prior for the spline coefficients instead. The underlying rational is that a larger number of knots should be located in areas with high variability while knots can be placed sparsely in areas with only moderate variability. To achieve such an adaptive knot placement, strategies inspired from variable selection are frequently employed.

In a Bayesian setting, the two most popular knot selection approaches either rely on reversible jump MCMC to move between model spaces of differing dimension, or the introduction of binary indicators that allow us to select or deselect basis functions from a rich set of candidate functions. In the latter case, we start

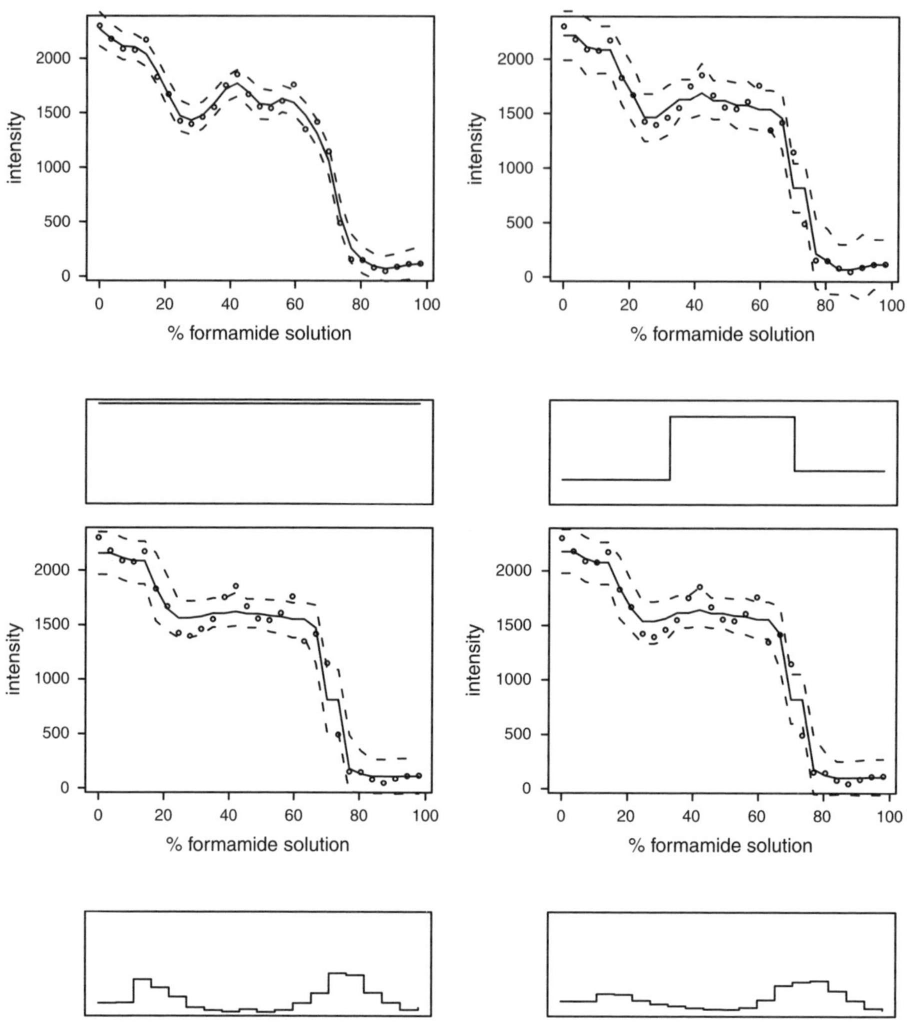

Fig. 4.29 Fractionation curves: Example of a spot with cross-hybridization. The upper left panel shows the results of a non-adaptive fit with global smoothing variance. The remaining three panels show results obtained for adaptive splines based on step functions for the variance with fixed changepoints (upper right), varying changepoint positions (lower left) and varying number and position of changepoints (lower right). The lower parts of the figures present the square-root of the estimated variance functions on the log-scale.

with an approximation of the unknown function $f(x)$ in terms of a basis function representation just as with penalized splines:

$$f(x) = \sum_{k=1}^{K} \gamma_k B_k(x).$$

However, instead of adding an informative prior/penalty to enforce smoothness of the resulting estimates, each of the basis functions is assigned an indicator variable $J_k \in \{0, 1\}$ that defines whether the corresponding basis function should actually be used in the reconstruction of f, that is,

$$f(x) = \sum_{k=1}^{K} J_k \gamma_k B_k(x).$$

Adding the binary indicators to the MCMC scheme allows for simultaneous inference on the relevant basis functions and the function f itself, see Smith and Kohn (1996) or Yau, Kohn and Wood (2003) for details. Similar approaches will be considered in Section 4.5.3 for variable, function and covariance selection.

Reversible jump MCMC approaches replace the idea of binary indicator variables by a more formal assessment of models of varying dimension. The principle idea is, however, related and relies on a set of candidate basis functions that are characterized in terms of locations. The reversible jump MCMC scheme then allows us to add or remove basis functions from the current model based on birth and death steps. Further steps corresponding to the movement of basis functions may additionally be included. Examples of such approaches, differing for example in the prior specifications for the knot configurations, can be found in Biller (2000), DiMatteo, Genovese and Kass (2001) or Denison, Mallick and Smith (1998).

4.5 Model choice and model checking

The previous sections have shown how advances in Bayesian computation make it possible to fit increasingly flexible models to longitudinal data and to complex real world data structures in general. Usually, several candidate models are fitted, and some form of model assessment and comparison is required to check model adequacy, to choose between competing models or to average over them. In the context of this chapter the following questions will typically be of interest: Has the effect of a covariate to be modelled nonparametrically, or is a linear parametric effect adequate? Should random effects be included? Are interactions or varying coefficient terms necessary? Are the model assumptions justified? How well do predictions from the model agree with observed data? Are there observations with a rather poor fit?

The general need for answering such questions has led to immense methodological research in comparing and assessing models, often with competing suggestions or strategies. This section begins with Bayes factors as a classical concept for model choice, but the focus is on model selection criteria, on predictive methods for model checking and on predictor selection. Presentation and choice of methods have been guided by the following aspects: relevance for empirical or fully Bayesian models considered in this book, and a balance between theoretical appeal and computational efforts. Although included in this chapter, concepts and methods apply to previous and subsequent chapters as well.

4.5.1 Bayes factors and model selection criteria

Bayes factors and marginal likelihoods

Suppose we have K (possibly non-nested) models $M_1, \ldots, M_K$ and we wish to compare them. In the context of SPMMs for longitudinal data, the models may have different covariates and sets of basis functions for modelling nonlinear effects, they may include random slopes or random intercepts only, the response variable may be transformed in a preprocessing step or not, etc. In generic notation the models are specified through conditional distributions $p(\boldsymbol{y}|\boldsymbol{\theta}_k, M_k)$ for the observed data $\boldsymbol{y}$, together with priors $p(\boldsymbol{\theta}_k|M_k)$ for the unknown parameter vector $\boldsymbol{\theta}_k$. In SPMMs, $\boldsymbol{\theta}_k$ comprises all regression and basis functions coefficients, random effects, variance and association parameters in model M_k. The marginal distribution

$$p(\boldsymbol{y}|M_k) = \int p(\boldsymbol{y}|\boldsymbol{\theta}_k, M_k)p(\boldsymbol{\theta}_k|M_k)d\boldsymbol{\theta}_k \tag{4.39}$$

is often called marginal likelihood of model M_k. A good survey on simulation-based approximations of the marginal likelihood is given in Frühwirth-Schnatter (2006, Ch. 5.4).

In the traditional formal approach, it is assumed that the set of models contains the true one (the 'M-closed' situation, see Bernardo and Smith, 2004). Then, under a 0-1 loss function, the expected posterior loss is minimized, if we choose the model with the largest posterior probability $p(M_k|\boldsymbol{y})$. With prior probability $p(M_k)$ of M_k being the true model, we have

$$p(M_k|\boldsymbol{y}) = \frac{p(\boldsymbol{y}|M_k)p(M_k)}{\sum_{j=1}^{K} p(\boldsymbol{y}|M_j)p(M_j)}.$$

For comparing just two models, M_k and M_l, this leads us to consider the posterior odds

$$\frac{p(M_k|\boldsymbol{y})}{p(M_l|\boldsymbol{y})} = \frac{p(\boldsymbol{y}|M_k)}{p(\boldsymbol{y}|M_l)} \cdot \frac{p(M_k)}{p(M_l)}.$$

This ratio is obtained by multiplying the prior odds $p(M_k)/p(M_l)$ with the marginal likelihood ratio or *Bayes factor*

$$B_{kl} = \frac{p(\boldsymbol{y}|M_k)}{p(\boldsymbol{y}|M_l)}.$$

Note that for equal prior model probabilities the Bayes factor coincides with the posterior odds. Thus, the marginal likelihoods $p(\boldsymbol{y}|M_k)$ are the key quantities for model choice within this formal Bayesian approach. However, apart from conjugate prior specifications, the integral in (4.39) cannot be solved analytically. Rearranging Bayes' theorem gives

$$p(\boldsymbol{y}|M_k) = \frac{p(\boldsymbol{y}|\boldsymbol{\theta}_k, M_k)p(\boldsymbol{\theta}_k|M_k)}{p(\boldsymbol{\theta}_k|\boldsymbol{y}, M_k)}. \tag{4.40}$$

This equivalent formula often provides a simpler way of deriving the marginal likelihood analytically, if possible at all, and it can also be useful for approximating marginal likelihoods from MCMC output.

A disadvantage of Bayes factors is that they are well-defined for proper priors only. Improper flat priors, for example $p(\boldsymbol{\beta}_k|M_k) \propto 1$ in a regression model, cause problems even if the posterior is proper. This is immediately seen from the definition of Bayes factors and marginal likelihoods, because we may also set $p(\boldsymbol{\beta}_k) = c_k$ with an arbitrary constant $c_k \neq 1$. In general, the problem cannot be circumvented by assuming vague but still proper priors $\boldsymbol{\beta} \sim \mathrm{N}(\boldsymbol{0}, c_k\boldsymbol{I})$ on the coefficients: Consider the two regression models

$$M_1\colon\ y = \sum_{k=1}^{K} \beta_k B_k(x) + \varepsilon, \quad M_2\colon\ y = \sum_{k=1}^{K+1} \beta_k B_k(x) + \varepsilon,$$

with basis functions $B_k(x)$, $k = 1, \ldots, K+1$, basis coefficients $\boldsymbol{\beta}_1 = (\beta_1, \ldots, \beta_K)$, $\boldsymbol{\beta}_2 = (\beta_1, \ldots, \beta_{K+1})$, and priors $\boldsymbol{\beta}_1 \sim \mathrm{N}(\boldsymbol{0}, c\boldsymbol{I}_K)$, $\boldsymbol{\beta}_2 \sim \mathrm{N}(\boldsymbol{0}, c\boldsymbol{I}_{K+1})$. As shown in Denison *et al.* (2002, Sec. 2.5), the Bayes factor B_{12} tends to infinity for $c \to \infty$, that is, model M_1 is favoured over M_2, irrespective of the data. This is an example of Lindley's paradox. For vague $\mathrm{N}(\boldsymbol{0}, c\boldsymbol{I})$ priors, with c large, we can expect that the simpler model M_1 will be favoured with high probability. This example, as well as others, lead to the following conclusions (Berger and Pericchi, 2001): When the models have parameter spaces with differing dimensions, use of improper noninformative priors yields indeterminate answers. This is seen by replacing improper priors $p_k = p(M_k)$ and $p_l = p(M_l)$ for models M_k and M_l, resulting in the 'formal' Bayes factor B_{kl}, with priors $c_k p_k$ and $c_l p_l$, in which case the Bayes factor would be $c_k/c_l B_{kl}$. Trying to avoid this indeterminacy by the use of 'vague proper priors' is just the 'devil in disguise': it usually gives unsatisfactory answers in model selection.

However, when the parameters $\boldsymbol{\theta}_k$ and $\boldsymbol{\theta}_l$ are close to each other, choosing improper noninformative priors with $c_k = c_l$ can be reasonable, see Berger *et al.* (1998) for situations when this is justified. Similarly, if both models contain a common subvector $\tilde{\boldsymbol{\theta}}$, an improper prior for $\tilde{\boldsymbol{\theta}}$ can be reasonable. In the context of SPMMs for longitudinal data such a situation arises, for example, if we wish to choose between a usual parametric linear or polynomial effect of a covariate x and a nonlinear effect $f(x)$ modelled through a Bayesian penalized spline. Then the vector $\boldsymbol{f}$ of function evaluations can be written as $\boldsymbol{f} = \boldsymbol{X}\boldsymbol{\gamma}$, where $\boldsymbol{X}$ is the design matrix constructed from the evaluations of basis functions and $\boldsymbol{\gamma}$ is the vector of basis function coefficients. Choosing a TPS basis or reparametrizing $\boldsymbol{\gamma} = \tilde{\boldsymbol{U}}\boldsymbol{\beta} + \tilde{\boldsymbol{Z}}\boldsymbol{\alpha}$ as in Section 4.2.1 leads to the representation

$$M_1: \ \boldsymbol{f} = \boldsymbol{U}\boldsymbol{\beta} + \boldsymbol{Z}\boldsymbol{\alpha},$$

where $\boldsymbol{U}\boldsymbol{\beta}$ represents the (unpenalized) parametric linear or polynomial part, with a flat prior $p(\boldsymbol{\beta}) \propto 1$ and a proper Gaussian prior $\boldsymbol{\alpha} \sim \mathrm{N}(\boldsymbol{0}, \tau^2\boldsymbol{I})$ for the 'random effects' part. The submodel

$$M_2: \ \boldsymbol{f} = \boldsymbol{U}\boldsymbol{\beta},$$

corresponds to the assumption of a linear or polynomial effect of x, again with flat prior $p(\boldsymbol{\beta}) \propto 1$. Comparison of M_1 and M_2 with 'formal' Bayes factors is analogous to the model choice problem considered by Zellner and Siow (1980) in linear regression and can be extended to SPMMs with general additive predictors.

For more than two, possibly non-nested models, Zellner and Siow (1980) suggest an 'encompassing' approach where each model M_k is compared to an encompassing model M_0 which contains all submodels. Then the Bayes factor of M_k to M_l is defined as $B_{kl} = B_{0l}/B_{0k}$, where B_{0k}, $k = 1, \ldots, K$, are the Bayes factors when comparing M_0 to M_k.

To avoid problems with improper noninformative priors at all, various 'default' Bayes factors for objective Bayesian model selection have been proposed, such as intrinsic Bayes factors (IBF, Berger and Pericchi, 1996) or fractional Bayes factors (FBF, O'Hagan, 1995). Motivated by the g-prior for regression coefficients in linear models (Zellner, 1986), IBFs and FBFs use the idea of 'training' samples to convert improper priors to proper priors. IBFs use a subset of the data as training sample, while FBFs are based on a fraction of the likelihood. Berger and Pericchi (2001) give a detailed review of these approaches in the context of linear models, see also the survey in Clyde and George (2004).

In any case, computation of Bayes factors or marginal likelihoods cannot be done in closed form for more complex models such as SPMMs and others considered in this book, but have to be approximated through asymptotic considerations, numerical or sampling methods. The following paragraph gives an account of various suggestions, focusing on approaches we found useful as a compromise between approximation quality and computational feasibility.

Methods for estimating Bayes factors and marginal likelihoods

There are two conceptual approaches: First, direct methods that involve separate sampling or approximations for each model and, second, methods based on simultaneous sampling from the joint parameter-model space, such as (modifications of) reversible jump MCMC.

In the following description of some direct methods, we suppress the dependence on the model indicator to simplify notation. Then $p(\boldsymbol{y})$ denotes a typical marginal model likelihood and $\hat{p}(\boldsymbol{y})$ an estimate. The computationally simplest estimate is the *harmonic mean estimator* (HM, Newton and Raftery, 1994)

$$\hat{p}_{\mathrm{HM}}(\boldsymbol{y}) = \left[\frac{1}{S} \sum_{s=1}^{S} \frac{1}{p(\boldsymbol{y}|\boldsymbol{\theta}^{(s)})} \right]^{-1},$$

where $p(\boldsymbol{y}|\boldsymbol{\theta})$ is the conditional observation density and $\{\boldsymbol{\theta}^{(s)},\ s = 1, \ldots, S\}$ refers to the MCMC output from the posterior $p(\boldsymbol{\theta}|\boldsymbol{y})$. The harmonic mean estimator is derived from rearranging Bayes theorem to

$$\frac{p(\boldsymbol{\theta})}{p(\boldsymbol{y})} = \frac{p(\boldsymbol{\theta}|\boldsymbol{y})}{p(\boldsymbol{y}|\boldsymbol{\theta})}$$

and integrating both sides with respect to $\boldsymbol{\theta}$. This result holds for a proper prior with $\int p(\boldsymbol{\theta})d\boldsymbol{\theta} = 1$, but also for a partially improper prior, with a flat prior $p(\tilde{\boldsymbol{\theta}}) \propto 1$ for a subvector $\tilde{\boldsymbol{\theta}}$ such as the regression coefficient $\boldsymbol{\beta}$ in SPMMs.

Unfortunately, the harmonic mean estimator can be very unstable because some of the likelihood terms $p(\boldsymbol{y}|\boldsymbol{\theta})$ will be close to zero. The *reciprocal importance estimator* (RI, Gelfand and Dey, 1994)

$$\hat{p}_{\mathrm{RI}}(\boldsymbol{y}) = \left[\frac{1}{S} \sum_{s=1}^{S} \frac{g(\boldsymbol{\theta}^{(s)})}{p(\boldsymbol{y}|\boldsymbol{\theta}^{(s)})p(\boldsymbol{\theta}^{(s)})} \right]^{-1},$$

where g is a density with known normalizing constant, generalizes the harmonic mean estimator. It is derived by multiplication with $g(\boldsymbol{\theta})/p(\boldsymbol{\theta})$. A reasonable choice for g is to match the posterior density as well as possible, for example through a Gaussian density with mean and covariance matrix estimated from the MCMC output or from an empirical Bayes estimate.

Bridge sampling (BS, Meng and Wong, 1996) is a further generalization. It is based on the identity

$$p(\boldsymbol{y}) = \frac{\int \alpha(\boldsymbol{\theta})p(\boldsymbol{y}|\boldsymbol{\theta})p(\boldsymbol{\theta})g(\boldsymbol{\theta})d\boldsymbol{\theta}}{\int \alpha(\boldsymbol{\theta})g(\boldsymbol{\theta})p(\boldsymbol{\theta}|\boldsymbol{y})d\boldsymbol{\theta}},$$

with appropriately chosen function $\alpha(\boldsymbol{\theta})$. With samples $\{\boldsymbol{\theta}^{(s)},\ s = 1, \ldots, S\}$ from the posterior $p(\boldsymbol{\theta}|\boldsymbol{y})$ and $\{\tilde{\boldsymbol{\theta}}^{(r)},\ r = 1, \ldots, R\}$ from $g(\boldsymbol{\theta})$ the BS estimator is

$$\hat{p}_{\mathrm{BS}} = \frac{\frac{1}{R}\sum_{r=1}^{R} \alpha(\tilde{\boldsymbol{\theta}}^{(r)})p(\boldsymbol{y}|\tilde{\boldsymbol{\theta}}^{(r)})p(\tilde{\boldsymbol{\theta}}^{(r)})}{\frac{1}{S}\sum_{s=1}^{S} \alpha(\boldsymbol{\theta}^{(s)})g(\tilde{\boldsymbol{\theta}}^{(s)})}.$$

The function $g(\boldsymbol{\theta})$ should again be chosen as a good approximation to the posterior, for example a matching Gaussian density. Different choices for $\alpha(\boldsymbol{\theta})$

define different BS estimators. The RI estimator is a special case with $\alpha(\boldsymbol{\theta}) = (p(\boldsymbol{y}|\boldsymbol{\theta})p(\boldsymbol{\theta}))^{-1}$. If

$$\alpha(\boldsymbol{\theta}) = (p(\boldsymbol{y}|\boldsymbol{\theta})p(\boldsymbol{\theta})g(\boldsymbol{\theta}))^{-1},$$

the BS estimator resembles the harmonic mean estimator. If

$$\alpha(\boldsymbol{\theta}) = (p(\boldsymbol{y}|\boldsymbol{\theta})p(\boldsymbol{\theta})g(\boldsymbol{\theta}))^{-1/2},$$

one gets the *geometric estimator* $\hat{p}_{\mathrm{G}}(\boldsymbol{y})$, see Lopes and West (2004).

Recursive schemes for obtaining an optimal estimate of $\alpha(\boldsymbol{\theta})$ are also available, see Kaufmann and Frühwirth-Schnatter (2002), Lopes and West (2004) and Miazhynskaia and Dorffner (2006).

Specifications of $g(\boldsymbol{\theta})$ and $\alpha(\boldsymbol{\theta})$ can be avoided by *Chib's candidate estimator* (CC, Chib, 1995) and its extension to output from MH algorithms (Chib and Jeliazkov, 2001). It is based on the identity

$$p(\boldsymbol{y}) = \frac{p(\boldsymbol{y}|\boldsymbol{\theta})p(\boldsymbol{\theta})}{p(\boldsymbol{\theta}|\boldsymbol{y})},$$

which is valid for any $\boldsymbol{\theta}$, in particular for a point $\boldsymbol{\theta}^*$ with high posterior density. With an estimate $\hat{p}(\boldsymbol{\theta}^*|\boldsymbol{y})$ of the unknown posterior, we obtain

$$\hat{p}_{\mathrm{CC}}(\boldsymbol{y}) = \frac{p(\boldsymbol{y}|\boldsymbol{\theta}^*)p(\boldsymbol{\theta}^*)}{\hat{p}(\boldsymbol{\theta}^*|\boldsymbol{y})}.$$

Chib (1995) and Chib and Jeliazkov (2001) partition the parameter vector into B blocks $\boldsymbol{\theta} = (\boldsymbol{\theta}_1, \ldots, \boldsymbol{\theta}_B)$, so that

$$p(\boldsymbol{\theta}^*|\boldsymbol{y}) = p(\boldsymbol{\theta}_1^*|\boldsymbol{y})p(\boldsymbol{\theta}_2^*|\boldsymbol{y}, \boldsymbol{\theta}_1^*) \cdot \ldots \cdot p(\boldsymbol{\theta}_B^*|\boldsymbol{y}, \boldsymbol{\theta}_1^*, \ldots, \boldsymbol{\theta}_{B-1}^*)$$

and estimate the factors from MCMC runs. For $B = 2$, the method is straightforward to implement but for $B > 2$ additional $B - 1$ Gibbs or MCMC runs are required, together with extensive bookkeeping.

Frühwirth-Schnatter and Wagner (2008) consider several new estimators of the marginal likelihood for non-Gaussian models, making use of the output of auxiliary mixture sampling. One of these estimators combines Chib's estimator with data augmentation, avoiding MH-algorithms.

The *Laplace–Metropolis estimator* (LM, Lewis and Raftery, 1997) combines (modified) Laplace approximation with MCMC output. It has the form

$$\hat{p}_{\mathrm{LM}}(\boldsymbol{y}) = (2\pi)^{d/2}|\hat{\boldsymbol{V}}|^{1/2}p(\boldsymbol{y}|\hat{\boldsymbol{\theta}})p(\hat{\boldsymbol{\theta}}),$$

where $\hat{\boldsymbol{\theta}}$ is the posterior mode or maximizes $p(\boldsymbol{y}|\boldsymbol{\theta})p(\boldsymbol{\theta})$ among the posterior draws, $\hat{\boldsymbol{V}}$ is the MCMC approximation to the posterior covariance matrix of $\boldsymbol{\theta}$, evaluated at $\hat{\boldsymbol{\theta}}$ and d is the dimension of $\boldsymbol{\theta}$. Modifications replace $\hat{\boldsymbol{\theta}}$ and $\hat{\boldsymbol{V}}$ with approximations obtained for example from empirical Bayes estimation.

As an alternative to direct methods, Carlin and Chib (1995) proposed a Gibbs sampler for stochastic search in the complete model-parameter product space. Computationally more efficient methods are based on (variations of) reversible jump MCMC algorithms, see Dellaportas, Forster and Ntzoufras (2002), Brooks, Giudici and Roberts (2003) and Green (2003). Han and Carlin (2001), Lopes and West (2004) and Miazhynskaia and Dorffner (2006) compare various direct and product space methods through simulation studies and applications to real data. A general conclusion is that product space methods work adequately, but, at least in current versions, may be difficult to tune and may only slowly converge to posterior model probabilities. The direct methods are much easier to program and to tune, and some of them, for example BS and LM estimators, turned out to be quite competitive.

Recently Congdon (2006) suggested a method for directly estimating posterior model probabilities and hence Bayes factors but not marginal likelihoods. The method uses only separate MCMC samples of different models and is therefore an easy postprocessing step for model choice. While the output is similar to joint model-parameter search methods, there is no mechanism proposing jumps between models. Robert and Marin (2008) show that the method produces a bias. Depending on the specific model, this bias may be small, but there are also situations where the divergence from the true posterior probability can become extreme. There is additional research needed to identify model classes where this approach works reliably.

Information criteria: AIC, BIC and DIC

Marginal likelihoods and Bayes factors are a formal approach to model selection in the M-closed situation, that is if one and only one of the competing models is true. In other circumstances, criteria based on prediction performance or on a trade-off between some measure of fit and complexity, measured through the effective number of parameters in the model, may be more appropriate. Information criteria such as Akaikes information criterion (AIC), the Bayesian information criterion (BIC), the deviance information criterion (DIC), and others are of the latter type, while cross validation and criteria based on predictive 'discrepancy measures' (Laud and Ibrahim, 1995; Gelfand and Ghosh, 1998) are motivated by prediction. Although derived under different perspectives, information criteria and predictive criteria are related: Stone (1974) showed the asymptotic equivalence of model comparison based on cross validation and AIC. The generalized cross validation criterion of Wahba (1990) sheds some light on the relation between the smoothing matrix in nonparametric regression and the effective number of parameters. Gelfand and Ghosh (1998) showed that their discrepancy measures have attractive interpretations as trade off between goodness of fit and prediction performance penalty terms. In the following, we focus on some selected information criteria.

BIC

The Bayesian (or Schwarz) information criterion (Schwarz, 1978) can be derived as an asymptotic first order approximation to the marginal likelihood given by

$$\text{BIC} = -2\log\{p(\boldsymbol{y}|\hat{\boldsymbol{\theta}})\} + \log(n)p,$$

for a model with p parameters and n observations. The derivation relies on the parametric Laplace approximation, assuming p is fixed while n tends to infinity, see for example, (Robert, 2001, Sec. 7.2.3). It is based on conventional Bayesian asymptotics (Bernardo and Smith, 2004), assuming that the prior is negligible compared to the likelihood when n becomes large. The BIC is a simple default method for model search within a parametric class of models, assuming independent observations. For longitudinal data models its use is problematic: A first difficulty is to determine the effective sample size in non-independent settings such as GLMMs (Pauler, Wakefield and Kass, 1999). Secondly, it is not appropriate in problems where the number of parameters increases with sample size n, such as in GAMMs, or in other non-standard asymptotic settings (Berger, Ghosh and Mukhopadhyay, 2003). Finally, the number of parameters is also not simple to assess when smoothness and regularization priors are involved. For example, the effective parameter count of a spline is strongly reduced when applying an informative smoothness prior such as a random-walk prior with a small variance. In this case, effective degrees of freedom should be used instead of a simple parameter count.

AIC

Although being a purely likelihood-based measure of model fit, Akaike's information criterion (AIC) may also be useful for model selection in particular in empirical Bayes contexts. The principal idea underlying the AIC is to select the model that leads to the best agreement with the true model. Assume that the data have been generated from the true model density $g(\boldsymbol{y})$ and that the true model should be approximated with a parametric model $p_{\boldsymbol{\theta}}(\boldsymbol{y}) = p(\boldsymbol{y}|\boldsymbol{\theta})$ where different values of $\boldsymbol{\theta}$ determine different models. The discrepancy between a candidate model $p(\boldsymbol{y}|\boldsymbol{\theta})$ and the true model $g(\boldsymbol{y})$ can be assessed in terms of the Kullback–Leibler-distance

$$\begin{aligned} K(p_{\boldsymbol{\theta}}, g) &= \int [\log(g(\boldsymbol{z})) - \log(p_{\boldsymbol{\theta}}(\boldsymbol{z}))]\, g(\boldsymbol{z}) d\boldsymbol{z} \\ &= \mathbb{E}_{\boldsymbol{z}} [\log(g(\boldsymbol{z})) - \log(p_{\boldsymbol{\theta}}(\boldsymbol{z}))]. \end{aligned}$$

where $\boldsymbol{z}$ is an independent replicate following the same distribution as $\boldsymbol{y}$. It can be shown that $K(f_{\psi}, g) \geq 0$ and $K(p_{\boldsymbol{\theta}}, g) = 0$ if and only if $p_{\boldsymbol{\theta}} = g$ almost everywhere and therefore the Kullback–Leibler-distance can in fact be interpreted as a distance measure.

Minimizing the Kullback–Leibler-distance leads to the model that best approximates the true model. However, in practice the parameters $\boldsymbol{\theta}$ of a model will have to be replaced by an estimate $\hat{\boldsymbol{\theta}} = \hat{\boldsymbol{\theta}}(\boldsymbol{y})$. To base the model decision on average properties of the models instead of a particular data realization, the Kullback–Leibler-distance is replaced by the expected Kullback–Leibler-distance

$$\mathbb{E}_{\boldsymbol{y}}[K(p_{\hat{\boldsymbol{\theta}}(\boldsymbol{y})}, g)] = \mathbb{E}_{\boldsymbol{y}}\left[\mathbb{E}_{\boldsymbol{z}}\left[\log(g(\boldsymbol{z})) - \log(p_{\hat{\boldsymbol{\theta}}(\boldsymbol{y})}(\boldsymbol{z}))\right]\right].$$

Moreover, the true model will not be known in practice but minimizing the expected Kullback–Leibler-distance is in fact equivalent to minimizing

$$-2\,\mathbb{E}_{\boldsymbol{y}}[\mathbb{E}_{\boldsymbol{z}}[\log(p_{\hat{\boldsymbol{\theta}}(\boldsymbol{y})}(\boldsymbol{z}))]] \tag{4.41}$$

(the expected *relative* Kullback–Leibler-distance) since $g(\boldsymbol{z})$ is independent of the model under consideration.

To actually implement model selection, the expectations in (4.41) have to be replaced by estimated quantities. The best estimate available is given by

$$-2\log(p_{\hat{\boldsymbol{\theta}}(\boldsymbol{y})}(\boldsymbol{y})) \tag{4.42}$$

but in contrast to (4.41) this estimate does no longer define a predictive quantity since both the estimate and the density of the candidate model are evaluated at the same data. This would induce a bias in model selection preferring more complex models. In fact, minimizing (4.42) simply leads to the model with the best fit in terms of the log-likelihood which will be the most complex model. Hence, a correction term penalizing complex models has to be introduced.

Under standard asymptotics, it can be shown that an asymptotically unbiased estimate for (4.41) is given by

$$\text{AIC} = -2\log(p_{\hat{\boldsymbol{\theta}}(\boldsymbol{y})}(\boldsymbol{y})) + 2d$$

where $d = \dim(\boldsymbol{\theta})$. This is exactly the commonly known definition of Akaike's information criterion (Akaike, 1973). Small sample corrected versions have also been proposed that modify the penalty term, see Sugiura (1978).

In particular in the context of empirical Bayes inference relying on mixed model methodology (as discussed in Section 4.2.1), it may be tempting to use AIC since it is also frequently delivered by statistical software packages. However, some difficulties arise: First of all, model selection can be based on either the conditional or the marginal perspective on the mixed model, leading to different expressions for the log-likelihood. In addition, the number of parameters is not simply given by the dimension of the model parameter vector since the random effects assumption induces a shrinkage effect. In particular, random effects parameters will typically not be counted as full parameters.

To make the discussion more concrete, consider the Gaussian mixed model

$$\boldsymbol{y} = \boldsymbol{U}\boldsymbol{\beta} + \boldsymbol{Z}\boldsymbol{\alpha} + \boldsymbol{\varepsilon}, \quad \boldsymbol{\alpha} \sim \text{N}(\boldsymbol{0}, \tau^2\boldsymbol{I}), \quad \boldsymbol{\varepsilon} \sim \text{N}(\boldsymbol{0}, \sigma^2\boldsymbol{I})$$

with fixed effects $\boldsymbol{\beta}$ and random effects $\boldsymbol{\alpha}$ arising either from the inclusion of subject-specific random effects or from reparameterizing penalized splines as random effects. Conditional on the random effects, this induces the distribution

$$\boldsymbol{y}|\boldsymbol{\alpha} \sim \mathrm{N}(\boldsymbol{U\beta} + \boldsymbol{Z\alpha}, \sigma^2 \boldsymbol{I})$$

for the responses. In contrast, the marginal distribution after integrating out the random effects $\boldsymbol{\alpha}$ is given by

$$\boldsymbol{y} \sim \mathrm{N}(\boldsymbol{U\beta}, \boldsymbol{V})$$

where $\boldsymbol{V} = \sigma^2 \boldsymbol{I} + \tau^2 \boldsymbol{ZZ}'$ denotes the marginal covariance matrix of $\boldsymbol{y}$.

Correspondingly, the conditional AIC (cAIC) is given by

$$\mathrm{cAIC} = -2l(\boldsymbol{y}|\hat{\boldsymbol{\alpha}}) + 2(\mathrm{df} + 1), \tag{4.43}$$

where

$$l(\boldsymbol{y}|\boldsymbol{\alpha}) = -\frac{N}{2}\log(\sigma^2) - \frac{1}{2\sigma^2}(\boldsymbol{y} - \boldsymbol{U\beta} - \boldsymbol{Z\alpha})'(\boldsymbol{y} - \boldsymbol{U\beta} - \boldsymbol{Z\alpha})$$

is the conditional log-likelihood, $N = \dim(\boldsymbol{y})$ denotes the total number of observations and

$$\mathrm{df} = \mathrm{tr}\left(\begin{pmatrix} \boldsymbol{U}'\boldsymbol{U} & \boldsymbol{U}'\boldsymbol{Z} \\ \boldsymbol{Z}'\boldsymbol{U} & \boldsymbol{Z}'\boldsymbol{Z} + \sigma^2/\tau^2 \boldsymbol{I} \end{pmatrix}^{-1} \begin{pmatrix} \boldsymbol{U}'\boldsymbol{U} & \boldsymbol{U}'\boldsymbol{Z} \\ \boldsymbol{Z}'\boldsymbol{U} & \boldsymbol{Z}'\boldsymbol{Z} \end{pmatrix}\right)$$

are the effective degrees of freedom of the model, i.e. the trace of the hat matrix. Hence, the parameter count in the conditional AIC is composed of a shrunken number of regression coefficients varying between $\dim(\boldsymbol{\beta})$ and $\dim(\boldsymbol{\beta}) + \dim(\boldsymbol{\alpha})$ depending on the value of the smoothing parameter σ^2/τ^2 and an additional count for the error variance σ^2.

As Vaida and Blanchard (2005) have shown, the conditional AIC is asymptotically unbiased for the expected relative Kullback–Leibler-distance if τ^2 is given. However, this is usually a very restrictive assumption and would preclude the use of the conditional AIC to decide on the inclusion of random effects or to test linear versus nonlinear modelling of a continuous covariate by comparing the two models

$$M_1 : \boldsymbol{y} \sim \mathrm{N}(\boldsymbol{U\beta} + \boldsymbol{Z\alpha}, \sigma^2 \boldsymbol{I}) \quad \text{and} \quad M_2 : \boldsymbol{y} \sim \mathrm{N}(\boldsymbol{U\beta}, \sigma^2 \boldsymbol{I})$$

that differ exactly in the value of τ^2. If τ^2 is replaced by a consistent estimate, one might hope that the conditional AIC would still be asymptotically unbiased but, as Greven and Kneib (2010) have shown, this is no longer the case. In contrast, the conditional AIC will select the more complex model including the random effect whenever $\hat{\tau}^2 > 0$. If the random effects variance is estimated to be zero, that is, if $\hat{\tau}^2 = 0$, the conditional AICs for both models coincide but the external criterion of parsimony would suggest selecting the simpler model in this case. Hence, choosing between the two models according to the conditional AIC would reduce to simply checking whether the estimated variance of the random effects is positive (use the model with random effects) or not (use the model excluding the random effects). In a simulation study, Greven and Kneib (2010) have shown that this unexpected behaviour leads to a false detection rate of random effects of approximately 20% for ML estimation and about 35% for REML estimation when the true model does not include random effects.

The reason why the conditional AIC as defined in (4.43) fails in practice is that uncertainty in estimating τ^2 is not taken appropriately into account in the determination of the degrees of freedom df. Liang, Wu and Zou (2008b) proposed a corrected version of the conditional AIC that rectifies this problem and adequately assesses the additional uncertainty in estimating df. While Liang, Wu and Zou (2008b) propose to compute the corrected effective parameter count based on numerical derivatives, Greven and Kneib (2010) derive an analytic expression that makes computation of the corrected conditional AIC feasible even for complex models and larger numbers of observations.

While the conditional AIC is suggested when the model under consideration will be used for predictions sharing the same random effects as the collected data (as for example in case of mixed model-based penalized spline smoothing), a marginal variant may be considered more appropriate when predicting observations with new random effects. The latter situation arises, for example, when predicting for new individuals in a longitudinal setting. From the marginal perspective on the mixed model, we obtain the marginal AIC

$$\text{mAIC} = -2l(\boldsymbol{y}) + 2(m+2)$$

where

$$l(\boldsymbol{y}) = -\frac{1}{2}\log(|\boldsymbol{V}|) - \frac{1}{2}(\boldsymbol{y} - \boldsymbol{U\beta})'\boldsymbol{V}^{-1}(\boldsymbol{y} - \boldsymbol{U\beta})$$

corresponds to the marginal log-likelihood and the parameter count consists of the fixed effects $m = \dim(\boldsymbol{\beta})$ as well as the two variance parameters τ^2 and σ^2. In this case, it is possible to simply count parameters since no shrinkage is present in determining $\hat{\boldsymbol{\beta}}$. Hence there seem to be no conceptual difficulties with the marginal AIC at first sight. However, when checking the assumptions underlying the asymptotic arguments for the AIC it becomes obvious that difficulties previously observed for likelihood ratio tests carry over to the marginal AIC. To be more specific, the parameter space of τ^2 is not an open interval but given by $[0, \infty)$ since we are interested in deciding whether $\tau^2 = 0$. Moreover, the observations $\boldsymbol{y}$ are not independent but marginally correlated when $\tau^2 > 0$.

To investigate the impact of these violations of assumptions, we take a closer look at the decision based on the marginal AIC. Model M_1 including the random effect ($\tau^2 > 0$) will be preferred over model M_2 ($\tau^2 = 0$) if $\text{mAIC}_1 < \text{mAIC}_2$. Inserting the definition of the marginal AIC yields

$$\begin{aligned}\text{mAIC}_1 < \text{mAIC}_2 &\Leftrightarrow -2l(\boldsymbol{y}|\hat{\boldsymbol{\beta}}_1, \hat{\tau}^2, \hat{\sigma}_1^2) + 2(m+2) < -2l(\boldsymbol{y}|\hat{\boldsymbol{\beta}}_2, 0, \hat{\sigma}_2^2) + 2(m+1) \\ &\Leftrightarrow 2l(\boldsymbol{y}|\hat{\boldsymbol{\beta}}_1, \hat{\tau}^2, \hat{\sigma}_1^2) - 2l(\boldsymbol{y}|\hat{\boldsymbol{\beta}}_2, 0, \hat{\sigma}_2^2) > 2 \qquad (4.44)\end{aligned}$$

where estimates in model M_1 are marked by subscript 1 and estimates from model M_2 are marked with subscript 2. In addition, we have explicitly added the parameter estimates to the log-likelihoods to stress the difference between the log-likelihoods from the two models. The left-hand side of (4.44) equals the test statistic of the likelihood ratio test of the hypothesis $H_0 : \tau^2 = 0$ versus $H_1 : \tau^2 > 0$. Under standard asymptotics and under the null hypothesis, this statistic would asymptotically follow a χ^2-distribution with one degree of freedom. In this case, the marginal AIC would have a type one error rate of approximately

$$\mathbb{P}(\chi_1^2 > 2) \approx 15.7\%$$

which has led to the common interpretation that AIC selects rather too many than too few parameters in variable selection problems since 15.7% is much larger than usual significance levels. However, as we have already seen in Section 4.2.1, page 207, the distribution of the likelihood ratio test statistic deviates from standard asymptotics since we are testing for a value on the boundary of the parameter space. This leads to a test distribution that is close to a mixture of a point mass in zero and a scaled χ_1^2-distribution. The point mass in zero corresponds to the probability to estimate $\hat{\tau}^2 = 0$ and is in many situations much larger than 50%. This high probability of estimating $\hat{\tau}^2$ to be exactly equal to zero induces a bias towards zero that turns into a positive bias of the marginal AIC for estimating the expected relative Kullback–Leibler-distance. As a consequence, the marginal AIC favours small models and in our example this would lead to a bias towards the model excluding the random effect (see Greven and Kneib, 2010, for more details and a rigorous proof). This is in contradiction to the classical perception that AIC selects rather too many than too few parameters and therefore using mAIC to select for example a confounder model will typically not lead to a useful solution. Moreover, the marginal AIC simply does not estimate the quantity of interest, that is, the expected relative Kullback–Leibler-distance.

In summary, both the conditional and the marginal AIC in their classical formulations do have conceptual flaws. While a correction is possible for the conditional AIC, such a correction is hard to imagine for the marginal AIC since the bias in mAIC depends on the true value of the random effects variance τ^2. Although we have exclusively dealt with the problem of selecting one single random effect, the conceptual difficulties carry over to more complex models and also do not disappear when replacing the asymptotic versions of AIC with finite sample corrected versions. However, both AIC variants should work fine for the selection of fixed effects.

DIC

Spiegelhalter *et al.* (2002) introduced a penalized fit criterion analogous to AIC and BIC, called *deviance information* criterion (DIC). The theoretical arguments supporting the definition of the DIC focus on generalized linear models, but the criterion is applicable to models with nonparametric components and random effects too. The DIC is based on the deviance

$$\mathrm{D}(\boldsymbol{\theta}) = -2\log p(\boldsymbol{y}|\boldsymbol{\theta}) + 2\log h(\boldsymbol{y}),$$

where $h(\boldsymbol{y})$ is some standardizing term, depending only on the data but not on unknown parameters. Therefore $h(\boldsymbol{y})$ does not affect model choice and may be set to 1 so that the second term vanishes. The *effective model dimension* or *number of parameters* is defined as

$$p_{\mathrm{D}} = \overline{\mathrm{D}(\boldsymbol{\theta})} - \mathrm{D}(\hat{\boldsymbol{\theta}}),$$

where $\overline{\mathrm{D}(\boldsymbol{\theta})}$ is the posterior mean deviance

$$\overline{\mathrm{D}(\boldsymbol{\theta})} = -2\,\mathbb{E}_{\boldsymbol{\theta}}\left[\log p(\boldsymbol{y}|\boldsymbol{\theta})|\boldsymbol{y}\right] + 2\log h(\boldsymbol{y}),$$

and $\hat{\boldsymbol{\theta}}$ is an estimate of $\boldsymbol{\theta}$ depending on $\boldsymbol{y}$. Mostly $\hat{\boldsymbol{\theta}}$ is taken to be the posterior mean but the posterior mode or median can also be justified. The posterior mean deviance can be regarded as a Bayesian measure of fit and p_{D} as a measure of model complexity. The DIC is then defined from this construction as

$$\begin{aligned}\mathrm{DIC} &= \overline{\mathrm{D}(\boldsymbol{\theta})} + p_{\mathrm{D}}\\ &= \mathrm{D}(\hat{\boldsymbol{\theta}}) + 2p_{\mathrm{D}}\\ &= 2\overline{\mathrm{D}(\boldsymbol{\theta})} - \mathrm{D}(\hat{\boldsymbol{\theta}}).\end{aligned}$$

If $D(\boldsymbol{\theta})$ is available in closed form, $\overline{\mathrm{D}(\boldsymbol{\theta})}$ can easily be estimated from MCMC output as the sample mean of simulated values $\mathrm{D}(\boldsymbol{\theta}^{(s)})$, $s = 1, \ldots, S$, and corresponding estimates for DIC and p_{D} can be obtained from their definition, plugging in the sample mean of the simulated values $\boldsymbol{\theta}^{(s)}$. Thus, an estimate of the DIC can be conveniently computed as an add-on of MCMC output, making the criterion quite popular in applications. Models with lower values of $\overline{\mathrm{D}(\boldsymbol{\theta})}$ or $\mathrm{D}(\hat{\boldsymbol{\theta}})$ combined with small values of p_{D} will be preferred by the DIC. The form $\mathrm{DIC} = \mathrm{D}(\hat{\boldsymbol{\theta}}) + 2p_{\mathrm{D}}$ shows that DIC can be seen as a Bayesian version of AIC and may underpenalize model complexity. Pourahmadi and Daniels (2002) therefore suggest

$$\mathrm{DIC}^* = \mathrm{D}(\hat{\boldsymbol{\theta}}) + \log(n)p_{\mathrm{D}}$$

as a criterion analogous to the BIC in an application to longitudinal data, where n denotes the number of individuals.

As already pointed out in Spiegelhalter *et al.* (2002), the fact that p_D depends on the parametrization of the model is one of the problems with the DIC in more complex hierarchical models. It can only be solved by specifying the level of the hierarchy the model focuses on. For example, when analysing probit models with linear or semiparametric predictors directly or using auxiliary latent Gaussian models, the complexity measures will obviously differ. This is directly seen from the relation

$$p(\boldsymbol{y}|\boldsymbol{\theta}) = \int p(\boldsymbol{y}, \boldsymbol{l}|\boldsymbol{\theta}) d\boldsymbol{l},$$

where $\boldsymbol{l}$ are auxiliary latent variables. Similar problems arise when $\boldsymbol{l}$ are missing data and, in principle, with all models involving latent variables. Celeux *et al.* (2006) discuss these issues with a view towards mixture models and models with random effects. Depending on the focus of models and analyses, they derive a number of DIC variants, depending on the parameterization and on which parameters are integrated out or not. Apart from models with auxiliary latent variables, we consider all 'random effects', in particular 'random' basis function coefficients for nonlinear effects, to be a focus of the models. Then, we choose $\boldsymbol{\theta}$ as the vector of all parameters, corresponding to the conditional DIC in Celeux *et al.* (2006), as our standard option.

In the discussion of Spiegelhalter *et al.* (2002) a nice interpretation of p_D and model complexity is given by Peter Green: Consider a Gaussian model

$$\boldsymbol{y} = \boldsymbol{f}_1 + \boldsymbol{f}_2 + \ldots + \boldsymbol{f}_r + \boldsymbol{\varepsilon},$$

where all effects $\boldsymbol{f}_l$ are Gaussian and $\text{Cov}(\boldsymbol{\varepsilon})$ is nonsingular. The effects represent all the various terms of the model: fixed effects with priors, random effects with different structures and errors. Then for any $l = 1, \ldots, r$, the sum of the first l terms may be considered as focus and the rest, including $\boldsymbol{\varepsilon}$, as noise. He shows that the corresponding effective model dimension is

$$p_D(l) = \text{tr}\left\{\text{Cov}(\boldsymbol{y})^{-1}\,\text{Cov}\left(\sum_{h=1}^{l} \boldsymbol{f}_h\right)\right\},$$

where

$$\text{Cov}(\boldsymbol{y}) = \text{Cov}\left(\sum_{h=1}^{l} \boldsymbol{f}_h + \boldsymbol{\varepsilon}\right)$$

is the marginal covariance matrix of $\boldsymbol{y}$. Thus, p_D generates a decomposition of the model complexity into nonnegative terms attributable to the levels $l = 1, \ldots, r$.

4.5.2 Predictive methods for model assessment

Bayes factors, marginal likelihoods and model selection criteria are global measures for choosing between models. Predictive methods offer a more flexible approach to assess or criticize various aspects of a model, ranging from graphical goodness-of-fit checks to model assessment via global discrepancy measures or scoring rules. They are used diagnostically to identify deficiencies of a model and can also help in model comparison. In the following we will describe posterior predictive goodness-of-fit checks as well as exact and approximate cross validatory assessment. The main diagnostic tools considered are PIT (probability integral transform) and BOT (Box ordinate transform) values for assessing probabilistic calibration, proper scoring rules and some custom test statistics or discrepancy measures.

Alternative predictive distributions

Predictive methods can be based on alternative predictive distributions. In generic notation, the marginal density

$$p(\boldsymbol{y}) = \int p(\boldsymbol{y}|\boldsymbol{\theta})p(\boldsymbol{\theta})d\boldsymbol{\theta}$$

of the observables $\boldsymbol{y}$ might be interpreted as a *prior predictive* density. Typically, however, some kind of posterior predictive distribution, given all or part of the data $\boldsymbol{y}$, will be employed.

The *posterior predictive* density

$$p(\boldsymbol{y}^{\text{pred}}|\boldsymbol{y}) = \int p(\boldsymbol{y}^{\text{pred}}|\boldsymbol{\theta})p(\boldsymbol{\theta}|\boldsymbol{y})d\boldsymbol{\theta} \tag{4.45}$$

is the predictive density of new independent random variables $\boldsymbol{y}^{\text{pred}}$ under the model, given the observed data $\boldsymbol{y}$. As a special case,

$$p(\boldsymbol{y}_i^{\text{pred}}|\boldsymbol{y}) = \int p(\boldsymbol{y}_i|\boldsymbol{\theta})p(\boldsymbol{\theta}|\boldsymbol{y})d\boldsymbol{\theta}$$

is the posterior predictive density for an individual scalar or multivariate observation. Comparing, for example, this posterior predictive density with the actual observation $\boldsymbol{y}_i$ introduces a form of conservatism into model checking, because the data $\boldsymbol{y}_i$ is used twice: For computing the posterior $p(\boldsymbol{\theta}|\boldsymbol{y})$ and for checking any divergence between $\boldsymbol{y}_i$ and its posterior predictive density.

Cross validation predictive densities aim at avoiding this conservatism. Let $\boldsymbol{y}_{\text{sub}}$ denote a subset of the data $\boldsymbol{y}$ (the 'validation data') and $\boldsymbol{y}_{\backslash\text{sub}}$ the remaining complementary subset of 'training data', excluding $\boldsymbol{y}_{\text{sub}}$ from $\boldsymbol{y}$. Then

$$p(\boldsymbol{y}_{\text{sub}}^{\text{pred}}|\boldsymbol{y}_{\backslash\text{sub}}) = \int p(\boldsymbol{y}_{\text{sub}}^{\text{pred}}|\boldsymbol{\theta}, \boldsymbol{y}_{\backslash\text{sub}})p(\boldsymbol{\theta}|\boldsymbol{y}_{\backslash\text{sub}})d\boldsymbol{\theta} \tag{4.46}$$

is a cross validation predictive density that can be compared to the observed data $\boldsymbol{y}_{\text{sub}}$. Dividing $\boldsymbol{y}$ repeatedly into k training and test data leads to k-fold cross validation. Leave-one-out cross validation is the special case where $\boldsymbol{y}_{\setminus \text{sub}} = \boldsymbol{y}_{\setminus i}$ consists of all observations except $\boldsymbol{y}_i$. The leave-one-out predictive densities

$$p(\boldsymbol{y}_i^{\text{pred}}|\boldsymbol{y}_{\setminus i}) = \int p(\boldsymbol{y}_i^{\text{pred}}|\boldsymbol{\theta}, \boldsymbol{y}_{\setminus i}) p(\boldsymbol{\theta}|\boldsymbol{y}_{\setminus i}) d\boldsymbol{\theta} \tag{4.47}$$

are sometimes called *conditional predictive ordinates* (CPOs). For conditionally independent observations $p(\boldsymbol{y}_{\text{sub}}^{\text{pred}}|\boldsymbol{\theta}, \boldsymbol{y}_{\setminus \text{sub}})$ and $p(\boldsymbol{y}_i^{\text{pred}}|\boldsymbol{\theta}, \boldsymbol{y}_{\setminus i})$ simplify to $p(\boldsymbol{y}_{\text{sub}}^{\text{pred}}|\boldsymbol{\theta})$ and $p(\boldsymbol{y}_i^{\text{pred}}|\boldsymbol{\theta})$, respectively. The predictive distributions $p(\boldsymbol{y}_i^{\text{pred}}|\boldsymbol{y}_{\setminus i})$ can be compared to the actually observed $\boldsymbol{y}_i$ in various ways (Gelfand, 1996).

Another special case are *intrinsic predictive* densities $p(\boldsymbol{y}_{\text{int}}^{\text{pred}}|\boldsymbol{y}_{\setminus \text{int}})$, where $\boldsymbol{y}_{\setminus \text{int}}$ is the smallest subset of $\boldsymbol{y}$ such that $p(\boldsymbol{\theta}|\boldsymbol{y}_{\setminus \text{int}})$ is proper. They have been suggested for constructing *intrinsic Bayes factors* to replace the Bayes factor when improper priors are used (Berger and Pericchi, 1996). Related ideas are formalized through partial and conditional predictive densities (Bayarri and Berger, 2000).

Because full cross validation predictive densities become computationally extremely intense for larger data sets, alternatives that imitate or approximate cross validation predictive densities are desirable, see Marshall and Spiegelhalter (2003, 2007) and the remainder of this section for suggestions in the context of hierarchical models.

Assessing calibration: PIT and BOT

Informally, calibration requires that the observed data are practically indistinguishable from random draws from the predictive density. This can be interpreted as some form of statistical consistency between the predictive distribution and the realized observations. Gneiting, Balabdaoui and Raftery (2007) define various modes of calibration more formally. Let F denote the cumulative distribution function (CDF) corresponding to some predictive density p, that is $\boldsymbol{y}^{\text{pred}} \sim F$ for some typical scalar or multivariate prediction random variable $\boldsymbol{y}^{\text{pred}}$. Assume now that a corresponding observation $\boldsymbol{y}$ is a realization from $\boldsymbol{y} \sim G$, where G is the CDF corresponding to the (true) data generating mechanism. In our framework and in many applications, there will be multiple conditionally independent prediction variables $\boldsymbol{y}_1^{\text{pred}}, \ldots, \boldsymbol{y}_n^{\text{pred}}$ and $\boldsymbol{y}_1, \ldots, \boldsymbol{y}_n$, with corresponding predictive CDFs $F_1, \ldots, F_n$ and data generating CDFs $G_1, \ldots, G_n$. For longitudinal data, $\boldsymbol{y}_1, \ldots, \boldsymbol{y}_n$ will usually consist of the vectors $\boldsymbol{y}_i = (y_{i1}, \ldots, y_{in_i})'$, $i = 1, \ldots, n$, of repeated individual observations. In the scalar case, $y_1, \ldots, y_n$ could be a set of individual scalar outcomes y_{ij} (after reindexing). Then Gneiting, Balabdaoui and Raftery (2007, p. 247) define (probabilistic) calibration as follows: If for all probabilities $\pi \in (0, 1)$

$$\frac{1}{n} \sum_{i=1}^{n} G_i(F_i^{-1}(\pi)) \to \pi$$

almost surely for $n \to \infty$, then the sequence $\{F_i, i = 1, 2, \ldots\}$ is calibrated relative to $\{G_i, i = 1, 2, \ldots\}$. Note, however, that this formal definition is only sensible for CDFs that are continuous and invertible. If $F_i \equiv G_i$ for all i, then the predictions are trivially calibrated. The probability integral transform (PIT) and the Box ordinate transform (BOT) are calibration checks for univariate and multivariate observations respectively that we will focus on in the following. Other tools are described in Gneiting, Balabdaoui and Raftery (2007) and, for multivariate observations, in Gneiting *et al.* (2008).

The *probability integral transform* was introduced by Dawid (1984) for scalar predictions $y^{\text{pred}} \sim F$ and observations y. It is defined as

$$\text{PIT}(F, y) = F(y), \tag{4.48}$$

which is simply the value that the predictive CDF attains at the observation y. If F and the data generating CDF G coincide, and the predictive distribution is continuous, then

$$\text{PIT} = F(y) \sim \text{U}(0, 1).$$

Given independent scalar observations $y_1, \ldots, y_n$, the empirical distribution of the PIT values $F_1(y_1), \ldots, F_n(y_n)$ can be compared against the standard uniform distribution, usually in form of a PIT histogram. If predictive CDFs are not available analytically, as for example in SPMMs for longitudinal data, they can be estimated by empirical CDFs $\hat{F}_i$, using MCMC samples $y_i^{\text{pred},s}$, $s = 1, \ldots, S$ from the predictive distributions. Gneiting, Balabdaoui and Raftery (2007) show that the convergence of the PIT histogram to the uniform density is equivalent to the formal definition of calibration given above.

The PIT histogram is typically used informally as a diagnostic tool. For (almost) perfectly calibrated predictive densities, that is $F_i \approx G_i$, an almost uniform PIT histogram is expected. Characteristic deviations from uniformity indicate uncalibrated predictive densities. For example, underdispersed predictive densities underestimate the variability of the data, that is the data generating density has heavier tails than the predictive density. This leads to a u-shaped form of the PIT histogram. Vice versa, overdispersed predictive densities lead to hump-shaped forms. Both deviations from uniformity are illustrated in Fig. 4.30 for simulated data.

For discrete observations, PIT values are no longer uniform even in case of perfect calibration. To remedy this, several authors have proposed a randomized PIT (e.g. Frühwirth-Schnatter, 1996; Brockwell, 2007) or a non-randomized modification for count data and ordered categorical data (Czado, Gneiting and Held, 2009). A simple modification is

$$\text{PIT}(F, x) = \mathbb{P}(y^{\text{pred}} < x) + 0.5\mathbb{P}(y^{\text{pred}} = x)$$

which has been suggested for example by Marshall and Spiegelhalter (2003).

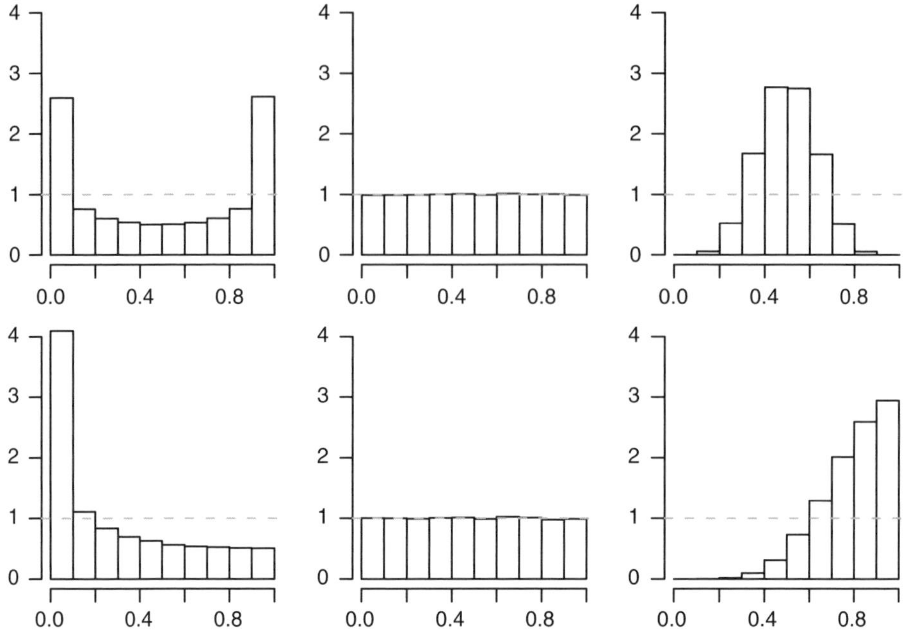

Fig. 4.30 PIT (top row) and BOT (bottom row) histograms for underdispersed (left), correct (middle) and overdispersed (right) forecasts.

For multivariate observations, such as response vectors $\boldsymbol{y}_i = (y_{i1}, \ldots, y_{in_i})'$, $i = 1, \ldots, n$, in longitudinal studies, the *Box ordinate transform* (BOT) has been proposed as a diagnostic tool for checking calibration, see for example Gneiting *et al.* (2008). For continuous data, it is defined as the tail probability

$$\mathrm{BOT}(p, \boldsymbol{y}) = \mathbb{P}(p(\boldsymbol{y}^{\mathrm{pred}}) \leq p(\boldsymbol{y})), \tag{4.49}$$

where p is the predictive density and $\mathbb{P}$ is the corresponding predictive distribution. It is easy to see that $\mathrm{BOT}(p, \boldsymbol{y})$ has a standard uniform distribution if $\boldsymbol{y}$ is distributed according to p.

Characteristic deviations of BOT histograms from uniformity indicate uncalibrated predictive densities as illustrated in (Gneiting *et al.*, 2008, Fig. 5) and Fig. 4.30. BOT histograms for underdispersed forecasts are typically right-skewed: the data generating distribution puts large probability mass on areas where the predictive density has very low values. On the other hand, BOT histograms for overdispersed forecasts are typically left-skewed.

For discrete observations a suitably randomized version of the BOT remains uniform (e.g. Brockwell, 2007) if predictive and data generating distributions coincide. However, power is reduced.

Proper scoring rules

Scoring rules can be used as omnibus performance measures for assessing both calibration and sharpness, where the latter refers to the concentration of predictive distributions. For a predictive CDF F and an observation $\boldsymbol{y}$, a score $S(F, \boldsymbol{y})$ is assigned as penalty. For $\boldsymbol{y} \sim G$,

$$S(F, G) = \mathbb{E}_G(S(F, \boldsymbol{y}))$$

is the expected score under the data generating distribution G. A scoring rule with

$$S(G, G) \leq S(F, G)$$

and $S(G, G) = S(F, G)$ if and only if $F = G$ is said to be strictly proper, see Gneiting and Raftery (2007) for definitions as well as examples of scoring rules. We will now present some common, strictly proper scoring rules.

The *continuous ranked probability score* (CRPS) can be used for a scalar observation y. It is defined as

$$\text{CRPS}(F, y) = \int_{-\infty}^{+\infty} \{F(y^{\text{pred}}) - \mathbb{1}_{[y,+\infty]}(y^{\text{pred}})\}^2 dy^{\text{pred}}, \tag{4.50}$$

which is the squared L_2-distance between the predictive distribution F and the CDF of the point mass δ_y. The CRPS is strictly proper for predictive distributions with finite expectation. The CRPS can also be written as

$$\text{CRPS}(F, y) = \mathbb{E}_F \left|y^{\text{pred}} - y\right| - \frac{1}{2}\,\mathbb{E}_F \left|y^{\text{pred}} - y^*\right|, \tag{4.51}$$

where y^{pred} and y^* are two independent copies from F and y is a fixed observation, see Gneiting and Raftery (2007). This representation of the CRPS can be interpreted as a decomposition into two terms representing deviations from calibration (consistency between the forecast y^{pred} and the observation y) and sharpness (concentration of the predictive distributions). This decomposition is analogous to the decomposition of the MSE in terms of bias and variance.

For discrete observations, the integral is replaced by the corresponding sum, see Czado, Gneiting and Held (2009) in the context of count data. For a categorical observation $y = c$, $c \in \{1, \dots, C\}$, the CRPS becomes the *quadratic* or *Brier score*

$$\text{BS}(\boldsymbol{p}, c) = \sum_{j=1}^{C} (\delta_{cj} - p_j)^2,$$

where $\boldsymbol{p} = (p_1, \dots, p_C)'$ is the predictive probability function for y^{pred}, and $\delta_{cj} = 1$ if $c = j$ and $\delta_{cj} = 0$ otherwise.

The representation (4.51) of the CRPS can be generalized to a multivariate observation $\boldsymbol{y}$, defined as the *energy score* (ES)

$$\mathrm{ES}(F, \boldsymbol{y}) = \mathbb{E}\left\|\boldsymbol{y}^{\mathrm{pred}} - \boldsymbol{y}\right\| - \frac{1}{2}\mathbb{E}\left\|\boldsymbol{y}^{\mathrm{pred}} - \boldsymbol{y}^*\right\|,$$

see Gneiting *et al.* (2008).

Perhaps most widely known is the *logarithmic score*

$$\mathrm{LogS}(F, \boldsymbol{y}) = -\log p(\boldsymbol{y}),$$

which can be used for continuous as well as discrete scalar or multivariate data. It is closely related to the *predictive deviance*

$$\mathrm{dev}(F, \boldsymbol{y}) = -2\log p(\boldsymbol{y}) + 2\log h(\boldsymbol{y}),$$

where $h(\boldsymbol{y})$ is a standardizing term, depending only on the data (Spiegelhalter *et al.*, 2002). The logarithmic score is also closely related to the Bayes factor

$$B_{kl} = \frac{p(\boldsymbol{y}|M_k)}{p(\boldsymbol{y}|M_l)}$$

for comparing two models M_k and M_l. Taking the marginal likelihoods (4.40) as predictive densities with CDFs F_k and F_l, then the log Bayes factor is the difference

$$\log B_{kl} = \mathrm{LogS}(F_k, \boldsymbol{y}) - \mathrm{LogS}(F_l, \boldsymbol{y})$$

of the two scores.

For conditionally independent observations $\boldsymbol{y}_1, \ldots, \boldsymbol{y}_n$ and corresponding predictive CDFs $F_1, \ldots, F_n$ the scores $S(F_i, \boldsymbol{y}_i)$, $i = 1, \ldots, n$, are combined into a *mean score*

$$\overline{S(F, \boldsymbol{y})} = \frac{1}{n}\sum_{i=1}^{n} S(F_i, \boldsymbol{y}_i). \tag{4.52}$$

Comparison with another model, producing predictive CDFs E_i, say, is based on comparing $\overline{S(F, \boldsymbol{y})}$ and $\overline{S(E, \boldsymbol{y})}$. This is theoretically well-founded because the mean score of single proper scores is again proper. For a formal comparison a permutation test may be used.

Custom summary statistics

There are many alternative proposals for omnibus measures as well as informal graphical tools and specific test quantities or discrepancy measures, tailored to check specific aspects of a working model, see Gelman *et al.* (1995) for an overview in the context of posterior predictive checking. Given the data $\boldsymbol{y}$, scalar test statistics $T(\boldsymbol{y})$ or discrepancy measures $T(\boldsymbol{y}; \boldsymbol{\theta})$, also depending on the unknown parameter $\boldsymbol{\theta}$, are chosen or designed to reflect aspects of the model that are relevant to the scientific purposes to which the inference will be applied (Gelman *et al.*, 1995, p. 172). For example, in a longitudinal setting, this could be the range of data points or some *ad hoc* check for symmetry, etc. The basic tools are *Bayesian p-values* or *tail probabilities* of the form

$$\mathbb{P}(T(\boldsymbol{y}^{\text{pred}}; \boldsymbol{\theta}) \geq T(\boldsymbol{y}; \boldsymbol{\theta})), \tag{4.53}$$

where $\mathbb{P}$ is the corresponding predictive distribution. In practice, samples from the predictive distributions are drawn to estimate these probabilities.

Regression-diagnostic type checks can be based on some form of (standardized or transformed) residuals $(y_l - \mathbb{E}(y_l^{\text{rep}}))/\sqrt{\text{Var}(y_l^{\text{rep}})}$, where $\mathbb{E}$ and Var refer to the predictive density and y_l refers to a selected scalar response. The sum of squared standardized residuals gives the χ^2-discrepancy

$$\chi^2 = \sum_l \frac{(y_l - \mathbb{E}(y_l^{\text{rep}}))^2}{\text{Var}(y_l^{\text{rep}})}.$$

Monte Carlo estimation of predictive measures

In practice, predictive densities and CDFs as well as corresponding expectations etc. appearing in the definition of PIT, BOT, scores and summary statistics are generally unknown and will be estimated from MCMC samples $\boldsymbol{y}_i^{\text{pred},s}$, $i = 1, \ldots, n$, $\boldsymbol{\theta}^{(s)}$, $s = 1, \ldots, S$.

For PIT values $\text{PIT}(F_i, \boldsymbol{y}_i) = F_i(y_i)$, the CDFs F_i are estimated by empirical CDFs $\hat{F}_i$ using samples $y_i^{\text{pred},1}, \ldots, y_i^{\text{pred},S}$ from the predictive CDFs F_i. In the discrete case, the probabilities in $\mathbb{P}(y_i^{\text{pred}} < y_i) + 0.5\mathbb{P}(y_i^{\text{pred}} = y_i)$ are estimated through relative frequencies.

For BOT values (4.49), we need estimates $\hat{p}_i(\boldsymbol{y})$ for predictive densities $p_i(\boldsymbol{y})$ evaluated at $\boldsymbol{y} = \boldsymbol{y}_i^{\text{pred},s}$ and $\boldsymbol{y}_i$, $i = 1, \ldots, n$. For full posterior predictive densities (4.45) as well as cross validation predictive densities (4.46) and (4.47) the predictive density is a mixture of the form

$$p_i(\boldsymbol{y}) = \int p_i(\boldsymbol{y}|\boldsymbol{\theta}) p_i(\boldsymbol{\theta}) d\boldsymbol{\theta},$$

where $p_i(\boldsymbol{y}|\boldsymbol{\theta})$ are known densities. Then the Monte Carlo estimate

$$\hat{p}_i(\boldsymbol{y}) = \frac{1}{S} \sum_{s=1}^{S} p_i(\boldsymbol{y}|\boldsymbol{\theta}^{(s)})$$

is a Rao–Blackwell estimate for $p_i(\boldsymbol{y})$ if the samples $\boldsymbol{\theta}^{(s)}$ are (at least approximately) from $p_i(\boldsymbol{\theta})$. Evaluation of $\hat{p}_i(\boldsymbol{y})$ for the samples $\boldsymbol{y}_i^{\text{pred},s}$ as well as observations $\boldsymbol{y}_i$ and estimating $\mathbb{P}$ through relative frequencies gives a BOT estimate.

The same procedure can be used to estimate log scores $-\log p_i(\boldsymbol{y}_i)$. For the CRPS, the predictive CDF F_i is replaced by its empirical counterpart $\hat{F}_i$ obtained from $\boldsymbol{y}_i^{\text{pred},1}, \dots, \boldsymbol{y}_i^{\text{pred},S}$. The evaluation of the integral in (4.50) is easy, because the integral is a step function. Alternatively, one can also use an efficient Monte Carlo estimate for the energy score proposed by Gneiting *et al.* (2008). It is given by

$$\widehat{\text{ES}}(F_i, \boldsymbol{y}_i) = \frac{1}{S}\sum_{s=1}^{S}\left\|\boldsymbol{y}_i^{\text{pred},s} - \boldsymbol{y}_i\right\| - \frac{1}{2(S-1)}\sum_{s=1}^{S-1}\left\|\boldsymbol{y}_i^{\text{pred},s+1} - \boldsymbol{y}_i^{\text{pred},s}\right\|,$$

where $\|\cdot\|$ denotes the Euclidian norm.

For custom summary statistics and discrepancy measures, tail probabilities or p-values (4.53) as well as residuals are estimated through their empirical counterparts obtained directly from MCMC output. For example, p-values are estimated as averages

$$\frac{1}{S}\sum_{s=1}^{S}\mathbb{1}(T(\boldsymbol{y}_i^{\text{pred},s}; \boldsymbol{\theta}^{(s)}) \geq T(\boldsymbol{y}_i; \boldsymbol{\theta}^{(s)})),$$

$i = 1, \dots, n$.

In the following, we describe sampling based on MCMC output for posterior predictive goodness-of-fit checking, exact cross validatory predictive assessment and an approximate cross validatory approach.

Posterior predictive goodness-of-fit assessment

Posterior predictive model checking based on replicate sampling—without omitting cases—is outlined in Gelman *et al.* (1995) and Gelman, Meng and Stern (1996). The general idea goes back to Rubin (1984) who suggested simulating replicate data sets $\boldsymbol{y}^{\text{rep}}$ from the posterior predictive distribution and comparing them with the observed data. Any significant difference indicates a possible deficiency of the model. Discrepancies between model and data can be assessed with more informal graphical tools, for example displaying data or residuals, or through custom summary statistics.

The basic tool for model assessment using test statistics or discrepancy measures are tail-area probabilities or p-values. The *classical p-value* is defined as

$$p_c = \mathbb{P}\{T(\boldsymbol{y}^{\text{rep}}, \boldsymbol{\theta}) \geq T(\boldsymbol{y}, \boldsymbol{\theta})|\boldsymbol{\theta}\},$$

where the probability is computed with respect to the distribution of $\boldsymbol{y}^{\text{rep}}$ with $\boldsymbol{\theta}$ fixed. This fixed parameter value may be $\boldsymbol{\theta} = \boldsymbol{\theta}_0$ corresponding to a 'point null hypothesis', or an estimate $\boldsymbol{\theta} = \hat{\boldsymbol{\theta}}$ obtained under a 'composite null hypothesis' specifying the model under consideration. In the latter case p_c may be termed a *plug-in p-value*.

The *Bayesian posterior predictive p-value* for a discrepancy measure is defined as

$$p_{\text{post}} = \mathbb{P}\{T(\boldsymbol{y}^{\text{rep}}, \boldsymbol{\theta}) \geq T(\boldsymbol{y}, \boldsymbol{\theta})|\boldsymbol{y}\},$$

where the probability is taken over the *joint* posterior distribution $p(\boldsymbol{y}^{\text{rep}}, \boldsymbol{\theta}|\boldsymbol{y})$ of $\boldsymbol{y}^{\text{rep}}$ and $\boldsymbol{\theta}$. Because $p(\boldsymbol{y}^{\text{rep}}|\boldsymbol{\theta}, \boldsymbol{y}) = p(\boldsymbol{y}^{\text{rep}}|\boldsymbol{\theta})$ we obtain

$$p_{\text{post}} = \iint \mathbb{1}_{T(\boldsymbol{y}^{\text{rep}}, \boldsymbol{\theta}) \geq T(\boldsymbol{y}, \boldsymbol{\theta})} p(\boldsymbol{y}^{\text{rep}}|\boldsymbol{\theta}) p(\boldsymbol{\theta}|\boldsymbol{y}) d\boldsymbol{y}^{\text{rep}} d\boldsymbol{\theta}.$$

This explicit form shows how to estimate p_{post} as a by-product of MCMC algorithms:

Algorithm 4.7: Bayesian posterior predictive p-values

i. Draw simulated values $\boldsymbol{\theta}^{(s)}$, $s = 1, \ldots, S$ from the posterior $p(\boldsymbol{\theta}|\boldsymbol{y})$, using some MCMC algorithm.
ii. Draw one $\boldsymbol{y}^{\text{rep},s}$ for each $\boldsymbol{\theta}^{(s)}$ from the predictive distribution $p(\boldsymbol{y}^{\text{rep}}|\boldsymbol{\theta}^{(s)})$, $s = 1, \ldots, S$.
iii. Compute $T(\boldsymbol{y}^{\text{rep},s}, \boldsymbol{\theta}^{(s)})$ and $T(\boldsymbol{y}, \boldsymbol{\theta}^{(s)})$, $s = 1, \ldots, S$.
iv. Estimate p_{post} as the relative frequency $\hat{p}_{\text{post}}$ of replicated discrepancy measures and realized discrepancies for which $T(\boldsymbol{y}^{\text{rep},s}, \boldsymbol{\theta}^{(s)}) \geq T(\boldsymbol{y}, \boldsymbol{\theta}^{(s)})$.

In addition to only computing $\hat{p}_{\text{post}}$, it may also be useful to compare replicated and realized discrepancies more informally, for example by displaying the pairs $\{T(\boldsymbol{y}, \boldsymbol{\theta}^{(s)}), T(\boldsymbol{y}^{\text{rep},s}, \boldsymbol{\theta}^{(s)})\}$ in a scatter plot.

Informal posterior predictive graphical model checks and posterior predictive p-values for custom summary statistics are a very flexible tool for model assessment. Such quantities could, for example, compare sample moments, quantiles or even just a part of the observations to their model-based counterparts. As an extreme case, one could estimate posterior predictive confidence intervals and compare with an individual observation. In the latter case it becomes quite apparent why posterior predictive checks may tend to being conservative: The full posterior $p(\boldsymbol{\theta}|\boldsymbol{y})$ is used to generate samples $\boldsymbol{\theta}^{(s)}$ which are likely under the model. Then the replicated data are generated from $p(\boldsymbol{y}^{\text{rep}}|\boldsymbol{\theta}^{(s)})$. The degree of this induced conservatism will depend on the specific aspects of a model and the discrepancy measures chosen for posterior predictive checking.

For linear mixed models, Kato and Hoijtink (2004) investigate the performance of plug-in p-values and posterior predictive p-values for checking homogeneity versus inhomogeneity of error variances, and fixed versus random slopes. They conclude that, with appropriately defined test statistics and discrepancy measures, both types of p-values are useful tools for checking the homogeneity assumptions for error variances or slopes. On the other hand, Sinharay and Stern (2003) demonstrate that it is difficult to detect non-normality of random effects distributions. Therefore posterior predictive checking is useful as a tool for goodness-of-fit assessment but, in a strict sense, not for predictive assessment.

Exact cross validatory predictive assessment

For avoiding conservatism when checking whether small subsets $\boldsymbol{y}_{\text{sub}}$ of the data or individual observations $\boldsymbol{y}_i$ are not well fitted or are outlying, it seems conceptually reasonable to compare them with their corresponding cross validation predictive densities (4.46) or (4.47). Gelfand (1996) suggests various ways for such comparisons.

In addition to residuals and custom summary statistics, one may also look at

$$\text{CPO}_i = p(\boldsymbol{y}_i|\boldsymbol{y}_{\backslash i}),$$

the so-called *conditional predictive ordinates* (CPOs, Gelfand, Dey and Chang, 1992; Chen, Shao and Ibrahim, 2000). Small CPOs indicate observations $\boldsymbol{y}_i$ that do not support the model. An interesting property of CPOs is the relation

$$\log \text{CPO}_i = \log p(\boldsymbol{y}) - \log p(\boldsymbol{y}_{\backslash i}).$$

It can be interpreted as the contribution of $\boldsymbol{y}_i$ to the marginal likelihood of all observations compared to the marginal likelihood $p(\boldsymbol{y}_{\backslash i})$ of the remaining observations. The CPOs can also be used to define pseudo Bayes factors

$$\begin{aligned}\text{PsBF}(M_1, M_2) &= \prod_{i=1}^{n} p(\boldsymbol{y}_i|\boldsymbol{y}_{\backslash i}, M_1) \,/\, p(\boldsymbol{y}_i|\boldsymbol{y}_{\backslash i}, M_2)\\ &= \prod_{i=1}^{n} \text{CPO}_i(M_1) \,/\, \text{CPO}_i(M_2).\end{aligned}$$

Pseudo Bayes factors are also well defined for models with improper priors as long as the posteriors $p(\boldsymbol{\theta}|\boldsymbol{y}_{\backslash i})$ are proper. One may also inspect plots of CPO_i values versus i, $i = 1, \ldots, n$, for each model and compare plots graphically.

Conceptually, generating samples $\boldsymbol{y}_i^{\text{rep},s}$ from $p(\boldsymbol{y}_i^{\text{rep}}|\boldsymbol{y}_{\backslash i})$ is straightforward: Rerunning an MCMC sampler using only $\boldsymbol{y}_{\backslash i}$, gives samples $\boldsymbol{\theta}^{(s)}$ from $p(\boldsymbol{\theta}|\boldsymbol{y}_{\backslash i})$ and $\boldsymbol{y}_i^{\text{rep},s}$ is then drawn from $p(\boldsymbol{y}|\boldsymbol{\theta}^{(s)})$. Obviously such resampling schemes will be computationally infeasible or extremely expensive for larger data sets. Therefore approximations, using only a single MCMC run for the full data $\boldsymbol{y}$, are desirable.

As an estimator for $\text{CPO}_i = p(\boldsymbol{y}_i|\boldsymbol{y}_{\backslash i})$, the harmonic mean estimator

$$\widehat{\text{CPO}_i} = \frac{S}{\sum_{s=1}^{S} 1 \,/\, p(\boldsymbol{y}_i|\boldsymbol{\theta}^{(s)})}$$

is suggested as such an approximation. The harmonic mean estimator may suffer from numerical instability, with extreme $\widehat{\text{CPO}}_i$ values if $p(\boldsymbol{y}_i|\boldsymbol{\theta}^{(s)}) \approx 0$ for some s. Congdon (2005) states that $\widehat{\text{CPO}}_i$ will be stable as long as individual log-likelihoods exceed -10. Ghosh and Tu (2009) use the harmonic mean approximation for model comparison in a longitudinal study.

Generally, the quality of such approximations is not well explored for the types of comparably complex hierarchical models considered in this book. Below we describe an approximate cross validatory approach which is designed for hierarchical models. To this end we first take a closer look at the exact leave-one-out predictive densities in SPMMs.

Following Marshall and Spiegelhalter (2003, 2007), we use a simplified generic notation and consider hierarchical models of the following form:

$$\begin{aligned} \boldsymbol{y}_i &\overset{\text{ind.}}{\sim} p(\boldsymbol{y}_i|\boldsymbol{\xi}, \boldsymbol{\alpha}_i), \quad i = 1, \ldots, n \\ \boldsymbol{\alpha}_i &\overset{\text{i.i.d.}}{\sim} p(\boldsymbol{\alpha}_i|\boldsymbol{\delta}) \\ (\boldsymbol{\xi}, \boldsymbol{\delta}, \boldsymbol{\tau}) &\sim p(\boldsymbol{\xi}, \boldsymbol{\delta}, \boldsymbol{\tau}) \end{aligned}$$

In the observation model, $\boldsymbol{\xi}$ comprises all parameters which are not individual-specific. For SPMMs with generic predictor $\boldsymbol{\eta}_i = \boldsymbol{U}_i\boldsymbol{\beta} + \boldsymbol{X}_{i1}\gamma_1 + \ldots + \boldsymbol{X}_{ir}\gamma_r + \boldsymbol{Z}_i\boldsymbol{\alpha}_i$, $\boldsymbol{\xi}$ consists of $\boldsymbol{\beta}$, $\gamma = (\gamma_1', \ldots, \gamma_r')'$ and, for Gaussian SPMMs, of variance parameters in the observation model for $\boldsymbol{y}_i$, $i = 1, \ldots, n$. For the random effects distribution, $\boldsymbol{\delta}$ contains all hyperparameters, that is, variance parameters for Gaussian random effects priors, and $\boldsymbol{\tau}$ is the vector of inverse smoothing parameters. Finally, $p(\boldsymbol{\xi}, \boldsymbol{\delta}, \boldsymbol{\tau})$ is the joint prior of regression coefficients and hyperparameters. For SPMMs it factorizes into $p(\boldsymbol{\beta})p(\gamma|\boldsymbol{\tau})p(\boldsymbol{\delta})p(\boldsymbol{\tau})$, the priors for regression and basis function coefficients and for variance parameters.

Then the cross validation predictive density for $\boldsymbol{y}_i^{\text{rep}}$ is

$$p(\boldsymbol{y}_i^{\text{rep}}|\boldsymbol{y}_{\setminus i}) = \iiiint p(\boldsymbol{y}_i^{\text{rep}}|\boldsymbol{\alpha}_i, \boldsymbol{\xi})p(\boldsymbol{\alpha}_i|\boldsymbol{\delta})p(\boldsymbol{\xi}, \boldsymbol{\delta}, \boldsymbol{\tau}|\boldsymbol{y}_{\setminus i})d\boldsymbol{\alpha}_i d\boldsymbol{\xi} d\boldsymbol{\delta} d\boldsymbol{\tau}.$$

Note that we have used conditional independence of observations $\boldsymbol{y}_i$ as well as $p(\boldsymbol{\alpha}_i|\boldsymbol{\delta}, \boldsymbol{\tau}, \boldsymbol{\xi}, \boldsymbol{y}_{\setminus i}) = p(\boldsymbol{\alpha}_i|\boldsymbol{\delta})$ to derive this representation. Thus, sampling from the exact leave-one-out predictive density proceeds as follows:

1. Draw $\boldsymbol{\xi}^{(s)}, \boldsymbol{\delta}^{(s)}, \boldsymbol{\tau}^{(s)}$ from the reduced posterior obtained from the reduced sample $\boldsymbol{y}_{\setminus i}$, $s = 1, \ldots, S$.
2. Draw random effects $\boldsymbol{\alpha}_i^{(s)}$ from $p(\boldsymbol{\alpha}_i|\delta^{(s)})$.
3. Draw $\boldsymbol{y}_i^{\text{rep},s}$ from $p(\boldsymbol{y}_i^{\text{rep}}|\boldsymbol{\alpha}_i^{(s)}, \boldsymbol{\xi}^{(s)})$.

This exact leave-one-out cross validation will usually be computationally extremely demanding. The following approximate sampling scheme has been suggested by Marshall and Spiegelhalter (2003).

Approximate cross validatory predictive assessment

The basic idea is to replace step 1 in the exact scheme by drawing only once from the full posterior instead of the reduced posterior, thus reducing conservatism to $\boldsymbol{\delta}, \boldsymbol{\tau}, \boldsymbol{\xi}$ and avoiding it, hopefully, for individual-specific effects $\boldsymbol{\alpha}_i$. Thus, approximate cross validatory sampling proceeds as follows:

1. Draw $\boldsymbol{\delta}, \boldsymbol{\tau}, \boldsymbol{\xi}$ from the full posterior based on the *complete* data $\boldsymbol{y}$.
2. Identical to exact sampling.
3. Identical to exact sampling.

Simulation studies in Sabanés Bové (2009) indicate that this approximate cross validation works quite well for Gaussian mixed models with normal random effects and avoids conservatism to a large extent when computing PIT, BOT and scores. Marshall and Spiegelhalter (2003) employ approximate sampling in a disease mapping application and Lewin *et al.* (2006) in a gene expression application. Additional empirical research is necessary to confirm this for a broader class of non-Gaussian SPMMs, however.

Example 4.12: Childhood obesity
We illustrate predictive model assessment by comparing four related but different additive mixed models. All models have the same additive structure

$$y_{ij} = \boldsymbol{u}'_{ij}\boldsymbol{\beta} + f(t_{ij}) + \boldsymbol{z}'_{ij}\boldsymbol{\alpha}_i + \varepsilon_{ij}, \quad \varepsilon_{ij} \overset{\text{i.i.d.}}{\sim} \mathrm{N}(0, \sigma^2),$$

with Gaussian random effects priors, considered already in Examples 3.2, 3.4, 4.2 and 4.4. They differ in the specification of the population trend $f(t)$ and the random effects component, while covariates in $\boldsymbol{u}$ are the same.

- Model 1 assumes a hockey-stick function

  $$f(t) = \beta_0 + \beta_1 t + \beta_2 (t-1)_+$$

  for the population trend and the nonlinear curve specification $g_i(t)$ defined in (3.7), Example 3.4.
- Model 2 assumes the nonlinear population trend (3.6) and the nonlinear random curve of Example 3.4.
- Model 3 assumes a cubic P-spline, with a second-order random-walk prior and 7 knots, and the nonlinear random curve of Example 3.4.
- Model 4 assumes the hockey-stick function but with a random line specification $\boldsymbol{z}'_{ij}\boldsymbol{\alpha}_i = \alpha_{i0} + \alpha_{i1} t_{ij}$.

The estimated population effects $f(t)$, $t =$ age, are shown in Fig. 4.31. For model 2 and 3, we see again the shapes already obtained in Examples 3.4 and 4.4. For model 1, we see a quite different form: Obviously, the major part of nonlinearity is absorbed by the random curves, so that the population trend falsely suggests an increasing effect of age on BMI! This is caused by a bad modelling strategy: The individual-specific effects $g_i(t)$ are more flexibly specified than the population effect. This artefact is avoided in model 4 with the simpler random lines specification.

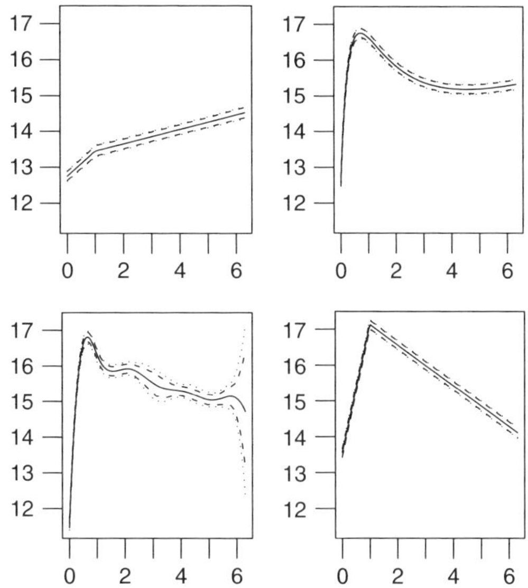

Fig. 4.31 Childhood obesity: Estimated fixed effects time trends (including the intercept) in the four models together with means, pointwise HPD and simultaneous credible intervals at the 0.95 level.

Posterior BOT histograms for assessing goodness-of-fit are shown in Fig. 4.32. Goodness-of-fit looks comparable for models 1,2 and 3. The corresponding BOT histograms are heavily left-skewed, indicating overdispersed predictive samples. Note that overdispersion is desirable here, since it means that posterior predictive distributions are centered around the original observations. For model 4, the fit of the data seems to be worse because more small posterior predictive BOT values are observed.

Table 4.6 contains the posterior predictive energy-score, the log-score and the posterior expected saturated deviance. This indicates the best fit for model 1, while model 4 has the worst fit.

These results illustrate that goodness-of-fit alone is not appropriate for model assessment: Model 1 is not plausible as discussed at the beginning of the example. We now turn to model assessment through approximate cross validation. Leave-one-out BOT histograms are shown in Figure 4.33.

It seems difficult to prefer one of the four models based on these histograms. This changes if we look at PIT histograms in Fig. 4.34: We see that the implausible model 1 is assessed much worse than models 2 and 3. Also, model 4 looks quite good now. This impression is confirmed by the approximate leave-one-out scores as well as DIC values in Table 4.7.

The parametric model 2, used in Example 3.4, is slightly preferred over model 3 by the scores, while model 3, used in Example 4.4, has the smaller DIC value.

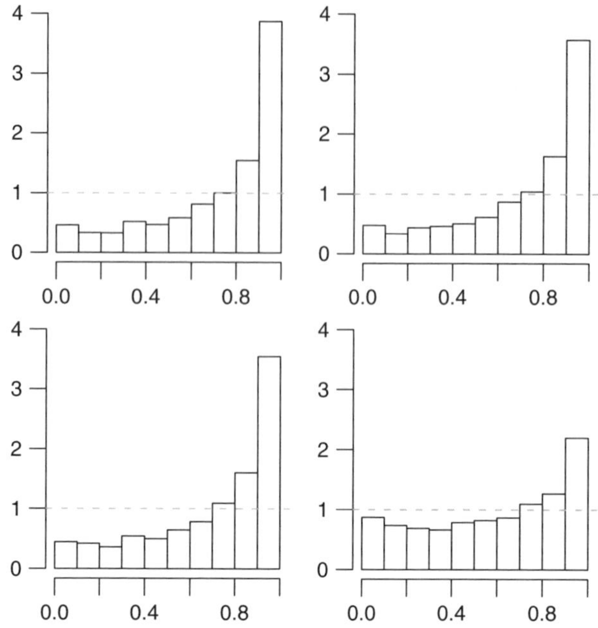

Fig. 4.32 Childhood obesity: Posterior predictive BOT histograms for goodness-of-fit assessment of the four models (from top left to bottom right).

Model 1 is worst for all three criteria, but the simple model 4 is best in terms of the DIC.

In summary, this justifies our model choice in Example 3.4: The P-spline specification of the population trend in Example 4.4 was useful as an exploratory tool to recognize the form of the trend and, then, to simplify through the parametric trend model (3.6) of Example 3.4. △

4.5.3 Predictor selection using spike and slab priors

Predictor selection in SPMMs comprises the following tasks: Variable selection, that is inclusion or exclusion of covariates u_l with linear effects $\beta_l u_l$ as part of the linear predictor $\boldsymbol{u}'\boldsymbol{\beta}$, selection of functions $f_l(u_l)$ and, for clustered or

Table 4.6 Childhood obesity: Posterior predictive mean energy- and log-scores of the four models as well as the posterior expected saturated deviances.

Fit criterion	Model 1	Model 2	Model 3	Model 4
energy-score	1.50	1.54	1.57	2.59
log-score	10.33	10.41	10.55	14.50
deviance	17308.38	17310.47	17311.83	17317.17

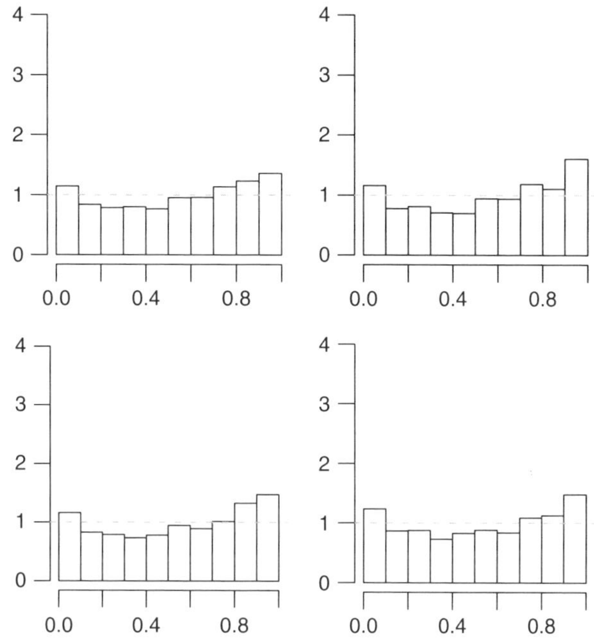

Fig. 4.33 Childhood obesity: Approximate BOT histograms for calibration assessment of the leave-one-out predictive distributions implied by the four models.

longitudinal data, selection of a parsimonious (Gaussian) random effects prior through a parsimonious choice of the covariance matrix of i. i. d. random effects. This section describes approaches and concepts based on spike and slab priors, allowing for jointly estimating and selecting the various components in a SPMM. Spike and slab priors are discrete mixture priors, formalizing the idea that a regression coefficient or a variance component is significant or (practically) non-significant. Following Fahrmeir, Kneib and Konrath (2010), we describe the three selection issues in turn, but they can be combined for jointly selecting and estimating components of interest.

Variable selection

Bayesian variable selection is most commonly based on spike and slab priors for regression coefficients, and it has been most extensively discussed for the classical linear Gaussian model $\boldsymbol{y} = \boldsymbol{U}\boldsymbol{\beta} + \boldsymbol{\varepsilon}$, $\boldsymbol{\varepsilon} \sim \mathrm{N}(\boldsymbol{0}, \sigma^2 \boldsymbol{I})$. We will consider extensions to Gaussian and non-Gaussian models with structured additive predictors such as

$$\boldsymbol{\eta} = \boldsymbol{U}\boldsymbol{\beta} + \boldsymbol{X}_1\boldsymbol{\gamma}_1 + \ldots + \boldsymbol{X}_r\boldsymbol{\gamma}_r$$

or, for longitudinal data,

$$\boldsymbol{\eta} = \boldsymbol{U}\boldsymbol{\beta} + \boldsymbol{X}_1\boldsymbol{\gamma}_1 + \ldots + \boldsymbol{X}_r\boldsymbol{\gamma}_r + \boldsymbol{Z}\boldsymbol{\alpha}.$$

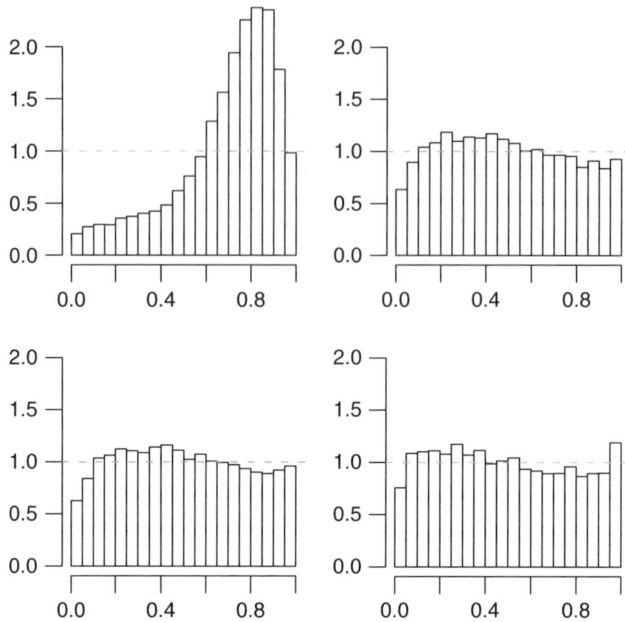

Fig. 4.34 Childhood obesity: Approximate scalar-PIT histograms for cross validated calibration assessment of the scalar predictive distributions in the four models.

The basic idea for a Gaussian linear model with $\boldsymbol{\beta} = (\beta_0, \beta_1, \ldots, \beta_m)'$, where β_0 is the intercept, is as follows (George and McCulloch, 1993, 1995, 1997): Each component β_l of $\boldsymbol{\beta}$ is modelled as having come either from a distribution with most (or all) of its mass concentrated around zero (the 'spike'), or from a comparably diffuse distribution with mass spread out over a large range of values (the 'slab'). Introducing binary latent indicators $J_l \in \{0, 1\}$, this prior can be formally expressed as a mixture

$$p(\beta_l | J_l) = (1 - J_l)\varphi(\beta_l; 0, v_{0l}^2) + J_l \varphi(\beta_l; 0, v_{1l}^2)$$

Table 4.7 Childhood obesity: Approximate mean energy- and logarithmic scores for the cross validated prediction of the four models and corresponding DIC based on the saturated deviance samples.

Model criterion	Model 1	Model 2	Model 3	Model 4
energy-score	4.69	2.62	2.63	3.11
log-score	14.07	12.99	13.07	15.38
DIC	21989.3	21389.4	21335.6	18958.9

of Gaussian densities with variances $v_{0l} \ll v_{1l}$, and

$$\mathbb{P}(J_l = 1|w_l) = 1 - \mathbb{P}(J_l = 0|w_l) = w_l.$$

The idea is that if $J_l = 0$ and v_{0l} is close to zero, then β_l is likely to be close to zero and the effect of the covariate u_l is 'practically' non-significant. Otherwise, if $J_l = 1$ and v_{1l} is large, then β_l is significant. George and McCulloch (1993, 1997) give guidelines for choosing the hyperparameters v_{0l}^2 and v_{1l}^2, including a semiautomatic approach. They also point out that computational problems can arise when the ratio v_{0l}^2/v_{1l}^2 is too large. To avoid such problems they recommend to restrict this ratio to 10,000. The hyperparameter $w_l = \mathbb{P}(J_l = 1|w_l)$ is the prior probability that u_l has a significant effect and should be included in the model. Usually, the intercept term β_0 is always included so that $w_0 = 1$. Excluding this degenerate case, the standard joint prior for $\boldsymbol{J} = (J_1, \ldots, J_m)$ is the independent Bernoulli prior

$$p(\boldsymbol{J}|\boldsymbol{w}) = \prod_{l=1}^{m} w_l^{J_l}(1 - w_l)^{1-J_l}.$$

A particularly simple choice is to assume $w_l = w = 0.5$ for all $l = 1, \ldots, m$, resulting in the seemingly uninformative, discrete uniform prior $p(\boldsymbol{J}) = 2^{-m}$. However, the number $m_{\boldsymbol{J}} = \sum_l J_l$ of significant effects has a binomial distribution $\mathrm{B}(m, 0.5)$, so that $\mathbb{E}(m_{\boldsymbol{J}}|w = 0.5) = m/2$. This may be unreasonable for high-dimensional $\boldsymbol{\beta}$. A more flexible approach is to place a hyperprior on w. A beta prior

$$p(w) \propto w^{a_w - 1}(1 - w)^{b_w - 1}$$

is a reasonable and convenient choice, including a uniform prior $w \sim \mathrm{U}(0, 1)$ as a special case for $a_w = b_w = 1$. Generally, a_w and b_w can be chosen so that they match desired values $\mathbb{E}(m_{\boldsymbol{J}})$ and $\mathrm{Var}(m_{\boldsymbol{J}})$, see Kohn, Smith and Chan (2001).

For joint selection of components of $\boldsymbol{\beta}$ and estimation of functional or random effects in structured additive regression models, we summarize as follows: The diffuse prior $p(\boldsymbol{\beta}) \sim \mathrm{N}(\boldsymbol{0}, \boldsymbol{C})$ is replaced by the hierarchy

$$\boldsymbol{\beta}|\boldsymbol{J} \sim \mathrm{N}(\boldsymbol{0}, \boldsymbol{D}_{\boldsymbol{J}}),$$

where $\boldsymbol{D}_{\boldsymbol{J}} = \mathrm{diag}((1 - J_l)v_{0l}^2 + J_l v_{1l}^2)$, setting $J_0 \equiv 1$ for always including the intercept term β_0,

$$p(\boldsymbol{J}|w) = \prod_{l=1}^{m} w^{J_l}(1 - w)^{1-J_l} = w^{m_J}(1 - w)^{m - m_J},$$

where $m_{\boldsymbol{J}} = \sum_l J_l$ is the number of significant effects, and

$$w \sim \mathrm{Be}(a_w, b_w).$$

Priors for functional or random effects are the same as in previous sections.

Extending Gibbs sampling for Gaussian models and MCMC algorithms with IWLS proposals for non-Gaussian exponential family models to include variable selection is conceptually straightforward:

Algorithm 4.8: Spike and slab priors for regression coefficients

i. To draw from full conditionals $\boldsymbol{\beta}|\cdot$ for regression coefficients (given $\boldsymbol{J}$, all other parameters and $\boldsymbol{y}$) the matrix $\boldsymbol{C}$ appearing in Gaussian full conditionals or MH-IWLS proposals is replaced by $\boldsymbol{D_J}$.

ii. The indicator variables J_l are drawn componentwise from Bernoulli distributions with probability

$$\mathbb{P}(J_l = 1|\beta_l) = \frac{A_l}{A_l + B_l},$$

where $A_l = p(\beta_l|J_l = 1)w$ and $B_l = p(\beta_l|J_l = 0)(1 - w)$.

iii. The weight w is drawn from

$$w|\cdot \sim \text{Be}(\tilde{a}_w, \tilde{b}_w)$$

with updated hyperparameters

$$\tilde{a}_w = a_w + m_{\boldsymbol{J}}, \quad \tilde{b}_w = b_w + (m - m_{\boldsymbol{J}}).$$

As a key advantage of MCMC inference, all other full conditionals remain untouched.

To avoid the choice of a suitably small hyperparameter v_{0l}^2 at all, it appears natural to consider the limiting case $v_{0l}^2 = 0$. This would lead to the prior

$$\beta_l|J_l \sim (1 - J_l)\delta_0(\cdot) + J_l\,\text{N}(0, v_{1l}^2),$$

where $\delta_0(\cdot)$ denotes a point mass at 0. In this formulation, $J_l = 0$ indicates that $\beta_l = 0$, that is, u_l is excluded from the linear predictor, and, vice versa, $J_l = 1$ indicates $\beta_l \neq 0$. However, then the Gibbs or MH sampling schemes as outlined above cannot be used anymore because they generate reducible, nonconvergent Markov chains. Without any modification, the sampler gets stuck when a value $\beta_l = 0$ is drawn. The method of Carlin and Chib (1995), reversible jump MCMC (Green, 1995), and the sampler of Geweke (1996) are conceptually feasible, but computationally very expensive alternatives.

For *Gaussian models*, a different strategy based on a conjugate hierarchical prior leads to computationally very efficient sampling schemes. Let $\boldsymbol{\beta_J}$ denote the subvector of components β_l of $\boldsymbol{\beta}$ for which $J_l = 1$, and $\boldsymbol{U_J}$ the corresponding submatrix of $\boldsymbol{U}$. Then the prior for $\beta_{\boldsymbol{J}}$, conditional on $\boldsymbol{J}$ and $\sigma^2 = \text{Var}(\varepsilon_i)$, is

$$p(\boldsymbol{\beta_J}|\boldsymbol{J}, \sigma^2) \sim \text{N}(\boldsymbol{0}, g\sigma^2(\boldsymbol{U}_{\boldsymbol{J}}'\boldsymbol{U_J})^{-1})$$

for a positive hyperparameter g. It is related to Zellner's g-prior formulation (Zellner, 1986) and has been suggested in the context of variable selection in Gaussian models by Smith and Kohn (1996) and George and McCulloch (1997). The priors for $\boldsymbol{J}$ and σ^2, for example an IG-prior or Jeffreys' prior, are assumed independent *a priori*. As a related data-dependent prior, it has also been suggested to assume the prior expectation $\mathbb{E}(\boldsymbol{\beta}_{\boldsymbol{J}}|\boldsymbol{J},\sigma^2) = \hat{\boldsymbol{\beta}}_{\gamma}$, the least squares estimate of $\boldsymbol{\beta}$, instead of $\mathbb{E}(\boldsymbol{\beta}|\boldsymbol{J},\sigma^2) = \boldsymbol{0}$, see for example Kohn, Smith and Chan (2001). A key advantage is that this conjugate prior setting yields rapidly computable closed form expressions for $p(\boldsymbol{y}|\boldsymbol{J})$ by analytically integrating out $\boldsymbol{\beta}$ and σ^2. This allows for efficient updating schemes and for direct computation of Bayes factors and posterior model probabilities $p(\boldsymbol{J}|\boldsymbol{y})$. Because the hyperparameter g acts as a complexity parameter, its choice can be critical. Various choices and hyperpriors for g have been suggested, see Liang *et al.* (2008a) for a review and suggestions for mixtures of g-priors. Avoiding the g-prior approach, Casella and Moreno (2006) propose intrinsic priors for objective Bayesian variable selection.

Conceptually, approaches based on this conjugate prior setting can be extended to non-Gaussian regression models that are formulated in terms of latent auxiliary Gaussian models in a further stage of the hierarchy, such as for binary or multicategorical probit models. For other non-Gaussian models, such as count data regression, analogous extensions could be based on auxiliary models with error distributions approximated through a finite mixture of normals (Frühwirth-Schnatter and Wagner, 2006; Frühwirth-Schnatter *et al.*, 2009). As an alternative, avoiding latent auxiliary models, Wang and George (2007) suggest a generalization of the g-prior approach to generalized linear models. Because the marginal $p(\boldsymbol{y}|\boldsymbol{J})$ cannot be integrated out analytically anymore, however, an approximation using integrated Laplace approximation is employed.

Instead of encouraging variable selection by placing priors directly on the regression coefficients, a different strategy is to place spike and slab priors on the variances $\tau^2_{\beta,l}$ in the conditional priors $p(\beta_l|\tau^2_{\beta,l})$ for regression coefficients. This concept is formalized through the spike and slab prior of Ishwaran and Rao (2003, 2005). Their hierarchical prior for $\beta_l|\tau^2_{\beta,l}$ is

$$\beta_l|\tau^2_{\beta,l} \overset{\text{ind.}}{\sim} \text{N}(0,\tau^2_{\beta,l})$$

where $\tau^2_{\beta,l} = J_l v_l^2$, with priors

$$\begin{aligned} J_l|w &\overset{\text{i.i.d.}}{\sim} (1-w)\delta_{\nu_0}(\cdot) + w\delta_{\nu_1}(\cdot), \\ v_l^2 &\overset{\text{i.i.d.}}{\sim} \text{IG}(a_v, b_v). \end{aligned}$$

The first component J_l is an indicator variable with point mass at the values $0 < \nu_0 \ll \nu_1$, with ν_0 close to zero. The parameter w controls how likely J_l equals ν_0 or ν_1, and therefore takes on the role of a complexity parameter. We assume a Beta prior

$$w \sim \text{Be}(a_w, b_w)$$

with the special case $w \sim \mathrm{U}[0,1]$ for $a_w = b_w = 1$. With an appropriate choice of a_w, b_w it is possible to favour more or less sparse models. Marginalizing over J_l and v_l^2, we have

$$\tau_{\beta,l}^2 | w \sim (1-w)\,\mathrm{IG}(a_v, \nu_0 b_v) + w\,\mathrm{IG}(a_v, \nu_1 b_v)$$

taht is, the prior for $\tau_{\beta,l}^2$ is a spike and slab prior for variances specified as a bimodal mixture of two inverse gamma distributions. Therefore, we call this spike and slab prior a NMIG (normal-mixture of inverse gamma) prior. Integrating out $\tau_{\beta,l}^2$, the marginal distribution of β_l is a mixture of two scaled t-distributions:

$$\beta_l | w \sim (1-w)\,\mathrm{t}(\mathrm{df}, \mathrm{scale}_0) + w\,\mathrm{t}(\mathrm{df}, \mathrm{scale}_1)$$

with

$$\mathrm{df} = 2a_\omega, \quad \mathrm{scale}_0 = \sqrt{\frac{\nu_0 b_\omega}{a_\omega}}, \quad \mathrm{scale}_1 = \sqrt{\frac{\nu_1 b_\omega}{a_\omega}}.$$

Thus, the inverse gamma spike and slab prior for variances induces a spike and slab prior for regression coefficients. The issue of specifying the variances v_{0l}^2 and v_{1l}^2 in the bimodal Gaussian mixture prior is now transferred to the choice of ν_0 and ν_1. For Gaussian models, Ishwaran and Rao (2005) suggest default options, in particular $\nu_1 = 1$, after standardizing all covariates and rescaling the model, see also Kneib, Konrath and Fahrmeir (2010b) for non-Gaussian models.

Extension of Gibbs sampling for Gaussian SPMMs and MCMC algorithms for exponential family SPMMs in Section 4.2.2 to include variable selection for regression parameters β_l proceeds in analogy to shrinkage priors in Section 4.4.2 and to spike and slab priors (4.54) of this section: The conditional prior for $\boldsymbol{\beta} = (\beta_1, \ldots, \beta_m)'$ is Gaussian,

$$\boldsymbol{\beta} | \boldsymbol{\tau}_\beta \sim \mathrm{N}(\mathbf{0}, \boldsymbol{D}_{\boldsymbol{\beta}}), \tag{4.54}$$

with $\boldsymbol{D}_{\boldsymbol{\beta}} = \mathrm{diag}(\tau_{\beta,0}^2, \tau_{\beta,1}^2, \ldots, \tau_{\beta,m}^2)$. Because variable selection does not include the intercept term, $\tau_{\beta,0}^2$ usually is a large (but fixed) value for a weakly informative prior for β_0, with the limiting case $\tau_{\beta,0}^2 \to \infty$ corresponding to a flat prior. (If some covariate u_l should be kept in the predictor, the same holds for $\tau_{\beta,l}^2$.) Then the update steps are as follows.

Algorithm 4.9: Normal mixture of inverse Gaussians

i. To draw from full conditionals $\boldsymbol{\beta}|\cdot$ for regression coefficients, given $\boldsymbol{\tau}_\beta = (\tau_{\beta,0}^2, \tau_{\beta,1}^2, \ldots, \tau_{\beta,m}^2)'$, all other parameters and $\boldsymbol{y}$, the matrix $\boldsymbol{C}$ in (4.19) or (4.26) is replaced by $\boldsymbol{D}_{\boldsymbol{\beta}}$ and $\boldsymbol{c}$ equals $\mathbf{0}$.

ii. The full conditionals for the binary indicator variables are

$$\mathbb{P}(J_l = \nu_0 | \cdot\,) = \frac{A_l}{A_l + B_l}, \quad \mathbb{P}(J_l = \nu_1 | \cdot\,) = \frac{B_l}{A_l + B_l},$$

where

$$A_l = (1-w)\frac{1}{\sqrt{\nu_0}}\exp\left\{-\frac{1}{2\nu_0 v_l^2}\beta_l^2\right\},$$

$$B_l = w\frac{1}{\sqrt{\nu_1}}\exp\left\{-\frac{1}{2\nu_1 v_l^2}\beta_l^2\right\}.$$

iii. The full conditionals for v_l^2 are inverse Gamma with

$$v_l^2 \sim \mathrm{IG}\left(a_v + \frac{1}{2}, b_v + \frac{\beta_l^2}{2J_l}\right).$$

iv. The full conditional for the weight w is

$$w \sim \mathrm{Be}(a_w + n_1, b_w + n_0),$$

with $n_0 = |\{l : J_l = \nu_0\}|$ and $n_1 = |\{l : J_l = \nu_1\}|$.

Implicitly, the NMIG prior approach provides a natural procedure for selecting covariates, using posterior samples of the indicator variables: The larger the estimated posterior probability of an indicator variable (i.e. the percentage of ν_1 values in the sample), the larger is the evidence that the corresponding covariate has non-negligible effects.

As with shrinkage priors, inclusion of NMIG priors instead of Gaussian priors $\boldsymbol{\beta} \sim \mathrm{N}(\boldsymbol{c}, \boldsymbol{C})$ is possible in all MCMC algorithms for (generalized) additive models and (semiparametric) mixed models in previous chapters and sections, setting $\boldsymbol{c} = \boldsymbol{0}$ and $\boldsymbol{C} = \boldsymbol{D}_{\boldsymbol{\beta}}$ in corresponding full conditionals and including steps ii and iii of Algorithm 4.9.

Example 4.13: Childhood obesity
We supplement the analysis in Example 4.10 (page 252) with results obtained with a NMIG spike and slab prior for population effects instead of lasso or ridge priors, after standardizing the continuous covariates. Otherwise the model is exactly the same as in Example 4.10. Posterior mean estimates and (estimated) inclusion probabilities $\mathbb{P}(J_l = 1|\boldsymbol{y})$, obtained from the MCMC output as relative frequencies of samples with $J_l = 1$, are displayed in Table 4.8. A comparison with lasso and ridge estimates shows that small effects are mostly shrunken closer to zero with the NMIG prior, while larger effects are less affected from the shrinkage. Using inclusion probabilities to select variables or to assess their influence leads to the same conclusions as in Example 4.10. △

Function selection

The following related issues are of interest in function selection: First, should a functional effect $f_l(x_l)$ be included in the predictor at all? Because we always include a common intercept β_0, we assume that all functions are centred around

Table 4.8 Childhood obesity: Posterior mean estimates for the fixed effects in the AMM with NMIG prior together with inclusion probabilities. wgain2y, outdoor, mBMI and mDiffBMI have been standardised for the analyses.

	mean	prob.
const	15.58	–
sex	0.0274	0.230
breast	0.1746	0.987
tvpc2	0.0000	0.028
tvpc3	0.0250	0.129
mSmoke	−0.0061	0.063
mEdu2	0.0915	0.359
mEdu3	0.3622	1.000
mEdu4	−0.2052	0.712
area	0.0054	0.064
wgain2y	0.1717	1.000
outdoor	0.0101	0.067
mBMI	0.1637	1.000
mDiffBMI	0.1914	1.000

zero. Therefore, we want to know whether $f_l(x_l)$ differs significantly from 0. Second, we are often more interested in the following question: Is the effect of a covariate x_l linear, that is $f_l(x_l) = \beta_l x_l$, or is it nonlinear, that is, does $f_l(x_l)$ differ significantly from a straight line? In this case, we may decompose the function into a linear and a nonlinear part, that is,

$$f_l(x_l) = \beta_l x_l + \tilde{f}_l(x_l),$$

where the nonlinear part $\tilde{f}_l(x_l)$ solely captures deviations from a linear effect. More generally, we may be interested whether a functional effect differs from a polynomial of higher degree. For identifiability, functions $\tilde{f}_l(x_l)$ have to be 'orthogonal' to the space of linear functions or, more generally, of polynomials. How to best approach these issues is also related to the choice of basis functions and smoothness priors for basis function coefficients.

This is most immediately seen when a function is represented by a linear combination of TPS basis functions. For a linear spline representation

$$f_l(x_l) = \beta_l x_l + \sum_{k=1}^{K} \gamma_{kl}(x_l - \kappa_k)_+$$

with knots κ_k the effect of x_l is exactly linear for $\boldsymbol{\gamma}_l = (\gamma_{1l}, \ldots, \gamma_{Kl})' = \mathbf{0}$, and it is 'practically' linear if $\boldsymbol{\gamma}$ is 'practically' zero. This leads immediately to the idea to transfer concepts from variable selection to function selection by introducing corresponding spike and slab priors directly for the basis function coefficients or, alternatively, for the variance parameters τ_l^2 in the conditional Gaussian prior

$$p(\gamma_l|\tau_l^2) \propto \exp\left(-\frac{1}{2\tau_l^2}\gamma_l'\boldsymbol{K}_l\gamma_l\right),$$

where $\boldsymbol{K}_l = \boldsymbol{I}$ for the TPS basis representation. If we want to explore if x_l has a significant effect $f_l(x_l) \neq 0$ at all, we have to combine this with a selection prior for β_l.

If, instead, we represent $f_l(x_l)$ through Bayesian P-splines or smoothing splines with a partially improper penalty matrix $\boldsymbol{K}_l$, choice of selection priors is less immediate: Improper priors should be avoided for variable selection problems, and function selection is also related to the null space of $\boldsymbol{K}_l$, that is, to functions that are not penalized through $\boldsymbol{K}_l$. Several strategies are possible for circumventing problems arising with (partially) improper priors for Bayesian P-splines: First, additional restrictions or proper priors can be imposed on initial values of parameters as in Chib and Jeliazkov (2006) or, more symmetrically, on initial and final values. This results in a modified, but proper penalty matrix. Second, the mixed model representation and reparametrization described in Section 4.2.1 may be used. For example for a penalized spline with second-order-random-walk prior for basis function coefficients, an additional linear effect $\beta_l x_l$ has to be included in the linear predictor while the vector $\boldsymbol{\alpha}_l$ of remaining reparametrized coefficients has a proper ridge-type prior $\boldsymbol{\alpha}_l \sim \mathrm{N}(\boldsymbol{0}, \tau_l^2\boldsymbol{I})$. Cottet, Kohn and Nott (2008) suggest such a strategy for smoothing splines with point mass at zero priors in (double) exponential families. To circumvent problems with high-dimensional parameters $\boldsymbol{\alpha}_l$ and to avoid mixing problems, they use a reduced rank representation of smoothing splines, effectively leading to a very small number of basis functions. As an alternative, Reich, Storlie and Bondell (2009) consider a smoothing spline ANOVA framework to perform variable and function selection in Gaussian regression models. Panagiotelis and Smith (2008) employ the mixed model representation to turn partially improper priors into proper priors, replacing the smoothness prior

$$p(\gamma_l|\tau_l^2) \propto \exp\left(-\frac{1}{2\tau_l^2}\gamma_l'\boldsymbol{K}_l\gamma_l\right)$$

through a linearly constrained smoothness prior

$$p(\gamma_l|\tau_l^2) \propto \exp\left(-\frac{1}{2\tau_l^2}\gamma_l'\boldsymbol{K}_l\gamma_l\right)\mathbb{1}(\boldsymbol{\Gamma}_{1l}'\gamma_l = \boldsymbol{0}). \tag{4.55}$$

The matrix $\boldsymbol{\Gamma}_{1l}$ is the submatrix of eigenvectors in the spectral decomposition $\boldsymbol{K}_l = (\boldsymbol{\Gamma}_{1l}, \boldsymbol{\Gamma}_{2l})\boldsymbol{\Omega}_l(\boldsymbol{\Gamma}_{1l}, \boldsymbol{\Gamma}_{2l})'$ corresponding to the null space of $\boldsymbol{K}_l$, see

Section 4.2.1, page 205. The indicator function $\mathbb{1}(\cdot)$ is 1 if the linear constraint is fulfilled, and 0 otherwise. The constraint turns the partially improper prior into a proper prior, but it also deletes the corresponding unpenalized part from the model. For a P-spline with second-order-random-walk, the constraint effectively removes the unpenalized linear effect from the model. Therefore, a linear effect $\beta_l x_l$ has to be added to the linear predictor. For a proper penalty matrix $\boldsymbol{K}_l$, we have $\boldsymbol{\Gamma}_1 = \boldsymbol{0}$, so that the constraint is always fulfilled and (4.55) reduces to the unconstrained smoothness prior.

Assuming from now on unconstrained or constrained proper smoothness priors, concepts for variable selection can be transferred to function selection. To define selection priors for the vector γ_l of basis function coefficients, an indicator variable J_l is introduced such that $J_l = 0$ indicates that $\gamma_l = \boldsymbol{0}$ (or is 'practically' zero) and $J_l = 1$ indicates that $\gamma_l \neq 0$ (or is significantly different from zero). Panagiotelis and Smith (2008) augment the (constrained) smoothness prior (4.55) with a point mass at 0:

$$\gamma_l | \tau_l^2, J_l \sim (1 - J_l)\delta_0(\cdot) + J_l \, \mathrm{N_{con}}(\boldsymbol{0}, \tau_l^2 \boldsymbol{K}_l^-), \tag{4.56}$$

where $\mathrm{N_{con}}$ denotes the (possibly constrained) Gaussian smoothness prior (4.55) and $\boldsymbol{K}_l^-$ is a (generalized) inverse of $\boldsymbol{K}_l$.

The spike and slab priors (4.56) for functions $f_l(x_l)$, $l = 1, \dots, r$, represented through $\boldsymbol{f}_l = \boldsymbol{Z}_l \gamma_l$ in the additive predictor, have to be supplemented with further priors for variance parameters τ_l^2, $l = 1, \dots, r$, and indicator variables $\boldsymbol{J} = (J_1, \dots, J_r)$. For variance parameters, inverse Gamma priors $\tau_l^2 \sim \mathrm{IG}(a_l, b_l)$ are a convenient choice, although Panagiotelis and Smith (2008) employ the prior $\log(\tau_l^2) \sim \mathrm{N}(a_l, b_l)$. For a small number r of functions, independent Bernoulli priors

$$J_l \stackrel{\text{ind.}}{\sim} \mathrm{B}(1, w_l)$$

are a standard choice, while for high-dimensional r an alternative prior is suggested in Cripps, Carter and Kohn (2005). Posterior inference is developed for *Gaussian additive models*

$$\boldsymbol{y} = \boldsymbol{U}\boldsymbol{\beta} + \boldsymbol{X}_1\gamma_1 + \ldots + \boldsymbol{X}_r\gamma_r + \boldsymbol{\varepsilon}, \quad \boldsymbol{\varepsilon} \sim \mathrm{N}(\boldsymbol{0}, \sigma^2\boldsymbol{I})$$

with possibly high-dimensional r and sampling schemes make essential use of the Gaussian assumption to circumvent possible pitfalls in connection with the point mass at zero in the mixture prior (4.56). Extensions of this approach using (mixtures of) Gaussian auxiliary latent variables is conceivable and would be interesting.

As an alternative, it seems natural to assume a Gaussian smoothness prior $\gamma_l \sim \mathrm{N}(\boldsymbol{0}, \tau_l^2 \boldsymbol{K}_l^-)$ and to impose a NMIG prior on the smoothing variance, i.e. to set $\tau_l^2 = J_l v_l^2$ with

$$\begin{aligned} J_l &\sim (1 - w_l)\delta_{\nu_0}(\cdot) + w_l \delta_{\nu_1}(\cdot), \quad 0 < \nu_0 \ll \nu_1, \\ v_l^2 &\sim \mathrm{IG}(a_l, b_l). \end{aligned}$$

Posterior probabilities for J_l would indicate whether the function f_l should be included in the predictor or not. However, as shown in Scheipl (2010b), it turns out that the resulting MCMC sampler suffers from bad mixing or may even get stuck already for moderate dimensions of $\boldsymbol{\gamma}_l$. To avoid these problems, Scheipl (2010b) combines the NMIG prior approach with the idea of parameter expansion in mixed models (Gelman *et al.*, 2008). In a first step, the mixed model reparametrization of P-splines is employed to represent the function $f_l(x_l)$ in the form $\beta_l x_l + \boldsymbol{\alpha}_l \boldsymbol{z}'(x_l)$, where $\boldsymbol{z}_l(x_l)$ is the $\tilde{K}$-dimensional vector of basis functions. Then parameter expansion is applied to $\boldsymbol{\alpha}_l$ in the multiplicative form

$$\boldsymbol{\alpha}_l = w_l \boldsymbol{\xi}_l, \quad \boldsymbol{\xi}_l = (\boldsymbol{\xi}_{l1}, \ldots, \boldsymbol{\xi}_{l\tilde{K}}).$$

A NMIG prior is now assumed for the scalar parameter w_l, that is $w_l \sim \mathrm{N}(0, J_l v_l^2)$ with priors for the indicator variables J_l and for v_l^2 as well as hyperparameters ν_0, w_l, a_l and b_l as before. The components of $\boldsymbol{\xi}_l$ are assumed to be independent (as the components of $\boldsymbol{\alpha}_l$) with Gaussian mixture priors

$$\xi_{lk} \overset{\text{ind.}}{\sim} 0.5\,\mathrm{N}(1,1) + 0.5\,\mathrm{N}(-1,1).$$

The scalar parameter w_l parametrizes the importance of $\boldsymbol{\alpha}_l$, that is of the nonlinear part $\boldsymbol{\alpha}_l \boldsymbol{b}'(x_l)$ of the function $f_l(x_l)$, while $\boldsymbol{\xi}_l$ spreads w_l across $\boldsymbol{\alpha}_l$. The effective dimension for updating the parameter coming from the NMIG prior is 1, thereby avoiding mixing problems. For the linear part $\beta_l x_l$ we may assume a flat prior, a shrinkage prior, or, again, a NMIG prior. In the latter case, inclusion probabilities can be estimated from MCMC output separately for the linear and nonlinear parts of the smooth function $f_l(x_l)$. Details of MCMC simulation for Gaussian and exponential family semiparametric (mixed) models are described in Scheipl (2010b) and implemented in the R-package `spikeSlabGAM`. Applications to simulated and many real data sets provide convincing evidence for good performance of this function selection method. The following example illustrates the method with an application to longitudinal data.

Example 4.14: Childhood obesity
We consider a Gaussian SPMM with the same covariate set as in Example 4.2 but include both the age of the child and the body mass index of the mother as covariates with potentially nonlinear effect. Therefore, we assume a predictor of the form

$$\eta_{ij} = \boldsymbol{u}_{ij}'\boldsymbol{\beta} + f_1(\text{age}) + f_2(\text{mBMI}) + \alpha_{i0}$$

where $\boldsymbol{u}$ comprises the remaining covariates. After the mixed model reparametrization, the predictor is

$$\eta_{ij} = \boldsymbol{u}_{ij}'\boldsymbol{\beta} + \beta_{\text{age}}\text{age} + \beta_{\text{mBMI}}\text{mBMI} + \tilde{f}_1(\text{age}) + \tilde{f}_2(\text{mBMI}) + \alpha_{i0}$$

where $\tilde{f}_1$ and $\tilde{f}_2$ denote the nonlinear parts of f_1 and f_2. For illustration purposes and because interest lies in the population effects part of the predictor, we have simplified the random effects part to a random intercept. As a consequence, estimates for fixed effects for some of the covariates are not comparable to previous results and are therefore omitted.

The posterior mean estimates for the linear effects of age and mother's BMI are $\hat{\beta}_{\text{age}} = -0.408$ (0.379) and $\hat{\beta}_{\text{mBMI}} = -0.056$ (0.051), with standard deviations in brackets. The corresponding inclusion probabilities 0.990 and 0.723 for the nonlinear parts $\tilde{f}_1(\text{age})$ and $\tilde{f}_2(\text{mBMI})$ show that it is reasonable to model the effects of both covariates as nonlinear. Figure 4.35 visualizes the (estimated) sums $f_1(\text{age})$ and $f_2(\text{mBMI})$ and the estimated population trend $f_1(\text{age})$ is in good agreement with previous results. △

Covariance matrix selection for random effects

Variable selection concepts can also be applied to variance component and covariance matrix selection in mixed models with Gaussian random effects $\boldsymbol{\alpha}_i \overset{\text{i.i.d.}}{\sim} \text{N}(\mathbf{0}, \boldsymbol{Q})$. If the components α_{il}, $l = 1, \ldots, q$ of $\boldsymbol{\alpha}_i$ are assumed to be independent *a priori*, then $\boldsymbol{Q} = \text{diag}(\delta_1^2, \ldots, \delta_q^2)$, with $\delta_1^2 = \text{Var}(\alpha_{i1}), \ldots, \delta_q^2 = \text{Var}(\alpha_{iq})$. To explore whether a random effect α_{il} should be included or not, a spike and slab prior for variances can be specified for δ_l^2 in complete analogy to the spike and slab prior of Ishwaran and Rao (2003, 2005) for variable selection. Then, the marginal prior for δ_l^2 is a bimodal mixture of two inverse gamma densities, replacing for example $\delta_l^2 \sim \text{IG}(a_l, b_l)$ as a standard conjugate prior for variances of Gaussian random effects.

Selection schemes for diagonal and offdiagonal elements of covariance matrices in case of correlated random effects have been developed by Smith and Kohn (2002) for the multivariate Gaussian distribution, Chen and Dunson (2003) and Frühwirth-Schnatter and Tüchler (2008) for Gaussian linear mixed models, and Cai and Dunson (2006) for generalized linear mixed models. All these approaches

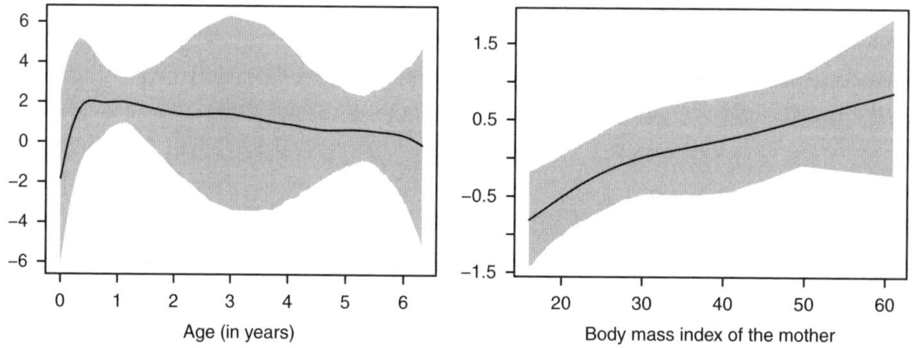

Fig. 4.35 Childhood obesity: Estimated nonlinear effect $f_1(\text{age})$ and $f_2(\text{mBMI})$ together with 90% pointwise credible intervals.

aim at parsimonious covariance matrix estimation to avoid unstructured highly parametrized specification through inverse Wishart priors. The common idea is to factorize $\boldsymbol{Q} = \text{Cov}(\boldsymbol{\alpha}_i)$ or its inverse through some type of Cholesky decomposition and to specify appropriate priors for Cholesky factors.

Cai and Dunson (2006) develop a covariance matrix selection method using the Cholesky decomposition $\boldsymbol{Q} = \boldsymbol{DLL'D}$, where $\boldsymbol{D}$ is a diagonal matrix and $\boldsymbol{L}$ is a lower triangular matrix with ones on the diagonal. For selection of elements in $\boldsymbol{D}$ and $\boldsymbol{L}$, they choose a mixture of a point mass at zero and a truncated normal density for the elements d_l^2, $l = 1, \ldots, q$, of $\boldsymbol{D}$, and a mixture of a point mass at zero and a normal density for the elements of $\boldsymbol{L}$. Posterior computation is based on MCMC algorithms in combination with a Laplace approximation of the marginal likelihood of $\boldsymbol{y}$, integrating out random effects. As an alternative, avoiding approximate Laplace integration, Kinney and Dunson (2007) extend a modified version of Chen and Dunson (2003) for LMMs to fixed and random effects selection in logistic models, using auxiliary latent LMMs.

Frühwirth-Schnatter and Tüchler (2008) use the Cholesky decomposition $\boldsymbol{Q} = \boldsymbol{CC'}$ to rewrite the random effects part $\boldsymbol{Z}_i\boldsymbol{\alpha}_i$ of the predictor as $\boldsymbol{Z}_i\boldsymbol{C}\boldsymbol{b}_i$, where $\boldsymbol{b}_i \sim \text{N}(\boldsymbol{0}, \boldsymbol{I})$. Stacking the elements of $\boldsymbol{C}$ in a vector $\boldsymbol{c}$, the random effects part can be rewritten as $\boldsymbol{W}_i\boldsymbol{c}$, where the design matrix $\boldsymbol{W}_i$ depends on $\boldsymbol{Z}_i$ and the latent random effects $\boldsymbol{b}_i$. The basic idea is: If the random effects $\boldsymbol{b}_i$ (and $\boldsymbol{W}_i$) *were known*, then the predictor $\boldsymbol{\eta}_i = \boldsymbol{X}_i\boldsymbol{\beta} + \boldsymbol{W}_i\boldsymbol{c}$ defines a classical linear Gaussian model, and selection of the matrix is transformed into a variable selection problem for the components of the vector $\boldsymbol{c}$. Given the random effects, Frühwirth-Schnatter and Tüchler (2008) follow the conjugate prior approach with a point mass at zero, assuming a conditional g-prior for $\boldsymbol{c}$. The MCMC scheme includes sampling the latent random effects from their (Gaussian) full conditionals, so that the complete MCMC sampler resembles samplers developed for Bayesian variable selection in Gaussian linear models. This approach may be extended to non-Gaussian mixed models that can be formulated in terms of latent auxiliary Gaussian mixed models, see Tüchler (2008) for logistic mixed models.

4.6 Notes and further reading

4.6.1 Individual-specific curves and functional mixed models

SPMMs may be extended to allow for nonparametric individual-specific curves. For example, Durban *et al.* (2005) consider an additive mixed model

$$y_{ij} = f(t_{ij}) + g_i(t_{ij}) + \varepsilon_{ij}, \quad \varepsilon_{ij} \sim \text{N}(0, \sigma^2),$$

where $f(t)$ is the population trend and $g_i(t_{ij})$ are individual-specific deviations. The ('fixed') population trend f and the random curves g_i are modelled through linear splines with a TPS basis, that is

$$g_i(t_{ij}) = \alpha_{i1} + \alpha_{i2}t_{ij} + \sum_{k=3}^{K} \alpha_{ik}(t_{ij} - \kappa_k)_+,$$

with individual-specific i. i. d. random effects $(\alpha_{i1}, \alpha_{i2}) \overset{\text{i.i.d.}}{\sim} \mathrm{N}(\mathbf{0}, \boldsymbol{Q})$, $\alpha_{ik} \overset{\text{i.i.d.}}{\sim} \mathrm{N}(0, \delta^2)$, $k = 1, \ldots, K$. For inference they suggest the EB appproach using mixed model technology. Extensions to more general predictors with several functional population and individual-specific effects are obvious. Inclusion of varying coefficient terms such as $g_{ij}(t_{ij})u_{ij}$ leads to so-called functional mixed models as an extension of functional data analysis to longitudinal data. Guo (2002) assumes smoothing splines for population and random functions and suggests EB as well as FB inference where the latter relies on a multivariate state-space formulation and corresponding MCMC techniques. Morris and Carroll (2006) develop wavelet-based functional mixed models with spike and slab priors for wavelet coefficients, with applications in proteomics (Morris *et al.*, 2006). A short review on functional mixed models is also given in Lin and Carroll (2009). Of course, SPMMs with individual-specific functional effects can also be based on B-splines and other bases. In any case, this leads to predictors of the form

$$\boldsymbol{\eta}_i = \boldsymbol{U}_i\boldsymbol{\beta} + \boldsymbol{X}_{i1}\boldsymbol{\gamma}_1 + \ldots + \boldsymbol{X}_{ir}\boldsymbol{\gamma}_r + \boldsymbol{Z}_i\boldsymbol{\alpha}_i + \boldsymbol{Z}_{i1}\boldsymbol{\alpha}_{i1} + \ldots + \boldsymbol{Z}_{im}\boldsymbol{\alpha}_{im},$$

where the terms $\boldsymbol{Z}_{ih}\boldsymbol{\alpha}_{ih}$, $h = 1, \ldots, m$, represent individual-specific functional effects. Gaussian priors for $(\boldsymbol{\alpha}_{i1}, \ldots, \boldsymbol{\alpha}_{im})$ can be appropriately structured to account for smoothness, i.e. correlation within functions, and for between-function correlation. The `R`-package `amer` (Scheipl, 2010a) has an option for mixed model-based EB inference in such models.

4.6.2 Approximate Bayesian inference

The aim of approximate Bayesian inference is to simplify or to avoid time-consuming full MCMC simulation in hierarchical models for complex data structures through analytical and numerical approximations. Let $p(\boldsymbol{y}|\boldsymbol{\xi}, \boldsymbol{\theta}_{\boldsymbol{y}})$ denote a generic observation model with possibly high-dimensional parameter vector $\boldsymbol{\xi}$ of primary interest and a low-dimensional hyperparameter $\boldsymbol{\theta}_{\boldsymbol{y}}$. Often $\boldsymbol{\xi}$ is split up in blocks, such as $\boldsymbol{\xi} = (\boldsymbol{\alpha}, \boldsymbol{\beta}, \boldsymbol{\gamma})$ with vectors of random effects $\boldsymbol{\alpha}$, fixed effects $\boldsymbol{\beta}$ and basis function coefficients $\boldsymbol{\gamma}$ in SPMMs. The prior $p(\boldsymbol{\xi}|\boldsymbol{\theta}_{\boldsymbol{\xi}})$ depends on a low-dimensional hyperparameter $\boldsymbol{\theta}_{\boldsymbol{\xi}}$, and $p(\boldsymbol{\theta})$ denotes a hyperprior for $\boldsymbol{\theta} = (\boldsymbol{\theta}_{\boldsymbol{y}}, \boldsymbol{\theta}_{\boldsymbol{\xi}})$.

Variational Bayes approaches

A generic approach for approximate inference developed in machine learning is the variational Bayes (VB) methodology (e.g. Bishop, 2006), with applications for a variety of complex data structures, see for example Jordan (2004) and Titterington (2004) for reviews. The principle of VB inference is to approximate the posterior distribution $p(\boldsymbol{\xi}, \boldsymbol{\theta}|\boldsymbol{y})$ through a simpler approximation $q(\boldsymbol{\xi}, \boldsymbol{\theta})$.

The quality of approximations is measured by the Kullback–Leibler-distance between the posterior and its approximation. Typically, $q(\boldsymbol{\xi}, \boldsymbol{\theta})$ is factorized into blocks of densities such as $q(\boldsymbol{\alpha})q(\boldsymbol{\beta})q(\boldsymbol{\gamma})q(\boldsymbol{\theta})$ trying to minimize the Kullback–Leibler-distance. Obviously, such a factorized approximation cannot capture dependencies between blocks of parameters but one hopes that the factors approximate well the true posterior marginals for blocks of parameters. A potential problem is that VB estimates may approximate well the posterior mode (Wang and Titterington, 2006) but the posterior variance may be severely underestimated, see Bishop (2006, Ch. 10) and Rue, Martino and Chopin (2009, Appendix A) for a simple example with a Gaussian prior for $\boldsymbol{\xi}$. Another approximation technique developed in machine learning is the expectation-propagation approach, see for example Kuss and Rasmussen (2005), Bishop (2006, Ch. 10), but it often overestimates posterior variances.

Integrated nested laplace approximation (INLA)

Rue and Martino (2007) and Rue, Martino and Chopin (2009) developed approximate inference for models with Gaussian priors $p(\boldsymbol{\xi}|\boldsymbol{\theta}_{\boldsymbol{\xi}})$, combining Laplace approximations with numerical integration methods. The main steps of their *integrated nested Laplace approximation* (INLA) are sketched in the following.

Basic assumptions are: Individual observations y_i are conditionally independent, the prior for $\boldsymbol{\xi}$ given $\boldsymbol{\theta}_{\boldsymbol{\xi}}$ is a Gaussian Markov random field (GMRF) with mean zero and (sparse) precision matrix $\boldsymbol{K}(\boldsymbol{\theta}_{\boldsymbol{\xi}})$, and the number of hyperparameters is small. The posterior is

$$\begin{aligned} p(\boldsymbol{\xi}, \boldsymbol{\theta}|\boldsymbol{y}) &\propto \left[\prod_i p(y_i|\boldsymbol{\xi}, \boldsymbol{\theta}_{\boldsymbol{y}})\right] p(\boldsymbol{\xi}|\boldsymbol{\theta}_{\boldsymbol{\xi}})p(\boldsymbol{\theta}) \\ &\propto \exp\left\{\sum_i p(y_i|\boldsymbol{\xi}, \boldsymbol{\theta}_{\boldsymbol{y}}) - \frac{1}{2}\boldsymbol{\xi}'\boldsymbol{K}(\boldsymbol{\theta}_{\boldsymbol{\xi}})\boldsymbol{\xi}\right\} p(\boldsymbol{\theta}). \end{aligned}$$

The main aim is to approximate posterior marginals $p(\xi_l|\boldsymbol{y})$ for components ξ_l of $\boldsymbol{\xi}$, $p(\boldsymbol{\theta}|\boldsymbol{y})$ and $p(\theta_j|\boldsymbol{y})$ for components of $\boldsymbol{\theta}$. The posterior marginals are

$$\begin{aligned} p(\xi_l|\boldsymbol{y}) &= \int p(\xi_l|\boldsymbol{\theta}, \boldsymbol{y})p(\boldsymbol{\theta}|\boldsymbol{y})d\boldsymbol{\theta} \\ p(\theta_j|\boldsymbol{y}) &= \int p(\boldsymbol{\theta}|\boldsymbol{y})d\boldsymbol{\theta}_{-j}. \end{aligned}$$

The key feature is to replace conditional densities $p(\cdot|\cdot)$ through approximate densities $\tilde{p}(\cdot|\cdot)$, combining Laplace approximations and numerical integrations. The marginal posterior of $\boldsymbol{\theta}$ is approximated by

$$\tilde{p}(\boldsymbol{\theta}|\boldsymbol{y}) \propto \left.\frac{p(\boldsymbol{\xi}, \boldsymbol{\theta}, \boldsymbol{y})}{\tilde{p}_G(\boldsymbol{\xi}|\boldsymbol{\theta}, \boldsymbol{y})}\right|_{\boldsymbol{\xi}=\boldsymbol{\xi}^*(\boldsymbol{\theta})},$$

where $\tilde{p}_G$ is the Gaussian approximation matching the posterior mode $\boldsymbol{\xi} = \boldsymbol{\xi}^*(\boldsymbol{\theta})$ and the posterior variance, given $\boldsymbol{\theta}$. This is equivalent to the Laplace approximation of Tierney and Kadane (1986). The main use of $\tilde{p}(\boldsymbol{\theta}|\boldsymbol{y})$ is to evaluate the integrals for $p(\xi_l|\boldsymbol{y})$ and $p(\theta_j|\boldsymbol{y})$ numerically, that is, it suffices to compute good evaluation points $\{\boldsymbol{\theta}_k\}$ for numerical integration. Approximate posterior marginals can then be obtained directly from $\tilde{p}(\boldsymbol{\theta}|\boldsymbol{y})$, using numerical integration. To obtain $\tilde{p}(\xi_l|\boldsymbol{y})$ by integrating out $\boldsymbol{\theta}$, an approximation $\tilde{p}(\xi_l|\boldsymbol{\theta}, \boldsymbol{y})$ is needed. Three options are suggested: A Gaussian approximation obtained from the marginals of $\tilde{p}_G(\boldsymbol{\xi}|\boldsymbol{\theta}, \boldsymbol{y})$ as the computationally cheapest but possibly crude approximation, another Laplace approximation, and a simplified Laplace approximation, see Rue, Martino and Chopin (2009) for details and applications to simulated and real data. INLA is implemented in the open source GMRFLib library and available in the `R` add-on package `INLA` (www.r-inla.org).

4.6.3 Further comments

In this chapter we focused on Bayesian P-splines for fitting nonlinear function effects in SPMMs. Of course, other approaches for smoothing such as locally adaptive variants or adaptive knot and basis function selection methods are possible, see for example Bigelow and Dunson (2007) for adaptive regression splines for hierarchical data. Another issue not touched upon is incomplete or missing data and covariate measurement error in nonparametric longitudinal regression. The handbook edited by Fitzmaurice *et al.* (2009) is a good source of information for such issues.

5 Spatial Smoothing, Interactions and Geoadditive Regression

Spatial information is generated or collected in diverse scientific fields and its availability is increasing in scientific and public data sources. In many applications such as environmental studies, disease mapping, or medical imaging, spatial data analysis and smoothing is of primary interest. In a very simple form, the problem of spatial smoothing can be formulated as follows: Let $\boldsymbol{s}$ denote the continuous spatial location in a subset or domain $D \subset \mathbb{R}^d$ and $y(\boldsymbol{s})$ a univariate spatial observation at $\boldsymbol{s}$. In most spatial applications, we will be faced with bivariate surfaces, that is, the special case $d = 2$, but most of the statistical methodology considered in the following can (at least conceptually) be extended to $d > 2$. For Gaussian responses, a purely spatial regression model assumes

$$y(\boldsymbol{s}) = f_{\text{spat}}(\boldsymbol{s}) + \varepsilon(\boldsymbol{s}),$$

where $f_{\text{spat}}(\boldsymbol{s})$ is an unknown spatial effect, image or surface observed with error $\varepsilon(\boldsymbol{s})$. A typical application appears in Example 5.1 for rainfall data in Switzerland where one is interested in inferring the amount of rainfall in all Switzerland based on observations from a number of measurement stations. Spatial smoothing then refers to estimating $f_{\text{spat}}(\boldsymbol{s})$ in a non- or semiparametric (Bayesian) fashion. If additional covariates $\boldsymbol{u}(\boldsymbol{s})$ carrying spatial information are available, the basic spatial model may be extended to

$$y(\boldsymbol{s}) = \boldsymbol{u}(\boldsymbol{s})'\boldsymbol{\beta} + f_{\text{spat}}(\boldsymbol{s}) + \varepsilon(\boldsymbol{s}).$$

For more general response types, a spatial regression model can analogously be defined by specifying a spatial predictor $\eta(\boldsymbol{s}) = f_{\text{spat}}(\boldsymbol{s})$ or $\eta(\boldsymbol{s}) = \boldsymbol{u}(\boldsymbol{s})'\boldsymbol{\beta} + f_{\text{spat}}(\boldsymbol{s})$ with $\mathbb{E}(y(\boldsymbol{s})) = h(\eta(\boldsymbol{s}))$. Again the aim is to recover the function $f_{\text{spat}}(\boldsymbol{s})$ from spatial data $y(\boldsymbol{s}_1), \ldots, y(\boldsymbol{s}_n)$ observed at locations $\boldsymbol{s}_1, \ldots, \boldsymbol{s}_n$. In the following sections, we will at some places focus on Gaussian models to fix ideas about spatial smoothing, but generalized responses can be treated with the same methods and the required extensions will be summarized at the end of the chapter.

For spatial smoothing, a basic differentiation is between continuous and discrete spatial information. In the continuous case, spatial information comes in the form of coordinates, that is, the location is represented as $\boldsymbol{s} = (s_1, s_2)$ with coordinates s_1, s_2, and $\boldsymbol{s}$ can vary continuously in $D \subset \mathbb{R}^2$. The location of the measurement stations in the Swiss rainfall data (Example 5.1) are typical for this situation. In the discrete case, $s \in \{1, \ldots, S\}$ is a spatial index, representing for example pixel s on a regular lattice of S pixels in imaging, or denoting region or area s within an irregular lattice of S regions such as in disease mapping (see Example 5.6), or in other epidemiological or geographical applications.

Spatial smoothing with continuous spatial information is obviously closely related to bivariate regression and to modelling interactions between two continuous covariates x_1, x_2. After inserting $\boldsymbol{x} = (x_1, x_2)$ and re-indexing observations,

$$y_i = f(x_{i1}, x_{i2}) + \varepsilon_i,$$

appears as the basic model of bivariate smoothing. The function $f(x_1, x_2)$ can now also be interpreted as a nonparametric interaction effect between x_1 and x_2. It is one of the aims of this chapter to highlight the close relationship between spatial smoothing and non- or semiparametric modelling of interactions. There are, however, some important differences between spatial smoothing and more general bivariate surface smoothing that require a separate treatment.

In many regression situations, spatial information in continuous or discrete form is available in addition to other covariates. Therefore, the incorporation of spatial effects into regression models has received a lot of attention in recent years. In particular, it has been recognized that ignoring spatial correlation in statistical analyses bears the risk of false conclusions about covariate effects, especially with regard to significance statements. The main reason is that spatial proximity of the collected data points typically leads to an effective reduction of the sample size. While each observation counts as one degree of freedom in an analysis based on independent data, positive correlations between observation points effectively reduce the number of observations. Ignoring this fact leads to over-optimistic statements regarding the efficiency and precision of statistical estimates. One aim of spatial regression models is therefore to correct this artificial behaviour by including the spatial correlation into the model. However, modelling spatial correlation should not be mistaken as the sole purpose of spatial regression models: Often the spatial effect (interpreted as a surface of regression effects) is also of interest in itself. It can for example provide information on unobserved covariates driving the data-generating process or can be the basis of a spatial interpolation strategy. Note also that although most spatial data situations will relate to positive spatial correlation, negative correlations are also sometimes encountered, for example in the case of disaggregation between animal populations.

Geoadditive regression models account for spatial effects by adding a spatial component $f_{\text{spat}}(\boldsymbol{s})$ to the additive predictor of semiparametric models considered in Chapter 2 for cross-sectional data as well as of semiparametric mixed models (Chapters 3 and 4). Methods for spatial smoothing and for modelling interactions are then another basic building block for inference in geoadditive regression models.

To summarize, this chapter provides an introduction to Bayesian spatial smoothing as a subject of interest in its own right, it points out the close relation between modelling interactions and spatial effects, and it extends smoothing and regression to geoadditive regression models. Section 5.1 introduces the different types of spatial data in more detail and provides first information on the corresponding modelling techniques. Section 5.2 describes Markov random fields as basic stochastic process models for discrete spatial data, Section 5.3 highlights relations between continuous spatial smoothing approaches and modelling of interactions, and Section 5.4 introduces Gaussian random fields as stochastic process models for continuous spatial data, including their use in classical geostatistics. Section 5.5 incorporates ideas from the previous sections in the general framework of geoadditive regression.

5.1 Spatial data structures

5.1.1 Point-referenced data: Continuous spatial information

When considering point-referenced data, the observations are indexed by spatial coordinates $\boldsymbol{s} = (s_1, \ldots, s_d)$ varying continuously in a suitable subspace of $\mathbb{R}^d$.

Example 5.1: Swiss rainfall data

In this example, taken from the R-package `geoR` accompanying the book by Diggle and Ribeiro Jr. (2007), information on the measured rainfall from 467 locations in Switzerland on 8 May, 1986 is given together with the the locations of the measurements in terms of coordinates. This information is visualized in Fig. 5.1 where each circle represents one single observation and the diameter of the circles is proportional to the amount of rain observed at this particular point. The main aim in this application will be to provide a smoothed surface representing the observed rainfall after denoising. In Section 5.4 we will also demonstrate the advantages of anisotropic (i.e. directional) modelling with this data. △

Example 5.2: Human brain mapping

Human brain mapping is concerned with the detection of regions in the brain that are associated with the completion of specific tasks such as the visual or the auditory cortex. The detection of these activated regions can be accomplished by functional magnetic resonance imaging (fMRI) techniques that rely on the increase of blood oxygenation in activated areas and the different magnetic properties of oxygenated and deoxygenated blood. In this example, a sequence of visual signals has been presented to the participants, separated by rest conditions.

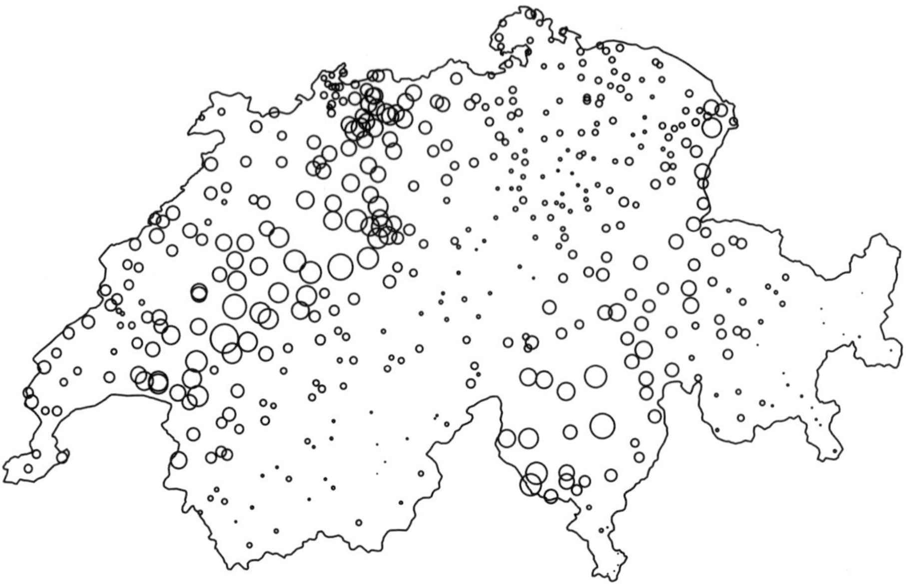

Fig. 5.1 Swiss rainfall: Location of observations. The diameter of the circles is proportional to the amount of rain observed at this point.

More precisely, the brain activity of a study participant is measured at a large number of volume elements (voxels) for a set of equidistant time points, yielding activation profiles $y(\boldsymbol{s}, t)$, where $\boldsymbol{s} \in \mathbb{R}^3$ is the location of the voxel in the brain (represented in terms of the 3D coordinates of the voxel centroid), and $t = 1, \ldots, T$, is the measurement time. In principle, we are therefore faced with a space-time data set but we will reduce the complexity in the following for the sake of illustration. Note also that voxel data are not necessarily treated within the framework of continuous spatial data since they can also be interpreted as discrete lattice data.

Usual analyses of fMRI data can be divided into three steps (see for example Gössl, Auer and Fahrmeir, 2001 and Fahrmeir and Gössl, 2002). The first step corresponds to data processing and comprises for example corrections for movements of the participant during the data acquisition. In the second step, voxelwise analyses are performed to remove temporal trends in the data and to relate the activation to the stimulus. The third step analyses spatial dependence in the results obtained in the voxelwise analyses. In this chapter, we will solely focus on the third step, that is, spatial smoothing of activation estimates obtained from voxelwise analyses.

Therefore, we use the same data as in Lang and Brezger (2004) with measurements on 70 time points and 2,950 pixels of a horizontal slice through the brain. To determine the activation profile at a specific pixel, Lang and Brezger (2004) utilize Bayesian P-splines to estimate the time-series model

$$y(\boldsymbol{s},t) = \beta_0 + f_1^{(\boldsymbol{s})}(t) + f_2^{(\boldsymbol{s})}(t)x_{\boldsymbol{s},t} + \varepsilon_{\boldsymbol{s},t}, \quad t = 1,\ldots,70,$$

separately for each pixel $\boldsymbol{s}$. While $f_1^{(\boldsymbol{s})}(t)$ is a general time trend, $f_2^{(\boldsymbol{s})}(t)$ corresponds to the time-varying activation due to the presented stimulus. To account for the smooth on- and offset of the activation in the brain, the zero-one-stimulus is folded with the so-called hemodynamic response function, transforming it into the continuous covariate $x_{\boldsymbol{s},t}$.

The final data for our spatial analyses are estimated effects $\hat{f}_2^{(\boldsymbol{s})}(t)$. To perform spatial smoothing, we fix a single point in time where the stimulus has been present. Figure 5.2 shows the corresponding estimates $\hat{f}_2^{(\boldsymbol{s})}(t)$ (as a function of the spatial location $\boldsymbol{s}$) which will serve as the responses later on in this chapter. While the visual cortex can be identified, there is quite a lot of noise and variation in the data. The purpose of spatial smoothing will then be both to denoise the data and to provide a clearer localization of the visual cortex. △

Example 5.3: Leukemia survival in northwest England
Henderson, Shimakura and Gorst (2002) present data on the survival times of adults after diagnosis of acute myeloid leukemia (AML). The data set contains information on all 1,043 cases of AML in northwest England collected between 1982 and 1998. In addition to the survival times and a number of covariates, the

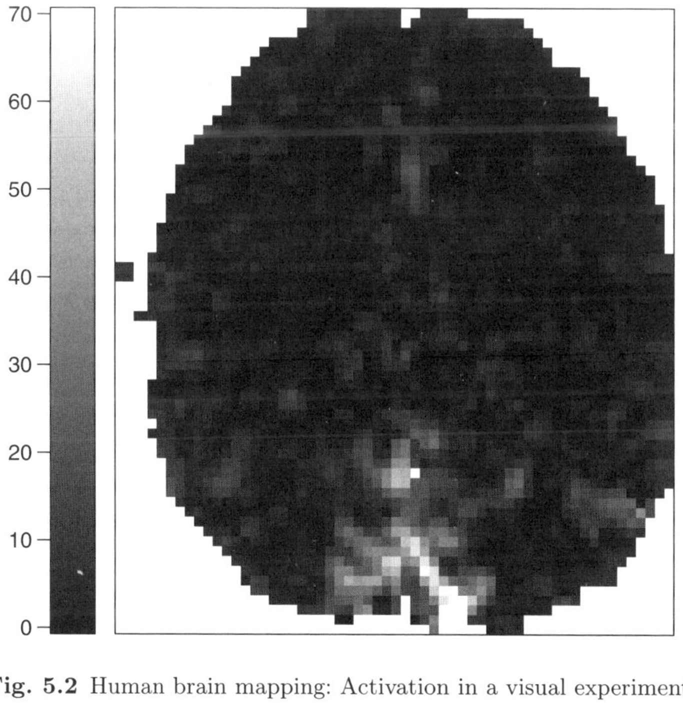

Fig. 5.2 Human brain mapping: Activation in a visual experiment.

exact location of the residence of the patients is available in terms of longitude and latitude as visualized in Fig. 5.3. In Chapter 6, we will incorporate this coordinate information in a geoadditive hazard regression model to take spatial correlations in the survival times into account. Hazard regression will provide us with a convenient framework for dealing with a common problem in the analysis of survival times: At the end of the study, several participants are still alive such that their survival times are censored at the end of the study period. Hazard regression models allow us to incorporate also these (right-) censored survival times as well as survival times with more general censoring mechanisms. Note that we are not interested in the spatial agglomeration of the data that is caused by the unequal population density, but in the distribution of survival times in different parts of northwest England. △

Fig. 5.3 Leukemia survival data in northwest England: Locations of the observations.

One of the main purposes of spatial regression models is to take spatial correlation between spatially aligned points into account. In case of point-referenced data, spatial Gaussian processes typically form the basis for such approaches, where the spatial correlation is modelled explicitly in terms of a spatially correlated process. In a simple linear model for normally distributed responses, this yields the regression model

$$y(\boldsymbol{s}) = \boldsymbol{u}(\boldsymbol{s})'\boldsymbol{\beta} + \gamma(\boldsymbol{s}) + \varepsilon(\boldsymbol{s}), \tag{5.1}$$

where $\boldsymbol{u}(\boldsymbol{s})'\boldsymbol{\beta}$ corresponds to linear effects with covariates $\boldsymbol{u}(\boldsymbol{s})$ indexed by their spatial location to stress the spatial allocation of the data and $\varepsilon(\boldsymbol{s})$ as the usual i.i.d. error. In contrast, $\gamma(\boldsymbol{s})$ is considered a spatially correlated, zero-mean Gaussian stochastic process, typically supplemented with additional assumptions such as stationarity to simplify the model. Stationary Gaussian processes are chosen mostly for convenience: They can be characterized in terms of their expectation and a covariance function, making the underlying spatial correlation evident in the model formulation. The Gaussian distribution also has computational advantages since it greatly facilitates inferential procedures. In particular, it readily fits into the general framework of Gaussian smoothness priors considered in Chapter 4.

Model (5.1) is traditionally known as a universal Kriging model after the South-African mining engineer D. G. Krige who invented models of this type for interpolating ore grades in gold mines (Krige, 1951). Krige also coined the term 'geostatistics' as a label for spatial approaches based on Gaussian process priors. Motivated from the original application, Kriging has traditionally been associated with interpolation and optimal prediction of the spatial process at unobserved locations. In the following, we will rather consider it as a spatial smoothing approach and discuss the classical approach only briefly. We will therefore also replace the notation in (5.1) that emphasizes the spatial nature of the data and the spatial process assumption with a variant that reveals the connection to smoothing and regression more clearly. We therefore consider $\boldsymbol{s}_i$ to be a covariate vector providing spatial information, index the observations by $i = 1, \ldots, n$, and write the spatial effect as a function $f_{\text{spat}}(\boldsymbol{s})$ of the coordinates, yielding

$$y_i = \boldsymbol{u}_i'\boldsymbol{\beta} + f_{\text{spat}}(\boldsymbol{s}_i) + \varepsilon_i$$

where the dependence of responses, covariates and errors on the spatial location is suppressed for simplicity. Supplementing appropriate prior or stochastic process assumptions to the function $f_{\text{spat}}(\boldsymbol{s})$ then makes the model equivalent to the original Kriging model. In the following, we will mostly suppress the index 'spat' to simplify notation.

As already noted above, a (stationary) Gaussian process with zero mean can be characterized in terms of a correlation function $\rho(\cdot)$. In spatial models, this correlation function is typically a function of the distance between two points in space, that is,

$$\mathrm{Cor}(f(\boldsymbol{s}_i), f(\boldsymbol{s}_j)) = \rho(||\boldsymbol{s}_i - \boldsymbol{s}_j||, \boldsymbol{\vartheta})$$

where $\boldsymbol{\vartheta}$ is a vector of parameters involved in the definition of the correlation function and $||.||$ denotes a general distance measure. Several types of correlation functions are available and will be discussed in more detail in Section 5.4. There we will also present theoretical characterizations that allow for the construction of new types of correlation functions.

Another source of diversity in spatial models arises from different distance measures that can be used in combination with a correlation function. The standard choice is of course the Euclidean distance in $\mathbb{R}^d$, where $|| \cdot ||$ denotes the Euclidean norm

$$||\boldsymbol{s}|| = \sqrt{s_1^2 + \ldots + s_d^2} = \sqrt{\boldsymbol{s}'\boldsymbol{s}}.$$

However, despite its immediate appeal, using the Euclidean distance comes with some implicit assumptions that will not necessarily be fulfilled for all applications. To be more specific, the Euclidean distance treats the distance between two points as the same regardless of the direction of the distance vector. In the context of geostatistics, this assumption is known as isotropy since in this case the contour lines of any correlation function defined upon the distance will be circles. In Section 5.4 we will present alternative distance measures that yield anisotropic correlation functions.

Though being introduced as spatial stochastic processes, it will turn out that geostatistical approaches are actually closely related to smoothing based on basis functions. This is in analogy to our considerations in the previous chapter on the relation between, for example, autoregressive priors, random walks and penalized splines in Section 4.3. As we will see later on in Section 5.4, the correlation functions then translate into basis functions and the penalty also depends on the correlation function. This connection also hints at a second possibility for including spatial effects in regression models with continuous coordinate information: Extending the concepts of penalized basis function smoothing, bivariate (or higher-dimensional) basis functions can be constructed and supplemented by appropriate penalties. This also opens the way to considering more general surface estimation, where the arguments are not necessarily coordinates but may be any type of continuous covariates.

5.1.2 Interaction surfaces

At first sight, modelling a nonparametric interaction surface $f(\boldsymbol{x})$ based on a d-dimensional covariate vector $\boldsymbol{x}$ seems to be exactly the same task as smoothing a d-dimensional spatial surface based on coordinates $\boldsymbol{s}$ since we are seeking a smooth d-dimensional surface estimate. However, the basic difference lies in the scaling of elements in $\boldsymbol{x}$: While the coordinates $\boldsymbol{s}$ will typically be measured in the same units such as metres or kilometres, general covariates $\boldsymbol{x}$ may be measured at extremely different scales. This makes the construction of suitable distance measures in the covariate space more difficult than in inherently spatial approaches. Therefore it seems more natural to extend basis function approaches in this situation, since these do not rely on distances between the points.

Example 5.4: Forest health data

In the forest health data introduced in Example 1.2, an interaction between age and calendar time has been hypothesized such that the effect of age is varying with time. This would require the estimation of a general surface $f(\text{age}, \text{year})$ that is smooth in both age and calendar time. To obtain a visual impression of whether the data support the necessity of an interaction, we computed temporal trends (with year of measurement as time unit) of the percentage of damaged trees for different age groups as visualized in Fig. 5.4. The differences in the three curves give at least some indication of a possible interaction. Specifying an interaction surface $f(\text{age}, \text{year})$ will allow us to investigate the hypothesis of an interaction in more detail and in particular does not rely on an arbitrary definition of age groups. △

Example 5.5: Munich rental guide

For the Munich rental guide data (Example 1.4), previous analyses have shown an interaction effect between floor size and year of construction. To explore the presence of such an interaction in our data set, we fitted separate univariate penalized spline smooths for the effect of floor size in subgroups defined by age classes. This yields the function estimates shown in Fig. 5.5. Obviously, there

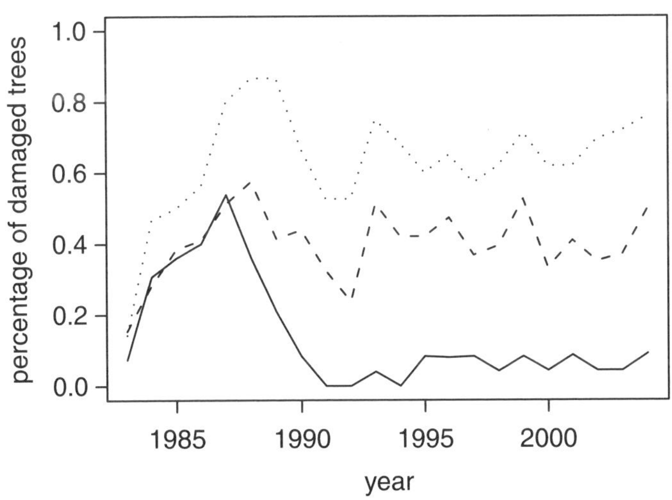

Fig. 5.4 Forest health: Temporal development of the percentage of damaged trees for different age groups. Young trees are represented by solid lines, middle-aged trees by dashed lines and old trees by dotted lines.

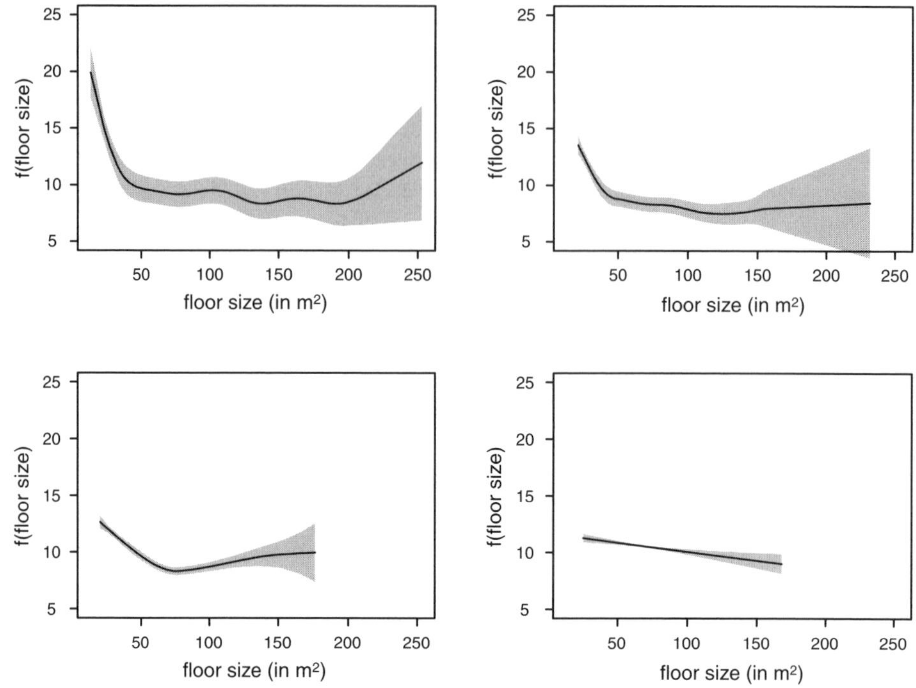

Fig. 5.5 Munich rental guide: Separate penalized spline fits for the effect of floor size in groups of buildings built before 1945 (top left), between 1946 and 1965 (top right), between 1966 and 1985 (bottom left) and after 1986 (bottom right).

are quite some differences between the effects in the subgroups. In particular, the effect for the newer buildings is only very moderate and close to linear while it is much more strongly expressed and highly nonlinear for older buildings.

The univariate fits also indicate that the range of the floor sizes observed for the flats varies considerably for the groups. Very large flats are relatively rare in newly constructed buildings, which is reflected both in the restricted range of the graphs on the floor size axis and the width of the confidence intervals for larger flats. Figure 5.6 visualizes the joint distribution of size and year of construction in terms of a scatter plot that further supports the change in the range of the floor size over time. It also reveals some clustering with respect to the year of construction that is due to limited information in the data acquisition. Several buildings are therefore assigned to the years 1918 and 1945 due to incomplete information.

Visual investigations of the joint distribution of covariates that should be included in terms of a nonparametric interaction surface such as $f(\text{size}, \text{year})$ are important to detect areas where estimation of surfaces is not well supported by the data. In the rental guide example, we might decide to exclude very

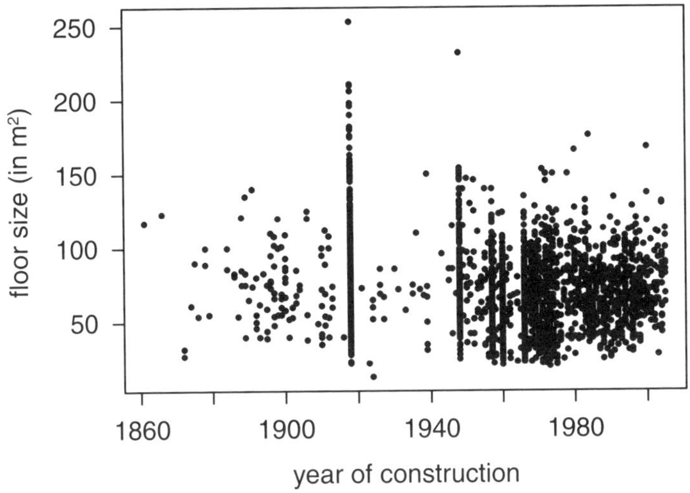

Fig. 5.6 Munich rental guide: Joint distribution of the covariates floor size and year of construction.

large flats when estimating an interaction effect to avoid artefacts for covariate combinations that are not well represented in the data. △

An easy possibility for constructing basis functions of higher dimension are tensor products. In this case, univariate bases for each of the elements in $\boldsymbol{x}$ are constructed first and the multivariate basis functions are then obtained by building all possible interaction terms, that is, by considering the products of the univariate basis functions as the multivariate basis. For example, in case of a bivariate surface, this yields the basis function representation

$$f(x_1, x_2) = \sum_{k_1=1}^{K_1} \sum_{k_2=1}^{K_2} \gamma_{k_1 k_2} B_{k_1 k_2}(x_1, x_2)$$

where the bivariate basis functions $B_{k_1 k_2}(x_1, x_2)$ are composed as products of univariate basis functions in x_1- and x_2-direction, that is, $B_{k_1 k_2}(x_1, x_2) = B_{k_1}(x_1) \cdot B_{k_2}(x_2)$. This approach is quite intuitive and is (at least conceptually) also applicable in higher dimensions. It is also possible to bring forward the penalties from the univariate bases to the multivariate case by considering suitable Kronecker products. We will discuss this extensively in Section 5.3.

By construction, tensor product P-splines of low degree have a relatively large amount of anisotropy. To overcome this, a second class of surface smoothers has been introduced: A set of points is chosen as spatial knots and basis functions are defined for each of these knots by inserting the distance between the point of interest and the knot into a univariate basis function. This yields so-called radial basis function approaches which will also be covered in more detail in Section 5.3.

5.1.3 Areal data: Discrete spatial information

Areal spatial data represent a different source of spatial information. In this case, no exact coordinate information on the observations is available but the observations are assigned to clusters with a specific spatial alignment. The most frequent example are census or survey data, where due to practical or confidentiality reasons, the place of residence is aggregated to administrative districts or regions. This spatial information is discrete by nature, since only a finite set of regions is available.

In principle, the clustering present in such spatial data is in complete analogy to the mechanisms considered in Chapter 3 where the clustering was induced, for example, by repeated measurements in longitudinal data. If the spatial effect is considered as a surrogate for unobserved covariates, usual i.i.d. random effects based on the regions will still be appropriate as long as the covariates do not obey a spatial structure. If this, however, is the case, regions that are spatially close by will tend to have similar covariate values, leading to correlations in the random effects.

In many spatial regression situations, spatial correlations will be present so that spatial closeness of the regions has to be taken into account. The difficulty arising from the discreteness of such spatial data is that no direct measure for the distance between two regions is available. As an indirect measure, spatial smoothing approaches based on Markov random fields consider the notion of neighbours and neighbourhoods. For example, regions that share a common boundary may be treated as neighbours. For regularly gridded regions, higher order neighbourhoods can also be defined which will be helpful in the construction of penalties for bivariate and higher order basis functions in the context of the estimation of interaction surfaces. In the following, we will typically consider first-order neighbours only.

Assuming a Gaussian Markov random field for the spatial effect now conceptualizes the idea that the effects in neighbouring regions should be similar. For a spatial regression model with predictor

$$\eta_i = \boldsymbol{u}_i'\boldsymbol{\beta} + f_{\text{spat}}(s_i),$$

we define the vector of spatial effects $\boldsymbol{\gamma}$ with elements $\gamma_s = f_{\text{spat}}(s)$ corresponding to the spatial effect in region s, $s = 1, \dots, S$. Then the conditional distribution of γ_s is *a priori* assumed to be a normal distribution where the expectation is given by the average of neighbouring values while the variance is inverse proportional to the number of neighbours. This yields the conditional distribution

$$\gamma_s | \boldsymbol{\gamma}_{-s} \sim \mathrm{N}\left(\frac{1}{N_s}\sum_{r\in\delta_s}\gamma_r, \frac{\tau^2}{N_s}\right) \tag{5.2}$$

where $\boldsymbol{\gamma}_{-s}$ denotes the vector of spatial effects excluding region s, δ_s contains the indices of all direct neighbours of s and $N_s = |\delta_s|$ is the number of such neighbours. The spatial regression model from above can then be rewritten as

$$\eta_i = \boldsymbol{u}_i'\boldsymbol{\beta} + \boldsymbol{x}_i'\boldsymbol{\gamma}$$

where $\boldsymbol{x}_i$ is a zero/one incidence vector that links observation i with the region it has been collected in and $\boldsymbol{\gamma}$ is supplemented with the Gaussian Markov random field prior (5.2). More details on this prior specification will be discussed in Section 5.2 including the derivation of a joint prior distribution for $\boldsymbol{\gamma}$ from the conditional distributions given in (5.2).

If both uncorrelated and correlated unobserved covariates are inducing the spatial effect, it may be useful to split up the overall spatial effect into a structured and an unstructured component. Unstructured effects allow for localized differences in effects of neighbouring regions which can not be captured by the smooth spatial trend induced by the Gaussian Markov random field prior. They can be represented by an i.i.d. random effect where the regions provide the clustering information. Note however, that it will typically not be possible to clearly separate between the structured and unstructured effects so that only the sum of both will be interpretable. Still, splitting up the effect may be helpful in reliably identifying the complete spatial pattern.

Example 5.6: Disease mapping of oral cavity cancer

The heading 'disease mapping' summarizes statistical approaches that analyse the geographical variation in disease risks, for example with respect to mortality from a particular disease reported on a regional level. Major aims in such analyses are the detection of unknown risk factors and the quantification of covariate impacts. If the disease is rare and non-infectious, the number of deaths or illnesses in a region can be approximated by a Poisson distribution with intensity λ_s, that is

$$y_s \sim \text{Po}(\lambda_s)$$

where both the response and the intensity of the Poisson distribution are indexed by $s \in \{1, \dots, S\}$ representing the region. To make the rates of the different regions comparable in magnitude, an additional quantity representing the expected risk is introduced. If, for example, the global risk for a single person averaged across all regions is given by a probability π and n_s denotes the population size in region s, one would expect $e_s = n_s\pi$ if no spatial differences exist. More generally, the expected cases e_s could also take special properties of a region into account, such as the age distribution or other known risk factors (although these may also be included as covariate information). Adding the expected number of cases to the model specification yields

$$y_s \sim \text{Po}(e_s\lambda_s).$$

In a general regression formulation, e_s is also called an offset.

In this example, we are dealing with mortality due to oral cavity cancer in Germany between 1985 and 1990. For a visual representation of the observed cases, it is advantageous to consider the standardized mortality ratios (SMR) which are given by the maximum likelihood estimates for λ_s

$$\hat{\lambda}_s = \frac{y_s}{e_s}.$$

The SMR are comparable in magnitude across districts and should deviate from the expected value 1 only due to Poisson variation if the observations in the regions are independent and risk homogeneous. Heterogeneity in the rates can also be assessed in terms of p-values for the hypotheses

$$H_0 : \lambda_s \leq 1 \quad \text{vs.} \quad H_1 : \lambda_s > 1.$$

Here we have used one-sided hypotheses since one is typically interested in regions that exceed the average mortality risk.

Figure 5.7 shows both the standardized mortality ratios and the p-values for the oral cavity cancer data. Note that both standardized mortality ratios

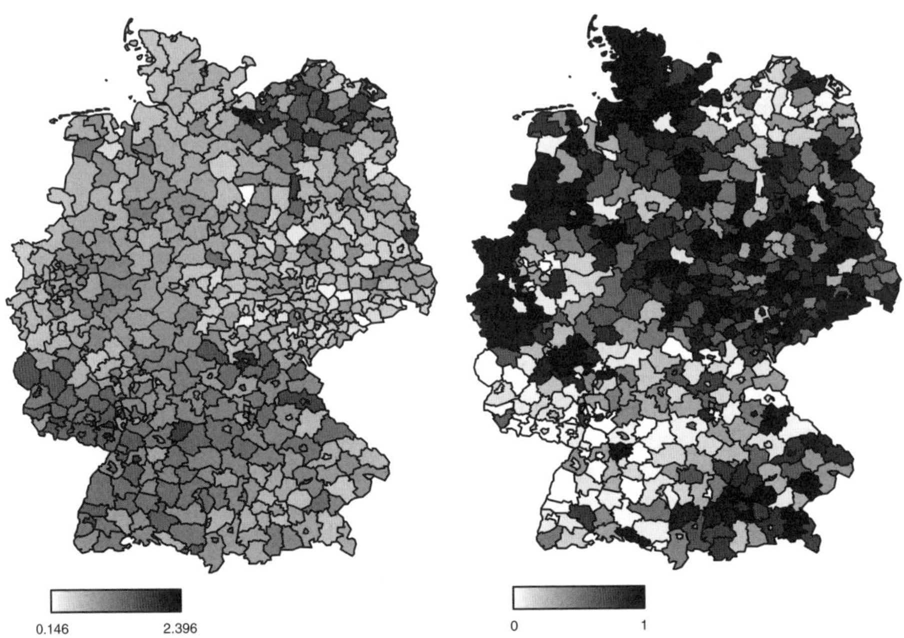

Fig. 5.7 Disease mapping: Standardized mortality ratios (left panel) and p-values (right panel) for mortality due to oral cavity cancer in Germany between 1985 and 1990.

and p-values are calculated independently and therefore ignore any spatial correlation that may be present in the data. Still, both figures show an increased risk in the northeastern part and also in the middle western part of Germany. However, the data also show a considerable amount of variation, and some cities such as Berlin show up with an increased risk despite the correction for expected no. of cases. In a spatial regression model, one would therefore specify a combination of a spatially structured and a spatially unstructured effect, for example

$$\lambda_s = \exp(\gamma_s + \alpha_s).$$

While the spatially structured effect $\gamma_s = f_{\text{spat}}(s)$ is considered a smooth function of the district variable and therefore captures spatial correlations, α_s represents an i.i.d. region-specific random effect that represents localised features.

For the Markov random field prior, we could simply assume that all regions with common boundary are treated as neighbours. Figure 5.8 indicates all neighbourhood relationships in terms of lines connecting the centroids of the regions. △

5.1.4 Continuous vs. discrete spatial information

Although we have described continuous and discrete spatial information as two separated concepts, the discrimination between the two is not that clear-cut in practice. For example, discrete information on regions may be turned into coordinate information by considering the centroids of the regions. This may be of practical relevance if the number of regions is very large, since in this case basis-function approaches allow us to reduce the computational burden.

On the other hand, continuous coordinate information can be treated within the framework of Markov random fields by defining neighbourhoods based on distance measures. An alternative discretization strategy is to overlay the observation area with a discrete grid and to assign observations to the grid cells. The grid cells can then again be treated as (regularly shaped) regions.

Example 5.7: Forest health

The forest health data set contains spatial information on the locations of the observation plots in terms of coordinates. Hence, spatial regression models including continuous spatial information would probably be the first choice. However, the number of distinct observation plots is relatively low and large parts of the rectangular observation area do not contain any data. Therefore, we may consider a discrete spatial model as an alternative. Figure 5.9 illustrates the definition of neighbours based on a distance of 1.2 km around an observation plot. All observation plots within this distance are defined to be neighbours of the point in the middle of the circle, leading to the neighbourhood structure visualized in Figure 5.10. △

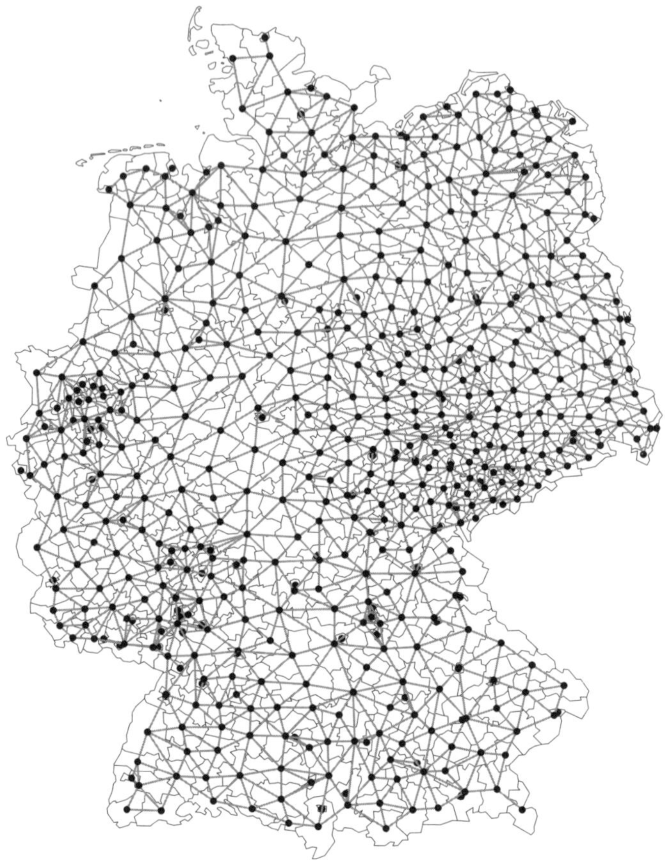

Fig. 5.8 Disease mapping: Neighbourhood relationships in the map of Germany.

In summary, the question whether continuous or discrete spatial modelling is more appropriate is not only related to the data provided, but also to statistical and computational feasibility.

5.1.5 Spatial regression models

The spatial modelling possibilities introduced in the remainder of this chapter will serve as additions to the regression toolbox, that is, they will be augmented to the different types of models considered in previous chapters. For example, we can define random effects models with an additional spatial component. More generally, we will consider geoadditive regression models that extend the semiparametric regression models from Chapter 2 and the semiparametric mixed models introduced in Chapter 4. For example, in case of point-referenced data, a geoadditive model extending the additive model (2.56) (page 91) and the generalized additive model (2.62) (page 100) is given by

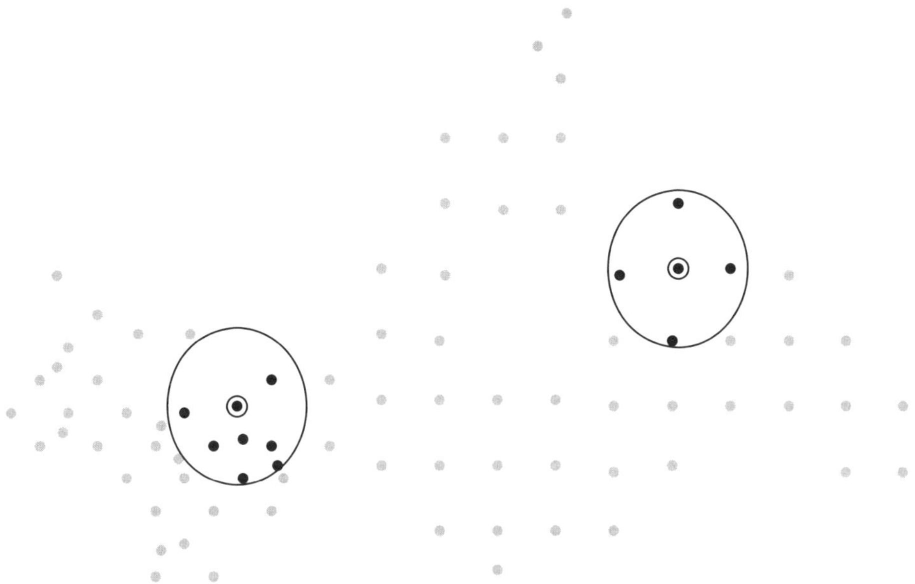

Fig. 5.9 Forest health: Neighbourhood definition based on a fixed distance.

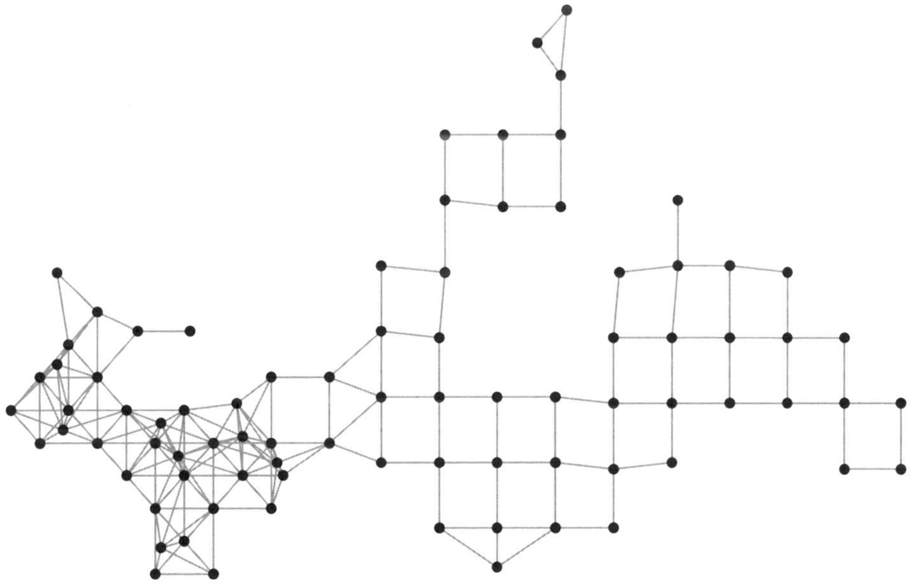

Fig. 5.10 Forest health: Neighbourhood structure obtained with a maximal distance of 1.2 km between adjacent observation plots.

$$\eta_i = \boldsymbol{u}_i'\boldsymbol{\beta} + f_1(x_{i1}) + \ldots + f_p(x_{ip}) + f_{\text{spat}}(\boldsymbol{s}_i)$$

where the only difference to a semiparametric regression model is the additional inclusion of the spatial effect $f_{\text{spat}}(\boldsymbol{s})$. For discrete spatial information, the surface $f_{\text{spat}}(\boldsymbol{s})$ is replaced with a spatial function $f_{\text{spat}}(s)$, where $s \in \{1, \ldots, S\}$ denotes a geographical index from a fixed set of S areas or regions. If longitudinal data have been collected in combination with spatial information as, for example, for the forest health data or the human brain mapping example, spatial effects can similarly be augmented to the predictor (4.2) of a semiparametric mixed model, yielding

$$\eta_{ij} = \boldsymbol{u}_{ij}'\boldsymbol{\beta} + f_1(x_{ij1}) + \ldots + f_p(x_{ijp}) + f_{\text{spat}}(\boldsymbol{s}_i) + \boldsymbol{z}_{ij}'\boldsymbol{\alpha}_i.$$

Example 5.8: Forest health
The forest health data are one example of spatially arranged data that also have a longitudinal component. Observations at the 83 observation plots are repeatedly collected between 1983 and 2004, leading to an ensemble of spatially correlated categorical time series. Figures 1.3 and 1.4 (page 8 and 9) show empirical estimates of the time trend and the spatial effect that indicate that both temporal and spatial effects should be included in the data. Since the same observation plots are considered over time, the additional inclusion of a plot-specific random effect seems to be recommended to adjust for the repeated measurements design. △

Extending the ideas of varying coefficient models, geoadditive models also comprise predictors with spatially varying effects. For example, if covariate u is suspected to have a spatially varying effect, the predictor could be augmented to

$$\eta_i = \ldots + u_i f_{\text{spat}}(\boldsymbol{s}_i) + \ldots$$

where now the spatial variable $\boldsymbol{s}$ is considered as the effect modifier. Such models can also be useful, if a small number of panel waves is available and if changes in the spatial pattern between the panel waves are of interest. With three panel waves, for example, the spatial effect could be represented as

$$\eta_i = \ldots + f_{1,\text{spat}}(\boldsymbol{s}_i) + u_{i2} f_{2,\text{spat}}(\boldsymbol{s}_i) + u_{i3} f_{3,\text{spat}}(\boldsymbol{s}_i) + \ldots$$

where u_2 and u_3 are binary indicators for the second and the third panel wave. In this case, $f_{1,\text{spat}}(\boldsymbol{s}_i)$ corresponds to the basic spatial effect in the first panel wave, whereas $f_{2,\text{spat}}(\boldsymbol{s}_i)$ and $f_{3,\text{spat}}(\boldsymbol{s}_i)$ represent the deviation from this basic effect in the second and the third panel wave. If both effects are small, this will reflect that no or only minor changes in the spatial pattern underlying the data have been present across panels. Note that similar identifiability problem will occur with spatially varying effects as we have discussed for additive and varying coefficient models in Section 4.2.1 (page 200).

When surfaces are employed for the analysis of interaction effects, the joint effect of a bivariate covariate $\boldsymbol{s} = (x_1, x_2)$ is frequently decomposed in an ANOVA-type model as

$$\eta_i = \ldots + f_1(x_{i1}) + f_2(x_{i2}) + f_{1,2}(x_{i1}, x_{i2}) + \ldots$$

where $f_1(x_1)$ and $f_2(x_2)$ represent additive main effects whereas $f_{1,2}(x_1, x_2)$ corresponds to the non-additive deviation from the main effects. In such models, the question of identifiability arises and it may frequently be the case that only the sum of the three effects, that is, the total interaction, is well identifiable while the decomposition into the separate terms cannot be completely recovered. In particular, the interaction is often only made identifiable by imposing informative prior distributions. Note that this typically does not cause numerical difficulties, that is, all model parameters are formally estimable but it is not always possible to interpret the estimated effects as straightforwardly as in the theoretical model formulation.

5.1.6 Other types of spatial data

The list of spatial data types we have considered so far is of course not exhaustive. In particular, similarly to the usual regression models, we take a conditional view on the generation of the spatial information, that is, we treat the spatial information as fixed. For point-referenced data, the location of the points may, however, also be random and may be of interest in itself. In this case, analyses frequently focus on models for the intensity at which observations arise at a particular spatial location (possibly depending on covariates in addition). The corresponding class of statistical models is formed by spatial point processes, see for example Schabenberger and Gotway (2005, Ch. 3), Møller and Waagepetersen (2004) or Illian, Penttinen, Stoyan and Stoyan (2008) for introductory texts. Both the focus on intensities and also the generated data of spatial point processes can be seen as generalizations of models that we will consider in Chapter 6, where the temporal development of the intensity for an event of interest will be modelled. A further extension, also contained in the references cited above, are marked point processes where a further response variable is associated with each location. This variable may for example represent different types of cells in biology, different types of trees in forestry, etc.

5.2 Discrete spatial data: Markov random fields

We start our considerations of spatial regression models with the case of discrete spatial information. The theoretical tool that will form the basis for setting up spatial functions $f_{\text{spat}}(s)$ based on areal data s are Markov random fields and in particular Gaussian Markov random fields as treated extensively in Rue and Held (2005). We will start with a special Gaussian Markov random field as a heuristic spatial smoothness prior to introduce the general ideas while avoiding theoretical issues and complexity for the moment. In the following sections, we will then thoroughly deal both with general Markov random fields and the special case of Gaussian Markov random fields.

5.2.1 A heuristic spatial smoothness prior

Consider spatially arranged data where the spatial information is given in terms of a regional index $s \in \mathcal{S}$, where $\mathcal{S} = \{s_1, \dots, s_S\}$ is the set of all regions. If no confusion can arise, we will also simply identify the S distinct regions with their indices leading to $\mathcal{S} = \{1, \dots, S\}$. For each of the regions, neighbourhood information is assumed to be given, that is, for each region s there exists a set of regions $\delta_s \subset \mathcal{S}$ which are considered to be 'close' to region s.

Our aim is to define suitable prior structures for imposing spatial smoothness and correlation in a regression model with predictor

$$\eta_i = \dots + \gamma_{s_i} + \dots$$

where $\gamma_{s_i} = f_{\text{spat}}(s_i)$ is the spatial effect in region s_i. Extended models will also allow for spatially varying effects of covariates via the inclusion of terms such as $u_i \gamma_{s_i}$. In both cases, a suitable prior distribution should express the assumption that neighbouring regions tend to have similar spatial effects γ_{s_i}. Such a prior distribution is given by

$$\gamma_s | \boldsymbol{\gamma}_{-s} \sim \mathrm{N}\left(\frac{1}{N_s} \sum_{r \in \delta_s} \gamma_r, \frac{\tau^2}{N_s} \right) \tag{5.3}$$

where $N_s = |\delta_s|$ is the number of neighbours for region s, τ^2 is a prior variance, and $\boldsymbol{\gamma}_{-s}$ is the vector comprising all spatial effects except the one for region s. The prior assumption (5.3) combines the following features:

- The (conditional) prior expectation for the spatial effect γ_s given the effects of all other regions equals the average of the neighbouring values. This relates to the assumption that—in tendency—neighbours share similar effects.
- The precision of the conditional prior assumption for a region s depends on the prior variance τ^2 and the number of neighbours N_s. The variance parameter can be interpreted in analogy to the smoothing variances in the context of nonparametric function estimation in Chapters 2 and 4, that is, a large variance allows for ample deviations from the prior expectation while a small value makes the prior assumption very strict. In (5.3), this translates to a flat surface estimate since for a small variance τ^2 all effects will tend to be the same. Of course, a suitable inferential procedure will include data-driven determination of τ^2 to avoid sensitivity with respect to extreme values chosen for τ^2. Making the prior variance inverse proportional to the number of neighbours indicates that regions with a very small number of neighbours are allowed to vary more strongly around the prior expectation than regions with a large number of neighbours.

- Since prior (5.3) is based on a normal distribution, the prior is completely characterized by expectation and variance. Both involve only neighbours of the region s and therefore implicitly induce a conditional independence assumption: The effect in each region is conditionally independent from all its non-neighbours given the values of the neighbouring regions. This results in a spatial Markov property that generalizes the temporal Markov property of random walks. The spatial Markov property will also form the basis for the more general treatment of smoothing priors for discrete spatial information based on Markov random fields in the following section.

Since (5.3) only defines full conditional distributions for the separate spatial effects, it is not clear at first sight, whether (5.3) actually defines a valid joint prior for the complete vector $\boldsymbol{\gamma} = (\gamma_1, \dots, \gamma_S)'$. However, in this particular case, it can be shown that no incompatibilities exist (see Section 5.2.2 for details) and that the joint prior is multivariate Gaussian, that is, of the form

$$p(\boldsymbol{\gamma}|\tau^2) \propto \exp\left(-\frac{1}{2\tau^2}\boldsymbol{\gamma}'\boldsymbol{K}\boldsymbol{\gamma}\right). \tag{5.4}$$

The precision matrix $\boldsymbol{K}$ is given by a kind of adjacency matrix with entries

$$\boldsymbol{K}[s,s] = N_s$$

$$\boldsymbol{K}[s,r] = \begin{cases} -1 & \text{if } s \text{ and } r \text{ are neighbours} \\ 0 & \text{otherwise.} \end{cases}$$

The precision matrix therefore contains the neighbourhood information in its non-zero entries. In addition, $\boldsymbol{K}$ is obviously rank-deficient since all rows and columns sum to one, making (5.4) a partially improper prior.

The logarithm of the joint prior distribution relates to a penalty term that again makes the spatial smoothing assumption obvious. The quadratic form in the prior can be rewritten as

$$\boldsymbol{\gamma}'\boldsymbol{K}\boldsymbol{\gamma} = \sum_{s \in \mathcal{S}} \sum_{r \in \delta_s, r<s} (\gamma_s - \gamma_r)^2$$

and therefore represents an overall measure for the deviations between parameters of adjacent regions. Similar as for penalized spline smoothing approaches, this penalty can also be employed to derive frequentist, penalized likelihood estimates where spatial smoothness is enforced by the penalty.

Since the prior distribution (5.4) is of the same form as the penalized spline smoothing priors, similar algorithms can be used for model estimation. Due to the larger dimensionality of the vector of regression coefficients $\boldsymbol{\gamma}$, numerical efficiency, however, becomes more important. Both sparse matrix methods and blocking strategies can be employed to reduce the computational burden, see for example Lang and Brezger (2004).

Although we have introduced an intuitive spatial smoothing approach that fits readily into the framework of the semiparametric mixed models considered in Chapter 4, prior (5.3) is actually only one of the easiest examples of a Markov random field prior. In more general situations, the questions already present in the simple example become more challenging and are, for example, as follows:

- Which specifications for full conditional prior distributions as in (5.3) are allowed? Which restrictions have to be imposed to obtain compatible full conditionals that actually correspond to a valid joint prior distribution?
- Can the impropriety of the joint prior be avoided by modifying the full conditional priors? In the temporal setting this would correspond, for example, to replacing the random-walk prior by an autoregressive prior with a suitable autoregressive parameter.
- What specific type of (marginal) dependence is induced by a Markov random field?
- Why are we using a latent formulation of Markov random fields where the spatial dependence structure is imposed on a regression coefficient in the predictor instead of directly assuming a Markov random field prior for the responses?

To answer these questions, we will now introduce Markov random fields in a theoretically sound framework. This will also allow us to generalize the heuristic spatial smoothing prior to include weights to account for differences in the dependency of region s on its neighbours, for example due to differences in the length of the common boundary or in the distance of region centroids.

5.2.2 Markov random fields

Definition of Markov random fields

As introduced in the previous section, Markov random fields are based on conditional independence assumptions applying to a random vector. Motivated from the spatial applications that will be the focus of this chapter, we consider the random vector $\boldsymbol{\gamma} = (\gamma_1, \ldots, \gamma_S)$ of spatial effects in regions $s = 1, \ldots, S$. Conditional independence is then translated into neighbourhood information, where two regions are assumed to be conditionally independent (given all other regions) if they are not neighbours. More formally, if we denote the neighbours of region s as δ_s, this assumption can be written as

$$p(\gamma_s|\boldsymbol{\gamma}_{-s}) = p(\gamma_s|\boldsymbol{\gamma}_{\delta_s}) \tag{5.5}$$

where $\boldsymbol{\gamma}_{\delta_s}$ is the collection of all neighboring parameter values. Then the joint distribution of $\boldsymbol{\gamma}$ is said to follow a Markov random field with respect to the given neighbourhood structure if exactly assumption (5.5) is fulfilled for all regions s. In extension to the example considered in the precious section, (5.5) defines a general spatial Markovian assumption that motivated the name Markov random field.

A simple example for a Markov random field has already been discussed extensively in Chapter 2 for the special case where the spatial domain is simply a line, i.e. if 'space' corresponds to the time axis. Then the first-order random-walk prior fulfils assumption (5.5) with the neighbourhood consisting of the preceding and the subsequent time point. For higher order random walks, the neighbourhood has to be extended to include also second or higher order neighbours.

A graphical description of the conditional independence properties holding in a Markov random field can be obtained from considering a graph $\mathcal{G} = (\mathcal{K}, \mathcal{E})$ with knots $\mathcal{K} = \{1, \ldots, S\}$ representing the regions and edges $\mathcal{E}$ corresponding to the dependency structure. An edge between two knots corresponds to conditional dependence while the absence of an edge relates to conditional independence. Figure 5.11 displays two examples for such graph structures based on regular and irregular regions.

Note that the notion of graphs and Markov random fields is not restricted to spatial or temporal applications, but can be directly extended to more general situations including higher-dimensional smoothing or spatio-temporal applications. The only necessary modification is the definition of the neighbourhood. In spatio-temporal models, for example, neighbours can be defined both in time and space. The more difficult question is how to specify the conditional distributions $p(\gamma_s|\gamma_{\delta_s})$ in such situations. A second example of generalized spatial Markov random fields arises when analysing for example the spread of infectious diseases among humans. Due to flight connections, not only regions that are

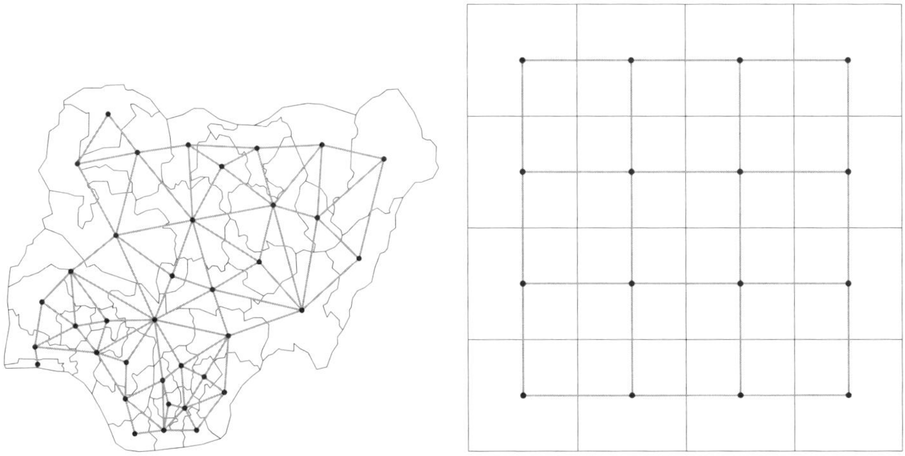

Fig. 5.11 Graph structures for a map of Nigeria and a regular map of rectangular regions with dots denoting the vertices of the graph. In both cases, two regions are treated as neighbours if they share a common boundary. The grey lines indicate conditional dependence between the connected regions.

geographically close to each other but also regions with direct flying connections will become 'close by' in the sense of the potential transfer of infections between the regions. In a Markov random field model, such additional neighbourhood relationships can easily be accommodated by adding edges between vertices representing airplane or other travel connections. Note however, that for modelling the spread of infectious diseases a purely spatial approach is usually not sufficient since one is interested in analysing the spread in space and time simultaneously.

In practice, the typical starting point for defining a Markov random field will not be the joint distribution of the full vector $\boldsymbol{\gamma}$, but the specification of the conditional priors $p(\gamma_s|\boldsymbol{\gamma}_{\delta_s})$. While these conditional distributions are easily derived from the joint distribution based on conditional probability rules, obtaining the joint distribution from the conditionals is more difficult. Moreover, and even more important, the derivation of a joint distribution from full conditionals is not always possible and certain restrictions on the specification of the full conditionals have to be imposed.

Consider the example of a two-state Markov random field $\boldsymbol{\gamma} = (\gamma_1, \gamma_2)'$ (taken from Banerjee, Carlin and Gelfand, 2003a) with the full conditional distributions

$$\gamma_1|\gamma_2 \sim \mathrm{N}(w_0 + w_1\gamma_2, \sigma_1^2) \quad \text{and} \quad \gamma_2|\gamma_1 \sim \mathrm{N}(v_0 + v_1\gamma_1^2, \sigma_1^2).$$

Based on iterated expectations, the unconditional expectations of γ_1 and γ_2 are obtained as

$$\begin{aligned}\mathbb{E}(\gamma_1) &= \mathbb{E}(\mathbb{E}(\gamma_1|\gamma_2)) = \mathbb{E}(w_0 + w_1\gamma_2) = w_0 + w_1\,\mathbb{E}(\gamma_2)\\ \mathbb{E}(\gamma_2) &= \mathbb{E}(\mathbb{E}(\gamma_2|\gamma_1)) = \mathbb{E}(v_0 + v_1\gamma_1^2) = v_0 + v_1\,\mathbb{E}(\gamma_1^2).\end{aligned}$$

Except for very specific choices of the parameter settings, these expectations are incompatible and the given full conditional distributions do not correspond to a valid joint distribution.

Following Besag (1974) we will now solve the aforementioned difficulties in two steps: First, we will show that (under certain assumptions) the joint distribution of $\boldsymbol{\gamma}$ is actually determined by the full conditional distributions up to proportionality. This result is known as Brook's lemma. Second, we will give a characterization of the form of the joint distribution in the Hammersley–Clifford theorem. This result will proof to be particularly useful in generating valid classes of Markov random field models. In fact, it will turn out that for exponential family Markov random fields only autoregressive specifications based on the natural link function are valid while more general autoregressive specifications are ruled out by the general Markov random field theory. Finally, we will return to Gaussian Markov random fields that will be used for spatial smoothing in the rest of this book.

Brook's lemma

Suppose that the support of the joint distribution of $\boldsymbol{\gamma}$ is the Cartesian product of the support of the marginal distributions of the elements $\gamma_1, \ldots, \gamma_s$, i.e. if

$p(\gamma_s) > 0$, $s = 1, \ldots, S$, then also $p(\boldsymbol{\gamma}) > 0$. This assumption, known as *positivity*, rules out situations as the one depicted in Fig. 5.12 where the marginal support for both γ_1 and γ_2 is given by the interval $[0, 1]$ but the joint support is a circle and not the unit square.

If positivity holds, Brook's lemma states that the joint density $p(\boldsymbol{\gamma})$ can be factorized in terms of the Brook's expansion as

$$p(\boldsymbol{\gamma}) = \frac{p(\gamma_1|\gamma_2,\ldots,\gamma_S)}{p(\gamma_1^{(0)}|\gamma_2,\ldots,\gamma_S)} \frac{p(\gamma_2|\gamma_1^{(0)},\gamma_3,\ldots,\gamma_S)}{p(\gamma_2^{(0)}|\gamma_1^{(0)},\gamma_3,\ldots,\gamma_S)} \frac{p(\gamma_3|\gamma_1^{(0)},\gamma_2^{(0)},\gamma_4,\ldots,\gamma_S)}{p(\gamma_3^{(0)}|\gamma_1^{(0)}\gamma_2^{(0)},\gamma_4,\ldots,\gamma_S)} \cdots$$

$$\frac{p(\gamma_S|\gamma_1^{(0)},\ldots,\gamma_{S-1}^{(0)})}{p(\gamma_S^{(0)}|\gamma_1^{(0)},\ldots,\gamma_{S-1}^{(0)})} p(\gamma_1^{(0)},\ldots,\gamma_S^{(0)}),$$

where $\boldsymbol{\gamma}^{(0)} = (\gamma_1^{(0)},\ldots,\gamma_S^{(0)})'$ is a fixed point from the support of $\boldsymbol{\gamma}$ (see Besag, 1974, for a proof).

Brook's lemma has the following implications:

- The joint distribution is determined by the full conditional distributions up to proportionality since, beside the constant $p(\gamma_1^{(0)},\ldots,\gamma_S^{(0)})$, only full

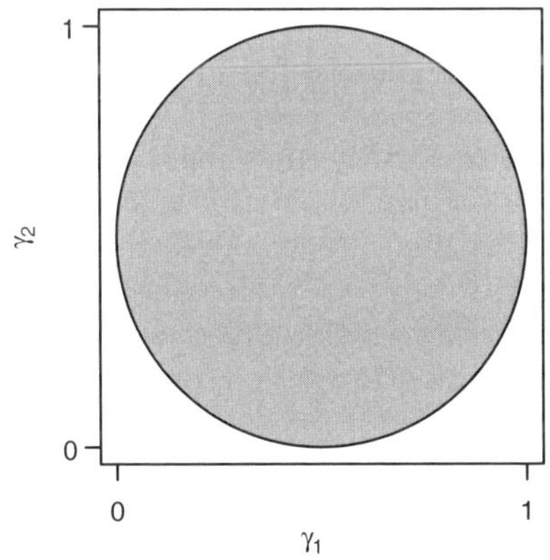

Fig. 5.12 Illustration of a situation where positivity does not hold. Suppose that the supports of γ_1 and γ_2 are given by the corresponding axes. Then, in case of positivity, the joint support should be square but is in fact given by the grey region, that is, a circle.

conditional distributions are involved in the right-hand side. This property becomes even more obvious when rewriting the above expression as

$$\frac{p(\boldsymbol{\gamma})}{p(\boldsymbol{\gamma}^{(0)})} = \prod_{s=1}^{S} \frac{p(\gamma_s|\gamma_1^{(0)}, \ldots, \gamma_{s-1}^{(0)}, \gamma_{s+1}, \ldots, \gamma_S)}{p(\gamma_s^{(0)}|\gamma_1^{(0)}, \ldots, \gamma_{s-1}^{(0)}, \gamma_{s+1}, \ldots, \gamma_S)}$$

where the normalizing constant in $p(\boldsymbol{\gamma})$ and $p(\boldsymbol{\gamma}^{(0)})$ cancels from the fractions. Provided that the joint distribution is proper, the normalizing constant can in principle be derived by integrating Brook's expansion, but this will be analytically infeasible in most cases.

- Brook's expansion also hints at a second problem in defining joint probability distributions in terms of conditionals: Propriety of the joint distribution cannot be deduced from propriety of the full conditionals since the joint distribution is only available upon proportionality. Banerjee, Carlin and Gelfand (2003a) give the simple example

$$p(\gamma_1, \gamma_2) \propto \exp\left(-\frac{1}{2}(\gamma_1 - \gamma_2)^2\right)$$

  where the joint distribution is improper while both conditionals are Gaussian distributions, that is, $\gamma_1|\gamma_2 \sim \mathrm{N}(\gamma_2, 1)$ and $\gamma_2|\gamma_1 \sim \mathrm{N}(\gamma_1, 1)$.
- Brook's expansion is defined with respect to a reference point $\boldsymbol{\gamma}^{(0)}$ since no natural ordering exists on a spatial lattice as compared to the time axis. In temporal conditional models, the first time point provides a natural starting point and a similar factorization without the necessity of a reference point is given by

$$p(\boldsymbol{\gamma}) = p(\gamma_1) \prod_{t=2}^{T} p(\gamma_t|\gamma_{t-1}, \ldots, \gamma_1).$$

  In contrast to Brook's expansion, the temporal expansion is directed in that it only contains preceeding time points in the condition.
- Brook's expansion is not unique. Since the elements in the reference point $\boldsymbol{\gamma}^{(0)}$ and in $\boldsymbol{\gamma}$ can be permuted arbitrarily, a total of $S!$ Brook's expansions can be derived for the same reference point. All these expansions have to define the same joint distribution so that restrictions on the full conditionals arise. These restrictions will be the subject of the next paragraph.
- Brook's expansion also has to be invariant to the choice of the reference point $\boldsymbol{\gamma}^{(0)}$.

Negpotential function and Hammersley–Clifford theorem

Similarly to the previous paragraph, we assume that positivity holds and in addition require that $\mathbf{0} = (0, \ldots, 0)'$ is in the support of $\boldsymbol{\gamma}$ (which can be achieved by a simple reparametrization if not fulfilled anyway). Under these conditions, we can define the *negpotential function*

$$Q(\boldsymbol{\gamma}) = \log\left(\frac{p(\boldsymbol{\gamma})}{p(\mathbf{0})}\right).$$

It can then be shown that $Q(\boldsymbol{\gamma})$ can be written as

$$Q(\boldsymbol{\gamma}) = \sum_{1\le s\le S} \gamma_s\phi(\gamma_s) + \sum\sum_{1\le r<s\le S} \gamma_r\gamma_s\phi(\gamma_r,\gamma_s) + \sum\sum\sum_{1\le r<s<t\le S} \gamma_r\gamma_s\gamma_t\phi(\gamma_r,\gamma_s,\gamma_t)$$
$$\ldots + \gamma_1\cdot\ldots\cdot\gamma_S\phi(\gamma_1,\ldots,\gamma_S)$$

where the functions $\phi(\cdot)$ are potentials, that is, they are invariant with respect to permutations of their arguments. This decomposition can be achieved by defining

$$\gamma_s\phi(\gamma_s) = Q(0,\ldots,0,\gamma_s,0,\ldots,0) - Q(\mathbf{0})$$

and similar expressions for the higher order potentials.

The essential statement of the Hammersley–Clifford theorem is now given by a characterization of non-null potentials, that is, potentials that are admissible in the sense that they lead to a well-defined joint distribution. This characterization is based on the notion of cliques (see Lauritzen, 1996, for a graph-theoretical treatment of cliques). A clique is a collection of regions $\mathcal{C}$, such that either $|\mathcal{C}| = 1$ (i.e. the clique consists of only one region) or all elements in $\mathcal{C}$ are neighbours, that is,

$$s \sim r, \quad \text{for all } s, r \in \mathcal{C}.$$

On a regular lattice structure with only first-order neighbours, there are only cliques of one and two elements. For irregular lattices, higher-order cliques may exist, as depicted in Fig. 5.13.

It turns out, that for a Markov random field with a given neighbourhood structure, only potentials defined upon cliques can be non-null, that is,

$$\phi(\gamma_{s_1},\ldots,\gamma_{s_p}) \neq 0$$

is admissible if and only if $s_1,\ldots,s_p$ form a clique. Proofs can be found in Cressie (1993, pp. 417–18) and Besag (1974). As a direct consequence, the joint density of $\boldsymbol{\gamma}$ takes the form

$$p(\boldsymbol{\gamma}) \propto \exp\left[\sum_{\mathcal{C}}\left(\prod_{s\in\mathcal{C}}\gamma_s\right)\phi_{\mathcal{C}}(\gamma_{\mathcal{C}})\right] \tag{5.6}$$

where the sum is over all possible cliques $\mathcal{C} = \{s_1,\ldots,s_p\}$ subject to $s_1 < \ldots < s_p$ to avoid redundancy. Distributions of this form (i.e. distributions defined only on potentials of cliques) are commonly referred to as Gibbs distributions. Therefore the *Hammersley–Clifford theorem* can briefly be restated as follows: The distribution of $\boldsymbol{\gamma}$ follows a Markov random field if and only if the corresponding joint

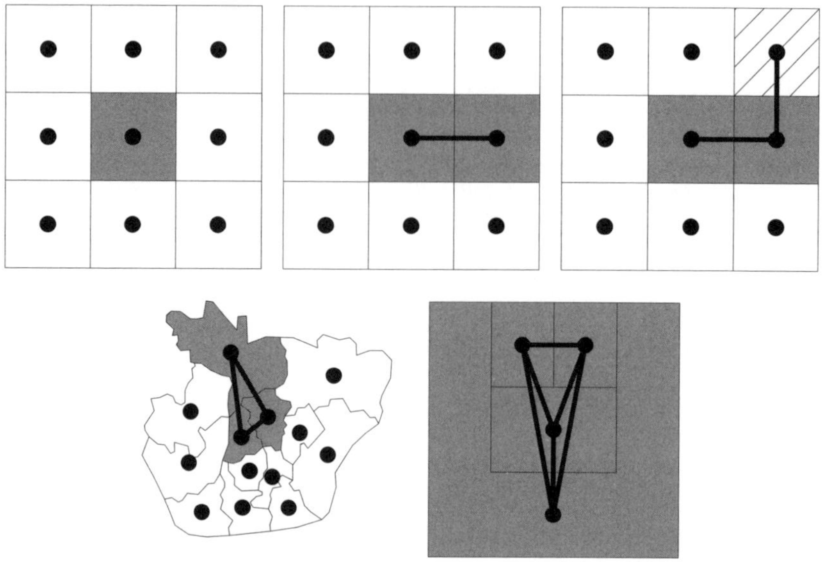

Fig. 5.13 Illustration of cliques for regular and irregular lattices: The top row shows cliques consisting of one and two regions. Trying to add the striped region as a third member does not define a valid clique since it is not a direct neighbour of the middle region. The bottom row shows examples of cliques with three and four members.

density defines a Gibbs distribution. This is also the origin of the term Gibbs sampler in Markov chain Monte Carlo simulation, termed by Geman and Geman (1984) in the context of image restoration.

The Hammersley–Clifford theorem gives us both a characterization of admissible MRF densities and a recipe for constructing valid MRFs based on suitable classes of potentials. Restrictions on the definition of MRFs arise from the invariance of the potentials with respect to permutations of the arguments. In the following, we will consider the class of auto-models derived for exponential family densities where only potentials of up to two arguments are involved. This leads to simplified validity conditions.

As an important by-product it will turn out that autoregressive regression model definitions have to be restricted to exponential family regression with the *natural* link function.

Auto-models

Auto-models are characterized by two assumptions:

- The conditional distributions $p(\gamma_s|\gamma_{-s})$ are of the univariate exponential family type.

- The negpotential function takes the form

$$Q(\boldsymbol{\gamma}) = \sum_{s=1}^{S} \alpha_s \gamma_s + \sum_{s \sim r, s<r} \beta_{sr} \gamma_s \gamma_r, \tag{5.7}$$

  that is, only cliques of up to two elements lead to nonzero contributions and the potentials are given by $\phi(\gamma_s) = \alpha_s$ and $\phi(\gamma_s, \gamma_r) = \beta_{rs}$. To obtain proper two-way potentials, the restriction $\beta_{rs} = \beta_{sr}$ has to be imposed on the coefficients.

It immediately follows from (5.6) that the conditional distributions are of the form

$$p(\gamma_s | \boldsymbol{\gamma}_{-s}) \propto \exp\left(\gamma_s \alpha_s + \gamma_s \sum_{r: r \sim s} \beta_{rs} \gamma_r \right).$$

For example, in case of binary variables $\gamma_s \in \{0, 1\}$, the density is given by

$$\mathbb{P}(\gamma_s = 1 | \boldsymbol{\gamma}_{-s}) = \frac{\exp\left(\alpha_s + \sum_{r: r \sim s} \beta_{rs} \gamma_r \right)}{1 + \exp\left(\alpha_s + \sum_{r: r \sim s} \beta_{rs} \gamma_r \right)}$$

$$\mathbb{P}(\gamma_s = 0 | \boldsymbol{\gamma}_{-s}) = \frac{1}{1 + \exp\left(\alpha_s + \sum_{r: r \sim s} \beta_{rs} \gamma_r \right)}$$

that is, takes the form of a classical logit model but with neighbouring coefficients γ_r included as 'explanatory variables'. Note that by convention region s is not a neighbour of itself such that the sum over all neighbours does not include index s. The binary auto-model is called auto-logistic model since the variables of interest are used as both 'responses' and 'covariates'.

It is important to realize that the definition of auto-probit models is ruled out by the Hammersley–Clifford theorem. More precisely, the auto-logistic model is the only valid MRF specification for binary variables with negpotential functions of the form (5.7). Hence, apart from the natural logistic link, no binary auto-models can be defined.

The famous *Ising model* (...) is obtained as a special auto-logistic model by taking $\alpha_s \equiv \alpha$ and $\beta_{rs} \equiv \beta$. In this case, the probability $\mathbb{P}(\gamma_s = 1 | \boldsymbol{\gamma}_{-s})$ simplifies to

$$\mathbb{P}(\gamma_s = 1 | \boldsymbol{\gamma}_{-s}) \propto \exp\left(\alpha + \beta \sum_{r: r \sim s} \mathbb{1}(\gamma_r = 1) \right),$$

that is, provided that $\beta > 0$, the probability $\mathbb{P}(\gamma_s = 1 | \boldsymbol{\gamma}_{-s})$ is high if several of γ_s's neighbours are equal to one and small otherwise. The Ising model is commonly used in image restoration of black and white pictures where $\gamma_s = 1$ refers to black and $\gamma_s = 0$ to white pixels. It can also be extended to colour or grey-scale images when considering multinomial auto-logit models. This corresponds to the Potts model, where a multinomial logit model is defined with probabilities relating to the number of neighbours of a particular colour, see Hurn, Husby and Rue (2003) for more details on the Ising and the Potts model as well as examples for applications in image restoration.

Auto-Poisson models can be defined in analogy to the auto-logit model, yielding

$$\mathbb{E}(\gamma_s|\boldsymbol{\gamma}_{-s}) = \exp\left(\alpha_s + \sum_{r:r\sim s} \beta_{rs}\gamma_s\right).$$

Again the auto-model employs the appropriate generalized linear model with natural link function, given by a loglinear Poisson model in this particular case.

5.2.3 Gaussian Markov random fields/auto-normal models

When considering geoadditive regression models with predictors of the form

$$\eta = \boldsymbol{u}'\boldsymbol{\beta} + f_1(x_1) + \ldots + f_r(x_r) + f_{\text{spat}}(s)$$

and a Markov random field prior for the spatial regression effects $\gamma_s = f_{\text{spat}}(s)$, the most natural choice is to assume conditional Gaussian structures. The *auto-normal model* obtained by setting $\phi(\gamma_s) = -\gamma_s/\tau_s^2$ and $\phi(\gamma_s,\gamma_r) = \beta_{rs}/\tau_s^2$ induces the conditional distribution

$$\gamma_s|\boldsymbol{\gamma}_{-s} \sim \mathrm{N}\left(\mu_s + \sum_{r:r\sim s} \beta_{rs}(\gamma_r - \mu_r), \tau_s^2\right).$$

This is a generalization of the model considered in Section 5.2.1, where $\mu_s \equiv 0$, $\beta_{rs} \equiv 1$ and $\tau_s^2 = \tau^2/|\delta_s|$. In particular, the conditional expectation of γ_s given all the neighbours, which was a simple average of the neighbours in Section 5.2.1, is now given by

$$\mathbb{E}(\gamma_s|\boldsymbol{\gamma}_{-s}) = \mu_s + \sum_{r:r\sim s} \beta_{rs}(\gamma_r - \mu_r).$$

Still, the expectation is formulated conditionally on the neighbours and therefore this model is also called a conditionally autoregressive (CAR) model.

The joint distribution of $\boldsymbol{\gamma}$ that corresponds to the auto-normal model can be calculated based on Brook's lemma. Given the reference point $\boldsymbol{\gamma}^{(0)} = \boldsymbol{\mu} = (\mu_1,\ldots,\mu_S)'$, the joint distribution is defined as

$$p(\boldsymbol{\gamma}) \propto \prod_{s=1}^{S} \frac{p(\gamma_s|\mu_1,\ldots,\mu_{s-1},\gamma_{s+1},\ldots,\gamma_S)}{p(\mu_s|\mu_1,\ldots,\mu_{s-1},\gamma_{s+1},\ldots,\gamma_S)}$$

To simplify notation, we furthermore define $\beta_{rs} = 0$ if $s \nsim r$, which allows us to rewrite Brook's expansion as

$$\log(p(\gamma)) = \sum_{s=1}^{S} \log\left(\frac{p(\gamma_s|\mu_1,\ldots,\mu_{s-1},\gamma_{s+1},\ldots,\gamma_S)}{p(\mu_s|\mu_1,\ldots,\mu_{s-1},\gamma_{s+1},\ldots,\gamma_S)}\right)$$

$$= \sum_{s=1}^{S} \frac{1}{2\tau_s^2}\left(\left[\sum_{r=s+1}^{S} \beta_{rs}(\gamma_r - \mu_r)\right]^2 - \left[\gamma_s - \mu_s - \sum_{r=s+1}^{S} \beta_{rs}(\gamma_r - \mu_r)\right]^2\right)$$

$$= \sum_{s=1}^{S} \frac{1}{2\tau_s^2}\left(-(\gamma_s - \mu_s)^2 + 2(\gamma_s - \mu_s)\sum_{r=s+1}^{S} \beta_{rs}(\gamma_r - \mu_r)\right)$$

$$= -\sum_{s=1}^{S} \frac{1}{2\tau_s^2}(\gamma_s - \mu_s)^2 + \sum_{s=1}^{S}\sum_{r=s+1}^{S} \frac{\beta_{rs}}{\tau_s^2}(\gamma_s - \mu_s)(\gamma_r - \mu_r)$$

where all equalities hold up to an additive constant. From this form of the density, we can deduce that

$$p(\boldsymbol{\gamma}) \propto \exp\left(-\frac{1}{2}(\boldsymbol{\gamma} - \boldsymbol{\mu})'\boldsymbol{\Sigma}^{-1}(\boldsymbol{I}_S - \boldsymbol{B})(\boldsymbol{\gamma} - \boldsymbol{\mu})\right) \tag{5.8}$$

where $\boldsymbol{\Sigma} = \text{diag}(\tau_1^2, \ldots, \tau_S^2)$ and

$$\boldsymbol{B}[s,r] = \begin{cases} \beta_{sr} & s \sim r, \\ 0 & \text{otherwise.} \end{cases}$$

Hence, $\boldsymbol{\gamma}$ is multivariate normal with expectation $\boldsymbol{\mu}$ and covariance $\boldsymbol{\Sigma}(\boldsymbol{I}_S - \boldsymbol{B})^{-1}$ provided that $(\boldsymbol{I}_S - \boldsymbol{B})$ is positive definite.

Typically, the coefficients β_{rs} are chosen to reflect some kind of weighting in the spatial dependence on the neighbours. On regular lattices, such weights can be used to differentiate between direct neighbours along the axes and diagonal neighbours which are farther apart. For regional data, the weights are frequently based on the length of the common boundary or the distance of the centroids of the regions. More specifically, we set $\mu_s = 0$ and

$$\beta_{sr} = \rho\frac{w_{sr}}{w_{s+}}, \qquad \tau_s^2 = \frac{\tau^2}{w_{s+}},$$

where

$$w_{s+} = \sum_{r:r\sim s} w_{sr}$$

and $\rho \in [0,1]$. This results in the conditional distribution

$$\gamma_s|\boldsymbol{\gamma}_{-s} \sim \text{N}\left(\rho\sum_{r:r\sim s}\frac{w_{sr}}{w_{s+}}\gamma_r, \frac{\tau^2}{w_{s+}}\right)$$

that extends the specification of a simple GMRF in (5.3).

The weights w_{sr} are for example calculated as follows:

- Apply the same weights for all neighbours, that is, without loss of generality $w_{sr} \equiv 1$ and $w_{s+} = |\delta_s|$. This results in the simple GMRF defined in (5.3).
- Choose the weights inverse proportional to the distance of the centroids of the regions, for example
$$w_{sr} \propto \exp(-d(s, r)),$$
where $d(s, r)$ denotes the distance between the centroids of regions s and r.
- Choose the weights proportional to the length of the common boundary between the neighbouring regions.

Note that all weight definitions are symmetric in s and r and therefore yield valid MRF specifications.

As long as the condition $0 \leq \rho < 1$ is fulfilled, the joint distribution in (5.8) is proper. If, however, the extreme case $\rho = 1$ is considered, we obtain an improper prior where $\boldsymbol{I} - \boldsymbol{B}$ is rank-deficient and not invertable (but still positive semidefinite). This special case is known as an intrinsic Gaussian Markov random field (IGMRF). The spatial prior (5.3) considered in Section 5.2.1 is of this type where the equal weights scheme defines the matrix $\boldsymbol{B}$. In the following, we will mostly focus on partially improper IGMRFs as the default choice for a spatial prior in case of area-coded spatial information. From our perspective, the improper version has the advantage that it does not penalize deviations from a constant level. While proper GMRF priors force the overall effect towards the prior expectation $\boldsymbol{\mu}$, this is not the case with the IGMRF version. Moreover, there is no necessity to estimate the hyperparameter ρ in the IGMRF. In a full Bayesian formulation, suitable hyperpriors would have to be assigned to ρ, see Gschlößl and Czado (2008) or Sun, Tsutakawa, Kim and He (2000) for examples. Finally, if a proper spatial prior is required, it may be advantageous to impose a linear constraint on the spatial coefficients $\boldsymbol{\gamma}$ (for example a sum to zero constraint) using conditional Gaussian Markov random fields as described in Rue and Held (2005).

The distinction between proper GMRFs and IGMRFs can conceptually be considered the same as the distinction between (first-order) temporally stationary autoregressive and (first-order) random-walk distributions. For the stationary autoregressive model, the sequence of parameter values is defined as

$$\gamma_t = \rho\gamma_{t-1} + u_t, \quad u_t \sim \mathrm{N}(0, \tau^2), \quad -1 < \rho < 1,$$

whereas the random-walk model assumes

$$\gamma_t = \gamma_{t-1} + u_t, \quad u_t \sim \mathrm{N}(0, \tau^2),$$

that is, corresponds to the non-stationary autoregressive model with $\rho = 1$. In fact, both the temporal autoregressive model and the temporal random-walk can be cast into the framework of Gaussian Markov random fields when considering time as the univariate spatial dimension and identifying the two nearest time points as neighbours. Unlike the temporally directed formulation of autoregressive and random-walk models, Markov random fields employ a symmetric perspective, since no natural ordering exists in the spatial domain. We already considered the undirected formulation of a random-walk in Section 2.2.1 and found that

$$\gamma_t \sim \mathrm{N}\left(\frac{1}{2}(\gamma_{t-1} + \gamma_{t+1}), \frac{\tau^2}{2}\right),$$

which exactly resembles the structure of an IGMRF. For the autoregressive model, analogous considerations yield

$$\gamma_t \sim \mathrm{N}\left(\frac{1}{1+\rho^2}(\rho\gamma_{t-1} + \rho\gamma_{t+1}), \frac{\tau^2}{1+\rho^2}\right).$$

While, so far, we derived the joint distribution induced by the full conditionals of a Gaussian Markov random field, it is of course also possible to proceed the other way round, that is, to obtain the full conditionals from the joint distribution. A vector $\boldsymbol{\gamma}$ is multivariate normal with expectation $\boldsymbol{\mu}$ and precision matrix $\boldsymbol{P}$ if and only if the conditional distribution of an element γ_s is given by

$$\gamma_s | \boldsymbol{\gamma}_{-s} \sim \mathrm{N}\left(\mu_s - \sum_{r \neq s} \frac{\boldsymbol{P}[s,r]}{\boldsymbol{P}[s,s]}(\gamma_r - \mu_r), \frac{1}{\boldsymbol{P}[s,s]}\right).$$

As a consequence, the conditional distribution of γ_s depends only on coefficients γ_r if the corresponding element in the precision matrix $\boldsymbol{P}[s,r] = \boldsymbol{P}[r,s]$ is different from zero. For a Markov random field, this is exactly the case if and only if the regions s and r are neighbours. Moreover, the conditional covariance structure of two elements γ_s and γ_r is also determined by the precision matrix:

$$\mathrm{Cov}(\gamma_s, \gamma_r | \boldsymbol{\gamma}_{-r,s}) = -\frac{\boldsymbol{P}[s,r]}{\boldsymbol{P}[s,s]\boldsymbol{P}[r,r] - \boldsymbol{P}[s,r]^2}.$$

With the conditional variance

$$\mathrm{Var}(\gamma_s | \boldsymbol{\gamma}_{-r,s}) = \frac{\boldsymbol{P}[s,s]}{\boldsymbol{P}[s,s]\boldsymbol{P}[r,r] - \boldsymbol{P}[r,s]^2}$$

this yields

$$\mathrm{Cor}(\gamma_s, \gamma_r | \boldsymbol{\gamma}_{-r,s}) = -\frac{\boldsymbol{P}[s,r]}{\sqrt{\boldsymbol{P}[s,s]\boldsymbol{P}[r,r]}}.$$

Again, from the properties of multivariate Gaussian distributions, we immediately obtain the result that non-neighbouring coefficients are not only conditionally uncorrelated but also conditionally independent.

All properties of Gaussian Markov random fields discussed in this section have been based on a conditional view on the prior distribution. Wall (2004) and Assunção and Krainski (2009) study the marginal properties induced by a Markov random field prior. In particular, Assunção and Krainski (2009) provide explanations for some unexpected results in Wall (2004) on the marginal distribution. For example, the marginal variances of regions with the same number of neighbours may be different although the conditional variance is the same. Assunção and Krainski (2009) develop a matrix series expansion that allows to study this behaviour and find the intuitive explanation that marginal properties do not only depend on the direct neighbours but also on higher order neighbours.

Basis function representation of GMRFs

To include Markov random fields in a spatial regression model, we need a representation of the spatial function $f_{\text{spat}}(s)$ in terms of the vector of spatial effects $\boldsymbol{\gamma}$. Since each function evaluation relates to exactly one value of the spatial effects vector, that is, $f_{\text{spat}}(s) = \gamma_s$, we only have to connect observations with the corresponding elements in $\boldsymbol{\gamma}$. This is achieved by defining the incidence $n \times S$ matrix $\boldsymbol{X}$ where

$$\boldsymbol{X}[i,s] = \begin{cases} 1 & \text{observation } i \text{ has been collected in region } s \\ 0 & \text{otherwise.} \end{cases}$$

Based on this definition, the vector of function evaluations $\boldsymbol{f}_{\text{spat}} = (f_{\text{spat}}(s_1), \ldots, f_{\text{spat}}(s_n))'$ can be written as

$$\boldsymbol{f}_{\text{spat}} = \boldsymbol{X}\boldsymbol{\gamma}$$

and the structure of $\boldsymbol{X}$ ensures that each observation is assigned the spatial effect of the region it pertains to.

For spatially varying effects $u\gamma_s$, the design matrix for the spatial effect has to pre-multiplied with a diagonal matrix containing the covariate values similar as for varying coefficient models. Therefore, the vector of function evaluations $\boldsymbol{f}_{\text{spat}} = (u_1 f_{\text{spat}}(s_1), \ldots, u_n f_{\text{spat}}(s_n))'$ can in this case be written as

$$\boldsymbol{f}_{\text{spat}} = \text{diag}(u_1, \ldots, u_n)\boldsymbol{X}\boldsymbol{\gamma} = \boldsymbol{X}^*\boldsymbol{\gamma}$$

where $\boldsymbol{X}^* = \text{diag}(u_1, \ldots, u_n)\boldsymbol{X}$.

In summary, GMRFs fit well into the framework of semiparametric regression as well as semiparametric mixed models with conditionally Gaussian priors and basis function representations. As a consequence, the inferential procedures described in Chapters 2 and 4 can be extended to include spatial effects based on GMRFs and we will present details in Section 5.5.

Example 5.9: Disease mapping of oral cavity cancer
As a first example on the application of GMRFs, we analyse the disease mapping data from Example 5.6. To differentiate between spatially correlated structures and uncorrelated, district-specific effects, we consider the model specification

$$y_s \sim \text{Po}(e_s \lambda_s), \quad \lambda_s = \exp(\gamma_s + \alpha_s)$$

where an intrinsic GMRF prior (5.3) is assigned to the spatially correlated effect γ_s while i.i.d. Gaussian effects $\alpha_s \sim \text{N}(0, \delta^2)$ are assumed for the unstructured effects α_s.

Figure 5.14 visualizes posterior mode estimates of the structured and unstructured effects obtained with mixed model-based empirical Bayes inference (see Section 5.5 for details). From the magnitude of the estimated effects, we can identify a clear domination of the spatially structured effect that explains a larger proportion of the variability in the data. In contrast, the unstructured effect is mostly negligible and only shows up for some of the larger cities that obviously have a different risk structure with an alleviated risk. The spatially structured effect reflects the behaviour observed in the exploratory analyses in Example 5.6 with particularly high risk in the northeastern and also in the southwestern part

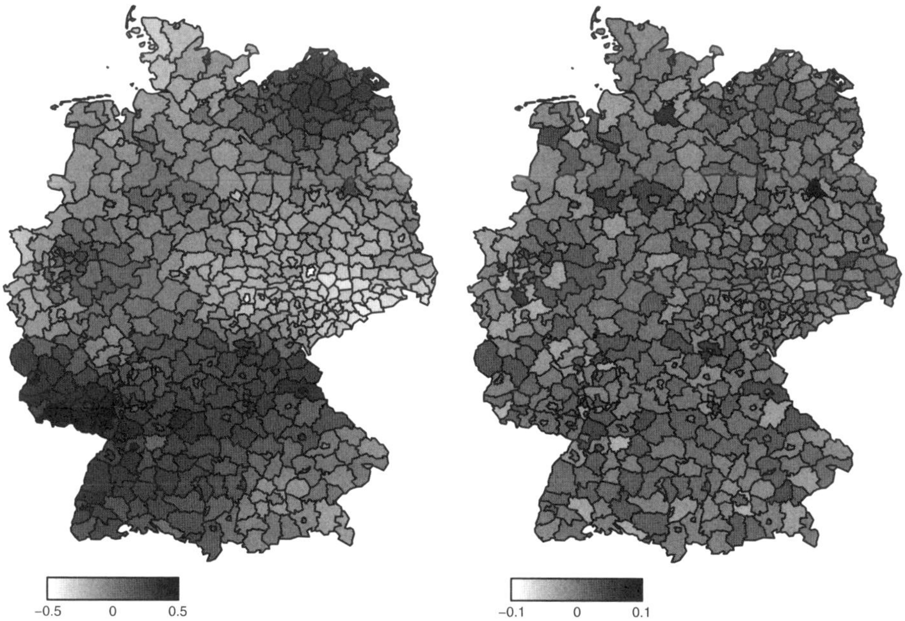

Fig. 5.14 Disease mapping: Estimated structured (left) and unstructured (right) spatial effects.

of Germany. However, since the effects are smoothed, it is now easier to identify spatial patterns.

The detection of such spatial patterns can be further enhanced by considering posterior probabilities as in Fig. 5.15. Here, the estimated spatial effect is classified into three categories according to the corresponding posterior credible interval. If the credible interval contains zero, the spatial effect is deemed to be insignificant and the region is coloured in grey. Similarly, black and white regions indicate significantly positive and negative effects, respectively. The resulting posterior significance map reveals clear patterns of reduced and alleviated risks that can be investigated further to detect reasons for these deviations from the average risk. Since the unstructured spatial effect is rather small, no significant effects have been identified in this case. △

Example 5.10: Forest health
In this example, we explore the spatial variation present in the forest health data introduced in Example 1.2. We use the three-categorical response variable representing no, moderate and severe defoliation and fit cumulative probit models to the data based on different specifications for the spatial effect.

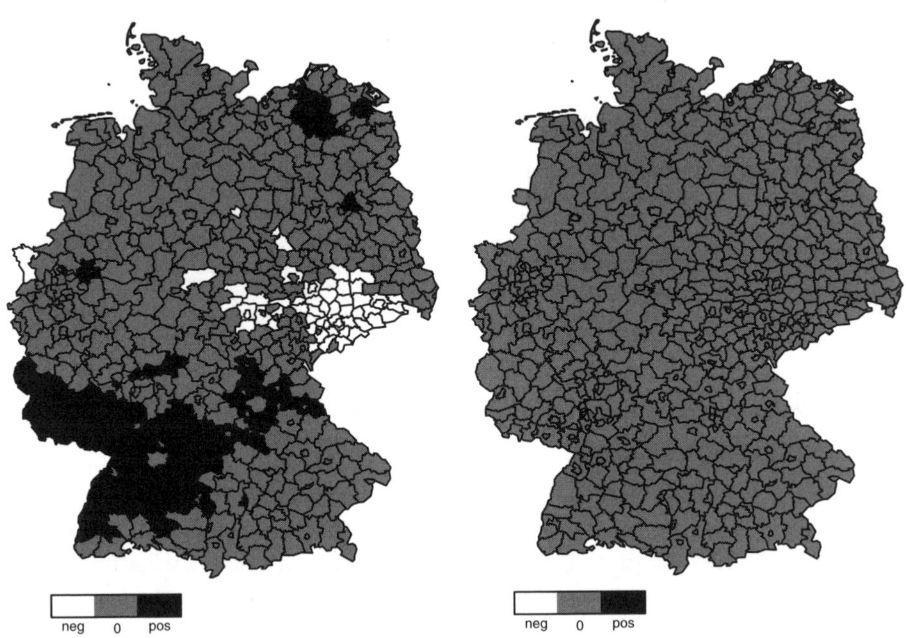

Fig. 5.15 Disease mapping: 95% posterior probabilities for the spatially structured (left) and unstructured (right) effects. Significantly positive effects are coloured in black, significantly negative effects are coloured in white. For regions coloured in grey, the 95% credible interval contained zero.

- Model 1: Spatially structured effect, that is

$$\eta_i = \theta^{(c)} - \gamma_{s_i}$$

where γ_{s_i} follows a GMRF prior.
- Model 2: Spatially unstructured effect, that is

$$\eta_i = \theta^{(c)} - \alpha_{s_i}$$

where $\alpha_{s_i} \sim \mathrm{N}(0, \delta^2)$ follows an i.i.d. Gaussian prior.
- Model 3: Combination of spatially structured and unstructured effect, that is

$$\eta_i = \theta^{(c)} - \gamma_{s_i} - \alpha_{s_i}$$

where γ_{s_i} follows a GMRF prior while α_{s_i} is i.i.d. Gaussian.

For the spatially correlated effects, two observation plots are treated as neighbours if their distance is less then 1.2 kilometres. Note that no further covariates are included in the model so that, for the moment, we restrict attention to the complete spatial variation present in the data.

Figures 5.16 and 5.17 show the resulting estimates (obtained from MCMC simulations in a full Bayesian specification). Surprisingly, both models comprising only one single effect produce almost the same estimates, although the GMRF prior enforces spatial smoothness while the random effects prior assumes prior independence. Obviously, the likelihood contains enough information to overwhelm the prior information at least to a certain extent.

The picture changes when including both the structured and the unstructured effect simultaneously in one model. Now the spatially structured part is much smoother and indicates a spatial trend underlying the defoliation probability while the unstructured effect catches more localized, plot-specific features. Still, it seems difficult to separate clearly between the two effects so that only the sum of both might be interpretable. However, including both the spatial effect and the random effect also allows us to reflect correlations in the data adequately, since the combination picks up both the spatial arrangement and the longitudinal component of the data.

To obtain an objective comparison of the model fit, we computed the deviance information criterion for the three models, see Table 5.1. The results for the posterior deviance $p(\hat{\boldsymbol{\theta}})$ confirm the visual impression that all three models provide comparable fits to the data. Also, the effective degrees of freedom p_{D} indicate similar complexity for the two models comprising only one spatial effect and a slightly increased complexity for the combined model. This increase in complexity results in a higher DIC value that would lead to a preference for the simpler models comprising only one spatial effect. The comparison of these two models yields a moderate preference for the GMRF model comprising a structured spatial effect. △

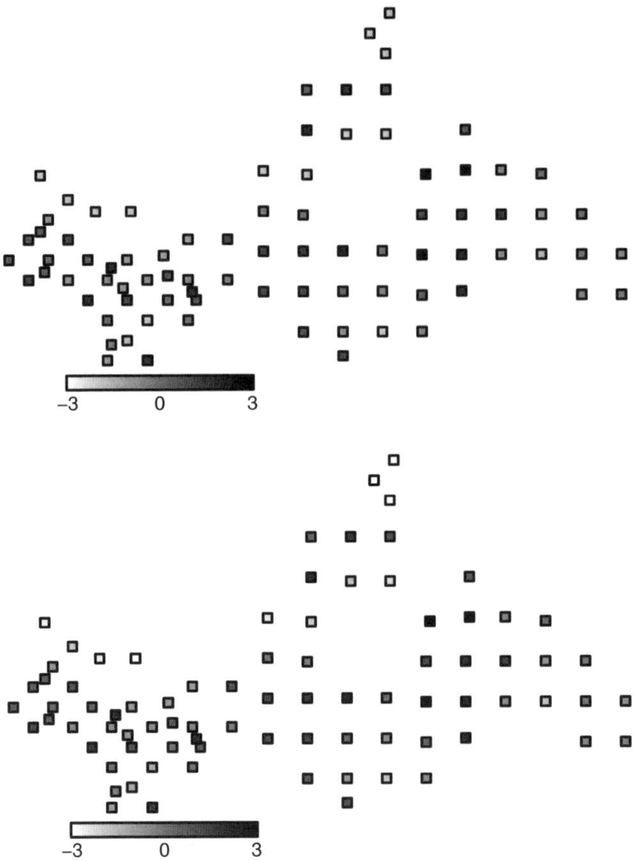

Fig. 5.16 Forest health: Unstructured (top) and structured (bottom) effects resulting from separate fits.

Direct vs. latent autoregressive models

All models we have discussed so far have been formulated as latent spatial models where γ_s is a part of the linear predictor of a spatial regression model. A direct approach to the inclusion of spatial correlations would be to assume spatial structures not for a part of the predictor but for the responses instead. For example, a Gaussian Markov random field specification for the responses would be given by

$$y_s | y_r, r \in \delta_s \sim \mathrm{N}\left(\frac{\rho}{N_s} \sum_{r:r \in \delta_s} y_r, \frac{\sigma^2}{N_s}\right)$$

where as before δ_s denotes the indices of neighbours of region s, $N_s = |\delta_s|$, and ρ is an autoregressive parameter. As we have seen in the previous sections, this

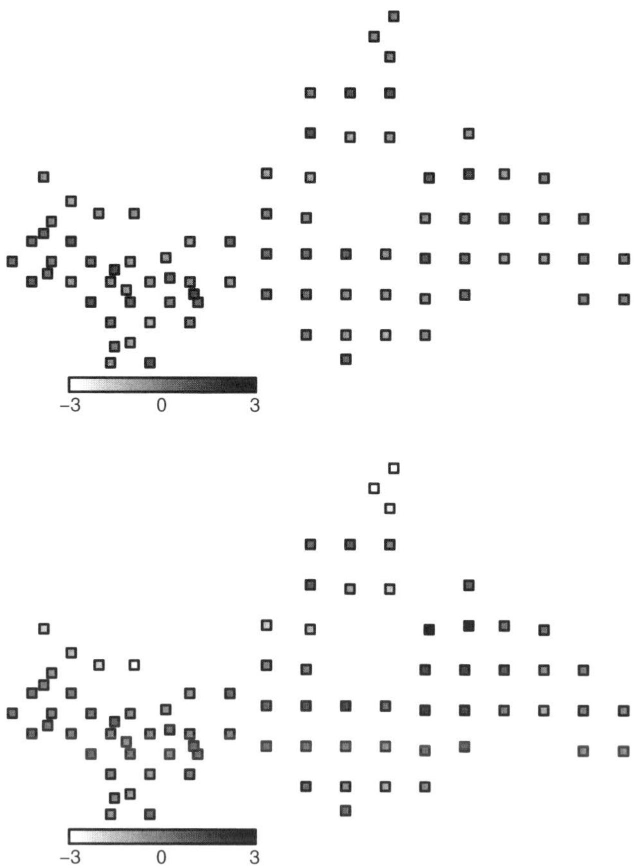

Fig. 5.17 Forest health: Unstructured (top) and structured (bottom) effects resulting from a model comprising both effects simultaneously.

conditional formulation is equivalent to the joint distribution

$$\boldsymbol{y} \sim \mathrm{N}\left(\mathbf{0}, \boldsymbol{\Sigma}(\boldsymbol{I}-\boldsymbol{B})^{-1}\right)$$

compare equation (5.8) with $\boldsymbol{\mu} = \mathbf{0}$. Similar specifications are possible for more general response types, for example based on the Ising or the Potts model considered in the section on auto-logistic models. These are standard tools in image reconstruction (see for example Møller, 2003) when being specified directly for binary/categorical responses.

Direct spatial models for the responses are very popular in spatial econometrics, see for example Anselin (1988), LeSage and Pace (2009) or Gamerman and Moreira (2004) (who also compare the direct and the latent specification of spatial regression models). We will, however, restrict our attention to latent models

Table 5.1 Forest health: Deviance, effective degrees of freedom and DIC for the three spatial models.

	$D(\hat{\boldsymbol{\theta}})$	p_D	DIC
Model 1	1616.13	70.60	1757.33
Model 2	1614.58	72.41	1759.39
Model 3	1614.50	75.15	1764.80

in the following since they have the particular advantage of being applicable with very general types of responses such as ordered categorical responses or survival times. These are typically difficult to handle in the direct approach. Moreover, there is the severe restriction imposed in the construction of direct auto-models that only autoregressive structures with natural link function are admissible. Defining the spatial effect as a part of the predictor therefore allows for considerably more flexibility concerning the model specification. For example, we can also consider spatial probit models or Gaussian models with other link functions than the identity when including the spatial effects in the predictor in latent formulation.

5.2.4 Extended Markov random field models

Extended versions of Markov random fields have been used in various situations. For example, an Ising prior can be used to achieve model selection for spatially correlated variables. In Chapter 4.5.3, binary indicators were introduced to differentiate between influential and non-influential covariates. In a related context, Smith, Pütz, Auer and Fahrmeir (2003) and Smith and Fahrmeir (2007) consider the differentiation between active and inactive voxels in the visual cortex of the brain. Since, in this case, voxels that are close by will tend to be activated simultaneously, the usual assumption of independent priors for the latent indicator variables clearly seems to be inappropriate. Smith *et al.* (2003) and Smith and Fahrmeir (2007) therefore employ an Ising prior on the indicators that achieves exactly a correlated prior structure.

Spatio-temporal and other combinations of Markov random fields can be achieved via tensor product priors. Such multivariate conditionally autoregressive (MCAR) prior structures have for example been considered in Mardia (1988), Gelfand and Vounatsou (2003), Jin, Carlin and Banerjee (2005) or Jin, Carlin and Banerjee (2007). The proposed specifications differ for example in the amount and precise definition of cross-correlation that is allowed between the CAR priors that have been used to set up the MCAR prior. Assunção (2003) considers Bayesian (multivariate) space varying coefficient models.

Another extension of Markov random field priors is to drop the assumption of fixed weights associated with the edges of the graph structure. Instead, an additional prior can be assumed for the weights, leading to adaptive Markov random fields where the strength of an edge is also determined by the data. Such priors are, for example, useful for brain imaging, where a sharp distinction between clusters of active voxels and inactive parts of the brain is desired for the sake of edge preservation. Using a fixed strength in this case may blur the estimated effect and induce artificial behaviour at the edges. Brezger, Fahrmeir and Hennerfeind (2007) propose such an adaptive Markov random field and apply it to the mapping of the visual cortex.

5.3 Spatial smoothing approaches and interactions

We will now turn to the case of point-referenced data, where the spatial information consists of continuous coordinate information, such as longitude and latitude in spatial examples. The immediate extension of univariate scatterplot smoothing is then to recover a smooth surface $f(\boldsymbol{s})$ from noisy data

$$y_i = f(\boldsymbol{s}_i) + \varepsilon_i, \quad i = 1, \dots, n, \tag{5.9}$$

where $\boldsymbol{s}_i = (s_{i1}, s_{i2}, \dots, s_{id}) \in \mathbb{R}^d$ represents continuous location information. Throughout the major part of this section, we will focus on the bivariate case with $d = 2$, which is the most common situation of spatial smoothing. In the first part of the section, we will extend the penalized spline approach to the estimation of surfaces based on the notion of tensor products. Tensor products can be interpreted as a natural extension of usual interaction modelling in parametric regression, where products of covariates are included as interaction terms. In case of tensor product penalized splines, products of basis functions will be considered as basis functions for more general, flexible interaction surfaces. The second part of this section will describe another spatial smoothing approach based on radial basis functions. In this case, we will not modify the basis functions but instead replace the d-dimensional argument of the basis functions with a univariate argument. Most commonly, this will be the Euclidean norm of the covariate vector, leading to basis functions with circular contour lines which are therefore called radial basis functions.

5.3.1 Tensor product penalized splines

A very simple approach to recover the surface in (5.9) that would enable application of univariate smoothing techniques is to assume that the surface is composed additively of two univariate functions, that is,

$$f(s_1, s_2) = f_1(s_1) + f_2(s_2). \tag{5.10}$$

In this case, f_1 and f_2 could again be approximated with some basis-functions approach leading to estimation schemes as considered in Chapters 2 and 4. However, in many situations and in particular in case of spatial effects, additivity is a very questionable assumption and more flexible solutions will be required.

In fact, the surface defined in (5.10) corresponds to a main effects model, where no interaction between the effects of s_1 and s_2 is present. In parametric regression models, interactions are commonly taken into account by including the product of certain design vectors as additional covariates. Since basis function approaches effectively convert the nonparametric smoothing problem into a large linear model (possibly augmented with prior information or a penalty), it seems reasonable to apply a similar idea to surface smoothing. This exactly leads to the notion of tensor product bases.

Tensor product bases

Starting from equation (5.10), we approximate both f_1 and f_2 in terms of spline bases. Let $B_{1,1}(s_1), \dots, B_{K_1,1}(s_1)$ and $B_{1,2}(s_2), \dots, B_{K_2,2}(s_2)$ be sets of polynomial spline bases of degrees D_1 and D_2 based on knots $\kappa_{0,1} < \kappa_{1,1} < \dots < \kappa_{M_1-1,1} < \kappa_{M_1,1}$ and $\kappa_{0,2} < \kappa_{1,2} < \dots < \kappa_{M_2-1,2} < \kappa_{M_2,2}$ such that $K_1 = M_1 + D_1$ and $K_2 = M_2 + D_2$. Then a tensor product polynomial spline basis is obtained by defining the bivariate basis functions

$$B_{k_1k_2}(s_1, s_2) = B_{k_1,1}(s_1)B_{k_2,2}(s_2), \quad k_1 = 1, \dots, K_1, k_2 = 1, \dots, K_2, \qquad (5.11)$$

that is, by taking all pairwise products of univariate basis functions. In most cases, the degrees of both univariate bases will be taken to be the same and we will refer to $D = D_1 = D_2$ as the degree of the tensor product polynomial spline in these cases.

As an illustration for the construction of tensor product basis functions, Fig. 5.18 shows linear tensor product truncated power series splines while Figs. 5.19 and 5.20 illustrate tensor product B-splines. The truncated power series leads to easily interpretable basis functions composed, for example, of polynomial effects in s_1- and s_2-direction, but again has the drawback of numerical instability. In fact, numerical problems tend to be even increased in the spatial setting. Tensor product B-splines, on the other hand, form a spatially localized basis with finite support. With increasing degree, they appear more and more Gaussian, that is they resemble the form of bivariate (independent) Gaussian distributions. Note, however, that regardless of the degree, tensor product B-splines do not have circular contour lines but show some anisotropic deviation from radiality as illustrated in Fig. 5.21 for different degrees. The amount of anisotropy is highest for zero degree B-splines and decreases with increasing degree. Still, first degree tensor product B-splines may induce artefacts in surfaces with single spikes dominating the surface.

Combining all bivariate basis functions yields a representation of the spatial surface $f(s_1, s_2)$ as

$$f(s_1, s_2) = \sum_{k_1=1}^{K_1} \sum_{k_2=1}^{K_2} \gamma_{k_1k_2} B_{k_1k_2}(s_1, s_2).$$

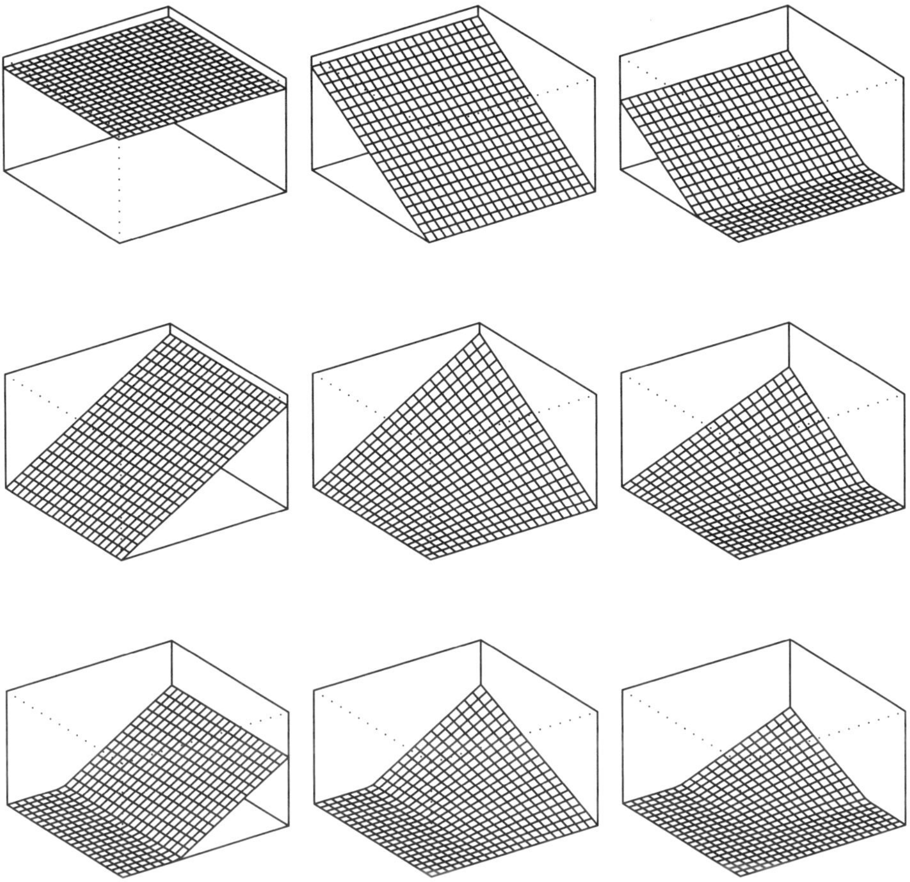

Fig. 5.18 Perspective plots of linear tensor product splines in truncated power series representation.

After re-indexing, we may also simply write

$$f(s_1, s_2) = \sum_{k=1}^{K} \gamma_k B_k(s_1, s_2)$$

where $K = K_1 \cdot K_2$ denotes the complete number of basis functions. The latter representation is easier if, for example, combinations of basis functions are considered and avoids the need for double indices. It also bridges the gap to basis-function approaches that rely on basis functions that are not necessarily aligned along a grid. Moreover, it reveals more clearly that even tensor product

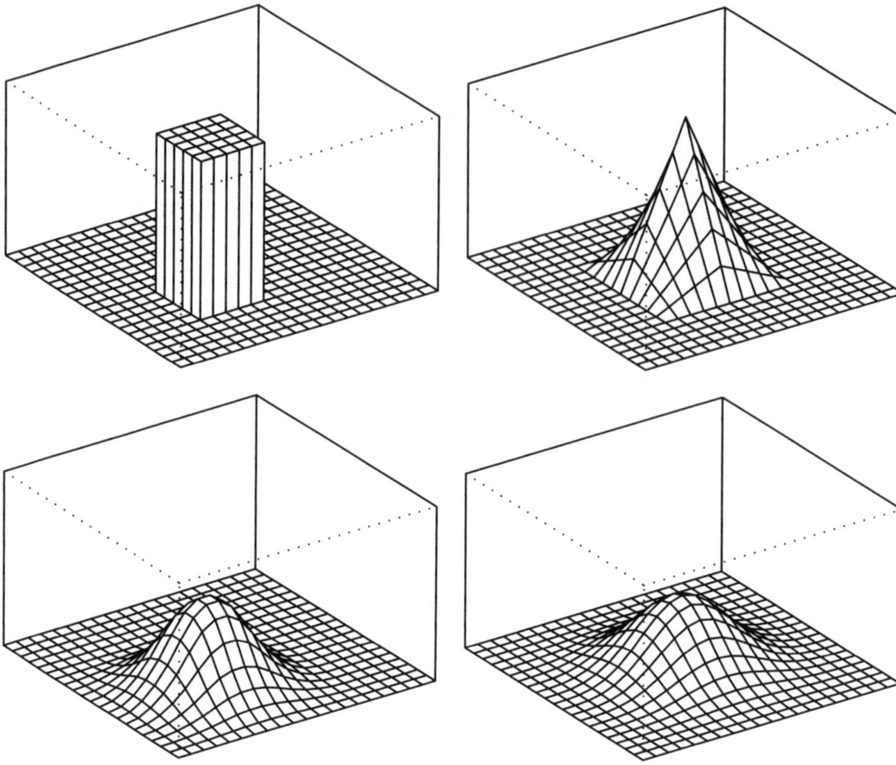

Fig. 5.19 Tensor product B-splines of degree $D = 0$, 1, 2, and 3.

basis function representations induce a large linear model with predictor

$$\boldsymbol{\eta} = \boldsymbol{X}\boldsymbol{\gamma}$$

where the design matrix is composed of basis function evaluations. More specifically, the i-th row of the design matrix is given by

$$\boldsymbol{x}_i' = (B_{11}(s_{i1}, s_{i2}), \ldots, B_{K_1 1}(s_{i1}, s_{i2}), \ldots, B_{1 K_2}(s_{i1}, s_{i2}), \ldots, B_{K_1 K_2}(s_{i1}, s_{i2}))$$

when using double indices and

$$\boldsymbol{x}_i' = (B_1(s_{i1}, s_{i2}), \ldots, B_K(s_{i1}, s_{i2}))$$

in the single index variant. The complete design matrix is then obtained by stacking all rows into the matrix $\boldsymbol{X}$. Correspondingly, the vector of regression coefficients is defined as

$$\boldsymbol{\gamma} = (\gamma_{11}, \ldots, \gamma_{K_1 1}, \ldots, \gamma_{1 K_2}, \ldots, \gamma_{K_1 K_2})'$$

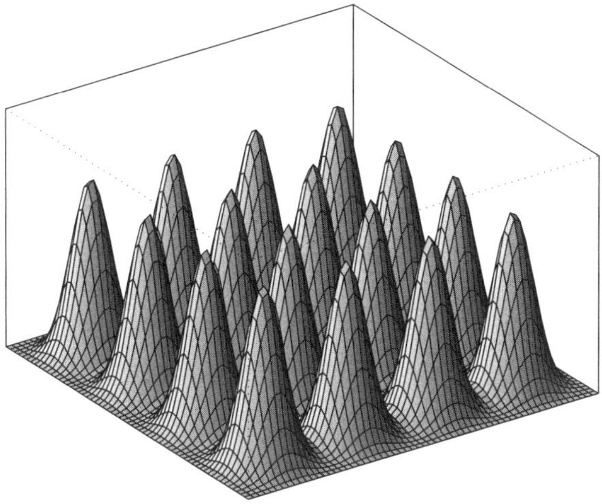

Fig. 5.20 Subset of a cubic tensor product B-spline basis.

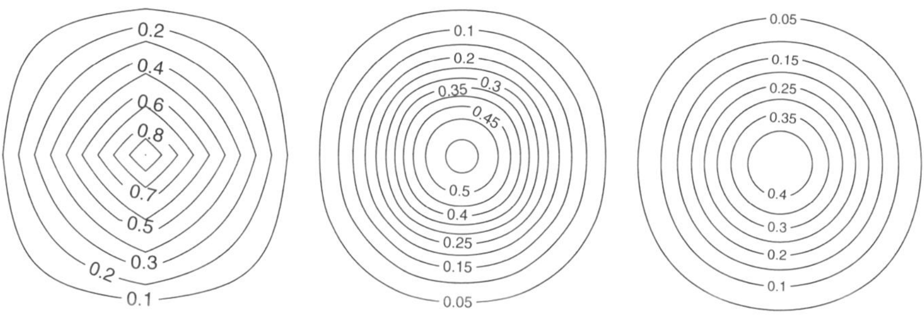

Fig. 5.21 Contour plots of tensor product B-splines of degree $D = 1$, 2 and 3.

in double index notation and as

$$\boldsymbol{\gamma} = (\gamma_1, \ldots, \gamma_K)'$$

when using single indices.

The elements of $\boldsymbol{X}$ can also be obtained from the vector-matrix representation of the additive model (5.10). Let $\boldsymbol{X}_1$ and $\boldsymbol{X}_2$ denote the design matrices corresponding to f_1 and f_2 respectively, such that

$$\boldsymbol{y} = \boldsymbol{X}_1 \boldsymbol{\gamma}_1 + \boldsymbol{X}_2 \boldsymbol{\gamma}_2 + \boldsymbol{\varepsilon}.$$

Than the columns of $\boldsymbol{X}$ are given by the (elementwise) products of all combinations of columns in $\boldsymbol{X}_1$ and $\boldsymbol{X}_2$.

Of course the idea of tensor product bases for bivariate smoothing can not only be applied in the context of polynomial splines but more generally based on any sets of univariate basis functions. We can also combine different types of bases in a tensor product if this seems to be required. For example, a penalized spline basis for a temporal effect could be combined with a Markov random field for a discrete spatial structure, leading to a spatio-temporal model.

One special feature of tensor product splines is that they can not only be justified in a heuristic manner but in fact form a class of bivariate polynomial splines that extends the definition of univariate polynomial splines. More specifically, a function $f : [a_1, b_1] \times [a_2, b_2] \to \mathbb{R}$ is called a bivariate polynomial spline of degree D with knots $\kappa_{0,1} < \kappa_{1,1} < \ldots < \kappa_{M_1-1,1} < \kappa_{M_1,1}$ and $\kappa_{0,2} < \kappa_{1,2} < \ldots < \kappa_{M_2-1,2} < \kappa_{M_2,2}$ if it satisfies the following conditions:

1. The partial derivatives
$$\frac{\partial^{i+j} f(s_1, s_2)}{\partial^i s_1 \partial^j s_2}, \quad 0 \leq i, j < D,$$
exist and are continuous and
2. $f(s_1, s_2)$ is a polynomial of degree D for $(s_1, s_2) \in [\kappa_{m_1,1}, \kappa_{m_1+1,1}) \times [\kappa_{m_2,2}, \kappa_{m_2+1,2})$, $m_1 = 0, \ldots, M_1 - 1$, $m_2 = 0, \ldots, M_2 - 1$.

The bivariate polynomial splines then form a $K_1K_2 = (M_1 + D)(M_2 + D)$-dimensional vector space and the tensor product basis functions defined in (5.11) can be shown to define a basis of this function space.

Kronecker product penalties for tensor product bases

After having defined bivariate basis functions, we will now consider extensions of the penalty concepts/prior specifications discussed in Chapter 2 for the bivariate case. Note that it is not straightforward to extend the idea of penalizing differences of adjacent parameter values since there is no natural ordering in space. As an alternative, concepts developed for discrete spatial information, and in particular Gaussian Markov random fields, will turn out to be useful for attributing spatial smoothness to the coefficient vector either in terms of a penalty or a spatial prior distribution. Therefore, we will for the moment no longer consider the vector of regression coefficients $\boldsymbol{\gamma}$ but will interpret the elements of this vector as a field of coefficients:

$$\begin{array}{cccc}
\gamma_{11} & \gamma_{21} & \cdots & \gamma_{K_1 1} \\
\gamma_{12} & \ddots & & \gamma_{K_1 2} \\
\vdots & & \ddots & \vdots \\
\gamma_{1K_2} & \gamma_{2K_2} & \cdots & \gamma_{K_1 K_2}
\end{array}$$

Hence, we can consider the regression coefficients as being spatially aligned on a discrete lattice laid out by the indices $k_1 = 1, \dots, K_1$ and $k_2 = 1, \dots, K_2$. Therefore we can now assign a spatial smoothness prior to the vector γ utilizing ideas for discrete lattices discussed in Section 5.2. Note that we can equivalently define penalties/joint priors of the form $\gamma' \boldsymbol{K} \gamma$ or conditional priors since these can be connected based on the rules for multivariate Gaussian distributions described on page 339.

Carrying over the idea of an intrinsic Gaussian Markov random field, we could assign the following prior to the coefficients in the interior of the field:

$$\gamma_{k_1 k_2} | \gamma_{-k_1 k_2} \sim \mathrm{N}\left(\frac{1}{4}[\gamma_{k_1-1,k_2} + \gamma_{k_1+1,k_2} + \gamma_{k_1,k_2-1} + \gamma_{k_1,k_2+1}], \frac{\tau^2}{4}\right) \quad (5.12)$$

that is, the prior is conditionally normal with the average of the four nearest neighbours as expectation. Prior distributions for the boundary coefficients can be obtained analogously by adjusting the number of neighbours appropriately.

While the intrinsic Gaussian Markov random field prior based on the four nearest neighbours intuitively makes sense, it may be useful to consider different, and in particular larger, neighbourhood structures. In the context of univariate smoothing, first-order random-walk priors are known to produce relatively wiggly estimates and therefore second-order random walks are commonly preferred. It would therefore be interesting to derive an analogous extension for the bivariate first-order random-walk prior (5.12). This leads to higher order Markov random fields which are, however, mathematically more involved in general. We have therefore restricted our attention to first-order Markov random fields in Section 5.2. Still, in the special case of regular lattices as the regularly arranged regression coefficients, higher-order Markov random fields are easier to obtain. To introduce such higher-order Markov random fields, we will start with penalization in a frequentist sense, where the difference penalty is to be extended to the bivariate smoothing case.

A natural extension from univariate to bivariate difference penalties would be to consider differences along the rows and the columns in the coefficient field. Hence, we start our considerations by constructing separate penalties for the rows and the columns, utilizing again the univariate bases that were used to set up the tensor product. Let $\boldsymbol{D}_1$ be a first order difference matrix for the univariate smooth in s_1-direction, that is, for a smooth along the rows of the coefficient vector. Since γ consists of K_2 vectors of regression coefficients corresponding to rows of basis functions in the spatial field, the univariate difference matrix has to be repeated K_2 times before applying it to the vector of regression coefficients. Mathematically, this repetition is achieved by taking the tensor product of a K_2-dimensional identity matrix and the difference matrix, yielding $\boldsymbol{I}_{K_2} \otimes \boldsymbol{D}_1 = \mathrm{blockdiag}(\boldsymbol{D}_1, \dots, \boldsymbol{D}_1)$. Multiplying the resulting matrix with the vector of regression coefficients then results in the vector of all row-wise differences

$$(\boldsymbol{I}_{K_2} \otimes \boldsymbol{D}_1)\boldsymbol{\gamma} = \begin{pmatrix} \gamma_{21} - \gamma_{11} \\ \vdots \\ \gamma_{K_1 1} - \gamma_{K_1 - 1, 1} \\ \vdots \\ \gamma_{2K_2} - \gamma_{1,K_2} \\ \vdots \\ \gamma_{K_1 K_2} - \gamma_{K_1 - 1, K_2} \end{pmatrix}$$

and, correspondingly, the quadratic form

$$\boldsymbol{\gamma}'(\boldsymbol{I}_{K_2} \otimes \boldsymbol{D}_1)'(\boldsymbol{I}_{K_2} \otimes \boldsymbol{D}_1)\boldsymbol{\gamma} = \sum_{k_2=1}^{K_2} \sum_{k_1=2}^{K_1} (\gamma_{k_1 k_2} - \gamma_{k_1 - 1, k_2})^2,$$

consists of all row-wise squared first order differences. In analogy, the sum of all squared first order differences along the columns is given by

$$\boldsymbol{\gamma}'(\boldsymbol{D}_2 \otimes \boldsymbol{I}_{K_1})'(\boldsymbol{D}_2 \otimes \boldsymbol{I}_{K_1})\boldsymbol{\gamma} = \sum_{k_1=1}^{K_1} \sum_{k_2=2}^{K_2} (\gamma_{k_1 k_2} - \gamma_{k_1, k_2 - 1})^2,$$

where $\boldsymbol{D}_2$ is the first order difference matrix for the univariate smooth in s_2-direction and $\boldsymbol{I}_{K_1}$ is the K_1-dimensional identity matrix. Note that the order in the Kronecker product has to be reversed to account for the ordering of coefficients in $\boldsymbol{\gamma}$.

Combining row-wise and column-wise squared differences finally yields the overall penalty term

$$\lambda \boldsymbol{\gamma}' \left[(\boldsymbol{I}_{K_2} \otimes \boldsymbol{D}_1)'(\boldsymbol{I}_{K_2} \otimes \boldsymbol{D}_1) + (\boldsymbol{D}_2 \otimes \boldsymbol{I}_{K_1})'(\boldsymbol{D}_2 \otimes \boldsymbol{I}_{K_1}) \right] \boldsymbol{\gamma} \tag{5.13}$$

which is of the form

$$\lambda \boldsymbol{\gamma}' \boldsymbol{K} \boldsymbol{\gamma}$$

with

$$\boldsymbol{K} = (\boldsymbol{I}_{K_2} \otimes \boldsymbol{D}_1)'(\boldsymbol{I}_{K_2} \otimes \boldsymbol{D}_1) + (\boldsymbol{D}_2 \otimes \boldsymbol{I}_{K_1})'(\boldsymbol{D}_2 \otimes \boldsymbol{I}_{K_1}).$$

An equivalent expression is given by

$$\boldsymbol{K} = \boldsymbol{I}_{K_2} \otimes \boldsymbol{K}_1 + \boldsymbol{K}_2 \otimes \boldsymbol{I}_{K_1},$$

with univariate penalty matrices $\boldsymbol{K}_1 = \boldsymbol{D}_1'\boldsymbol{D}_1$ und $\boldsymbol{K}_2 = \boldsymbol{D}_2'\boldsymbol{D}_2$. This representation reveals more clearly that we cannot only construct difference penalties for tensor product basis functions but that any univariate penalty concept can be inserted to obtain a valid tensor product penalty.

It now turns out that for first-order differences for both rows and columns, the difference penalty induced by penalty matrix$\boldsymbol{K}$ corresponds exactly to the first-order IGMRF (5.12) that we considered in the beginning of this section. To be more specific, we interpret $\boldsymbol{K}$ as the precision matrix of the Gaussian distribution

$$p(\boldsymbol{\gamma}) \propto \exp\left(-\frac{1}{2\tau^2}\boldsymbol{\gamma}'\boldsymbol{K}\boldsymbol{\gamma}\right)$$

and, utilizing the rules for deriving full conditional distributions from joint Gaussians introduced in Section 5.2.2, page 339, we obtain the conditional dependence structure

$$\gamma_{k_1 k_2}|\cdot \sim \mathrm{N}\left(\frac{1}{4}(\gamma_{k_1-1,k_2} + \gamma_{k_1+1,k_2} + \gamma_{k_1,k_2-1} + \gamma_{k_1,k_2+1}), \frac{\tau^2}{4}\right)$$

which exactly coincides with (5.12). This prior structure will also be our standard choice for interaction surfaces in the remainder of this book.

Example 5.11: Forest health

Continuing Example 5.4, we investigate the presence of an interaction between the temporal trend and average age at the observation plot using bivariate penalized splines. More specifically, we will consider the following three predictor specifications

$$\eta_{it}^{(c)} = \theta^{(c)} - f_1(\text{age}_{it}) - f_2(t) \tag{5.14}$$

$$\eta_{it}^{(c)} = \theta^{(c)} - f(\text{age}_{it}, t) \tag{5.15}$$

$$\eta_{it}^{(c)} = \theta^{(c)} - f_1(\text{age}_{it}) - f_2(t) - f_{1,2}(\text{age}_{it}, t) \tag{5.16}$$

in a three-categorical cumulative probit model with thresholds $\theta^{(c)}$. The first model comprises an additive combination of univariate penalized splines and therefore reduces to a main effects model. The second model contains only the interaction surface while the third model combines main effects with an interaction surface. The main advantage of the last formulation is that it allows for an easier interpretation since $f_{1,2}$ really refers to an additional interaction effect. However, as with the separation of structured and unstructured spatial effects, the separation between main effects and interaction surface may be difficult. A further advantage of the third model is that it offers increased flexibility since separate smoothing parameters are available for the main effects and for the interaction.

For all nonparametric functions and surfaces, we employ cubic B-splines based on 20 equidistant knots in both covariate directions. A second-order random-walk prior is assumed for the univariate splines while a first-order random-walk (5.12) has been chosen for the interaction. The latter choice ensures that the model is formally identifiable (subject to a centring condition). With a second-order random-walk for the interaction effect, both the univariate main effects and the interaction leave linear effects in age and time unpenalized.

To make such a model formally identifiable, separate linear effects would have to be included in the model, while both univariate effects and main effects are centred around the linear effect. For MCMC simulations, this can be achieved by substracting the linear part of the fit in each iteration. For mixed model-based estimation, duplicated rows in the fixed effects parts of the functions would have to be deleted.

Figure 5.22 and 5.23 show posterior mode estimates for the main effects and the interaction obtained with mixed model-based empirical Bayes inference. The differences between the main effects estimated in models (5.14) and (5.16) are relatively small although the effects appear to be somewhat smoother when the interaction effect is included in addition. This also intuitively makes sense since localized variability is picked up by the interaction effect. The differences between the estimated interaction effects are much larger, but this is simply due to the fact that the interaction in model (5.15) also contains the main effects. As a consequence, the estimated surface is largely dominated by the main effects and it is hard to detect the precise form of the interaction from the estimated surface. In contrast, the surface estimated in model (5.15) really represents an interaction

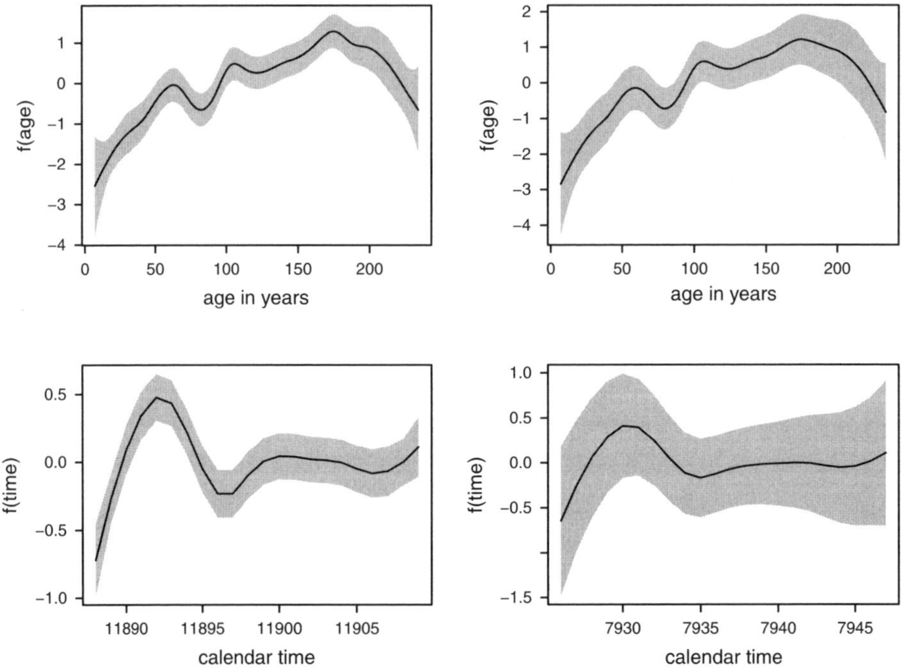

Fig. 5.22 Forest health: Estimated main effects of calendar time and age together with 95% credible intervals. The left panel contains results for the pure main effects model (5.14), the right panel results for the interaction model (5.16).

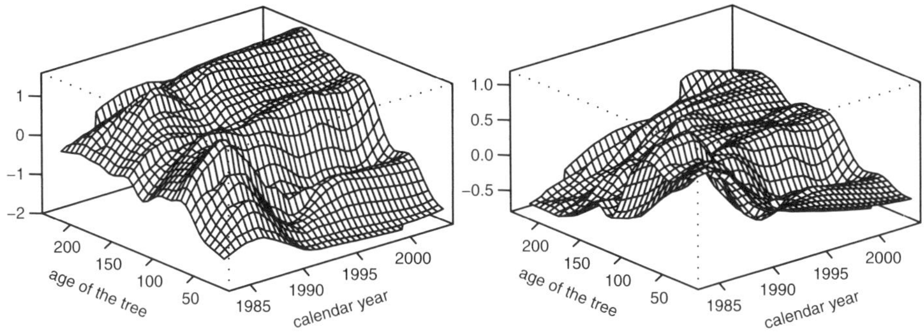

Fig. 5.23 Forest health: Estimated interaction effect for the pure interaction model (5.15) (left) and the interaction model including main effects (5.16) (right).

and leads to the conclusion that young trees had an alleviated defoliation risk early in the observation period but recovered later on. Vice versa, older trees started better off but their defoliation probability increased over the observation period. △

The advantage of the Kronecker product penalty (5.13) is that it offers large flexibility in deriving other types of penalties and therefore corresponding Markov random field structures. For example, we can of course employ higher-order difference matrices for the univariate fits, yielding also higher-order Markov random field equivalents. For example, with second order differences, we obtain a precision matrix that induces the following conditional prior mean and conditional prior variance for an element γ_{k_1,k_2}:

$$\begin{aligned}\mathbb{E}(\gamma_{k_1,k_2}|\cdot) = \frac{1}{12}\,[&4\gamma_{k_1-1,k_2} + 4\gamma_{k_1+1,k_2} + 4\gamma_{k_1,k_2-1} + 4\gamma_{k_1,k_2+1}\\ &-\gamma_{k_1-2,k_2} - \gamma_{k_1+2,k_2} - \gamma_{k_1,k_2-2} - \gamma_{k_1,k_2+2}]\end{aligned}$$

and

$$\mathrm{Var}(\gamma_{k_1,k_2}|\cdot) = \frac{\tau^2}{12}$$

in the interior of the coefficient field. This can now be interpreted as a second-order Markov random field on a regular lattice. More compactly, the dependence structure of such a higher-order Markov random field can be summarized in the graphical representation shown in Figure 5.24. The weights represented in the graph have to be interpreted as follows: The weight in the middle circle corresponds to the diagonal element of $\boldsymbol{K}$ that specifies the sum of weights assigned to the neighbours of the regression coefficient. The weights in the neighbouring circles specify the weights assigned to the neighbouring coefficients in the weighted average that forms the conditional expectation of $\gamma_{k_1 k_2}$. Hence, when comparing

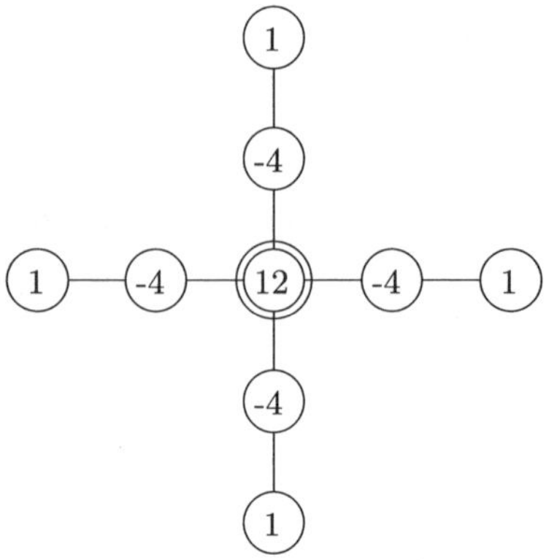

Fig. 5.24 Kronecker sum of two univariate second-order random walks: Coefficients of the precision matrix.

the weights with those in the conditional expectation above, we find the expected equivalence. The graphical representation will be valuable in the following when discussing different types of penalty structures, to give a compact summary of the induced dependence structure. In particular, the graph structure also reveals more clearly one of the distinct features of the higher order MRFs derived from Kronecker product penalties: Conditional dependence is only induced along the rows and columns while no dependence is assumed along the diagonals. This property also holds for Markov random fields derived from tensor products of higher-order differences.

Generalized penalty concepts

The definition of bivariate penalties we have considered so far assumes a global smoothing parameter λ. In particular this means that the induced penalization has the same strength along columns and rows. More flexibility can be achieved by introducing separate smoothing parameters which leads to

$$\lambda_1 \gamma' \left[\boldsymbol{I}_{K_2} \otimes \boldsymbol{K}_1\right] \gamma + \lambda_2 \gamma' \left[\boldsymbol{K}_2 \otimes \boldsymbol{I}_{K_1}\right] \gamma.$$

However, especially for spatial effects, the usefulness of this extension crucially depends on the the choice of the coordinate axes. If different smoothing is needed along the diagonals or some other angle, the penalty will not be able to accomodate this even with separate smoothing parameters. An additional rotation parameter might be introduced to overcome this limitation, leading to a more data-driven determination of the axes.

The difference penalty for univariate penalized splines was motivated as an approximation to integrated squared derivative penalties. Replacing the difference approximation with the integral penalty (as for example in Wood, 2006) therefore leads to a closely related penalization approach. For penalties in the bivariate case, we could similarly consider a bivariate differential operator like the biharmonic differential operator

$$\frac{\partial^2}{\partial^2 s_1} + 2\frac{\partial^2}{\partial s_1 \partial s_2} + \frac{\partial^2}{\partial^2 s_2} \tag{5.17}$$

that represents a bivariate extension of the second-order derivative considered in the context of smoothing splines. Seeking the solution of the penalized log-likelihood criterion

$$l(f) - \frac{1}{2}\lambda \int\int \left[\left(\frac{\partial^2}{\partial^2 s_1} + 2\frac{\partial^2}{\partial s_1 \partial s_2} + \frac{\partial^2}{\partial^2 s_2}\right) f(s_1, s_2)\right]^2 ds_1 ds_2 \rightarrow \max_f$$

over the space of all twice continuously differentiable bivariate functions leads to the thin plate spline which is the two-dimensional analogon to the natural cubic smoothing spline. The natural boundary conditions in this case resemble the behaviour of a thin metal plate that is fixed at the function evaluations at the knots but returns to its natural behaviour at the boundaries. We will discuss thin plate splines in somewhat more detail in the following section.

Instead of determining the thin plate spline from the biharmonic differential operator penalty, we can also simply apply it as a general penalty concept whenever using bivariate basis-function approaches. If the bivariate function $f(s_1, s_2)$ is approximated as

$$f(s_1, s_2) = \sum_{k=1}^{K} \gamma_k B_k(s_1, s_2),$$

it turns out that the penalty resulting from the biharmonic differential operator can be written as a quadratic form $\boldsymbol{\gamma}'\boldsymbol{K}\boldsymbol{\gamma}$ where the elements of $\boldsymbol{K}$ correspond to

$$\int\int B''_{k_1}(s_1, s_2) B''_{k_2}(s_1, s_2) ds_1 ds_2$$

where

$$B''_k(s_1, s_2) = \left(\frac{\partial^2}{\partial^2 s_1} + 2\frac{\partial^2}{\partial s_1 \partial s_2} + \frac{\partial^2}{\partial^2 s_2}\right) B_k(s_1, s_2).$$

Similarly to the case of univariate smoothing, it is easier to work with an approximation based on difference operators than to evaluate the derivatives explicitly. Such an approximation can be derived by considering difference operators based on the twelve nearest neighbours. This results in a precision matrix with non-zero elements defined by

$$-\left(\Delta^2_{(1,0)} + \Delta^2_{(0,1)}\right)^2 = -\left(\Delta^4_{(1,0)} + 2\Delta^2_{(1,0)}\Delta^2_{(0,1)} + \Delta^4_{(0,1)}\right), \tag{5.18}$$

where $\Delta^2_{(1,0)}$ and $\Delta^2_{(0,1)}$ denote second-order difference operators along the s_1- and s_2-direction, respectively. It now turns out (Yue and Speckman, 2010) that the corresponding penalty matrix can be written as

$$\boldsymbol{K} = \boldsymbol{I}_{K_2} \otimes \boldsymbol{K}_1^{(2)} + \boldsymbol{K}_2^{(2)} \otimes \boldsymbol{I}_{K_1} + 2\boldsymbol{K}_2^{(1)} \otimes \boldsymbol{K}_1^{(1)}$$

where $\boldsymbol{K}_1^{(d)}$ and $\boldsymbol{K}_2^{(d)}$ are univariate d-th order difference penalty matrices constructed along the columns and the rows, respectively.

Figure 5.25 visualizes the structure of the penalty matrix for an interior point of the coefficient lattice. The main difference to a second-order difference Kronecker product penalty is that the biharmonic differential operator also induces dependencies along the diagonals. This is also reflected in the null space of the penalties: The null space of the second-order Kronecker product penalty leaves the interaction effect $s_1 \cdot s_2$ unpenalized while the biharmonic differential operator also penalizes this interaction (see the next paragraph for details).

An alternative perspective on the approximation of the biharmonic differential operator is obtained by considering the squares of increments

$$(\gamma_{k_1+1,k_2} + \gamma_{k_1-1,k_2} + \gamma_{k_1,k_2+1} + \gamma_{k_1,k_2-1}) - 4\gamma_{k_1 k_2}$$

which can be regarded as extensions of the increments $\gamma_{k+1} - 2\gamma_k + \gamma_{k-1}$ defined for univariate second-order random walks. Rue and Held (2005) moreover discuss refined ways to approximate the biharmonic differential operator using larger neighbourhoods or including differences along the diagonals in approximation (5.18).

As a last possibility to derive bivariate penalties, we will reconsider the interpretation of the random-walk prior for penalized spline coefficients as local

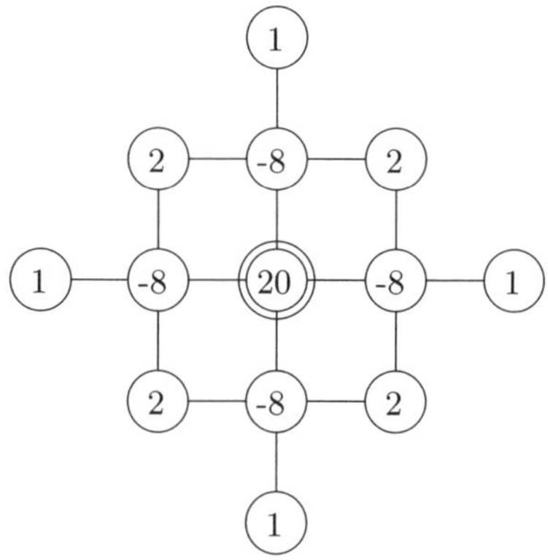

Fig. 5.25 Approximation to the biharmonic differential operator: Coefficients of the precision matrix.

polynomial fits based on the neighbouring coefficients (Section 2.2.1, page 50). For example, the expectation of a first-order random-walk turned out to be equivalent to a local linear fit based on the preceding and the subsequent time point (Fig. 2.11). For the bivariate case, we could analogously consider a local linear fit through the four nearest neighbours γ_{k_1-1,k_2}, γ_{k_1,k_2-1}, γ_{k_1+1,k_2} and γ_{k_1,k_2+1} to obtain the conditional prior expectation of γ_{k_1,k_2}. From the weights of the linear combination of the neighbours, we could then deduce the structure of the corresponding penalty by reversing the ideas for obtaining conditional expectations from the precision matrix of a joint Gaussian distribution. The weights of the local linear fit are obtained by computing the prediction for γ_{k_1,k_2} from a regression model of the form

$$\begin{pmatrix} \gamma_{k_1-1,k_2} \\ \gamma_{k_1,k_2-1} \\ \gamma_{k_1+1,k_2} \\ \gamma_{k_1,k_2+1} \end{pmatrix} = \begin{pmatrix} 1 & k_1-1 & k_2 \\ 1 & k_1 & k_2-1 \\ 1 & k_1+1 & k_2 \\ 1 & k_1 & k_2+1 \end{pmatrix} \begin{pmatrix} \beta_1 \\ \beta_2 \\ \beta_3 \end{pmatrix} + \varepsilon.$$

It turns out that all four nearest neighbours are assigned equal weight so that we obtain the same conditional prior as with a first-order difference Kronecker product penalty, that is, (5.12).

Analogously, Besag and Higdon (1999) propose to perform a local quadratic fit through the twelve next neighbours to obtain a second-order random-walk penalty. In this case, the coefficients associated with the conditional expectation of $\gamma_{k_1k_2}$ are obtained from predicting $\gamma_{k_1k_2}$ from the regression model of the form

$$\begin{pmatrix} \gamma_{k_1-2,k_2} \\ \gamma_{k_1-1,k_2-1} \\ \gamma_{k_1-1,k_2} \\ \gamma_{k_1-1,k_2+1} \\ \gamma_{k_1,k_2-2} \\ \gamma_{k_1,k_2-1} \\ \gamma_{k_1,k_2+1} \\ \gamma_{k_1,k_2+2} \\ \gamma_{k_1+1,k_2-1} \\ \gamma_{k_1+1,k_2} \\ \gamma_{k_1+1,k_2+1} \\ \gamma_{k_1+2,k_2} \end{pmatrix} = \begin{pmatrix} 1 & k_1-2 & k_2 & (k_1-2)^2 & k_2^2 \\ 1 & k_1-1 & k_2-1 & (k_1-1)^2 & (k_2-1)^2 \\ 1 & k_1-1 & k_2 & (k_1-1)^2 & k_2^2 \\ 1 & k_1-1 & k_2+1 & (k_1-1)^2 & (k_2+1)^2 \\ 1 & k_1 & k_2-2 & k_1^2 & (k_2-2)^2 \\ 1 & k_1 & k_2-1 & k_1^2 & (k_2-1)^2 \\ 1 & k_1 & k_2+1 & k_1^2 & (k_2+1)^2 \\ 1 & k_1 & k_2+2 & k_1^2 & (k_2+2)^2 \\ 1 & k_1+1 & k_2-1 & (k_1+1)^2 & (k_2-1)^2 \\ 1 & k_1+1 & k_2 & (k_1+1)^2 & k_2^2 \\ 1 & k_1+1 & k_2+1 & (k_1+1)^2 & (k_2+1)^2 \\ 1 & k_1+2 & k_2 & (k_1+2)^2 & k_2^2 \end{pmatrix} \begin{pmatrix} \beta_1 \\ \beta_2 \\ \beta_3 \\ \beta_4 \\ \beta_5 \end{pmatrix} + \varepsilon.$$

This yields the dependence structure and coefficients shown in Fig. 5.26. Similarly to the biharmonic differential operator, there is a dependence also along the diagonal that results from employing the twelve next neighbours.

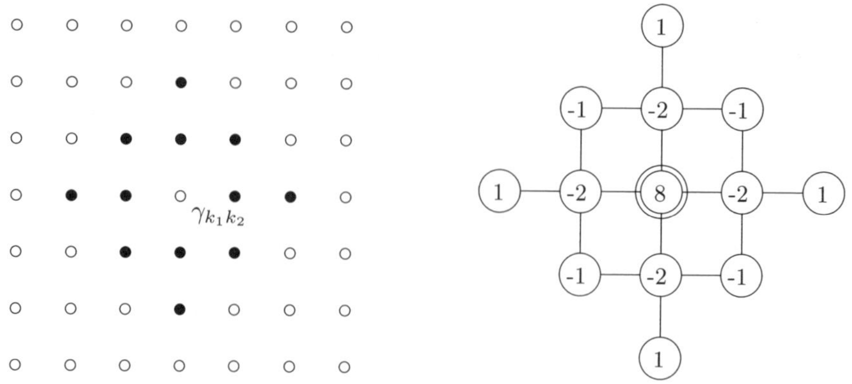

Fig. 5.26 Dependence structure and coefficients of the precision matrix for a two-dimensional second-order random-walk based on a local quadratic fit through the twelve next neighbours.

The main difficulty with local polynomial approximations for the derivation of prior distributions on lattices is the consideration of boundary modifications, since the presented coefficients are only valid for the interior points of the lattice. Besag and Higdon (1999) propose to extend the lattice at the boundaries by a number of additional knots and then to work on a torus, resulting in a theoretically infinite lattice. However, this approach is not very intuitive since observations of extremely high distance become close by when folding the surface to a torus. An alternative is to restrict an infinite lattice to the finite case without correcting for the boundaries, as proposed for example in Rue and Held (2005. page 115). While this approach may be used with MCMC simulation techniques, it has the drawback of enlarging the null space of the precision matrix, leading to a noninterpretable basis of this null space. A further argument against the local quadratic fit is given by Rue and Held (2005. page 119): The penalty derived from the precision matrix cannot be represented in terms of increments depending only on the same neighbourhood of twelve nearest neighbours. In contrast, the increments depend on a much larger neighbourhood, which is not a desirable property.

In summary, the local polynomial fit priors do not seem to be recommended while both Kronecker sums and the biharmonic differential operator yield intuitive and interpretable prior distributions.

Null spaces of bivariate penalties

Besides comparing the spatial dependence structure imposed by a prior distribution or penalty, it is also interesting to study the null space of the penalty matrix involved. First and foremost, this null space tells us what kind of functions remain unpenalized by a specific penalty but the null space will also be important when considering mixed model-based inference for spatial effects in analogy to Section 4.2.1. Therefore, we will not only give a qualitative description of the null spaces but will also provide bases that will be required as a building block when coming back to mixed model-based inference in Section 5.5.

For tensor product penalized splines with penalties derived from Kronecker products in (5.13), it can be shown that the null space of the penalty equals the tensor product of the univariate null spaces. If we consider difference penalties of orders d_1 and d_2 for rows and columns, this corresponds to null spaces of polynomials

$$\beta_{0,1} + \beta_{1,1}s_1 + \beta_{2,1}s_1^2 + \ldots + \beta_{d_1-1,1}s_1^{d_1-1}$$

and

$$\beta_{0,2} + \beta_{1,2}s_2 + \beta_{2,2}s_2^2 + \ldots + \beta_{d_2-1,2}s_2^{d_2-1}.$$

When taking the tensor product of these bases, we have to consider all twoway combinations of polynomials. For example, in case of first-order differences for both rows and columns, both null spaces reduce to a constant and the corresponding tensor product is also a constant. For second-order differences, both univariate null spaces are linear effects and the tensor product yields the four-dimensional null space spanned by the linear effects of directions s_1 and s_2 plus an interaction effect, that is, the space

$$\beta_0 + \beta_1 s_1 + \beta_2 s_2 + \beta_3 s_1 s_2.$$

A basis of this null space (for the function evaluations of a given data set) can be constructed based on any combination of equidistant values in the s_1- and s_2-domain. For example, when employing the knots for simplicity, we obtain the basis

$$\begin{pmatrix} 1 & \kappa_{1,1} & \kappa_{1,2} & \kappa_{1,1}\cdot\kappa_{1,2} \\ 1 & \kappa_{2,1} & \kappa_{1,2} & \kappa_{2,1}\cdot\kappa_{1,2} \\ \vdots & \vdots & \vdots & \vdots \\ 1 & \kappa_{K_1,1} & \kappa_{1,2} & \kappa_{K_1,1}\cdot\kappa_{1,2} \\ 1 & \kappa_{1,1} & \kappa_{2,2} & \kappa_{1,1}\cdot\kappa_{2,2} \\ \vdots & \vdots & \vdots & \vdots \\ 1 & \kappa_{K_1,1} & \kappa_{2,2} & \kappa_{K_1,1}\cdot\kappa_{2,2} \\ \vdots & \vdots & \vdots & \vdots \\ 1 & \kappa_{1,1} & \kappa_{K_2,2} & \kappa_{1,1}\cdot\kappa_{K_2,2} \\ \vdots & \vdots & \vdots & \vdots \\ 1 & \kappa_{K_1,1} & \kappa_{K_2,2} & \kappa_{K_1,1}\cdot\kappa_{K_2,2} \end{pmatrix}, \tag{5.19}$$

where $\kappa_{1,1},\ldots,\kappa_{K_1,1}$ and $\kappa_{1,2},\ldots,\kappa_{K_2,2}$ are the knots in s_1- and s_2-direction, respectively.

For the approximation of the biharmonic differential operator (5.18), the null space is only three-dimensional, containing functions of the form

$$\beta_0 + \beta_1 s_1 + \beta_2 s_2$$

resulting from the additional penalty along the diagonal. Consequently, the last column has to be deleted from (5.19) to obtain the corresponding basis.

Higher-order Markov random fields on regular grids

The considerations about penalties in the previous sections are not only valuable in the context of bivariate smoothing but actually also extend the concepts discussed in the context of Gaussian Markov random fields in Section 5.2. There we restricted our attention to first-order neighbourhoods, mainly because we had to deal with irregular lattices resulting from geographical data. As a byproduct, higher-order penalties for surfaces allow us to consider higher-order Gaussian Markov random fields on regular lattices. As a consequence, the higher-order prior distributions are not only useful as penalties in basis-function approaches but can also be used to model spatial functions directly. This provides an extension of random-walk smoothing discussed in Sections 2.1 and 2.2 which is the univariate analogon to modelling discrete spatial data on a regular lattice. There we obtained random walks as a limiting case of zero degree P-splines with knots at each distinct observation point.

Similarly, we can consider spatial data arranged on a regularly spaced grid or discretize the observation area to rectangles. Then we can assign a separate coefficient to each of the rectangles, which essentially corresponds to zero degree tensor product splines with knots placed at the unique observed locations. Instead of writing $f(s_1, s_2)$ we can then also simply write $f(k_1, k_2) = \gamma_{k_1,k_2}$. Assigning one of the previously considered priors to this spatial surface results in a higher-order Gaussian Markov random field for the regular grid of function values γ_{k_1,k_2}.

Higher-order interactions

Conceptually, it is quite simple to extend the idea of tensor product penalized splines to interactions of more than two variables. For example, to determine a three-way interaction surface $f(s_1, s_2, s_3)$, we can set up univariate bases along all three directions and then define the trivariate tensor product basis functions

$$B_{k_1,k_2,k_3}(s_1, s_2, s_3) = B_{k_1}(s_1)B_{k_2}(s_2)B_{k_3}(s_3)$$

yielding the representation

$$f(s_1, s_2, s_3) = \sum_{k_1=1}^{K_1} \sum_{k_2=1}^{K_2} \sum_{k_3=1}^{K_3} \gamma_{k_1,k_2,k_3} B_{k_1,k_2,k_3}(s_1, s_2, s_3).$$

Analogously, the penalty matrix is constructed as

$$\boldsymbol{K} = \boldsymbol{I}_{K_3} \otimes \boldsymbol{I}_{K_2} \otimes \boldsymbol{K}_1 + \boldsymbol{I}_{K_3} \otimes \boldsymbol{K}_2 \otimes \boldsymbol{I}_{K_1} + \boldsymbol{K}_3 \otimes \boldsymbol{I}_{K_2} \otimes \boldsymbol{I}_{K_1},$$

where $\boldsymbol{K}_1$, $\boldsymbol{K}_2$ and $\boldsymbol{K}_3$ are univariate penalty matrices in directions s_1, s_2 and s_3.

Hence, at first sight there do not seem to be larger difficulties associated with estimating general interaction surfaces $f(s_1, \ldots, s_d)$ of dimension d. However, two problems render higher order interaction surfaces cumbersome. The first difficulty simply relates to the number of regression coefficients involved in specifying a tensor product spline. When starting with 20 basis functions for each of the univariate fits, the dimension of γ is already 400 for a bivariate surface and grows to 8,000 for the trivariate effect. In general, a number of 20^d regression coefficients would have to be estimated. The exponential growth has two immediate consequences: First of all, estimation procedures have to be able to deal with a large number of regression coefficients. When estimating all parameters simultaneously, a system of equations with dimension $20^d \times 20^d$ has to be solved, which rapidly becomes numerically intractable even when using sparse matrix methods or efficient representations such as array regression (Currie, Durban and Eilers, 2006). Moreover, although penalization will effectively reduce the number of free parameters, the large number of coefficients requires a large amount of data in order to be well identified.

This becomes even clearer when considering the second difficulty that is known as the curse of dimensionality. The idea behind penalized splines with B-splines was to consider local polynomials with appropriate smoothness conditions to approximate unknown functions. Hence, flexibility in the function estimate is introduced in terms of local polynomial coefficients. However, the intuition of locally defined basis functions does not carry over to higher-order interactions. To see this, consider a d-dimensional unit hypercube containing uniformly distributed data. Now we want to construct a subcube with edges of length l that should contain a fixed fraction p of the data. For $d = 1$, an interval of length $l = p$ would on average contain the desired amount of data. For a square, that is, for $d = 2$, we already require $l = \sqrt{p}$ and in general we have

$$l = p^{\frac{1}{d}}.$$

When considering for example $p = 0.1$, this yields

$$\begin{array}{ll} d = 1 & l = 0.1 \\ d = 2 & l \approx 0.3 \\ d = 3 & l \approx 0.47 \\ \vdots & \vdots \\ d = 10 & l \approx 0.8 \end{array}$$

as illustrated in Fig. 5.27 for $d = 1$, 2 and 3. Hence, if we want to define a local basis function that spans over an area containing approximately 10% of the

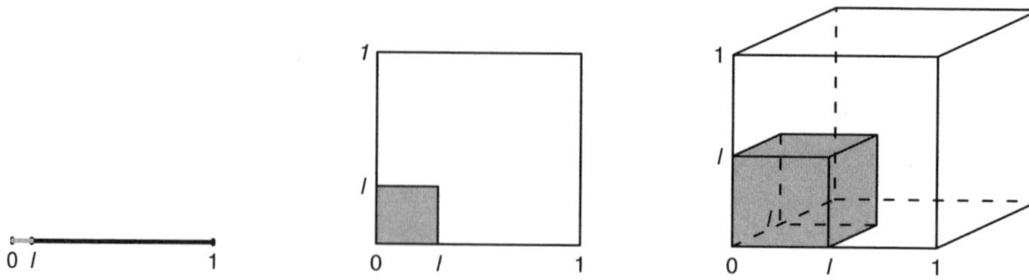

Fig. 5.27 Curse of dimensionality: Increasing the dimension of the interaction makes the basis functions less and less local.

data, it will be defined on a cube with edges of length 0.47 for a trivariate basis. Although our aim was to obtain a local basis by considering only 10% of the data points, such a basis function would hardly be considered local anymore. Put the other way round, we can also deduce that using a fixed edge length for the basis functions (as would be the case by simply considering 20 knots in each of the univariate directions regardless of the dimension of the fit) would lead to basis functions that are less and less supported by the data as the dimensionality increases. As a consequence, interactions of order larger than $d = 2$ are only rarely used in practice and models are mostly restricted to follow an additive structure.

Still, tensor products can be useful as a tool for constructing effects that go beyond surfaces. The most important example of such tensor product approaches are spatio-temporal models where separate bases are set up for the spatial and the temporal effect that can then be combined into a flexible interaction in terms of a tensor product. Note that the spatial and the temporal bases do not have to be penalized splines and this opens up the possibility of defining a more parsimonious basis in particular for the spatial part of the function to reduce the dimensionality of the parameter vector. For example, Augustin *et al.* (2009) consider a radial basis (see the next section) for the spatial effect, combined with a penalized spline basis for the temporal effect in a spatio-temporal tensor product.

5.3.2 Radial bases

While tensor product splines aim at combining univariate bases into a multidimensional basis function, the basic idea of radial bases is to modify the multivariate *argument* of the basis functions such that it becomes univariate. Therefore, we consider a set of knots $\boldsymbol{\kappa}_1, \ldots, \boldsymbol{\kappa}_K$ in space, where the basis functions should be located. Note that these knots can be placed arbitrarily and in particular are not restricted to forming a regular grid. This will turn out to be one of the major advantages of radial bases since it allows us to place more knots in areas with dense data. In most cases, the knots will be located at the observation points or a representative subset. We will discuss suitable knot placement strategies later on in this section.

For a given knot configuration, separate basis functions are assigned to these knots, that is, in contrast to tensor product B-splines each basis function should be associated with exactly one knot. This is formalized by defining the basis functions

$$B_k(\boldsymbol{s}) = B(||\boldsymbol{s} - \boldsymbol{\kappa}_k||)$$

where $B(\cdot)$ is a univariate function and $||\boldsymbol{s} - \boldsymbol{\kappa}_k||$ measures the distance between the point of interest and the knot $\boldsymbol{\kappa}_k$ defining the basis function. A spatial function $f(\boldsymbol{s})$ is then approximated by

$$f(\boldsymbol{s}) = \sum_{k=1}^{K} \gamma_k B_k(\boldsymbol{s}). \tag{5.20}$$

Obviously, different types of radial bases are obtained by varying either the function $B(\cdot)$ or the distance measure $||\cdot||$. In the context of radial basis functions, the distance is typically restricted to being Euclidean since this is exactly the distance measure that corresponds to circular contour lines and therefore radial bases. Still, other non-Euclidean distance measures could be employed and we will see in the following section on Gaussian random fields that there is a close connection to anisotropic correlation functions in this case.

Some of the most common radial basis functions are given by

$$\begin{aligned}
B(r) &= \exp(-cr^2) \quad \text{with } c > 0,\\
B(r) &= r^2 \log(r) \quad \text{(thin plate spline)},\\
B(r) &= r^k, \quad k \text{ even},\\
B(r) &= \sqrt{r^2 + c^2} \quad \text{with } c > 0,
\end{aligned}$$

where r denotes the distance between a location $\boldsymbol{s}$ and a knot $\boldsymbol{\kappa}$. Figure 5.28 shows a perspective plot of the first two examples of radial basis functions. Figure 5.29 indicates that in fact all radial basis functions in the representation (5.20) have exactly the same shape since they are all defined based on the same function $B(\cdot)$, and that each basis function is centred around one knot. It also stresses again that radial basis functions may be unevenly distributed across the observation area. In contrast, tensor product spline basis functions are necessarily aligned on a grid due to their construction based on univariate bases. This may have two undesirable consequences: First of all, several basis functions may be placed in areas that do not contain any data. For example, if the observation locations arise from a bivariate Gaussian density with considerable correlation, as illustrated in the left part of Fig. 5.30, a large portion of knots for the bivariate spline would be placed in areas without any data. Though the placement of

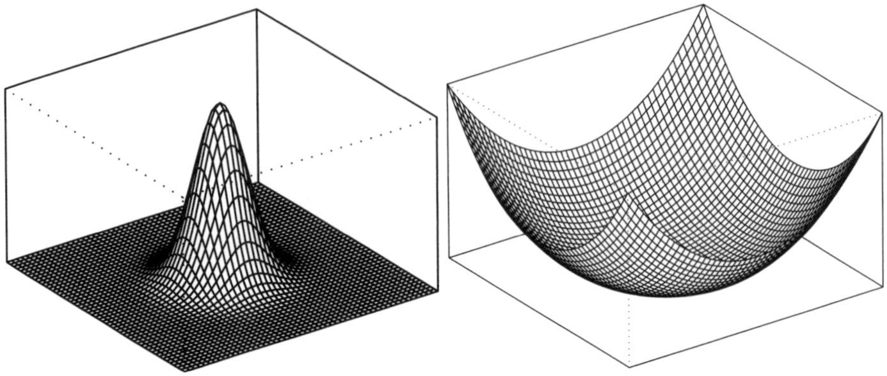

Fig. 5.28 Perspective plots of a Gaussian and a thin plate spline radial basis function.

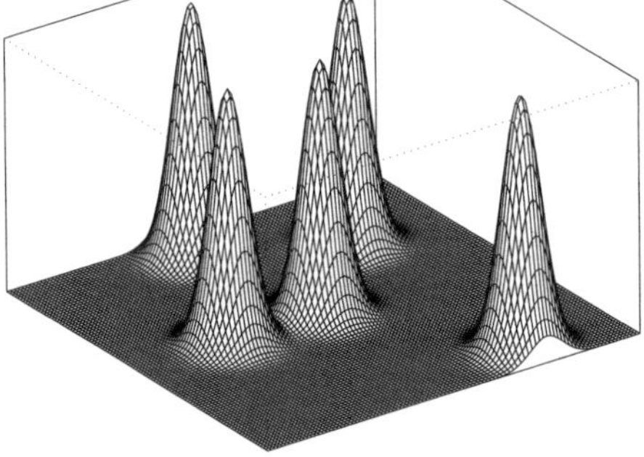

Fig. 5.29 Perspective plot of five randomly placed radial basis functions.

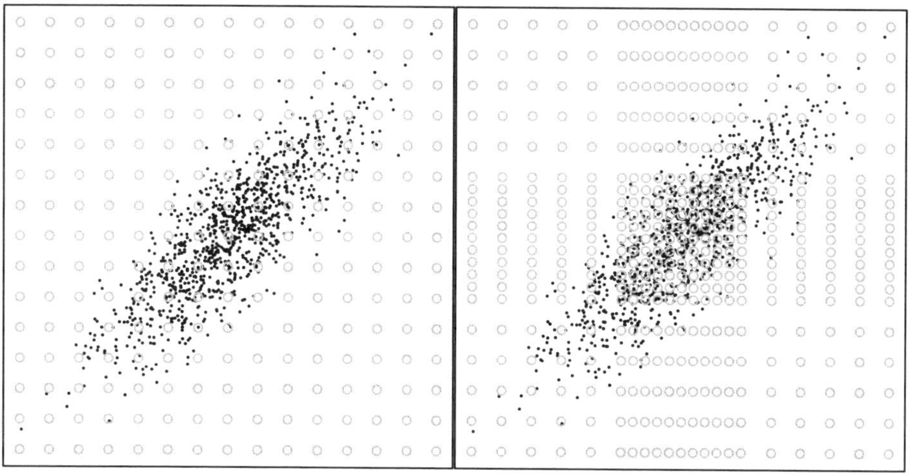

Fig. 5.30 Knot locations for tensor product splines. In the left panel, an equidistant grid is set up while in the right panel an increased number of knots is located in the middle part of the observation area.

knots in areas without observations does not typically cause estimation problems due to the regularization imposed by the penalty of tensor product P-splines, it unnecessarily increases the number of regression coefficients to be estimated. By restricting the basis functions to a specific area, considerable savings in model complexity could be achieved.

A second drawback of the knot placement in tensor product splines arises when certain areas are of particular interest and should therefore be supplemented with a higher number of basis functions. For example, if higher variability in the spatial effect is expected in a certain area, one might try to compensate for this fact by adding more basis functions in this area. With tensor product splines, the only possibility to achieve this is to increase the number of knots in the univariate bases as illustrated in Fig. 5.30. However, this not only increases the number of knots in the designated area of interest but also in all other parts of the observation area that are connected to the additional knots in the univariate bases.

Radial basis functions allow us to overcome both difficulties by allowing arbitrary placement of the knots $\boldsymbol{\kappa}$ (although observation points are mostly preferred as knot locations). In situations with a relatively small number of distinct locations, a useful default is to simply use all unique observation points. However, with a moderate number of locations this already leads to quite a considerable number of regression coefficients that have to be estimated. It would therefore be desirable to obtain knot locations that still represent the data well but allow for a reduction of the required number of basis functions.

Space filling algorithm

To achieve this goal, we will derive measures for the agreement between a set of locations and a set of knots. Following Johnson, Moore and Ylvisaker (1990) and Nychka and Saltzman (1998), we will describe a space filling algorithm that provides both such a measure and also a possibility for achieving an optimal knot configuration. Let $\mathcal{D} = \{\boldsymbol{\kappa}_1, \dots, \boldsymbol{\kappa}_K\}$ be a subset of the observation points $\mathcal{C} = \{\boldsymbol{s}_1, \dots, \boldsymbol{s}_n\}$ defining the knots of a radial basis and define the distance measure

$$d(\boldsymbol{s}, \mathcal{D}) = \left(\sum_{\boldsymbol{\kappa} \in \mathcal{D}} ||\boldsymbol{s} - \boldsymbol{\kappa}||^p \right)^{\frac{1}{p}} \tag{5.21}$$

between an arbitrary location $\boldsymbol{s} \in \mathcal{C}$ and the candidate set of knots $\mathcal{D}$. The distance measure is characterized by the coefficient p and the restriction $p < 0$ ensures that the distance between the set of knots and a knot contained in this set is zero, that is,

$$d(\boldsymbol{\kappa}, \mathcal{D}) = 0 \quad \text{for all } \boldsymbol{\kappa} \in \mathcal{D}.$$

The overall coverage of the knots (or the agreement between knots and data points) can then be defined as

$$\left(\sum_{\boldsymbol{s} \in \mathcal{C}} d(\boldsymbol{s}, \mathcal{D})^q \right)^{\frac{1}{q}} \tag{5.22}$$

with $q > 0$. The parameter q has to be strictly positive in order to ensure large values of $d(\boldsymbol{s}, \mathcal{D})$ for observation points far from all knots $\mathcal{D}$. Standard choices for p and q advocated in Johnson, Moore and Ylvisaker (1990) are $p = -20$ and $q = 20$, respectively.

The optimal set of knots (for a given number of knots) would now be given by the knot configuration minimizing (5.22). However, global minimization of the criterion is not straightforward and a heuristical minimization approach is typically employed. Note also that it might be less important to obtain the optimal knot configuration—in most practical cases it will be sufficient to obtain a reasonable solution. Nychka and Saltzman (1998) propose a simple swapping algorithm for the minimization of (5.22):

Algorithm 5.1: Space filling algorithm

i. Randomly choose a starting design of candidate knots $\mathcal{D}$ and compute the coverage criterion (5.22) for these knots.
ii. Loop over the set of remaining observation points $\boldsymbol{s} \in \mathcal{C} \setminus \mathcal{D}$ and the current set of candidate knots $\boldsymbol{\kappa} \in \mathcal{D}$. For each pair $(\boldsymbol{s}, \boldsymbol{\kappa})$, compute the coverage criterion if $\boldsymbol{\kappa}$ is replaced by $\boldsymbol{s}$. Note that this computation can be performed efficiently, since only a small part of the addends in sum (5.21) computed in step (i) has to be changed.

iii. Perform the substitution (or swap) that yields the largest reduction of (5.22) and replace the coverage criterion with its new value.
iv. If no further improvement can be achieved, stop the algorithm. Otherwise, return to (ii).

The algorithm will always terminate at a local minimum but will not necessarily find the globally optimal knot configuration.

A reduction of computing time can be achieved if the algorithm does not work with the points themselves but swaps only their indices. An implementation of the swapping algorithm that works in the described manner is available in the R-package `fields` (see Nychka, Haaland, O'Connel and Ellner, 1998, for a description of its S-Plus predecessor FUNFITS).

Figure 5.31 illustrates the knot selection procedure with two examples. Grey points indicate a randomly chosen starting design for the given number of knots. Arrows illustrate the progress towards the final solution by connecting two points that are interchanged in one of the swapping steps. The selected, optimal set of knots is indicated by black points. While the upper panel shows results for a set of regular points on a two-dimensional lattice, the lower panel illustrates the same information for an irregular set of points.

Penalties for radial bases

Having chosen a suitable set of basis functions (characterized by a suitable set of knots), penalties for radial bases have to be developed. Obviously, the concept of row-wise and column-wise penalties can no longer be applied since the basis functions are no longer aligned on a grid. However, it is possible to employ bivariate extensions of the squared second-order derivative such as the biharmonic differential operator discussed for tensor product splines in the previous section. This penalty is also closely related to the most well-known radial basis, the thin plate spline. Given observations at distinct locations $\boldsymbol{s}_1, \ldots, \boldsymbol{s}_n$, minimizing the penalized least-squares criterion

$$\sum_{i=1}^{n}(y_i - f(\boldsymbol{s}_i))^2 + \lambda \int\int \left[\left(\frac{\partial^2}{\partial^2 s_1} + 2\frac{\partial^2}{\partial s_1 \partial s_2} + \frac{\partial^2}{\partial^2 s_2}\right) f(s_1, s_2)\right]^2 ds_1 ds_2 \to \min_f$$

within the space of all twice continuously differentiable bivariate functions $f(\boldsymbol{s})$ yields the bivariate equivalent to the smoothing spline discussed in Section 2.2.4. It turns out that the resulting function can be represented as

$$f(\boldsymbol{s}) = \beta_0 + \beta_1 s_1 + \beta_2 s_2 + \sum_{k=1}^{n} \gamma_k B_k(\boldsymbol{s})$$

where

$$B_k(\boldsymbol{s}) = B(||\boldsymbol{s} - \boldsymbol{s}_k||) = ||\boldsymbol{s} - \boldsymbol{s}_k||^2 \log(||\boldsymbol{s} - \boldsymbol{s}_k||).$$

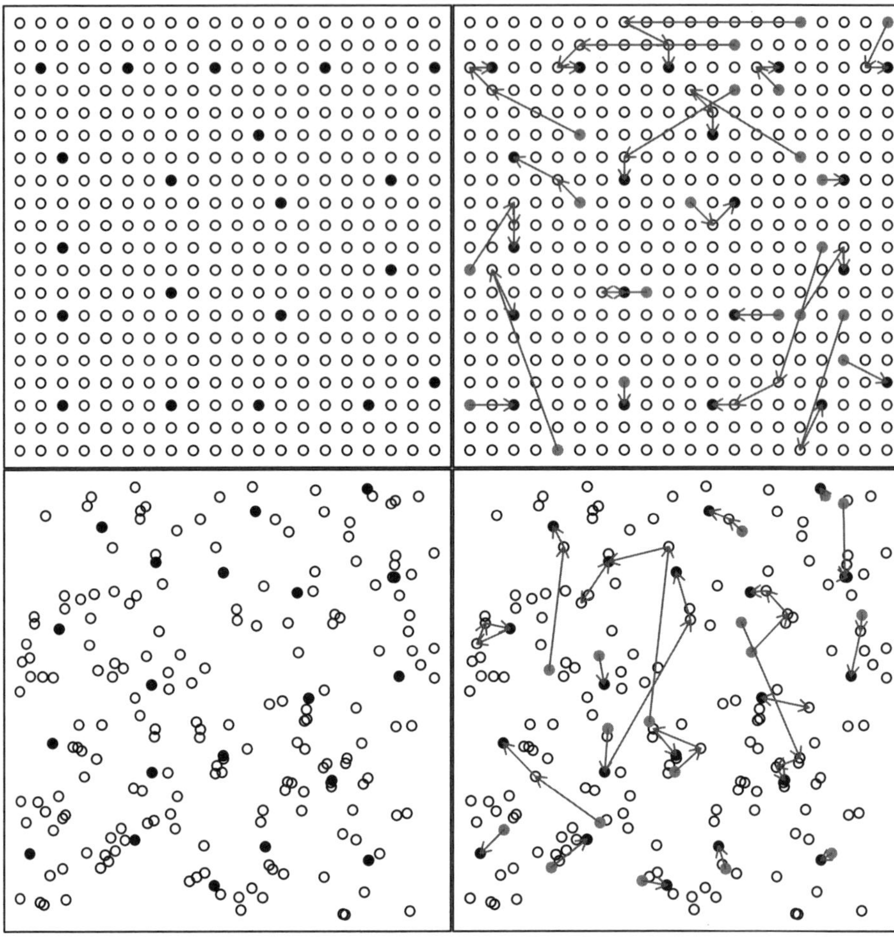

Fig. 5.31 Low-rank approximation to the full radial basis based on space filling algorithm for regular and irregular data points. The left panel shows the resulting 20 knots (black points). The right panel shows the way from the starting configuration (grey points) to the final configuration (black points).

The function is referred to as the thin plate spline since it fulfils the natural boundary conditions, that is, it is linear in the area of the boundary knots, and therefore resembles the behaviour of a thin metal plate fixed at the observation points and the corresponding function estimates. Apart from the linear parts, the thin plate spline is constructed from a radial basis with basis functions

$$B(r) = r^2 \log(r).$$

and knots at each distinct observation location. Similarly to the univariate smoothing spline, the thin plate spline has the drawback that despite its optimality property it can be hardly used in practice when the number of observation points is large since it involves all distinct observation points as knots. Again it is common to reduce the number of knots based on space filling approaches but then the optimality property of the thin plate spline is lost.

Still, the penalty involved in deriving the thin plate spline can be generally applied, also when using other types of basis functions or when considering a reduced number of knots. This results in a quadratic penalty

$$\lambda \boldsymbol{\gamma}' \boldsymbol{K} \boldsymbol{\gamma}$$

where the elements of the penalty matrix are given by

$$\int B_j''(\boldsymbol{s}) B_k''(\boldsymbol{s}) d\boldsymbol{s}$$

and

$$B_k''(\boldsymbol{s}) = \left(\frac{\partial^2}{\partial^2 s_1} + 2\frac{\partial^2}{\partial s_1 \partial s_2} + \frac{\partial^2}{\partial^2 s_2} \right) B_k(\boldsymbol{s})$$

is used as a short-hand notation for the biharmonic differential operator (5.17) applied to the basis functions.

An alternative penalty concept arises when establishing a connection between bivariate smoothing with radial basis functions and reproducing kernel Hilbert spaces (RKHS, Section 2.2.4). If the radial basis functions are positive definite, that is, if

$$\sum_{i=1}^{m} \sum_{j=1}^{m} a_i a_j B(||\boldsymbol{s}_i - \boldsymbol{s}_j||) > 0$$

for all $m \in \mathbb{N}$, arbitrary constants $a_1, \dots, a_m$ (with at least one a_i being different from zero), and arbitrary locations $\boldsymbol{s}_1, \dots, \boldsymbol{s}_m$, they can be considered as the kernel functions of a RKHS. From the theory of RKHS we then again obtain a quadratic penalty but now with the elements of the penalty matrix obtained via

$$B(||\boldsymbol{\kappa}_j - \boldsymbol{\kappa}_k||)$$

that is, by evaluating the radial basis function at the distance between the corresponding knots $\boldsymbol{\kappa}_j$ and $\boldsymbol{\kappa}_k$. We will reconsider this type of penalty in the context of Gaussian random fields in the following section where the notion of positive definiteness of the basis functions will be linked to the validity of covariance functions.

5.3.3 Tensor products vs. radial bases

After introducing tensor product splines and radial bases as two variants of bivariate (or even higher dimensional) smoothing, we will now compare both and discuss their advantages and disadvantages with respect to practical applicability.

- *Scaling of the covariates*: Tensor products do not make any assumption on the scaling of the univariate covariates and can therefore be applied to any combination of covariates. In particular, this enables the construction of interaction surfaces also for covariates that are measured on completely different units, such as year of construction and floor space in the Munich rental guide example. In contrast, the application of the Euclidean distance for radial bases makes these more suitable in situations where both covariates are measured on the same scale. The most common example will be spatial coordinates that are both measured in, for example, kilometres. Since the Euclidean difference treats s_1 and s_2 with exactly the same weight, it will in general be difficult to combine covariates such as year of construction and floor space. Usually, scaling to the unit square $[0,1] \times [0,1]$ is considered as a remedy but the choice is arbitrary and results depend on the interval of the scaling. From a mathematical perspective, tensor products are invariant under linear transformations of the covariates while radial basis functions are invariant under rotation of the covariate space.
- *Knot placement*: While knots are placed on the univariate covariate axes for the tensor product spline approach, they are allowed to be placed freely for the radial bases. In particular, the placement of knots can be adapted to non-rectangular observation areas or to varying curvature in the surface to be estimated when using radial bases.
- *Higher-dimensional smoothing*: Both approaches can be extended to smoothing in higher dimensions based on data $\boldsymbol{s} \in \mathbb{R}^d$. While higher-order tensor products have to be considered on the one hand, radial bases in higher dimensions are immediately available since the Euclidean distance is defined for arbitrary dimensions. Especially in higher dimensions, the ability to reduce the number of knots and to adapt knots to the area containing the data will be valuable and therefore indicates a preference for radial bases. Still, the curse of dimensionality discussed in the context of tensor products of course also applies for radial bases and higher dimensional interactions should be used with care. In addition, the tensor product approach has the advantage that it allows us to combine, for example, spatial and temporal bases to a trivariate smoother which would be difficult with radial bases due to the different scaling in time and space.

5.4 Continuous spatial data: Stationary Gaussian random fields

After having discussed bivariate smoothing based on point-referenced data with a focus on modelling interactions, we will now return more explicitly to the case of spatial data and discuss approaches that have been summarized under the heading of geostatistics or Kriging. The basic model has already been introduced in Section 5.1.1 and adds a spatially correlated process $\gamma(\boldsymbol{s})$ to the predictor of a regression model, yielding for example

$$y(\boldsymbol{s}) = \boldsymbol{u}(\boldsymbol{s})'\boldsymbol{\beta} + \gamma(\boldsymbol{s}) + \varepsilon(\boldsymbol{s}) \tag{5.23}$$

in the case of a linear model, where $\boldsymbol{s}$ denotes points within a spatial domain, that is, $\boldsymbol{s} \in \mathbb{R}^2$ or more generally $\boldsymbol{s} \in \mathbb{R}^d$. To make the formulation more general and to emphasize the connection to the two former smoothing approaches discussed in this chapter, we will, however, in the following use the notation $f_{\text{spat}}(\boldsymbol{s})$ and discuss later on why it actually is possible to interpret the spatial process $\gamma(\boldsymbol{s})$ as a spatial function $f_{\text{spat}}(\boldsymbol{s})$. Note that from a Bayesian perspective this interpretation intuitively makes sense since all functions are in fact considered random quantities and therefore there is no conceptual difference between a smooth function and a spatial process. However, it will turn out that the spatial processes considered in this section can in fact be rewritten utilizing a basis-function approach and therefore the notation $f_{\text{spat}}(\boldsymbol{s})$ will also be useful in general. Note that we will suppress the spatial index of $f(\boldsymbol{s})$ in the following for the sake of notational simplicity.

The assumption of a Gaussian process prior for $\gamma(\boldsymbol{s})$ enables the explicit inclusion of correlations between spatial locations in terms of a correlation function. A large part of this section will therefore be devoted to the discussion of issues arising in the specification of spatial correlation functions.

Model formulation

In most classical introductions to Kriging and geostatistics, it is common to differentiate between different types of Kriging models. For example, ordinary Kriging considers the model

$$y(\boldsymbol{s}) = \beta_0 + f(\boldsymbol{s}) + \varepsilon(\boldsymbol{s})$$

where β_0 denotes a fixed constant level, that is, no covariates are included in the model formulation. Interest is then mostly in predicting a spatial surface $\hat{f}(\boldsymbol{s})$ or the conditional expectation of the response $\hat{y}(\boldsymbol{s}) = \hat{\beta}_0 + \hat{f}(\boldsymbol{s})$ at arbitrary locations. More generally, universal Kriging includes (spatial) covariates, leading to the model equation

$$y(\boldsymbol{s}) = \boldsymbol{u}(\boldsymbol{s})'\boldsymbol{\beta} + f(\boldsymbol{s}) + \varepsilon(\boldsymbol{s}).$$

Again, predictions $\hat{f}(\boldsymbol{s})$ or $\hat{y}(\boldsymbol{s}) = \boldsymbol{u}(\boldsymbol{s})'\hat{\boldsymbol{\beta}} + \hat{f}(\boldsymbol{s})$ are the quantities of interest in classical approaches. In contrast, we want to consider spatial processes $f(\boldsymbol{s})$ as a further component to be added to the predictor of semiparametric (mixed) models. This is in analogy to Gaussian Markov random fields and surface smoothers considered in the previous sections of this chapter. This also means that we will not restrict our attention to Gaussian response models but will be interested in general semiparametric (mixed) models, for example with responses from univariate or multivariate exponential families. These will be formulated in terms of one or several predictors including $f(\boldsymbol{s})$ as a further modelling component. For the moment, we will mainly focus on models having only a spatial effect, that is, with predictor

$$\eta = \beta_0 + f(\boldsymbol{s}),$$

in order to present the modelling ideas for Gaussian random fields. In Section 5.5, we will then describe how to include such Gaussian random fields into models with general semiparametric (mixed) predictors.

A Gaussian random field is a stochastic process $\{f(\boldsymbol{s}),\ \boldsymbol{s} \in \mathbb{R}^d\}$ (possibly restricted to a suitable subset of $\mathbb{R}^d$), where all finite dimensional distributions are Gaussian, that is, for arbitrary combinations $\boldsymbol{s}_1, \ldots, \boldsymbol{s}_m$, $m \geq 1$, the vector $\boldsymbol{f} = (f(\boldsymbol{s}_1), \ldots, f(\boldsymbol{s}_m))'$ of process evaluations follows a multivariate Gaussian distribution. A Gaussian random field can then be characterized in terms of the expectation function

$$\mu(\boldsymbol{s}) = \mathbb{E}(f(\boldsymbol{s}))$$

and the covariance function

$$c(\boldsymbol{s}, \tilde{\boldsymbol{s}}) = \text{Cov}(f(\boldsymbol{s}), f(\tilde{\boldsymbol{s}}))$$

since the first two moments define the Gaussian distribution. Alternatively, we can decompose the covariance into variance and correlation and obtain the variance function

$$\tau^2(\boldsymbol{s}) = \text{Var}(f(\boldsymbol{s}))$$

and the correlation function

$$\rho(\boldsymbol{s}, \tilde{\boldsymbol{s}}) = \text{Cor}(f(\boldsymbol{s}), f(\tilde{\boldsymbol{s}}))$$

to supplement the expectation function.

For most situations, general Gaussian random fields are too flexible and attention is restricted to the class of stationary random fields where the finite dimensional distributions are invariant under spatial translation, that is,

$$f(\boldsymbol{s}_1), \ldots, f(\boldsymbol{s}_m) \stackrel{\mathcal{D}}{=} f(\boldsymbol{s}_1 + \boldsymbol{h}), \ldots, f(\boldsymbol{s}_m + \boldsymbol{h})$$

where $\boldsymbol{h} \in \mathbb{R}^d$ is a shift vector and $\stackrel{\mathcal{D}}{=}$ denotes equality in terms of distributions. This generalizes the definition of temporal stationarity where the finite dimensional distributions of the process for a set of time points is invariant under temporal translation. While stationarity is hard to establish for most types of distributions, it reduces to conditions on the first two moments for Gaussian random fields: The expectation function has to be constant in space, and since there is an intercept in the model anyway, we will assume

$$\mu(\boldsymbol{s}) = \mathbb{E}(f(\boldsymbol{s})) = 0 \quad \text{for all } \boldsymbol{s}$$

in the following. In addition, the covariance between two points $\boldsymbol{s}$ and $\tilde{\boldsymbol{s}}$ under stationarity only depends on their difference $\boldsymbol{s} - \tilde{\boldsymbol{s}}$, that is, on the direction and length of the vector that connects the points:

$$c(\boldsymbol{s}, \tilde{\boldsymbol{s}}) = \text{Cov}(f(\boldsymbol{s}), f(\tilde{\boldsymbol{s}})) = c(\boldsymbol{s} - \tilde{\boldsymbol{s}}).$$

More generally, constant expectation and difference-dependent covariance function define weak stationarity, while translation invariance defines strong stationarity (with strong stationarity being the more restrictive concept). While both concepts do not coincide for arbitrary spatial processes, they exactly match for Gaussian random fields and we will therefore simply refer to Gaussian random fields fulfilling any of the two as being stationary.

Stationarity of a Gaussian random field implies homoscedasticity, that is, the variance function is also constant over space:

$$\text{Var}(f(\boldsymbol{s})) = \tau^2 \quad \text{for all } \boldsymbol{s}.$$

This also leads to the conclusion that the correlation function depends only on the distance vector $\boldsymbol{s} - \tilde{\boldsymbol{s}}$ just as the covariance function, that is,

$$\rho(\boldsymbol{s}, \tilde{\boldsymbol{s}}) = \text{Cor}(f(\boldsymbol{s}), f(\tilde{\boldsymbol{s}})) = \rho(\boldsymbol{s} - \tilde{\boldsymbol{s}}).$$

If we treat the correlation function as known for the moment, we can derive the covariance matrix of a vector of process evaluations $\boldsymbol{f} = (f(\boldsymbol{s}_1), \ldots, f(\boldsymbol{s}_n))'$ as

$$\text{Cov}(\boldsymbol{f}) = \tau^2 \boldsymbol{R}$$

where the elements of the correlation matrix $\boldsymbol{R}$ are given by

$$\boldsymbol{R}[i, j] = \text{Cor}(f(\boldsymbol{s}_i), f(\boldsymbol{s}_j)) = \rho(\boldsymbol{s}_i - \boldsymbol{s}_j).$$

We can now include $\boldsymbol{f}$ into the predictor of an additive regression model and interpret the Gaussian process assumption as a prior that enforces spatial correlation. From a frequentist or empirical Bayes perspective, we can equivalently interpret the spatial effect as a random effect that is spatially correlated.

To make the formulation more general, we have to allow for replicated observations at the same locations that induce specific artificial correlations of magnitude 1 in the vector $\boldsymbol{f}$. To circumvent this difficulty, we summarize the process evaluations at all K *distinct* observation points in the vector $\boldsymbol{\gamma}$, that is, if $\boldsymbol{\kappa}_1, \ldots, \boldsymbol{\kappa}_K$, $K \leq n$, denote all distinct locations in the data, then $\boldsymbol{\gamma} = (f(\boldsymbol{\kappa}_1), \ldots, f(\boldsymbol{\kappa}_K))'$ The complete vector of process evaluations for all observations $\boldsymbol{f}$ can then be written as

$$\boldsymbol{f} = \boldsymbol{X}\boldsymbol{\gamma}$$

where $\boldsymbol{X}$ is a $n \times K$ zero-one incidence matrix similar as in the case of Markov random fields, that is,

$$\boldsymbol{X}[i,k] = \begin{cases} 1 & \text{if observation } i \text{ is located at point } \boldsymbol{\kappa}_k, \text{ and} \\ 0 & \text{otherwise,} \end{cases}$$

for $i = 1, \dots, n$, and $j = 1, \dots, K$. The distributional assumption for the Gaussian process translates to a distribution for the vector of coefficients $\boldsymbol{\gamma}$ where

$$\boldsymbol{\gamma} \sim \mathrm{N}(\mathbf{0}, \tau^2 \boldsymbol{R})$$

and $\boldsymbol{R}$ is a $K \times K$ dimensional matrix containing the correlations of the observed locations, that is,

$$\boldsymbol{R}[j,k] = \rho(\boldsymbol{\kappa}_j - \boldsymbol{\kappa}_k), \quad j,k = 1, \dots, K.$$

Hence, the joint density of the coefficients $\boldsymbol{\gamma}$ is given by

$$p(\boldsymbol{\gamma}|\tau^2) \propto \exp\left(-\frac{1}{2\tau^2}\boldsymbol{\gamma}\boldsymbol{R}^{-1}\boldsymbol{\gamma}\right)$$

which resembles the form of prior distributions employed in semiparametric (mixed) models in Chapters 2 and 4 with penalty matrix $\boldsymbol{K} = \boldsymbol{R}^{-1}$.

As a consequence, both empirical Bayes and full Bayes inference can be utilized for the estimation of spatial regression models including a Gaussian random field prior for the spatial component. Still, estimation is complicated by the fact that the correlation function is typically unknown in practice. It can either be estimated nonparametrically or parametrized by assuming a specific class of correlation functions that is characterized by additional parameters $\boldsymbol{\vartheta}$. We will focus on parametric classes of correlation functions in the following paragraph and briefly comment on nonparametric estimation subsequently.

Correlation functions

The construction of correlation functions for general d-dimensional distance vectors $\boldsymbol{h}$ is a nontrivial undertaking since certain restrictions have to be fulfilled. For example, valid correlation functions are symmetric since $\mathrm{Cor}(f(\boldsymbol{s}), f(\tilde{\boldsymbol{s}})) = \mathrm{Cor}(f(\tilde{\boldsymbol{s}}), f(\boldsymbol{s}))$ and therefore $\rho(\boldsymbol{h}) = \rho(-\boldsymbol{h})$. In addition, a valid correlation function has to be positive definite, that is,

$$\sum_{i=1}^{m}\sum_{j=1}^{m} a_i a_j \rho(\boldsymbol{s}_i, \boldsymbol{s}_j) > 0$$

for all $m \in \mathbb{N}$, arbitrary constants $a_1, \dots, a_m$ (with at least one a_i being different from zero), and arbitrary locations $\boldsymbol{s}_1, \dots, \boldsymbol{s}_m$. This property ensures that the variance of linear combinations of process evaluations is positive since

$$\operatorname{Var}\left(\sum_{i=1}^{m} a_i f(\boldsymbol{s}_i)\right) = \tau^2 \sum_{i=1}^{m}\sum_{j=1}^{m} a_i a_j \rho(\boldsymbol{s}_i, \boldsymbol{s}_j).$$

Unfortunately and unlike symmetry, positive definiteness is typically hard to establish in practice. It is therefore common to consider correlation functions that take only a norm of the distance vector as argument, such that

$$\rho(\boldsymbol{s}, \tilde{\boldsymbol{s}}) = \rho(||\boldsymbol{s} - \tilde{\boldsymbol{s}}||) = \rho(h)$$

for a suitable norm $h = ||\boldsymbol{s} - \tilde{\boldsymbol{s}}||$ measuring distance between $\boldsymbol{s}$ and $\tilde{\boldsymbol{s}}$ in $\mathbb{R}^d$. The assumption that the correlation function only depends on a scalar distance measure is introduced mainly for the sake of parsimony. It is much easier to parametrize a correlation function with a scalar argument than a general correlation function with bivariate or even d-dimensional argument. It will then also be possible to give an easy characterization of positive definite correlation functions that enables to ensure positive definiteness when constructing a new class of correlation functions.

When employing the Euclidean distance

$$h = ||\boldsymbol{h}|| = \sqrt{h_1^2 + \ldots + h_d^2},$$

the exact direction of the distance vector $\boldsymbol{h} = \boldsymbol{s} - \tilde{\boldsymbol{s}}$ is no longer important but only the length h. In this case, the correlation function is said to be isotropic since it has radial contour lines. In all other cases, that is, when the contour lines deviate from circles, the correlation function is called anisotropic. In the following, we will focus on isotropic correlation functions and discuss more general choices at the end of this section. Note already here that the definition of the correlation function in terms of a distance and in particular in terms of the Euclidean distance relates to the radial basis functions considered in Section 5.3.2. We will make the exact connection clearer later on in this section.

Before introducing suitable parametric classes of correlation functions, we will now define some quantities that allow us to characterize the behaviour of correlation functions relating to different types of anisotropy. The *nugget* describes the behaviour of the correlation function in the origin $\boldsymbol{h} = \boldsymbol{0}$. Of course, $\rho(\boldsymbol{0}) = 1$ has to hold for any valid correlation function, but the transition from $\rho(\alpha\boldsymbol{h})$ to $\rho(\boldsymbol{0})$ as $\alpha \to 0$ is not necessarily continuous. If it is continuous, we will see in the following that this induces specific smoothness properties of the paths (or surfaces to be more specific) of the process. For general, possibly anisotropic correlation functions, the nugget for direction $\boldsymbol{h}/||\boldsymbol{h}||$ is defined as

$$N_{\boldsymbol{h}} = \lim_{\alpha \to 0} \rho(\alpha\boldsymbol{h}).$$

If the nugget depends on the direction $\boldsymbol{h}$, the correlation function is said to be nugget anisotropic, while for isotropic correlation functions, there is only one single nugget

$$N = \lim_{\boldsymbol{h} \to 0} \rho(\boldsymbol{h}).$$

Assuming a nugget different from $\rho(\mathbf{0}) = 1$ may seem strange at first sight but makes sense when micro-variation is present in the spatial effect. For example, the correlation function for the responses in the ordinary Kriging model (5.23) is given by

$$\begin{aligned} \operatorname{Cor}(y(\boldsymbol{s}), y(\tilde{\boldsymbol{s}})) &= \frac{\operatorname{Cov}(y(\boldsymbol{s}), y(\tilde{\boldsymbol{s}}))}{\sqrt{\operatorname{Var}(y(\boldsymbol{s}))\operatorname{Var}(y(\tilde{\boldsymbol{s}}))}} \\ &= \frac{\operatorname{Cov}(f(\boldsymbol{s}), f(\tilde{\boldsymbol{s}}))}{\sqrt{[\operatorname{Var}(f(\boldsymbol{s})) + \operatorname{Var}(\varepsilon(\boldsymbol{s}))] \cdot [\operatorname{Var}(f(\tilde{\boldsymbol{s}})) + \operatorname{Var}(\varepsilon(\tilde{\boldsymbol{s}}))]}} \\ &= \frac{\tau^2 \rho(\boldsymbol{s} - \tilde{\boldsymbol{s}})}{\tau^2 + \sigma^2}. \end{aligned}$$

In the limit $\boldsymbol{s} - \tilde{\boldsymbol{s}} \to \mathbf{0}$ we therefore obtain

$$\operatorname{Cor}(y(\boldsymbol{s}), y(\tilde{\boldsymbol{s}})) \to \frac{\tau^2}{\tau^2 + \sigma^2} < 1,$$

where the nugget is introduced by the additional error variation $\varepsilon(\boldsymbol{s})$. For spatial smoothing, that is, when considering the correlation $\operatorname{Cor}(f(\boldsymbol{s}), f(\tilde{\boldsymbol{s}}))$, assuming correlation functions with nugget $N \neq 1$ does not make too much sense since smoothness of $f(\boldsymbol{s})$ is one of the inherent assumptions. In contrast, correlation functions with nugget have been considered in classical geostatistics, in particular when predicting the response at unobserved locations.

The distance $R_{\boldsymbol{h}}$ at which $\rho(\alpha \boldsymbol{h})$ reaches its limiting value

$$\lim_{\alpha \to \infty} \rho(\alpha \boldsymbol{h}) \tag{5.24}$$

in direction $\boldsymbol{h}/||\boldsymbol{h}||$ is called the corresponding *range* of the correlation function. In most cases of practical relevance, the limit in (5.24) will be zero since the correlation decays to zero with increasing distance. In this case, the notion of the range of a correlation function intuitively makes sense since the range defines the distance at which two observations will be uncorrelated. For many correlation functions, however, the range is infinite since $\rho(\alpha \boldsymbol{h}) > 0$ for all α. Therefore, it is common practice to consider the *effective range*, that is, the distance $R_{\boldsymbol{h}}$ where the correlation function becomes smaller than a small (but still positive) value. By construction, the effective range provides a measure for the spread of a correlation function, that is, the effective range is small for correlation functions that are concentrated around zero while it becomes large for correlation functions that decay to zero very slowly.

Range (or more generally effective range) anisotropy is the type of anisotropy considered most frequently in practice. The rationale is that correlations will decay faster in certain directions than in others. This may for example be the case for meteorological phenomena where correlations are larger along the major wind directions and smaller in directions orthogonal to these directions. Possibilities for incorporating range anisotropy by modifying the Euclidean distance or the correlation function will be considered later on and we will focus on isotropic correlation functions for the moment.

We have already stated qualitatively that it is possible to read smoothness properties of a Gaussian process from the behaviour of the correlation function in the origin, that is, by studying the nugget. Since we are concerned with the smoothness of a random quantity or more precisely the paths (or surfaces) of a stochastic process, we require appropriate definitions of smoothness. The concept of continuity of deterministic functions is generalized as follows: A stochastic process f is called L^2-continuous, if

$$\lim_{\alpha \to 0} \mathbb{E}\left(|f(\boldsymbol{s}+\alpha\boldsymbol{h}) - f(\boldsymbol{s})|^2\right) = 0 \quad \text{for all} \quad \boldsymbol{h}.$$

Similarly, the process f is called L^2-differentiable, if a process f' exists such that

$$\lim_{\alpha \to 0} \mathbb{E}\left(\left|\frac{f(\boldsymbol{s}+\alpha\boldsymbol{h}) - f(\boldsymbol{s})}{\alpha||\boldsymbol{h}||} - f'(\boldsymbol{s})\right|^2\right) = 0 \quad \text{for all} \quad \boldsymbol{h}.$$

It is of course also possible to consider direction specific derivatives.

Both smoothness properties are difficult to validate for a given process but fortunately the correlation function contains all the information required. It can be shown (see for example Stein, 1999) that for stationary processes f,

- f is L^2-continuous if and only if $\rho(\boldsymbol{h})$ is continuous in $\boldsymbol{h} = \boldsymbol{0}$, that is, if the nugget equals 1 for all directions $\boldsymbol{h}$.
- f is k times L^2-differentiable if $\rho(\boldsymbol{h})$ is $2k$ times differentiable in $\boldsymbol{h} = \boldsymbol{0}$.

Parametric classes of correlations functions

Let us now turn to some well-known classes of parametric correlation functions. We will define all correlation functions in isotropic versions, that is, for a scalar argument h.

Some popular correlation functions are contained in the power exponential class (De Oliveira, Kedem and Short, 1997) given by

$$\rho(h; \phi, \nu) = \exp\left(-|h/\phi|^{\nu}\right)$$

where $\phi > 0$ determines the effective range by specifying the rate of the exponential decay of the correlation. The second parameter $0 < \nu \leq 2$ determines the smoothness of the corresponding process since it determines the smoothness and differentiability properties in the origin of the correlation function. Figure 5.32 shows some examples of power exponential correlation functions.

For $\nu = 2$ we obtain the special case of the Gaussian correlation function

$$\rho(h; \phi, \nu = 2) = \exp\left(-h^2/\phi^2\right)$$

which is infinitely often differentiable in zero and therefore yields very smooth process realizations as illustrated in Fig. 5.33 for one simulated example. The simple exponential correlation function

$$\rho(h; \phi, \nu = 1) = \exp\left(-|h|/\phi\right)$$

results for $\nu = 1$ and resembles the exponential decay of the well-known correlation function of first-order autoregressive temporal processes. It is continuous in zero but not differentiable and, as a consequence, the resulting paths are much more variable than in case of the Gaussian correlation function (see again the example realizations in Fig. 5.33).

A very rich collection of correlation functions is available via the Matérn class (Matérn, 1986; Handcock and Stein, 1993) defined by

$$\rho(h; \phi, \nu) = \frac{1}{2^{\nu-1}\Gamma(\nu)} \left(\frac{h}{\phi}\right)^{\nu} \mathcal{K}_\nu\left(\frac{h}{\phi}\right)$$

where $\phi > 0$ again determines the effective range, $\nu > 0$ is a parameter specifying overall smoothness properties and $\mathcal{K}_\nu(\cdot)$ is the modified Bessel function of the third kind and order ν (Abramowitz and Stegun, 1974). The Matérn class is often recommended for practical analyses (see for example Stein, 1999) due to its flexibility. However, since the Bessel function is not available in closed form for general ν, the numerical computation of estimates may be time-consuming and requires numerical optimization. In contrast, for smoothness parameters $\nu = 0.5, 1.5, 2.5, \ldots$, a closed form representation of the Matérn correlation functions exists. For example, the Matérn correlation functions visualized in Fig. 5.32 are given by

$$\begin{aligned}
\rho(h; \phi, \nu = 0.5) &= \exp(-|h/\phi|), \\
\rho(h; \phi, \nu = 1.5) &= \exp(-|h/\phi|)(1 + |h/\phi|), \\
\rho(h; \phi, \nu = 2.5) &= \exp(-|h/\phi|)(1 + |h/\phi| + \tfrac{1}{3}|h/\phi|^2).
\end{aligned}$$

The simple exponential correlation function is again included as a special case for $\nu = 0.5$. Increasing the value of ν leads to correlation functions that are smoother and have larger effective ranges, and in the limit $\nu \to \infty$ the Matérn correlation function converges to the Gaussian correlation function. Figure 5.33 also shows realizations from Gaussian processes with Matérn correlation functions that further support the increased amount of smoothness observed for larger values of ν.

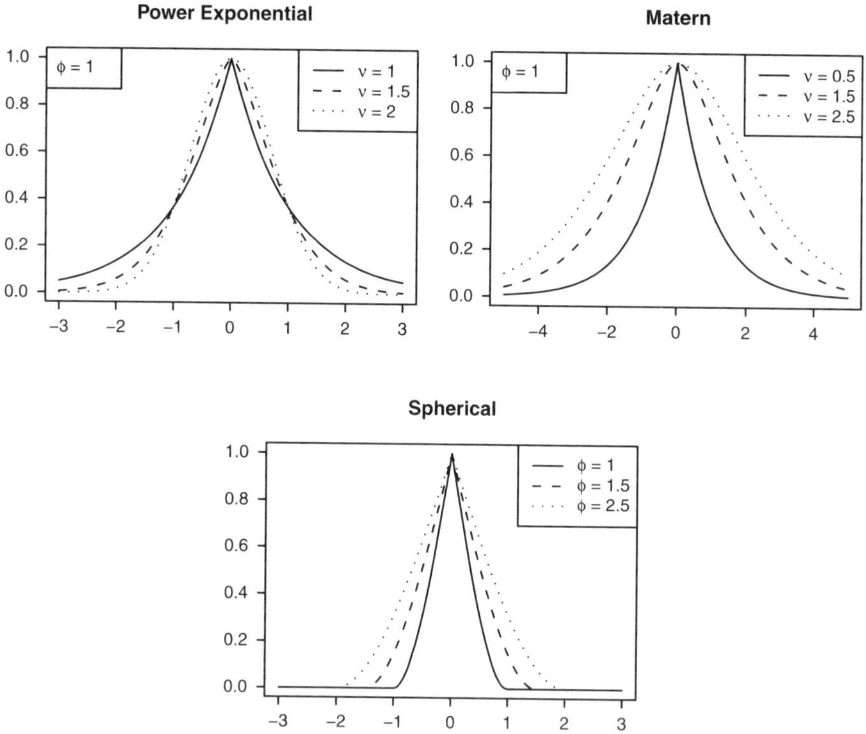

Fig. 5.32 Examples of correlation functions: The upper left panel shows power exponential correlation functions for $\nu = 1$ (solid), $\nu = 1.5$ (dashed) and $\nu = 2$ (dotted). The upper right panel shows Matérn correlation functions for $\nu = 0.5$ (solid), $\nu = 1.5$ (dashed) and $\nu = 2.5$ (dotted). For both classes, the range parameter is fixed at $\phi = 1$. The lower panel shows spherical correlation functions for $\phi = 1$ (solid), $\phi = 1.5$ (dashed) and $\phi = 2.5$ (dotted).

Another example of a parametric correlation function is the rational quadratic correlation function (Yaglom, 1987), given by

$$\rho(h; \phi, \nu) = \left(1 + \left(\frac{h}{\phi}\right)^2\right)^{-\nu}$$

with range parameter $\phi > 0$ and smoothness parameter $\nu > 0$.

Power exponential, Matérn and rational quadratic correlation functions have infinite range and were therefore characterized in terms of an effective range parameter ϕ. The last correlation function we will consider is an example with finite range that actually is also the only parameter. The spherical correlation function (Wackernagel, 1995) is given by

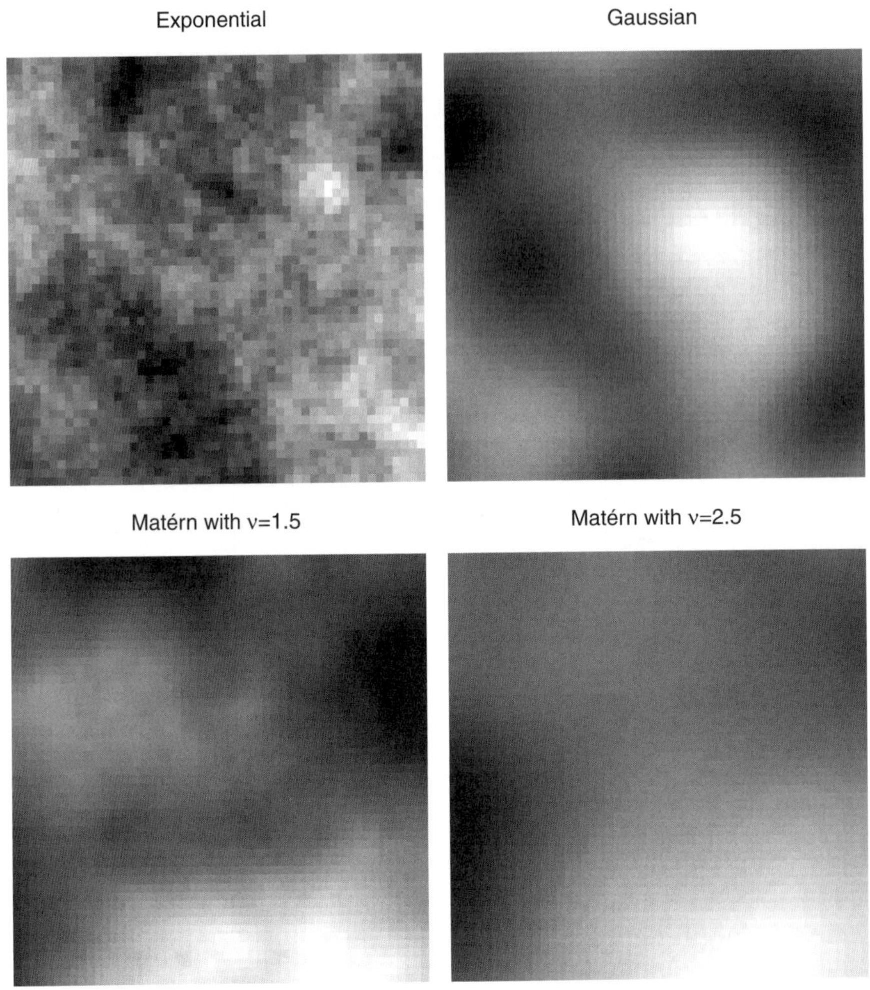

Fig. 5.33 Sample realizations of Gaussian processes based on different correlation functions (top left: exponential, top right: Gaussian, bottom left: Matérn with $\nu = 1.5$, bottom right: Matérn with $\nu = 2.5$).

$$\rho(h;\phi) = \begin{cases} 1 - \frac{3}{2}|h/\phi| + \frac{1}{2}|h/\phi|^3 & 0 \leq h \leq \phi, \\ 0 & h > \phi, \end{cases}$$

and obviously the range is given by ϕ. The spherical correlation function is not used too much in practice since it is only once differentiable in ϕ which results in difficulties in maximum likelihood estimation of the range. The spherical correlation function is visualized in Fig. 5.32 for different values of the range parameter.

Bochner's theorem

All correlation functions introduced in the previous section are of course valid and therefore positive definite. However, as already noted before, it is actually quite difficult to validate positive definiteness for a given class of functions. A criterion that is easier to verify is given in Bochner's theorem (Bochner, 1955) that provides a characterization of isotropic stationary correlation functions. It basically states that a correlation function $\rho(h)$ is positive definite if and only if it is the characteristic function of a symmetric random variable Z, that is, if it can be written as

$$\rho(h) = \int_{-\infty}^{\infty} \exp(ihz) f_Z(z) dz \tag{5.25}$$

where $i = \sqrt{-1}$ and $f_Z(z)$ is the density of Z with $f_Z(z) = f_Z(-z)$ since Z is assumed to be symmetric. Employing the representation of the complex exponential function as

$$\exp(ihz) = \cos(hz) + i \sin(hz)$$

equation (5.25) can be rewritten as

$$\rho(h) = 2 \int_{0}^{\infty} \cos(hz) f_Z(z) dz \tag{5.26}$$

avoiding the need to work with complex arguments.

The benefit of Bochner's theorem is actually two-fold. First of all, criterion (5.26) can be inverted to

$$f_Z(z) = \frac{1}{\pi} \int_{0}^{\infty} \cos(-hz) \rho(h) dh. \tag{5.27}$$

This allows us to check whether a given candidate function $\rho(h)$ is a valid correlation function by deriving the corresponding quantity $f_Z(z)$ via (5.27). If (and only if) $f_Z(z)$ is the density of a random variable Z that is symmetric about zero, $\rho(h)$ is a positive definite and therefore valid correlation function.

The second advantage arising from Bochner's theorem is that it allows us to construct new classes of correlation functions by inserting symmetric densities. We can start by plugging any type of symmetric density into (5.26), deriving the corresponding correlation function and then studying its further properties. For example, the Gaussian correlation function (with range parameter $\phi = 1$) arises from the density

$$f_Z(z) = \frac{1}{\sqrt{2\pi}} \exp\left(-\frac{1}{2}z^2\right),$$

that is, from a standard normal density. The exponential correlation function is obtained from the Cauchy density

$$f_Z(z) = \frac{1}{\pi(1+z^2)}.$$

Note that similar considerations are common in time series analysis, to characterize valid temporal correlation functions (see, for example, Diggle, 1990). In the context of time series, Bochner's theorem corresponds to the discrete Fourier and inverse Fourier transform of the correlation function and the density $f_Z(z)$ is called the spectral density. The spectral density (and its empirical analogon the spectrum) are particularly useful for studying cyclic behaviour in time series, a phenomenon that is less common in spatial data.

A further advantage of Bochner's theorem is that it allows us to prove that products of valid correlation functions are again valid. Consider two independent and symmetric random variables Z_1 and Z_2 with densities $f_1(z_1)$ and $f_2(z_2)$. Then the sum $Z = Z_1 + Z_2$ is also symmetric and the characteristic function of the sum Z is simply the product of the characteristic functions of Z_1 and Z_2. Hence, for given correlation functions $\rho_1(h)$ and $\rho_2(h)$, Bochner's theorem states that symmetric random variables Z_1 and Z_2 exist. Moreover, the product of the correlation functions $\rho(h) = \rho_1(h)\rho_2(h)$ corresponds to the characteristic function of the sum $Z = Z_1 + Z_2$ which is also symmetric. In summary, $\rho(h) = \rho_1(h)\rho_2(h)$ is a valid correlation function.

This property can be used in a variety of settings to generate useful new classes of correlation functions. For example, if we want to modify a correlation function such that it has finite range, we can simply multiply it with a second correlation function of finite range. This is commonly referred to as tapering of correlation functions, see for example Furrer, Genton and Nychka (2006) or Stein (1999). A second application is the construction of range anisotropic correlation functions.

Range anisotropic correlation functions

The assumption of isotropy may for example be violated when considering environmental data where the spatial effect is induced by some metereological phenomenon such as wind. Typically there will be a predominant direction of the wind that will induce directional correlations. In this case, the isotropic Euclidean distance can be replaced by the anisotropic version

$$||\boldsymbol{s}||_B = \sqrt{\boldsymbol{s}'\boldsymbol{B}\boldsymbol{s}}$$

where $\boldsymbol{B}$ is any symmetric, positive definite matrix. In this case, known as *geometric anisotropy*, the circular contour lines are replaced by elliptic contour lines and the prolongation and the rotation of the ellipses correspond to the strength and the direction of anisotropy, respectively. In the special case $d = 2$, a constructive definition of $\boldsymbol{B}$ that reveals these features is based on the representation

$$\boldsymbol{B} = \boldsymbol{R}(\psi_A)'\boldsymbol{P}(\psi_R)\boldsymbol{R}(\psi_A),$$

where $\boldsymbol{R}(\psi_A)$ is a rotation matrix with anisotropy angle $\psi_A \in [0, 2\pi]$ of the form

$$\boldsymbol{R}(\psi_A) = \begin{pmatrix} \cos(\psi_A) & \sin(\psi_A) \\ -\sin(\psi_A) & \cos(\psi_A) \end{pmatrix},$$

and $\boldsymbol{P}(\psi_R)$ is a prolongation matrix with anisotropy ratio $\psi_R \geq 1$ given by

$$\boldsymbol{P}(\psi_R) = \begin{pmatrix} \psi_R^{-1} & 0 \\ 0 & 1 \end{pmatrix},$$

see Chiles and Delfiner (1999, Ch. 2.5.2). The two parameters ψ_A and ψ_R exhibit an intuitive interpretation: With $\psi_R = 1$, we obtain isotropic correlation functions, while $\psi_R > 1$ leads to elliptically shaped correlation functions. More precisely, ψ_R depends on the ratio of the ranges of the correlation function along the two principal axes. The anisotropy angle ψ_A causes a rotation of the correlation function around the origin. Figure 5.34 visualizes the contours of some anisotropic correlation functions for selected choices of anisotropy angle and ratio.

Example 5.12: Swiss rainfall
To illustrate the impact of assuming an anisotropic correlation function, we present spatial fits for the Swiss rainfall data from Example 5.1 obtained with either an isotropic or an anisotropic exponential correlation function (and parameter estimation based on restricted maximum likelihood as detailed later on in this section). Figure 5.35 shows the two resulting surfaces that indicate a clear difference. While localized features with circular contour lines appear in the isotropic fit, these are connected to smoother and elliptically shaped areas in the anisotropic fit. The anisotropy angle of about 71° reflects the orientation of the ellipses approximately in northwestern direction, while the estimated anisotropy ratio of 2.9 represents a considerable amount of anisotropy. △

Two further types of distances are the Manhattan-distance (or block-metric)

$$||\boldsymbol{s}|| = \sum_{i=1}^{d} |s_i|$$

which is motivated from measuring the distance between two points along blocks in a city with regular streets, and the geodesic distance that measures the shortest distance along the Earth's surface taking the curvature of the globe into account (i.e. the arc length on a sphere with radius R corresponding to the Earth's radius). For the latter case, let $\boldsymbol{s}_i = (s_{i1}, s_{i2})'$ denote a point on the Earth's surface in terms of latitude s_{i1} and longitude s_{i2}. If, furthermore, R is the radius of the Earth, the distance between two points $\boldsymbol{s}_i$ and $\boldsymbol{s}_j$ is given by

$$d(\boldsymbol{s}_i, \boldsymbol{s}_j) = R \arccos(\phi_{ij}),$$

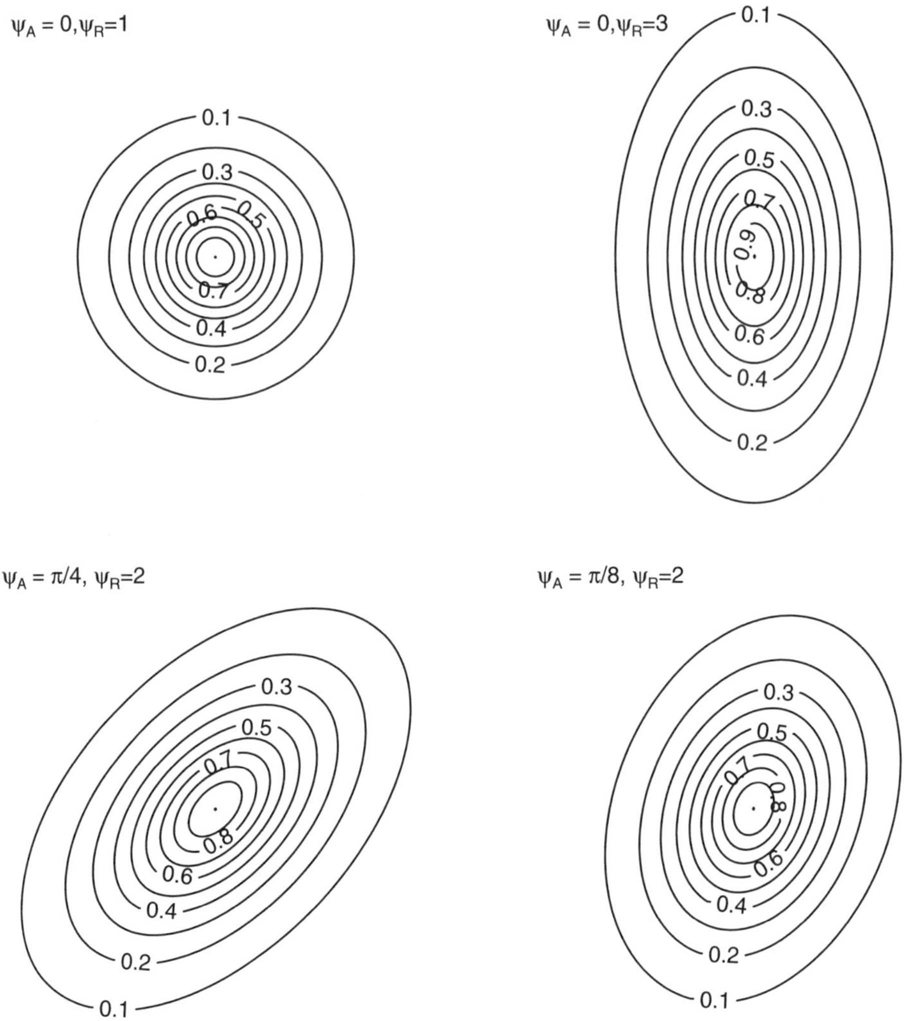

Fig. 5.34 Contour lines of some geometrically range anisotropic correlation functions.

where

$$\phi_{ij} = \sin(s_{i1})\sin(s_{j1}) + \cos(s_{i1})\cos(s_{j1})\cos(s_{i2} - s_{j2})$$

is the corresponding radian, see Banerjee, Carlin and Gelfand (2003a, page 17) for a derivation.

Picking the right type of distance measure is mainly a question of the application at hand. If, for example, the area considered is relatively small, the curvature of the Earth's surface can be safely ignored and the Euclidean distance or an

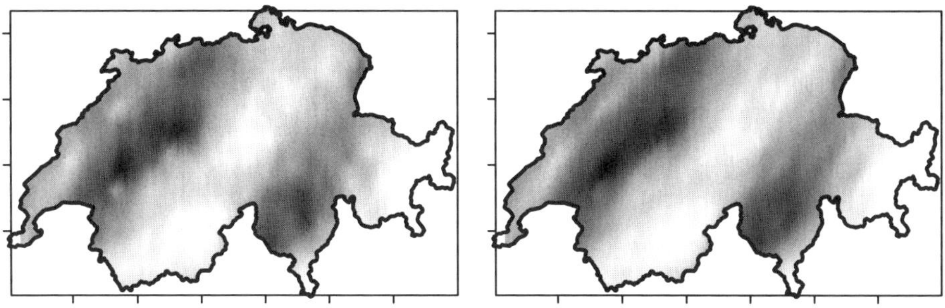

Fig. 5.35 Swiss rainfall: Surface estimates obtained with an isotropic (left) and an anisotropic (right) exponential correlation function.

anisotropic version might be a good starting point. The Manhattan distance will probably only be useful in very specialized applications.

A more general form of non-geometric range anisotropy has been introduced in Ecker and Gelfand (2003) based on a sequence of univariate (isotropic) correlation functions $\rho_1(h), \dots, \rho_k(h)$ and positive definite matrices $\boldsymbol{B}_1, \dots, \boldsymbol{B}_k$. It follows from Bochner's theorem that the correlation function

$$\rho(\boldsymbol{h}) = \rho_1(\sqrt{\boldsymbol{h}'\boldsymbol{B}_1\boldsymbol{h}}) \cdot \ldots \cdot \rho_k(\sqrt{\boldsymbol{h}'\boldsymbol{B}_k\boldsymbol{h}})$$

is also valid. Moreover, it is in general range anisotropic (unless $\boldsymbol{B}_1 = \ldots = \boldsymbol{B}_k = \boldsymbol{I}_d$) and has contour lines that deviate from the elliptical form of geometric range anisotropy (provided that $k > 2$). Of course, the matrices $\boldsymbol{B}_1, \dots, \boldsymbol{B}_k$ can again be represented in terms of anisotropy angle and anisotropy ratio although their interpretation is more complicated in the product of correlation functions. In most cases, the number of correlation functions k will be small and the same simple parametric correlation function type will be used for all of them.

Figure 5.36 illustrates the types of anisotropy achieved by combining two exponential correlation functions. Of course, there is some redundancy in the parameters employed in the definition of $\rho(\boldsymbol{h})$ and several parameter combinations may lead to exactly the same fit. Ecker and Gelfand (2003) provide a parametrization with unique parameters by simultaneous codiagonalization of the matrices $\boldsymbol{B}_1, \dots, \boldsymbol{B}_k$.

Variogram

In classical geostatistics, it is common practice to use a different quantity than the correlation function to characterize spatial dependence. The variogram for a stochastic process f is defined as

$$v(\boldsymbol{s}, \tilde{\boldsymbol{s}}) = \frac{1}{2}\,\mathrm{Var}(f(\boldsymbol{s}) - f(\tilde{\boldsymbol{s}})).$$

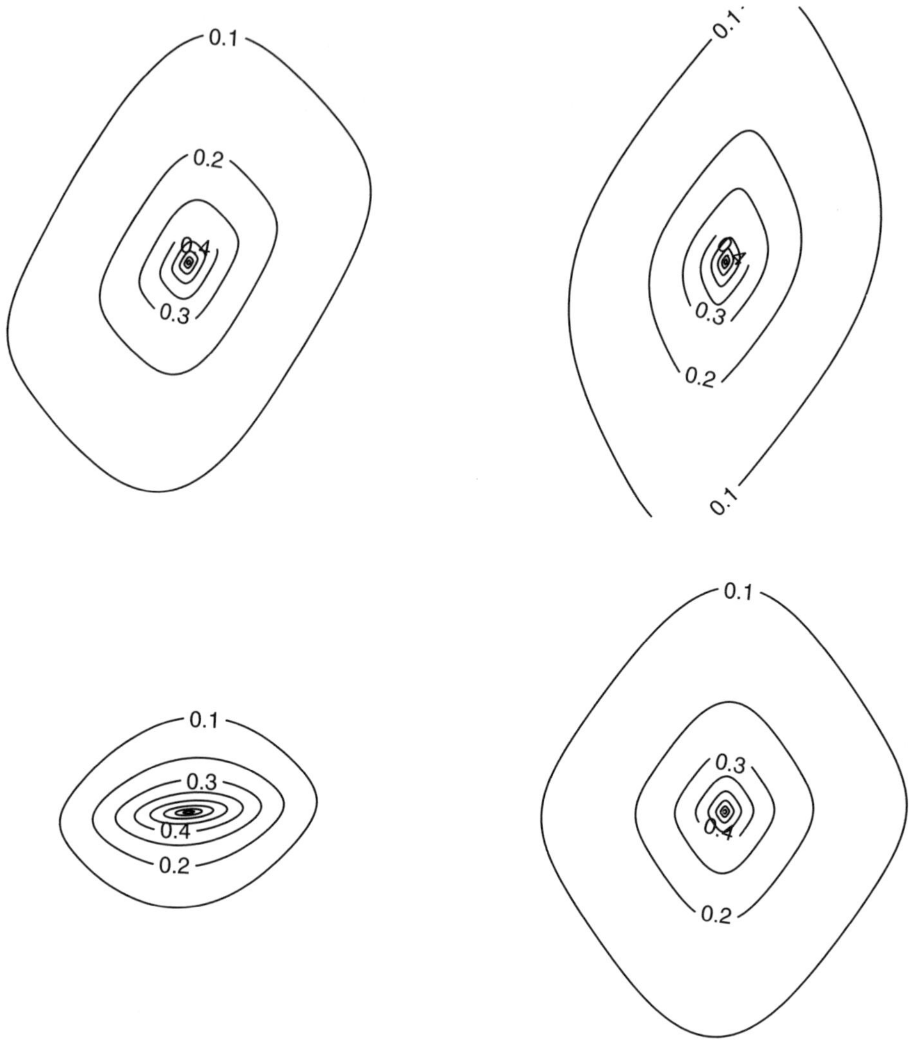

Fig. 5.36 Contour lines of some nongeometrically range anisotropic correlation functions.

Based on the variogram, there is also an alternative definition of stationarity. A stochastic process is called intrinsically stationary if it has constant expectation

$$\mathbb{E}(f(\boldsymbol{s})) = \mu$$

and the variogram only depends on the distance vector $\boldsymbol{h} = \boldsymbol{s} - \tilde{\boldsymbol{s}}$, that is,

$$v(\boldsymbol{s}, \tilde{\boldsymbol{s}}) = v(\boldsymbol{s} - \tilde{\boldsymbol{s}}).$$

If, furthermore, the variogram only depends on the Euclidean distance such that

$$v(\boldsymbol{s}, \tilde{\boldsymbol{s}}) = v(||\boldsymbol{s} - \tilde{\boldsymbol{s}}||)$$

the process is again said to be isotropic.

It can be shown that stationarity in terms of the covariance or correlation function is actually a stronger property than intrinsic stationarity, that is, a process that is stationary is also intrinsically stationary but the reverse implication does not hold in general. Moreover, for stationary processes the variogram can be derived from the correlation function via

$$v(\boldsymbol{h}) = \tau^2(1 - \rho(\boldsymbol{h})). \tag{5.28}$$

In analogy to parametric correlation functions, it is also possible to define parametric classes of variograms. In particular, we can transform any of the classes of correlation functions to the corresponding variogram class. However, in practice nonparametric estimation of the variogram is much more common, as outlined at the end of the following paragraph.

Estimation of covariance and correlation parameters

Although we have already seen that Gaussian random fields fit into the inferential frameworks introduced in Chapters 2 and 4 and Section 3.3 regardless of the specification of the covariance or correlation function, estimation is of course complicated by the additional parameters $\boldsymbol{\vartheta}$ contained in the correlation matrix $\boldsymbol{R} = \boldsymbol{R}(\boldsymbol{\vartheta})$ such as the effective range ϕ or a smoothness parameter ν, and in addition possibly anisotropy parameters such as the anisotropy angle ψ_A and the anisotropy range ψ_R. When considering the Gaussian random field as a spatially correlated random effect, it is actually possible to estimate the unknown parameters $\boldsymbol{\vartheta}$ via maximum likelihood or restricted maximum likelihood just as in Section 3.3. For example, for a Gaussian model

$$\boldsymbol{y} = \boldsymbol{U}\boldsymbol{\beta} + \boldsymbol{X}\gamma + \varepsilon,$$

the restricted log-likelihood for $\boldsymbol{\vartheta}$ is given by

$$l^*(\boldsymbol{\vartheta}) = -\frac{1}{2}\log(|\boldsymbol{V}(\boldsymbol{\vartheta})|) - \frac{1}{2}\log(|\boldsymbol{U}'\boldsymbol{V}(\boldsymbol{\vartheta})^{-1}\boldsymbol{U}|) - \frac{1}{2}(\boldsymbol{y} - \boldsymbol{U}\hat{\boldsymbol{\beta}})'\boldsymbol{V}(\boldsymbol{\vartheta})^{-1}(\boldsymbol{y} - \boldsymbol{U}\hat{\boldsymbol{\beta}})$$

where $\boldsymbol{V}(\boldsymbol{\vartheta}) = \sigma^2\boldsymbol{I} + \tau^2\boldsymbol{X}\boldsymbol{R}\boldsymbol{X}'$ is the marginal covariance matrix of the observation vector $\boldsymbol{y}$. Estimates obtained from maximizing the restricted likelihood will then be inserted into the penalized least-squares estimates that can be derived for $\boldsymbol{\beta}$ and $\boldsymbol{\gamma}$. For non-Gaussian responses, repeated REML estimation in working models

$$\tilde{\boldsymbol{y}} \sim \mathrm{N}(\boldsymbol{U}\boldsymbol{\beta} + \boldsymbol{X}\gamma, \boldsymbol{W}^{-1})$$

with working responses and weights as in usual GLMs yields estimates $\hat{\boldsymbol{\vartheta}}$. Correspondingly, the estimates for $\boldsymbol{\beta}$ and $\boldsymbol{\gamma}$ are obtained by iteratively weighted least-squares fitting. In a Bayesian interpretation, REML estimation leads to empirical Bayes estimates where the estimated parameters $\hat{\boldsymbol{\vartheta}}$ are inserted into the posterior mode estimates for $\boldsymbol{\gamma}$ and $\boldsymbol{\beta}$.

For full Bayesian analyses, suitable prior specifications have to be supplemented for the unknown parameters $\boldsymbol{\vartheta}$. The range parameter present in all correlation functions that we considered should be positive, and therefore priors also considered for precisions or variances (such as Gamma or inverse Gamma) can be employed in principle. Diggle and Ribeiro Jr. (2007) suggest using a discrete prior to circumvent difficulties in the prior specification of the range parameter, where positive probability mass is only assigned to a user-specified grid of range values. This has the advantage that the full conditional for the range is also discrete, allowing us to draw random samples from the full conditionals without the need of Metropolis–Hastings steps. Berger, de Oliveira and Sansó (2001) consider flat priors for the range parameter.

For the Matérn class, the smoothness parameter ν is often fixed in advance to avoid additional estimation complexity. We will also argue in the following that the exact specification of the correlation function does not make too much of a difference in the context of spatial smoothing, as long as the range parameter is set to a suitable value. The rationale is that Kriging can actually be considered a radial basis-function smoothing approach where the form of the basis functions corresponds exactly to the correlation function. Hence, it is important to choose a correlation function that fits to the assumptions about the underlying spatial effect (e.g. in terms of smoothness and range) but often small variations in the parameters do not impact the results too much. Moreover, the Gaussian process assumption is only an approximation to the true data generating mechanism anyway, and provides only the prior assumptions in our Bayesian perspective on spatial smoothing.

Nonparametric covariogram and variogram estimation

In order to choose a specific class of correlation functions, it is useful to also have a nonparametric estimate available. Such nonparametric estimates have been derived for Gaussian models but are generally not available in generalized models. To detect the form of spatial dependence in the presence of further covariate effects $\boldsymbol{\beta}$, we will study the residuals $\varepsilon(\boldsymbol{s}) = y(\boldsymbol{s}) - \boldsymbol{u}(\boldsymbol{s})'\boldsymbol{\beta}$ in the following. Of course, in practice the parameters $\boldsymbol{\beta}$ have to be replaced by estimates obtained when ignoring the spatial process in the model specification leading to the estimated residuals $\hat{\varepsilon}(\boldsymbol{s})$. Note that it is sufficient to have an unbiased estimate such as $\hat{\boldsymbol{\beta}} = (\boldsymbol{U}'\boldsymbol{U})^{-1}\boldsymbol{U}'\boldsymbol{y}$ for the determination of residuals.

The basis for nonparametric estimation of either the covariance function or the variogram are the empirical covariogram or variogram cloud. For each pair of observations (i, j) derive the distance

$$d_{ij} = ||\boldsymbol{s}_i - \boldsymbol{s}_j||, \quad i, j = 1, \ldots, n,$$

in combination with

$$c_{ij} = (\hat{\varepsilon}(\boldsymbol{s}_i) - \bar{\varepsilon})(\hat{\varepsilon}(\boldsymbol{s}_j) - \bar{\varepsilon})$$

and

$$v_{ij} = \frac{1}{2}[\hat{\varepsilon}(\boldsymbol{s}_i) - \hat{\varepsilon}(\boldsymbol{s}_j)]^2$$

where

$$\bar{\varepsilon} = \frac{1}{n}\sum_{i=1}^{n} \varepsilon(\boldsymbol{s}_i)$$

is the average residual. The quantities c_{ij} and v_{ij} define estimates for the covariance and the variogram at a given distance d_{ij}. For exploratory purposes, plots with coordinates (d_{ij}, c_{ij}) or (d_{ij}, v_{ij}) can be drawn, yielding the so-called covariogram or variogram cloud (see the top row of Fig. 5.37 for an example obtained from data simulated with a Matérn covariance function with $\nu = 1.5$). Both

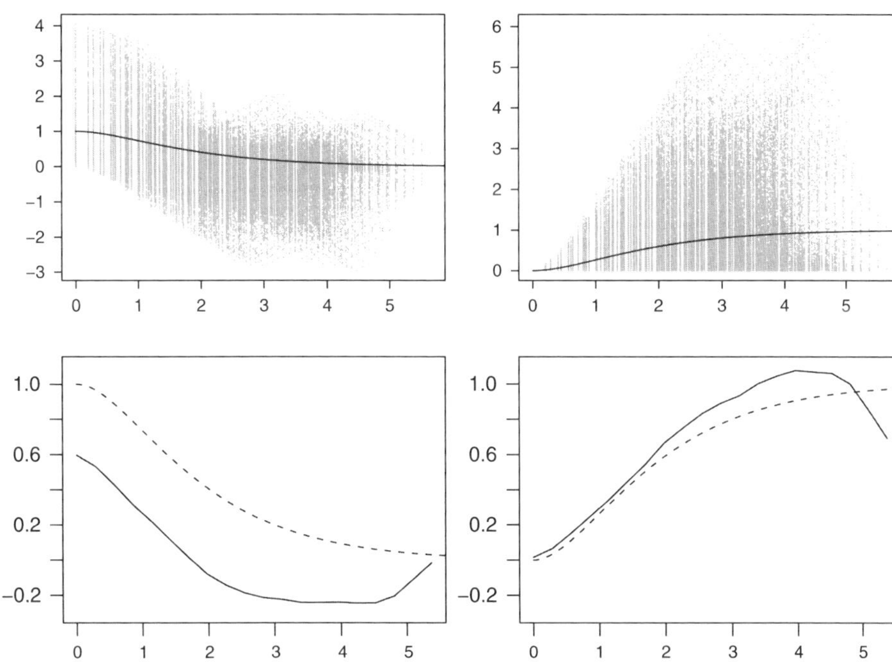

Fig. 5.37 Top row: Covariogram (left) and variogram (right) cloud with the true covariance function and variogram superimposed. Bottom row: Covariogram (left) and variogram (right) estimates obtained by averaging over a grid of 20 equidistant values as solid lines with true covariance function and variogram superimposed as dashed lines.

provide a raw impression of the form of underlying covariance structure but are typically not very suitable deciding on a specific class of parametric covariance (or variogram) functions. The underlying reason is at least two-fold: On the one hand, all entries in the cloud depend on only two observations, inducing high variation and large uncertainty about the precise form of the covariance function. On the other hand, the cloud consists of $n(n-1)$ points derived from only n observations which is ignored in the raw figures.

To deal with the high variation, it seems suitable to combine all observation pairs at a certain distance h and to average over the corresponding covariogram and variogram ordinates, leading to

$$\hat{c}(h) = \frac{1}{|N(h)|} \sum_{N(h)} (\hat{\varepsilon}(\boldsymbol{s}_i) - \bar{\varepsilon})(\hat{\varepsilon}(\boldsymbol{s}_j) - \bar{\varepsilon})$$

and

$$\hat{v}(h) = \frac{1}{2|N(h)|} \sum_{N(h)} [\hat{\varepsilon}(\boldsymbol{s}_i) - \hat{\varepsilon}(\boldsymbol{s}_j)]^2$$

where $N(h) = \{(i,j) : ||\boldsymbol{s}_i - \boldsymbol{s}_j|| = h\}$ contains the pairs (i,j) which have distance $h = ||\boldsymbol{s}_i - \boldsymbol{s}_j||$ and $|N(h)|$ denotes the number of such pairs, that is, the number of addends in the sum. While the derived empirical variogram is an unbiased estimate for the true variogram (note that $\mathbb{E}(v_{ij}) = v(||\boldsymbol{s}_i - \boldsymbol{s}_j||)$), the covariance estimate is biased since the estimation of the average $\bar{\varepsilon}$ is not taken into account. This may partially explain why the variogram estimate is more frequently considered in practice than the covariogram.

Unfortunately, the number of pairs contained in $N(h)$ will typically be small for most or all distances, leaving still quite large variation also for the averaged estimates. As a remedy, the domain of distances can be divided into intervals defined by cutpoints $0 = h_0 < h_1 < \ldots < h_M$ to obtain estimates only for a prespecified grid of values. Define $N(h_m) = \{(i,j) : h_{m-1} < ||\boldsymbol{s}_i - \boldsymbol{s}_j|| \leq h_m\}$ as the set of observation pairs with distance falling into the m-th interval $(h_{m-1}, h_m]$ and consider

$$\hat{c}(h_m) = \frac{1}{|N(h_m|)} \sum_{(i,j) \in N(h_m)} (\hat{\varepsilon}(\boldsymbol{s}_i) - \bar{\varepsilon})(\hat{\varepsilon}(\boldsymbol{s}_j) - \bar{\varepsilon}), \quad m = 1, \ldots, M,$$

as an estimate for the covariance function (and a similar expression for the variogram). Figure 5.37 illustrates the process of estimating covariance function and variogram based on averaging the values in the covariogram and variogram cloud for a simulated data set.

Both nonparametric estimates can also be adapted to situations with anisotropic covariance by defining the intervals as not only distance- but also direction-specific. However, the number of intervals increases at a high rate, requiring quite large amounts of data to obtain reliable estimates.

Gaussian random fields as radial basis function smoothers

So far, we have treated the spatial Gaussian random field as a stochastic process similar to a random effect but including spatial correlation. Although it is obvious from a Bayesian perspective that Gaussian random fields also provide a surface estimate and therefore a spatial smoother, we will now make this connection more explicit and show that Gaussian random fields can in fact be interpreted as a radial basis function approach as introduced in Section 5.3.2.

Recall that the vector of spatial process evaluations can be written as $\boldsymbol{f} = \boldsymbol{X}\boldsymbol{\gamma}$ where $\boldsymbol{X}$ is a zero/one incidence matrix for the distinct spatial locations $\boldsymbol{\kappa}_1, \dots, \boldsymbol{\kappa}_K$ while

$$\boldsymbol{\gamma} \sim \mathrm{N}(\mathbf{0}, \tau^2 \boldsymbol{R})$$

is the corresponding prior distribution. Let us now consider the reparametrization

$$\boldsymbol{f} = \boldsymbol{X}\boldsymbol{R}\boldsymbol{R}^{-1}\boldsymbol{\gamma} = \tilde{\boldsymbol{X}}\tilde{\boldsymbol{\gamma}}$$

where $\tilde{\boldsymbol{X}} = \boldsymbol{X}\boldsymbol{R}$ and $\tilde{\boldsymbol{\gamma}} = \boldsymbol{R}^{-1}\boldsymbol{\gamma}$. It then follows for the regression coefficients that

$$\tilde{\boldsymbol{\gamma}} \sim \mathrm{N}(\mathbf{0}, \tau^2 \boldsymbol{R}^{-1})$$

while the vector corresponding to the i-th row of the matrix $\tilde{\boldsymbol{X}}$ is given by

$$\tilde{\boldsymbol{x}}_i = (\rho(||\boldsymbol{s}_i - \boldsymbol{\kappa}_1||), \dots, \rho(||\boldsymbol{s}_i - \boldsymbol{\kappa}_K||)'$$

and $f(\boldsymbol{s}_i) = \tilde{\boldsymbol{x}}_i'\tilde{\boldsymbol{\gamma}}$. This may be interpreted as follows: Consider the distinct observation locations $\boldsymbol{\kappa}_1, \dots, \boldsymbol{\kappa}_K$ to be knots and define the basis functions

$$B_k(\boldsymbol{s}) = \rho(||\boldsymbol{s} - \boldsymbol{\kappa}_k||).$$

Then the function $f(\boldsymbol{s})$ can be represented in terms of these basis functions as

$$f(\boldsymbol{s}) = \tilde{\boldsymbol{x}}'\tilde{\boldsymbol{\gamma}} = \sum_{k=1}^{K} \tilde{\gamma}_k B_k(\boldsymbol{s}).$$

In summary, the reparametrization allows us to express the spatial process $f(\boldsymbol{s})$ as a basis function smoother with radial basis functions defined by the correlation function. Especially from this perspective it is no longer surprising that the behaviour of the correlation function translates into the smoothness of the process, although we are now considering path smoothness while previously we have considered smoothness of a stochastic process. The basis function representation also allows us to apply the same knot reduction strategies as outlined in Section 5.3 for radial basis functions to reduce the dimensionality of $\tilde{\boldsymbol{\gamma}}$. Moreover, it provides a link to smoothing based on reproducing kernel Hilbert spaces (RKHS) as outlined in Section 2, page 87: The positive definite correlation functions define valid, positive definite kernels for setting up a RKHS.

The interpretation as a basis-function approach also supports our conjecture that the exact form of the correlation function is of less importance when reproducing a spatial surface. In particular, several correlation functions have quite similar form once they are adjusted for the effective range. In fact, the range seems to be the most important quantity in determining the spatial surface since it specifies how far the basis functions spread over the observation area. In our experience, it is typically safe to fix all parameters of the correlation function except the range according to subject matter knowledge. For example, when working with the Matérn class, the smoothness parameter ν can be chosen from the sequence $0.5, 1.5, 2.5, \dots$ according to the desired smoothness of the surface (in analogy to the degree for spline estimates). As a rule of thumb, choosing the range parameters ϕ as

$$\hat{\phi} = \max_{j,k} ||\boldsymbol{\kappa}_j - \boldsymbol{\kappa}_k||/c,$$

ensures scale invariance of the basis functions. The constant $c > 0$ is chosen such that $\rho(c; \phi = 1)$ is small (e.g. 0.001), that is, c determines the desired effective range. As a consequence, the different values of $||\boldsymbol{s} - \boldsymbol{\kappa}_j||/\hat{\phi}$ are spread out over the distance axis of the correlation function.

Example 5.13: Human brain mapping

As an illustration for the application of Gaussian random fields in spatial smoothing, we reconsider the human brain mapping data introduced in Example 5.2. We specify a Gaussian random field prior for the spatial effect, utilizing Matérn correlation functions with smoothness parameters $\nu = 0.5$, 1.5 and 2.5 favouring an increasing amount of smoothness. To reduce the complexity of the problem, we apply the space filling algorithm (page 370) to select 100 representative points as knots for the basis function representation of the Gaussian process. Estimation is based on restricted maximum likelihood, that is, we will consider empirical Bayes estimates.

Figure 5.38 displays the corresponding posterior mode estimates for the activation profiles. As expected, the Matérn correlation function with $\nu = 2.5$ leads to the smoothest estimates and therefore does a better job in reducing background noise. However, it also smoothes the peak of the activation profile in the visual cortex that is better preserved when using a smaller value of the smoothness parameter. Obviously, the brain mapping data would require a more adaptive method to obtain a clearer separation between background with considerable smoothing and activation where not too much smoothness should be imposed. △

Identifiability in geostatistical models

While it is sometimes recommended to include low order polynomials of the coordinates into the fixed effects part $\boldsymbol{u}(\boldsymbol{s})'\boldsymbol{\beta}$ of the model specification, the basis

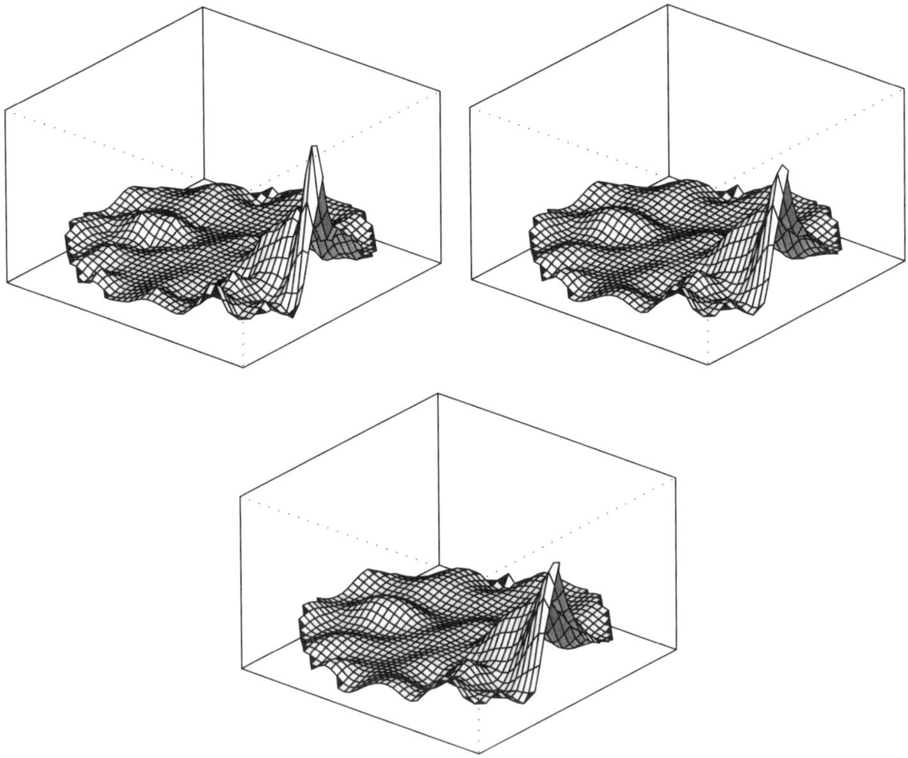

Fig. 5.38 Human brain mapping: Estimated activation based on Matérn correlation functions with $\nu = 0.5$ (top left), $\nu = 1.5$ (top right) and $\nu = 2.5$ (bottom).

function perspective of Gaussian random fields also indicates that identifiability problems may arise when making the polynomial part overly flexible or when replacing it with a (deterministic) nonparametric spatial smoother such as a tensor product penalized spline. For illustration purposes, consider the model

$$y(\boldsymbol{s}) = \boldsymbol{u}(\boldsymbol{s})'\boldsymbol{\beta} + g(\boldsymbol{s}) + f(\boldsymbol{s}) + \varepsilon(\boldsymbol{s})$$

where, as before, $f(\boldsymbol{s})$ denotes the Gaussian random field, while $g(\boldsymbol{s})$ denotes a further spatial surface estimate. From a frequentist perspective, $g(\boldsymbol{s})$ and $f(\boldsymbol{s})$ are then often separated by considering the former as deterministic (a surface) while the latter is a random quantity that should capture spatial correlations. This distinction no longer makes sense in the Bayesian interpretation and in particular the radial basis function representation of $f(\boldsymbol{s})$ reveals that two types of surfaces are specified in the model. This is comparable to our considerations in Section 4.3, where we found similar identifiability restrictions in the context of univariate smoothing.

Classical geostatistics

The perspective we have taken on geostatistics, in particular in the last paragraphs, is quite different from what is conventionally considered as geostatistics. From a classical perspective, $f(\boldsymbol{s})$ is treated as a stochastic process and its randomness is essential in the analysis of spatial data. In contrast, we have shown that the assumption of a stationary Gaussian process corresponds to a basis-function approach with a specific prior distribution. Moreover, in classical geostatistics the focus lies on prediction of either the spatial effect or the responses at unobserved locations. We will now outline classical geostatistical estimation and prediction in Gaussian random field models of the form

$$\boldsymbol{y} = \boldsymbol{U}\boldsymbol{\beta} + \boldsymbol{X}\boldsymbol{\gamma} + \boldsymbol{\varepsilon}$$

where $\boldsymbol{U}\boldsymbol{\beta}$ captures covariate effects that are treated as fixed while $\boldsymbol{\gamma}$ represents the random spatial effect. Similarly to the mixed models, it is common to assume

$$\begin{pmatrix}\boldsymbol{\gamma}\\ \boldsymbol{\varepsilon}\end{pmatrix} \sim \mathrm{N}\left(\begin{pmatrix}\boldsymbol{0}\\ \boldsymbol{0}\end{pmatrix}, \begin{pmatrix}\tau^2\boldsymbol{R} & \boldsymbol{0}\\ \boldsymbol{0} & \sigma^2\boldsymbol{I}\end{pmatrix}\right),$$

that is, joint normality and independence of error terms and spatial effect.

Note that in classical geostatistics $\boldsymbol{\varepsilon}$ is used both to correct for measurement error in the response and to account for small-scale local variation (the nugget effect). The rationale is that spatially correlated variation may also be present at a micro-level of very short-ranged distances. In principle, a second spatially correlated Gaussian random field could be included in the model formulation to account for this fact, but if all distances between observations exceed a certain threshold, it will be hard or even impossible to separate the small scale spatially correlated process from the error term $\boldsymbol{\varepsilon}$. As a consequence, interest in classical geostatistics often lies in interpolating the observed data instead of smoothing it. Implicitly, one is assuming that the small scale spatial variation overweighs the measurement error, which is plausible in classical applications of geostatistics such as the spatial interpolation of ore grades.

To derive predictions at unobserved locations, we start with the marginal distribution of the responses after integrating out the spatial effect. Similarly to mixed models, we obtain

$$\boldsymbol{y} \sim \mathrm{N}(\boldsymbol{U}\boldsymbol{\beta}, \sigma^2\boldsymbol{I} + \tau^2\boldsymbol{X}\boldsymbol{R}\boldsymbol{X}').$$

The term $\sigma^2\boldsymbol{I}$ in the covariance matrix represents the nugget effect in the covariance function for the responses $\boldsymbol{y}$. We are now interested in predictions at unobserved locations $\boldsymbol{s}_0$ for the sum of deterministic effects and the spatial effect, that is, for $\boldsymbol{u}(\boldsymbol{s}_0)'\boldsymbol{\beta}+\gamma(\boldsymbol{s}_0)$. Let $\boldsymbol{r} = (\rho(\boldsymbol{s}_1 - \boldsymbol{s}_0), \dots, \rho(\boldsymbol{s}_n - \boldsymbol{s}_0))'$, and $y_0 = y(\boldsymbol{s}_0)$, then the joint distribution of $\boldsymbol{y}$ and y_0 is given by

$$\begin{pmatrix}\boldsymbol{y}\\ y_0\end{pmatrix} \sim \mathrm{N}\left(\begin{bmatrix}\boldsymbol{U}\boldsymbol{\beta}\\ \boldsymbol{u}(\boldsymbol{s}_0)'\boldsymbol{\beta}\end{bmatrix}, \begin{bmatrix}\sigma^2\boldsymbol{I} + \tau^2\boldsymbol{X}\boldsymbol{R}\boldsymbol{X}' & \tau^2\boldsymbol{r}'\\ \tau^2\boldsymbol{r} & \tau^2\end{bmatrix}\right)$$

From the joint Gaussian distribution, the mean squared error optimal prediction can be derived as the conditional expectation

$$\hat{y}_0 = \mathbb{E}(y_0|\boldsymbol{y}) = \boldsymbol{u}(\boldsymbol{s}_0)'\boldsymbol{\beta} + \tau^2\boldsymbol{r}'(\tau^2\boldsymbol{X}\boldsymbol{R}\boldsymbol{X}' + \sigma^2\boldsymbol{I})^{-1}(\boldsymbol{y} - \boldsymbol{U}\boldsymbol{\beta})$$

with prediction variance

$$\text{Var}(y_0|\boldsymbol{y}) = \tau^2 - \tau^2\boldsymbol{r}'(\tau^2\boldsymbol{X}\boldsymbol{R}\boldsymbol{X}' + \sigma^2\boldsymbol{I})^{-1}\boldsymbol{r}\tau^2$$

(see Cressie, 1993). The same prediction rule is also obtained without the assumption of Gaussian errors when restricting attention to the best linear unbiased predictor in case of i.i.d. error terms $\boldsymbol{\varepsilon}$.

To make the prediction rule applicable in practice, estimated variance parameters and regression coefficients have to be inserted. In classical geostatistics, these are typically obtained either based on empirical variograms and corresponding least-squares estimation or (restricted) maximum likelihood estimation.

The main difficulty with the classical approach to geostatistics is that it is hard to generalize to non-Gaussian response types. In this case, optimal prediction rules are much harder to find and it is easier to consider penalized estimation within the basis function interpretation of geostatistical models. In particular, the basis function representation perfectly fits into the general framework of SPMMs, making the inferential procedures derived in Chapter 4 available.

5.5 Geoadditive regression

As already noted at various places throughout this chapter, the considered types of spatial functions and associated prior structures readily fit into the frameworks of semiparametric regression and semiparametric mixed models considered in Chapters 2 and 4. To be more specific, all types of spatial effects could be expressed in terms of a design matrix and a vector of regression coefficients, that is,

$$\boldsymbol{f}_{\text{spat}} = \boldsymbol{X}_{\text{spat}}\boldsymbol{\gamma}_{\text{spat}} \tag{5.29}$$

where $\boldsymbol{f}_{\text{spat}}$ denotes the vector of evaluations of the spatial function. The vector of regression coefficients is again supplemented with a multivariate Gaussian prior

$$p(\boldsymbol{\gamma}_{\text{spat}}) \propto \exp\left(-\frac{1}{2\tau^2_{\text{spat}}}\boldsymbol{\gamma}'_{\text{spat}}\boldsymbol{K}_{\text{spat}}\boldsymbol{\gamma}_{\text{spat}}\right)$$

where the precision matrix $\boldsymbol{K}_{\text{spat}}$ can be rank-deficient. The specific details depend on the type of spatial effect employed:

- *Markov random fields*: For Gaussian Markov random fields, the design matrix $\boldsymbol{X}_{\text{spat}}$ is simply a zero/one incidence matrix that relates an observation to the spatial effect of the region the observation pertains to. The precision matrix represents spatial neighbourhoods and has the form of an adjacency matrix.

- *Interactions*: For tensor product P-splines or general basis functions, the design matrix contains the basis function evaluations while the precision matrix enforces smoothness of the estimated surface. This may be achieved either in terms of appropriate difference penalties along rows and columns, extending concepts introduced for univariate P-splines, or by considering higher-dimensional extensions of differential operators.
- *Gaussian random fields*: For stationary Gaussian random fields, the design matrix typically will again be an incidence matrix and the precision is given by the inverse correlation matrix of the random field. When representing Gaussian random fields as radial basis function smoothers, the design matrix contains basis function evaluations while the inverse correlation matrix has to be replaced by the correlation matrix itself as penalty matrix.

The spatial function is now simply augmented to the semiparametric (mixed) predictor yielding *geoadditive regression models* where the predictor is given for example by

$$\eta_i = \boldsymbol{u}_i'\boldsymbol{\beta} + f_1(x_{i1}) + \ldots + f_r(x_{ir}) + f_{\text{spat}}(s_i)$$

for an additive model for cross-sectional data or by

$$\eta_{ij} = \boldsymbol{u}_{ij}'\boldsymbol{\beta} + f_1(x_{ij1}) + \ldots + f_r(x_{ijr}) + f_{\text{spat}}(s_i) + \boldsymbol{z}_{ij}'\boldsymbol{\alpha}_i$$

for a semiparametric mixed model for longitudinal or cluster data, where s_i stands for any type of spatial information to be included in the model and $\boldsymbol{\alpha}_i$ denotes cluster specific random effects. In vector-matrix representation, the full model can then always be written as

$$\boldsymbol{\eta} = \boldsymbol{U\beta} + \boldsymbol{X}_1\boldsymbol{\gamma}_1 + \ldots + \boldsymbol{X}_r\boldsymbol{\gamma}_r + \boldsymbol{X}_{\text{spat}}\boldsymbol{\gamma}_{\text{spat}} + \boldsymbol{Z\alpha}$$

using (5.29). However, there are some subtle differences or additions that will be discussed in the following paragraphs. For example, the vector of regression coefficients $\boldsymbol{\gamma}_{\text{spat}}$ typically has higher dimension compared to the nonparametric effects vectors $\boldsymbol{\gamma}_1, \ldots, \boldsymbol{\gamma}_r$ such that efficient implementations of the inferential procedures will be more important.

Full Bayes inference

For inference based on Markov chain Monte Carlo simulations, similar algorithms to those discussed in Section 4.2.2 arise. If the model can be expressed in terms of (latent) Gaussian responses the Gibbs sampler for Gaussian SPMMs discussed on page 213 can be adapted to include spatial effects. To be more specific, a Gaussian full conditional with parameters as in (4.20) results for the spatial component by inserting the design matrix and precision matrix for $\boldsymbol{X}_l$ and $\boldsymbol{K}_l$. In the case of inherently non-Gaussian responses, methods for exponential family

SPMMs developed on page 219 can be utilized. In particular, IWLS proposals from a Gaussian distribution with parameters as in (4.27) result for the spatial effect.

What makes spatial models more demanding when implementing them with MCMC is that the vector of regression coefficients is typically of much higher dimension and that penalty and design matrix are not necessarily sparse. For Gaussian Markov random fields, the design matrix is simply an incidence matrix (and therefore very sparse) while the structure of the penalty matrix depends on the neighbourhoods represented in the geographical information. In particular, the sparseness of $\boldsymbol{K}$ depends on the order of the spatial regions in the vector γ and efficient ordering may considerably improve the sparseness of $\boldsymbol{K}$. Corresponding methods such as the Cuthill–McKey algorithm for reordering and efficient matrix calculations for the resulting sparse matrices are for example described in George and Liu (1981). Brezger and Lang (2006) consider the application of such methods in the context of geoadditive regression models.

For bivariate penalized splines, at least part of the sparseness of the univariate splines carries over to the bivariate case. However, in case of radial bases and Gaussian random fields, the precision matrix is typically full, that is, it only has a very small number of non-zero entries. Trying to circumvent this by reparametrizing $\boldsymbol{K}$ yields a full design matrix such that the MCMC algorithm is again faced with high-dimensional non-sparse systems of equations. As a consequence, computations will typically be too slow when treating the complete vector γ as one block in the MCMC algorithm. On the other hand, decomposing it may yield samplers with insufficient mixing behaviour since correlations of the elements in γ are no longer taken into account. More details on Bayesian estimation of Gaussian random field models can be found, for example, in Banerjee, Carlin and Gelfand (2003a) or Diggle and Ribeiro Jr. (2007, Ch. 7).

Example 5.14: Childhood malnutrition in India

As a first example of a geoadditive regression model, we consider the childhood malnutrition data introduced in Example 1.3. The predictor

$$\begin{aligned}\eta_i = \boldsymbol{u}_i'\boldsymbol{\beta} &+ f_1(cage) + f_2(breastfeeding) + f_3(mbmi) + f_4(mage) + f_5(medu)\\ &+ f_6(edupartner) + f_{\text{spat}}(s_i) + \alpha_{s_i}\end{aligned}$$

for the stunting score comprises six nonparametric functions of continuous covariates modelled as cubic penalized B-splines with twenty inner knots and second-order random-walk penalty. In addition, we supplement a structured spatial effect $f_{\text{spat}}(s_i)$ based on a Markov random field prior that captures smoothly varying effects on the district level and an unstructured spatial random effect α_{s_i} that represents localized spatial features. The latter is assigned an i.i.d. Gaussian prior and corresponds to a random intercept based on clusters defined by regions. Assuming a Gaussian distribution for the error terms leads to a Gibbs sampling algorithm.

Figure 5.39 shows the posterior mean estimates for nonparametric effects together with pointwise 95% credible intervals. In particular the effect of the age of the child turns out to be highly nonlinear with a steady decline until an age of about two years. The increasing effect that is observed afterwards is due to a change in the reference population that is used to define the Z-score for stunting. The body mass index of the mother as well as the variables capturing the educational level of the mother and the partner of the mother show the expected behaviour, where a higher BMI and higher educational level have a beneficial effect on the nutritional status of the child. All three effects are not too far from linear but high uncertainty is attached to the effects of larger covariate values as represented in the wide credible intervals. The effect of duration of breastfeeding appears to be counterintuitive since a longer period of breastfeeding seems to be associated with a higher risk for malnourishment. A possible explanation is that large durations of breastfeeding are associated with the absence of alternative possibilities to feed the child and are therefore an indicator for poorer families. It is also worth noting that the estimated effect is rising somewhat for the first months, giving a slight indication that no breastfeeding does not have a beneficial effect. However, again uncertainty is quite large and no definite conclusions about this effect can be drawn.

Figure 5.40 visualizes both point estimates (posterior means) and posterior probabilities for the structured and the unstructured spatial effect. The latter are obtained by classifying the regions into groups with significantly positive and negative effects based on the 80% credible intervals. From the range of the estimated effects, we find a domination of the spatially structured effect that also translates into a number of significant effects. In contrast, for the unstructured effect there are almost no credible intervals that do not contain zero. A better 'nutritional status' is observed in the southern part of India and along the coasts while the situation is worse in the northern part of India and at the border to Pakistan.

Parametric effects are summarized in Table 5.2 in terms of the posterior mean, standard deviation and 95% credible interval. Some of the variables representing assets of the household (motorcycle, refrigerator, television, electricity) show the expected, significantly positive effect on the Z-score, that is, children in households with assets representing wealth are in general also better nourished. A large positive effect is also associated with children being the firstborn while the nutritional status worsens for children that come later in the birth order. The largest effect (in magnitude) is observed for the twin variable that is associated with a large negative effect. △

A major advantage of full Bayesian inference based on MCMC is that the extensions discussed in Section 4.4 can be included in exactly the same way, since a spatial effect only adds one further step to the sampling algorithm. We will demonstrate this possibility by analysing the Munich rental guide where a geoadditive model specification is combined with Bayesian regularization priors

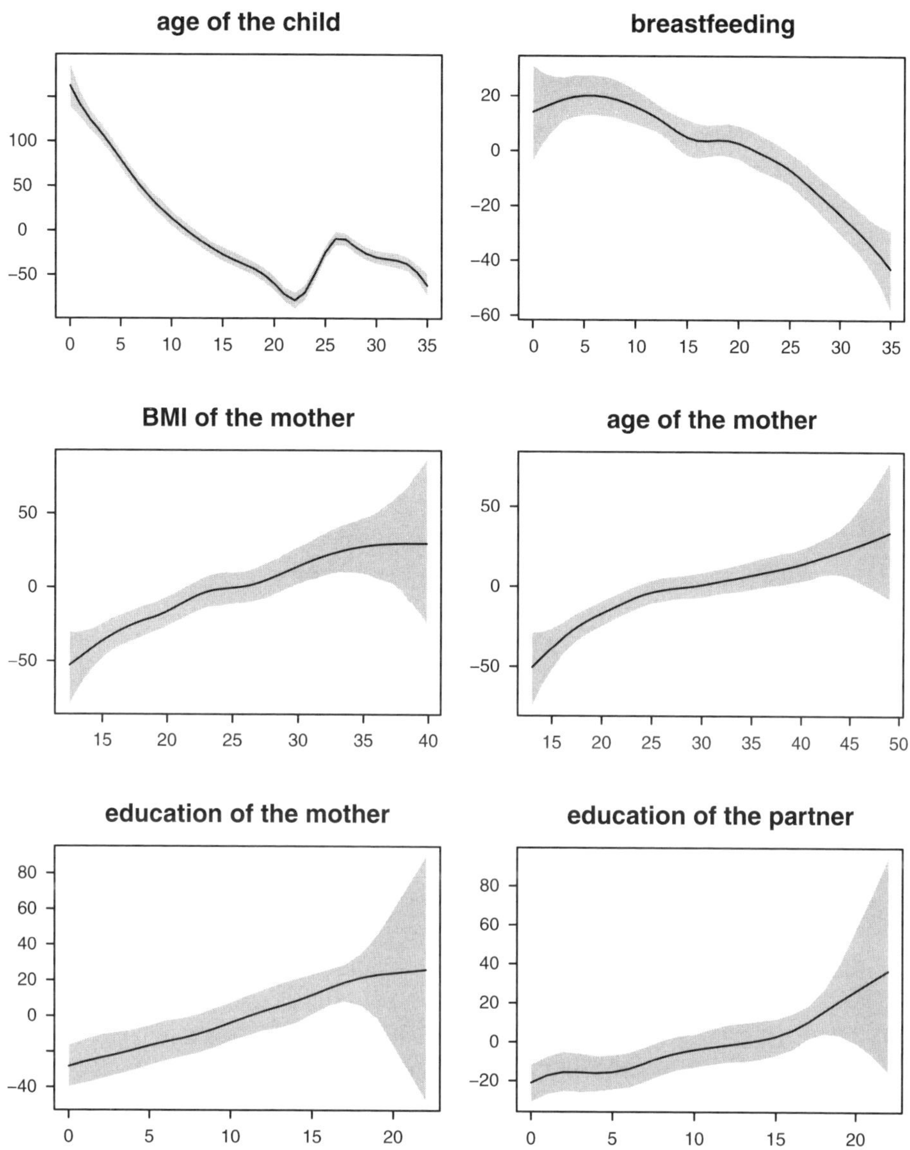

Fig. 5.39 Childhood malnutrition: Estimated nonparametric effects together with pointwise 95% credible intervals.

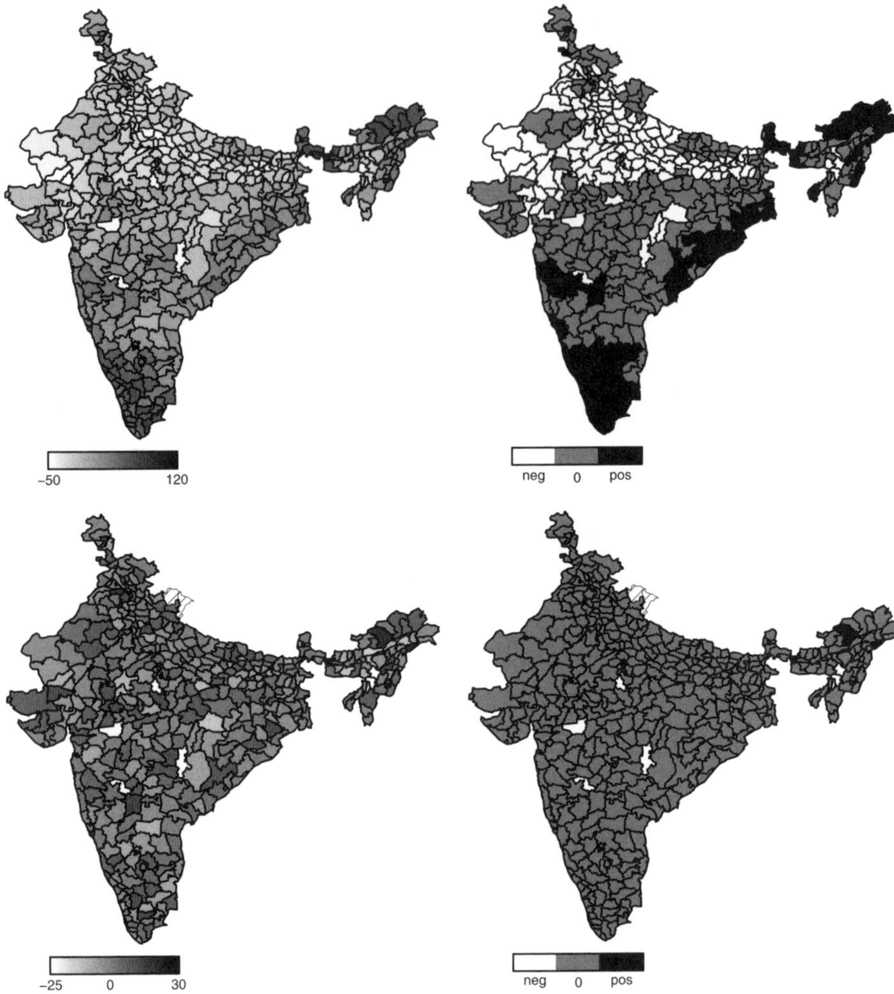

Fig. 5.40 Childhood malnutrition: Estimated structured (top) and unstructured (bottom) spatial effects (left) and corresponding 80% posterior probabilities (right).

discussed in Section 4.4.2. A more comprehensive analysis of this example can be found in Kneib, Konrath and Fahrmeir (2010b).

Example 5.15: Munich rental guide
Our second example on geoadditive regression deals with the Munich rental guide data described in detail in Example 1.4. Again, the predictor combines nonparametric effects of continuous covariates modelled as Bayesian penalized splines with a spatial effect based on a Markov random field prior, yielding

Table 5.2 Childhood malnutrition: Point estimates, standard deviations and 95% credible interval for the parametric effects.

variable	estimate	sd	95% credible	interval
intercept	−127.468	9.215	−146.800	−109.269
csex	−2.324	1.841	−6.111	1.282
ctwin	−69.866	9.953	−88.335	−50.415
munemployed	−2.890	2.209	−7.008	1.357
cbirthorder (5+)	−27.622	4.792	−36.863	−18.323
cbirthorder (4)	−11.861	4.061	−20.029	−3.901
cbirthorder (3)	−15.273	3.309	−21.946	−8.793
cbirthorder (2)	−8.628	2.544	−13.359	−3.479
cbirthorder(1)	63.384			
deadchildren (3+)	−13.839	7.565	−28.837	0.694
deadchildren (2)	−2.130	5.310	−12.444	8.368
deadchildren (1)	−3.094	3.103	−9.107	2.698
deadchildren (0)	19.063			
mreligion (Other)	7.194	6.773	−5.747	21.140
mreligion (Sikh)	13.061	8.284	−3.285	29.698
mreligion (Christian)	17.390	5.521	6.361	28.278
mreligion (Muslim)	−4.594	3.230	−10.904	1.367
mreligion (Hindu)	−33.051			
mresidence	−0.914	2.680	−6.417	4.487
electricity	9.838	2.655	4.690	14.883
radio	4.171	2.172	−0.369	8.448
television	15.302	2.702	10.148	20.550
refrigerator	11.493	3.784	4.216	19.062
bicycle	−3.273	2.136	−7.334	0.926
motorcycle	10.054	3.431	3.561	17.102
car	2.636	7.551	−12.272	17.337

$$\eta_i = \boldsymbol{u}_i'\boldsymbol{\beta} + f_1(size) + f_2(year) + f_{\text{spat}}(s_i)$$

where s_i indexes subquarters in Munich. In contrast to the childhood malnutrition example, the vector of covariates $\boldsymbol{u}$ has a much higher dimension and comprises 238 mostly categorical covariates representing different features of the flat, such as equipment in the bathroom or the kitchen, or variables for floor material. To ensure stable estimation of these effects even in situations where data are sparse and to improve the predictive ability of our model, we assign informative priors to the parametric effects $\boldsymbol{\beta}$ (excluding the intercept) utilizing regularization and variable selection priors discussed in Sections 4.4.2 and 4.5.3. To be more specific, we will consider the Bayesian analogues of ridge and lasso regression, corresponding to i.i.d. Gaussian and i.i.d. Laplace priors, and a normal mixture of inverse gamma (NMIG) prior. As a consequence, we enforce that small coefficients are shrunk towards zero such that they effectively drop out of the model. For the NMIG prior, we can additionally determine the inclusion probability for a specific covariate based on the corresponding inclusion indicator.

Figure 5.41 displays posterior mean estimates for the nonparametric effects of size in square metre and the year of construction on the net rent per square metre obtained with either the ridge or the lasso prior for parametric effects (results for the NMIG prior are very similar and therefore omitted). Obviously, the impact of the different prior specifications on the estimated nonparametric effects is rather small. In both cases, the effect of the size shows an elevated effect for small flats while the effect remains on a much smaller and almost constant level for larger flats. For the effect of year of construction, we find a mostly constant curve for flats built before 1945, followed by a moderate decrease during the 1950s and 1960s. Afterwards, the effect starts to increase corresponding to higher rents in flats built only recently.

Although the categorical covariates included in the regression model already capture several aspects concerning the location and the quality of the flat, there remains a strongly expressed spatial effect with both the lasso and the ridge

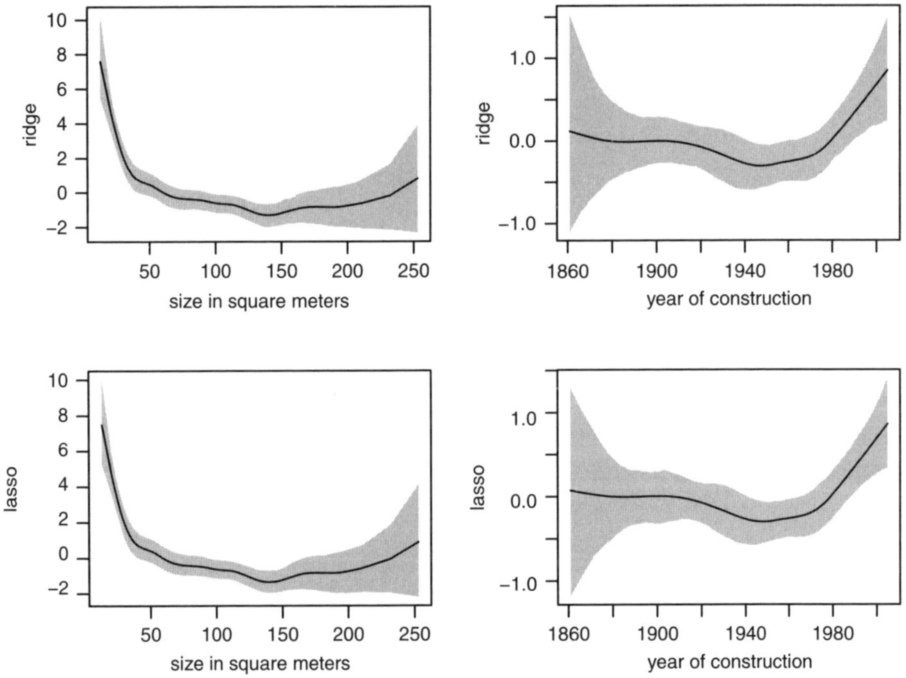

Fig. 5.41 Munich rental guide: Estimated nonparametric effects together with pointwise 95% credible intervals for a model with lasso prior (bottom) and a ridge prior (top) for the parametric effects.

prior (see Figure 5.42). In particular, rents are increased in the centre of Munich as well as along the River Isar that separates the city from south to north. In contrast, reduced rents are observed in some areas at the border of Munich.

Instead of displaying all categorical effects, we selected those that achieved a posterior probability (defined as $2\min\{\mathbb{P}(\beta_j < 0); \mathbb{P}(\beta_j > 0)\}$) smaller than 0.05 with either the Bayesian lasso, the Bayesian ridge or the NMIG prior (Table 5.3). While there is a general agreement in the sign of coefficients, some differences in the absolute magnitude can be observed. In particular, larger coefficients are shrunk less while smaller coefficients are shrunk more strongly when estimated with the lasso prior as compared to the ridge prior. This is exactly what was expected for the lasso prior that has heavier tails than the ridge and a more strongly expressed peak in zero. Hence, it induces more regularization for small coefficients while larger coefficients remain almost untouched. For effects with inclusion probability larger than 0.5 in the NMIG model, coefficients tend to be larger under this prior than with the Bayesian lasso. Vice versa, for effects with inclusion probability smaller than 0.5 the effect estimates tend to be smaller than compared with the Bayesian lasso. This more adaptive amount of shrinkage results from the ability of the NMIG prior to select or deselect covariates that are then estimated almost unpenalized or extremely penalized, respectively. Still, there is a general agreement on the importance of covariates across all three prior structures. △

Empirical Bayes inference

For empirical Bayes inference as discussed in detail in Section 4.2.1, we require a decomposition of the spatial effect as

$$\boldsymbol{\gamma}_{\text{spat}} = \tilde{\boldsymbol{U}}\boldsymbol{\beta}_{\text{spat}} + \tilde{\boldsymbol{Z}}\boldsymbol{\alpha}_{\text{spat}}$$

where $\boldsymbol{\beta}_{\text{spat}}$ corresponds to the fixed effects part with flat prior $p(\boldsymbol{\beta}_{\text{spat}}) \propto \text{const}$ and $\boldsymbol{\alpha}_{\text{spat}}$ corresponds to the random effects part with $\boldsymbol{\alpha}_{\text{spat}} \sim \text{N}(\mathbf{0}, \tau^2_{\text{spat}}\boldsymbol{I})$. Such a representation can always be achieved based on the eigenvalue decomposition of the penalty matrix $\boldsymbol{K}$ as defined in equation (4.13), page 193. Although the eigenvalue decomposition provides us with a basis for the null space of the penalty, too, it is advantageous to have an interpretable characterization in addition. For intrinsic Gaussian Markov random fields including only first-order neighbourhoods, the rank-deficiency of the penalty matrix is typically equal to one, provided that the regions are not disconnected. In the latter case, for example if there are regions without neighbours or if the map consists of two separated parts, the rank-deficiency will increase. If the regions are connected, the rank-deficiency reflects the fact that only differences in parameters enter the penalty and, as a consequence, the level of the spatial effect remains unpenalized. Therefore, a basis of the null space is simply obtained as a vector of ones. The reparametrization of the intrinsic Gaussian Markov random field into a constant

Table 5.3 Munich rental guide: Posterior means and posterior probabilities for regularized effects with posterior probability lower than 0.05 in either the Bayesian lasso, ridge or NMIG model. For the NMIG prior, the last column shows the estimated inclusion probabilities obtained from the binary inclusion indicators.

	lasso		ridge		NMIG		
variable	mean	prob.	mean	prob.	mean	prob.	ind.
company housing	−1.318	0.000	−0.977	0.000	−1.455	0.000	1.000
fridge	0.436	0.000	0.414	0.000	0.589	0.008	0.992
intercom	0.375	0.000	0.362	0.000	0.484	0.000	1.000
drying area	−0.429	0.000	−0.427	0.000	−0.497	0.000	1.000
playground	−0.474	0.000	−0.455	0.000	−0.566	0.000	1.000
attic	−0.565	0.000	−0.540	0.000	−0.632	0.000	1.000
no. of occupants	0.119	0.000	0.125	0.006	0.057	0.076	0.188
side costs	−0.005	0.000	−0.005	0.000	−0.004	0.000	0.001
floor covering (PVC)	0.545	0.000	0.489	0.000	0.954	0.000	1.000
floor covering (felt files)	0.441	0.000	0.389	0.004	0.825	0.000	1.000
rent start date before 2002	−0.916	0.000	−0.847	0.000	−1.031	0.000	1.000
tiled bathroom	0.239	0.002	0.250	0.002	0.185	0.220	0.542
average residential area	−0.459	0.002	−0.429	0.004	−1.509	0.000	1.000
no waterheating	−1.052	0.004	−0.532	0.004	−2.127	0.000	1.000
no additional heating	−0.375	0.014	−0.400	0.002	−0.379	0.238	0.610
special windows	0.268	0.014	0.272	0.010	0.087	0.480	0.235
grassy area	−0.260	0.016	−0.294	0.004	−0.044	0.498	0.127
kitchen size > 12 m^2	−0.207	0.022	−0.233	0.006	−0.022	0.540	0.048
drying room	−0.182	0.022	−0.202	0.006	−0.083	0.296	0.279
bath width	0.139	0.024	0.160	0.018	0.033	0.334	0.072
no. roof terraces	0.506	0.026	0.406	0.024	1.132	0.014	0.985
floor covering (simple wood)	0.337	0.026	0.312	0.010	0.793	0.000	1.000
modernized windows	−0.161	0.028	−0.189	0.024	−0.011	0.628	0.008
modernized floor	0.183	0.030	0.197	0.024	0.036	0.456	0.094
bathtub	0.260	0.032	0.297	0.018	0.014	0.736	0.028
room 3 facing garden	0.219	0.040	0.247	0.054	0.082	0.418	0.229
built-in kitchen	0.215	0.044	0.236	0.030	0.044	0.542	0.109
2nd bathroom	0.327	0.048	0.327	0.030	0.514	0.140	0.769
shower	0.216	0.054	0.254	0.006	0.011	0.672	0.014
no balcony	−0.207	0.056	−0.242	0.036	−0.014	0.632	0.028
satellite dish	−0.253	0.060	−0.273	0.022	−0.052	0.554	0.120
room 4 facing street	−0.290	0.060	−0.300	0.032	−0.109	0.550	0.227
cooking hobs	0.231	0.064	0.262	0.026	0.051	0.590	0.132
room 1 facing south	0.173	0.064	0.219	0.016	0.012	0.622	0.009
stairwell renovation	−0.175	0.066	−0.205	0.020	−0.014	0.636	0.022
penthouse appartment	0.279	0.076	0.297	0.024	0.011	0.882	0.027
electric/gas stove	−0.220	0.078	−0.240	0.054	−0.237	0.188	0.678
floor covering (linoleum)	−0.248	0.088	−0.313	0.010	0.001	0.986	0.014
2nd toilet	0.187	0.100	0.199	0.048	0.053	0.582	0.140
video interphone system	0.339	0.154	0.296	0.132	0.798	0.230	0.717
ground floor flat	−0.342	0.268	−0.262	0.246	−2.090	0.064	0.937
gas furnace	−0.195	0.274	−0.190	0.294	−0.353	0.304	0.611
floor covering (felt)	0.196	0.314	0.194	0.264	0.852	0.108	0.894
good address	−0.146	0.394	−0.125	0.382	−1.155	0.000	1.000
basic address	−0.218	0.400	−0.212	0.324	−2.585	0.008	0.991

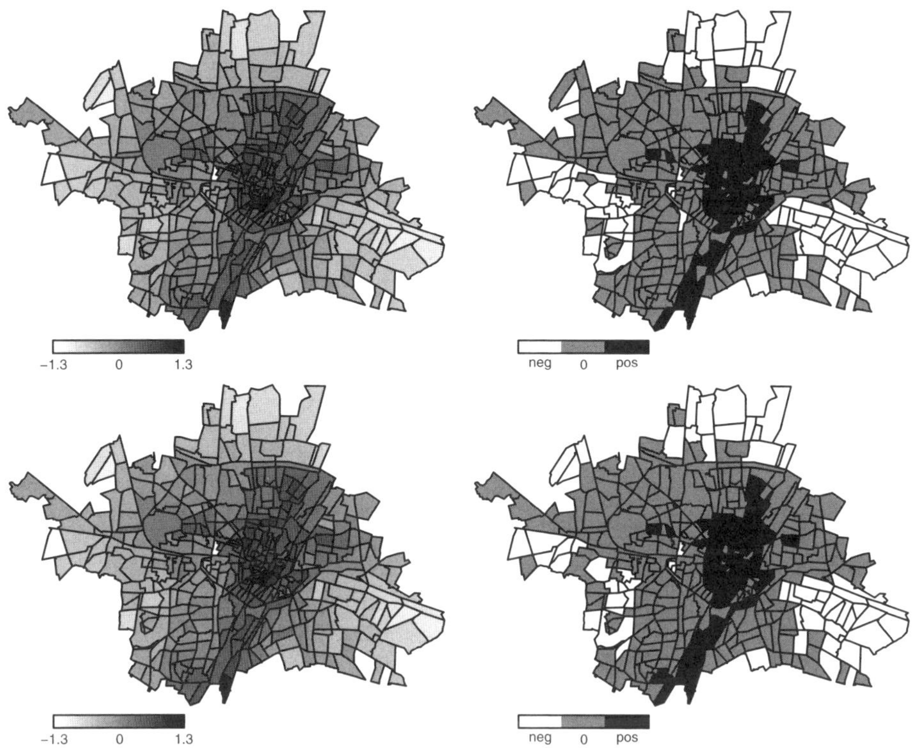

Fig. 5.42 Munich rental guide: Estimated spatial effects with lasso (top) and ridge (bottom) prior for the fixed effects. The left column shows posterior means, the right column corresponding 80% posterior probabilities.

fixed effect and i.i.d. random effects representing the deviation from the constant, can be performed exactly as outlined in Section 4.2.1 based on the spectral decomposition of the adjacency matrix $\boldsymbol{K}$.

For tensor product splines with penalty based on the Kronecker product, the null space of the penalty matrix is obtained as the Kronecker product of the univariate spline bases and penalties. For example, in case of second-order differences, a linear effect remains unpenalized for both univariate effects.

Hence, the null space of the penalty matrix for the bivariate effect consists of a constant, two univariate effects in direction of s_1 and s_2 and the interaction effect $s_1 s_2$. When using the biharmonic differential operator to construct the penalty, the interaction effect disappears and the null space is only three-dimensional consisting of the terms β_0, $\beta_1 s_1$ and $\beta_2 s_2$.

Finally, for radial basis functions and Gaussian random fields, the penalty matrix has full rank and therefore the mixed model representation is not necessary. Still, it may be advantageous to transform the correlated vector of regression coefficients to an i.i.d. vector to yield a comparable structure for all effects included in the regression model. Again, this reparametrization can be based on the spectral decomposition but the fixed effects part drops from the model since there is no rank-deficiency.

After augmenting the spatial effect in mixed model representation, we can use the algorithm described in Section 4.2.1 on page 198.

Example 5.16: Forest health

As a last example of geoadditive regression, we consider a non-Gaussian regression model for the defoliation degree in the forest health data introduced in Example 1.2. We use a categorical response variable with three categories and specify the following predictor for a cumulative probit model:

$$\eta_{it}^{(c)} = \theta^{(c)} - \boldsymbol{u}_{it}'\boldsymbol{\beta} - f_1(age_{it}) - f_2(t) - f_{1,2}(age_{it}, t) - f_{\text{spat}}(s_i) - \alpha_{s_i}$$

Since the forest health data combine longitudinal and spatial information, we include both an unstructured random effect α_{s_i} that accounts for the repeated measurements design and a spatially structured function $f_{\text{spat}}(s_i)$ that captures spatial correlation. The former is assigned an i.i.d. Gaussian prior while the latter is included as a Markov random field with neighbourhoods defined by distances smaller than 1.8 km in between two observation plots.

In addition, nonparametric effects of age and calendar time have been included as well as a nonparametric interaction surface between the two. Cubic penalized B-splines with 20 inner knots and second-order random-walk penalty are employed for the univariate nonparametric effects while a cubic tensor product spline with 20 inner knots along each of the axes and a first-order spatial random-walk prior is specified for the interaction effect. We consider both full Bayesian and empirical Bayes inference.

Figure 5.43 compares posterior mode estimates obtained from the empirical Bayes procedure and posterior mean estimates obtained with MCMC for the nonparametric effects. MCMC yields somewhat more variable estimates but reduced uncertainty reflected in the narrower credible intervals. Still, the general conclusions drawn about the underlying effects remain qualitatively the same with an increased tendency towards defoliation for older trees and at the end of the 1980s. In particular, the temporal effect reflects the empirical trends observed in Fig. 1.3 (page 8).

For the interaction effect displayed in Fig. 5.44, the behaviour observed for the univariate nonparametric effects is reversed: Posterior mode estimation yields a more wiggly surface while posterior mean estimation produces an interaction surface with overall smoother appearance. Both approaches identify an alleviated risk for defoliation in young trees early in the 1980s and also for older trees after 2000. This might reflect a recovery of younger trees over time while older trees might suffer from long-term consequences of air pollution.

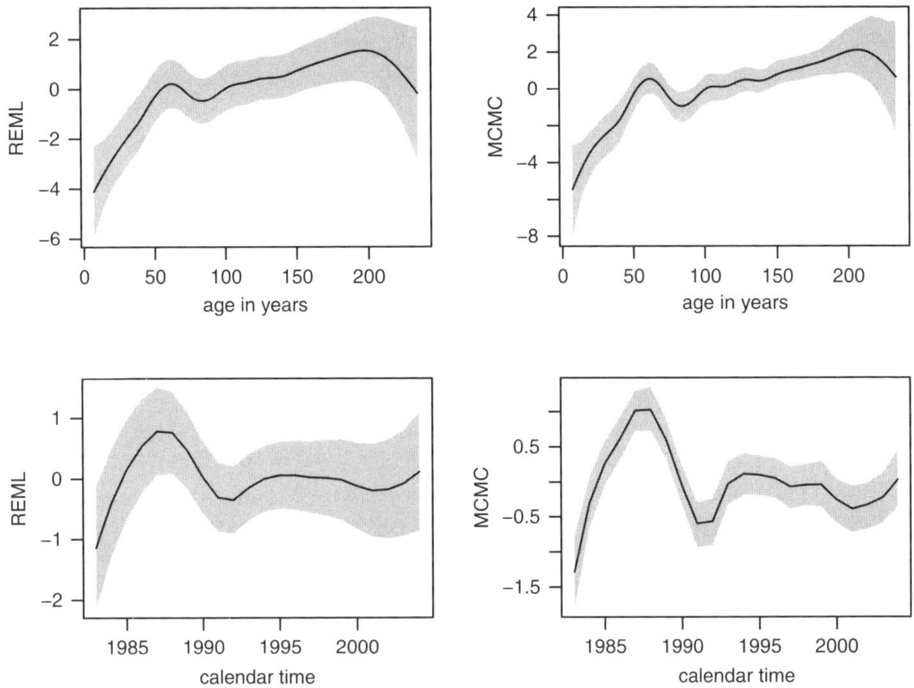

Fig. 5.43 Forest health: Estimated nonparametric effects together with pointwise 95% credible intervals.

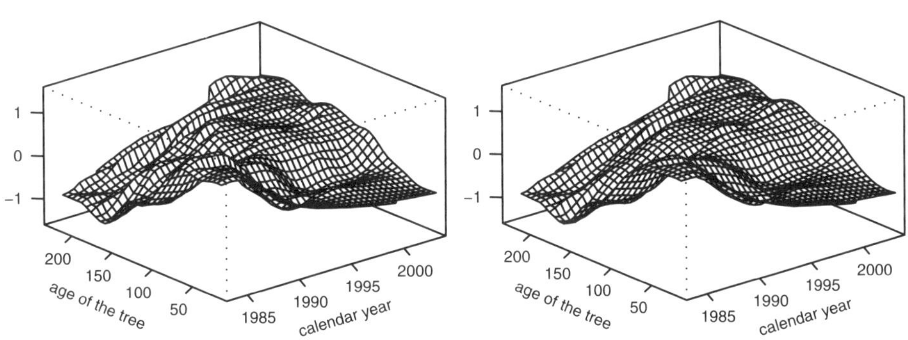

Fig. 5.44 Forest health: Posterior mode (left) and mean (right) estimates for the interaction effect.

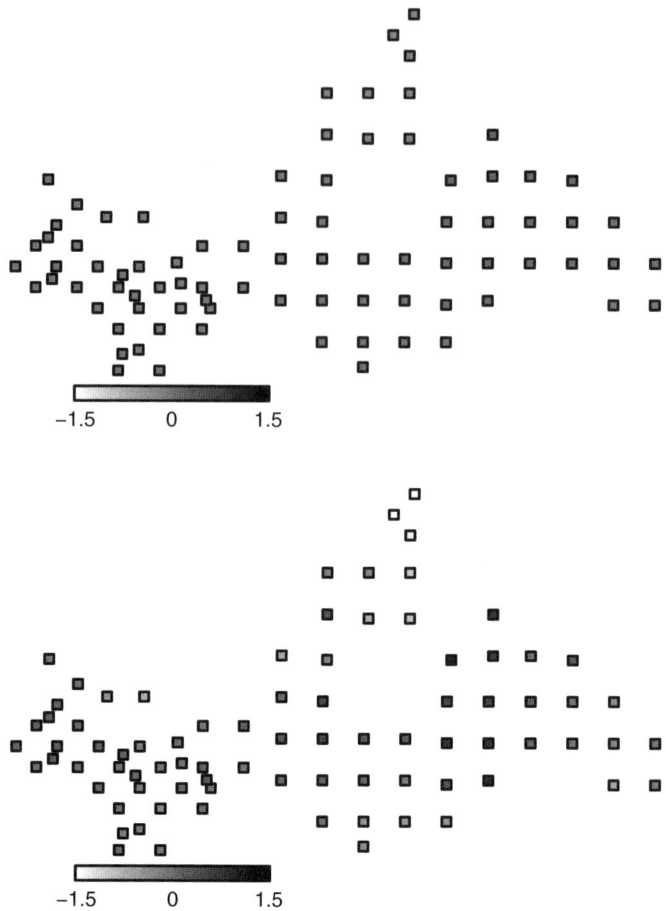

Fig. 5.45 Forest health: Posterior mode (top) and mean (bottom) estimates for the structured spatial effect.

Structured and unstructured spatial effects are represented in Figs. 5.45 and 5.46, respectively. The largest differences are observed for the structured effect where REML estimation identifies only very minor variation while the effect is much more strongly expressed for MCMC-based estimation. In contrast, the spatially structured effect is recovered to a comparable extent by both approaches.

Table 5.4 summarizes estimates for the parametric effects. Again, we find a close agreement between both empirical and full Bayes inference. Further details on this example can be found in Kneib and Fahrmeir (2010). △

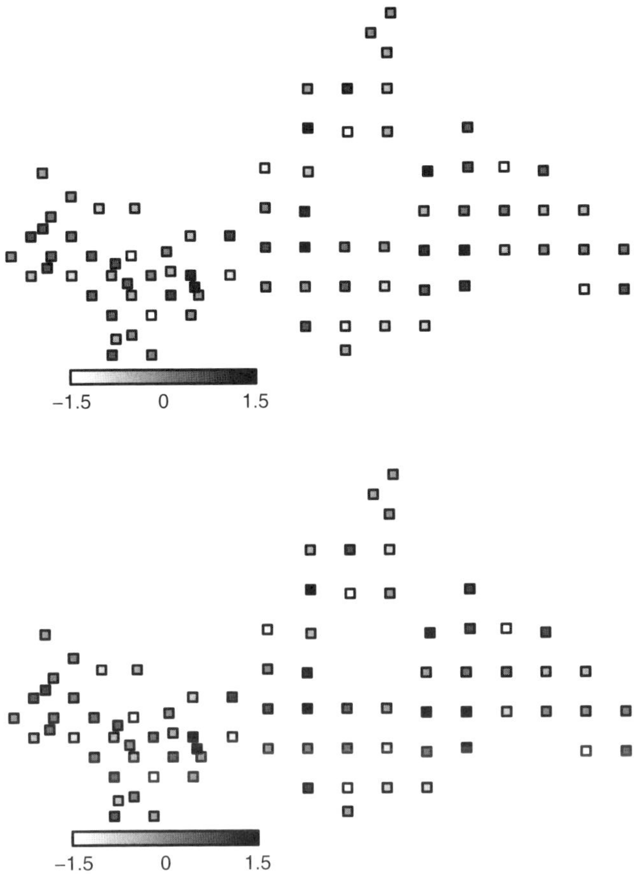

Fig. 5.46 Forest health: Posterior mode (top) and mean (bottom) estimates for the unstructured spatial effect.

5.6 Notes and further reading

A lot of the recently growing interest in spatial statistics is related to the advent of geographical information systems (GIS) that provide a convenient way to store and process spatially oriented data. We have deliberately left out any relation to GIS software in this chapter and focused on the principles of statistical models for spatial data. A statistical perspective on GIS software in combination with a discussion of `R`-packages that provide an interface between GIS and `R` can be found in Bivand, Pebesma and Gómez-Rubio (2008).

An important extension of concepts considered in this chapter are spatio-temporal models that combine temporal and spatial smoothing in one regression model. While this is conceptually simple if spatial and temporal effects are

Table 5.4 Forest health: Posterior mode and posterior mean estimates together with 95% credible intervals for parametric effects.

	REML			MCMC		
variable	post. mode	95% ci		post. mean	95% ci	
elevation	0.001	−0.005	0.007	0.002	−0.004	0.011
inclination	0.007	−0.022	0.036	0.018	−0.013	0.053
soil	−0.021	−0.050	0.008	−0.019	−0.058	0.016
ph	−0.048	−0.462	0.366	−0.023	−0.439	0.393
canopy	−2.032	−2.765	−1.300	−2.223	−3.074	−1.453
standcat	0.228	−0.058	0.515	0.140	−0.145	0.493
fertilization	−0.434	−0.963	0.094	−0.728	−1.381	−0.076
humus0	−0.259	−0.470	−0.049	−0.263	−0.481	−0.026
humus1	−0.133			−0.137		
humus2	0.140	−0.028	0.309	0.141	−0.034	0.306
humus3	0.132	−0.067	0.331	0.150	−0.055	0.340
humus4	0.120	−0.158	0.398	0.109	−0.173	0.404
moisture1	−0.591	−1.220	0.037	−0.667	−1.366	0.032
moisture2	0.181			0.225		
moisture3	0.410	−0.041	0.860	0.442	−0.012	0.915
alkali1	0.415	−0.228	1.058	0.105	−0.743	0.895
alkali2	−0.407			−0.607		
alkali3	−0.368	−1.044	0.308	−0.186	−1.052	0.602
alkali4	0.360	−0.519	1.239	0.688	−0.225	1.560

combined additively, more complex models are required in case of spatio-temporal interactions of a more complex form. Finkenstädt, Held and Isham (2007) collect recent approaches for spatio-temporal statistics and provide references for further reading.

A completely different type of spatial data that has not been considered in this chapter are spatial point processes. While we have kept the location of observations as fixed, the placement of observations is exactly the quantity of interest in the analysis of spatial point processes. Most models in this area focus on the estimation of, and models for, intensity functions that basically measure the instantaneous probability for an observation at a particular point in space. We will consider related models for temporal intensities in the next chapter. The mathematical background and basic statistical approaches for spatial point processes are provided in Møller and Waagepetersen (2004). More applied introductions are for example Illian, Penttinen, Stoyan and Stoyan (2008), Diggle (2003) or the corresponding chapter in Schabenberger and Gotway (2005). Møller and Waagepetersen (2007) provide a recent survey on modern statistics for spatial point processes, including full Bayes approaches.

6 Event History Data

This chapter extends semiparametric Bayesian approaches for modelling and analysing longitudinal and spatial data to regression models for survival and event history data. Generally speaking, such regression models are needed in studies where analysing the influence of covariates on times until the occurrence of certain events or on the duration times between events, observed as time goes by, is of primary interest.

Survival analysis appears as a special case, most popular in medical and epidemiological applications. The quantity of main interest is, for example, the time between diagnosis and death from a certain disease. A typical application appears in Example 6.1, where death from leukemia after diagnosis is the event of interest. The data set contains about one thousand cases of acute myeloid leukemia (AML) that have been diagnosed between 1982 and 1998 in northwest England. As a typical feature of such studies, not all survival times are completely available: Almost 16% of the cases are right censored because death was not observed within the study period or because patients dropped out of the study. Covariates include age and sex of the patient, the white blood cell count at diagnosis, a district-specific deprivation index, and location of residence of a patient, given in terms of longitude and latitude.

Another application described in more detail in Examples 1.6 and 6.2 is childhood mortality in developing countries, with Nigeria as a case study. The data have been collected within the 2003 Demographic and Health Survey (DHS) for Nigeria. Survival times of children are obtained from retrospective interviews with their mothers. Due to memory effects, survival times are often not known exactly, that is, in days, but only in months. This can be considered as interval-censoring, because the exact days of death are only known to lie between the first and the last day of the corresponding month. In addition, a large number of personal, socio-economic and public health factors with a possible influence on survival times is contained in the DHS data.

There are many other fields of application where the analysis of times until a certain event occurs is of interest, but the term 'survival time' may not always be appropriate. For example, terminology like 'life times', 'failure times' or 'durations' are also quite common. In principle, survival or duration times could be analysed with regression models introduced in previous chapters, accounting for non-negativity of survival times. However, special models have been developed to incorporate certain features of survival data, in particular the fact that survival times are often not completely observed, as in the examples just described.

Moreover, the inherent time-dependence of the process that shall be analysed requires a more explicit treatment of time-varying effects and covariates. Probably the most popular class of models in survival analysis are continuous-time hazard rate models, following the seminal work of Cox (1972). The influence of covariates $\boldsymbol{u}$ on survival times T is specified through a regression model for the hazard rate

$$\lambda(t|\boldsymbol{u}) = \lim_{\Delta t \to 0} \frac{1}{\Delta t} \mathbb{P}(t \leq T < t + \Delta t \mid T \geq t, \boldsymbol{u}). \tag{6.1}$$

The hazard rate can be interpreted as the instantaneous rate of an event in the interval $[t, t + \Delta t)$ given survival up to time t. In the Cox proportional hazards model, a multiplicative structure of the form

$$\lambda(t|\boldsymbol{u}) = \lambda_0(t) \exp(\boldsymbol{u}'\boldsymbol{\beta}) \tag{6.2}$$

is assumed for the hazard rate where $\lambda_0(t)$ is an unspecified baseline hazard rate and $\boldsymbol{u}'\boldsymbol{\beta}$ is a linear predictor. This model separates time-dependence (in the baseline hazard $\lambda_0(t)$) and covariate effects (in the multiplicative factor $\exp(\boldsymbol{u}'\boldsymbol{\beta})$) into two model components, making the model easy to interpret but also quite restrictive. After describing survival data and characteristic features in more detail in Section 6.1, we will basically extend the Cox model to Cox-type hazard rate models in Section 6.2, replacing the linear predictor through a structured additive predictor comprising time-varying covariates and effects as required for the applications just sketched.

Time can often not be observed continuously but measured only on a discrete scale. For example, in market labour studies, duration of unemployment is mostly given in months, see Example 6.3. In this situation, discrete-time survival (or duration) models are more appropriate. They can also be useful as an alternative to continuous-time hazard rate models with interval censoring as for the DHS mortality study, with survival times measured in months only (see Example 6.2). It will turn out that discrete-time survival models are closely related to regression models for binary longitudinal data (Section 6.1.4).

As a third alternative, seemingly closer to conventional regression models, we will also discuss transformation or accelerated failure time (AFT) models. To account for nonnegativity a transformation is applied. The most common models are based on a logarithmic transformation yielding a regression model

$$\log(T) = \boldsymbol{u}'\boldsymbol{\beta} + \sigma\varepsilon.$$

Different models are characterized by the distribution of the error term ε. Exponentiating the loglinear model shows that covariates act multiplicatively on the survival time. Section 6.4 will consider semiparametric extensions of conventional AFT models, including smoothing and shrinkage.

Survival data can also be seen as a special case of data generated from a stochastic process with a finite state space: individuals or objects, initially in a transient state ('alive') are observed over time until a transition to an absorbing state ('death') occurs. For right-censored data, this transition cannot be observed within the observation period. Multi-state models (Section 6.5) are flexible tools of event history analysis when more transitions between several states are the events of interest. Some common examples are illustrated in Fig. 6.1 in form of reachability graphs. For recurrent events (Fig. 6.1(a)), individuals or objects are moving repeatedly between a fixed set of states. The application on sleep research presented in Example 6.13 is of this type: states are given by sleep states such as awake, REM and NON-REM. Figure 6.2 shows two typical realizations of such a multi-state sleep process, together with nocturnal cortisol level measurements as time-varying covariates. Similar data structures arise in marketing studies when individuals are repeatedly buying products, choosing between different brands, or in unemployment studies, where, for example, unemployment, full-time job and

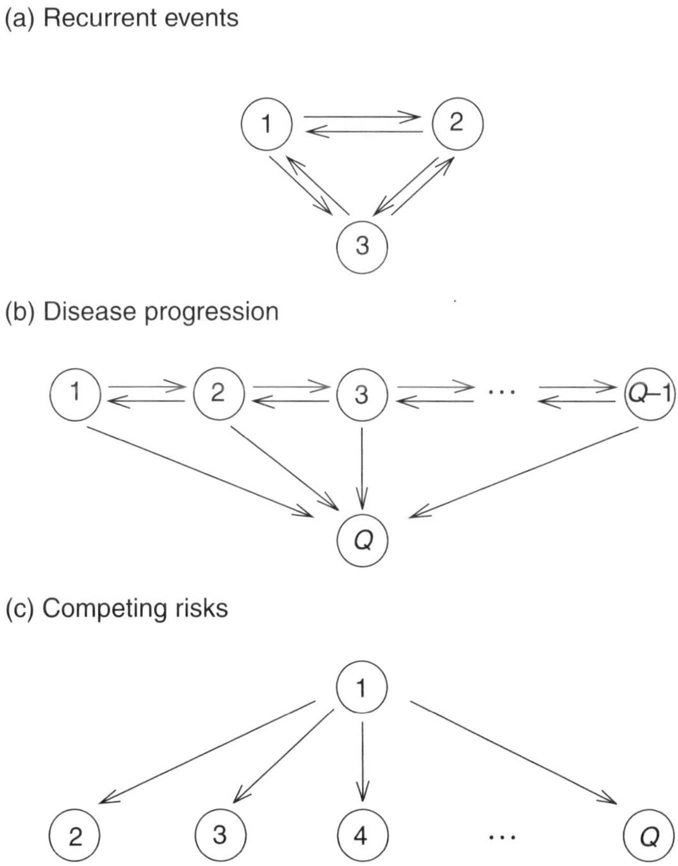

Fig. 6.1 Reachability graphs for three common types of multi-state models.

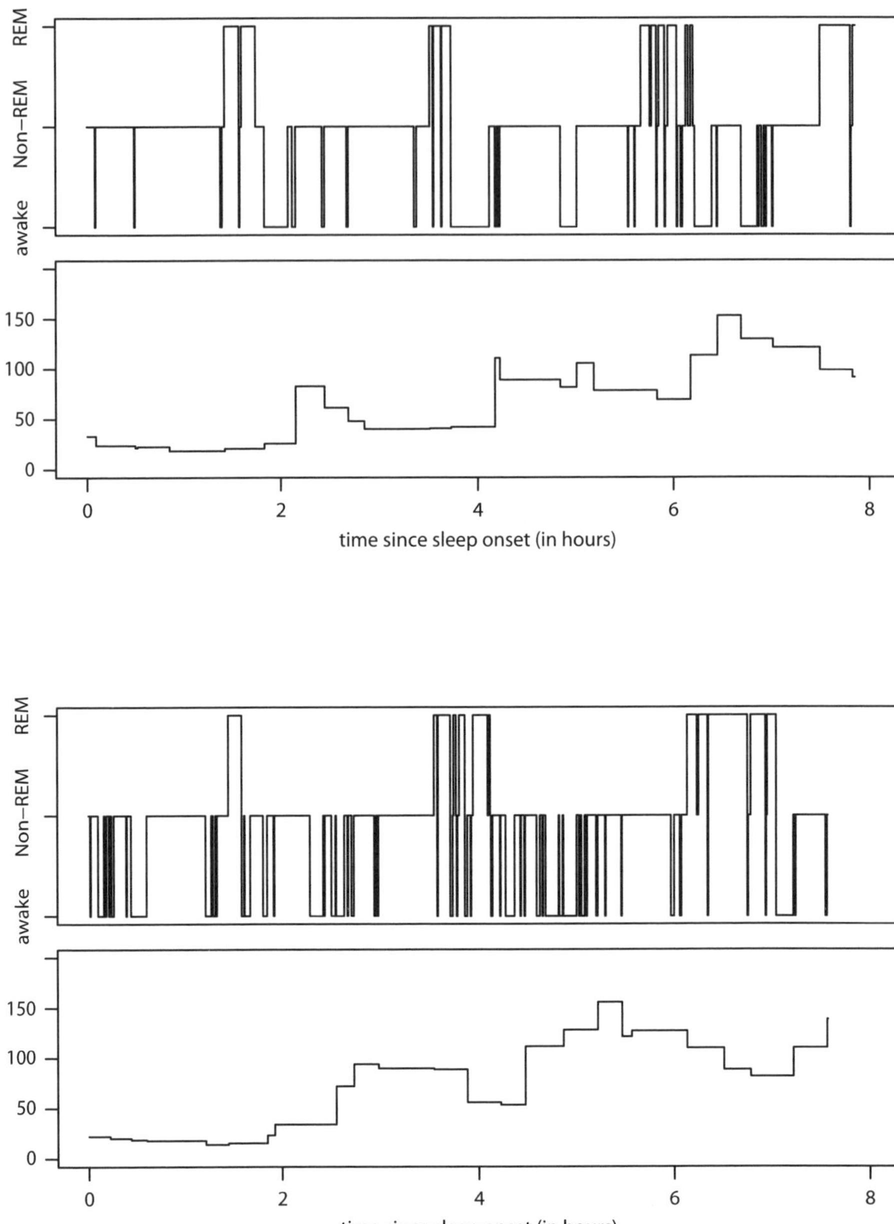

Fig. 6.2 Human sleep: Realizations of two sleep processes (first and third panel) together with the corresponding nocturnal secretion of cortisol (second and fourth panel).

part-time job define the recurrent states. Disease progression models (Fig. 6.1(b)) describe the chronological development of a certain disease: If an individual is currently in a (transient) disease state $q = 1, \dots, Q-1$, it can move to one of the contiguous disease states or to the absorbing state Q ('death'). A model with several absorbing states is the competing risks model (Fig. 6.1(c)) where, for example, different causes of death are analysed simultaneously.

Multi-state hazard rate models (see Section 6.5) extend hazard rate models for survival data by specifying the various rates $\lambda^h(t|\mathcal{F}_t)$ for transitions $h = 1, \dots, H$, of interest given the history of the process $\mathcal{F}_t$ in analogy to hazard rates $\lambda(t|\boldsymbol{u})$ for survival times. These transition rates are defined as

$$\lambda^h(t|\mathcal{F}_t) = \lim_{\Delta t \to 0} \frac{1}{\Delta t} \mathbb{P}(\text{a transition of type } h \text{ occurs in } [t, t + \Delta t) \mid \mathcal{F}_t),$$

where $\mathcal{F}_t$ denotes the information given by the history of the multi-state process up to time t, including time-constant and, under certain conditions, time-varying covariates. The theory of counting processes and marked point processes provides a general framework for developing likelihood-based and Bayesian models, including a variety of censoring schemes. Transformation and AFT models are less suited to model the (time-varying) influence of information $\mathcal{F}_t$ in flexible form since temporal information cannot be provided both in terms of the response and as covariate information.

In summary, this chapter will extend Bayesian approaches for smoothing and regression developed in previous chapters to regression models for survival and event history data with structured additive predictors. This allows us to include nonlinear time-varying effects and flexible covariate effects, spatial effects, and random effects in addition to common linear predictors and to estimate them simultaneously based on full or empirical Bayes inference.

Alternative approaches and other model types will be outlined in Section 6.6.

6.1 Survival data

This section describes some basic concepts for survival analysis, in particular various censoring mechanisms, and presents resulting likelihoods needed for Bayesian modelling and inference.

6.1.1 Basic notions for continuous survival times

Let T denote a non-negative continuous random variable with density function $f(t)$ and distribution function $F(t) = \mathbb{P}(T \leq t)$. In survival analysis, it is often more common to consider the *survivor function*

$$S(t) = 1 - F(t) = \mathbb{P}(T > t),$$

that is the probability of surviving time t. Defining the *hazard rate*

$$\lambda(t) = \lim_{\Delta t \to 0} \frac{\mathbb{P}(t \le T < t + \Delta t | T \ge t)}{\Delta t}$$

as in the introduction and the *cumulative hazard rate* as

$$\Lambda(t) = \int_0^t \lambda(u) du,$$

the following equations relate these quantities to each other:

$$\begin{aligned} \lambda(t) &= \frac{f(t)}{S(t)}, \\ S(t) &= \exp(-\Lambda(t)), \\ f(t) &= \lambda(t)\exp(-\Lambda(t)) = \lambda(t)\exp\left(-\int_0^t \lambda(u)du\right). \end{aligned}$$

The first equation follows directly from the definition of the hazard rate, and the second equation from

$$\Lambda(t) = \int_0^t \lambda(u)du = \int_0^t f(u)/(1 - F(u))du = -\log(1 - F(t)) = -\log S(t).$$

The distribution of T is completely determined by any of these quantities. As an example, the Weibull distribution has hazard rate

$$\lambda(t) = \lambda\alpha(\lambda t)^{\alpha-1},$$

with scale parameter $\lambda > 0$ and shape parameter $\alpha > 0$. The survivor function is thus given by

$$S(t) = \exp(-(\lambda t)^\alpha),$$

leading to

$$f(t) = \lambda(t)S(t) = \lambda\alpha(\lambda t)^{\alpha-1}\exp(-(\lambda t)^\alpha).$$

The exponential distribution with constant hazard rate $\lambda(t) = \lambda > 0$ is included as a special case with $\alpha = 1$.

6.1.2 Censoring and truncation

The most common type of incomplete survival data are *right-censored* observations. This censoring scheme can be formalized by a censoring time C. Instead of observing the true survival time T, only $\tilde{T} = \min(T, C)$ is observed together with the censoring indicator

$$D = \begin{cases} 1, & T \le C \\ 0, & \text{else.} \end{cases}.$$

Right-censored survival times occur in Example 6.1.

Example 6.1: Leukemia survival times
In this example, we analyse data on survival times for adults that are diagnosed with acute myeloid leukemia (AML). The data (a total of 1,043 cases) have been collected between 1982 and 1998 in northwest England and comprise all AML cases that have been diagnosed in this time interval. Approximately 16% of the patients were still alive at the end of the study and are therefore treated as right-censored. Continuous covariates include the age of the patient, the white blood cell count (wbc) at diagnosis and the Townsend deprivation index (tpi) which measures the deprivation for the enumeration district of residence. Gender is the only categorical covariate and will be used in dummy-coding (1=female, 0=male).

In addition, spatial information is available in two resolutions: On the one hand, the exact location of a patient's residence is given in terms of longitude and latitude. On the other hand, this exact information can be traced back to the district level by assigning each observation to the region that contains the exact location. Figure 6.3 illustrates both types of spatial information and shows the district boundaries in combination with the coordinate information of the observations. The rare situation that both point-referenced and district-referenced spatial information is available will allow us to determine the loss of information caused by aggregating the exact data to the district level, which is often performed for confidentiality reasons. △

Other applications confront the analyst with more general types of incomplete data. As already mentioned in the introduction, *interval-censored* survival times T are only known to fall in the interval $[T_{\mathrm{lo}}, T_{\mathrm{up}}]$. If $T_{\mathrm{lo}} = 0$, these survival times are *left-censored*: it is only known that the event of interest occurred before T_{up}.

Example 6.2: Childhood mortality in Nigeria
In this example, we analyse the survival times of children obtained from the 2003 Nigeria Demographic and Health Survey. The data are collected in retrospective questionnaires and, as a consequence, only survival times up to two months are recorded exactly while all other survival times are in fact only given in months. In contrast, right-censoring times for children that were still alive at the time of the interview can be computed exactly as the difference between the date of birth and the date of the interview. Instead of discretizing the complete information to monthly data, we will consider survival times exceeding two months as being interval censored with intervals given by the first and last day of the corresponding month.

We will also extend the interval censoring approach to account for memory effects induced by the retrospective collection of the data. From the histograms of uncensored and censored survival times shown in Fig. 6.4, we can conclude that actually a large number of survival times is heaped at 12, 18, 24, 36, and 48 months since the exact survival times can not be retrieved in the retrospective questionnaire but are instead rounded to multiples of half or full years. Ignoring

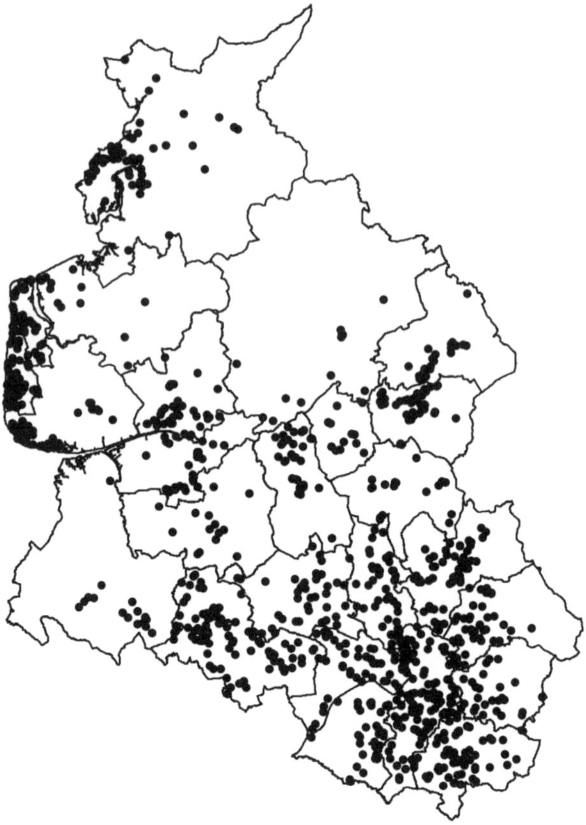

Fig. 6.3 Leukemia survival: Spatial distribution of the observations in northwest England. Each point corresponds to one observation.

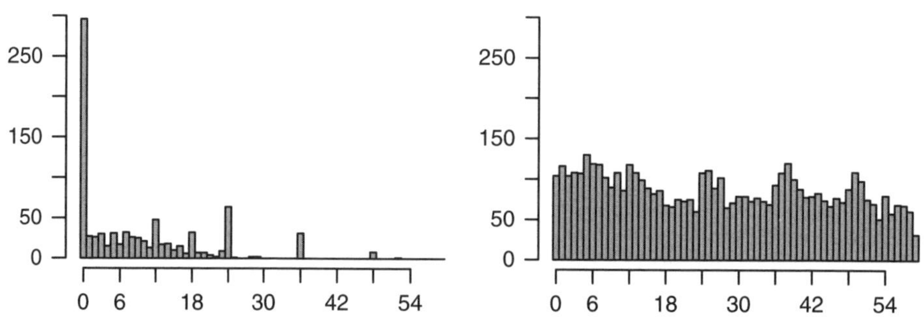

Fig. 6.4 Childhood mortality: Histograms for uncensored (left) and right-censored (right) survival times.

this heaping effect could potentially bias estimation results so that we consider larger intervals for heaped survival times. More specifically, non-overlapping, symmetric intervals of 6 or 12 months length are assigned to the heaped survival times. For example, an observation that is assigned to the value 18 in the original data will be treated as interval-censored with an interval ranging from 15 to 21 months.

After removing observations with incomplete covariate information, $n = 5,323$ observations remain in the data set. A total of 117 children died within the first two months, leading to uncensored survival times. The remaining 474 children that died within the study time are treated as interval censored, accounting also for the heaping effects as just described. △

In addition, each of the censoring schemes may appear in combination with *left truncation*. This means, that the survival time can only be observed if it exceeds the truncation time T_{tr}. Accordingly, some survival times are not observable and the likelihood has to be adjusted appropriately. Figure 6.5 illustrates the different censoring schemes we will consider in the following: The true survival time is denoted by T which is observed for individuals 1 and 2. While individual 1 is not truncated, individual 2 is left truncated at time T_{tr}. Similarly, individuals 3 and 4 are right-censored at time C and individuals 5 and 6 are interval-censored with interval $[T_{lo}, T_{up}]$ and the same pattern of left truncation.

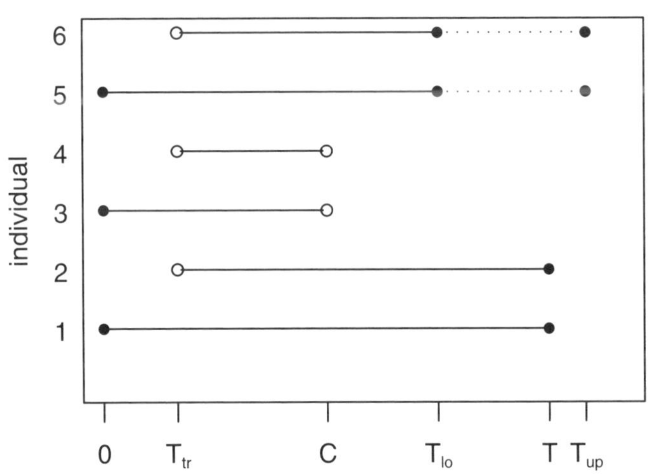

Fig. 6.5 Illustration of different censoring schemes: For individuals 1 and 2, the true survival time T is observed, individuals 3 and 4 are right-censored at time C, and individuals 5 and 6 are interval-censored, where the interval is given by $[T_{lo}, T_{up}]$. Individuals 2, 4 and 6 are left truncated at time T_{tr}.

In a general framework combining all the aforementioned issues, an observed survival time can be completely described by the quadruple $(T_{\text{tr}}, T_{\text{lo}}, T_{\text{up}}, D)$, with

$$\begin{array}{ll} T_{\text{lo}} = T_{\text{up}} = T,\ D = 1 & \text{if the observation is uncensored,} \\ T_{\text{lo}} = T_{\text{up}} = C,\ D = 0 & \text{if the observation is right-censored,} \\ T_{\text{lo}} < T_{\text{up}},\ D = 0 & \text{if the observation is interval-censored.} \end{array}$$

For left truncated observations we have $T_{\text{tr}} > 0$, and $T_{\text{tr}} = 0$ for observations which are not truncated.

6.1.3 Likelihood contributions for different types of censoring

Under the assumption of conditional independence for the survival times given the covariates, the likelihood $L(\cdot)$ is the product of individual likelihood contributions. For a right-censored observation the concept of random censoring is often assumed to hold: The survival time T_i and the (random) censoring time C_i of each individual $i = 1, \ldots, n$ are assumed to be conditionally independent random variables given the covariates. Let t_i denote a realization of $\tilde{T}_i = \min(T_i, C_i)$, that is the observed, possibly censored survival time and d_i the observed censoring indicator. Under the assumption of noninformative censoring, that means the distribution of C_i is not determined by parameters of interest contained in the distribution of T_i, the likelihood contribution of an uncensored observation $(T_i = t_i, d_i = 1)$ is given by

$$L_i = f_i(t_i) = \lambda_i(t_i) S_i(t_i) = \lambda_i(t_i) \exp\left(-\int_0^{t_i} \lambda_i(t)dt\right). \tag{6.3}$$

For a right-censored observation $(T_i > t_i, d_i = 0)$ it is given by

$$L_i = \mathbb{P}(T_i > t_i) = S_i(t_i) = \exp\left(-\int_0^{t_i} \lambda_i(t)dt\right). \tag{6.4}$$

In these expressions, $f_i(t)$, $S_i(t)$ and $\lambda_i(t)$ are the density, survivor function and hazard rate corresponding to survival time T_i of individual i. Dependence on i usually comes in through covariates. For a time-varying covariate $u_i(t)$, say, additional assumptions are needed: Either the covariate is 'external', i.e. known in advance, such as the age of a patient, or, in case of an 'internal' covariate influenced by the survival process itself, it has to be known shortly before t. More formally, such covariates are called *predictable*, see Kalbfleisch and Prentice (2002) for a detailed discussion.

If an observation is left truncated, the integrals start from the observed truncation time $t_{\text{tr},i} > 0$ instead of 0, for example

$$L_i = S_i(t_i)/S_i(t_{\text{tr},i}) = \exp\left(-\int_{t_{\text{tr},i}}^{t_i} \lambda_i(t)dt\right)$$

for a right-censored observation.

For an interval-censored observation, let $[t_{\mathrm{lo},i}, t_{\mathrm{up},i}]$ denote the interval that contains the true, unobservable survival time T_i. Then,

$$\begin{aligned} L_i &= \mathbb{P}(t_{\mathrm{lo},i} \leq T_i \leq t_{\mathrm{up},i}) = S_i(t_{\mathrm{lo},i}) - S_i(t_{\mathrm{up},i}) \\ &= \exp\left(-\int_0^{t_{\mathrm{lo},i}} \lambda_i(t)dt\right) - \exp\left(-\int_0^{t_{\mathrm{up},i}} \lambda_i(t)dt\right) \\ &= \exp\left(-\int_0^{t_{\mathrm{lo},i}} \lambda_i(t)dt\right)\left(1 - \exp\left(-\int_{t_{\mathrm{lo},i}}^{t_{\mathrm{up},i}} \lambda_i(t)dt\right)\right). \end{aligned} \tag{6.5}$$

Again, the first integral starts at $t_{\mathrm{tr},i}$ for left truncated observations.

All likelihood expressions presented in this paragraph are special cases of a general likelihood derived from counting process theory, see Andersen *et al.* (1993) and Section 6.5. This general approach also accounts for more general censoring mechanisms in a unified form.

6.1.4 Discrete-time survival data

Models for discrete-time survival data are basically used in two situations. First, survival or duration times may be truly discrete, as for example durations of unemployment, since employments usually end and begin with monthly allowance. Second, true survival times are continuous but are only known to lie within intervals $[\kappa_0, \kappa_1), [\kappa_1, \kappa_2), \ldots, [\kappa_{q-1}, \kappa_q), [\kappa_q, \infty)$, as in internal censoring. Such *grouped survival times* may be caused through observations that are only made at successive follow-ups or for reasons of data protection. Note that the interval boundaries are assumed to be independent of the individual index i such that a common structure underlies the discretization process.

Let now $T \in \{1, 2, \ldots, t, \ldots\}$ denote the discrete observed survival time, where $T = t$ means end of duration at the discrete time point t or within interval $[\kappa_{t-1}, \kappa_t)$, that is, the discrete time index t is identified with the interval. Often $\kappa_0 = 0$ may be assumed and κ_q denotes the final follow-up. Then the hazard rate is replaced through the *discrete hazard function*

$$\lambda(t) = \mathbb{P}(T = t | T \geq t), \quad t = 1, 2, \ldots,$$

which is the conditional probability for end of duration at time point t or in interval $[\kappa_{t-1}, \kappa_t)$ given that t (or the corresponding interval) is reached.

Let the *discrete survivor function* be given by

$$S(t) = \mathbb{P}(T \geq t) = \prod_{s=1}^{t-1}(1 - \lambda(s)),$$

then the unconditional probability for failure at $T = t$ is given by

$$\mathbb{P}(T = t) = \lambda(t) \prod_{s=1}^{t-1} (1 - \lambda(s)) = \lambda(t) S(t), \quad t = 1, 2, \ldots.$$

These expressions are the discrete-time analoga of the survivor function and the density $f(t)$ for a continuous survival time.

Censoring and truncation can also be defined analogously. For example, right-censoring means that $\tilde{T} = \min(T, C)$ is observed together with the censoring indicator

$$D = \begin{cases} 1, & \text{failure at } \tilde{T} = t \text{ or in interval } [\kappa_{t-1}, \kappa_t), \\ 0, & \text{censoring at } \tilde{T} = t \text{ or in interval } [\kappa_{t-1}, \kappa_t). \end{cases}$$

The discrete hazard function can be related to covariates using sequential models for ordinal responses as described in Section 2.2.3. Each survival time corresponds to a sequence of binary decisions indicating whether the corresponding individual survived the t-th interval or not.

Although sequential models exactly fit the needs of a discrete-time survival model without censoring, it is easier to model the binary decisions directly if right-censoring is to be included. Let right-censored discrete-time survival data be given by observed survival times $t_i = \min(T_i, C_i)$, censoring indicators d_i, and time-constant covariates $\boldsymbol{u}_i$, $i = 1, \ldots, n$. Then binary indicators y_{is}, $s = 1, \ldots, t_i$, are introduced with

$$y_{is} = \begin{cases} 1, & \text{if } t_i = s \text{ and } d_i = 1, \\ 0, & \text{otherwise.} \end{cases}$$

Hence, for a right-censored observation all binary indicators equal zero, while for an uncensored observation the last indicator equals one and all other indicators are zero. Corresponding to the data augmentation of the response variable, all covariates have to be replicated as illustrated in the following example. Suppose that discrete-time survival data

t_i	d_i	u_{i1}	u_{i2}
4	0	0	2
3	1	1	0
⋮	⋮	⋮	⋮

are given. The first observation is censored in the fourth interval. The second observation is uncensored and has an event in the third interval. Augmenting the data as described above yields

y_{is}	i	s	d_i	u_{i1}	u_{i2}
0	1	1	0	0	2
0	1	2	0	0	2
0	1	3	0	0	2
0	1	4	0	0	2
0	2	1	1	1	0
0	2	2	1	1	0
1	2	3	1	1	0
⋮	⋮	⋮	⋮	⋮	⋮

Note that the data augmentation allows for the routine inclusion of time-varying covariates, as u_{i1} and u_{i2} do not necessarily have to be constant over time and can adopt different values for each interval.

Formally, discrete-time survival data are thus rewritten as longitudinal data, and the discrete hazard function can be equivalently expressed as

$$\lambda(t|\boldsymbol{u}_i) = \mathbb{P}(y_{it} = 1|\boldsymbol{u}_i) = h(\eta_{it}),$$

with a suitable response function leading to a logit, probit or complementary log-log model with predictors η_{it} as a function of (discrete) time t and covariates $\boldsymbol{u}_i$.

Example 6.3: Unemployment duration

Discrete-time survival or duration models are a natural choice for analysing unemployment durations because a new job usually starts on the first day of a month, so that unemployment spells as well as calendar dates are reported in months. Our example is based on a subset of data from the (West) German Federal Employment Office for the years 1980–1995 and on the analysis in Fahrmeir *et al.* (2003). For each individual in the sample, the data provide information about the calendar dates when unemployment starts and ends, the district where the individual worked prior to unemployment, and a number of personal characteristics like age, gender, nationality, education or vocational training, periods for which a claimant receives unemployment benefits, and the economic sector in which the individual works. To adequately use this detailed temporal, spatial and personal information, discrete-time duration models with structured additive predictors including nonlinear effects of duration, calendar time and age, spatial effects as well as linear effects of other covariates will be applied. Men and women are analysed separately, and we only consider persons with transitions back to a full time job. Censoring means that no transition occured within the observation period or that durations were longer than 36 months, which is only the case for a very small fraction of individuals. Sample sizes were approximately 16,000 men and 20,000 women, excluding persons with other transitions, for example to a part-time job. We will report results only for women, see Fahrmeir *et al.* (2003) for the complete results and comparisons with men.

Let t denote duration of unemployment (in months), *date* the date (in months) of calendar time, *age* the age at the beginning of unemployment, and s the labour market region where the unemployed has his home. Furthermore, the following partly time-varying categorical covariates are included in the vector $\boldsymbol{u}$ (in effect coding):

- Nationality N:
  - German
  - foreign (reference category)
- Education E:
  - no vocational training
  - vocational training (reference category)
  - university degree
- Unemployment benefits B_t (in month t of current unemployment period):
  - unemployment insurance benefit (reference category)
  - unemployment assistance benefit
- Previous unemployment periods P_d (in month *date* of calendar time):
  - 0 (reference category)
  - 1 or 2
  - 3 or more
- Economic sectors:
  - agriculture
  - manufacturing (reference category)
  - construction
  - trade
  - transport and communications
  - financial sector
  - service industry
  - non-profit organizations and private households
  - territorial authorities and social insurance △

Example 6.4: Childhood mortality in Nigeria
As a second example on discrete-time survival models, we will analyse the childhood mortality data from Example 6.2. In this case, we will discretize both uncensored and right-censored survival times to months to obtain a data structure that fits the definition of a discrete-time model. To reduce the difficulties with survival times heaped at 18, 24, 36 and 48 months, we restrict all observations to 18 months. This also leads to a significant reduction of the sample size resulting from the data augmentation for the discrete-time analysis. △

6.2 Continuous-time hazard regression

This section considers extensions of the Cox model (6.2) to hazard rate models with more general predictors. Throughout, we will assume that survival or duration times T_i, $i = 1, \ldots, n$, are conditionally independent, given covariates and parameters. Note that this assumption does not necessarily exclude analyses of multivariate, correlated survival or duration times: inclusion of common individual- or group-specific random effects leads to marginally correlated times after integrating out such random effects.

For time-constant covariates $\boldsymbol{u}_i$, $i = 1, \ldots, n$, the *Cox model* specifies the hazard rate of individual i as the product

$$\lambda_i(t) = \lambda(t|\boldsymbol{u}_i) = \lambda_0(t) \exp(\boldsymbol{u}_i'\boldsymbol{\beta}), \tag{6.6}$$

where $\lambda_0(t)$ is the baseline hazard that remains unspecified and is independent of the covariates, so that it depends only on time. The linear predictor $\boldsymbol{u}_i'\boldsymbol{\beta}$ is independent of time and does not contain an intercept term, which is already included in $\lambda_0(t)$. Through the exponential link function, the covariates act multiplicatively on the hazard rate. The Cox model with time constant covariates is also called *proportional hazards (PH) model*, since the ratio of the hazard rates for two individuals with covariate vectors $\boldsymbol{u}_1$ and $\boldsymbol{u}_2$ is given by

$$\frac{\lambda(t|\boldsymbol{u}_1)}{\lambda(t|\boldsymbol{u}_2)} = \exp\left[(\boldsymbol{u}_1 - \boldsymbol{u}_2)'\boldsymbol{\beta}\right].$$

Therefore, the hazard rates for different covariate combinations are proportional and the ratio of two hazard rates is independent of time t. This implicit assumption is likely to fail in complex data examples. It is, however, convenient if primary interest is on estimating the covariate effect $\boldsymbol{\beta}$ while considering the baseline hazard rate $\lambda_0(t)$ as a 'nuisance parameter'. Estimation of $\boldsymbol{\beta}$ can then be based on a partial likelihood instead of the full likelihood of the data. For right-censored survival data $(t_i, d_i, \boldsymbol{u}_i)$, $i = 1, \ldots, n$, it is defined as

$$\mathrm{PL}(\boldsymbol{\beta}) = \prod_{i=1}^{I} \frac{\exp(\boldsymbol{u}_i'\boldsymbol{\beta})}{\sum_{j \in R(t_{(i)})} \exp(\boldsymbol{u}_j'\boldsymbol{\beta})}, \tag{6.7}$$

where $t_{(i)}$, $i = 1, \ldots, I \leq n$ are the ordered survival times $t_{(1)} < \ldots < t_{(i)} < \ldots < t_{(I)}$ of uncensored observations, and $R(t_{(i)})$ is the set of individuals still at risk immediately before $t_{(i)}$, that is, $R(t_{(i)})$ contains the indices of all individuals that are still alive and uncensored immediately before $t_{(i)}$. The partial likelihood can be derived, for example, from a profile likelihood approach, assuming $\lambda_0(t)$ is a step function with jumps at the survival times $t_{(i)}$. Maximizing $\mathrm{PL}(\boldsymbol{\beta})$ gives the MPL estimate $\hat{\boldsymbol{\beta}}$, having the same asymptotic properties as full ML estimators under regularity conditions. The step function for $\lambda_0(t)$ may be estimated in the second phase of the profile likelihood approach, with a correction for tied survival times (Breslow, 1974).

In this section, Bayesian inference for the Cox model and its extensions will usually be based on the full likelihood as derived in the previous section instead of a partial likelihood. One reason is that the Bayesian paradigm requires that a full likelihood is combined with priors to derive a proper full posterior. It seems tempting to work with a partial instead of a full likelihood in a Bayesian analysis of the Cox model. Sinha, Ibrahim and Chen (2003) provide a rigorous justification for Bayesian inference based on the partial likelihood when the prior for the (cumulative) baseline hazard function is specified through a Gamma process. However, the loss of information when working with a 'partial posterior' instead of the full posterior is generally not clear for example when the (log-) baseline hazard rate is modelled through a Gaussian smoothness prior. A second reason is that we will consider extensions of the Cox model where it is not possible or at least difficult to work with a partial likelihood. Two typical situations where such extensions are needed are the following:

- In applications where prediction of survival times is of interest, smooth nonparametric estimation of the baseline hazard, simultaneously with estimation of other effects, is required.
- Effects of some covariates may smoothly vary over time. Including such time-varying effects as well as time-varying covariates will relax the proportional hazards assumption.

Furthermore, extended Cox-type hazard rate models are needed in many applications, such as in our examples, to account for the following issues:

- Effects of continuous covariates as well as interaction between covariates may be of unknown nonlinear form.
- If spatial information is available, explicit modelling of spatial effects as in Chapter 5 will not only account for spatial correlation of survival times but will often be of interest itself.
- Unobserved individual- or cluster-specific heterogeneity may be present and must be considered appropriately.

To take these issues into account within a joint model, we introduce Cox-type hazard regression models.

6.2.1 Observation models, priors and likelihoods

Observation models

Reparametrizing the baseline hazard rate through the log-baseline hazard rate $g_0(t) = \log(\lambda_0(t))$ and incorporating additional terms, we extend the Cox model (6.6) to a semiparametric hazard rate model

$$\lambda_i(t) = \exp(\eta_i(t)), \quad i = 1, \ldots, n, \tag{6.8}$$

with structured additive predictor

$$\eta_i(t) = \boldsymbol{u}_i'(t)\boldsymbol{\beta} + g_0(t) + \sum_{l=1}^{m} g_l(t)v_{il}(t) + \sum_{j=1}^{p} f_j(x_{ij}(t)) + f_{\text{spat}}(s_i) + \alpha_{g_i}. \tag{6.9}$$

Including the baseline hazard in form of the log-baseline effect $g_0(t)$ into the predictor will allow us to specify $g_0(t)$ as a P-spline without having to deal with the non-negativity restriction $\lambda_0(t) \geq 0$. The functions $g_l(t)$ are time-varying effects of covariates $v_l(t)$, while $f_j(x_j(t))$ and $f_{\text{spat}}(s)$ are nonlinear effects of continuous covariates and spatial effects as in geoadditive regression models of Chapter 5. All covariates are allowed to be (piecewise constant) time-varying. The unit- or group-specific random effects α_g, $g = 1, \ldots, G$, account for unobserved heterogeneity or 'frailty', with $\alpha_{g_i} = \alpha_g$ if individual i belongs to group g. For $G = n$, we obtain individual-specific random effects. For $G < n$, α_g might be the effect of centre g in a multicentre study or the unstructured (i.e. uncorrelated) effect of an area. Several extensions of the models, such as inclusion of interaction terms or random slopes are possible but we omit this here to simplify the presentation. As in previous structured additive regression models, all unknown functions are centred about zero and an intercept is included in the linear parametric part $\boldsymbol{u}_i'(t)\boldsymbol{\beta}$.

Priors

Priors for the different components in the predictor are specified as in Chapters 4 and 5: As long as $\boldsymbol{\beta}$ is low-dimensional, we will assume a flat prior $p(\boldsymbol{\beta}) \propto \text{const}$ or a weakly informative Gaussian prior. For high-dimensional $\boldsymbol{\beta}$ we may assume shrinkage priors, such as the Bayesian ridge or lasso, or spike and slab priors with implicit variable selection as considered in Sections 4.4.2 (page 245) and 4.5.3 (page 290).

Nonparametric effects f_j as well as time-varying effects g_0 and g_l are modelled as penalized splines with (conditionally) Gaussian smoothness priors for basis function coefficients. In principle, representations or approximations through linear combinations of other basis functions are possible.

Priors for structured spatial effects f_{spat} are specified through Gaussian Markov random fields or stationary Gaussian random fields as in Chapter 5. For the vector $\boldsymbol{\alpha}$ of i. i. d. random effects α_g, $g = 1, \ldots, G$, we assume a Gaussian prior as in Chapter 3.3.

As in previous chapters, all different terms in (6.9) can be cast into one generic form. We represent the log-baseline hazard through

$$g_0(t) = \boldsymbol{x}_0'(t)\boldsymbol{\gamma}_0,$$

where $\boldsymbol{x}_0'(t) = (B_1(t), \ldots, B_K(t))$ is a vector of basis functions evaluated at t and $\boldsymbol{\gamma}_0$ is the vector of the basis function coefficients. For time-varying effects, we similarly assume

$$g_l(t) = \boldsymbol{x}_l^{*\prime}(t)\boldsymbol{\gamma}_l, \quad l = 1, \ldots, m,$$

with $\boldsymbol{x}_l^*(t)$ as a vector of basis functions—possibly the same as in $\boldsymbol{x}_0(t)$—and $\boldsymbol{\gamma}_l$ as the corresponding vector of basis function coefficients. Defining $\boldsymbol{x}_{il}'(t) = \boldsymbol{x}_l^{*\prime}(t)v_{il}(t)$, we obtain the generic representation

$$g_l(t)v_{il}(t) = \boldsymbol{x}'_{il}(t)\boldsymbol{\gamma}_l.$$

In complete analogy, nonlinear effects of possibly time-dependent covariates $x_j(t)$, $j = 1, \dots, p$, are expressed as

$$f_j(x_{ij}(t)) = \boldsymbol{x}'_{ij}(t)\boldsymbol{\gamma}_j.$$

Because we always assume that spatial effects $f_{\text{spat}}(s_i)$ and random effects α_{g_i} are time-independent, design vectors $\boldsymbol{x}_{i,\text{spat}}$ and $\boldsymbol{x}_{i,\text{ran}}$ are time-independent 0-1 incidence vectors. After reindexing, the predictor (6.9) can be rewritten in generic notation as

$$\eta_i(t) = \boldsymbol{u}'_i(t)\boldsymbol{\beta} + \boldsymbol{x}'_0(t)\boldsymbol{\gamma}_0 + \boldsymbol{x}'_{i1}(t)\boldsymbol{\gamma}_1 + \ldots + \boldsymbol{x}'_{ir}(t)\boldsymbol{\gamma}_r. \tag{6.10}$$

If present, spatial effects and random effects are identified with the last two terms, for example $\boldsymbol{\alpha} = \boldsymbol{\gamma}_r$, and the corresponding design vectors do not depend on t.

Priors for $\boldsymbol{\gamma}_l$, $l = 0, \dots, r$, are (possibly partially improper) Gaussian

$$p(\boldsymbol{\gamma}_l|\tau_l^2) \propto \frac{1}{(\tau_l^2)^{\text{rank}(\boldsymbol{K}_l)/2}} \exp\left(-\frac{1}{2\tau_l^2}\boldsymbol{\gamma}'_l\boldsymbol{K}_l\boldsymbol{\gamma}_l\right),$$

with appropriately defined precision matrices, for example with $\boldsymbol{K}_l = \boldsymbol{D}'\boldsymbol{D}$ for P-splines as in (4.8) (page 185), or $\boldsymbol{K}_r = \boldsymbol{I}$ for the vector $\boldsymbol{\gamma}_r = \boldsymbol{\alpha}$ of i.i.d. random effects.

As in previous chapters, unknown smoothing or variance parameters τ_l^2 are considered as unknown constants for empirical Bayes (EB) inference. For full Bayes (FB) inference, we assume (possibly improper) inverse gamma priors as in Chapter 4 (see Section 4.1.3).

Likelihoods

Let $\boldsymbol{\beta}$ and $\boldsymbol{\gamma} = (\boldsymbol{\gamma}'_0, \dots, \boldsymbol{\gamma}'_r)'$ denote the vectors of all regression parameters. Under the assumptions of conditional independence and noninformative censoring, the likelihood is the product $L(\boldsymbol{\beta}, \boldsymbol{\gamma}) = \prod_i L_i(\boldsymbol{\beta}, \boldsymbol{\gamma})$ of individual likelihood contributions. The likelihood contributions $L_i(\boldsymbol{\beta}, \boldsymbol{\gamma})$ are obtained by inserting $\lambda_i(t) = \exp(\eta_i(t))$ into the expressions (6.3), (6.4) and (6.5) of Section 6.1 for the different censoring schemes. Apart from special cases, however, some numerical integration technique has to be employed, since integrals of the form

$$\int_l^u \exp(\eta_i(t))dt$$

appearing in these expressions can in general not be solved analytically. In our experience, the simple trapezoidal rule is sufficiently accurate. It approximates an integral of the form

$$\int_l^u g(t)dt$$

by the sum

$$\sum_{k=1}^{K}(t_k - t_{k-1})\frac{g(t_k) + g(t_{k-1})}{2}$$

based on integration points $l = t_0 < \ldots < t_K = u$. This corresponds to approximating g with a piecewise linear function. The number of knots should not be too small, and locations of the knots can be based on an equidistant grid or on quantiles of observed survival times. Time-varying covariates can be easily incorporated by splitting up integrals into sums of integrals over piecewise constant parts of a covariate.

A special case, where numerical integration can be avoided, is a piecewise exponential model for right-censored observations, see the following section.

6.2.2 Full Bayes inference for right-censored observations

For noninformative right-censoring, the log-likelihood for regression parameters $\boldsymbol{\beta}, \boldsymbol{\gamma}$ is given by

$$l(\boldsymbol{\beta}, \boldsymbol{\gamma}) = \sum_{i=1}^{n} d_i \eta_i(t_i) - \sum_{i=1}^{n} \int_0^{t_i} \exp(\eta_i(t))dt = \sum_{i=1}^{n} \left[d_i \eta_i(t_i) - \Lambda(t_i)\right],$$

where d_i are the censoring indicators, $t_i = \min(T_i, C_i)$ are right-censored survival times, and $\eta_i(t)$ is the generic predictor (6.10). As already pointed out, the integrals for $\Lambda(t_i)$, as well as others appearing in expressions for first and second derivatives, can in general not be solved analytically.

Piecewise exponential model

The piecewise exponential model, originally introduced for flexible estimation of the baseline hazard $\lambda_0(t)$ in the Cox model (see for example Fahrmeir and Tutz, 2001, Ch. 9.1; or Ibrahim, Chen and Sinha, 2001, Section 3.1) assumes a special form for time-varying quantities which allows us to circumvent numerical integrations. Similarly as for discrete-time survival data, suppose that the time axis is divided into successive intervals

$$[\kappa_0, \kappa_1), [\kappa_1, \kappa_2), \ldots, [\kappa_{q-1}, \kappa_q), [\kappa_q, \infty).$$

The knots $\kappa_0, \ldots, \kappa_q$ may be equidistant or based on quantiles of survival times. In the original piecewise exponential model, $\lambda_0(t)$ is assumed to be piecewise constant on that grid, that is, $\lambda_0(t) = \lambda_{0k}$ for $t \in [\kappa_{k-1}, \kappa_k)$. Reparametrizing through $\gamma_{0k} = \log(\lambda_{0k})$ allows us to avoid the non-negativity restriction $\lambda_{0k} \geq 0$

and to estimate the log-baseline $g_0(t) = \log(\lambda_0(t))$ as a step function jointly with the regression parameters in the Cox model.

More generally, assuming that all time-varying quantities are piecewise constant on the intervals, we can rewrite the predictor (6.9) as

$$\eta_i(t) = \eta_{ik} = \boldsymbol{u}_{ik}'\boldsymbol{\beta} + \gamma_{0k} + \sum_{l=1}^{m} \gamma_{lk} v_{ilk}(t) + \sum_{j=1}^{p} f_j(x_{ijk}) + f_{\text{spat}}(s_i) + \alpha_{g_i}$$

where $\eta_i(t) = \eta_{ik}$, $\boldsymbol{u}_i(t) = \boldsymbol{u}_{ik}$, $g_l(t) = \gamma_{lk}$ for $t \in [\kappa_{k-1}, \kappa_k)$, $k = 1, \ldots, q$, and similar definitions hold for the other time-varying quantities. Defining event indicators

$$y_{ik} = \begin{cases} 1, & \text{individual } i \text{ has an event in } [\kappa_{k-1}, \kappa_k), \\ 0, & \text{individual } i \text{ survives or is censored in } [\kappa_{k-1}, \kappa_k), \end{cases}$$

the log-likelihood can be written as

$$\sum_{i=1}^{n} d_i \eta_i(t) - \sum_{i=1}^{n} \Lambda(t_i) = \sum_{k=1}^{q} \sum_{i \in R_k} (y_{ik}\eta_{ik} - \Delta_{ik} \exp(\eta_{ik}))$$

where R_k is the risk set for $[\kappa_{k-1}, \kappa_k)$, i.e. R_k contains all individuals that had no event or were not censored prior to κ_{k-1}. The Δ_{ik} are offsets defined by

$$\Delta_{ik} = \max\left\{0, \min\left\{\kappa_k - \kappa_{k-1}, t_i - \kappa_{k-1}\right\}\right\}.$$

The likelihood is formally equivalent to the likelihood of a loglinear Poisson model with responses y_{ik} and predictors η_{ik}, including an additional offset term Δ_{ik}. Therefore a piecewise exponential model can be estimated using methodology described in previous chapters for exponential family models after appropriate data augmentation. To avoid too rough estimates for the step function, we use a comparably large number q for knots, combined with random-walk priors for the sequences $\{\gamma_{lk}, k = 1, \ldots, q\}$, $l = 0, \ldots, m$, as introduced in Section 2.1.1 for smoothing of time series. As the definition of (penalized) B-splines in Section 2.2.1 shows, this is equivalent to P-spline smoothing with B-splines of degree zero.

Models with general structured additive predictor

One drawback of the piecewise exponential model is that the data augmentation trick strongly increases the effective number of observations. This may lead to long execution times as well as to numerical problems due to the large matrices involved. A second drawback is that the likelihood cannot be transformed for other censoring schemes than right-censoring. Last but not least, step functions are still an unsmooth approximation of underlying smooth time-varying effects.

These disadvantages can be avoided by using P-splines of higher order for smoothing, and by evaluating the likelihoods using some numerical integration technique such as the trapezoidal rule.

In addition to likelihoods, score functions and observed information matrices corresponding to the regression vectors $\boldsymbol{\beta}, \boldsymbol{\gamma}_0, \ldots, \boldsymbol{\gamma}_r$ in generic model (6.9) will be needed for defining Gaussian proposals for MH algorithms that are closely related to MH algorithms with IWLS proposals in Section 4.2.2. Individual score function contributions are obtained as first derivatives of individual log-likelihoods

$$l_i(\boldsymbol{\beta}, \boldsymbol{\gamma}) = d_i \eta_i(t_i) - \int_0^{t_i} \exp(\eta_i(t)) dt.$$

Inserting the generic predictor (6.10) and differentiating gives the score function contributions

$$\begin{aligned} \boldsymbol{s}_{i,\boldsymbol{\beta}}(\boldsymbol{\beta}, \boldsymbol{\gamma}) &= \frac{\partial l_i(\boldsymbol{\beta}, \boldsymbol{\gamma})}{\partial \boldsymbol{\beta}} = d_i \boldsymbol{u}_i(t_i) - \int_0^{t_i} \boldsymbol{u}_i(t) \exp(\eta_i(t)) dt, \\ \boldsymbol{s}_{il}(\boldsymbol{\beta}, \boldsymbol{\gamma}) &= \frac{\partial l_i(\boldsymbol{\beta}, \boldsymbol{\gamma})}{\partial \boldsymbol{\gamma}_l} = d_i \boldsymbol{x}_{il}(t_i) - \int_0^{t_i} \boldsymbol{x}_{il}(t) \exp(\eta_i(t)) dt, \end{aligned} \tag{6.11}$$

and the score functions

$$\begin{aligned} \boldsymbol{s}_{\boldsymbol{\beta}}(\boldsymbol{\beta}, \boldsymbol{\gamma}) &= \sum_{i=1}^n \boldsymbol{s}_{i,\boldsymbol{\beta}}(\boldsymbol{\beta}, \boldsymbol{\gamma}), \\ \boldsymbol{s}_l(\boldsymbol{\beta}, \boldsymbol{\gamma}) &= \sum_{i=1}^n \boldsymbol{s}_{il}(\boldsymbol{\beta}, \boldsymbol{\gamma}), \quad l = 0, \ldots, r. \end{aligned}$$

Observed information matrices $\boldsymbol{H}_{\boldsymbol{\beta}}$ and $\boldsymbol{H}_l$ are obtained by summing up individual (negative) second derivatives

$$\begin{aligned} \boldsymbol{H}_{i,\boldsymbol{\beta}}(\boldsymbol{\beta}, \boldsymbol{\gamma}) &= -\frac{\partial^2 l_i(\boldsymbol{\beta}, \boldsymbol{\gamma})}{\partial \boldsymbol{\beta} \partial \boldsymbol{\beta}'} = \int_0^{t_i} \boldsymbol{u}_i(t) \boldsymbol{u}_i'(t) \exp(\eta_i(t)) dt, \\ \boldsymbol{H}_{il}(\boldsymbol{\beta}, \boldsymbol{\gamma}) &= -\frac{\partial^2 l_i(\boldsymbol{\beta}, \boldsymbol{\gamma})}{\partial \boldsymbol{\gamma}_l \partial \boldsymbol{\gamma}_l'} = \int_0^{t_i} \boldsymbol{x}_{il}(t) \boldsymbol{x}_{il}'(t) \exp(\eta_i(t)) dt. \end{aligned} \tag{6.12}$$

Note that these expressions simplify considerably for models with time-constant covariates $\boldsymbol{u}_i(t) \equiv \boldsymbol{u}_i$, $\boldsymbol{x}_l(t) = \boldsymbol{x}_l$ and without time-varying effects $g_l(t)$, $l = 1, \ldots, m$. Then the baseline function $g_0(t)$ is the only time-varying quantity, so that $\eta_i(t) = g_0(t) + \tilde{\eta}_i$, where $\tilde{\eta}_i$ is the time-constant part of the predictor. The integrals in (6.11), (6.12) simplify to

$$\boldsymbol{u}_i \Lambda(t_i), \quad \boldsymbol{x}_{il} \Lambda(t_i), \quad \boldsymbol{u}_i \boldsymbol{u}_i' \Lambda(t_i), \quad \boldsymbol{x}_{il} \boldsymbol{x}_{il}' \Lambda(t_i),$$

with cumulative hazard rate

$$\Lambda(t_i) = \int_0^{t_i} \exp(g_0(t) + \tilde{\eta}_i)dt = \Lambda_0(t_i)\exp(\tilde{\eta}_i). \tag{6.13}$$

Then only the cumulative baseline hazard rate $\Lambda_0(t)$ has to be computed by numerical integration. Similar simplifications are possible if time-varying quantities are piecewise constant, that is, step functions, with only a small number of different function values. Otherwise, computational efforts in evaluating the integrals can strongly increase.

Let $\boldsymbol{y} = (t_i, d_i, i = 1, \ldots, n)$ denote the vector of observed survival times and censoring indicators, $\boldsymbol{\tau} = (\tau_0^2, \ldots, \tau_r^2)$ the vector of variance parameters, $p(\boldsymbol{\beta}, \boldsymbol{\gamma}, \boldsymbol{\tau}|\boldsymbol{y})$ the posterior, and $p(\boldsymbol{\beta}|\cdot)$, $p(\boldsymbol{\gamma}_l|\cdot)$, $p(\tau_l^2|\cdot)$ the corresponding full conditionals. Suppose we want to update $\boldsymbol{\gamma}_l$ with current state $\boldsymbol{\gamma}_l^c$ of the Markov chain. Then the full conditional is

$$p(\boldsymbol{\gamma}_l|\cdot) \propto \exp\left(l(\boldsymbol{\beta}, \boldsymbol{\gamma}) - \frac{1}{2\tau_l^2}\boldsymbol{\gamma}_l'\boldsymbol{K}_l\boldsymbol{\gamma}_l\right).$$

To simplify notation, let $l(\boldsymbol{\eta}^c)$ denote the log-likelihood, $\boldsymbol{s}_l(\boldsymbol{\eta}^c)$ the score function, and $\boldsymbol{H}_l(\boldsymbol{\eta}^c)$ the observed information matrix evaluated at current states $\boldsymbol{\beta}^c, \boldsymbol{\gamma}^c$ of the chain, and $l(\boldsymbol{\eta}^p)$ the log-likelihood where $\boldsymbol{\gamma}_l^c$ is replaced through a proposed value $\boldsymbol{\gamma}_l^p$ while holding other parameters fixed at their current values. Following the idea of IWLS proposals as considered in Section 4.2.2, the log-likelihood $l(\boldsymbol{\eta}^p)$ is approximated through a second-order Taylor expansion around $\boldsymbol{\eta}^c$:

$$l(\boldsymbol{\eta}^p) \approx l(\boldsymbol{\eta}^c) + (\boldsymbol{\gamma}_l^p - \boldsymbol{\gamma}_l^c)'\boldsymbol{s}_l(\boldsymbol{\eta}^c) - \frac{1}{2}(\boldsymbol{\gamma}_l^p - \boldsymbol{\gamma}_l^c)'\boldsymbol{H}_l(\boldsymbol{\eta}^c)(\boldsymbol{\gamma}_l^p - \boldsymbol{\gamma}_l^c).$$

Inserting this approximation into the full conditional and completing squares, leads to the Gaussian approximation

$$p(\boldsymbol{\gamma}_l^p|\cdot) \approx \varphi(\boldsymbol{\gamma}_l^p; \boldsymbol{\mu}_l^c, \boldsymbol{\Sigma}_l^c), \tag{6.14}$$

with covariance matrix and mean

$$\boldsymbol{\Sigma}_l^c = \left(\boldsymbol{H}_l(\boldsymbol{\eta}^c) + \frac{\boldsymbol{K}_l}{(\tau_l^2)^c}\right)^{-1}, \quad \boldsymbol{\mu}_l^c = \boldsymbol{\Sigma}_l^c\left[\boldsymbol{s}_l(\boldsymbol{\eta}^c) + \boldsymbol{H}_l(\boldsymbol{\eta}^c)\boldsymbol{\gamma}_l^c\right].$$

Then a new value $\boldsymbol{\gamma}_l^p$ is proposed by drawing a random number from the Gaussian proposal density $\varphi(\boldsymbol{\gamma}_l; \boldsymbol{\mu}_l^c, \boldsymbol{\Sigma}_l^c)$. It is accepted with probability

$$\alpha(\boldsymbol{\gamma}_l^c, \boldsymbol{\gamma}_l^p) = \frac{p(\boldsymbol{y}|\boldsymbol{\eta}^p)p(\boldsymbol{\gamma}_l^p|(\tau_l^2)^c)\varphi(\boldsymbol{\gamma}_l^c; \boldsymbol{\mu}_l^p, \boldsymbol{\Sigma}_l^p)}{p(\boldsymbol{y}|\boldsymbol{\eta}^c)p(\boldsymbol{\gamma}_l^c|(\tau_l^2)^c)\varphi(\boldsymbol{\gamma}_l^p; \boldsymbol{\mu}_l^c, \boldsymbol{\Sigma}_l^c)}, \tag{6.15}$$

compare (4.29) in Section 4.2.2 (page 220) The observation likelihood $p(\boldsymbol{y}|\boldsymbol{\eta}^c) = L(\boldsymbol{\eta}^c)$, the prior for $\boldsymbol{\gamma}_l$ and the proposal density in the denominator are evaluated at current values of parameters and predictors. In the numerator, $\boldsymbol{\gamma}_l^c$ is replaced through the proposal value $\boldsymbol{\gamma}_l^p$, holding the remaining parameters at their current values. The MH step for updating a current value $\boldsymbol{\beta}^c$ of the vector $\boldsymbol{\beta}$ of linear effects is quite similar: For a Gaussian prior $\boldsymbol{\beta} \sim \mathrm{N}(\boldsymbol{0}, \boldsymbol{C})$, the proposal density is $\varphi(\boldsymbol{\beta}^p|\boldsymbol{\mu}_{\boldsymbol{\beta}}^c, \boldsymbol{\Sigma}_{\boldsymbol{\beta}}^c)$, with

$$\boldsymbol{\Sigma}_{\boldsymbol{\beta}}^c = (\boldsymbol{H}_{\boldsymbol{\beta}}(\boldsymbol{\eta}^c) + \boldsymbol{C}^{-1})^{-1}, \quad \boldsymbol{\mu}_{\boldsymbol{\beta}}^c = \boldsymbol{\Sigma}_{\boldsymbol{\beta}}^c \left[\boldsymbol{s}_{\boldsymbol{\beta}}(\boldsymbol{\eta}^c) + \boldsymbol{H}_l(\boldsymbol{\eta}^c)\boldsymbol{\beta}_l^c\right],$$

and a flat prior $p(\boldsymbol{\beta}) \propto \text{const}$ corresponds to the limiting case $\boldsymbol{C}^{-1} = \boldsymbol{0}$.

For time-constant covariates $\boldsymbol{u}$, nonlinear effects $f_j(x_j)$ with time constant covariates x_j, spatial effects and random effects, the Gaussian proposal densities for corresponding parameters can be rewritten in IWLS form in similar fashion as for exponential family models in previous chapters. In generic notation, let $\boldsymbol{X}_l$ denote the corresponding design matrix with time-constant design vectors $\boldsymbol{x}_{il}'$, $i = 1, \ldots, n$, as rows, and $\boldsymbol{\gamma}_l$ the corresponding parameter. As already described, the integrals appearing in the score functions (6.11) and information matrices (6.12) simplify to $\boldsymbol{x}_{il}\Lambda(t_i)$ and $\boldsymbol{x}_{il}\boldsymbol{x}_{il}'\Lambda(t_i)$. Summing up, the information matrix simplifies to

$$\boldsymbol{H}_l(\boldsymbol{\beta}, \boldsymbol{\gamma}) = \boldsymbol{X}_l'\boldsymbol{W}\boldsymbol{X}_l,$$

where $\boldsymbol{W} = \boldsymbol{W}(\boldsymbol{\beta}, \boldsymbol{\gamma}) = \text{diag}(w_1, \ldots, w_i, \ldots, w_n)$ with weights $w_i = \Lambda(t_i)$. The score function can be written in the form

$$\boldsymbol{s}_l(\boldsymbol{\beta}, \boldsymbol{\gamma}) = \boldsymbol{X}_l'\boldsymbol{d} - \boldsymbol{X}_l'\boldsymbol{w},$$

where $\boldsymbol{d} = (d_1, \ldots, d_n)'$ is the vector of censoring indicators and $\boldsymbol{w} = (w_1, \ldots, w_i, \ldots, w_n)'$ the vector of weights. After some algebraic manipulations, this results in the IWLS form

$$\boldsymbol{\Sigma}_l^c = \left(\boldsymbol{X}_l'\boldsymbol{W}^c\boldsymbol{X}_l + \frac{\boldsymbol{K}_l}{(\tau_l^2)^c}\right)^{-1}, \quad \boldsymbol{\mu}_l^c = \boldsymbol{\Sigma}_l^c\boldsymbol{X}_l'\boldsymbol{W}^c(\tilde{\boldsymbol{y}}^c - \boldsymbol{\eta}_{-l}^c), \tag{6.16}$$

where $\boldsymbol{\eta}_{-l}^c$ is the current predictor after removing $\boldsymbol{X}_l\boldsymbol{\gamma}_l^c$, and the current working observations $\tilde{\boldsymbol{y}}^c = (\tilde{y}_1^c, \ldots, \tilde{y}_i^c, \ldots, \tilde{y}_n^c)'$ are given by

$$\tilde{y}_i^c = \eta_i^c(t_i) + \frac{d_i}{w_i^c} - 1.$$

As already discussed, computation of the weights $w_i = \Lambda(t_i)$ greatly simplifies for models where the log-baseline hazard $g_0(t)$ is the only time-varying quantity. In this case w_i is given by (6.13).

For regression coefficients corresponding to time-dependent effects or time-dependent covariates, a computationally attractive alternative to the Gaussian proposals (6.14) are conditional prior proposals, see Knorr-Held (2000) and Fahrmeir and Lang (2001*a*) for details, requiring only likelihoods but no derivatives.

With (possibly improper) inverse gamma priors for variance parameters, full conditionals are (proper) inverse gamma densities as in previous chapters.

In summary, updating steps resemble closely the corresponding steps of the MH algorithm with IWLS proposals for full Bayes inference in semiparametric (mixed) models (Section 4.4.2) for exponential family responses.

Algorithm 6.1: MH algorithm for right-censored survival data

i. *Initialization:* as in Section 4.2.2.

ii. *Update basis coefficients $\boldsymbol{\gamma}_l$ for time-constant terms:*
   - Compute the likelihood $p(\boldsymbol{y}|\boldsymbol{\eta}^c)$, using the trapezoidal rule to evaluate the cumulative hazard function $\Lambda^c(t_i)$.
   - Draw a proposal $\boldsymbol{\gamma}_l^p$ from the Gaussian proposal density with mean $\boldsymbol{\mu}_l^c$ and covariance matrix $\boldsymbol{\Sigma}_l^c$ given in (6.16).
   - Compute the likelihood $p(\boldsymbol{y}|\boldsymbol{\eta}^p)$.
   - Compute the acceptance probability (6.15) and accept $\boldsymbol{\gamma}_l^p$ with this probability.

iii. *Update fixed effects parameters $\boldsymbol{\beta}$* by similar steps.

iv. *Update basis coefficients $\boldsymbol{\gamma}_l$ for the baseline hazard $g_0(t)$ and time-varying effects $g_1(t), \ldots, g_m(t)$:*
   - Proceed either as in step ii., but drawing a proposal $\boldsymbol{\gamma}_l^p$ from (6.14), or
   - update $\boldsymbol{\gamma}_l^c$ through a MH step with conditional prior proposals as in Fahrmeir and Lang (2001*a*) for GAMMs, requiring only evaluation of likelihoods.

v. *Update variance parameters τ_l^2:* Draw from inverse gamma full conditionals given in (4.23), page 215.

Remarks:

1. If the predictor contains a Gaussian frailty term α_g, for example for observations coming from clusters $g = 1, \ldots, G$, then proceed as in steps iv and vi of the MH-IWLS algorithm of Section 4.2.2.
2. To include shrinkage or variable selection for regression parameters β_l, the algorithm can be extended in complete analogy to Section 4.4.2, specifying ridge, lasso or other shrinkage priors, and to Section 4.5.3, imposing spike and slab priors, in particular the NMIG variance for $\beta_l|\tau_l^2$. As in these sections, only full conditionals for regression parameters and parameters related to shrinkage or spike and slab priors have to be modified with the formulae given there. The remaining full conditionals remain completely untouched.

Example 6.5: Coronary artery bypass graft

We illustrate full Bayesian inference for both piecewise exponential models and models based on P-splines with an application analysing waiting times to coronary artery bypass graft (CABG) in London and Essex (Example 1.5).

Information on waiting times (in days), the area of residence (in terms of one of 488 electoral wards) and some further individual-specific covariates are available for a total of 3,015 patients with definite coronary artery disease who were referred to a single cardiothoracic unit from five contiguous health authorities. Given covariate information includes gender of the patient, age in years, and the number of diseased vessels (1, 2 or 3). The data were originally analysed by Crook, Knorr-Held and Hemingway (2003) based on discrete-time methodology and Hennerfeind, Brezger and Fahrmeir (2006) using geoadditive hazard regression.

Our basic model equation is

$$\eta_i(t) = \beta_0 + \beta_1 \mathrm{sex}_i + g_0(t) + \mathrm{DV}_{i2} g_1(t) + \mathrm{DV}_{i3} g_2(t) + f_1(\mathrm{age}_i) + f_{\mathrm{spat}}(s_i) + \alpha_{s_i}$$

where $g_0(t)$ represents the log-baseline, $g_1(t)$ and $g_2(t)$ are time-varying effects corresponding to the indicators DV_2 and DV_3 specifying the number of diseased vessels (two for DV_2 and three for DV_3), $f_1(\mathrm{age})$ is a nonparametric effect of age, $f_{\mathrm{spat}}(s_i)$ and α_{s_i} correspond to spatially structured and unstructured effects included based on a Markov random field and an i.i.d. Gaussian random effect, respectively, and β_1 is the gender difference.

All time-dependent quantities were approximated based on an equidistant grid with grid length of 50 days for the estimation of a piecewise exponential model. Alternatively, cubic P-splines with 20 inner knots and second-order random-walk penalty were assigned to the log-baseline and the time-varying effects to allow for a full likelihood approach with smooth time-dependent effects. In any case, the age effect was included as an additional cubic P-spline with 20 inner knots and second-order random-walk penalty.

Figure 6.6 shows posterior mean estimates for the nonparametric effects arising from full Bayes inference. Posterior mean estimates obtained with the penalized spline specification for the time-dependent effects are shown as solid black lines with 95% pointwise credible intervals indicated as shaded areas around the point estimates. Posterior means from the piecewise exponential model are superimposed as dashed black lines. Obviously, results are visually quite close to each other, at least for the effect of age and the log-baseline $g_0(t)$, if ignoring the principle difference between the smooth function and the step function. For the time-varying effects, differences are somewhat more expressed with less variability for results from the piecewise exponential model. Still, effects are comparable to each other, in particular if uncertainty in the point estimates is taken into account.

The log-baseline effect in Fig. 6.6 shows an initially high, but strongly decreasing likelihood of CABG immediately after diagnosis, followed by a slow increase between 150 and 450 days. Later, the likelihood of being operated decreases. A higher number of diseased vessels leads to a significant shift upwards of the intensity such that the general likelihood of a surgery increases. Moreover, there is an additional shift in the peak observed at about 450 days. For longer waiting times, uncertainty increases and it is hard to decide whether there still is a difference between patients with different numbers of diseased vessels. The effect

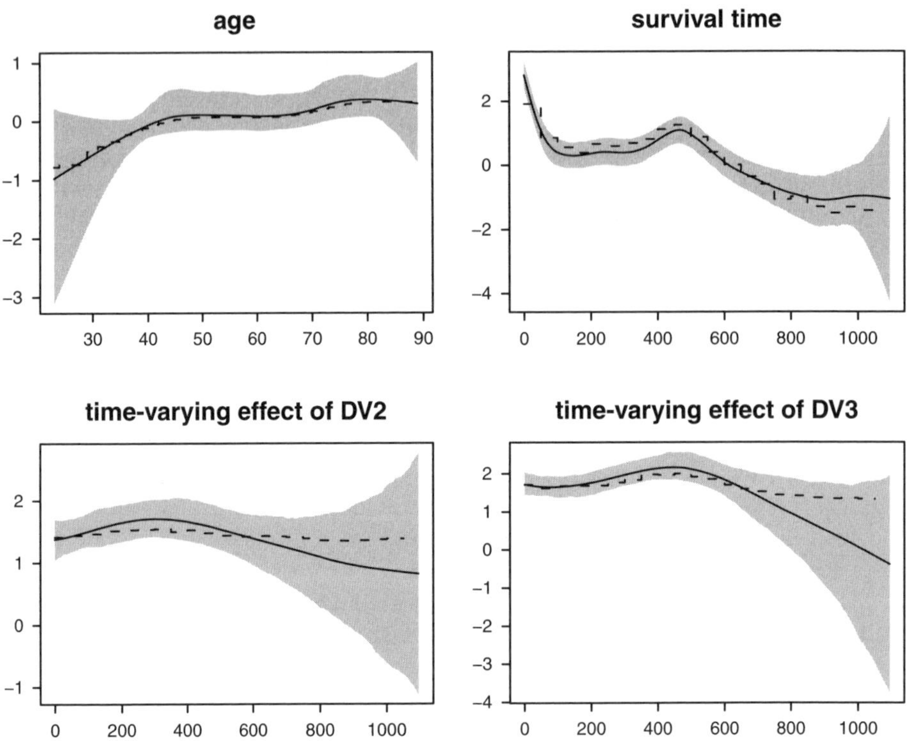

Fig. 6.6 Coronary artery bypass graft: Posterior mean estimates for nonparametric effects together with 95% pointwise credible intervals. Posterior means from a piecewise exponential model are superimposed as dashed black lines.

of age is almost constant between 40 and 80 years and does not seem to have significant effect on the waiting time.

Figure 6.7 shows posterior mean estimates for the structured and unstructured spatial effects as well as 80% posterior probabilities that give an impression of the spatially varying likelihood of CABG, with light (dark) areas indicating a decreased (increased) effect. Since results from the piecewise exponential model and the P-spline model are visually indistinguishable, we only show results from the P-spline model. Areas with increased likelihoods are Chelmsford and Malden in North Essex, whereas in areas around Harlow in North Essex and Walthamstow and Chingford in northeast London likelihoods are lower, that means patients have to wait longer for surgery. No pointwise significant effects can be identified for the unstructured spatial effect. △

Example 6.6: Primary biliary cirrhosis of the liver

To illustrate the inclusion of regularization priors in Bayesian survival time regression, we consider a data set collected between 1974 and 1984 at Mayo

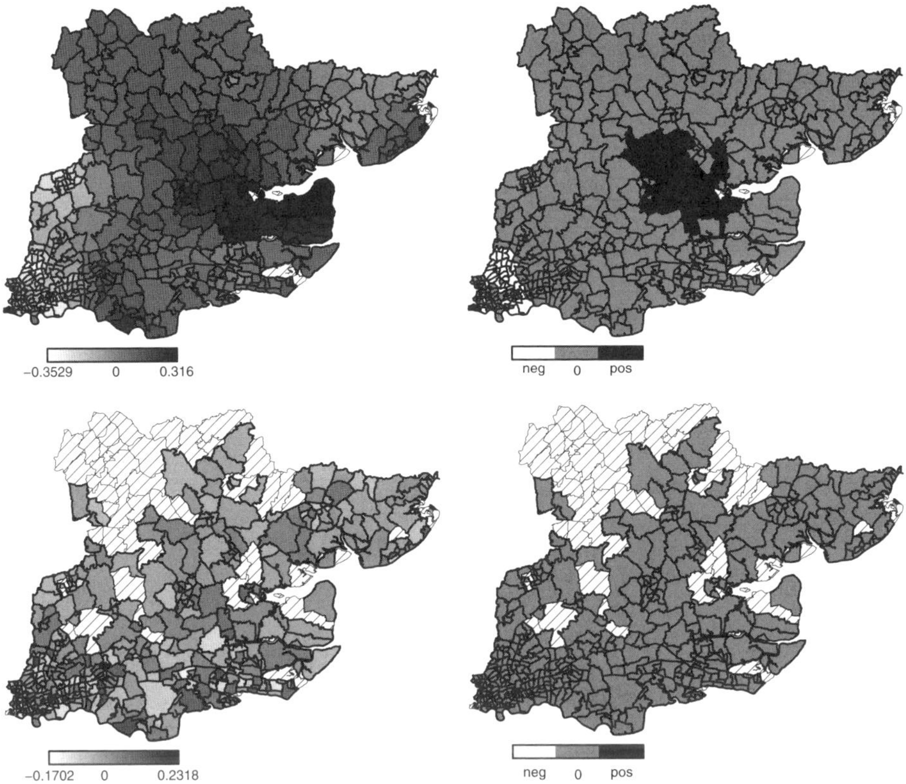

Fig. 6.7 Coronary artery bypass graft: Posterior mean estimates (left) for structured (top) and unstructured (bottom) spatial effects together with 80% pointwise posterior probabilities (right).

Clinic in a trial on primary biliary cirrhosis (PBC) of the liver (see Therneau and Grambsch, 2000, and Tibshirani, 1997, for previous analyses of this data set). PBC is an autoimmune disease of the liver marked by a progressive destruction of the small bile ducts that leads to scarring, fibrosis, cirrhosis and ultimately liver failure and death of the patient. After excluding observations with incomplete information, a total of 276 observations remain for the analysis with a rate of 58.4% for right-censoring. We consider the following variables in our analysis:

- age: age in years,
- alb: albumin in gm/dl,
- alkphos: alkaline phosphatase in U/litre,
- ascites: presence of ascites, 0 = no, 1 = yes,
- bili: serum bilirubin in mg/dl,
- chol: serum cholesterol in mg/dl,

- copper: urine copper in ug/day,
- edtrt: presence of edema, 0.0 = no edema and no diuretic therapy for edema, 0.5 = edema present without diuretics or edema resolved by diuretics, 1.0 = edema despite diuretic therapy,
- hepmeg: presence of hepatomegaly, 0 = no, 1 = yes,
- platelet: platelets per cubic ml/1000,
- protime: standardized blood clotting time, prothrombin time in seconds,
- sex: 0 = male, 1 = female,
- sgot: liver enzyme SGOT in U/ml,
- spiders: presence of spiders, 0 = no, 1 = yes,
- stage: histologic stage of disease,
- trig: triglicerides in mg/dl,
- trt: treatment/drug, 1 = D-penicillamine, 2 = Placebo.

All covariates were standardized to have zero mean and unit variance and were then fed into a hazard regression model

$$\lambda_i(t) = \exp(g_0(t) + \boldsymbol{x}_i'\boldsymbol{\beta})$$

where $\boldsymbol{x}_i$ is the covariate vector and $\boldsymbol{\beta}$ denotes the corresponding regression coefficients. We considered a standard Cox model based on the partial likelihood as well as Bayesian hazard regression models with lasso or normal mixture of inverse Gamma (NMIG) priors for the regression coefficients (see Sections 4.4.2 and 4.5.3 for details). For the latter two models, the baseline hazard function $g_0(t)$ was modelled as a cubic penalized spline with second order difference penalty and 20 inner knots such that inferences can be based on the full instead of the partial likelihood.

The resulting coefficient estimates together with one standard deviation uncertainty intervals are shown in Fig. 6.8 as solid lines. Since the number of regression coefficients as compared to the sample size is not very large, the shrinkage effect induced by the regularization priors is not terribly large but still visible for most of the coefficients. Since the Bayesian lasso as well as the NMIG prior do not have built in variable selection properties, we performed variable selection based on the one standard deviation intervals (for the lasso) and the inclusion frequency obtained from the latent binary indicators with a cut off value of 0.5 (for the NMIG prior). The models were then re-estimated with the reduced variable set. For comparison, we also ran a stepwise procedure based on the Cox model and the partial likelihood. The resulting estimates are also contained in Fig. 6.8 as dashed lines. Mostly the results are pretty consistent among the three different methods applied to the PBC liver data.

Figure 6.9 illustrates some of the additional information that can be extracted from the Bayesian model with the NMIG prior. It shows the estimated inclusion probability obtained from the latent binary indicators of the NMIG prior. For most of the variables, the estimated probabilities are well-separated from the cut off value 0.5 and either quite close to one or relatively close to zero. The sole

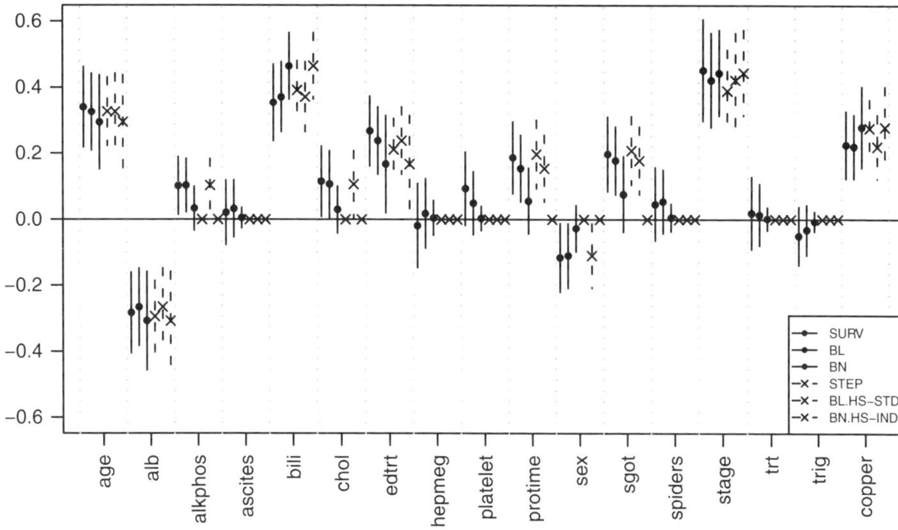

Fig. 6.8 PBC liver data: Estimated covariate effects from a standard Cox model (SURV), a Bayesian hazard regression model with lasso prior (BL) or NMIG prior (BN). Results including variable selection have been obtained with a stepwise procedure (STEP), hard shrinkage based on a one standard deviation interval (BL.HS-STD) and the inclusion indicators (BN.HS-IND) of the NMIG prior.

exception is the variable representing the presence of edema where one might be in doubt whether the variable should actually be included in the model or not.

Finally, to show that the Bayesian lasso also enables the construction of coefficient paths, Fig. 6.10 shows such coefficient paths for both a frequentist lasso estimate obtained from a regularized Cox model and from the Bayesian lasso based on the full likelihood. Both Figures show qualitatively similar paths for most of the variables but also reveal some of the distinct differences between frequentist and Bayesian lasso. Most notably, the Bayesian lasso does not allow us to set coefficients exactly to zero. Therefore all paths already move away from zero for large values of the smoothing parameter (corresponding to a small value of the norm of the coefficient vector). In contrast, some of the paths stay at zero for a while in the case of the frequentist lasso and only move away from zero once a certain threshold for the smoothing parameter is reached. Both figures also include estimated smoothing parameters obtained either from cross validation (for the frequentist lasso) or from sampling from the corresponding full conditional (for the Bayesian lasso). It turns out that the Bayesian lasso is more generous, leading to a larger estimate of the smoothing parameter. One of the advantages of the Bayesian lasso is, however, that it also enables proper inference about the smoothing parameter based on the posterior samples. △

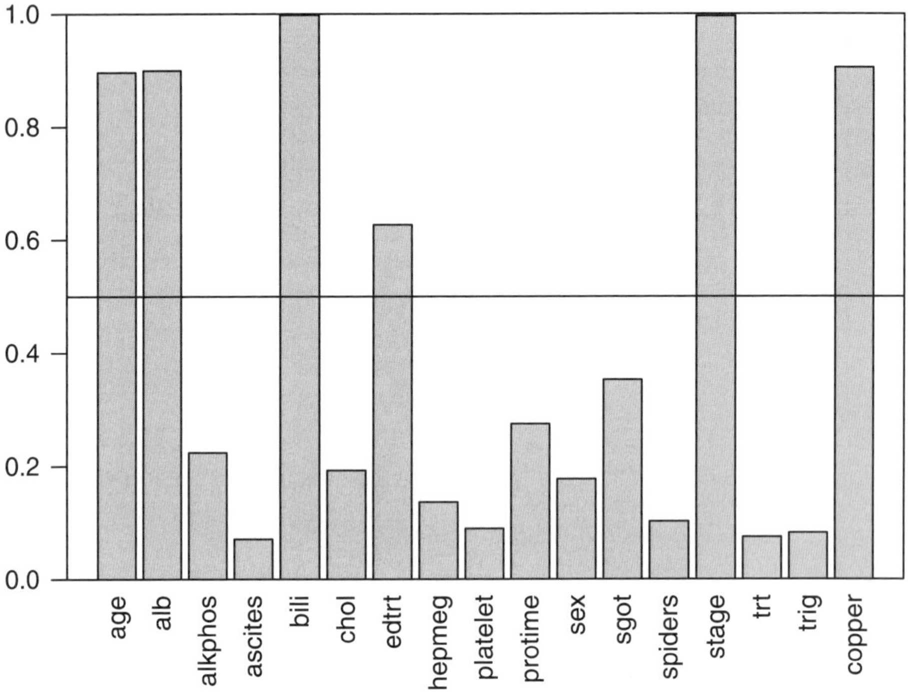

Fig. 6.9 PBC liver data: Inclusion frequencies for the binary coefficient indicators in the Bayesian hazard regression model with NMIG prior.

6.2.3 Empirical Bayes (EB) inference for right-censored observations

For EB inference, we assume a flat prior $p(\boldsymbol{\beta}) \propto$ const. Proceeding conceptually as in Section 4.2.1, we decompose the other coefficient vectors in the generic predictor (6.10) into

$$\boldsymbol{\gamma}_l = \tilde{\boldsymbol{U}}_l \boldsymbol{\beta}_l + \tilde{\boldsymbol{Z}}_l \boldsymbol{\alpha}_l, \quad l = 0, \dots, r,$$

where $\boldsymbol{\beta}_l$ corresponds to the unpenalized part of $\boldsymbol{\gamma}_l$, with flat prior $p(\boldsymbol{\beta}_l) \propto$ const and $\boldsymbol{\alpha}_l \sim \mathrm{N}(\boldsymbol{0}, \tau_l^2 \boldsymbol{I})$ captures the penalized part of $\boldsymbol{\gamma}_l$. Defining $\boldsymbol{\beta}_{\text{fix}} = (\boldsymbol{\beta}', \boldsymbol{\beta}_0', \dots, \boldsymbol{\beta}_r')'$ and $\boldsymbol{\alpha}_{\text{ran}} = (\boldsymbol{\alpha}_0', \dots, \boldsymbol{\alpha}_r')'$, the predictor (6.10) can be rewritten as

$$\eta_i(t) = \boldsymbol{u}_{i,\text{fix}}'(t)\boldsymbol{\beta}_{\text{fix}} + \boldsymbol{z}_{i,\text{ran}}'(t)\boldsymbol{\alpha}_{\text{ran}}. \tag{6.17}$$

The vector $\boldsymbol{u}_{i,\text{fix}}'(t)$ contains the subvector $\boldsymbol{u}_i'(t)$ as well as (transformed) subvectors $\boldsymbol{x}_{il}'(t)\tilde{\boldsymbol{U}}_l$, and $\boldsymbol{z}_{i,\text{ran}}'(t)$ contains the subvectors $\boldsymbol{x}_{il}'(t)\tilde{\boldsymbol{Z}}_l$. Priors for the unpenalized and penalized coefficients are then

$$p(\boldsymbol{\beta}_{\text{fix}}) \propto \text{const}, \quad \boldsymbol{\alpha}_{\text{ran}} \sim \mathrm{N}(\boldsymbol{0}, \boldsymbol{G}),$$

with $\boldsymbol{G} = \text{blockdiag}(\tau_0^2 \boldsymbol{I}, \dots, \tau_r^2 \boldsymbol{I})$.

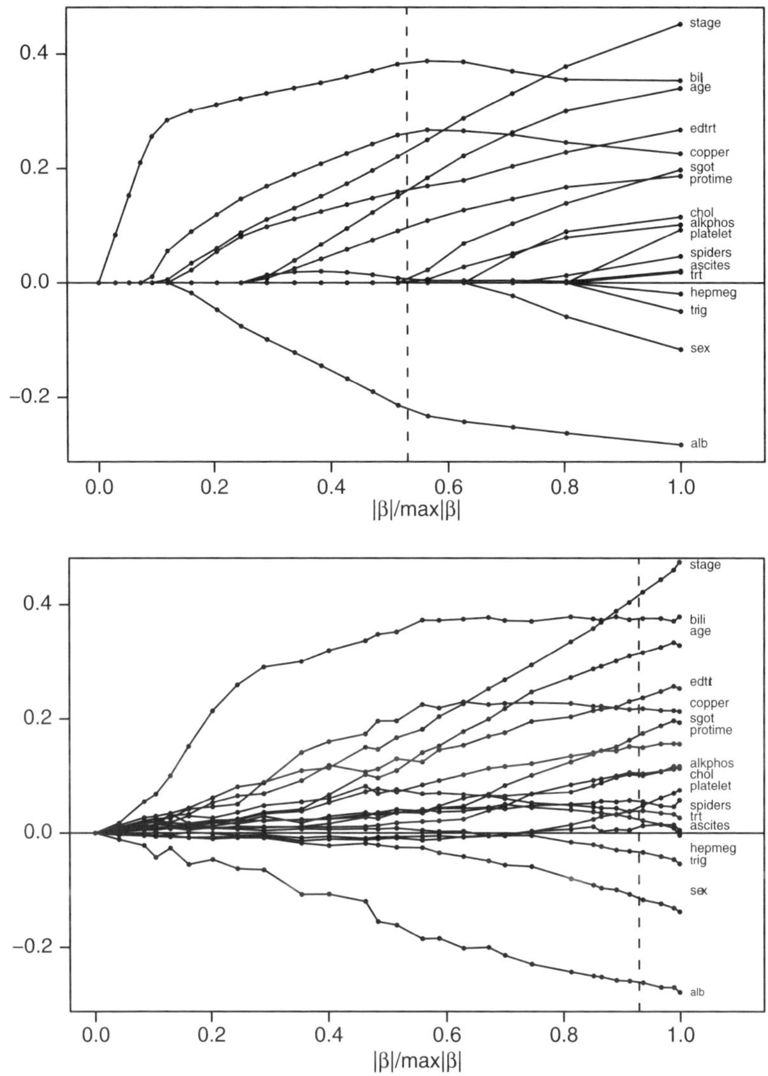

Fig. 6.10 PBC liver data: Coefficient paths for the frequentist lasso based on the partial likelihood (top) and the Bayesian hazard regression model with lasso prior for the regression coefficients (bottom). The vertical dashed line indicates the location of the smoothing parameter chosen by cross validation (top) and the MCMC sampler (bottom).

Defining $\boldsymbol{\eta} = (\eta_1(t_1), \ldots, \eta_i(t_i), \ldots, \eta_n(t_n))'$ as the vector of predictors evaluated at the observed survival times t_i, we obtain

$$\boldsymbol{\eta} = \boldsymbol{U}_{\text{fix}}\boldsymbol{\beta}_{\text{fix}} + \boldsymbol{Z}_{\text{ran}}\boldsymbol{\alpha}_{\text{ran}},$$

with design matrices $\boldsymbol{U}_{\text{fix}}$ and $\boldsymbol{Z}_{\text{ran}}$ defined by their rows $\boldsymbol{u}'_{i,\text{fix}}(t_i)$ and $\boldsymbol{z}'_{i,\text{ran}}(t_i)$, $i = 1 \ldots, n$. The penalized log-likelihood is

$$l_{\text{pen}}(\boldsymbol{\beta}_{\text{fix}}, \boldsymbol{\alpha}_{\text{ran}}) = l(\boldsymbol{\beta}_{\text{fix}}, \boldsymbol{\alpha}_{\text{ran}}) - \frac{1}{2}\boldsymbol{\alpha}'_{\text{ran}}\boldsymbol{G}^{-1}\boldsymbol{\alpha}_{\text{ran}}$$

where

$$l(\boldsymbol{\beta}_{\text{fix}}, \boldsymbol{\alpha}_{\text{ran}}) = \boldsymbol{d}'\boldsymbol{\eta} - \mathbf{1}'\boldsymbol{\Lambda}$$

with

$$\boldsymbol{\Lambda} = (\Lambda(t_1), \ldots, \Lambda(t_i), \ldots, \Lambda(t_n))', \quad \Lambda(t_i) = \int_0^{t_i} \exp(\eta_i(t))dt$$

is the log-likelihood $l(\boldsymbol{\beta}, \boldsymbol{\gamma})$ in reparametrized notation.

The score vector can be partitioned into two parts defined by the derivatives with respect to the unpenalized and the penalized vector of regression coefficients:

$$\boldsymbol{s}(\boldsymbol{\beta}_{\text{fix}}, \boldsymbol{\alpha}_{\text{ran}}) = \begin{pmatrix} \frac{\partial l_{\text{pen}}}{\partial \boldsymbol{\beta}_{\text{fix}}} \\ \frac{\partial l_{\text{pen}}}{\partial \boldsymbol{\alpha}_{\text{ran}}} \end{pmatrix} = \begin{pmatrix} \boldsymbol{U}'_{\text{fix}}\boldsymbol{d} - \sum_{i=1}^n \frac{\partial \Lambda(t_i)}{\partial \boldsymbol{\beta}_{\text{fix}}} \\ \boldsymbol{Z}'_{\text{ran}}\boldsymbol{d} - \sum_{i=1}^n \frac{\partial \Lambda(t_i)}{\partial \boldsymbol{\alpha}_{\text{ran}}} - \boldsymbol{G}^{-1}\boldsymbol{\alpha}_{\text{ran}} \end{pmatrix}. \tag{6.18}$$

Similarly, the observed Fisher information is partitioned into four blocks

$$\boldsymbol{H}(\boldsymbol{\beta}_{\text{fix}}, \boldsymbol{\alpha}_{\text{ran}}) = \begin{pmatrix} \sum_{i=1}^n \frac{\partial^2 \Lambda(t_i)}{\partial \boldsymbol{\beta}_{\text{fix}} \partial \boldsymbol{\beta}'_{\text{fix}}} & \sum_{i=1}^n \frac{\partial^2 \Lambda(t_i)}{\partial \boldsymbol{\beta}_{\text{fix}} \partial \boldsymbol{\alpha}'_{\text{ran}}} \\ \sum_{i=1}^n \frac{\partial^2 \Lambda(t_i)}{\partial \boldsymbol{\alpha}_{\text{ran}} \partial \boldsymbol{\beta}'_{\text{fix}}} & \sum_{i=1}^n \frac{\partial^2 \Lambda(t_i)}{\partial \boldsymbol{\alpha}_{\text{ran}} \partial \boldsymbol{\alpha}'_{\text{ran}}} + \boldsymbol{G}^{-1} \end{pmatrix}. \tag{6.19}$$

The derivatives appearing in these expressions are calculated in analogy to the expressions (6.11) and (6.12) for the score functions and observed Fisher information matrices in the original $(\boldsymbol{\beta}, \boldsymbol{\gamma})$-parametrization, including numerical integration. For given variance parameters $\boldsymbol{\tau} = (\tau_0^2, \ldots, \tau_r^2)$ they are used in Newton–Raphson steps for computing updated estimates for the regression coefficients.

Estimation of variance parameters is based on maximizing the marginal likelihood

$$L^{\text{marg}}(\boldsymbol{\tau}) = \int L_{\text{pen}}(\boldsymbol{\beta}_{\text{fix}}, \boldsymbol{\alpha}_{\text{ran}}, \boldsymbol{\tau})d\boldsymbol{\beta}_{\text{fix}}d\boldsymbol{\alpha}_{\text{ran}}$$

where L_{pen} denotes the penalized likelihood corresponding to l_{pen}. As shown by Harville (1974), maximizing the marginal likelihood is equivalent to REML estimation in Gaussian linear mixed models and allows us to extend REML estimation to general, non-Gaussian responses. However, direct maximization of $L^{\text{marg}}(\boldsymbol{\tau})$ is intractable, since the high-dimensional integrals cannot be evaluated, neither analytically nor numerically. We apply a Laplace approximation to the marginal likelihood yielding the approximate log-likelihood

$$l^{\text{marg}}(\boldsymbol{\tau}) \approx l(\hat{\boldsymbol{\beta}}_{\text{fix}}, \hat{\boldsymbol{\alpha}}_{\text{ran}}) - \frac{1}{2}\log|\boldsymbol{G}| - \frac{1}{2}\hat{\boldsymbol{\alpha}}_{\text{ran}}'\boldsymbol{G}^{-1}\hat{\boldsymbol{\alpha}}_{\text{ran}} - \frac{1}{2}\log|\boldsymbol{H}|.$$

Assuming that $l(\hat{\boldsymbol{\beta}}_{\text{fix}}, \hat{\boldsymbol{\alpha}}_{\text{ran}})$ and $\hat{\boldsymbol{\alpha}}_{\text{ran}}$ vary only slowly when changing the variance parameters, we can further reduce the marginal log-likelihood to

$$l^{\text{marg}}(\boldsymbol{\tau}) \approx -\frac{1}{2}\log|\boldsymbol{G}| - \frac{1}{2}\log|\boldsymbol{H}| - \frac{1}{2}\boldsymbol{\alpha}_{\text{ran}}'\boldsymbol{G}^{-1}\boldsymbol{\alpha}_{\text{ran}}, \qquad (6.20)$$

where $\boldsymbol{\alpha}_{\text{ran}}$ denotes a fixed value not depending directly on the variances, for example a current estimate. Both approximations are in the spirit of the penalized likelihood approach for generalized linear mixed models proposed in Breslow and Clayton (1993).

First and second derivatives of (6.20) can be easily derived, based on differentiation rules for matrices. Since expressions for general covariance matrices $\boldsymbol{G}$ become quite lengthy, we make use of the block diagonal structure $\boldsymbol{G} = \text{blockdiag}(\tau_0^2\boldsymbol{I}, \ldots, \tau_r^2\boldsymbol{I})$ in variance components models to obtain simpler formulae. For the score function this yields

$$\boldsymbol{s}_j^* = \frac{\partial l^{\text{marg}}(\boldsymbol{\tau})}{\tau_j^2} = -\frac{R_j}{2\tau_j^2} + \frac{1}{2\tau_j^4}\,\text{tr}\left(\tilde{\boldsymbol{H}}_{jj}^{pp}\right) + \frac{1}{2\tau_j^4}\boldsymbol{\alpha}_{\text{ran},j}'\boldsymbol{\alpha}_{\text{ran},j}, \qquad (6.21)$$

where $R_j = \text{rank}(\boldsymbol{K}_j)$, $\tilde{\boldsymbol{H}} = \boldsymbol{H}^{-1}$ denotes the inverse Fisher information (for the regression coefficients) and $\tilde{\boldsymbol{H}}_{jj}^{pp}$ is the diagonal block of $\boldsymbol{H}^{-1}$ corresponding to $\boldsymbol{\alpha}_{\text{ran},j}$. The elements of the expected Fisher information for $\boldsymbol{\tau}$ are given by

$$\boldsymbol{F}_{jk}^* = \mathbb{E}\left(-\frac{\partial^2 l^{\text{marg}}(\boldsymbol{\tau})}{\partial\tau_j^2\partial\tau_k^2}\right) = \frac{1}{2\tau_j^4\tau_k^4}\,\text{tr}\left(\tilde{\boldsymbol{H}}_{jk}^{pp}\tilde{\boldsymbol{H}}_{kj}^{pp}\right). \qquad (6.22)$$

Here, $\tilde{\boldsymbol{H}}_{jk}^{pp}$ denotes the off-diagonal block of $\boldsymbol{H}^{-1}$ corresponding to $\boldsymbol{\alpha}_{\text{ran},j}$ and $\boldsymbol{\alpha}_{\text{ran},k}$. Both expressions are numerically simple to evaluate since $\boldsymbol{H}^{-1}$ and $\boldsymbol{\alpha}_{\text{ran}}$ are direct byproducts from the estimation of the regression coefficients. Based on the score function and the Fisher information we can compute updated variances via a Fisher-scoring step.

In summary, we obtain the following algorithm for empirical Bayes inference in structured hazard regression:

Algorithm 6.2: Empirical Bayes inference for structured hazard regression

i. Initialize all parameters with suitable starting values.
1. Update the regression coefficients in a Newton–Raphson step based on the score function and Fisher information given in (6.18) and (6.19) given the current variance parameters.
ii. Update the variance parameters in a Fisher-scoring step based on the score function and Fisher information in (6.21) and (6.22) given the current regression coefficients.
iii. Iterate steps ii. and iii. until both regression coefficients and variance parameters no longer change.

Example 6.7: Leukemia survival

Continuing Example 6.1, we specify geoadditive predictors

$$\eta_i(t) = \beta_0 + \beta_1 \mathrm{sex}_i + g_0(t) + f_1(\mathrm{age}_i) + f_2(\mathrm{wbc}_i) + f_3(\mathrm{tpi}_i) + f_{\mathrm{spat}}(s_i)$$

for the leukemia survival data where $g_0(t)$ is the log-baseline, f_1, f_2 and f_3 are smooth functions of the continuous covariates age, white blood cell count and Townsend deprivation index and f_{spat} is a spatial effect. Both g_0 and the f_j will be modelled as cubic P-splines with second order difference penalty and 20 knots. In an individual-level analysis, we employ the coordinate information $s_i = (s_{i1}, s_{i2})$ on the exact location of the patient's residence while in a district-level analysis the spatial information is given in terms of the district s_i the patient lives in.

In the district-level analysis, the natural choice to model the spatial effect is a Gaussian Markov random field with neighbourhoods determined by common boundaries. For an individual-level analysis, we replace the GMRF with either a low rank Kriging term corresponding to a stationary Gaussian random field prior or with a tensor product penalized spline. For the low rank Kriging term we computed the knot locations based on the space filling algorithm described in Section 5.3.2 (page 370). Comparing results with 50, 100 and 200 knots yielded essentially the same results so we restrict our attention to the model with 100 knots. For the bivariate spline, we used cubic tensor product B-splines, a first-order random-walk penalty and 12 inner knots in each of the coordinate directions. Estimates for the baseline effect as well as the nonparametric effects were rather insensitive to the choice of the spatial effect so we only show results obtained from the individual-level analysis with GRF prior in the following. A more detailed analysis of the Leukemia data can be found in Kneib and Fahrmeir (2007).

As depicted in Fig. 6.11, the log-baseline shows a steady monotonic decrease over most of the observation period, alternating between periods with a relatively steep decrease and almost constant periods. At the very end of the observation period, the baseline shows a strong increase, but since this increase is

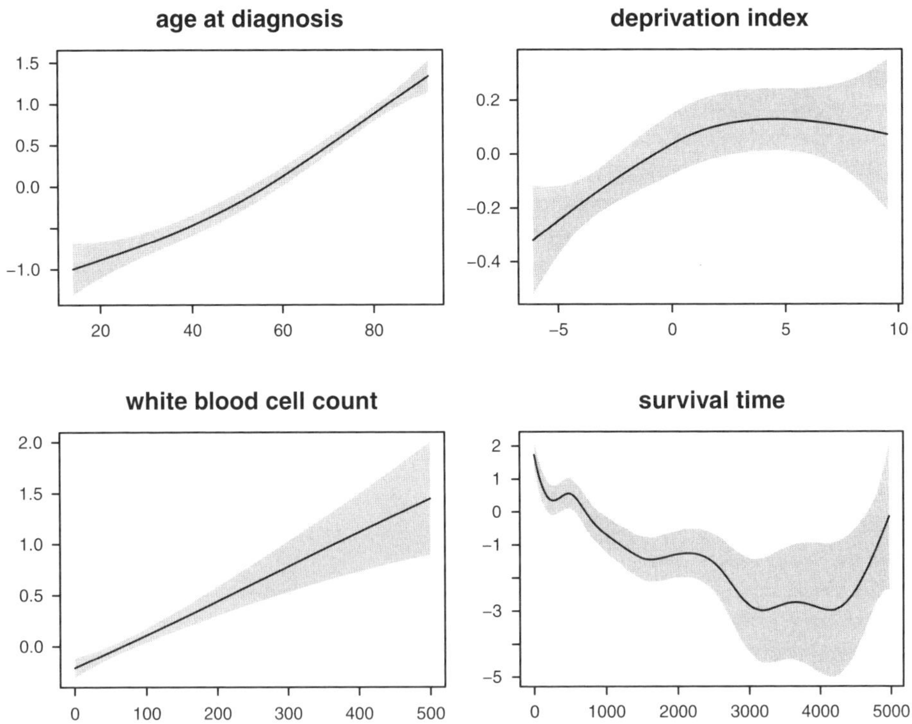

Fig. 6.11 Leukemia survival: Estimated nonparametric effects together with 95% pointwise credible intervals.

determined by only 26 individuals that survived longer then 10 years, it is associated with rather large uncertainty. The effects of age and white blood cell count are approximately linear and may therefore be replaced by linear terms in a simplified model specification. In contrast, the effect of the deprivation index is at least moderately nonlinear. Interestingly, the effect is almost constant for larger values corresponding to higher deprivation. Obviously, there is a gain in survival time for richer regions while poorer regions are mostly comparable with respect to the survival time.

Figure 6.12 shows the estimated spatial effects obtained with either a GMRF prior, a low-rank Kriging specification or a bivariate tensor product spline. In general, all variants identify a comparable structure and in particular both individual-level models are visually indistinguishable. Both also reveal a more detailed picture of the spatial effects simply because the effect is also allowed to vary smoothly within districts instead of restricting it to district-specific effects. In all models, we find low risks in the western part of the map, surrounded by areas of increased risk. In the southern part of the map there are also some areas with lower risk but the spatial effect is less pronounced there. △

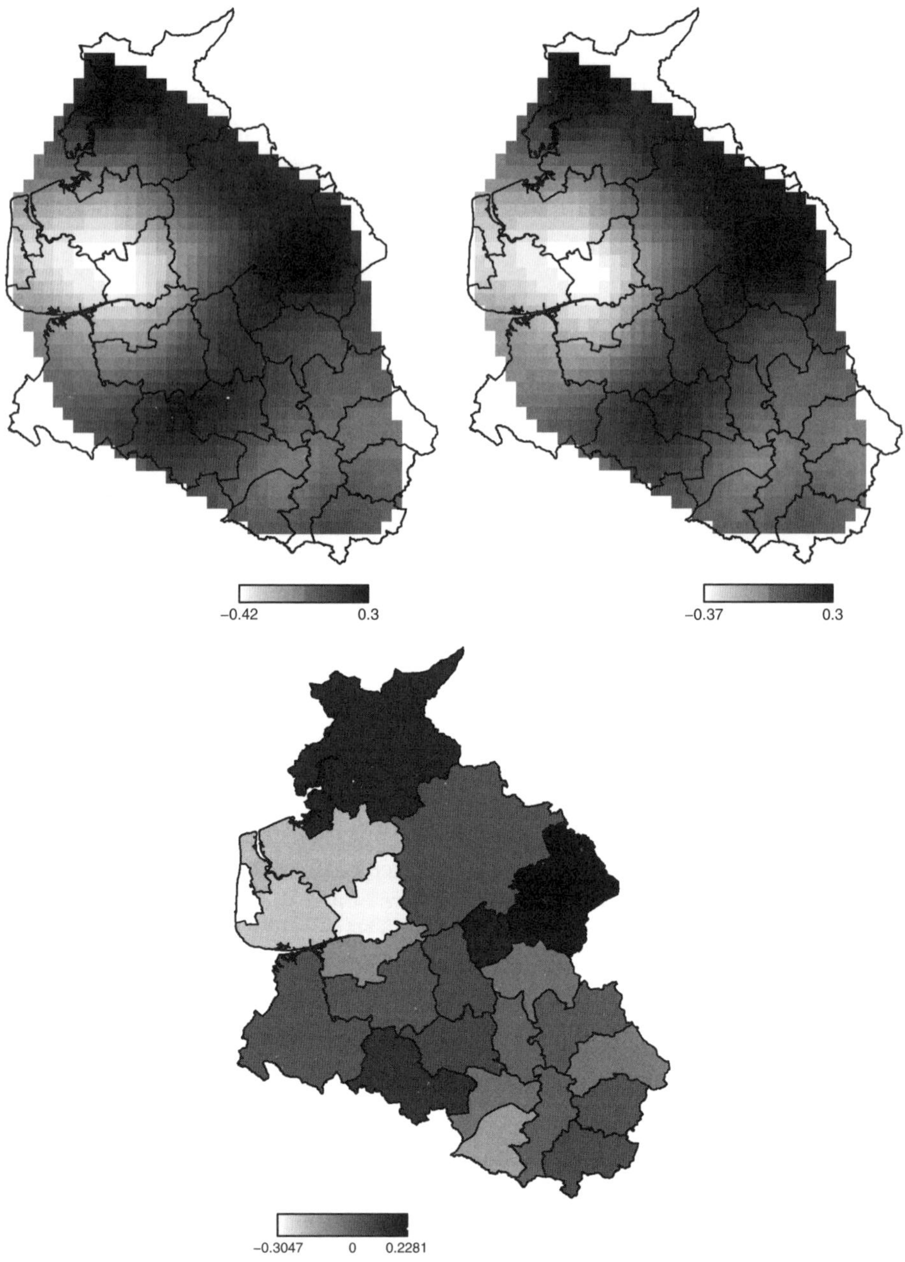

Fig. 6.12 Leukemia survival: Estimated spatial effect when the spatial function is represented as a Gaussian random field (top left) as a bivariate P-spline (top right) or as a Markov random field (bottom).

6.2.4 Inference for interval-censored observations

For interval-censored survival times, the basic concepts for EB and FB inference remain the same, but the formulae for first and second derivatives of the log-likelihood contributions become more complicated. To ease notation, consider for the moment a hazard rate of the form $\lambda(t) = \exp(\boldsymbol{x}(t)'\boldsymbol{\beta})$ which essentially reflects the structure of our hazard regression models. Defining

$$E_j(t) = -\frac{\partial}{\partial \beta_j} \int_0^t \lambda(u)du = -\int_0^t x_j(u)\lambda(u)du$$

and

$$F_{jk}(t) = -\frac{\partial^2}{\partial \beta_j \partial \beta_k} \int_0^t \lambda(u)du = -\int_0^t x_j(u)x_k(u)\lambda(u)du,$$

first and second derivatives of the log-likelihood contributions for uncensored and right-censored observations are given by (compare (6.3) and (6.4), in combination with left truncation)

$$d \cdot x_j(t_{\text{up}}) + E_j(t_{\text{up}}) - E_j(t_{\text{tr}}) \quad \text{and} \quad F_{jk}(t_{\text{up}}) - F_{jk}(t_{\text{tr}}).$$

For interval-censored survival times, formulae become more complicated. Here, first and second derivatives of the log-likelihood contributions can be shown to equal

$$E_j(t_{\text{lo}}) - E_j(t_{\text{tr}}) - \frac{\exp\left[\Lambda(t_{\text{lo}}) - \Lambda(t_{\text{up}})\right]\left[E_j(t_{\text{lo}}) - E_j(t_{\text{up}})\right]}{1 - \exp\left[\Lambda(t_{\text{lo}}) - \Lambda(t_{\text{up}})\right]}$$

and

$$\begin{aligned} & F_{jk}(t_{\text{lo}}) - F_{jk}(t_{\text{tr}}) \\ & - \frac{\exp\left[\Lambda(t_{\text{lo}}) - \Lambda(t_{\text{up}})\right]^2 \left[E_j(t_{\text{lo}}) - E_j(t_{\text{up}})\right]\left[D_k(t_{\text{lo}}) - D_k(t_{\text{up}})\right]}{\left\{1 - \exp\left[\Lambda(t_{\text{lo}}) - \Lambda(t_{\text{up}})\right]\right\}^2} \\ & - \frac{\exp\left[\Lambda(t_{\text{lo}}) - \Lambda(t_{\text{up}})\right]\left[D_k(t_{\text{lo}}) - D_k(t_{\text{up}})\right]\left[E_j(t_{\text{lo}}) - E_j(t_{\text{up}})\right]}{1 - \exp\left[\Lambda(t_{\text{lo}}) - \Lambda(t_{\text{up}})\right]} \\ & - \frac{\exp\left[\Lambda(t_{\text{lo}}) - \Lambda(t_{\text{up}})\right]\left[F_{jk}(t_{\text{lo}}) - F_{jk}(t_{\text{up}})\right]}{1 - \exp\left[\Lambda(t_{\text{lo}}) - \Lambda(t_{\text{up}})\right]}, \end{aligned}$$

see Section 6.1.2 (page 424) for the definition of the time variables t_{lo}, t_{up} and t_{tr}. Note that for $t_{\text{tr}} = 0$ these results are equivalent to those presented in Kooperberg and Clarkson (1997). To evaluate the derivatives $E_j(t)$ and $F_{jk}(t)$, we again employ the trapezoidal rule or some comparable numerical integration technique.

These expressions are used for calculating score functions and observed information matrices needed in MH proposals for FB and in Newton-Raphson steps for EB inference. Apart from likelihood contributions, modified for interval-censoring, variance parameters are again estimated by maximizing a Laplace-approximated marginal log-likelihood in case of EB inference.

Example 6.8: Childhood mortality in Nigeria
For analysing childhood mortality in Nigeria (Examples 1.6, 6.2 and 6.4) using EB estimation, we consider a hazard rate model with predictor

$$\eta(t) = g_0(t) + f_1(\text{bmi}) + f_2(\text{age}) + f_3(\text{bord}) + f_4(\text{size}) + f_{\text{spat}}(s) + \boldsymbol{u}(t)'\boldsymbol{\beta}$$

where $g_0(t)$ denotes the log-baseline hazard rate, $f_1, \ldots, f_4$ are functions of the continuous covariates body mass index of the mother (bmi), age of the mother at birth (age), number of the child in the birth order (bord) and number of household members (size), f_{spat} models a spatial effect based on the district s the mother lives in and $\boldsymbol{u}(t)$ comprises fixed effects of numerous categorical covariates including variables describing the economic situation of the household or the breastfeeding behaviour of the mother. The latter is a time-varying covariate that equals one as long as the child is breastfed and drops to zero afterwards. All nonparametric effects as well as the log-baseline are modelled as cubic P-splines with 20 inner knots and second-order random-walk prior. The spatial effect is specified via a Gaussian Markov random field and estimation is based on empirical Bayes inference.

In the following, we will restrict ourselves to a discussion of nonparametric and spatial effects and will not present results for the fixed effects part of our model. Estimates of all nonparametric effects (except the log-baseline) are shown in Fig. 6.13. It turns out that the effect of the maternal body mass index is very close to linear and shows only a minor increase with increasing body mass index. Moreover, the confidence bands for this effect are rather wide, indicating that the effect may be negligible after all. The effect of the mother's age is also not too far from linear but clearly has a much stronger impact on the hazard rate (with higher risks for older women). Both a high number of the child in the birth order and larger households are associated with a decreased hazard rate. This may be related to an increased knowledge about childcare for children that are born later and the fact that well-endowed households tend to attract additional household members.

Figure 6.14 shows the estimated log-baseline hazard rate for different approaches dealing with the interval-censoring present in the data (see again Example 6.2). In the first model, we account for both rounding of the survival times to months and the additional heaping to specific months such as 12, 18, 24 and so on. The second model ignores the heaping effect and only treats all survival times as interval censored with intervals of exactly one month width. In the third model, all survival times are randomly spread across the intervals of the model including the heaping effect. This allows us to treat the corresponding observations as uncensored but of course relies on the assumption that events

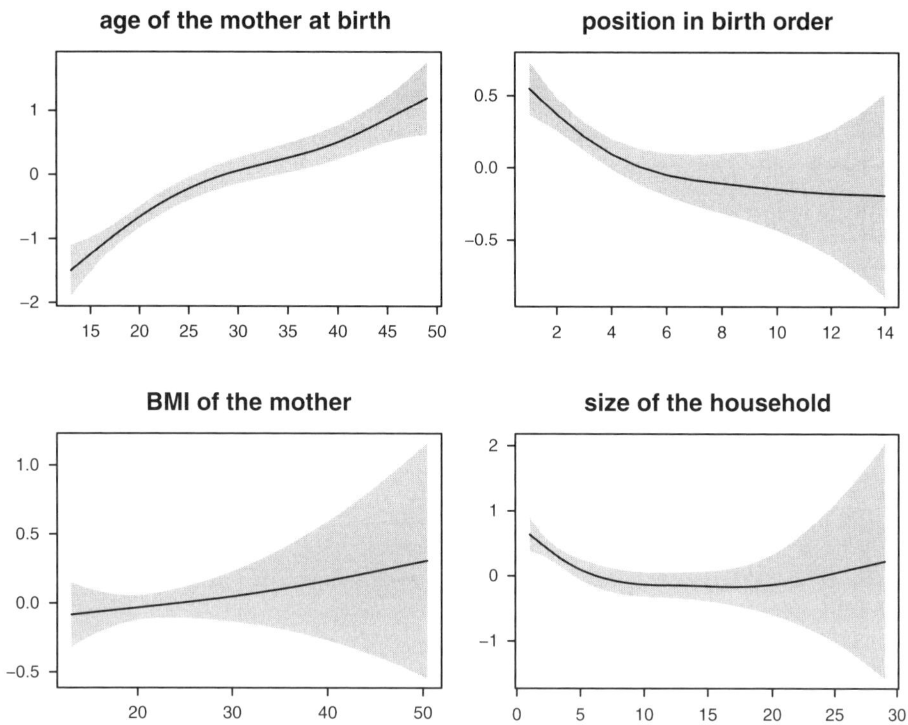

Fig. 6.13 Childhood mortality: Estimated nonparametric effects together with 95% pointwise credible intervals in a continuous-time analysis accounting for interval censoring and heaping.

are uniformly distributed within the intervals and ignores uncertainty about the exact placement of the events.

From the results we can conclude that ignoring the heaping effect leads to biased and highly implausible estimates for the baseline hazard. In particular, the log-baseline deviates to minus infinity in areas where no events have been observed relating to a hazard rate that is basically equal to zero. When the heaping effect is incorporated, this effect is strongly reduced, although some fluctuations remain in the estimate that are not expected to reflect the true temporal development of the hazard rate. The model based on a random relocation of event times leads to the most plausible, very smooth effect estimate. Possibly this additional smoothness results from the additional information assumed in the model. Since all observed death times are treated as being exactly observed, the model contains much more information than the corresponding model based on interval-censoring. This makes the interval-censoring approach more susceptible to small variations in the data inducing some of the variability in the function estimate.

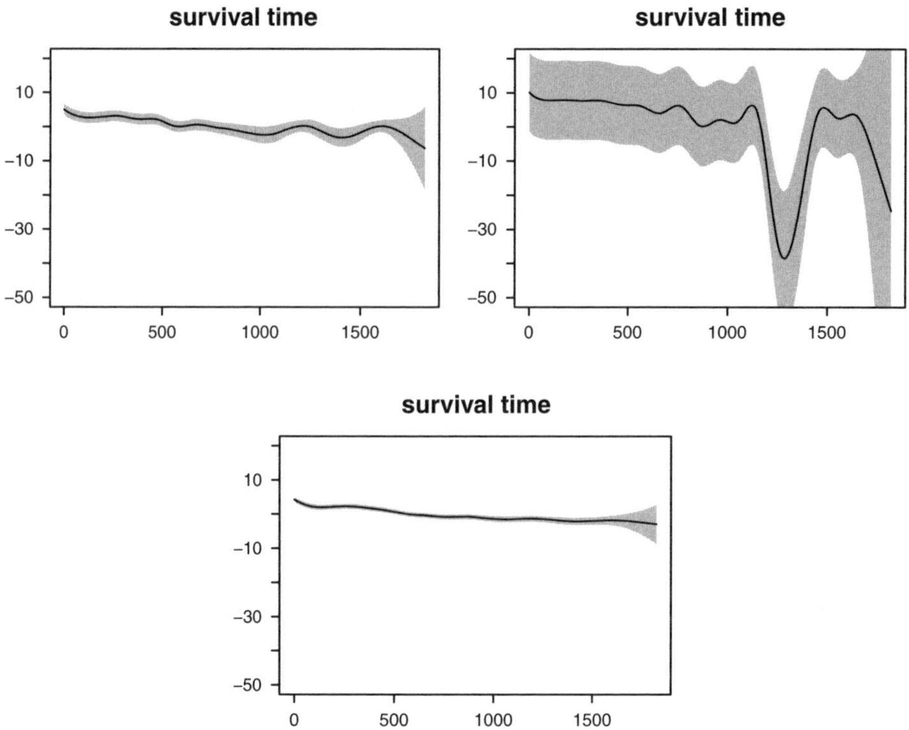

Fig. 6.14 Childhood mortality: Estimated log-baselines together with 95% pointwise credible intervals in continuous-time analyses accounting for interval-censoring and heaping (top left), accounting for interval-censoring only (top right) and with event times distributed randomly across intervals (bottom).

Figure 6.15 shows estimated spatial effects for two models including interval-censoring and heaping. The first model contains all further covariates and results in a very flat spatial effect that does not show any significant deviations from zero (bottom row of Fig. 6.15). To check whether any spatial correlation has been present in the raw data, we additionally computed a model that does not include any further covariates apart from the log-baseline and the spatial effect (upper row of Fig. 6.15). The corresponding effect estimate shows strong spatial variation and in particular leads to some pointwise significant posterior mode estimates. A decreased risk is observed in the southwestern part of Nigeria while an increased risk is observed in the northeastern part. As a consequence, spatial correlations can be concluded to be present in the data at hand but actually these spatial correlations are completely accounted for by the (spatially varying) covariates. △

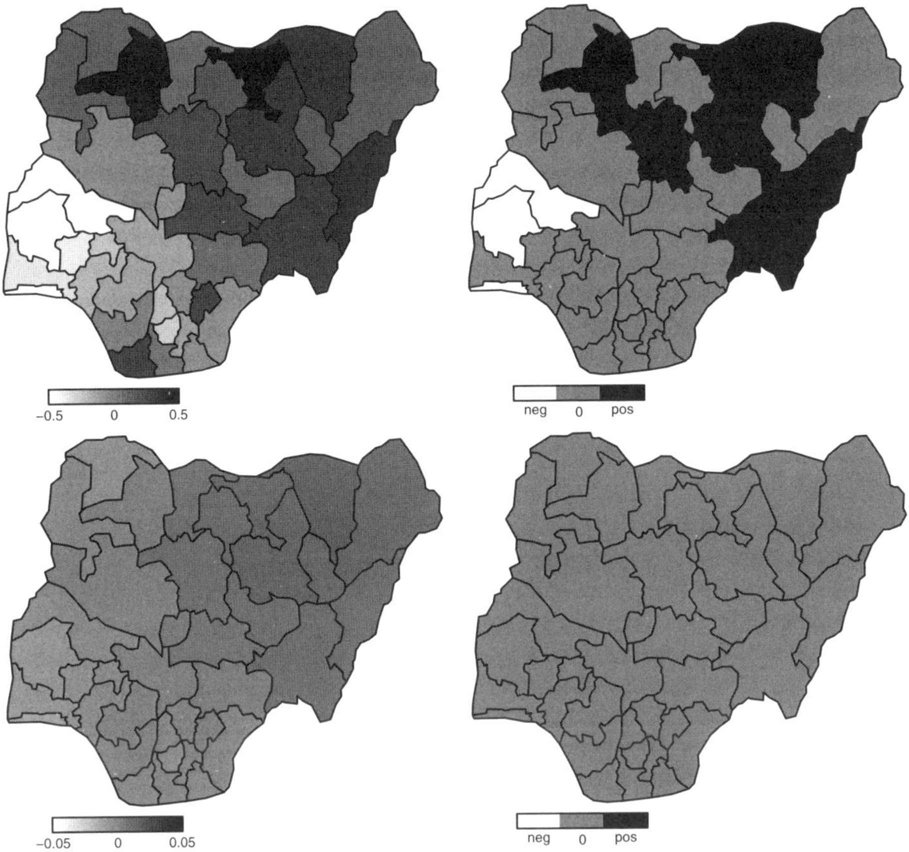

Fig. 6.15 Childhood mortality: Estimated spatial effect (left) together with 95% pointwise posterior probabilities (right) in continuous-time analyses accounting for interval-censoring and heaping. The top row shows estimates from a model that only comprises the spatial effect and the log-baseline while the lower row shows estimates from the full model.

6.3 Discrete-time hazard regression

As described in Section 6.1.1, regression models for the discrete hazard function $\lambda(t|\boldsymbol{u}_i) = \mathbb{P}(T = t|T \geq t, \boldsymbol{u}_i)$, where $\boldsymbol{u}_i$ is a time-constant vector of covariate values, can be expressed as a sequence of binary regression models

$$\lambda(t|\boldsymbol{u}_i) = \mathbb{P}(y_{it} = 1|\boldsymbol{u}_i) = h(\eta_{it})$$

with response function h and predictors

$$\eta_{it} = \boldsymbol{u}_i'\boldsymbol{\beta} + \gamma_{0t}, \quad t = 1, \ldots, q,$$

where γ_{0t} corresponds to the baseline effect. The predictor and the data augmentation concept can be extended to structured additive predictors, including time-varying covariates that are constant within intervals. The discrete-time analogue to the continuous-time predictor (6.9) is

$$\eta_{it} = \boldsymbol{u}_{it}'\boldsymbol{\beta} + \gamma_{0t} + \sum_{l=1}^{m} \gamma_{lt} v_{ilt} + \sum_{j=1}^{p} f_j(x_{ijt}) + f_{\text{spat}}(s_i) + \alpha_{gi},$$

similarly as for the piecewise exponential model. As for longitudinal and spatio-temporal data, both random walks and penalized splines can be used as priors to derive smooth estimates for the functions (or sequences) $\boldsymbol{\gamma}_l = (\gamma_{l1}, \ldots, \gamma_{lt}, \ldots, \gamma_{lq})'$, $l = 0, \ldots, m$, of the baseline effects and time-varying effects. Priors for f_j, $j = 1, \ldots, p$, f_{spat} and the random effects α_g, $g = 1, \ldots, G$, are specified as usual. EB and FB inference then proceeds as for longitudinal or spatio-temporal data with binary responses and will not be repeated here.

For models without time-varying effects $\boldsymbol{\gamma}_l$, $l = 1, \ldots, m$, and with time-constant covariates, there is a very close relationship between (extended) Cox models and the complementary log-log model for binary responses. Consider a continuous-time Cox-type model of the form

$$\lambda_c(t|\tilde{\eta}_i) = \lambda_0(t)\exp(\tilde{\eta}_i),$$

where $\tilde{\eta}_i$ is a time-constant structured additive predictor. Introducing the parameters

$$\gamma_{0t} = \log\left(\int_{\kappa_{t-1}}^{\kappa_t} \lambda_0(t)dt\right)$$

leads to the discrete-time hazard function

$$\lambda_d(t|\tilde{\eta}_i) = 1 - \exp(-\exp(\gamma_{0t} + \tilde{\eta}_i)) = \mathbb{P}(y_{it} = 1|\eta_{it}), \quad t = 1, \ldots, q,$$

that is, to a complementary log-log model. Note that the predictor $\tilde{\eta}_i$ remains unchanged when switching between the continuous-time and the discrete-time view. Therefore the complementary log-log model is also called *grouped proportional hazards model.*

Fleming and Harrington (1991, page 126) show a similarly close relationship between the logit model for discrete-time survival data and the Cox model, if time-intervals become small. Because logit and probit models usually lead to parameter estimates which are close to each other (up to a common factor), probit models are also a good choice: introducing latent Gaussian variables $l_{it} = \eta_{it} + \varepsilon_{it}$, $\varepsilon_{it} \sim \text{N}(0, 1)$, and assuming

$$\mathbb{P}(y_{it} = 1) \Leftrightarrow l_{it} \geq 0$$

leads to $\mathbb{P}(y_{it}|\eta_{it}) = \Phi(\eta_{it})$. As described in Section 2.2.2, this latent variable approach allows us to construct Gibbs samplers for probit models avoiding MH algorithms with IWLS proposals. We use such a latent variable approach in the following examples.

Example 6.9: Unemployment duration

Our analyses of unemployment duration (Example 6.3) are based on probit models

$$\mathbb{P}(y_{it}|\eta_{it}) = \Phi(\eta_{it}),$$

so that Gibbs sampling in combination with latent Gaussian variables $l_{it} = \eta_{it} + \varepsilon_{it}$ can be used for full Bayes inference. In the basic model, the predictor contains main effects of all covariates, that is, we assume

$$\eta_{it} = \boldsymbol{u}_{it}'\boldsymbol{\beta} + g_0(t) + f_1(age_i) + f_2^T(date_{it}) + f_3^S(date_{it}) + f_{\text{spat}}(s_i),$$

where $g_0(t)$ is the baseline effect, $f_1(age)$ a nonlinear effect of age, and $f_{\text{spat}}(s)$ the effect of labour market regions. The effect of calendar date is split up into a smooth trend $f_2^T(date_t)$ and a seasonal component $f_3^S(date_t)$. Note that calendar date is formally treated like a time-varying covariate. We assume a flat prior for $\boldsymbol{\beta}$, cubic P-splines with second-order random-walk priors for g_0, f_1 and f_2, and a Markov random field prior for f_{spat}. For the seasonal component f_3^S we assume the Gaussian smoothness prior

$$f_3^S(date) - f_3^S(date-1) - \ldots - f_3^S(date-11) = u_d^S \sim \text{N}(0, \tau_3^2),$$

a common model for a monthly seasonal component in Gaussian time series analysis and state space models. Defining $\boldsymbol{\gamma}_3 = (f_3^S(d), d = 1, \ldots, D)'$ as the vector of all seasonal effects and an appropriate precision matrix $\boldsymbol{K}^S$, the seasonal prior can be written in our generic form of Gaussian smoothness priors (see also Section 2.1.2, page 27). In contrast to conventional parametric modelling of seasonal effects by dummy variables for months, this prior allows for a flexible seasonal pattern that may change over the years.

Table 6.1 contains the estimation results for the effects of the categorical covariates in $\boldsymbol{u}$. They mainly confirm some facts already known from labour market research in that period. Chances for reemployment are better for German women. Women with a university degree have significantly higher reemployment chances. Positive effects of unemployment insurance benefits are in agreement with findings in previous studies, even with data from the German Socio-Economic Panel, see for example Fahrmeir and Knorr-Held (1997). One possible explanation is that individuals with unemployment benefits had regular jobs earlier, and therefore get offers for a job more easily.

The positive effect of a higher number of previous unemployment periods seems strange at first view. Additional analyses, not reported here, provide evidence that the effect is highly associated with seasonally unemployed workers.

Table 6.1 Unemployment duration: Posterior means, standard deviations and 80% credible intervals for parametric effects of categorical covariates.

variable	estimate	sd	80% ci	
German	0.07	0.01	0.06	0.08
foreign	−0.07	0.01	−0.08	−0.06
no vocational training	−0.04	0.01	−0.05	−0.03
vocational training	−0.04	0.01	−0.05	−0.02
university	0.07	0.02	0.05	0.10
unemployment assistance	−0.09	0.01	−0.10	−0.09
unemployment benefit	0.09	0.01	0.09	0.10
$P_d = 0$	−0.09	0.01	−0.10	−0.08
$P_d = 1, 2$	−0.03	0.01	−0.04	−0.02
$P_d \geq 3$	0.12	0.01	0.11	0.13
agriculture	0.19	0.03	0.15	0.22
manufacturing	−0.15	0.01	−0.16	−0.14
construction	−0.02	0.03	−0.06	0.02
trade	−0.01	0.01	−0.02	0.01
transport	0.09	0.02	0.06	0.12
financial sector	−0.18	0.03	−0.22	−0.15
service industry	0.06	0.01	0.05	0.07
private	0.00	0.02	−0.02	0.03
public	0.02	0.02	0.00	0.04

The effects of economic sectors differ significantly. In particular, the financial sector has a strong negative effect.

The nonlinear effects of the three time scales, that is, unemployment duration, age and calendar time are shown in Fig. 6.16. Panel a) provides clear evidence of negative duration effects: The log-baseline hazard peaks in the first two to three months, then chances for reemployment tend to fall with duration.

The age effect in panel b) declines fairly rapidly up to the age of 30. Between 30 and 50 years, the age function is flat. This pattern is in line with the hypothesis that the age effect captures the role of family and children. Institutional features of the unemployment system and rules for early retirement contribute to the pattern beyond 50 years of age.

The estimated calendar time trend (panel c) reflects fairly well and without much lag the German business cycle during the observation period. We find a sharp decline in the transition rates to (full time) employment of unemployed women at the beginning of the 1980s. They reach their trough at the beginning of 1983, the year in which the economy started its upturn. During the first phase of the upturn from 1983 to 1987 GDP growth never exceeded 3%. In that period

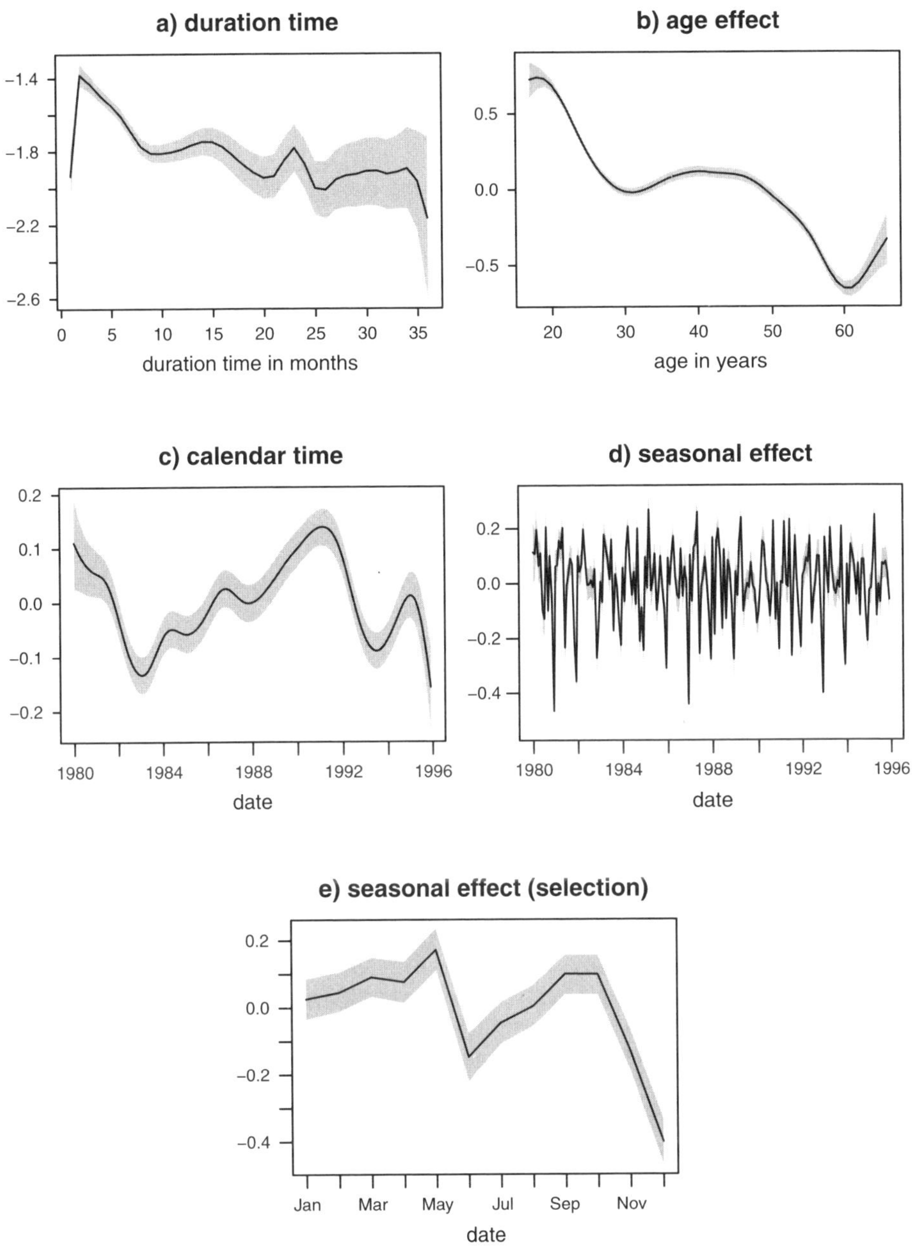

Fig. 6.16 Unemployment duration: Posterior mean estimates and 80% pointwise credible intervals for nonparametric and seasonal effects.

the hazards increase somewhat. In 1988 GDP growth accelerated. It reached even more than 5% at the beginning of the 90s as a consequence of the demand shock due to German unification. By that time, the female transition rates increased further and reached their pre-recession levels. The next downturn started in 1992 and again led to a sharp fall in the transition rates until mid-1993. The following upturn did not last for much longer than one year, so that the transition rates increased only temporarily.

Panel d) displays the seasonal effects over the entire observation period, and panel e) shows seasonal effects for 1992 in more detail. The seasonal effects correspond to commonly expected patterns, but are relatively small compared to those of men (not shown).

Estimated spatial effects in Fig. 6.17 show a strong spatial pattern. Employment chances are highest in the south and are lower in the middle and

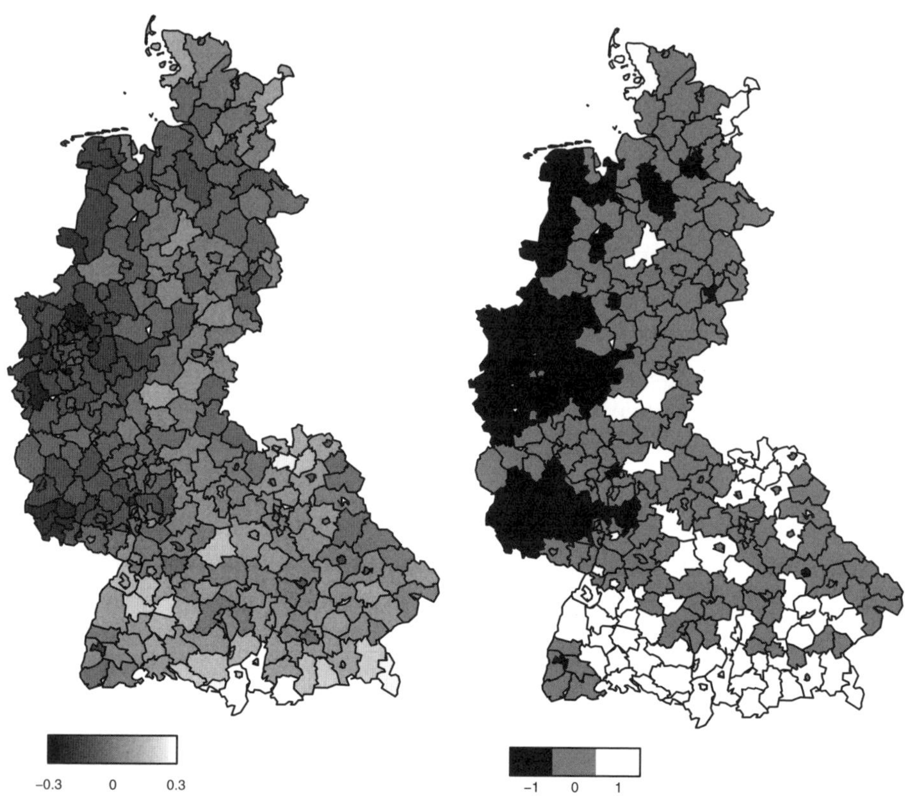

Fig. 6.17 Unemployment duration: Posterior mean estimates and 80% posterior probabilities for spatial effects.

the northern regions of West Germany. Particularly striking are the negative effects in the west, corresponding to the Saar and the Ruhr area. These regions are known for the massive structural adjustments during the observation period.

By assumption, the basic additive main effects model only allows us to estimate spatial effects averaged over the entire observation period. To investigate whether or not spatial effects are varying over time, interaction terms between calendar time and regions have to be included. One possibility is to work with a space-varying coefficient model after dividing the observation period into (a small number of) intervals. More specifically, we subdivide calendar time into the three distinct intervals 1980 to 1985, 1986 to 1990, and 1991 to 1995, and define indicator variables d^{86} and d^{91} whose values are one if an observation falls into the interval 1986 to 1990 or 1991 to 1995 respectively, and zero otherwise. Time-space interactions are then modelled by the predictor

$$\begin{aligned}\eta_{it} = \boldsymbol{u}_{it}'\boldsymbol{\beta} + g_0(t) + f_1(age_i) + f_2^T(date_{it}) + f_3^S(date_{it}) + \\ f_{1,\text{spat}}(s_i) + f_{2,\text{spat}}(s_i)d_{it}^{86} + f_{3,\text{spat}}(s_i)d_{it}^{91}.\end{aligned}$$

The function $f_{1,\text{spat}}$ corresponds now to the spatial effect for the time period between 1980 and 1985, and the two space-varying coefficient terms $f_{2,\text{spat}}$ and $f_{3,\text{spat}}$ can be considered as deviations from this reference effect in the following two time-periods. In this model, the sum

$$f_2^T(date) + f_{1,\text{spat}}(s), \qquad date \in [1980, 1985]\,,$$

is the effect of time and space for the period 1980 to 1985. The sums

$$\begin{aligned}f_2^T(date) + f_{2,\text{spat}}(s), \qquad date \in [1986, 1990]\,,\\ f_2^T(date) + f_{3,\text{spat}}(s), \qquad date \in [1991, 1995]\,,\end{aligned}$$

correspond to the effects of the second and third period respectively. To visualize possible space-time interactions, we compute differences of effects between a certain year, for example 1986, to a reference year, for example 1980. Figures 6.18 and 6.19 display such differences together with the 'baseline' effect for 1980. We can conclude that reemployment chances for females became worse in all districts in the second time period. In the third time period, there is a distinct positive development in western and northern regions. We also see some negative effects in border regions in the northeast and mideast close to the former (East) German Democratic Republic. A possible explanation is that the unemployment rate for women in East Germany was quite high, so that they competed with West Germans for vacancies in the former border region.

Estimates of remaining effects are practically the same as for the main effects model and are, therefore, not presented. △

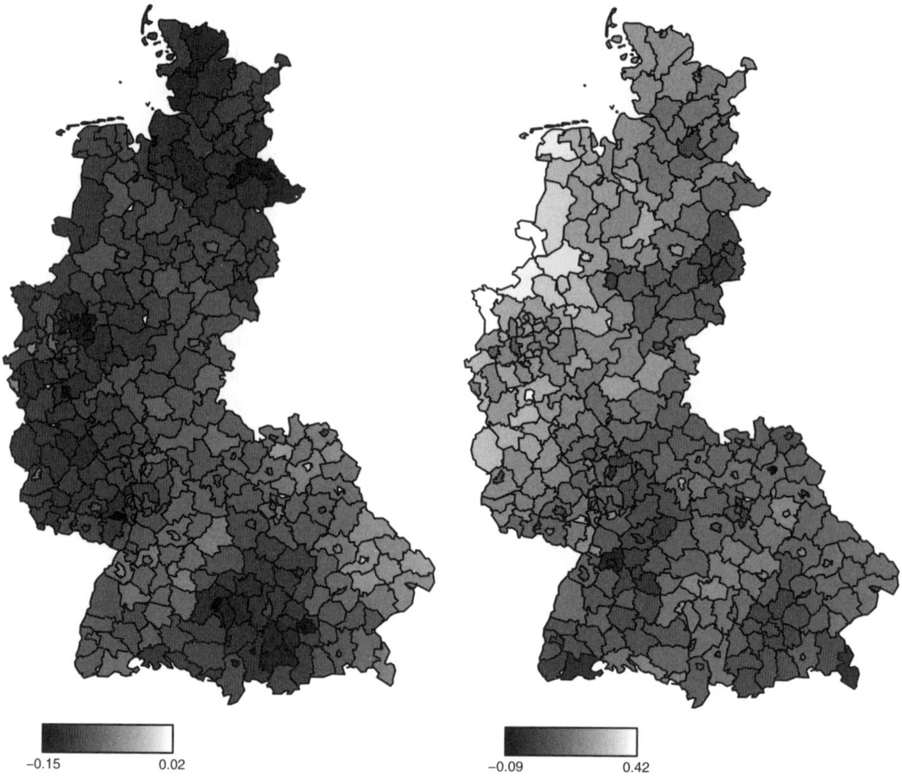

Fig. 6.18 Unemployment duration: Posterior mean estimates for the differences in spatial effects in years 1986 (left) and 1991 (right) as compared to the reference year 1980.

Example 6.10: Childhood mortality in Nigeria
As already noted in Section 6.1.1, discrete-time models also allow us to account for special cases of interval-censoring. In the example on childhood mortality in Nigeria, interval-censoring occurred in two places: Survival times were rounded to months and in addition heaping at half years and years was present for larger survival times. The first problem can be accounted for in a discrete-time model where all durations are measured in months instead of days. To avoid difficulties with heaping, we restricted the following analysis to duration times up to 18 months. For the analyses, we considered a fully Bayesian geoadditive probit model that resembles the structure of the model in Example 6.8 with estimation being based on MCMC simulations. We also used the same specifications for nonparametric and spatial effects as in Example 6.8.

Figure 6.20 shows the estimated nonparametric effects and the estimated log-baseline effect. For the nonparametric effects of continuous covariates, the

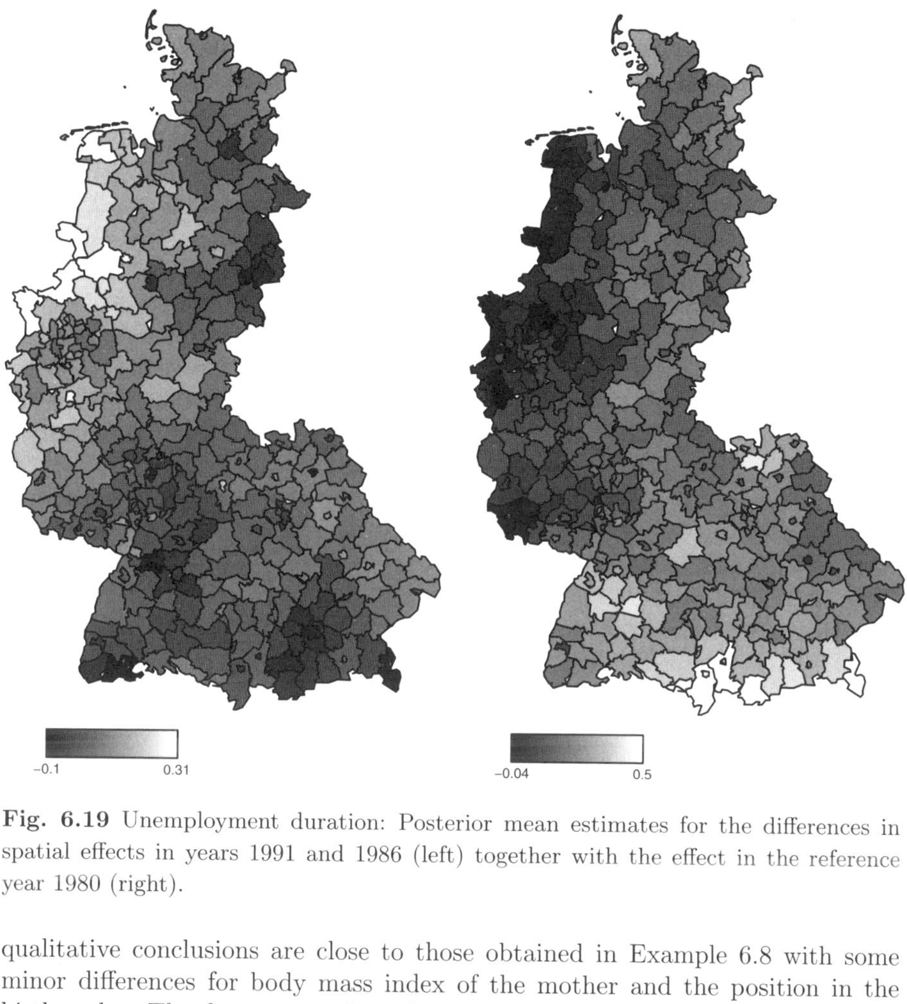

Fig. 6.19 Unemployment duration: Posterior mean estimates for the differences in spatial effects in years 1991 and 1986 (left) together with the effect in the reference year 1980 (right).

qualitative conclusions are close to those obtained in Example 6.8 with some minor differences for body mass index of the mother and the position in the birth order. The former now has almost constant effect for low values of the BMI and shows a stronger increase for large values. However, note again that we are not only using a different model specification but also a different time horizon that might be responsible for the changes in estimated covariate effects. Moreover, we also do not account for heaping effects that already appear for 18 and 12 months. These lead to the peaks in the estimated log-baseline.

In the discrete-time analysis, the spatial effect (Fig. 6.21) is much more expressed than in the analysis based on interval-censored observations. The reason is two-fold: On the one hand, there seems to be more spatial variability in the durations up to 18 months than in the complete data. On the other hand, we have used a fully Bayesian approach based on MCMC in the discrete-time analysis that typically leads to somewhat more variable spatial effect estimates. △

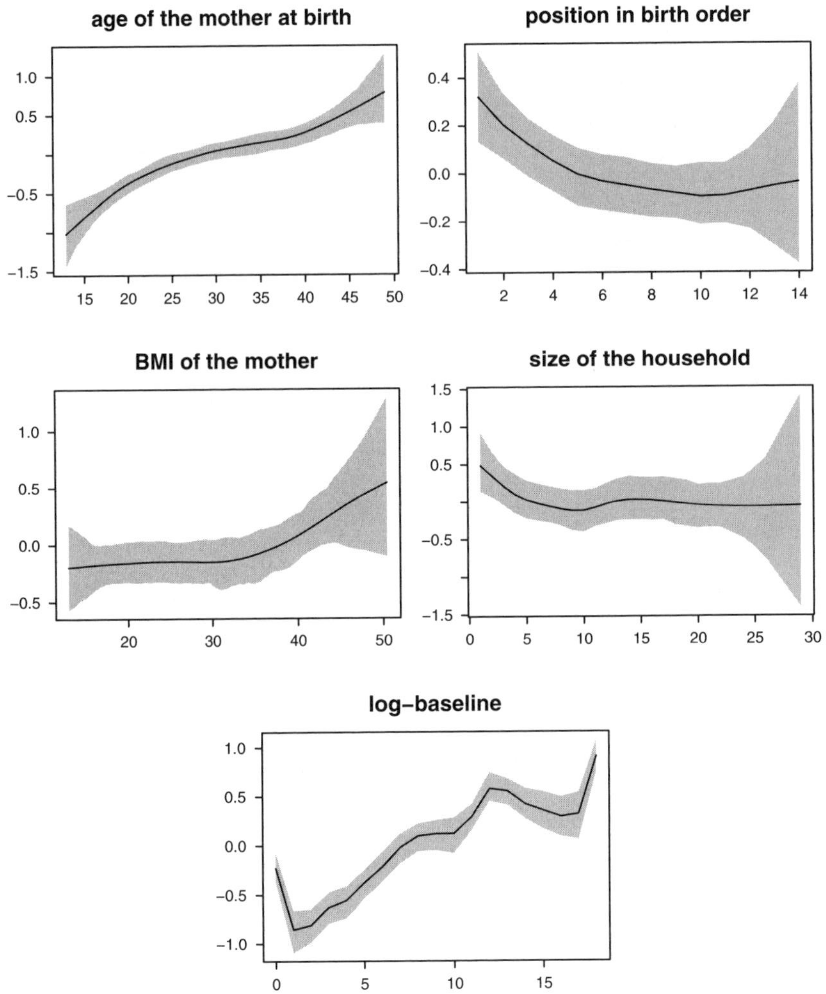

Fig. 6.20 Childhood mortality: Estimated nonparametric effects together with 95% pointwise credible intervals in a discrete-time analysis with time horizon restricted to 18 months.

6.4 Accelerated failure time models

Accelerated failure time models are a useful alternative to continuous-time Cox-type hazard rate models, in particular when the assumption of proportional hazards is questionable or not fulfilled. For a vector $\boldsymbol{u}$ of time-constant covariates, the AFT model conceptualizes the effects of covariates on survival time T via

$$T = \exp(\boldsymbol{u}'\boldsymbol{\beta})V, \tag{6.23}$$

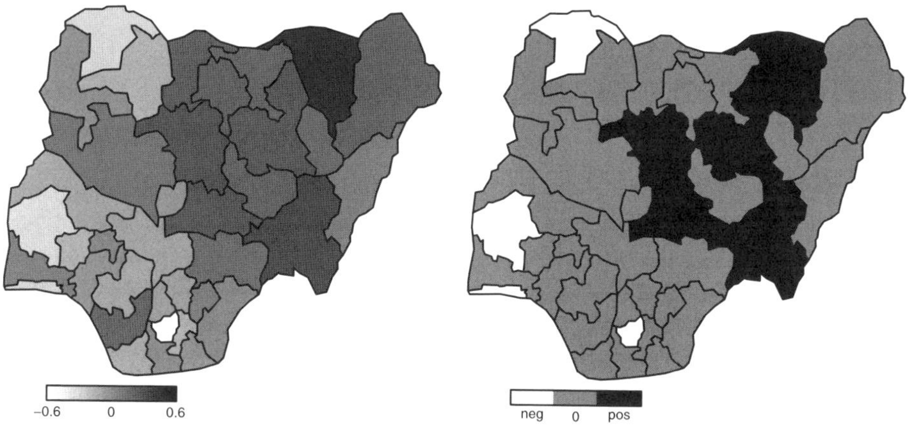

Fig. 6.21 Childhood mortality: Estimated spatial effect (left) together with 95% pointwise posterior probabilities (right) in a discrete-time analysis with time horizon restricted to 18 months.

where V is a baseline survival time. The factor $\exp(\boldsymbol{u}'\boldsymbol{\beta})$ accelerates (or decelerates) the baseline survival time. On the logarithmic scale, the model becomes a simple linear regression model

$$\log(T) = \boldsymbol{u}'\boldsymbol{\beta} + \sigma\varepsilon \tag{6.24}$$

with $\sigma\varepsilon = \log(V)$, where σ is a scale parameter and ε has density $f_\varepsilon(\varepsilon)$, with known location and scale parameter. Absorbing the location parameter into the intercept term, we may without loss of generality assume that $\mathbb{E}(\varepsilon) = 0$. Instead of transforming T to $\log(T)$, other transformations are conceivable. Therefore, AFT models are also called transformation models. We will, however, only consider the logarithmic transformation.

Parametric AFT models assume that $f_\varepsilon(\varepsilon)$ is a density of a specific type, for example standard-normal, standard-logistic or (standard-) minimum-extreme-value. Then the parameters $\boldsymbol{\beta}$ and σ are commonly estimated by maximum likelihood techniques. To avoid such parametric assumptions, a possible approach is to leave $f_\varepsilon(\varepsilon)$ unspecified, similarly as the baseline hazard $\lambda_0(t)$ in the Cox model, and to estimate only the regression parameters $\boldsymbol{\beta}$, based on linear rank tests (Kalbfleisch and Prentice, 2002, Ch. 7). However, this approach is practically applicable only with low-dimensional covariate vectors $\boldsymbol{u}$. Furthermore, there is no possibility for estimating the density or hazard rate of ε or the baseline survival time V, which would be needed for prediction.

This section considers two extensions of AFT models. First, parametric assumptions on $f_\varepsilon(\varepsilon)$ are replaced by flexible semiparametric distributions such as the penalized Gaussian mixture (Komárek, Lesaffre and Hilton, 2005) or a DPM prior (Kuo and Mallick, 1997). Second, the linear predictor $\boldsymbol{u}'\boldsymbol{\beta}$ can be extended to (high-dimensional) structured additive predictors as in Cox-type hazard rate models. High-dimensional predictors can be regularized by imposing shrinkage or spike and slab priors, and nonlinear or spatial effects may be included as before.

6.4.1 Observation models, likelihoods and priors

Consider first the basic AFT model (6.24) for independent, right-censored survival times T_i and time constant covariates $\boldsymbol{u}_i$, $i = 1, \ldots, n$. The density and the survivor function of $Y = \log(T)$ are given by

$$f_Y(y) = \frac{1}{\sigma} f_\varepsilon(\varepsilon), \quad S_Y(y) = S_\varepsilon(\varepsilon) = 1 - F_\varepsilon(\varepsilon),$$

where $\varepsilon = (y - \boldsymbol{u}'\boldsymbol{\beta})/\sigma$. Under non-informative right-censoring and observed log-survival times $y_i = \log(t_i)$ and censoring indicators d_i, $i = 1, \ldots, n$, the likelihood function is

$$L(\boldsymbol{\beta}, \sigma) = \prod_{i=1}^{n} \left(\frac{1}{\sigma} f_\varepsilon(\varepsilon_i) \right)^{d_i} S_\varepsilon(\varepsilon_i)^{1-d_i}, \tag{6.25}$$

with $\varepsilon_i = (y_i - \boldsymbol{u}_i'\boldsymbol{\beta})/\sigma$. For a *known*, parametric density f of ε, ML estimators, score functions and information matrices can then be derived in the usual way, see for example Kalbfleisch and Prentice (2002).

Replacing the linear predictor $\eta_i = \boldsymbol{u}_i'\boldsymbol{\beta}$ through a structured additive predictor in generic form

$$\eta_i = \boldsymbol{u}_i'\boldsymbol{\beta} + \boldsymbol{x}_{i1}'\boldsymbol{\gamma}_1 + \ldots + \boldsymbol{x}_{ir}'\boldsymbol{\gamma}_r$$

and imposing priors on $\boldsymbol{\beta}$ and $\boldsymbol{\gamma}_1, \ldots, \boldsymbol{\gamma}_r$ as in previous chapters, we obtain semiparametric Bayesian AFT models, but still under the restriction that f is *known*, for example standard normal, logistic, and so an. This assumption can be relaxed in different ways through flexible semiparametric error distributions. One approach is based on penalized Gaussian mixtures as already outlined for non-Gaussian random effects in Section 3.2.1.

Example 6.11: Primary biliary cirrhosis of the liver

We complement the analyses conducted in Example 6.6 based on Cox-type hazard regression models with some results from an AFT model with Gaussian error distribution and log-transformed survival times. Again, we consider Bayesian lasso and normal mixture of inverse gamma (NMIG) priors in case of the Bayesian AFT models and a standard parametric AFT model for comparison. The resulting parameter estimates are shown in Fig. 6.22. When comparing the estimates with the results from the hazard regression model, it is important to note that the sign of the coefficients is reversed since we are now considering effects directly on the survival time and not on the hazard rate. While positive effects in a hazard regression model lead to an increased hazard and therefore shorter survival

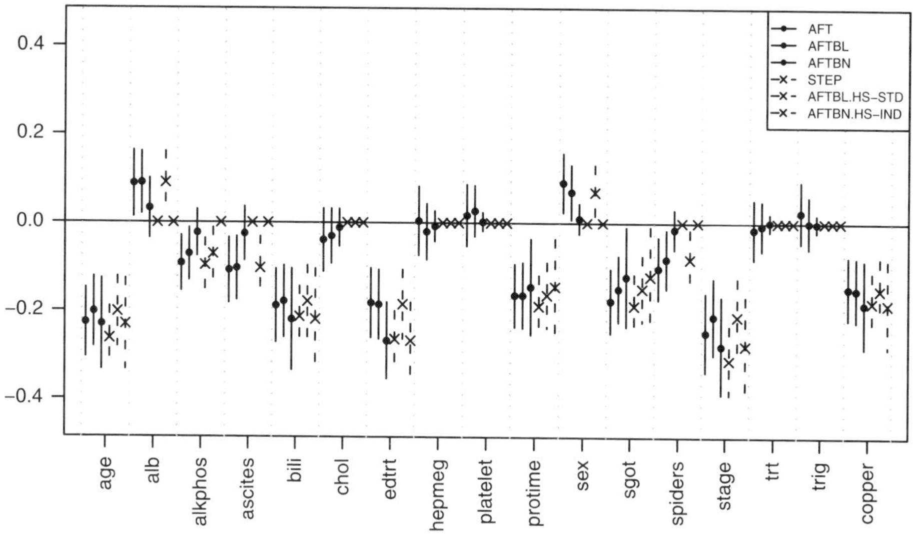

Fig. 6.22 PBC liver data: Estimated covariate effects from a standard AFT model (AFT), a Bayesian AFT model with lasso prior (AFTBL) or NMIG prior (AFTBN). Results including variable selection have been obtained with a stepwise procedure (STEP), hard shrinkage based on a one standard deviation interval (AFTBL.HS-STD) and the inclusion indicators (AFTBN.HS-IND) of the NMIG prior.

times, the reverse implication holds in AFT models where positive effects indicate larger survival times. Taking this change in signs into account, results from the AFT model and the Cox-type models are mostly consistent with each other. Figure 6.22 also includes results obtained from applying the same variable selection rules as in Example 6.6 (one standard deviation intervals for the lasso and a cut off of 0.5 for the NMIG prior) as well as results from a stepwise AFT analysis.

In contrast, the estimated inclusion probabilities in Fig. 6.23 show some differences to the results from the hazard regression models. The separation between important and irrelevant covariates is clearer in this picture and there are also some other effects selected. Note also that the variable corresponding to the presence of edema is clearly selected in the AFT model. Finally, Fig. 6.24 presents a complement to the coefficient paths from the hazard regression models. △

Penalized Gaussian mixture (PGM)

Komárek *et al.* (2005) express the density of the error term in the basic AFT model (6.24) as

$$f_\varepsilon(\varepsilon|\boldsymbol{c}) = \sum_{k=1}^{K} c_k \varphi(\varepsilon; \mu_k, \sigma_0^2), \tag{6.26}$$

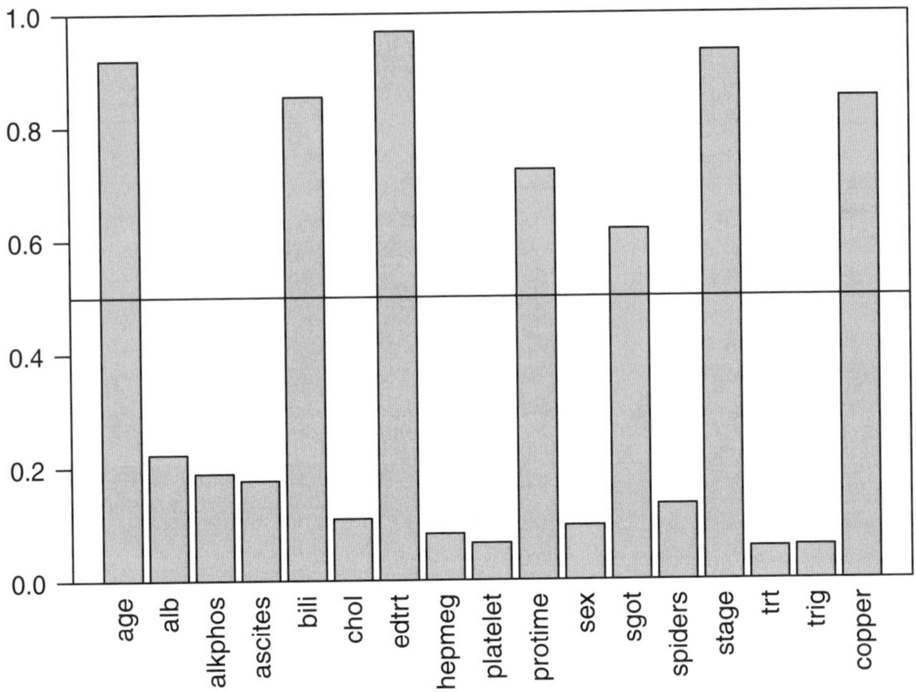

Fig. 6.23 PBC liver data: Inclusion frequencies for the binary coefficient indicators in the Bayesian AFT model with NMIG prior.

where $\varphi(\varepsilon; \mu_k, \sigma_0^2)$ are Gaussian densities with means $\mu_1 < \ldots < \mu_K$, variance σ_0^2 fixed by design, and $\boldsymbol{c} = (c_1, \ldots, c_k)'$ are mixture coefficients to be estimated. The approach is motivated by P-spline representations of regression functions: B-splines are replaced through the basis functions $\varphi(\varepsilon; \mu_k, \sigma_0^2)$, $k = 1, \ldots, K$, and c_k are the basis coefficients. As with P-splines, K is chosen to be comparably large to achieve flexibility, and a smoothness prior is placed on the coefficient vector $\boldsymbol{c}$. The means $\mu_1 < \ldots < \mu_K$ play the role of knots. To choose the knots $\mu_1 < \ldots < \mu_K$ and the basis standard deviation σ_0^2, the following is recommended as a standard option: The knots are placed on an equidistant grid from -6 to $+6$, with a distance of 0.3 between consecutive knots. With $\sigma_0^2 = 2/3(\mu_k - \mu_{k-1})$, a Gaussian basis density is practically zero outside (μ_{k-2}, μ_{k+2}).

To ensure that $f_\varepsilon(\varepsilon|\boldsymbol{c})$ in (6.26) is a density, the coeficients c_k have to be positive and must sum up to 1. These constraints are automatically fulfilled by reparametrizing through coefficients $a_1, \ldots, a_{K-1}$ related to $c_1, \ldots, c_K$ by

$$c_k = \frac{\exp(a_k)}{1 + \sum_{l=1}^{K-1} \exp(a_l)}, k = 1, \ldots, K-1; \quad c_K = \frac{1}{1 + \sum_{l=1}^{K-1} \exp(a_l)}.$$

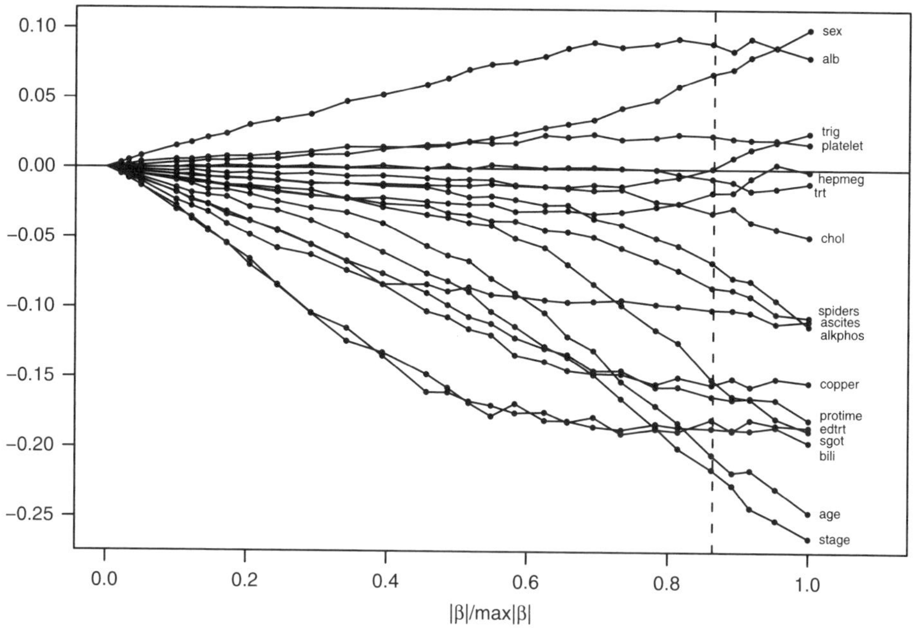

Fig. 6.24 PBC liver data: Coefficient paths for the Bayesian AFT model with lasso prior for the regression coefficients. The vertical dashed line indicates the location of the estimated smoothing parameter.

Two further restrictions on the (reparametrized) coefficients are required to guarantee that ε is normalized, that is,

$$\mathbb{E}(\varepsilon) = \sum_{k=1}^{K} c_k \mu_k = 0, \quad \text{Var}(\varepsilon) = \sum_{k=1}^{K} c_k(\mu_k^2 + \sigma_0^2) = 1.$$

This can be achieved by expressing, for example, a_{K-2} and a_{K-1} through the remaining coefficients $\boldsymbol{a} = (a_1, \dots, a_{K-3})'$. It can be shown that

$$a_k = \log\left\{\omega_{K,k} + \sum_{l=1}^{K-3} \omega_{l,k} \exp(a_l)\right\}, \quad k = K-2, K-1,$$

with

$$\omega_{l,K-2} = -\frac{\mu_l - \mu_{K-1}}{\mu_{K-2} - \mu_{K-1}} \cdot \frac{1 - \sigma_0^2 + \mu_{K-1}\mu_l}{1 - \sigma_0^2 + \mu_{K-1}\mu_{K-2}},$$

$$\omega_{l,K-1} = -\omega_{l,K-2} \cdot \frac{\mu_{K-2}}{\mu_{K-1}} - \frac{\mu_l}{\mu_{K-1}}, \quad l = 1, \dots, K-3, K.$$

To enforce smoothness of $f_\varepsilon(\varepsilon)$, a random-walk prior is assumed for the sequence $a_1, a_2, \ldots$ of coefficients, in complete analogy to P-splines. Thus the prior is (conditionally) Gaussian

$$p(\boldsymbol{a}|\tau_a^2) \propto \exp\left(-\frac{1}{2\tau_a^2}\boldsymbol{a}'\boldsymbol{K}\boldsymbol{a}\right),$$

with the usual penalty matrix $\boldsymbol{K} = \boldsymbol{D}'\boldsymbol{D}$. The log-prior corresponds to the quadratic penalty function $\lambda\boldsymbol{a}'\boldsymbol{K}\boldsymbol{a}$ used in penalized likelihood or empirical Bayes inference. For full Bayes inference a hyperprior, for example an $\mathrm{IG}(\varepsilon, \varepsilon)$ prior is specified for τ_a^2.

AFT models with DP(M) priors

As an alternative to the PGM method, AFT models with DPM priors for the distribution of V in (6.23) or ε in (6.24) have been proposed. Kuo and Mallick (1997) consider a general hierarchical model for (6.23) as follows:

$$\begin{aligned} T_i &= \exp(\boldsymbol{u}_i'\boldsymbol{\beta})V_i, \quad i = 1, \ldots, n, \\ V_i|\theta_i &\overset{\text{ind.}}{\sim} f_V(\cdot|\theta_i), \\ \theta_i|G &\overset{\text{i.i.d.}}{\sim} G, \\ G &\sim \mathrm{DP}(\alpha, G_0). \end{aligned}$$

In their application, they assume a Gaussian density for $f_V(\cdot|\theta_i)$ and an exponential distribution for G_0. To avoid negative values for V_i when using a normal density, Ghosh and Ghosal (2006) propose a scale mixture of Weibull densities instead, and Hanson (2006) suggests a DPM model of gamma densities, mixing over both the scale and the slope. A different nonparametric prior in form of a normalized inverse-Gaussian process has been recently proposed by Argiento, Guglielmi and Pievatolo (2009).

6.5 Multi-state models

As discussed in the introductory section, multi-state models extend survival models to event history data that are observed when individuals or objects move to or between different types of states as time proceeds. Multi-state models can be described in terms of transitions between states. A state structure specifies the diverse states, that might be transient, recurrent or absorbing, and defines which transitions are possible. The simplest models of this type are stationary Markov processes with a discrete state space where each of the transitions is associated with a time-constant transition rate $\lambda^{(h)} > 0$, with $h = 1, \ldots, H$, indexing the set of possible transitions. Even if transition rates are generalized to transition-specific Cox models

$$\lambda_i^{(h)}(t) = \lambda_0^{(h)}(t) \exp(\boldsymbol{u}_i'(t)\boldsymbol{\beta}^{(h)}),$$

relaxing the Markov assumption and accounting for population heterogeneity through covariates, such models are often still too restrictive in many real data applications.

As for survival data, multi-state processes may be observed on a continuous or a discrete time-scale. For example, the analysis on unemployment duration (Example 6.9) may be extended to include part time jobs, continuation training and retirement as further states. Also, original time may be grouped into coarser intervals, as in the application to human sleep data (Example 6.13).

In all cases, there is a need for extending parametric predictors to structured additive predictors. Section 6.5.1 considers continuous-time multi-state hazard regression, and Section 6.5.2 extends concepts for discrete-time survival data to multi-state models.

6.5.1 Continuous-time transition rate models

Multi-state models for transitions of types h, $h = 1, \ldots, H$, and individuals i, $i = 1, \ldots, n$, are fully described by individual-specific transition rates

$$\lambda_i^{(h)}(t) = \lim_{\Delta t \to 0} \frac{1}{\Delta t} \mathbb{P}(\text{obs. } i \text{ has type-}h \text{ transition in } [t + \Delta t) \mid \mathcal{F}_t,\ i \in R^{(h)}(t)),$$

where $\mathcal{F}_t$ denotes the information given through covariates and the history of the multi-state process up to time t, and risk set $R^{(h)}(t)$ is the set of individuals under risk for a type-h transition immediately before t. In terms of the multi-state process $\{y_i(t), t \geq 0\}$ of individual i, where $y_i(t)$ is a discrete response with states $\{1, \ldots, Q\}$ observed over time, let h denote a typical transition between states j and k, say. Then transition rates can also be defined by

$$\lambda_i^{(h)}(t) = \lim_{\Delta t \to 0} \frac{1}{\Delta t} \mathbb{P}(y_i(t + \Delta t) = k \mid y_i(t) = j,\ \mathcal{F}_t). \tag{6.27}$$

Assuming that the information $\mathcal{F}_t$ can be expressed in the form of a structured additive predictor $\eta_i^{(h)}(t)$, transition rates are modelled in analogy to survival hazard rates through

$$\lambda_i^{(h)}(t) = \exp(\eta_i^{(h)}(t)), \quad h = 1, \ldots, H,$$

where

$$\eta_i^{(h)}(t) = \boldsymbol{u}_i'(t)\boldsymbol{\beta}^{(h)} + g_0^{(h)}(t) + \sum_{l=1}^{m} g_l^{(h)}(t) v_{il}(t) + \sum_{j=1}^{p} f_j^{(h)}(x_{ij}(t)) + \alpha_{g_i}^{(h)}. \tag{6.28}$$

The components in this additive predictor are essentially the same as in Section 6.2: Parametric effects $\boldsymbol{\beta}^{(h)}$ of covariates $\boldsymbol{u}_i(t)$, a time-varying, nonparametric baseline effect $g_0^{(h)}(t)$ common for all individuals, covariates $v_{il}(t)$ with time-varying effects $g_l^{(h)}(t)$, nonparametric effects of continuous covariates $x_{ij}(t)$, and group-specific frailty or random effects $\alpha_{g_i}^{(h)}$. Modifications or extensions of the predictor (6.28) could be individual-specific frailties α_i, nonparametric interaction terms between time t and a continuous covariate or between two continuous covariates, and spatial effects if spatial information is present. We will not incorporate such effects in our examples, but the methodology can be extended to include such components in a straightforward manner. Covariates may be time-constant or time-varying. In the latter case it is assumed that they are predictable, that is, the value at time t is already known immediately before t. A sufficient condition is that the path of such a covariate is continuous from the left.

As in structured additive predictors (6.9) for survival models, all functions are represented through B-splines or, in principle, other basis function expansions, and random effects are i. i. d. Gaussian random variables. Therefore (6.28) can be rewritten in the same generic form as the predictor (6.10) after reindexing:

$$\eta_i^{(h)}(t) = \boldsymbol{u}_i'(t)\boldsymbol{\beta}^{(h)} + \boldsymbol{x}_0'(t)\boldsymbol{\gamma}_0^{(h)} + \boldsymbol{x}_{i1}'(t)\boldsymbol{\gamma}_1^{(h)} + \ldots + \boldsymbol{x}_{ir}'(t)\boldsymbol{\gamma}_r^{(h)}. \tag{6.29}$$

In principle, covariates as well as design vectors might also be transition-specific, that is, they may depend on h. We omit this for notational simplicity.

Priors for parameters are the same as for survival models but transition-specific. All parameters are assumed to be *a priori* independent.

6.5.2 Counting process representation and likelihood contributions

For each individual, we assume that a full transition path is observed consisting of the following parts: A set of ordered time points $0 = S_{i0} < S_{i1} < \ldots < S_{ir} < \ldots < S_{i,m_i}$ indicating the time points of transitions for individual i, the states $y_i(S_{ir})$, $r = 1, \ldots, m_i$, the individual moves to at the end of a time interval, and indicators $d_i^{(h)}(t)$ for type h transitions at time t. From the time points of transitions, we can deduce the duration time in the r-th state as $T_{ir} = S_{ir} - S_{i,r-1}$, $r = 1, \ldots, m_i$.

The likelihood contributions can be derived from a counting process representation of a multi-state model. Let $N_i^{(h)}(t)$, $h = 1, \ldots, H$, be a set of counting processes, counting transitions of type h for individual i up to time t. Consequently, h indexes the observable transitions in the model under consideration and the jumps of the counting processes $N_i^{(h)}(t)$ are defined by the transition times S_{ir}, $r = 1, \ldots, m_i$, of the corresponding multi-state process for individual i.

From classical counting process theory (see for example Andersen *et al.*, 1993, Ch. VII.2), the intensity processes $\alpha_i^{(h)}(t)$ of the counting processes $N_i^{(h)}(t)$ are derived as the product of the hazard rate for type h transitions $\lambda_i^{(h)}(t)$ and a predictable at-risk indicator process $I_i^{(h)}(t)$ (a sufficient condition for $I_i^{(h)}(t)$ to be predictable are paths continuous from the left), that is,

$$\alpha_i^{(h)}(t) = I_i^{(h)}(t)\lambda_i^{(h)}(t),$$

where the hazard rates are constructed in terms of covariates as described in the introduction of this section. The at-risk indicator $I_i^{(h)}(t)$ takes the value one if individual i is at risk for a type h transition immediately before time t and zero otherwise. For example, in the multi-state model of Fig. 6.1(a), an individual in state 2 is at risk for both transitions to state 1 and state 3. Hence, the at-risk indicators for both the transitions '2 to 1' and '2 to 3' will be equal to one as long as the individual remains in state 2.

Under mild regularity conditions, the individual log-likelihood contributions can now be obtained from counting process theory as

$$l_i = \sum_{h=1}^{H} \left(\int_0^{S_{i,m_i}} \log(\lambda_i^{(h)}(t)) dN_i^{(h)}(t) - \int_0^{S_{i,m_i}} \lambda_i^{(h)}(t) I_i^{(h)}(t) dt \right), \qquad (6.30)$$

where S_{i,m_i} denotes the time until which individual i has been observed. The likelihood contributions can be interpreted similarly to hazard rate models for survival times (and in fact coincide with these in the case of a multi-state process with only one transition to an absorbing state). The first term corresponds to contributions at the transition times since the integral with respect to the counting process in fact equals a simple sum over the transition times. Each of the summands is then given by the log-intensity for the observed transition evaluated at this particular time point. In survival models, this term simply equals the log-hazard evaluated at the survival time for uncensored observations. The second term reflects cumulative intensities integrated over accordant waiting periods between two successive transitions. The integral is evaluated for all transitions the corresponding person is at risk at during the current period. In survival models there is only one such transition (the transition from 'alive' to 'dead') and the integral is evaluated from the time of entrance to the study to the survival or censoring time.

These considerations yield an alternative representation of the likelihood, where each of the individual contributions is expressed in terms of the transition indicators $d_i^{(h)}(t)$ and observed transition times S_{ir}, $r = 0, \ldots, m_i$. The indicators $d_i^{(h)}(t)$ take the value one if a transition of type h is observed at time t for individual i and zero otherwise, while the S_{ir} are defined by the times at which the corresponding individual experiences a transition. This leads to the alternative log-likelihood formula

$$l_i = \sum_{r=1}^{m_i} \sum_{h=1}^{H} \left(d_i^{(h)}(S_{ir}) \log(\lambda_i^{(h)}(S_{ir})) - I_i^{(h)}(S_{ir}) \int_{S_{i,r-1}}^{S_{ir}} \lambda_i^{(h)}(t) dt \right), \qquad (6.31)$$

which reveals more clearly the connection to the commonly known likelihood of hazard rate models in case of continuous survival times.

Under the usual assumption of conditional independence, the complete log-likelihood is given by the sum of the individual contributions.

6.5.3 Empirical and full Bayes inference

Inserting $\lambda_i^{(h)}(t) = \exp(\eta_i^{(h)}(t))$, with the predictor $\eta_i^{(h)}(t)$ in generic form (6.29), the log-likelihood contributions become

$$l_i = \sum_{h=1}^{H} l_i^{(h)}$$

with

$$l_i^{(h)} = \sum_{r=1}^{m_i} \left(d_i^{(h)}(S_{ir})\eta_i^{(h)}(S_{ir}) - I_i^{(h)}(S_{ir}) \int_{S_{i,r-1}}^{S_{ir}} \exp(\eta_i^{(h)}(t))dt \right).$$

As long as the predictors $\eta_i^{(h)}(t)$ contain no common parameters, and that is what we assume, the individual log-likelihood contributions $l_i(\boldsymbol{\beta}, \boldsymbol{\gamma})$ decompose into a sum of transition-specific log-likelihood contributions:

$$l_i(\boldsymbol{\beta}, \boldsymbol{\gamma}) = \sum_{h=1}^{H} l_i^{(h)}(\boldsymbol{\beta}^{(h)}, \boldsymbol{\gamma}^{(h)}).$$

Summing up over individual contributions, the log-likelihood decomposes into the sum

$$l(\boldsymbol{\beta}, \boldsymbol{\gamma}) = \sum_{h=1}^{H} l^{(h)}(\boldsymbol{\beta}^{(h)}, \boldsymbol{\gamma}^{(h)}),$$

of transition-specific log-likelihoods $\sum_{i=1}^{n} l_i^{(h)}(\boldsymbol{\beta}^{(h)}, \boldsymbol{\gamma}^{(h)})$. Moreover, each summand

$$l_{ir}^{(h)} = d_i^{(h)}(S_{ir})\eta_i^{(h)}(S_{ir}) - I_i^{(h)}(S_{ir}) \int_{S_{i,r-1}}^{S_{ir}} \exp(\eta_i^{(h)}(t))dt$$

in $l_i^{(h)}$ has essentially the same structure as an individual log-likelihood contribution in Cox-type survival models (subject to an appropriate adjustment of the risk set). Therefore, inference is closely related to inference in survival models, so that we do not repeat details (which may be found in Kneib and Hennerfeind, 2008).

For EB inference, the penalized log-likelihood decomposes into separate penalized log-likelihoods

$$l_{\text{pen}}^{(h)}(\boldsymbol{\beta}^{(h)}, \boldsymbol{\gamma}^{(h)}) = l^{(h)}(\boldsymbol{\beta}^{(h)}, \boldsymbol{\gamma}^{(h)}) - \sum_{l=0}^{r} \frac{1}{2\tau_{lh}^2} \gamma_l^{(h)\prime} \boldsymbol{K}_l^{(h)} \gamma_l^{(h)}.$$

Therefore, after reparametrizing in mixed model representation, maximization of $l_{\text{pen}}(\boldsymbol{\beta}, \boldsymbol{\gamma})$ reduces to separate maximization problems which can be achieved by a

Newton–Raphson-type algorithm that extends the algorithm for survival models. Estimation of variance parameters is based on (approximate) maximization of marginal likelihoods $L^{(h)}(\boldsymbol{\tau}^{(h)})$, using the same Laplace approximations as for survival data.

Example 6.12: Nosocomial infection

In this example, we analyse data on 1,876 intensive care patients from the Charité University Hospital in Berlin collected from February 2000 until July 2001. Among these 1,876 patients, 158 acquired nosocomial pneumonia and our main interest is on the effect of this intermediate event on the time until discharge. We therefore consider a multi-state model where each patient starts in the state 'intensive care' immediately after delivery to the intensive care unit. From this initial state, patients can either move directly to 'discharge' or perform an intermediate transition to the state 'infection'. The resulting structure of the multi-state model is depicted in the left panel of Fig. 6.25 while the right panel shows a common misspecification where the temporal alignment of intermediate event is ignored. Instead of taking the time of infection into account, the group of patients is separated into a group with and a group without infection as if the infection had already been present at the time of delivery to the intensive care unit. We will see later on that this misspecification has a very specific impact on the resulting transition intensities.

In this application, no further covariates are considered in the analyses and therefore each transition intensity is only represented by a single penalized spline that corresponds to the baseline intensity (on the log scale). We use cubic penalized splines with 20 inner knots and second-order random-walk penalty and estimate the intensities based on EB inference. The resulting posterior mode estimates for the discharge rates are shown in Fig. 6.26 and indicate that an intermediate infection has the expected effect of prolonging stays at the intensive care unit since the discharge hazard is much lower for patients in state 'infection'. This difference is particularly expressed early at the time scale but disappears after about 20 days in intensive care.

Fig. 6.25 Nosocomial infection: Structure of the correctly specified (left) and a misspecified (right) multi-state model.

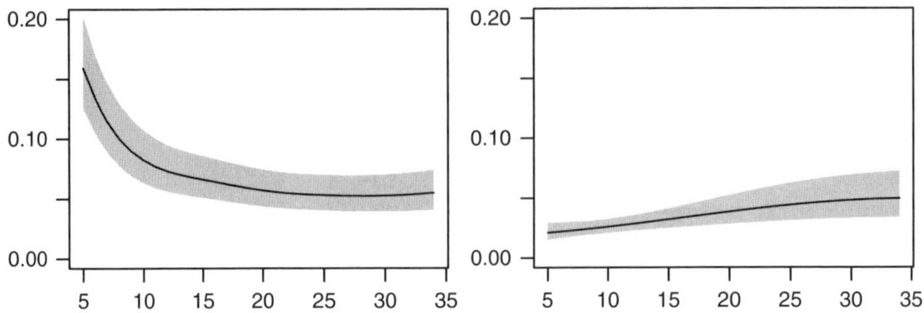

Fig. 6.26 Nosocomial infection: Posterior mode estimates together with pointwise 95% credible intervals for the discharge hazard without infection (left) and with infection (right).

Figure 6.27 depicts the impact of misspecifying intermediate infection as a time-constant factor. Interestingly, the discharge hazard for infected patients is estimated to be even lower while the discharge hazard for patients without infection is higher than for the correct model specification. Hence the impact of infections would be overestimated in the misspecified model. This is not only the case in our application but can be shown more generally, see Beyersmann, Kneib, Schumacher and Gastmeier (2009) for more details both on this application and the underlying reasons for biased estimation. △

After minor modifications, FB inference also proceeds similarly as in in Section 6.2.2. Predictor vectors are extended to

$$\boldsymbol{\eta}^{(h)} = (\eta_1^{(h)}(S_{1,0}), \dots, \eta_1^{(h)}(S_{1,m_1}), \dots, \eta_n^{(h)}(S_{n,0}), \dots, \eta_n^{(h)}(S_{n,m_n}))'$$

and can be represented in generic notation as

$$\boldsymbol{\eta}^{(h)} = \boldsymbol{U}^{(h)}\boldsymbol{\beta}^{(h)} + \boldsymbol{X}_0^{(h)}\boldsymbol{\gamma}_0^{(h)} + \dots + \boldsymbol{X}_l^{(h)}\boldsymbol{\gamma}_l^{(h)} + \dots + \boldsymbol{X}_r^{(h)}\boldsymbol{\gamma}_r^{(h)},$$

with design matrices constructed as usual. To update regression parameters we use MH algorithms with Gaussian proposal densities having a similar structure as survival models. For example, for updating coefficients $\boldsymbol{\gamma}_l^{(h)}$ corresponding to time-independent functions or random effects, we draw proposals $\boldsymbol{\gamma}_{lh}^p$ from $\varphi(\boldsymbol{\gamma}_{lh}^p; \boldsymbol{\mu}_l^{(h)}, \boldsymbol{\Sigma}_l^{(h)})$, with

$$\boldsymbol{\Sigma}_l^{(h)} = \left(\boldsymbol{X}_l^{(h)\prime}\boldsymbol{W}_h^c\boldsymbol{X}_l^{(h)} + \frac{\boldsymbol{K}_l}{(\tau_{lh}^2)^c}\right)^{-1}, \quad \boldsymbol{\mu}_l^{(h)} = \boldsymbol{\Sigma}_l^{(h)}\boldsymbol{X}_l^{(h)\prime}\boldsymbol{W}_h^c\tilde{\boldsymbol{y}}_h^c.$$

The weight matrix is $\boldsymbol{W}_h^c = \text{diag}(w_{h11}^c, \dots, w_{h1,m_1}^c, \dots, w_{hn,m_n}^c)'$ with weights

$$w_{hir} = \int_{S_{i,r-1}}^{S_{ir}} I_i(t)\lambda_i^{(h)}(t)dt,$$

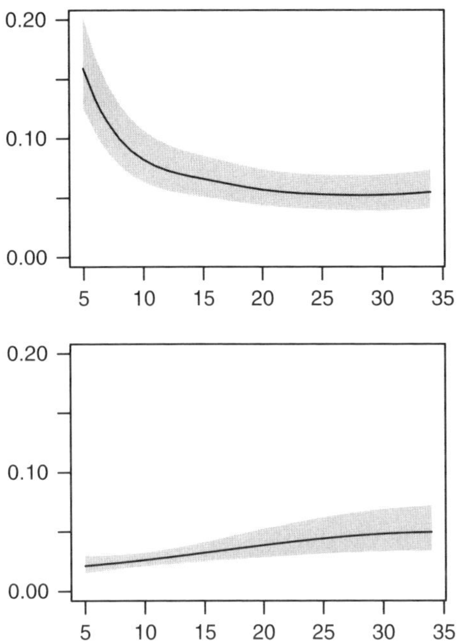

Fig. 6.27 Nosocomial infection: Posterior mode estimates together with pointwise 95% credible intervals for the discharge hazard without infection (upper row) and with infection (lower row) in case of the correctly specified model (left column) and in case of the misspecified model (right column).

evaluated at current values of parameters, and working observations

$$\tilde{\boldsymbol{y}}_h^c = \boldsymbol{X}_l^{(h)} \gamma_{lh}^c + (\boldsymbol{W}_h^c)^{-1} \boldsymbol{d} - \mathbf{1}.$$

For parameters corresponding to time-dependent effects, we use the computationally faster MH algorithm with conditional prior proposals, requiring no derivatives of the log-likelihoods.

Assuming (possibly improper) inverse gamma distributions for variance parameters, full conditionals are again (proper) inverse gamma distributions with updated parameters.

Example 6.13: Human sleep

In this application, we demonstrate both EB and FB inference for the human sleep data already discussed in the introduction to this chapter and in Example 1.7. In particular, we will differentiate between the sleep states 'awake', 'REM' and 'Non-REM', using the multi-state structure depicted in Fig. 6.28. We simplified the structure by aggregating 'REM' and 'Non-REM' to one joint

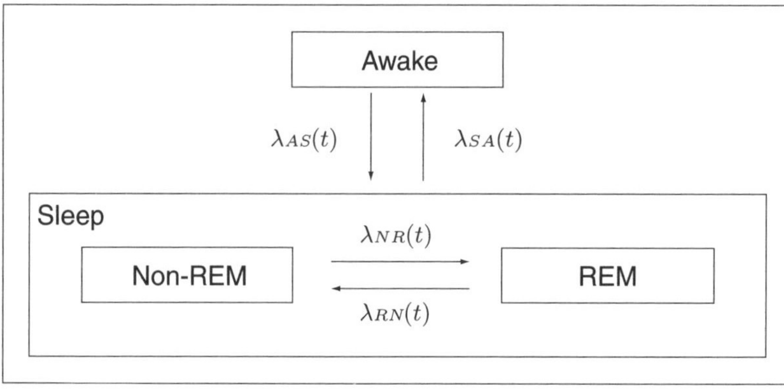

Fig. 6.28 Human sleep: Schematic representation of sleep stages and the transitions of interest.

state 'sleep' for the transitions between sleep and awake (and vice versa) but keep the original differentiation for transitions within the 'sleep' state.

In addition to the baseline intensities, we account for possible effects of nocturnal cortisol secretion and individual-specific frailties. The latter allow us to account for individual-specific sleeping behaviour with, for example, an increased tendency for some individuals to wake up more frequently throughout the night. Similarly, the inclusion of cortisol secretion allows to investigate whether an increased level of cortisol affects the transition intensities between Non-REM and REM-sleep phases, a relationship that has been found in exploratory correlation and variance analyses.

In summary, we chose the following specification for the four transition hazards:

$$\begin{aligned}
\lambda_{AS,i}(t) &= \exp\left[g_0^{(AS)}(t) + \alpha_i^{(AS)}\right], \\
\lambda_{SA,i}(t) &= \exp\left[g_0^{(SA)}(t) + \alpha_i^{(SA)}\right], \\
\lambda_{NR,i}(t) &= \exp\left[g_0^{(NR)}(t) + u_i(t)g_1^{(NR)}(t) + \alpha_i^{(NR)}\right], \\
\lambda_{RN,i}(t) &= \exp\left[g_0^{(RN)}(t) + u_i(t)g_1^{(RN)}(t) + \alpha_i^{(RN)}\right].
\end{aligned}$$

Each of the transitions is described in terms of a baseline effect $g_0^{(h)}(t)$ and a frailty term $\alpha_i^{(h)}$. Time-varying effects $g_1^{(h)}(t)$ of high cortisol secretion were included for the transition rates between Non-REM and REM, where $u_i(t)$ is a dichotomized binary indicator for a high level of cortisol, that is, $u_i(t)$ takes the value one if the cortisol level exceeds 60 nmol/l at time t and zero otherwise. Therefore $g_0^{(h)}(t)$ represents the log-baseline intensity in case of a low cortisol level while $g_0^{(h)}(t) + g_1^{(h)}(t)$ represents the log-baseline intensity for high levels of cortisol. Cubic P-splines with second-order random-walk prior and 40 inner knots were assigned to all time-varying effects $g_l^{(h)}(t)$, $l = 0, 1$, where the high number of knots was chosen to allow for sufficient flexibility. The individual-specific random effects $\alpha_i^{(h)}$ are assumed to be i.i.d. Gaussian $\alpha_i^{(h)} \sim \mathrm{N}(0, \delta_h^2)$ with transition-specific variances δ_h^2.

Posterior mode and posterior mean estimates for the different log-intensities and time-varying effects are shown in Figs. 6.29 (transitions from awake to sleep and vice versa), 6.30 (transition from Non-REM to REM) and 6.31 (transition from REM to Non-REM). In general, EB and FB inference yield results that are consistent with each other while the EB estimates for the transition from awake to sleep are notably lower than their FB counterparts. Overall we find strong evidence for temporal variability in the transition intensities with cyclic patterns for transitions between awake and sleep and the transition from Non-REM to REM. Low intensities are for example found for the transition from awake to sleep for patients who wake up more than seven hours after sleep onset. For the

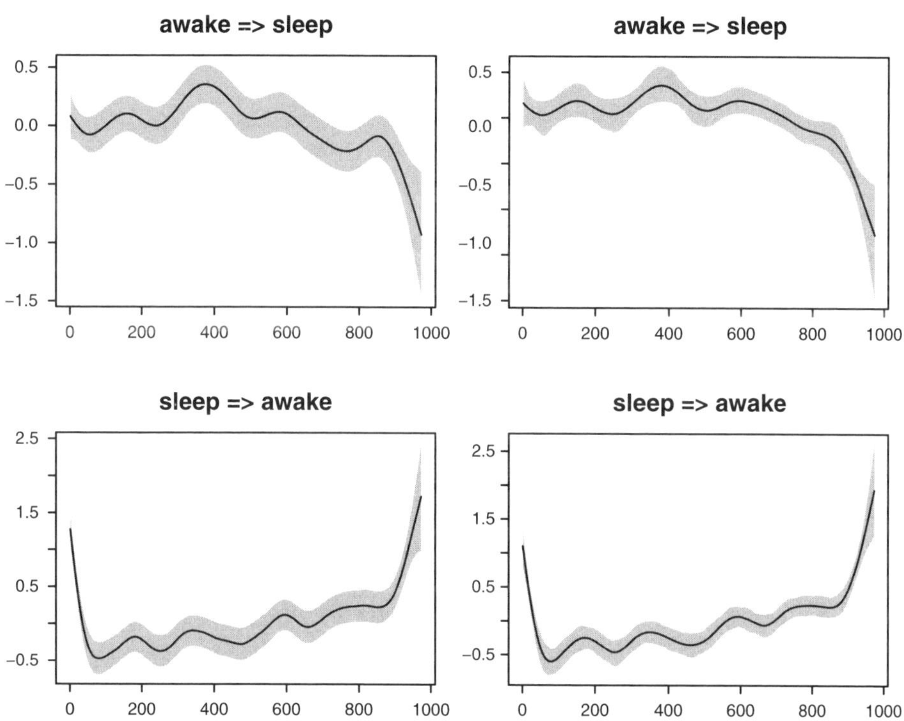

Fig. 6.29 Human sleep: Posterior mode (left) and posterior mean (right) estimates for the transition intensities from awake to sleep (top) and from sleep to awake (bottom) together with pointwise 95% credible intervals.

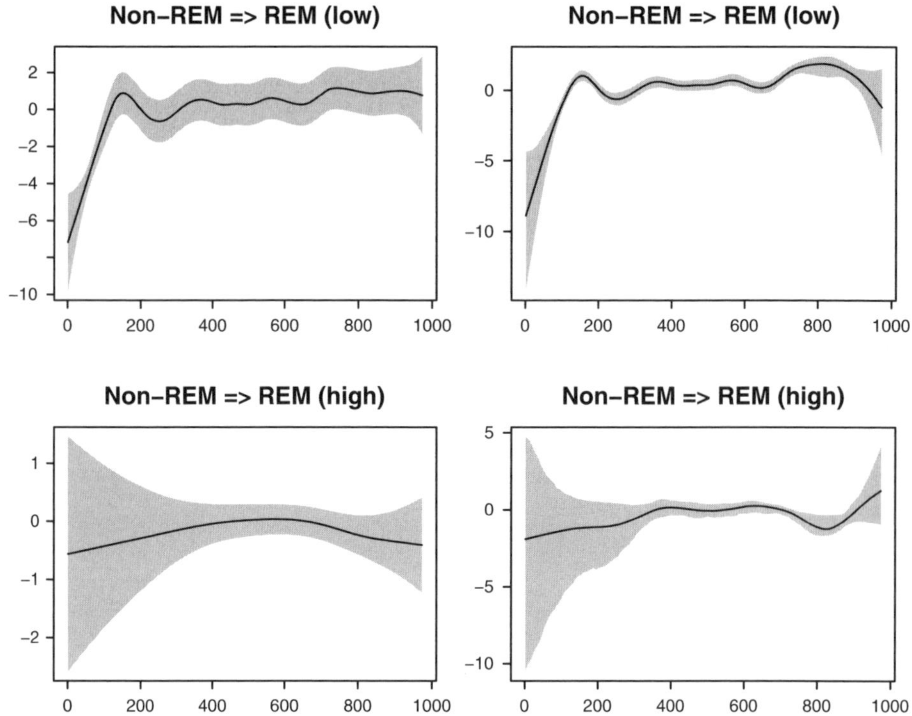

Fig. 6.30 Human sleep: Posterior mode (left) and posterior mean (right) estimates for the transition intensities from Non-REM to REM in case of low (top) and high (bottom) levels of cortisol together with pointwise 95% credible intervals.

reverse transition from sleep to awake, we find a u-shaped pattern with high transition intensities both early in the night and at the end of the night.

For the transitions between REM and Non-REM sleep within the sleep state, the log-baseline effects $g_0^{(h)}(t)$ represent the effects for a low level of cortisol, while the $g_1^{(h)}(t)$ capture deviations from these effects if the level of cortisol is high. Both EB and FB estimates yield a strong increase early in the night followed by an approximately constant effect for the transition from Non-REM to REM and a low cortisol level. For the complementary transition from REM to Non-REM the transition intensity is initially high immediately after sleep onset and decreases approximately linearly afterwards. While some additional temporal variability can be found for a high cortisol level and the transition from Non-REM to REM, the additional effect is negligible for the reverse transition (see Fig. 6.30). FB estimation identifies the presence of individual-specific frailties for all four transitions in our model while EB inference only finds frailties for the transition from REM to Non-REM. △

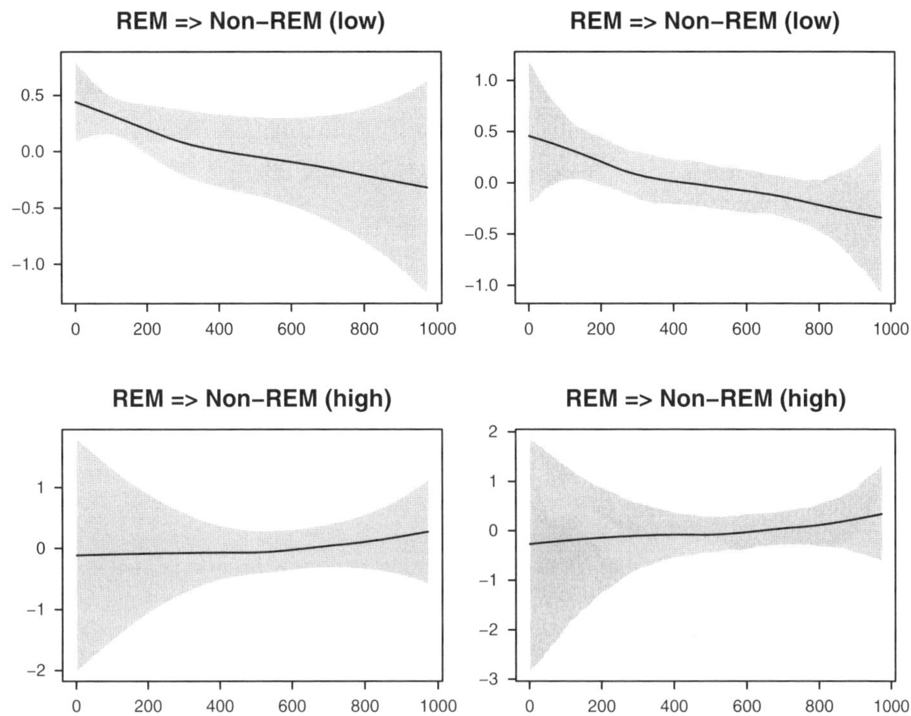

Fig. 6.31 Human sleep: Posterior mode (left) and posterior mean (right) estimates for the transition intensities from REM to Non-REM in case of low (top) and high (bottom) levels of cortisol together with pointwise 95% credible intervals.

6.5.4 Model checking based on martingale residuals

The counting process representation of multi-state models is not only useful to determine the associated likelihood but can also be used as a means of model checking. Therefore we follow the idea of Aalen, Fosen, Weedon-Fekjær, Borgan and Husebye (2004) and Kneib and Hennerfeind (2008) who considered martingale residual processes based on the Doob–Meyer decomposition. Since every counting process $N_i^{(h)}(t)$ is a submartingale by definition, it can be decomposed as

$$\begin{aligned} N_i^{(h)}(t) &= A_i^{(h)}(t) + M_i^{(h)}(t) \\ &= \int_0^t \alpha_i^{(h)}(u)du + M_i^{(h)}(t), \end{aligned}$$

where $M_i^{(h)}(t)$ is a martingale and $A_i^{(h)}(t)$ is a predictable process called the compensator of $N_i^{(h)}(t)$. The compensator process can be written as the integral over the intensity process of the model under consideration and is therefore also called the cumulative intensity process. In summary, the Doob–Meyer decomposition yields a decomposition into a predictable trend determined by the model and a noise process. Replacing the compensator process with an estimate $\hat{A}_i^{(h)}(t)$ yields estimated noise processes $N_i^{(h)}(t) - \hat{A}_i^{(h)}(t)$. If the model specification is valid, the estimated noise process should (at least approximately) have martingale properties. For example, its expectation should be zero and increments in non-overlapping intervals should be uncorrelated (Hall and Heyde, 1980, Sec. 1.6). These properties can now be visually assessed to decide whether a given model appropriately describes the given data.

Example 6.14: Human sleep

We continue the example on the human sleep process and evaluate the model fit of our semiparametric models with respect to a simple Markov model that only consists of a time-constant intensity not depending on any covariates. The resulting maximum likelihood estimates are given by the inverse of the average waiting time for the corresponding transition.

Figure 6.32 exemplarily displays martingale residuals for the transition from Non-REM to REM as a time series plot. From this graphical representation, we can already draw the conclusion that the Markov model tends to overestimate the number of transitions especially early in the night. In contrast, the semiparametric models yield mostly symmetric distributions around zero and also reduce the overall magnitude of the residual process at least in case of FB inference. This is related to the larger variability of the frailty term that has been obtained in the FB model.

A more detailed analysis of the residual processes is possible when displaying kernel density estimates of the residual processes at given fixed time points as in Fig. 6.33. In particular, this representation allows us to disentangle the cluttered picture resulting from the graphical display of 70 individual processes in one single figure. The kernel density estimates further support our conclusion that semiparametric modelling of the transition intensities leads to an improved model fit as compared to the simple Markov model. Overall, results from FB inference are rated best since the corresponding residual distributions are mostly symmetric around zero. In contrast, the residual distributions for the parametric model are considerably shifted to either a positive or negative value, indicating under- and overestimation of the expected number of transitions, respectively. Results obtained with the empirical Bayes approach are somewhere in between with the transition between REM and Non-REM as a notable exception. Especially for this transition, individual-specific frailties are also identified by EB inference such that differences to the FB results are rather small.

A further possibility to assess martingale residuals are empirical autocorrelation functions as shown in Fig. 6.34 for 30 second increments of the residual processes. For a valid model, these increments should be approximately uncorrelated according to martingale theory. Although none of the fitted models achieves

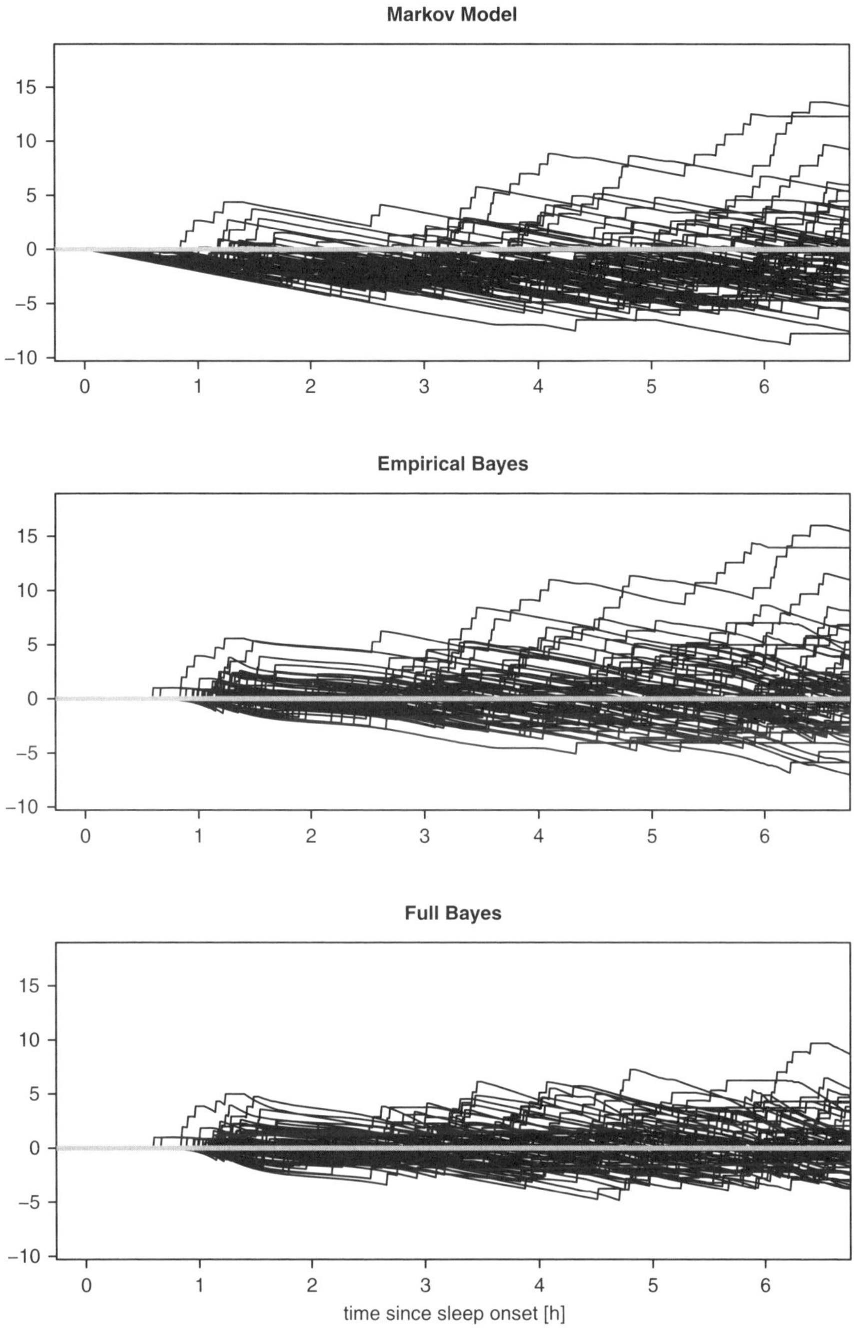

Fig. 6.32 Human sleep: Martingale residuals for the transition from Non-REM to REM.

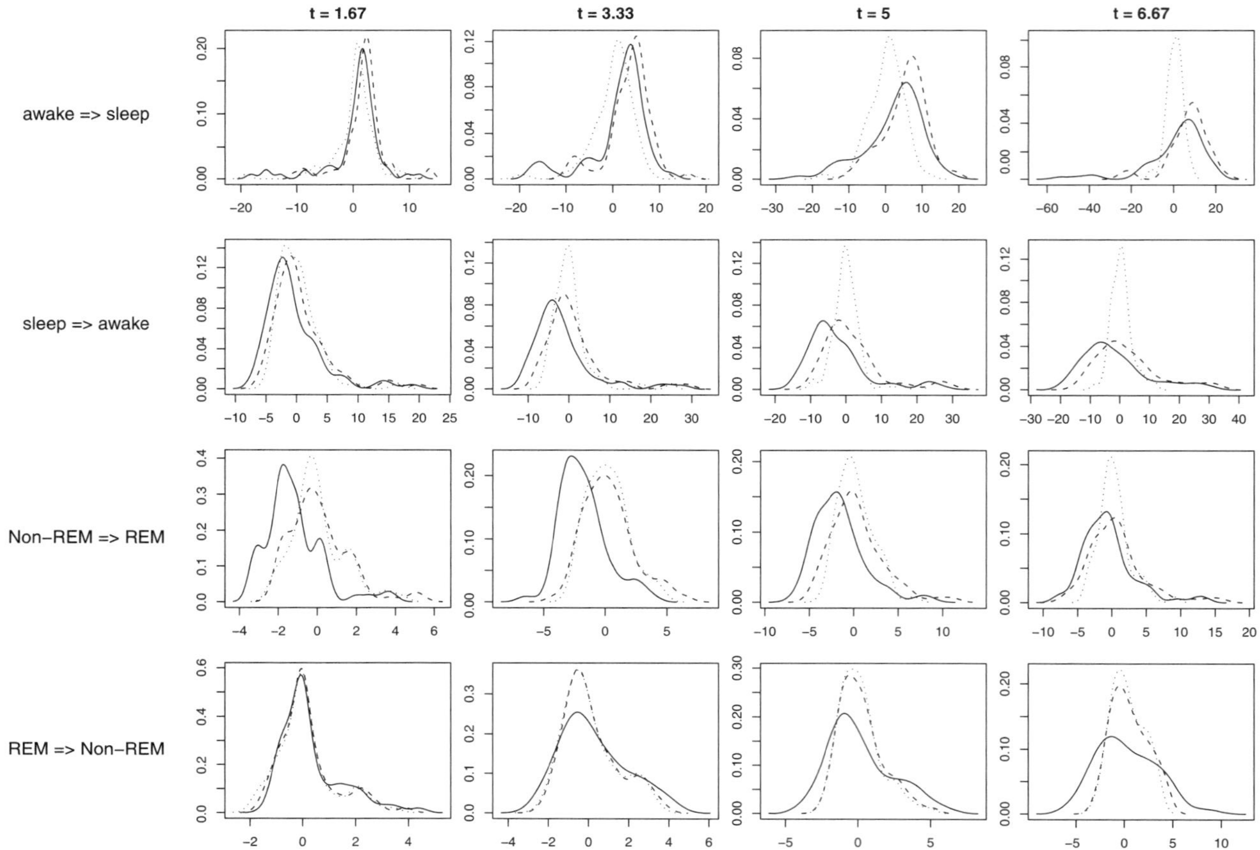

Fig. 6.33 Human sleep: Kernel density estimates of martingale residuals at selected time points for the Markov model (solid lines), empirical Bayes inference (dashed lines) and fully Bayesian inference (dotted lines).

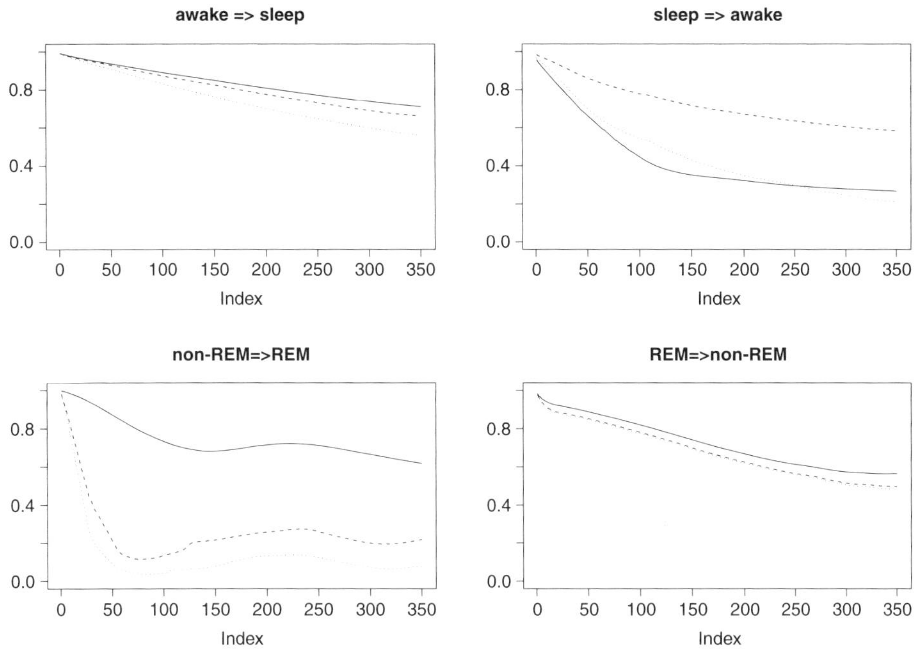

Fig. 6.34 Human sleep: Autocorrelation functions for the Markov model (solid lines), empirical Bayes inference (dashed lines) and full Bayes inference (dotted lines).

independent or uncorrelated increments, there is a clear improvement when moving from the Markov model to semiparametric model specifications and in particular to results from the FB model. For the transition from Non-REM to REM, autocorrelations die out relatively quickly while the transitions from awake to sleep and from REM to Non-REM still exhibit long-time autocorrelation and show the smallest differences between the three models.

Based on the inspection of residual processes, we can conclude that semiparametric models improve upon Markov models but are still not able to capture all of the essential features influencing the sleep process. Moreover, the preference for the FB results indicate the relevance of individual-specific variation in our data set. This might also hint at an explanation why all models fail to fully explain the variation in the data: The data set only contains very little information on the participants of the study such that a lot of unobserved heterogeneity remains unexplained even after including the frailty terms. △

6.5.5 Discrete-time multi-state models

In Section 6.3 we saw how discrete-time models in survival analysis can be recast as binary regression models via data augmentation. Similarly, it is possible to relate discrete-time multi-state models to categorical regression models and more specifically multinomial logit models. For example, in case of the human sleep data, we considered the 30 second epochs as the basic time scale up to now, leading naturally to continuous-time models. However, the hormone measurements are taken on a much coarser time scale (approximately every 20 minutes) and we may also be interested in analyses on this time level. In this case, discrete-time models may be deemed more appropriate, see Kalus, Kneib, Steiger, Holsboer and Yassouridis (2009) and the following Example 6.15.

The basic idea underlying discrete-time approximations to continuous multi-state models is as follows: The transition intensity between two states j and k is defined as

$$\lambda_{jk}(t) = \lim_{\Delta t \to 0} \frac{\mathbb{P}(Y_{t+\Delta t} = k \mid Y_t = j, \mathcal{F}_t)}{\Delta t}, \tag{6.32}$$

where Y_t represents the state of the process at time t. The transition intensities are then related to predictors $\eta_{jk}(t)$ via

$$\lambda_{jk}(t) = \exp(\eta_{jk}(t)).$$

For small time increments Δt, (6.32) can be approximated by

$$\mathbb{P}(Y_{t+\Delta t} = k \mid Y_t = j, \mathcal{F}_t) \approx \lambda_{jk}(t)\Delta t \propto \exp(\eta_{jk}(t)) \tag{6.33}$$

leading to a regression model for the transition probabilities in terms of the same predictor as in the continuous-time multi-state model. The regression model corresponding to (6.33) is exactly the multinomial logit model, where the probability $\pi_{jk}(t)$ for the transition from j to k at time t is given by

$$\pi_{jk}(t) = \mathbb{P}(Y_{t+1} = k \mid Y_t = j, \mathcal{F}_t) = \frac{\exp(\eta_{jk}(t))}{\sum_{l=1}^{L} \exp(\eta_{jl}(t))}$$

and $l = 1, \ldots, L$ denotes all sleep states that are reachable from sleep state j (including sleep state j itself for the case when no transition into another stage occurs). Of course, discrete-time models can also be based on data that is inherently discrete, like in the example on unemployment durations with multiple transitions between states like 'full employment', 'part time employment' and 'unemployed'.

To use the link to multinomial logit models for estimating discrete-time multi-state models, we employ a similar data augmentation strategy as with discrete-time survival models. As a very simple example, consider a data set where individual processes start in state '0' and move to one out of two absorbing states '1' or '2' afterwards. A small part of such a data set might look like this:

t	d	Y	u_1	u_2
4	0	1	0	2
3	1	2	1	0
5	1	1	0	3
⋮	⋮	⋮	⋮	⋮

The first individual is censored ($d = 0$) and the observed duration time is $t = 4$. The second individual is uncensored with duration time $t = 3$ and terminating transition to $Y = 2$. The third individual is again uncensored with duration time $t = 5$ and terminating event $Y = 1$. For all observations, additional information is available on covariates u_1 and u_2. We augment the data set as follows:

y	id	t	d	u_1	u_2
0	1	1	0	0	2
0	1	2	0	0	2
0	1	3	0	0	2
0	1	4	0	0	2
0	2	1	1	1	0
0	2	2	1	1	0
2	2	3	1	1	0

y	id	t	d	u_1	u_2
0	3	1	1	0	3
0	3	2	1	0	3
0	3	3	1	0	3
0	3	4	1	0	3
1	3	5	1	0	3
⋮	⋮	⋮	⋮	⋮	⋮

For the first individual we create four observations because the observed duration time is $t = 4$. The event indicator y always equals 0 because the observation is censored. For the second individual, we obtain three observations and the event indicator jumps at time $t = 3$ from 0 to 2 to indicate the corresponding transition. For the third individual the event indicator jumps at time $t = 5$ from 0 to 1. Now we can estimate a multinomial logit or probit model with y as the response, reference category 0, and covariates t, u_1, and u_2. If more transitions and recurrent events are to be analysed, the data augmentation scheme has to be extended accordingly by splitting the data according to the current state. Another extension concerns time-varying covariates that can easily be included by letting u_1 and / or u_2 vary over the spells of the augmented data set.

Example 6.15: Human sleep
Following Kalus *et al.* (2009), we again analyse the human sleep data but now take the 20 minute time intervals defined by the hormonal measurements as the basis for a discrete-time model. Moreover, we consider a somewhat finer differentiation between sleep states, where the Non-REM state is splitted into two further states corresponding to light sleep (LS) and slow wave sleep (SWS). The resulting model structure is represented in Table 6.2 that shows all possible transitions given a specific current state.

In the following, we will focus on models that investigate interactions between hormonal secretion and time based on the predictor

Table 6.2 Human sleep: Possible transitions given the current state together with the number of corresponding transitions observed in the data set. The transition from SWS to REM was excluded due to its rare occurrence.

Initial State	Transitions
Wake	Wake→Sleep (1,374), no transition (8,160)
Sleep	Sleep→Wake (1,372), no transition (35,395)
REM	REM→NREM (191), REM→Wake (301), no transition (6,779)
NREM	NREM→REM (441), NREM→Wake (1,071), no transition (27,984)
LS	LS→Wake (969), LS→SWS (1,019), LS→REM (440), no transition (21,734)
SWS	SWS→Wake (102), SWS→LS (927), no transition (4,304)

$$\eta_{jk}(t) = f_{0,jk}(t) + f_{1,jk}(\text{cortisol}) + f_{2,jk}(t, \text{cortisol}) + f_{3,jk}(\text{age}) + \beta_{jk}\text{gender}.$$

where $f_{0,jk}(t)$ and $f_{1,jk}(\text{cortisol})$ represent main effects of time and cortisol while $f_{2,jk}(t, \text{cortisol})$ captures an interaction effect between them. Note that it is much easier to include such interactions in the discrete-time model since we do not have to integrate over time-depending quantities to obtain the likelihood. In continuous-time models this would imply integration over the interaction surface with respect to only one of the two covariates, making an efficient implementation rather difficult. In addition to hormonal and temporal effects, we include a gender effect and a nonparametric age effect. As competitors, we consider models that do not include the interaction surface but are restricted to purely additive predictors. All models are estimated based on EB inference.

Figures 6.35 and 6.36 display some exemplary results for the transitions from LS to SWS and from REM to wake or Non-REM, respectively. Both figures show interaction effects (obtained as the sum of main effects and interaction surface) and results from the purely additive model for comparison. When comparing the surfaces in Fig. 6.35, the explicit incorporation of an interaction effect seems to yield a refined surface structure. For low cortisol levels earlier in the night, peaks for the transition probability can be found, which do not exactly coincide with the peaks of the curves from the model without interaction. Hence, structures emerge which cannot be detected if the analysis is restricted to additive main effects. △

6.6 Notes and further reading

In this chapter, we have considered event history data and survival analysis in the context of Bayesian structured additive regression. More general (and basic) introductory text books on survival analysis are for example Klein and Moeschberger (2005), Kleinbaum and Klein (2005) or Hosmer, Lemeshow

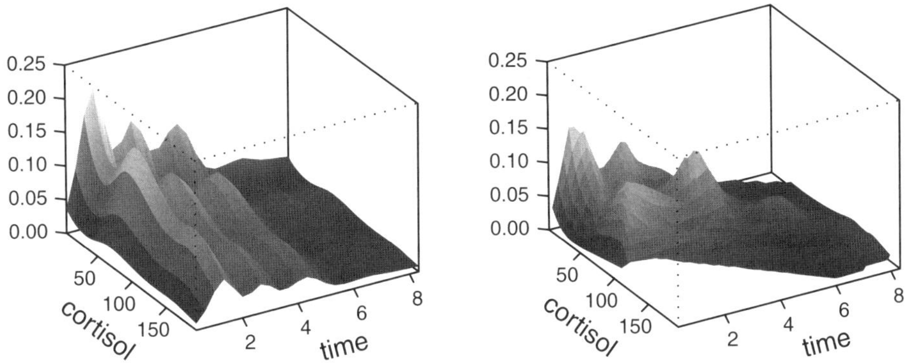

Fig. 6.35 Human sleep: Influence of time and cortisol level on the transition probability from LS to SWS. The left panel shows the combination of additive main effects, the right panel shows the sum of main effects and interaction.

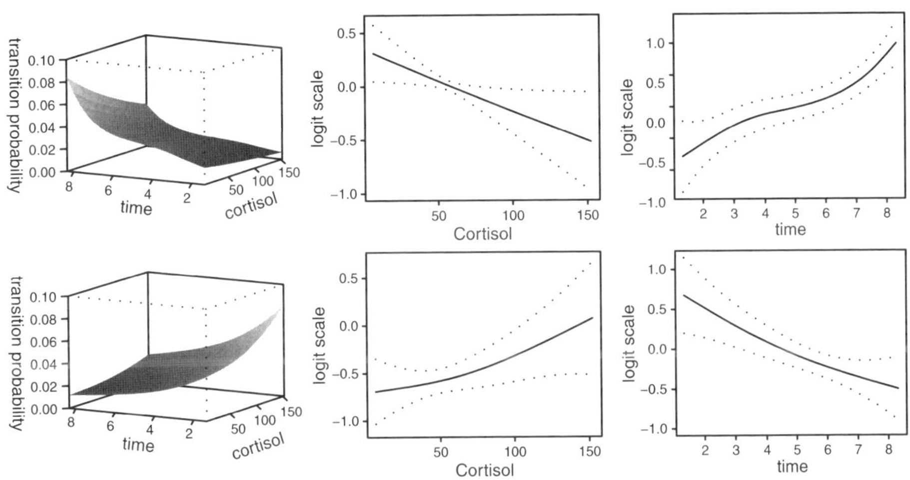

Fig. 6.36 Human sleep: Influence of time and cortisol level on the transitions from REM to wake (upper row) and from REM to Non-REM (lower row). The leftmost panel shows estimated interaction surfaces while the middle and the rightmost panel show estimated nonparametric main effects for cortisol and time together with 95% pointwise credible intervals.

and May (2008). These text books provide a from scratch introduction to survival analysis (and also some extensions for event history analysis) including for example a completely nonparametric treatment of estimating cumulative hazard rates or survivor functions based on the famous Nelson–Aalen and Kaplan–Meier estimates. Therneau and Grambsch (2000) also provide a general introduction to survival analysis but with a particular focus on the Cox proportional hazards model and its extensions. An extensive treatment of multi-state models within the framework of counting processes can be found in Andersen *et al.* (1993). Commenges (1999), Andersen and Keiding (2002) and Meira-Machado *et al.* (2009) are good introductions to multi-state models. Fahrmeir and Klinger (1998) deal with penalized likelihood inference in event history models, with an application to sleep data.

An alternative class of regression models for analysing duration times is Aalen's additive risk model where the hazard rate is specified as

$$\lambda_i(t) = \boldsymbol{u}_i(t)'\boldsymbol{\beta}(t).$$

The model inherently permits time-varying covariates and effects but estimation of the regression coefficients $\boldsymbol{\beta}(t)$ has to be restricted to ensure nonnegativity of the hazard rate. Such estimation schemes rely on counting process and martingale theory and provide estimates for the cumulative effects. An introduction to Aalen's additive risk models and extensions such as partially multiplicative models can be found in Martinussen and Scheike (2009).

6.6.1 Models for correlated survival data

Correlated survival or duration times arise in various applications. Dependence between survival times may be induced by *clustering*, as for example in familial studies, dental research or multi-centre studies, or if *multiple duration times* are observed for individuals, as, for example, in unemployment studies with multiple spells inducing within-subject correlation. For general expositions on multivariate survival times we refer to Hougaard (2000) and Therneau and Grambsch (2000), focusing on frequentist and empirical Bayes approaches.

Let T_{ij}, $j = 1, \dots, n_i$, denote survival times or repeated duration times for cluster or subject $i = 1, \dots, n$. To account for within-cluster or within-subject correlation, models based on random effects or *frailties* in analogy to mixed models for longitudinal data are most appropriate for empirical or full Bayesian inference.

The simplest extension of the Cox model are frailty models with hazard rates

$$\lambda_{ij}(t) = \lambda_0(t)\nu_i \exp(\boldsymbol{u}_{ij}'\boldsymbol{\beta}) = \lambda_0(t)\exp(\boldsymbol{u}_{ij}'\boldsymbol{\beta} + \alpha_i), \tag{6.34}$$

where ν_i or $\alpha_i = \log \nu_i$ are i. i. d. frailty or random effects. The most common distributional assumptions for ν_i are gamma priors with $\mathbb{E}(\nu_i) = 1$ or normal priors for α_i with $\mathbb{E}(\alpha_i) = 0$. Bayesian frailty models for multivariate survival data have been studied in various contexts and under varying specifications of frailty priors (Clayton, 1991; Sinha, 1993; Sahu *et al.*, 1997; Aslanidou, Dey and Sinha, 1998; Qiou, Ravishanker and Dey, 1999; among others). Yin and Ibrahim (2005) extend the proportional hazards model with gamma frailty to a more general class of hazard rate models, encompassing the PH model as well as an additive hazard rate model of the form $\lambda(t_{ij}) = \lambda_0(t)\nu_i + \boldsymbol{u}_{ij}'\boldsymbol{\beta}$ as special cases. The second specification in (6.34) makes the close relation to generalized linear mixed models evident: The (log-)frailty α_i corresponds to the random intercept term. Actually, such a cluster-specific random intercept term has already been included in the semiparametric hazard rate model (6.8), (6.9) in Section 6.2.1. If we split up the spatial effect f_{spat} into a structured (spatially correlated) effect and unstructured (i. i. d.) effects for regions $s = 1, \ldots, S$, as in Example 6.5 then the unstructured effect corresponds to α_i in (6.34) after reindexing. A natural extension of the proportional hazards model (6.34) is to extend the predictor $\eta_{ij} = \boldsymbol{u}_{ij}'\boldsymbol{\beta} + \alpha_i$ to linear mixed model predictors with Gaussian i. i. d. random effects $\boldsymbol{\alpha}_i$

$$\eta_{ij}^{\text{lin}} = \boldsymbol{u}_{ij}'\boldsymbol{\beta} + \boldsymbol{z}_{ij}'\boldsymbol{\alpha}_i$$

as for example in Vaida and Xu (2000). More generally, predictors including time-varying effects and other nonparametric terms as in (6.9) may be added to η_{ij}^{lin}, leading to models with a structured additive predictor. Kauermann, Xu and Vaida (2008) develop empirical Bayes inference using a Laplace approximation for models with time-varying effects $\boldsymbol{\beta}(t)$. Extensions of FB inference for models with a structured additive predictors considered in this chapter are—at least conceptually—straightforward. As for (semiparametric) mixed models for longitudinal data in Sections 3.2, 3.3.5 and 4.4.1, it might also be of interest to replace the Gaussian (or any other parametric) random effects prior by a more flexible prior. Naskar, Das and Ibrahim (2005) have considered an extension of the frailty model (6.34) in the more general framework of competing risks where the random effects distribution is specified through a DP prior. Dunson and Chen (2004) assume a zero-inflated inverse-gamma random effects prior to explicitly include the hypothesis of homogeneity with positive probability.

Accelerated failure time (AFT) models for correlated data can be formulated in mixed model form by

$$\log(T_{ij}) = \boldsymbol{u}_{ij}'\boldsymbol{\beta} + \boldsymbol{z}_{ij}'\boldsymbol{\alpha}_{ij} + \varepsilon_{ij}, \quad i = 1, \ldots, n, \ j = 1, \ldots, n_i. \tag{6.35}$$

For the special case of a random intercept model, we have $\boldsymbol{z}_{ij} \equiv 1$. Such a mixed model has been studied for right-censored data, Gaussian random effects and Gaussian errors by Pan and Louis (2000) and Pan and Connett (2001). Bayesian inference for the general mixed effects AFT model (6.35) with interval-censored data has been developed by Komárek and Lesaffre (2007), assuming Gaussian random effects and a Gaussian mixture prior for the error distribution, and in Komárek and Lesaffre (2008) with penalized Gaussian mixture priors for random effects as well as errors.

The models considered so far account for correlation through i. i. d. random effects. In the predictor (6.9), the spatial term f_{spat} accounts for *spatial correlation* through Markov random field or stationary Gaussian random field priors. In other words, correlation is induced through (spatially) correlated random effects. Related survival models for otherwise linear predictors have also been suggested under various assumptions on priors for the baseline hazard and spatial components. Li and Ryan (2002) added a spatial component in form of a stationary Gaussian process to the linear predictor of the Cox model. Treating the baseline hazard as a nuisance parameter, inference for the linear predictor and for correlation function parameters is based on a marginal rank likelihood. Henderson *et al.* (2002) proposed a Cox model with conditionally independent spatial gamma frailties. Banerjee, Wall and Carlin (2003b) assumed a parametric Weibull baseline hazard and geostatistical or MRF priors for the spatial component. Banerjee and Carlin (2003) developed Bayesian spatio-temporal survival models, modelling baseline hazard functions nonparametrically through a beta mixture approach and assuming MRF or conditionally autoregressive (CAR) priors for spatial effects, and Carlin and Banerjee (2003) extended this approach to multivariate MRF models. A good overview is given by Banerjee and Carlin (2004). Gamerman, Moreira and Rue (2003) present space-varying regression models based on CAR priors, and Nathoo and Dean (2008) develop spatial multi-state models with piecewise exponential or cubic B-spline specifications for baseline transition intensities, linear predictors for covariate effects and multivariate CAR priors for spatial effects.

6.6.2 Joint modelling of longitudinal and event history data

In many longitudinal studies it has become increasingly common that both longitudinal data and event history data are observed. A well-known example that has stimulated much methodological development in joint modelling is AIDS research, where CD4 counts are collected longitudinally and the time to AIDS or death is also recorded. Depending on the specific application, the following three objectives of joint modelling may be distinguished:

a) modelling the joint evolution of the longitudinal data and the survival or event history data processes;
b) inference about terminating or recurrent events, considering longitudinal data as a time-varying covariate process measured with error;
c) correcting for the informative, outcome-dependent drop-out in longitudinal studies.

Objective a) treats longitudinal and event history data symmetrically, while in b) and c) they are considered asymmetrically. Most approaches for relating the two components, symmetrically and asymmetrically, are based on joint random effects or latent processes. Therefore, empirical and full Bayesian inference is quite common in joint modelling.

The general structure of *symmetric* joint models is as follows. For longitudinal data y_{ij}, $i = 1, \ldots, n$, $j = 1, \ldots, n_i$ observed at time points t_{ij}, a mixed model is assumed with a linear predictor

$$\boldsymbol{\eta}_{ij} = \boldsymbol{u}'_{ij}\boldsymbol{\beta} + \boldsymbol{z}'_{ij}\boldsymbol{\alpha}_i \tag{6.36}$$

as in Chapter 3, or with an additive predictor

$$\boldsymbol{\eta}_{ij} = \boldsymbol{u}_{ij}'\boldsymbol{\beta} + \boldsymbol{z}_{ij}'\boldsymbol{\alpha}_i + f(t_{ij}) + g_i(t_{ij}) \tag{6.37}$$

as in Chapter 4. With a linear predictor, population trends and individual profiles are modelled parametrically, for example through (piecewise) polynomials. In additive predictors, the population trend $f(t)$, and possibly also individual profiles $g_i(t_{ij})$, are modelled through penalized splines, (stationary) Gaussian processes, etc. Sometimes $f(t)$ and $g_i(t)$ are nonparametric 'stochastic' deviations from simple 'deterministic' curves. Survival data are mostly modelled through mixed model type extensions of Cox models

$$\lambda_i(t) = \lambda_0(t)\exp(\tilde{\boldsymbol{u}}_i(t)'\tilde{\boldsymbol{\beta}} + \tilde{\boldsymbol{z}}_i(t)'\tilde{\boldsymbol{\alpha}}_i) \tag{6.38}$$

as discussed in the previous section, often with the special case $\tilde{\boldsymbol{z}}_i(t) \equiv 1$ of simple frailty models. More generally, hazard rate models for recurrent events can be modelled by assuming additive predictors with time-varying (random) effects, similarly as in (6.37). Covariates or design vectors $\boldsymbol{u}$, $\boldsymbol{z}$ and $\tilde{\boldsymbol{u}}$, $\tilde{\boldsymbol{z}}$ in both components may be the same, at least partly. Special joint models differ in the way in which they are linked together through joint effects and in the specific inferential approaches. For example, Henderson, Diggle and Dobson (2000) propose a Gaussian linear mixed model with random intercepts and slopes,

$$\boldsymbol{z}_{it}'\boldsymbol{\alpha}_i = \alpha_{i0} + \alpha_{i1}t, \quad (\alpha_{i0}, \alpha_{i1})' \sim \mathrm{N}(\boldsymbol{0}, \boldsymbol{Q})$$

for the longitudinal component, and

$$\tilde{\boldsymbol{z}}_{it}\tilde{\boldsymbol{\alpha}}_i = \gamma_1\alpha_{i0} + \gamma_2\alpha_{i1} + \gamma_3(\alpha_{i0} + \alpha_{i1}t) + \tilde{\alpha}_{i3},$$

where $\tilde{\alpha}_{i3} \sim \mathrm{N}(0, \delta_3^2)$ is independent of α_{i0}, α_{i1}. In this model, the parameters γ_1, γ_2 and γ_3 measure the association induced through the intercept, slope and individual-specific trend. As a more flexible extension, a random function $g_i(t)$ is added to the linear random part, resulting in an additive predictor (6.37). The functions $g_i(t)$ are specified as realizations from a (one-dimensional) stationary Gaussian random field. The random part of the hazard rate model (6.38) is specified in similar form as just described, replacing $\alpha_{i0} + \alpha_{i1}t$ through $\alpha_{i0} + \alpha_{i1}t + g_i(t)$. Estimation is based on empirical Bayes inference, involving an EM algorithm. Diggle, Henderson and Philipson (2009) embed other special models into the general forms (6.37), (6.38) and outline empirical Bayes inference using the EM algorithm.

Elashoff, Li and Li (2008) suggest a more symmetric way of linking the two components in the more general setting of multiple failure types. For the special case of survival data with only one type of failure their joint model consists of a Gaussian linear mixed model (6.36) and a frailty model (6.38) with $\tilde{\boldsymbol{z}}_i(t) \equiv 1$. The models are joined by assuming that $\boldsymbol{\alpha}_i$ and $\tilde{\boldsymbol{\alpha}}_i$ have a joint multinormal distribution

$$(\boldsymbol{\alpha}_i, \tilde{\boldsymbol{\alpha}}_i) \sim \mathrm{N}(\boldsymbol{0}, \boldsymbol{Q})$$

Again, an empirical Bayes approach is developed for inference.

Rizopoulos and Ghosh (2009) consider a multivariate joint model for multiple longitudinal outcomes and survival time. The longitudinal outcomes are modelled through generalized additive mixed models to flexibly capture possibly nonlinear shapes of subject specific evolutions. Instead of Gaussian priors, DPM priors are used for joining the models. Liu and Huang (2009) develop joint analysis of correlated repeated measures and recurrent events processes in the presence of death. Estimation is based on the EM algorithm.

A prominent application where the focus is on the *asymmetric* objective b) are HIV clinical trials, where some time-constant covariates are recorded at baseline and markers for immunologic and virologic status, such as CD4 count and viral RNA copy number, are taken as repeated, longitudinal measurements $y_i(t_{ij}) = y_{ij}$ at times t_{ij}, $i = 1, \ldots, n$, $j = 1, \ldots, n_i$. Times to progression to AIDS or death, possibly censored, are also recorded for each participant. In a naive approach, one would perhaps use the longitudinal data $y_i(t)$ as a time-varying covariate in a Cox or AFT model for survival times. However, $y_i(t)$ is measured with noise, leading to biased estimates. Instead, $y_i(t)$ is modelled through a mixed model

$$y_i(t) = \eta_i(t) + \varepsilon_i(t) \tag{6.39}$$

with linear or additive mixed model predictor $\eta_i(t)$. Then, instead of using $y_i(t)$ as a covariate, the latent covariate process $\eta_i(t)$ is inserted, for example, in a Cox model

$$\lambda_i(t) = \lambda_0(t) \exp(\boldsymbol{u}_i' \tilde{\boldsymbol{\beta}} + \gamma \eta_i(t)). \tag{6.40}$$

Thus, the two models are joined through the common latent mixed model predictor $\eta_i(t)$. Tsiatis and Davidian (2004) provide a good overview on various concepts for inference in such joint models. Again, approaches for modelling differ in the way how the submodels (6.39) and (6.40) are specified. For example, Wang and Taylor (2001) specify an additive predictor

$$\eta_i(t) = \alpha_{i0} + \beta_1 t + \boldsymbol{u}_i' \boldsymbol{\beta} + g_i(t),$$

where $g_i(t)$ is specified through a Gaussian integrated Ornstein–Uhlenbeck process and α_{i0} are Gaussian random intercepts. The baseline hazard rate is assumed as a piecewise exponential model. Brown and Ibrahim (2003) assume a parametric predictor

$$\eta_i(t) = \tilde{\alpha}_{i0} + \tilde{\alpha}_{i1} t + \tilde{\alpha}_{i2} t^2$$

for individual latent profiles, but a DP prior for the random effects. Brown, Ibrahim and DeGruttola (2005) use cubic B-splines to specify $\eta_i(t)$ more flexibly, with a Gaussian prior for basis function coefficients. Inferences are fully Bayesian using MCMC techniques.

As an alternative to Cox-type models, Tseng, Hsieh and Wang (2005) have employed an AFT model with time-varying covariates for the survival component in combination with a likelihood-based approach for inference.

Bibliography

Aalen, O. O., Fosen, J., Weedon-Fekjær, H., Borgan, Ø., and Husebye, E. (2004). Dynamic analysis of multivariate failure time data. *Biometrics*, **60**, 764–73.

Abramovich, F., Sapatinas, T., and Silverman, B. W. (1998). Wavelet thresholding via a Bayesian approach. *Journal of the Royal Statistical Society, Series B*, **60**, 725–749.

Abramowitz, M. and Stegun, I. A. (1974). *Handbook of Mathematical Functions.* Dover.

Adebayo, S. and Fahrmeir, L. (2005). Analysing child mortality in Nigeria with geoadditive discrete-time survival models. *Statistics in Medicine*, **24**, 709–28.

Agresti, A. (2002). *Categorical Data Analysis.* Wiley.

Akaike, H. (1973). Information theory and an extension of the maximum likelihood principle. In *2nd International Symposium on Information Theory* (ed. B. N. Petrov and F. Csaki), pp. 267–81. Akademiai Kiado.

Albert, J. and Chib, S. (1993). Bayesian analysis of binary and polychotomous response data. *Journal of the American Statistical Association*, **88**, 669–79.

Andersen, P. K., Borgan, Ø., Gill, R., and Keiding, N. (1993). *Statistical Models Based on Counting Processes.* Springer.

Andersen, P. K. and Keiding, N. (2002). Multi-state models for event history analysis. *Statistical Methods in Medical Research*, **11**, 91–115.

Anselin, L. (1988). *Spatial Econometrics: Methods and Models.* Kluwer Academic Publishers.

Argiento, R., Guglielmi, A., and Pievatolo, A. (2009). A comparison of nonparametric priors in hierarchical mixture modelling for AFT regression. *Journal of Statistical Planning and Inference*, **139**, 3989–4005.

Aslanidou, H., Dey, D. K., and Sinha, D. (1998). Bayesian analysis of multivariate survival data using Monte Carlo methods. *Canadian Journal of Statistics*, **26**, 33–48.

Assunção, R. (2003). Space varying coefficient models for small area data. *Environmetrics*, **14**, 453–73.

Assunção, R. and Krainski, E. (2009). Neighborhood dependence in Bayesian spatial models. *Biometrical Journal*, **51**, 851–69.

Augustin, N., Musio, M., von Wilpert, K., Kublin, E., Schumacher, M., and Wood, S. (2009). Modeling spatio-temporal forest health monitoring data. *Journal of the American Statistical Association*, **104**, 899–911.

Baladandayuthapani, V., Mallick, B. K., and Carroll, R. J. (2005). Spatially adaptive Bayesian penalized regression splines (P-splines). *Journal of Computational and Graphical Statistics*, **14**, 378–394.

Banerjee, S. and Carlin, B. P. (2003). Semiparametric spatio-temporal frailty modeling. *Environmetrics*, **14**, 523–35.

Banerjee, S. and Carlin, B. P. (2004). Parametric spatial cure rate models for interval-censored time-to-relapse data. *Biometrics*, **60**, 268–75.

Banerjee, S., Carlin, B. P., and Gelfand, A. E. (2003a). *Hierarchical Modeling and Analysis for Spatial Data.* Chapman & Hall/CRC.

Banerjee, S., Wall, M. M., and Carlin, B. P. (2003b). Frailty modeling for spatially correlated survival data, with application to infant mortality in Minnesota. *Biostatistics*, **4**, 123–42.

Baumgartner, B., Steiner, W. J., and Kneib, T. (2009). Estimating time-dependent effects in brand choice models: A semiparametric approach. Technical report.

Bayarri, M. and Berger, J. (2000). P-values for composite null models. *Journal of the American Statistical Association*, **95**, 1127–42.

Berger, J. O., de Oliveira, V., and Sansó, B. (2001). Objective Bayesian analysis of spatially correlated data. *Journal of the American Statistical Association*, **96**, 1361–74.

Berger, J. O. and Pericchi, L. R. (1996). The intrinsic Bayes factor for model selection and prediction. *Journal of the American Statistical Association*, **91**, 109–122.

Berger, J. O. and Pericchi, L. R. (2001). Objective Bayesian methods for model selection: Introduction and comparison (with discussion). *Model Selection* (P. Lahiri, ed.), **38**, 135–207.

Berger, J. O., Pericchi, L. R., and Varshavsky, J. A. (1998). Bayes factors and marginal distributions in invariant situations. *The Indian Journal of Statistics, Series A*, **60**, 307–21.

Berger, J. O., Ghosh, J. K., and Mukhopadhyay, N. (2003). Approximations and consistency of Bayes factors as model dimension grows. *Journal of Statistical Planning and Inference*, **112**, 241–258.

Bernardo, J. M. and Smith, A. F. M. (2004). *Bayesian Theory.* Wiley.

Besag, J. (1974). Spatial interaction and the statistical analysis of lattice systems (with discussion). *Journal of the Royal Statistical Society, Series B*, **36**, 192–236.

Besag, J. and Higdon, D. (1999). Bayesian analysis of agricultural field experiments. *Journal of the Royal Statistical Society B*, **61**, 691–746.

Besag, J., Green, P., Higdon, D., and Mengersen, K. (1995). Bayesian computation and stochastic systems (with discussion). *Statistical Science*, **10**, 3–66.

Beyersmann, J., Kneib, T., Schumacher, M., and Gastmeier, P. (2009). Nosocomial infection, length of stay and time-dependent bias. *Infection Control & Hospital Epidemiology*, **30**, 273–76.

Bigelow, J. L. and Dunson, D. B. (2007). Bayesian adaptive regression splines for hierarchical data. *Biometrics*, **63**, 724–32.

Biller, C. (2000). Adaptive Bayesian regression splines in semiparametric generalized linear models. *Journal of Computational and Graphical Statistics*, **9**, 122–140.

Biller, C. and Fahrmeir, L. (1997). Bayesian spline-type smoothing in generalized regression models. *Computational Statistics*, **12**, 135–51.

Biller, C. and Fahrmeir, L. (2001). Bayesian varying-coefficient models using adaptive regression splines. *Statistical Modelling*, **1**, 195–211.

Bishop, C. M. (2006). *Pattern Recognition and Machine Learning.* Springer.

Bivand, R. S., Pebesma, E. J., and Gómez-Rubio, V. (2008). *Applied Spatial Data Analysis with R.* Springer.

Bochner, S. (1955). *Harmonic Analysis and the Theory of Probability.* University of California Press.

Bondell, H. D. and Reich, B. J. (2008). Simultaneous regression shrinkage, variable selection, and supervised clustering of predictors with OSCAR. *Biometrics*, **64**, 115–23.

Breslow, N. (1974). Covariance analysis of censored survival data. *Biometrics*, **30**, 89–99.

Breslow, N. E. and Clayton, D. G. (1993). Approximate inference in generalized linear mixed models. *Journal of the American Statistical Association*, **88**, 9–25.

Brezger, A. and Lang, S. (2006). Generalized additive regression based on Bayesian P-splines. *Computational Statistics & Data Analysis*, **50**, 967–91.

Brezger, A., Fahrmeir, L., and Hennerfeind, A. (2007). Adaptive Gaussian Markov random fields with applications in human brain mapping. *Applied Statistics*, **56**, 327–45.

Brockwell, A. E. (2007). Universal residuals: A multivariate transformation. *Statistics & Probability Letters*, **77**, 1473–78.

Brooks, S. P., Giudici, P., and Roberts, G. O. (2003). Efficient construction of reversible jump Markov Chain Monte Carlo proposal distributions (with discussion). *Journal of the Royal Statistical Society, Series B*, **65**, 3–55.

Brown, E. R. and Ibrahim, J. G. (2003). A Bayesian semiparametric joint hierarchical model for longitudinal and survival data. *Biometrics*, **59**, 221–28.

Brown, E. R., Ibrahim, J. G., and DeGruttola, V. (2005). A flexible B-spline model for multiple longitudinal biomarkers and survival. *Biometrics*, **61**, 64–73.

Browne, W. J. and Draper, D. (2000). Implementation and performance issues in the Bayesian and likelihood fitting of multilevel models. *Computational Statistics*, **15**, 391–420.

Browne, W. J. and Draper, D. (2006). A comparison of Bayesian and likelihood-based methods for fitting multilevel models. *Bayesian Analysis*, **1**(3), 473–514.

Browne, W. J., Draper, D., Goldstein, H., and Rasbash, J. (2002). Bayesian and likelihood methods for fitting multilevel models with complex level-1 variation. *Computational Statistics & Data Analysis*, **39**, 203–25.

Bryk, A. S. and Raudenbush, S. W. (1992). *Hierarchical Linear Models.* Sage.

Buja, A., Hastie, T., and Tibshirani, R. (1989). Linear smoothers and additive models. *Annals of Statistics*, **17**, 453–510.

Cai, B. and Dunson, D. (2006). Bayesian covariance selection in generalized linear mixed models. *Biometrics*, **62**, 446–57.

Cantoni, E. and Hastie, T. (2002). Degrees-of-freedom tests for smoothing splines. *Biometrika*, **89**, 251–63.

Carlin, B. P. and Chib, S. (1995). Bayesian model choice via Markov Chain Monte Carlo methods. *Journal of the Royal Statistical Society B*, **57**, 473–484.

Carlin, B. P. and Banerjee, S. (2003). Hierarchical multivariate CAR models for spatio-temporally correlated data. In *Bayesian Statistics 7* (ed. J. M. Bernardo, M. J. Bayarri, J. O. Berger, A. P. Dawid, D. Heckerman, A. F. M. Smith, and M. West), pp. 45–63. Oxford University Press.

Casella, G. and Moreno, E. (2006). Objective Bayesian variable selection. *Journal of the American Statistical Association*, **101**, 157–67.

Caulfield, L. E., de Onis, M., Blössner, M., and Black, R. E. (2004). Undernutrition as an underlying cause of child deaths associated with diarrhea, pneumonia, malaria and measles. *American Journal of Clinical Nutrition*, **80**, 193–98.

Celeux, G., Forbes, F., Robert, C. P., and Titterington, D. M. (2006). Deviance information criteria for missing data models. *Bayesian Analysis*, **1**(4), 651–74.

Chen, M. H. and Dey, D. K. (2000). Bayesian analysis for correlated ordinal data models. In *Generalized Linear Models: A Bayesian Perspective* (ed. D. K. Dey, S. K. Ghosh, and B. K. Mallick), pp. 133–57. Marcel Dekker.

Chen, Z. and Dunson, D. (2003). Random effects selection in linear mixed models. *Biometrics*, **59**, 762–69.

Chen, M.-H., Shao, Q.-M., and Ibrahim, J. G. (2000). *Monte Carlo methods in Bayesian computation*. Springer.

Chen, L., Qin, Z. S., and Lius, J. S. (2001). Exploring Hybrid Monte Carlo in Bayesian computation. In *Proceedings of ISBA 2000 - The Sixth World Meeting of the International Society for Bayesian Analysis*. International Society for Bayesian Analysis.

Chernoff, H. (1954). On the distribution of the likelihood ratio. *Annals of Mathematical Statistics*, **25**, 573–78.

Chib, S. (1995). Marginal likelihood from the Gibbs output. *Journal of the American Statistical Association*, **90**, 1313–21.

Chib, S. and Jeliazkov, I. (2001). Marginal likelihood from the Metropolis-Hastings output. *Journal of the American Statistical Association*, **96**, 270–81.

Chib, S. and Jeliazkov, I. (2006). Inference in semiparametric dynamic models for binary longitudinal data. *Journal of the American Statistical Association*, **101**, 685–700.

Chiles, J.-P. and Delfiner, P. (1999). *Geostatistics: Modeling Spatial Uncertainty*. Wiley.

Claeskens, G., Krivobokova, T., and Opsomer, J. D. (2009). Asymptotic properties of penalized spline estimators. *Biometrika*, **96**, 529–44.

Clayton, D. G. (1991). A Monte Carlo method for Bayesian inference in frailty models. *Biometrics*, **47**, 467–85.

Clyde, M. and George, E. I. (2000). Flexible empirical Bayes estimation for wavelets. *Journal of the Royal Statistical Society, Series B*, **62**, 681–98.

Clyde, M. and George, E. I. (2004). Model uncertainty. *Statistical Science*, **19**(1), 81–94.

Clyde, M., Parmigiani, G., and Vidakovic, B. (1998). Multiple shrinkage and subset selection in wavelets. *Biometrika*, **85**, 391–402.

Commenges, D. (1999). Multi-state models in epidemiology. *Lifetime Data Analysis*, **5**, 315–27.

Congdon, P. (2005). *Bayesian Models for Categorical Data*. John Wiley & Sons.

Congdon, P. (2006). Bayesian model choice based on Monte Carlo estimates of posterior model probabilities. *Computational Statistics & Data Analysis*, **50**, 346–57.

Cottet, R., Kohn, R. J., and Nott, D. J. (2008). Variable selection and model averaging in semiparametric overdispersed generalized linear models. *Journal of the American Statistical Association*, **103**, 661–71.

Cox, D. R. (1972). Regression models and life tables (with discussion). *Journal of the Royal Statistical, Series B*, **34**, 187–220.

Crainiceanu, C. and Ruppert, D. (2004a). Likelihood ratio tests for goodness-of-fit of a nonlinear regression model. *Journal of Multivariate Analysis*, **91**, 35–52.

Crainiceanu, C. M. and Ruppert, D. (2004b). Likelihood ratio tests in linear mixed models with one variance component. *Journal of the Royal Statistical Society B*, **66**, 165–85.

Crainiceanu, C., Ruppert, D., Claeskens, G., and Wand, M. P. (2005). Exact likelihood ratio tests for penalised splines. *Biometrika*, **92**, 91–103.

Crainiceanu, C. M., Ruppert, D., Carroll, R. J., Joshi, A., and Goodner, B. (2007). Spatially adaptive Bayesian penalized splines with heteroscedastic errors. *Journal of Computational and Graphical Statistics*, **16**, 265–88.

Cressie, N. A. C. (1993). *Statistics for Spatial Data*. John Wiley.

Cripps, E., Carter, C., and Kohn, R. (2005). Variable selection and covariance selection in multivariate regression models. In *Handbook of Statistics 25, Bayesian Thinking: Modeling and Computation* (ed. D. Dey and C. Rao), pp. 519–52. North Holland.

Crook, A., Knorr-Held, L., and Hemingway, H. (2003). Measuring spatial effects in time to event data: a case study using months from angiography to coronary artery bypass graft (CABG). *Statistics in Medicine*, **22**, 2943–61.

Currie, I. D., Durban, M., and Eilers, P. H. C. (2006). Generalized linear array models with applications to multidimensional smoothing. *Journal of the Royal Statistical Society, Series B*, **68**, 259–80.

Czado, C., Gneiting, T., and Held, L. (2009). Predictive model assessment for count data. *Biometrics*, **65**, 1254–61.

Daniels, M. and Hogan, J. (2008). *Missing Data In Longitudinal Studies: Strategies for Bayesian Modeling and Sensitivity Analysis*. Chapman & Hall.

Dawid, A. P. (1984). Statistical theory: The prequential approach. *Journal of the Royal Statistical Society, Series A*, **147**(2), 278–92.

de Boor, C. (2001). *A Practical Guide to Splines*. Springer.

de Leeuw, J. and Meijer, E. (ed.) (2007). *Handbook of Multilevel Analysis*. Springer.

De Oliveira, V., Kedem, B., and Short, D. A. (1997). Bayesian prediction of transformed Gaussian random fields. *Journal of the American Statistical Association*, **92**, 1422–33.

Dellaportas, P., Forster, J. J., and Ntzoufras, I. (2002). On Bayesian model and variable selection using MCMC. *Statistical Computation*, **12**, 27–36.

Demidenko, E. (2004). *Mixed Models*. Wiley.

Denison, D. G. T., Holmes, C. C., Mallick, B. K., and Smith, A. F. M. (2002). *Bayesian Methods for Nonlinear Classification and Regression*. Wiley.

Denison, D. G. T., Mallick, B. K., and Smith, A. F. M. (1998). Automatic Bayesian curve fitting. *Journal of the Royal Statistical Society B*, **60**, 333–50.

Dey, D., Müller, P., and Sinha, D. (1998). *Practical Nonparametric and Semiparametric Bayesian Statistics*. Springer.

Dey, D., Gosh, S. K., and Mallick, B. K. (2000). *Generalized Linear Models: A Bayesian Perspective*. Dekker.

Dierckx, P. (1993). *Curve and Surface Fitting with Splines*. Oxford University Press.

Diggle, P., Henderson, R., and Philipson, P. (2009). Random-effects models for joint analysis of repeated-measurement and time-to-event outcomes. In *Longitudinal Data Analysis: A Handbook of Modern Statistical Methods* (ed. G. Fitzmaurice, M. Davidian, G. Verbeke, and G. Molenberghs), pp. 349–66. Chapman & Hall.

Diggle, P. J. (1990). *Time Series*. Oxford University Press.

Diggle, P. J. (2003). *Statistical Analysis of Spatial Point Patterns*. Arnold.

Diggle, P. J., Heagerty, P., Liang, K.-L., and Zeger, S. L. (2002). *Analysis of Longitudinal Data*. Clarendon Press.

Diggle, P. J. and Ribeiro Jr., P. J. (2007). *Model-based Geostatistics*. Springer.

Dimatteo, I., Genovese, C. R., and Kass, R. E. (2001). Bayesian curve-fitting with free-knot splines. *Biometrika*, **88**, 1055–71.

Donoho, D. L. and Johnstone, I. M. (1994). Ideal spatial adaptation by wavelet shrinkage. *Biometrika*, **81**, 425–55.

Donoho, D. L. and Johnstone, I. M. (1995). Adapting to unknown smoothness via wavelet shrinkage. *Journal of the American Statistical Association*, **90**, 1200–24.

Drobyshev, A., Machka, C., Horsch, M., Seltmann, M., Liebscher, V., de Angelis, M., and Beckers, J. (2003). Specificity assessment from fractionation experiments (SAFE): a novel method to evaluate microarray probe specificity based on hybridisation stringencies. *Nucleic Acids Research*, **31**, e1.

Dunson, D. (2007). Nonparametric Bayes applications to biostatistics. Technical report, National Institute of Environmental Health Sciences, NIH.

Dunson, D. and Chen, Z. (2004). Selecting factors predictive of heterogeneity in multivariate event time data. *Biometrics*, **60**, 352–58.

Dunson, D., Yang, M., and Baird, D. (2007). Semiparametric hierarchical models with mean and variance constraints. Technical report, Department of Statistical Science, Duke University.

Dunson, D., Yang, M., and Baird, D. (2010). Semiparametric Bayes hierarchical models with mean and variance constraints. *Computational Statistics & Data Analysis*, **54**, 2172–86.

Durban, M., Harezlak, J., Wand, M. P., and Carroll, R. J. (2005). Simple fitting of subject-specific curves for longitudinal data. *Statistics in Medicine*, **24**, 1153–67.

Durbin, J. and Koopman, S. J. (2001). *Time Series Analysis by State Space Methods*. Oxford University Press.

Ecker, M. D. and Gelfand, A. E. (2003). Spatial modelling and prediction under stationary non-geometric range anisotropy. *Environmental and Ecological Statistics*, **10**, 165–78.

Efron, B., Hastie, T., Johnstone, I., and Tibshirani, R. (2004). Least angle regression. *Annals of Statistics*, **32**, 407–99.

Eilers, P. H. C. and Marx, B. D. (1996). Flexible smoothing using B-splines and penalized likelihood. *Statistical Science*, **11**, 89–121.

Elashoff, R. M., Li, G., and Li, N. (2008). A joint model for longitudinal measurements and survival data in the presence of multiple failure types. *Biometrics*, **64**, 762–71.

Escobar, M. D. and West, M. (1995). Bayesian density estimation and inference using mixtures. *Journal of the Americal Statistical Association*, **90**, 577–88.

Escobar, M. D. and West, M. (1998). Computing Bayesian nonparametric hierarchical models. In *Practical Nonparametric and Semiparametric Bayesian Statistics* (ed. D. Dey, P. Müller, and D. Sinha), pp. 1–22. Springer.

Evgeniou, T., Pontil, M., and Poggio, T. (2000). Regularization networks and support vector machines. *Advances in Computational Mathematics*, **13**, 1–50.

Fahrmeir, L. and Knorr-Held, L. (1997). Dynamic discrete time duration models. *Sociological Methodology*, **27**, 417–52.

Fahrmeir, L. and Klinger, A. (1998). A nonparametric multiplicative hazard model for event history analysis. *Biometrika*, **85**, 581–92.

Fahrmeir, L. and Gössl, C. (2002). Bayesian models for human brain mapping. *Statistical Modelling*, **2**, 235–49.

Fahrmeir, L. and Lang, S. (2001a). Bayesian inference for generalized additive mixed models based on Markov random field priors. *Applied Statistics*, **50**, 201–20.

Fahrmeir, L. and Lang, S. (2001b). Bayesian semiparametric regression analysis of multicategorical time-space data. *Annals of the Institute of Statistical Mathematics*, **53**, 11–30.

Fahrmeir, L. and Osuna, L. (2006). Structured additive regression for overdispersed and zero-inflated count data. *Applied Stochastic Models in Business and Industry*, **22**, 351–69.

Fahrmeir, L. and Tutz, G. (2001). *Multivariate Statistical Modelling Based on Generalized Linear Models*. Springer.

Fahrmeir, L. and Kneib, T. (2008). On the identification of trend and correlation in temporal and spatial regression. In *Recent advances in Linear Models and Related Areas* (ed. Shalab and C. Heumann), pp. 1–27. Springer.

Fahrmeir, L. and Kneib, T. (2009). Propriety of posteriors in structured additive regression models: theory and empirical evidence. *Journal of Statistical Planning and Inference*, **39**, 843–59.

Fahrmeir, L., Kneib, T., and Lang, S. (2004). Penalized structured additive regression for space-time data: a Bayesian perspective. *Statistica Sinica*, **14**, 731–61.

Fahrmeir, L., Kneib, T., and Konrath, S. (2010). Bayesian regularisation in structured additive regression: A unifying perspective on shrinkage, smoothing and predictor selection. *Statistics and Computing*, **20**, 203–19.

Fahrmeir, L., Lang, S., Wolff, J., and Bender, S. (2003). Semiparametric Bayesian time-space analysis of unemployment duration. *Allgemeines Statistisches Archiv*, **87**, 281–307.

Fan, J. and Gijbels, I. (1996). *Local Polynomial Modelling and Its Applications*. Chapman & Hall.

Fan, J. and Li, R. (2001). Variable selection via nonconcave penalized likelihood and its oracle properties. *Journal of the American Statistical Association*, **96**, 1348–60.

Fenske, N., Fahrmeir, L., Rzehak, P., and Höhle, M. (2008). Detection of risk factors for obesity in early childhood with quantile regression methods for longitudinal data. Technical Report 38, Department of Statistics, Ludwig-Maximilians-University Munich.

Fenske, N., Kneib, T., and Hothorn, T. (2009). Identifying risk factors for severe childhood malnutrition by boosting additive quantile regression. Technical Report 52, Department of Statistics, Ludwig-Maximilians-University Munich.

Ferguson, T. S. (1973). A Bayesian analysis of some nonparametric problems. *Annals of Statistics*, **1**, 209–230.

Fernández, C. and Steel, M. (1998). On Bayesian Modelling of fat tails and skewness. *Journal of the American Statistical Association*, **93**, 359–71.

Finkenstädt, B., Held, L., and Isham, V. (2007). *Statistical Methods for Spatio-Temporal Systems*. Chapman & Hall/CRC.

Fitzmaurice, G., Laird, N., and Ware, J. (2004). *Applied Longitudinal Analysis*. John Wiley & Sons.

Fitzmaurice, G., Davidian, M., Verbeke, G., and Molenberghs, G. (2009). *Longitudinal Data Analysis: A Handbook of Modern Statistical Methods*. Chapman & Hall.

Fleming, T. R. and Harrington, D. P. (1991). *Counting Processes and Survival Analysis*. John Wiley & Sons.

Friedman, J. H. (1991). Multivariate adaptive regression splines. *The Annals of Statistics*, **19**, 1–141.

Frühwirth-Schnatter, S. (1996). Recursive residuals and model diagnostics for normal and non-normal state space models. *Environmental and Ecological Staistics*, **3**, 291–309.

Frühwirth-Schnatter, S. (2006). *Finite Mixture and Markov Switching Models.* Springer.

Frühwirth-Schnatter, S. and Wagner, H. (2006). Auxiliary mixture sampling for parameter-driven models of time series of small counts with applications to state space modelling. *Biometrika*, **93**, 827–41.

Frühwirth-Schnatter, S. and Wagner, H. (2008). Marginal likelihoods for non-gaussian models using auxiliary mixture sampling. *Computational Statistics & Data Analysis*, **52**, 4608–24.

Frühwirth-Schnatter, S. and Wagner, H. (2010). Stochastic model specification search for Gaussian and partially non-Gaussian state space models. *Journal of Econometrics*, **154**, 85–100.

Frühwirth-Schnatter, S. and Frühwirth, R. (2007). Auxiliary mixture sampling with applications to logistic models. *Computational Statistics & Data Analysis*, **51**, 3509–28.

Frühwirth-Schnatter, S. and Frühwirth, R. (2010). Data augmentation and MCMC for binary and multinomial logit models. In *Statistical Modelling and Regression Structures – Festschrift in Honour of Ludwig Fahrmeir* (ed. T. Kneib and G. Tutz), pp. 111–32. Physica-Verlag.

Frühwirth-Schnatter, S., Tüchler, R., and Otter, T. (2004). Bayesian analysis of the heterogeneity model. *Journal of Business & Economic Statistics*, **22**, 2–15.

Frühwirth-Schnatter, S. and Tüchler, R. (2008). Bayesian parsimonious covariance estimation for hierarchical linear mixed models. *Statistics and Computing*, **18**, 1–13.

Frühwirth-Schnatter, S., Frühwirth, R., Held, L., and Rue, H. (2009). Improved auxiliary mixture sampling for hierarchical models of non-Gaussian data. *Statistics and Computing*, **19**, 479–92.

Fu, W. J. (1998). Penalized regressions: The bridge versus the LASSO. *Journal of Computational and Graphical Statistics*, **7**, 397–416.

Furrer, R., Genton, M. G., and Nychka, D. (2006). Covariance tapering for interpolation of large spatial datasets. *Journal of Computational and Graphical Statistics*, **15**, 502–23.

Gamerman, D. (1997). Sampling from the posterior distribution in generalized linear mixed models. *Statistics and Computing*, **7**, 57–68.

Gamerman, D. and Moreira, A. R. B. (2004). Multivariate spatial regression models. *Journal of Multivariate Analysis*, **91**, 262–81.

Gamerman, D., Moreira, A. R. B., and Rue, H. (2003). Space-varying regression models: specifications and simulation. *Computational Statistics & Data Analysis*, **42**, 513–33.

Gelfand, A. (1996). Model determination using sampling based methods. In *Markov Chain Monte Carlo in Practice* (ed. W. Gilks, S. Richardson, and D. Spiegelhalter), pp. 145–61. Chapman & Hall/CRC, Boca Raton.

Gelfand, A. and Dey, D. (1994). Bayesian model choice: Asymptotic and exact calculations. *Journal of the Royal Statistical Society, Series B*, **56**, 501–14.

Gelfand, A. E. and Ghosh, S. K. (1998). Model choice: A minimum posterior predictive loss approach. *Biometrika*, **85**, 1–11.

Gelfand, A. and Vounatsou, P. (2003). Proper multivariate conditional autoregressive models for spatial data analysis. *Biostatistics*, **4**, 11–25.

Gelfand, A. E., Dey, D. K., and Chang, H. (1992). Model determination using predictive distributions with implementation via sampling-based methods. In *Bayesian Statistics 4* (ed. J. M. Bernardo, J. O. Berger, A. P. Dawid, and A. F. M. Smith), pp. 147–67. Oxford University Press, Oxford.

Gelman, A. (2006). Prior distributions for variance parameters in hierarchical models. *Bayesian Analysis*, **1**, 515–534.

Gelman, A. and Hill, J. (2006). *Data Analysis Using Regression and Multilevel/Hierarchical Models.* Cambridge University Press.

Gelman, A., Carlin, J. B., Stern, H. S., and Rubin, D. B. (1995). *Bayesian Data Analysis.* Chapman & Hall.

Gelman, A., Meng, X.-L., and Stern, H. S. (1996). Posterior predictive assessment of model fitness via realized discrepancies. *Statistica Sinica*, **6**, 733–60.

Gelman, A., van Dyk, D. A., Huang, Z., and Boscardin, J. W. (2008). Using redundant parameterizations to fit hierarchical models. *Journal of Computational and Graphical Statistics*, **17**(1), 95–122.

Geman, S. and Geman, D. (1984). Stochastic relaxation, Gibbs distributions, and the Bayesian restoration of images. *IEEE Transactions on Pattern Analysis and Machine Intelligence*, **6**, 721–41.

Gençay, R., Selçuk, F., and Whitcher, B. (2002). *An Introduction to Wavelets and Other Filtering Methods in Finance and Economics.* Academic Press.

George, A. and Liu, J. W. (1981). *Computer Solution of Large Sparse Positive Definite Systems.* Prentice-Hall.

George, E. I. and McCulloch, R. E. (1993). Variable selection via Gibbs sampling. *Journal of the American Statistical Society*, **88**(423), 881–89.

George, E. I. and McCulloch, R. E. (1995). Stochastic search variable selection. In *Markov Chain Monte Carlo in Practice* (ed. W. R. Gills, S. Richardson and D. J. Spiegethalter), pp. 203–14. Chapman & Hall.

George, E. I. and McCulloch, R. E. (1997). Approaches for Bayesian variable selection. *Statistica Sinica*, **7**(2), 339–74.

Geweke, J. (1991). Efficient simulation from the multivariate normal and student-t distributions subject to linear constraints. In *Computer Science and Statistics; Proceedings of the 23rd Symposium on the Interface*, pp. 571–78. Interface Foundation of North America.

Geweke, J. (1996). Variable selection and model comparison in regression. In *Bayesian Statistics 5* (ed. J. M. Bernardo, J. O. Berger, A. P. Dawid and A. F. M. Smith), pp. 609–20. Oxford University Press.

Ghosh, S. K. and Ghosal, S. (2006). Semiparametric accelerated failure time models for censored data. In *Bayesian Statistics and its Applications* (ed. S. K. Upadhyay, U. Singh, and D. K. Dey), pp. 213–29. Anamaya Publishers.

Ghosh, P. and Tu, W. (2009). Assessing sexual attitudes and behaviors of young women: A joint model with nonlinear time effects, time varying covariates, and dropouts. *Journal of the American Statistical Association*, **104**, 474–85.

Gneiting, T., Balabdaoui, F., and Raftery, A. E. (2007). Probabilistic forecasts, calibration and sharpness. *Journal of the Royal Statistical Society, Series B*, **69**, 243–68.

Gneiting, T. and Raftery, A. E. (2007). Strictly proper scoring rules, prediction, and estimation. *Journal of the American Statistical Association*, **102**, 359–78.

Gneiting, T., Stanberry, L. I., Grimit, E. P., Held, L., and Johnson, N. A. (2008). Assessing probabilistic forecasts of multivariate quantities, with an application to ensemble predictions of surface winds. *Test*, **17**(2), 211–35.

Goldstein, H. (1979). *The Design and Analysis of Longitudinal Studies*. Academic Press.

Goldstein, H. (2002). *Multilevel Statistical Models*. Hodder Arnold Publications.

Gössl, C., Auer, D. P., and Fahrmeir, L. (2001). Bayesian spatio-temporal inference in functional magnetic resonance imaging. *Biometrics*, **57**, 554–62.

Green, P. J. (1995). Reversible jump Markov chain Monte Carlo computation and Bayesian model determination. *Biometrika*, **82**, 711–32.

Green, P. J. (2003). Trans-dimensional Markov Chain Monte Carlo. In *Highly Structured Stochastic Systems* (ed. P. Green, N. Hjort, and S. Richardson), pp. 179–96. Oxford University Press.

Greven, S. (2007). *Non-Standard Problems in Inference for Additive and Linear Mixed Models*. Cuvillier Verlag.

Greven, S. and Kneib, T. (2010). On the behaviour of marginal and conditional Akaike information criteria in linear mixed models. *Biometrika*, **97**, 773–89.

Greven, S., Crainiceanu, C. M., Küchenhoff, H., and Peters, A. (2008). Restricted likelihood ratio testing for zero variance components in linear mixed models. *Journal of Computational and Graphical Statistics*, **870**, 870–91.

Griffin, J. E. and Brown, P. J. (2005). Alternative prior distributions for variable selection with very many more variables than observations. Technical report, University of Warwick, Department of Statistics.

Griffin, J. E. and Brown, P. J. (2007). Bayesian adaptive lassos with non-convex penalization. Technical report, University of Warwick, Department of Statistics.

Gschlößl, S. and Czado, C. (2008). Does a Gibbs sampler approach to spatial Poisson regression models outperform a single site MH sampler? *Computational Statistics & Data Analysis*, **52**, 4184–202.

Guo, W. (2002). Functional mixed effects models. *Biometrics*, **58**, 121–128.

Hall, P. and Heyde, C. C. (1980). *Martingale Limit Theory and Its Application.* Academic Press.

Han, C. and Carlin, B. P. (2001). Markov Chain Monte Carlo methods for computing bayes factors: A comparative review. *Journal of the American Statistical Association*, **96**, 1122–32.

Handcock, M. S. and Stein, M. L. (1993). A Bayesian analysis of kriging. *Technometrics*, **35**, 403–410.

Hanson, T. E. (2006). Modeling censored lifetime data using a mixture of gammas baseline. *Bayesian Analysis*, **1**, 575–94.

Harvey, A. (1989). *Forecasting, Structural Time Series Models and the Kalman Filter.* Cambridge University Press.

Harville, D. A. (1974). Bayesian inference for variance components using only error contrasts. *Biometrika*, **61**, 383–85.

Harville, D. A. (1976). Extensions of the Gauss-Markov theorem to include the estimation of random effects. *The Annals of Statistics*, **4**, 384–95.

Harville, D. A. (1977). Maximum likelihood approaches to variance component estimation and to related problems. *Journal of the American Statistical Association*, **72**, 320–38.

Harville, D. A. (1997). *Matrix Algebra From a Statistician's Perspective.* Springer.

Hastie, T. J. (1996). Pseudosplines. *Journal of the Royal Statistical Society, Series B*, **58**, 379–96.

Hastie, T. J. and Tibshirani, R. J. (1990). *Generalized Additive Models.* Chapman & Hall/CRC.

Hastie, T. J. and Loader, C. (1993). Local regression: Automatic kernel carpentry. *Statistical Science*, **8**(2), 120–43.

Hastie, T. J. and Tibshirani, R. J. (2000). Bayesian backfitting. *Statistical Science*, **15**, 193–23.

Hastie, T. J., Tibshirani, R. J., and Friedman, J. (2003). *The Elements of Statistical Learning.* Springer.

Heinzl, F., Kneib, T., and Fahrmeir, L. (2009). Additive mixed models with dirichlet process mixture and P-spline priors. Technical Report 68, Department of Statistics, Ludwig-Maximilians-University Munich.

Henderson, R., Diggle, P., and Dobson, A. (2000). Joint modelling of longitudinal measurements and event time data. *Biostatistics*, **1**, 465–80.

Henderson, R., Shimakura, S., and Gorst, D. (2002). Modeling spatial variation in leukemia survival data. *Journal of the American Statistical Association*, **97**, 965–72.

Hennerfeind, A., Brezger, A., and Fahrmeir, L. (2006). Geoadditive survival models. *Journal of the American Statistical Association*, **101**, 1065–75.

Hobert, J. P. and Casella, G. (1996). The effect of improper priors on Gibbs sampling in hierarchical linear mixed models. *Journal of the Americal Statistical Association*, **91**, 1461–73.

Holmes, C. C. and Held, L. (2006). Bayesian auxiliary variable models for binary and multinomial regression. *Bayesian Analysis*, **1**, 145–68.

Hosmer, D. W. and Lemeshow, S. (2000). *Applied Logistic Regression.* Wiley.

Hosmer, D. W., Lemeshow, S., and May, S. (2008). *Applied Survival Analysis: Regression Modeling of Time to Event Data.* Wiley.

Hougaard, P. (2000). *Analysis of Multivariate Survival Data.* Springer.

Hurn, M., Husby, O. K., and Rue, H. (2003). A tutorial on image analysis. In *Spatial Statistics and Computational Methods* (ed. J. Møller), pp. 87–117. Springer.

Hutchinson, M. F. and de Hoog, F. R. (1985). Smoothing noisy data with spline functions. *Numerische Mathematik*, **47**, 99–106.

Ibrahim, J. and Kleinman, K. (1998). Semiparametric Bayesian methods for random effects models. In *Practical nonparametric and semiparametric Bayesian Statistics* (ed. D. Dey, P. Müeller, and D. Sinha), Volume 133 of *Lecture Notes in Statistics.* Springer.

Ibrahim, J., Chen, M.-H., and Sinha, D. (2001). *Bayesian Survival Analysis.* Springer.

Illian, J., Penttinen, A., Stoyan, H., and Stoyan, D. (2008). *Statistical Analysis and Modelling of Spatial Point Patterns.* Wiley.

Imai, K. and van Dyk, D. A. (2005). A Bayesian analysis of the multinomial probit model using marginal data augmentation. *Journal of Econometrics*, **124**, 311–34.

Ishwaran, H. and James, L. F. (2001). Gibbs sampling methods for stick-breaking priors. *Journal of the American Statistical Association*, **96**, 161–73.

Ishwaran, H. and Rao, J. S. (2003). Detecting differentially expressed genes in microarrays using Bayesian model selection. *Journal of the American Statistical Association*, **98**, 438–55.

Ishwaran, H. and Rao, J. S. (2005). Spike and slab variable selection: frequentist and Bayesian strategies. *The Annals of Statistics*, **33**, 730–73.

Jara, A. (2007). Applied Bayesian non- and semi-parametric inference using DPpackage. *R News*, **7**(3), 17–26.

Jin, X., Carlin, B. P., and Banerjee, S. (2005). Generalized hierarchical multivariate CAR models for areal data. *Biometrics*, **61**, 950–61.

Jin, X., Carlin, B. P., and Banerjee, S. (2007). Order-free co-regionalized areal data lattice models with application to multiple-disease mapping. *Journal of the Royal Statistical Society, Series B*, **69**, 817–38.

Johnson, M. E., Moore, L. M., and Ylvisaker, D. (1990). Minimax and maximin distance designs. *Journal of Statistical Planning and Inference*, **26**, 131–48.

Jordan, M. I. (2004). Graphical models. *Statistical Science*, **19**, 140–155.

Kalbfleisch, J. D. and Prentice, R. L. (2002). *The Statistical Analysis of Failure Time Data.* Wiley.

Kalus, S., Kneib, T., Steiger, A., Holsboer, F., and Yassouridis, A. (2009). A new strategy to analyze possible association structures between dynamic nocturnal hormone activities and sleep alterations in humans. *American Journal of Physiology – Regulatory, Integrative and Comparative Physiology*, **296**, R1216–R1227.

Karatzoglou, A. and Feinerer, I. (2007). Text clustering with string kernels in R. In *Advances in Data Analysis* (ed. R. Decker and H.-J. Lenz), pp. 91–98. Springer.

Kato, B. and Hoijtink, H. (2004). Testing homogeneity in a random intercept model using asymptotic, posterior predictive and plug-in p-values. *Statistica Neerlandica*, **58**, 179–96.

Kauermann, G., Xu, R., and Vaida, F. (2008). Stacked Laplace-EM algorithm for duration models with time-varying and random effects. *Computational Statistics & Data Analysis*, **52**, 2514–28.

Kauermann, G., Krivobokova, T., and Fahrmeir, L. (2009). Some asymptotic results on generalized penalized spline smoothing. *Journal of the Royal Statistical Society, Series B*, **71**, 487–503.

Kaufmann, S. and Frühwirth-Schnatter, S. (2002). Bayesian analysis of switching ARCH models. *Journal of Time Series Analysis*, **23**(4), 425–58.

Kinney, S. K. and Dunson, D. B. (2007). Fixed and random effects selection in linear and logistic models. *Biometrics*, **63**, 690–98.

Klein, J. P. and Moeschberger, M. L. (2005). *Survival Analysis*. Springer.

Kleinbaum, D. G. and Klein, M. (2005). *Survival Analysis: A Self-Learning Text*. Springer.

Kleinman, K. and Ibrahim, J. (1998a). A semiparametric Bayesian approach to generalized linear mixed models. *Statistics in Medicine*, **17**, 2579–96.

Kleinman, K. and Ibrahim, J. (1998b). A semiparametric Bayesian approach to the random effects model. *Biometrics*, **54**, 921–38.

Kneib, T. (2006). Mixed model-based inference in geoadditive hazard regression for interval censored survival times. *Computational Statistics & Data Analysis*, **51**, 777–92.

Kneib, T. and Fahrmeir, L. (2006). Structured additive regression for categorical space-time data: A mixed model approach. *Biometrics*, **62**, 109–18.

Kneib, T. and Fahrmeir, L. (2007). A mixed model approach for geoadditive hazard regression. *Scandinavian Journal of Statistics*, **34**, 207–228.

Kneib, T. and Hennerfeind, A. (2008). Bayesian semiparametric multi-state models. *Statistical Modelling*, **8**, 169–98.

Kneib, T. and Fahrmeir, L. (2010). A space-time study on forest health. In *Statistical Methods for Trend Detection and Analysis in the Environmental Sciences* (ed. R. E. Chandler and M. Scott). Wiley. To appear.

Kneib, T. and Fahrmeir, L. (2011). A space–time study on forest health. In *Statistical Methods for Trend Detection and Analysis in the Environmental Sciences* (ed. R. E. Chandler and M. Scott). Wiley.

Kneib, T., Baumgartner, B., and Steiner, W. J. (2007). Semiparametric multinomial logit models for analysing consumer choice behaviour. *AStA Advances in Statistical Analysis*, **91**, 225–44.

Kneib, T., Knauer, F., and Küchenhoff, H. (2010a). A general approach for the analysis of habitat selection. *Environmental and Ecological Statistics*. To appear.

Kneib, T., Konrath, S., and Fahrmeir, L. (2010b). High-dimensional structured additive regression models: Bayesian regularisation, smoothing and predictive performance. *Applied Statistics*. To appear.

Kneib, T., Knauer, F., and Küchenhoff, H. (2011a). A general approach for the analysis of habitat selection. *Environmental and Ecological Statistics*. To appear.

Kneib, T., Konrath, S., and Fahrmeir, L. (2011b). High-dimensional structured additive regression models: Bayesian regularisation, smoothing and predictive performance. *Applied Statistics*. To appear.

Knorr-Held, L. (2000). Bayesian modelling of inseparable space-time variation in disease risk. *Statistics in Medicine*, **19**, 2555–67.

Knorr-Held, L. and Rue, H. (2002). On block updating in Markov random field models for disease mapping. *Scandinavian Journal of Statistics*, **29**, 597–614.

Kohn, R. and Ansley, C. F. (1988). Equivalence between Bayesian smoothness priors and optimal smoothing for function estimation. In *Bayesian Analysis of Time Series and Dynamic Models* (ed. J. C. Spall), pp. 393–430. Dekker, M.

Kohn, R., Smith, M., and Chan, D. (2001). Nonparametric regression using linear combinations of basis functions. *Statistics and Computing*, **11**, 313–22.

Komárek, A. and Lesaffre, E. (2007). Bayesian accelerated failure time model for correlated interval-censored data with a normal mixture as error distribution. *Statistica Sinica*, **17**, 549–69.

Komárek, A. and Lesaffre, E. (2008). Generalized linear mixed model with a penalized Gaussian mixture as a random effects distribution. *Computational Statistics & Data Analysis*, **52**, 3441–58.

Komárek, A., Lesaffre, E., and Hilton, J. F. (2005). Accelerated failure time model for arbitrarily censored data with smoothed error distribution. *Journal of Computational and Graphical Statistics*, **14**, 726–45.

Kooperberg, C. and Clarkson, D. B. (1997). Hazard regression with interval-censored data. *Biometrics*, **53**, 1485–94.

Krige, D. G. (1951). A statistical approach to some basic mine valuation problems on the Witwatersrand. *Journal of the Chemical, Metallurgical & Mining Society of South Africa*, **52**, 119–39.

Krivobokova, T., Crainiceanu, C. M., and Kauermann, G. (2008). Fast adaptive penalized splines. *Journal of Computational and Graphical Statistics*, **17**, 1–20.

Krivobokova, T., Kneib, T., and Claeskens, G. (2010). Simultaneous confidence bands for penalized spline estimators. *Journal of the American Statistical Association*, **105**, 852–63.

Kuo, L. and Mallick, B. (1997). Bayesian semiparametric inference for the accelerated failure-time model. *Canadian Journal of Statistics*, **25**, 457–72.

Kuss, M. and Rasmussen, C. E. (2005). Assessing approximate inference for binary Gaussian process classification. *Journal of Machine Learning Research*, **6**, 1679–1704.

Laird, N. M. and Ware, J. H. (1982). Random-effects models for longitudinal data. *Biometrics*, **38**(4), 963–74.

Lang, S. and Brezger, A. (2004). Bayesian P-splines. *Journal of Computational and Graphical Statistics*, **13**, 183–212.

Lang, S., Fronk, E.-M., and Fahrmeir, L. (2002). Function estimation with locally adaptive dynamic models. *Computational Statistics*, **17**, 479–99.

Laud, P. W. and Ibrahim, J. (1995). Predictive model selection. *Journal of the Royal Statistical Society, Series B*, **57**, 247–62.

Lauritzen, S. L. (1996). *Graphical Models.* Oxford University Press.

Lee, K. and Thompson, S. (2008). Flexible parametric models for random-effects distributions. *Statistics in Medicine*, **27**, 418–34.

Lee, Y. and Nelder, J. A. (1996). Hierarchical generalized linear models. *Journal of the Royal Statistical Society B*, **58**(4), 619–78.

Lee, Y., Nelder, J. A., and Pawitan, Y. (2006). *Generalized Linear Models with Random Effects.* Chapman & Hall.

Lenk, P. J. and DeSarbo, W. S. (2000). Bayesian inference for finite mixtures of generalized linear models with random effects. *Psychometrika*, **65**, 93–119.

LeSage, J. and Pace, R. K. (2009). *Introduction to Spatial Econometrics.* Chapman & Hall.

Lewin, A., Richardson, S., Marshall, C., Glazier, A., and Aitman, T. (2006). Bayesian modelling of differential gene expression. *Biometrics*, **62**, 1–9.

Lewis, S. M. and Raftery, A. E. (1997). Estimating Bayes factors via posterior simulation with the Lapace-Metropolis estimator. *Journal of the American Statistical Association*, **92**, 648–55.

Li, Q. and Lin, N. (2010). The Bayesian elastic net. *Bayesian Analysis*, **5**, 847–66.

Li, Y. and Ryan, L. (2002). Modeling spatial survival data using semiparametric frailty models. *Biometrics*, **58**, 287–97.

Li, Y. and Ruppert, D. (2008). On the asymptotics of penalized splines. *Biometrika*, **95**, 415–36.

Li, Y., Lin, X., and Müller, P. (2010). Bayesian inference in semiparametric mixed models for longitudinal data. *Biometrics*, **66**, 70–78.

Liang, F., Paulo, R., Molina, G., Clyde, M. A., and Berger, J. O. (2008a). Mixtures of g priors for Bayesian variable selection. *Journal of the American Statistical Association*, **103**(481), 410–23.

Liang, H., Wu, H., and Zou, G. (2008b). A note on conditional AIC for linear mixed-effects models. *Biometrika*, **95**, 773–78.

Lin, X. and Zhang, D. (1999). Inference in generalized additive mixed models by using smoothing splines. *Journal of the Royal Statistical Society B*, **61**, 381–400.

Lin, X. and Carroll, R. (2009). Non-parametric and semi-parametric regression methods for longitudinal data. In *Longitudinal Data Analysis: A Handbook of Modern Statistical Methods* (ed. G. Fitzmaurice, M. Davidian, G. Verbeke, and G. Molenberghs). Chapman & Hall.

Lindgren, F. and Rue, H. (2008). On the second-order random-walk model for irregular locations. *Scandinavian Journal of Statistics*, **35**, 691–700.

Liu, L. and Huang, X. (2009). Joint analysis of correlated repeated measures and recurrent events processes in the presence of death, with application to a study on acquired immune deficiency syndrome. *Applied Statistics*, **58**(1), 65–81.

Loader, C. (1999). *Local Regression and Likelihood.* Springer.

Lopes, H. F. and West, M. (2004). Bayesian model assessment in factor analysis. *Statistica Sinica*, **14**, 41–67.

MacEachern, S. N. and Müller, P. (1998). Estimating mixture of Dirichlet process models. *Journal of Computational and Graphical Statistics*, **7**, 223–38.

Mardia, K. V. (1988). Multi-dimensional multivariate Gaussian Markov random fields with application to image processing. *Journal of Multivariate Analysis*, **24**, 265–84.

Marshall, E. C. and Spiegelhalter, D. J. (2003). Approximate cross-validatory predictive checks in disease mapping models. *Statistics in Medicine*, **22**, 1649–60.

Marshall, E. C. and Spiegelhalter, D. J. (2007). Identifying outliers in Bayesian hierarchical models: A simulation-based approach. *Bayesian Analysis*, **2**, 409–44.

Martinussen, T. and Scheike, T. H. (2009). *Dynamic Regression Models for Survival Data.* Springer.

Matérn, B. (1986). *Spatial Variation.* Springer.

McCullagh, P. and Nelder, J. A. (1989). *Generalized Linear Models.* Chapman & Hall/CRC.

McCulloch, C. E. and Searle, S. R. (2001). *Generalized, Linear, and Mixed Models.* John Wiley.

McCulloch, R. and Rossi, P. E. (1994). An exact likelihood analysis of the multinomial probit model. *Journal of Econometrics*, **64**, 207–40.

McFadden, D. (1973). Conditional logit analysis of qualitative choice behaviour. In *Frontiers in Econometrics* (ed. P. Zarembka). Academic Press.

Meira-Machado, L. F., de Uña Álvarez, J., Cadarso-Suárez, C., and Andersen, P. K. (2009). Multi-state models for the analysis of time-to-event data. *Statistical Methods in Medical Research*, **18**, 195–222.

Meng, X.-L. and Wong, W. H. (1996). Simulating ratios of normalizing constants via a simple identity: A theoretical exploration. *Statistica Sinica*, **6**, 831–60.

Miazhynskaia, T. and Dorffner, G. (2006). A comparison of Bayesian model selection based on MCMC with an application to GARCH-type models. *Statistical Papers*, **47**, 525–49.

Moberg, A., Sonechkin, D. M., Holmgren, K., Datsenko, N. M., and Karlén, W. (2005). Highly variable Northern Hemisphere temperatures reconstructed from low- and high-resolution proxy data. *Nature*, **433**, 613–17. Data available from http://www.ncdc.noaa.gov/paleo/pubs/moberg2005/moberg2005.html.

Molenberghs, G. and Verbeke, G. (2001). A review on linear mixed models for longitudinal data, possibly subject to dropout. *Statistical Modelling*, **1**, 235–69.

Molenberghs, G. and Verbeke, G. (2005). *Models for Discrete Longitudinal Data.* Springer.

Møller, J. (2003). *Spatial Statistics and Computational Methods.* Springer.

Møller, J. and Waagepetersen, R. P. (2004). *Statistical Inference and Simulation for Spatial Point Processes.* Chapman & Hall/CRC.

Møller, J. and Waagepetersen, R. P. (2007). Modern statistics for spatial point processes. *Scandinavian Journal of Statistics*, **34**, 643–84.

Moran, P. (1971). Maximum likelihood estimation in non-standard conditions. *Proceedings of the Cambridge Philosophical Society – Mathematical and Physical Sciences*, **70**, 441–50.

Morris, J. S., Arroyo, C., Coull, B. A., Ryan, M. L., Herrick, R., and Gortmaker, S. L. (2006). Using wavelet-based functional mixed models to characterize population heterogeneity in accelerometer profiles: A case study. *Journal of the American Statistical Association*, **101**(476), 1352–64.

Morris, J. S. and Carroll, R. J. (2006). Wavelet-based functional mixed models. *Journal of the Royal Statistical Society, Series B*, **68**(2), 179–99.

Müller, P. and Quintana, F. A. (2004). Nonparametric Bayesian data analysis. *Statistical Science*, **19**(1), 95–110.

Naskar, M., Das, K., and Ibrahim, J. G. (2005). A semiparametric mixture model for analyzing clustered competing risks data. *Biometrics*, **61**, 729–37.

Natarajan, R. and Kass, R. (2000). Reference Bayesian methods for generalized linear mixed models. *Journal of the American Statistical Association*, **95**, 227–37.

Nathoo, F. S. and Dean, C. B. (2008). Spatial multistate transitional models for longitudinal event data. *Biometrics*, **64**, 271–79.

Neal, R. M. (2000). Markov chain sampling methods for dirichlet process mixture models. *Journal of Computational and Graphical Statistics*, **9**, 249–65.

Newton, M. A. and Raftery, A. E. (1994). Approximate Bayesian inference with the weighted likelihood bootstrap. *Journal of the Royal Statistical Society, Series B*, **56**, 3–48.

Nobile, A. (1998). A hybrid Markov chain for the Bayesian analysis of the multinomial probit model. *Statistics and Computing*, **8**, 229–42.

Nychka, D. and Saltzman, N. (1998). Design of air-quality monitoring networks. In *Case Studies in Environmental Statistics* (ed. D. Nychka, W. W. Piegorsch, and L. H. Cox), pp. 51–76. Springer.

Nychka, D., Haaland, M., O'Connel, M., and Ellner, S. (1998). FUNFITS, data analysis and statistical tools for estimating functions. In *Case Studies in*

Environmental Statistics (ed. D. Nychka, W. W. Piegorsch, and L. H. Cox), pp. 159–79. Springer.

Ogden, R. T. (1997). *Essential Wavelets for Statistical Applications and Data Analysis*. Birkhäuser.

O'Hagan, A. (1995). Fractional Bayes factors for model comparison (with discussion). *Journal of the Royal Statistical Society, Series B*, **57**, 99–138.

Ohlssen, D. I., Sharples, L. D., and Spiegelhalter, D. J. (2007). Flexible random-effects models using Bayesian semi-parametric models: applications to institutional comparisons. *Statistics in Medicine*, **26**, 2088–112.

Pan, W. and Louis, T. A. (2000). A linear mixed-effects model for multivariate censored data. *Biometrics*, **56**, 160–166.

Pan, W. and Connett, J. E. (2001). A multiple imputation approach to linear regression with clustered censored data. *Lifetime Data Analysis*, **7**, 111–23.

Panagiotelis, A. and Smith, M. (2008). Bayesian identification, selection and estimation of semiparametric functions in high-dimensional additive models. *Journal of Econometrics*, **143**, 291–316.

Park, T. and Casella, G. (2008). The Bayesian lasso. *Journal of the American Statistical Association*, **103**, 681–86.

Patterson, H. D. and Thompson, R. (1971). Recovery of inter-block information when block sizes are unequal. *Biometrika*, **58**, 545–54.

Pauler, D. K., Wakefield, J. C., and Kass, R. E. (1999). Bayes factors and approximations for variance component models. *Journal of the American Statistical Association*, **94**, 1242–53.

Pourahmadi, M. and Daniels, M. J. (2002). Dynamic conditionally linear mixed models for longitudinal data. *Biometrics*, **58**, 225–31.

Qiou, Z., Ravishanker, N., and Dey, D. K. (1999). Multivariate survival analysis with positive stable frailties. *Biometrics*, **55**, 637–44.

Reich, B. J., Storlie, C. B., and Bondell, H. D. (2009). Variable selection in Bayesian smoothing spline ANOVA models: application to deterministic computer codes. *Technometrics*, **51**, 110–20.

Reinsch, C. H. (1967). Smoothing by spline functions. *Numerische Mathematik*, **10**, 177–83.

Rizopoulos, D. and Ghosh, P. (2009). A Bayesian semiparametric multivariate joint model for multiple longitudinal outcomes and a time-to-event. Technical report, Department of Biostatistics, Erasmus Medical Center, Rotterdam.

Robert, C. and Marin, J.-M. (2008). On some difficulties with a posterior probability approximation technique. *Bayesian Analysis*, **3**(2), 427–42.

Robert, C. P. (1995). Simulation of truncated normal variables. *Statistics and Computing*, **5**, 121–25.

Robert, C. P. (2001). *The Bayesian Choice*. Springer.

Rossi, P. E., Allenby, G. M., and McCulloch, R. (2005). *Bayesian Statistics and Marketing*. Wiley.

Rubin, D. B. (1984). Bayesian justifiable and relevant frequency calculations for the applied statistician. *Annals of Statistics*, **12**, 1151–72.

Rue, H. (2001). Fast sampling of Gaussian Markov random fields with applications. *Journal of the Royal Statistical Society, Series B*, **63**, 325–38.

Rue, H. and Held, L. (2005). *Gaussian Markov Random Fields.* Chapman & Hall/CRC.

Rue, H. and Martino, S. (2007). Approximate Bayesian inference for hierarchical Gaussian Markov random fields models. *Journal of Statistical Planning and Inference*, **137**, 3177–92.

Rue, H., Martino, S., and Chopin, N. (2009). Approximate Bayesian inference for latent Gaussian models by using integrated nested Laplace approximations. *Journal of the Royal Statistical Society, Series B*, **71**, 319–92.

Rummel, D. (2006). *Correction for covariate measurement error in nonparametric regression.* PhD thesis, available from http://edoc.ub.uni-muenchen.de/6436/.

Ruppert, D. and Carroll, R. J. (2000). Spatially-adaptive penalties for spline fitting. *Australian & New Zealand Journal of Statistics*, **42**, 205–23.

Ruppert, D., Wand, M. P., and Carroll, R. J. (2003). *Semiparametric Regression.* Cambridge University Press.

Ruppert, D., Wand, M. P., and Carroll, R. J. (2009). Semiparametric regression during 2003–2007. *Electronic Journal of Statistics*, **3**, 1193–56.

Sabanés Bové, D. (2009). Predictive assessment of Bayesian hierarchical models. Master's thesis, Ludwig-Maximilians-University Munich, Department of Statistics.

Sahu, S. K., Dey, D. K., Aslanidou, H., and Sinha, D. (1997). A weibull regression model with gamma frailties for multivariate survival data. *Lifetime Data Analysis*, **3**, 123–37.

Schabenberger, O. and Gotway, C. A. (2005). *Statistical Methods for Spatial Data Analysis.* Chapman & Hall/CRC.

Scheipl, F. (2010a). amer: additive mixed models with lme4. http://CRAN.R-project.org/package=amer. R-package.

Scheipl, Fabian (2010b). Normal-mixture-of-inverse-gamma priors for Bayesian regularization and model selection in generalized additive models. Technical Report 84, Department of Statistics, LMU München.

Scheipl, F. (2010c). RLRsim: exact (restricted) likelihood ratio tests for mixed and additive models. http://CRAN.R-project.org/package=RLRsim. R-package.

Scheipl, F. and Kneib, T. (2009). Locally adaptive Bayesian P-splines with a Normal-Exponential-Gamma prior. *Computational Statistics & Data Analysis*, **53**, 3533–52.

Scheipl, F., Greven, S., and Küchenhoff, H. (2008). Size and power of tests for a zero random effect variance or polynomial regression in additive and linear mixed models. *Computational Statistics & Data Analysis*, **52**, 3283–99.

Schwarz, G. (1978). Estimating the dimension of a model. *Annals of Statistics*, **6**, 461–64.

Self, S. G. and Liang, K. Y. (1987). Asymptotic properties of maximum likelihood estimators and likelihood ratio tests under nonstandard conditions. *Journal of the American Statistical Association*, **82**, 605–10.

Sethuraman, J. (1994). A constructive definition of Dirichlet priors. *Statistica Sinica*, **2**, 639–650.

Sinha, D. (1993). Semiparametric Bayesian analysis of multiple event time data. *Journal of the American Statistical Association*, **88**, 979–83.

Sinha, D., Ibrahim, J., and Chen, M.-H. (2003). A Bayesian justification of Cox's partial likelihood. *Biometrika*, **90**, 629–41.

Sinharay, S. and Stern, H. (2003). Posterior predictive model checking in hierarchical models. *Journal of Statistical Planning and Inference*, **111**, 209–21.

Skrondal, A. and Rabe-Hesketh, S. (2004). *Generalized Latent Variable Modeling: Multilevel, Longitudinal, and Structural Equation Models.* Chapman & Hall/CRC.

Smith, M. and Kohn, R. (1996). Nonparametric regression using Bayesian variable selection. *Journal of Econometrics*, **75**, 317–43.

Smith, M. and Kohn, R. (2002). Parsimonious covariance matrix estimation for longitudinal data. *Journal of the American Statistical Association*, **97**, 1141–53.

Smith, M. and Fahrmeir, L. (2007). Spatial Bayesian variable selection with application to functional magnetic resonance imaging. *Journal of the American Statistical Association*, **102**, 417–31.

Smith, M., Pütz, B., Auer, D., and Fahrmeir, L. (2003). Assessing brain activity through spatial Bayesian variable selection. *NeuroImage*, **20**, 802–15.

Smola, A., Gretton, A., Song, L., and Schölkopf, B. (2007). A Hilbert space embedding for distributions. In *Algorithmic Learning Theory: 18th International Conference (ALT 2007)* (ed. M. Hutter, R. A. Servedio, and E. Takimoto), pp. 13–31. Springer.

Spiegelhalter, D. J., Best, N. G., Carlin, B. P., and van der Linde, A. (2002). Bayesian measures of model complexity and fit. *Journal of the Royal Statistical Society, Series B*, **64**, 583–639.

Sprott, D. A. (1975). Marginal and conditional sufficiency. *Biometrika*, **62**, 599–605.

Stein, M. L. (1999). *Interpolation of Spatial Data: Some Theory for Kriging.* Springer.

Stone, C. J., Hansen, M. H., Kooperberg, C., and Truong, Y. K. (1997). Polynomial splines and their tensor products in extended linear modeling. *Annals of Statistics*, **25**, 1371–1470.

Stone, M. (1974). Cross-validatory choice and assessment of statistical predictions (with discussion). *Journal of the Royal Statistical Society*, **36**, 111–47.

Stram, D. O. and Lee, J. W. (1994). Variance components testing in the longitudinal mixed effects model. *Biometrics*, **50**, 1171–77.

Sugiura, N. (1978). Further analysis of the data by Akaike's information criterion and the finite corrections. *Communications in Statistics, Theory and Methods*, **7**, 13–26.

Sun, D., Tsutakawa, R. K., Kim, H., and He, Z. (2000). Spatio-temporal interaction with disease mapping. *Statistics in Medicine*, **19**, 2015–35.

Sun, D., Tsutakawa, R., and He, Z. (2001). Propriety of posteriors with improper priors in hierarchical linear mixed models. *Statistica Sinica*, **11**, 77–95.

Sun, J. and Loader, C. R. (1994). Simultaneous confidence bands for linear regression and smoothing. *Annals of Statistics*, **22**, 1328–45.

Therneau, T. M. and Grambsch, P. M. (2000). *Modeling Survival Data: Extending the Cox Model.* Springer.

Tibshirani, R. (1996). Regression shrinkage and selection via the lasso. *Journal of the Royal Statistical Society, Series B*, **58**, 267–88.

Tibshirani, R. (1997). The lasso method for variable selection in the Cox model. *Statistics in Medicine*, **16**, 385–95.

Tierney, L. and Kadane, J. B. (1986). Accurate approximations for posterior moments and marginal densities. *Journal of the American Statistical Association*, **81**, 82–86.

Tipping, M. E. (2001). Sparse Bayesian learning and the relevance vector machine. *Journal of Machine Learning Research*, **1**, 211–44.

Titterington, D. M. (2004). Bayesian methods for neural networks and related models. *Statistical Science*, **19**, 128–39.

Tjelmeland, H. and Hegstad, B. K. (2001). Mode jumping proposals in MCMC. *Scandinavian Journal of Statistics*, **28**, 205–23.

Tseng, Y., Hsieh, F., and Wang, J. (2005). Joint modelling of accelerated failure time and longitudinal data. *Biometrika*, **92**, 587–603.

Tsiatis, A. A. and Davidian, M. (2004). Joint modeling of longitudinal and time-to-event data: An overview. *Statistica Sinica*, **14**, 809–34.

Tüchler, R. (2008). Bayesian variable selection for logistic models using auxiliary mixture sampling. *Journal of Computational and Graphical Statistics*, **17**, 76–94.

Tutz, G. (2003). Generalized semiparametrically structured ordinal models. *Biometrics*, **59**, 263–73.

Vaida, F. and Xu, R. (2000). Proportional hazards model with random effects. *Statistics in Medicine*, **19**, 3309–24.

Vaida, F. and Blanchard, S. (2005). Conditional Akaike information for mixed-effects models. *Biometrika*, **92**, 351–70.

van Dyk, D. and Meng, X.-L. (2001). The art of data augmentation. *Journal of Computational and Graphical Statistics*, **10**, 1–50.

Verbeke, G. and Lesaffre, E. (1996). A linear mixed-effects model with heterogeneity in the random-effects population. *Journal of the American Statistical Association*, **91**, 217–21.

Verbeke, G. and Molenberghs, G. (2000). *Linear Mixed Models for Longitudinal Data*. Springer.

Vishwanathan, S. V. N., Schraudolph, N., Kondor, R., and Borgwardt, K. M. (2009). Graph kernels. Technical report. Available from http://arxiv.org/abs/0807.0093.

Wackernagel, H. (1995). *Multivariate Geostatistics*. Springer.

Wahba, G. (1978). Improper priors, spline smoothing and the problem of guarding against model errors in regression. *Journal of the Royal Statistical Society, Series B*, **40**, 364–72.

Wahba, G. (1990). *Spline Models for Observational Data*. SIAM.

Wall, M. (2004). A close look at the spatial structure imploed by the CAR and SAR models. *Journal of Statistical Planning and Inference*, **121**, 311–24.

Wang, B. and Titterington, D. M. (2006). Convergence properties of a general algorithm for calculating variational Bayesian estimates for a normal mixture model. *Bayesian Analysis*, **1**, 625–50.

Wang, X. and George, E. I. (2007). Adaptive Bayesian criteria in variable selection for generalized linear models. *Statistica Sinica*, **17**, 667–90.

Wang, Y. and Taylor, J. M. G. (2001). Jointly modeling longitudinal and event time data with application to acquired immunodeficiency syndrome. *Journal of the American Statistical Association*, **96**, 895–905.

Welham, S. J., Cullis, B. R., Kenward, M. G., and Thompson, R. (2006). The analysis of longitudinal data using mixed model L-splines. *Biometrics*, **62**, 392–401.

West, M. and Harrison, P. J. (1997). *Bayesian Forecasting and Dynamic Models (2nd ed.)*. Springer.

Whittaker, E. T. (1923). On a new method of graduation. *Proceedings of the Edinburgh Mathematical Society*, **41**, 63–75.

Wood, S. N. (2006). *Generalized Additive Models: An Introduction with R*. Chapman & Hall/CRC.

Yaglom, A. (1987). *Correlation Theory of Stationary and Related Random Functions I: Basic Results*. Springer.

Yau, P., Kohn, R., and Wood, S. (2003). Bayesian variable selection and model averaging in high-dimensional multinomial nonparametric regression. *Journal of Computational and Graphical Statistics*, **12**, 23–54.

Yin, G. and Ibrahim, J. G. (2005). A class of Bayesian shared Gamma frailty models with multivariate failure time data. *Biometrics*, **61**, 208–16.

Yue, Y. and Speckman, P. L. (2010). Nonstationary spatial Gaussian Markov random fields. *Journal of Computational and Graphical Statistics*, **19**, 96–116.

Zellner, A. (1986). On assessing prior distributions and Bayesian regression analysis with g-prior distributions. In *Bayesian Inference and Decision Techniques* (ed. P. Goel and A. Zellner), pp. 233–43. North-Holland.

Zellner, A. and Siow, A. (1980). Posterior odds ratios for selected regression hypotheses. In *Bayesian Statistics* (ed. J. M. Bernardo, M. H. DeGroot, D. V. Lindley, and A. F. M. Smith), pp. 585–603. Valencia University Press.

Zhao, Y., Staudenmayer, J., Coull, B. A., and Wand, M. P. (2006). General design Bayesian generalized linear mixed models. *Statistical Science*, **21**(1), 35–51.

Zou, H. and Hastie, T. (2005). Regularization and variable selection via the elastic net. *Journal of the Royal Statistical Society, Series B*, **67**, 301–20.

Index